McGRAW-HILL
YEARBOOK OF
SCIENCE &
TECHNOLOGY

2010

Comprehensive coverage of recent events and research as compiled by
the staff of the McGraw-Hill Encyclopedia of Science & Technology

New York   Chicago   San Francisco   Lisbon   London   Madrid   Mexico City

Milan   New Delhi   San Juan   Seoul   Singapore   Sydney   Toronto

On the front cover
A bee visits a rose flower in Indiana.
(*Photo: David L. Dilcher*)

ISBN   978-007-163928-6
MHID   0-07-163928-4
ISSN   0076-2016

1 2 3 4 5 6 7 8 9 0   DOW/DOW   0 5 4 3 2 1 0 9

This book was printed on acid-free paper.

*It was set in Garamond Book and Neue Helvetica Black Condensed
by Aptara, New Delhi, India. The art was prepared by Aptara.
The book was printed and bound by RR Donnelley.*

Contents

iv

Consulting Editors

Article Titles and Authors

Preface

As we approach the completion of the first decade of the twenty-first century, it is apparent that the rapidly increasing pace of research and development in the sciences and technology offers not only novel opportunities for solving many of humankind's problems, such as feeding the hungry and curing once-incurable diseases, but also poses new challenges, such as thwarting the threat of bioterrorism or the misuse of widely available powerful computing and communications technologies. Thus our need for scientific literacy remains stronger than ever to help us make informed decisions in matters ranging from our personal lives to society as a whole. The 2010 *McGraw-Hill Yearbook of Science & Technology* continues its nearly 50-year mission of keeping professionals and nonspecialists alike abreast of important research and development with a broad range of concise reviews invited by a distinguished panel of consulting editors and written by leaders in international science and technology.

In this edition, for example, we report on the rapid advances in cell biology and genetics with articles on topics such as community genetics; the programmed death of cells in normal embryo development but also in controlled response to cell damage; epigenetics: the study of heritable changes in gene expression controlled by factors and processes other than changes in DNA sequence; and the remarkable packaging of DNA and the wide-ranging involvement of modifications of packaging in normal and potentially pathological processes. Reviews in topical areas of biomedicine cover recent research on the detection of respiratory viruses; the development of electronic medical records; a new drug to control tuberculosis; a new malaria vaccine; the emerging field of personalized medicine made possible by rapid advances in the basic sciences; and stimulating innate immunity as cancer therapy. In psychiatry, psychology, and the neurosciences, we review bipolar disorder; body self-perception; exercise and cognitive functioning; poverty, stress, and cognitive functioning; schizophrenia and prenatal infections; and sleep-dependent memory processes. Major developments in the plant sciences are covered, for example, in articles on the origin of the flowering plants; freezing tolerance and cold acclimation in plants; and plant phylogenomics. In the environmental sciences we discuss biotechnology and the environment; carbon capture and storage; the treatment of diesel engine exhaust; green computing; marine transportation and the environment; and transportation efficiency and smart

growth. Advances in computing and communication are documented, for example, in articles on CMOS technology; computational photography; intervehicle communications; optical Ethernet; picocells and femtocells (wireless technologies); and the Wide-Area Augmentation System (WAAS), a new system used in aviation safety to monitor GPS performance. In the earth and atmospheric sciences, we look at how ancient zircons provide a new picture of the early Earth; EarthScope, a network of geodetic instrumentation revealing information about structures and processes beneath the Earth's surface and providing new insights into earthquake and volcanic activity; precious element resources; satellite detection of thunderstorm intensity; and how wave processes shape shoreline changes. In chemistry, we review cluster ion mass spectrometry; Fischer-Tropsch synthesis; molecular modeling of polymers and biomolecules; and stimuli-responsive polymers. In physics and astronomy, we examine advances in the study of cosmic acceleration and galaxy cluster growth; the *Fermi Gamma-ray Space Telescope*; Large Kuiper Belt objects; lattice quantum chromodynamics; the experimental search for gluonic hadrons; strong-interaction theories based on gauge/gravity duality; and science on the *International Space Station*. A broad array of articles in engineering review, for example, arctic engineering; active noise control in vehicles; airplane wing design; the smart grid; cell phone cameras; deployable systems; the structural design of high-rise towers; and digital ultrasonics for materials science. All in all, more than 150 articles chronicle advances in sciences from agriculture to zoology.

Each contribution to the *Yearbook* is a concise yet authoritative article authored by one or more authorities in the field. The topics are selected by our consulting editors, in conjunction with our editorial staff, based on present significance and potential applications. McGraw-Hill strives to make each article as readily understandable as possible for the nonspecialist reader through careful editing and the extensive use of specially prepared graphics.

Librarians, students, teachers, the scientific community, journalists and writers, and the general reader continue to find in the *McGraw-Hill Yearbook of Science & Technology* the information they need in order to follow the rapid pace of advances in science and technology and to understand the developments in these fields that will shape the world of the twenty-first century.

Mark D. Licker
PUBLISHER

A–Z

Acoustic cloaking

Camouflage techniques have been employed by both humans and animals trying to become invisible to their predators or to their prey. Human fascination with invisibility can be traced from ancient times to modern literature. Materials that have the property of making objects or bodies invisible to the eye or undetectable by the ears or other sensing devices have been a dream of military engineers and scientists for centuries. While camouflage has been the solution adopted to achieve optical invisibility, for acoustic invisibility the reduction of noise generated when the object or body moves was formerly the more feasible solution.

Optically invisible regions are commonly obtained in the optical phenomenon of mirage, where the light rays are bent by the gradient of the refractive index that naturally builds up in the air above a hot surface, as in the desert. The air above the heated surface has a temperature that depends of the distance to the surface. Since higher temperature rarefies the air, making it less dense, the index of refraction gradually changes from the surface to the upper regions above. An observer looking at the surface sees objects placed above it because of the bending of light rays. As a consequence, a region on the surface becomes invisible to the observer. Invisibility is then obtained by a naturally produced light-guiding mechanism. Acoustic mirage is also a natural phenomenon, creating regions that are acoustically invisible, that is, regions that sound does not reach. As for light waves, the refractive index n describes the ratio of the local phase velocity of the sound wave to the bulk value. Since the sound speed in the atmosphere depends on the local temperature, sound waves are deflected and lost by the gradient of n created in the atmosphere on warm days (**Fig. 1**).

Guiding of sound waves is the physical mechanism behind the phenomenon of acoustic cloaking, which defines the behavior of some particularly designed materials that guide the flow of acoustic energy around the region that is being cloaked. As a result, the impinging sound wavefront is perfectly reconstructed on the opposite side, implying acoustic undetectability of objects inside the cloaked region (**Fig. 2**).

Anisotropic mass density and acoustic cloaking. Sound propagation is a phenomenon that occurs naturally in fluid materials such as water. The material parameters governing the sound propagation are the dynamical mass density ρ, which measures the resistance to propagation of sound waves, and the bulk modulus B, which measures the resistance of the fluid to compression. Because of the disordered structure of fluids, both ρ and B are scalar magnitudes; that is, they are the same in all directions. This property of fluids is also called isotropy. However, cloaking theory has shown that the phenomenon of acoustic cloaking is possible only in certain generalized fluids having a scalar modulus but with an anisotropic mass density. This extraordinary property, which does not exist in natural fluids, means that the mass density should be dependent on the direction of sound propagation. In fact, acoustic cloaking is obtained as an application of transformation optics, an exciting branch of modern optics. It is applied in acoustics to render a volume effectively invisible to incident radiation. The design process involves a coordinate transformation that squeezes

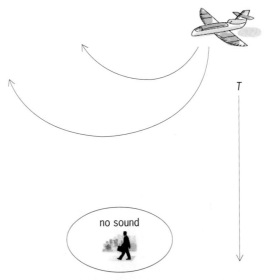

Fig. 1. Acoustic mirage. Acoustically invisible regions can be created on the Earth's surface when a temperature (*T*) gradient exists in the atmosphere. Guiding of sound waves produced by the resulting gradient of the acoustic refractive index is the mechanism explaining "deaf" regions.

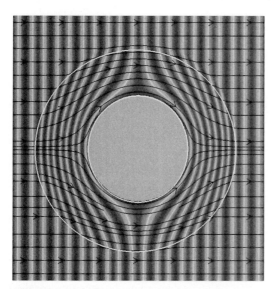

Fig. 2. Acoustic cloaking. The white circle defines the border of the cloaking shell. Black lines indicate the flow of acoustic energy through space. The energy paths bend inside the cloak and surround the invisible region (the inner gray circle). Note how the sound wavefront is perfectly reconstructed on the right side.

space from a volume into a shell surrounding the concealment volume. For the cylindrical cloaking shell considered in Fig. 2, the required parameters depend on the distance r to the origin according to Eqs. (1)–(3), where R_1 is the radius of the invisible

$$\frac{\rho_r}{\rho_0} = \frac{r}{r - R_1} \tag{1}$$

$$\frac{\rho_\theta}{\rho_0} = \frac{r - R_1}{r} \tag{2}$$

$$\frac{B}{B_0} = \left(\frac{R_2 - R_1}{R_2}\right)^2 \frac{r}{r - R_1} \tag{3}$$

region and R_2 is the radius of the outer edge of the cloaking shell. Moreover, $\rho_r(r)$ and $\rho_\theta(r)$ are the mass density along the radial distance and along its perpendicular direction at every value of r, respectively. The quantities with subscript zero are the properties of the fluid outside the shell.

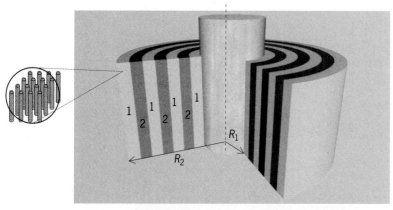

Fig. 3. Schematic view of the cloaking shell for a cylindrical region of radius R_1. It consists of alternating layers of two isotropic metamaterials, 1 and 2. Each metamaterial is made of a specially designed periodic distribution of sound scatterers.

Metamaterials. A new paradigm emerged in the 1960s when Victor Veselago studied electromagnetic materials with dielectric permittivity and magnetic permeability both negative. To engineer these materials, which do not exist in nature, became a challenge to scientists since their astonishing electromagnetic properties, such as negative refractive index, were of extraordinary interest for device applications. A medium with a negative index of refraction bends light to a negative angle with the surface normal, a property not observed in natural optical materials, which always bend light to a positive angle. The negative index was used by John Pendry to design superresolution lenses. But more importantly, he also realized that negative parameters can be achieved simultaneously by a lattice of linear wires in combination with a structure containing loops of conducting wires, provided that the working wavelength of the lens is large in comparison with lattice separation between individual units. The metamaterial concept arises from that simple idea; a metamaterial is a composite made of a periodic distribution of subwavelength units that behaves with extraordinary properties resulting from its structural arrangement rather than the material composition of its individual units.

Acoustic metamaterials can be simply made by setting up a periodic distribution of sonic scatterers, such as solid cylinders or metal spheres, in a fluid or a gas. When the cylinders are arranged in structures having a square or hexagonal symmetry, the resulting solid system effectively behaves like a homogeneous and isotropic fluid. This behavior is obtained when the wavelength λ of the propagating sound is larger than about four times the lattice separation between individual scatterers a; that is, $\lambda \geq 4a$. Materials behaving with an effective anisotropic mass density can be engineered by arranging cylinders on nonsymmetric lattices, that is, other than square or hexagonal. Acoustic parameters depending on the local coordinates, $\rho(r)$ and $B(r)$, can be also achieved by simply changing the dimensions of the scatterers.

Sonic crystals. Any periodic distribution of sound scatterers in a fluid or a gas is called a sonic crystal. A sonic crystal can be one-dimensional, two-dimensional (2D), or three-dimensional according to the periodicity of the material's parameters along the Cartesian directions. They can be periodic along only one direction, in two spatial directions, or in all three spatial directions, respectively. For the simple case of 2D sonic crystals, when the structures behave as isotropic fluidlike materials, their effective parameters can be tailored largely by changing the lattice separation between scatterers or by mixing scatterers of two different solid materials. In principle, 2D sonic crystals made of nonisotropic lattices could be used to build acoustic cloaks. However, no proposal has been reported to engineer acoustic cloaks based on nonisotropic lattices of sonic crystals because of the limited feasibility of tailoring them. A solution has been proposed to construct an acoustic cloak. It consists of two alternating isotropic metamaterials, layers 1 and 2, whose parameters must fulfill two main

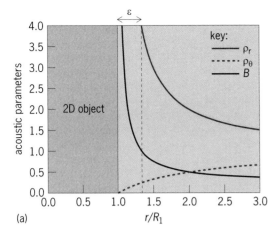

(a)

(b)

Fig. 4. Imperfect cloak. (*a*) Dependence of parameters on distance *r* to the origin. The singular values of parameters near the cloaked region are impossible to achieve. The vertical broken line defines the region that is "imperfectly" cloaked by the finite parameters $\rho_r = 4\rho_0$, $\rho_\theta = 0.25\rho_0$, and $B = B_0$. (*b*) Pressure field produced by the scattering of sound waves impinging an object surrounded by an imperfect cloak. Note that, in comparison with the ideal cloak shown in Fig. 2, there is not a perfect reconstruction of the sound wavefront at both sides of the cloak.

conditions: (1) the density of metamaterial 1 is the reciprocal of that of metamaterial 2, that is, $\rho_2(r) = \rho_0/\rho_1(r)$; and (2) the sound speeds in both materials are the same at the same radial distance, that is, $c_2(r) = c_1(r)$ [**Fig. 3**]. An interesting aspect of this proposal is that the cloak thickness is determined only by technology, not by any other constraint. This means that an extremely thin cloak could be constructed based on the proposed solution.

It should be mentioned that the ideal cloak, as defined by Eqs. (1)–(3) [for the 2D case], is impossible to achieve since it implies that at the border of the cloaked region, $r = R_1$, the values required for the metamaterial parameters are $\rho_r = \infty$, $\rho_\theta = 0$, and $B = \infty$. These values are not achievable by any material distribution. Therefore, real acoustic cloaks will be necessarily imperfect since only finite values for the density and bulk modulus are physically feasible. The finite value parameters define a region slightly larger than the ideally designed cloaked region (**Fig. 4a**). The imperfect cloak will produce a non-

negligible scattering that should be observed by detecting devices only if these devices have sufficient sensitivity (Fig. 4b).

For background information *see* ATMOSPHERIC ACOUSTICS; METEOROLOGICAL OPTICS; MIRAGE; REFRACTION OF WAVES in the McGraw-Hill Encyclopedia of Science & Technology.

José Sánchez-Dehesa; Daniel Torrent

Bibliography. S. A. Cummer and D. Schurig, One path to acoustic cloaking, *New. J. Phys.*, 9:45, 2007; G. W. Milton, M. Briane, and J. R. Willis, On cloaking for elasticity and physical equations with a transformation invariant form, *New. J. Phys.*, 8:248, 2006; J. B. Pendry, Negative refraction makes a perfect lens, *Phys. Rev. Lett.*, 85:3966–3969, 2000; D. Torrent and J. Sánchez-Dehesa, Acoustic cloaking in two-dimensions: A feasible approach, *New. J. Phys.*, 10:063015, 2008; D. Torrent and J. Sánchez-Dehesa, Anisotropic mass density by two dimensional acoustic metamaterials, *New. J. Phys.*, 10:023004, 2008; V. G. Veselago, Electrodynamics of substances with simultaneously negative values of sigma and mu, *Sov. Phys. Usp.*, 10:509–514 1968.

Active noise control in vehicles

An important way in which vehicles can be made more fuel-efficient is to make them lighter. Unfortunately, lighter vehicle structures inevitably transmit more low-frequency noise than heavier structures, and to make light-weight vehicles acceptable to passengers, this low-frequency noise has to be controlled. Conventional passive treatments for low-frequency noise tend to add weight to the structure, reducing the environmental benefits. Active noise control, in which the sound generated by an array of loudspeakers or shakers reduces vehicle noise by destructive interference, can, however, be achieved with much smaller increases in weight. There is thus a growing interest in reducing the low-frequency noise inside vehicles using active noise control.

Application to aircraft. The principle of active noise control is perhaps most widely known because of its use in active headphones, such as those that are supplied to premier-class passengers by several airlines.

Fig. 1. Turboprop aircraft of the type fitted with active noise control systems (Bombardier Q400).

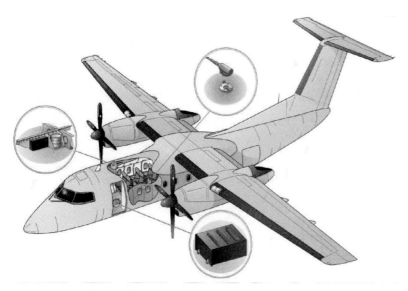

Fig. 2. Components of the active control system fitted in an aircraft, including an electromagnetic shaker and associated amplifier fitted to the fuselage (upper left), a microphone fitted behind the trim panel (upper right), and the electronic controller (lower right).

These headphones use local feedback loops with microphones and miniature loudspeakers inside each earcup to attenuate the ambient noise by about 10 decibels (dB), corresponding to a subjective halving of the perceived noise, up to a frequency of about 1 kHz. It is not always convenient to wear headphones, however, and a more efficient solution would be to control the sound in the whole aircraft cabin. Active control systems that achieve this objective in propeller aircraft were originally developed some time ago, and are now being fitted as standard into aircraft such as the Bombardier Q400 (**Fig. 1**).

Turboprop aircraft such as the Q400 use significantly less fuel than aircraft with jet turbine engines on short-haul flights of a few hundred miles (1 mi = 1.6 km). Historically, however, the passenger cabins of turboprop aircraft have been very noisy, with a loud booming sound generated by the wake of the propeller blades exciting tonal vibration in the aircraft fuselage. The active sound system reduces this booming sound using about 40 small shakers attached to the fuselage, which are driven at the same frequencies as the propeller tones and have their amplitudes and phases adjusted to act out of phase with the propeller excitation (**Fig. 2**). This delicate balancing act is maintained using an electronic controller that uses the outputs from about 90 microphones hidden in the trim panels of the aircraft. Impressive reductions in interior noise levels, with a 10-dB reduction in perceived sound pressure level (measured in adjusted decibels or dBA) throughout the passenger cabin, can be achieved with only about a 50-kg (110-lb) increase in weight (**Fig. 3**).

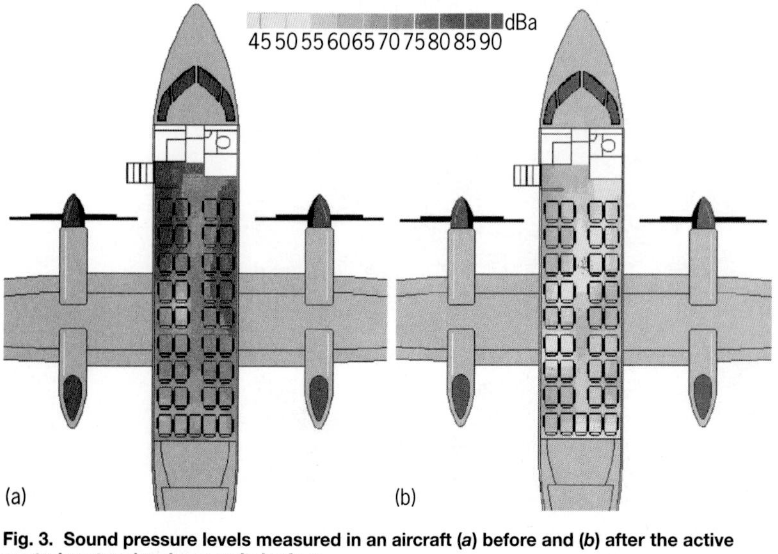

Originally, the systems used to demonstrate active noise control in propeller aircraft used loudspeakers to cancel the noise. There are practical disadvantages in aircraft to fitting loudspeakers in the small gap between the trim panel and the fuselage, but there are also fundamental restrictions on the frequency range of such a system. This is because the number of loudspeakers required for good control, which depends on the number of significantly excited acoustic modes in the aircraft cabin, rises with the cube of the excitation frequency. The number of shakers needed for good control of the vibration of the structure, however, which depends on the number of significantly excited structural modes, rises only linearly with frequency. Even so, practical limitations on the complexity of the system limit the frequencies that can be controlled to those below about 400 Hz. There are also benefits in reducing the low-frequency vibration in these aircraft, and for this reason the control system in the Q400 also controls the outputs of a few accelerometers fitted to seat rails.

More than 1000 active noise control systems of at least six different types are now fitted to turboprop aircraft. Generally, they are arranged to control the sound at the aircraft's blade-passing frequency (BPF), which is somewhere between 70 and 100 Hz, and the second and third harmonics of this fundamental frequency (2 and 3 BPF). The **table** shows the range of the reduction in average sound pressure level throughout the cabin that has been achieved in all these aircraft.

The very significant reductions achieved at the fundamental blade-passing frequency are due to the good matching that can be achieved between the sound field generated by the propellers and that generated by the shakers at this low excitation frequency, for which the acoustic wavelength is of the order of 4 m (13 ft). At the higher harmonics, the acoustic wavelength is proportionally reduced and the number of acoustic modes in the passenger cabin rises significantly, so that smaller attenuations in sound pressure levels can be achieved with a given number of shakers. The performance of passive noise control methods, such as the use of sound absorbers, is very poor at the fundamental but begins to improve at the higher harmonics. The active

45 50 55 60 65 70 75 80 85 90 ⬛dBa

(a)   (b)

Fig. 3. Sound pressure levels measured in an aircraft (a) before and (b) after the active control system has been switched on.

Attenuation in sound pressure level achieved in a number of turboprop aircraft at different frequencies	
Sound component	Attenuation in overall level, dB
BPF	12–20
2 BPF	8–13
3 BPF	2–4

systems thus complement the performance of conventional passive noise control methods, providing good control of the sound at low frequencies with only a modest weight penalty but reduced performance at higher frequencies.

Current developments in this area include the design of active control systems for significantly larger propeller aircraft, such as four-engined military transport vehicles. Research into future active noise control systems in aircraft includes the development of different structural actuators and investigation of the performance of large arrays of modular control systems, each of which includes an actuator, sensor, and local controller. If such an active control module could be shown to give good global control, large numbers could be mass-produced for use in many different applications.

The use of feedback control within these modules would also allow the control of the low-frequency roar due to the random excitation in jet aircraft, caused by the turbulent boundary layer and engine noise, for example. It is not yet clear, however, whether such a module would be best placed on the fuselage, the trim panels, or perhaps between the two, and the issues of integration also become more interesting as aircraft structures move to a mostly composite construction.

Application to road vehicles. Active control has also found use in controlling the interior noise in road vehicles such as cars, particularly the low-frequency engine noise. In this case, the size of the passenger compartment is considerably smaller than that of a passenger aircraft, and only about four loudspeakers are needed to control the sound at frequencies up to a few hundred hertz. Although some active control systems have appeared on Japanese cars, the real advantages of such systems will probably be seen when very light-weight cars are brought into production in order to achieve the mileage per gallon of fuel required in proposed legislation. An interesting trend in automotive active systems is the control of overall sound quality, rather than just sound level. This may be important in hybrid vehicles or those with variable cylinder management, where the sound characteristics of the car may change abruptly and disconcertingly when moving from one mode of propulsion to another. It is known that a smoothly changing sound profile with road speed provides a good sound quality, and this can be achieved with an active noise control system by reducing the sound when it is louder than expected, but also enhancing the sound if it suddenly becomes much quieter than expected. This electronic manipulation of the acoustic environment in cars has seen resistance from some commentators, being seen as cheating compared with mechanical modifications. In future vehicles with electronic control of braking, steering, and handling, however, this does not seem an unreasonable step and may be part of a wider strategy. The sound provided by an active control system could also, for example, be used to encourage the owner to adopt a more fuel-efficient driving behavior.

Another potential advantage of active noise control systems in cars that is still to be fully exploited is integration. Many of the components required for an active system are already present in the audio entertainment and communication systems of modern vehicles, particularly cars. Apart from the loudspeakers and amplifiers, most vehicle entertainment systems now contain at least one digital signal processor, part of whose power could be used to implement active control algorithms. There are also acoustic synergies between the audio and the noise control systems, by ensuring that the low-frequency noise is always sufficiently well controlled so as not to mask the music or speech from the audio system, for example. Responsibility for controlling the noise in cars and providing the audio system has historically been split between different manufacturers, but there is a clear opportunity for integration at the electronic and the acoustic levels.

In summary, active noise control has an important part to play in making the acoustic environment acceptable in lighter, more fuel-efficient, vehicles. This contribution has already been recognized in turboprop aircraft such as the Q400, which would be unacceptably noisy for many passengers without active noise control, despite using significantly less fuel than a jet turbine aircraft on short-haul flights. In the future, it is believed that the control of low-frequency noise with light-weight active systems will be more widely used to achieve fuel savings, particularly on road vehicles.

For background information *see* ACCELEROMETER; ACOUSTIC NOISE; ACTIVE SOUND CONTROL; AERODYNAMIC SOUND; HARMONIC (PERIODIC PHENOMENA); MECHANICAL VIBRATION; SIGNAL PROCESSING in the McGraw-Hill Encyclopedia of Science & Technology.

Stephen J. Elliott; Ian M. Stothers

Bibliography. S. J. Elliott, *Signal Processing for Active Control.* Academic Press, 2001; P. A. Nelson and S. J. Elliott, *Active Control of Sound*, Academic Press, 1996.

Airplane wing design

The design of the airplane wing is fundamental to efficient flight. The choice of the wing shape for a particular airplane is dictated by the application, the plane's speed (subsonic or supersonic), and whether it is a transport or a fighter. Recent interest in drones (customarily termed unmanned aerial vehicles or UAVs) has led to unusual wing shapes. The next important requirement is the takeoff and landing distance. In the case of military aircraft, the maneuver

Fig. 1. Typical transport wing, the Boeing Dash 80, forerunner of the Boeing 707, at the Smithsonian Air and Space Museum. (*Photograph by William H. Mason*)

and acceleration requirements also influence the design.

Multidisciplinary design optimization (MDO). Several disciplines are important, but the two key disciplines are aerodynamics and structures. Aerodynamic design focuses on reducing airplane drag, leading to efficient use of fuel and resulting in lower operating costs. The structural design focuses on making the wing as light as possible, and that is roughly related to the manufacturing cost of the airplane.

The inherent conflict between the two technologies is evident because a wing that is too thick may have a high drag. Yet to handle the loads on the wing with the least amount of structural material, the distance between the top and bottom of the wing should be large, in the same way that I-beams separate the top and bottom flanges. Thus, a wing with a low thickness to achieve low drag may weigh too much.

The proper integration of structures and aerodynamics requires that both components of the design be developed together, in contrast to traditional design approaches where the aerodynamic and structures groups worked essentially independently. This process is known as multidisciplinary design optimization (MDO), and is part of modern computational design. The idea is to use very accurate, high-fidelity computational simulations for both the structures and the aerodynamics simultaneously. This approach was not possible until the required computing power became available. Research on MDO techniques started in earnest in the 1990s, and it is now starting to become available for actual design applications. Other technologies such as propulsion, stability and control, materials, emissions, noise, and manufacturing complexity also need to be considered.

Characteristics of modern wings. Figures 1 and **2** illustrate typical transport and fighter wings. The two wings are quite different. The Boeing transport (Fig. 1) illustrates the use of a large wingspan to achieve efficient cruise flight. The X-35B (Fig. 2) has a much shorter span because its wing must be able to make rapid maneuvers at very high loads. Under these conditions, the wing weight becomes too high if the span is large.

Engines are installed on pods hanging from the wing on the Dash 80 (Fig. 1). It was discovered that engines could be mounted below the wing without a large drag penalty. This practice allows the engine weight to counteract the wing lift, so that the wing structural weight can be reduced, and it also allows for ease of access to the engines for maintenance.

Nomenclature. Wings are defined in terms of the wing area S, wing chord c, span b, thickness t, and sweepback angle Λ (**Fig. 3**). Aerodynamicists speak in terms of the aspect ratio (AR $= b^2/S$), taper ratio ($\lambda = c_t/c_r$, the tip chord divided by the root chord), and thickness-to-chord ratio (t/c). Wings are also defined in terms of a twist distribution, airfoil sections, and camber shapes. Finally, to produce high lift at takeoff and landing, most wings will have so-called devices on the leading edge and trailing edges of the wing. These devices are commonly described as flaps, and come in a wide variety of shapes.

Flight requirements. The basic understanding of the requirements for efficient flight can be identified by examining the Breguet range equation,

$$R = \frac{V\left(\dfrac{L}{D}\right)}{\text{sfc}}\ln\left(\frac{W_{\text{zero fuel}} + W_{\text{fuel}}}{W_{\text{zero fuel}}}\right)$$

where R is the airplane range, V is the flight velocity, L/D is the lift-over-drag ratio, sfc is the specific fuel consumption (the ratio of the weight flow of fuel to the thrust), $W_{\text{zero fuel}}$ is the zero-fuel weight of the airplane, and W_{fuel} is the fuel weight of the airplane. This equation shows that range increases with increasing velocity until the speed reaches a value where adverse transonic effects are encountered due to the formation of shock waves, and the airframe efficiency, L/D, drops rapidly. Most transports cruise at the speed where the drag starts to increase rapidly. Similarly, the propulsion fuel efficiency should be

Fig. 2. Typical fighter wing, the Lockheed Martin X-35B, which was the concept demonstrator leading to the F-35. It is capable of vertical takeoff and landing, using a shaft-driven lift fan located just behind the canopy. (*Photograph by William H. Mason*)

high (that is, sfc should be low) and the weight of the basic airplane without fuel should be low. To maximize L/D, the wing is sized so that the basic drag of the airplane, the so-called parasite drag, is slightly larger than the drag produced by generating lift, the drag due to lift. This requires the proper choice of wing loading, W/S, the airplane weight divided by the wing area. For transport aircraft, typical values would be in the range of 140 lbf/ft^2 (680 kgf/m^2), whereas a typical value for an air superiority fighter might be 65 lbf/ft^2 (320 kgf/m^2). Thus, the size of the wing relative to the airplane weight is a fundamental consideration in sizing the wing.

Physics driving the wing design. The wing planform and thickness will be dictated primarily by the desired flight speed, specified in terms of the Mach number, which is the flight speed divided by the speed of sound. Jet transports typically fly at high subsonic speeds, that is, 0.80–0.85 Mach. To fly efficiently at these speeds, the wings are swept, typically 30–35°. This reduces the adverse compressibility effects that produce extra drag (wave drag) as the airplane speed approaches the speed of sound. However, wing sweep also increases the wing weight. The wing is tapered to reduce wing weight. As aerodynamic design capability has advanced, airfoils have become more efficient and the required wing sweep can be reduced. In addition, wings designed to fly where compressibility effects are important need to be relatively thin (a low t/c, say 7–10%), although this increases the wing weight. If the plane flies much slower than the speed of sound, there is no reason to sweep the wing, and an unswept wing will be the best choice. For example, the piston-powered transports and fighters of the World War II era had unswept wings. In the case of long-range transports, the wing design emphasizes cruise efficiency, and the most efficient wing area requires sophisticated high-lift systems for takeoff and landing.

Airfoil design. For the wing concepts described above, the flow over the wing should be attached; that is, the flow should move over the wing in a smooth, streamlined pattern. This is achieved by the use of an airfoil shape that promotes streamlined flow at the design conditions. With the choice of planform made, the designer selects, or designs, an airfoil to achieve the desired conditions. In making the planform choice, the designer makes an assumption, based on experience, regarding the availability of an appropriate airfoil. To achieve an efficient lift distribution across the wing, the wing is twisted, typically a few degrees up at the root (washin) and a few degrees down at the tip (washout).

Supersonic aircraft. If the application is for a supersonic transport or a fighter, the concept will be different. In the case of a supersonic transport, the wings will tend to be highly swept—so-called slender wings—of which the Concorde is an example. However, it is possible for a supersonic airplane to have very thin unswept wings. Examples include the Lockheed F-104 and the proposed Aerion supersonic business jet. Fighters are required to demonstrate performance at numerous design points, and they

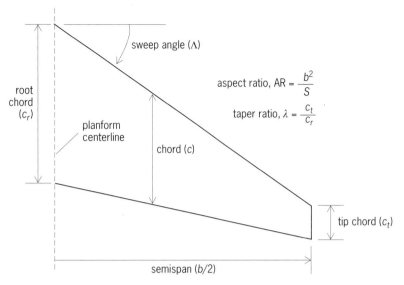

Fig. 3. Wing planform, showing the basic nomenclature of a wing.

may use a combination of flow concepts and planform shapes to meet the diverse requirements. The F-16 and F-18 use relatively unswept wings combined with a "strake" or "leading edge extension" (LEX). In this case the inboard portions of the wing act as a slender wing and provide lift for maneuvering.

Computational methods. Detailed design of today's wings is a critical aspect of competitiveness, so that few details are available for the latest designs. The airfoil and wing designers use computational fluid dynamics (CFD) and wind tunnels to develop their designs. The most important portions of the flight are the transonic cruise (and related buffet boundary and overspeed condition) and the low-speed landing condition. Both of these cases are still difficult to compute. Modern CFD methods are still being developed for both accuracy and efficiency, but for an entire airplane the equations still employ an element of empiricism, known as turbulence models, to simulate the flow field immediately adjacent to the surface. Progress is continuing, and designers use the latest CFD methodology and the most powerful available supercomputers available. The structural design is done using finite-element methods (FEM), which also use supercomputers. The wings are optimized both within individual disciplines and then together using MDO to arrive at the final, fully-integrated design. The fully coupled approach has not quite reached the state of development where it is relied upon completely for very large, expensive projects such as the B-787.

New concepts. Modern computational methods also allow designers to seriously consider new concepts. Two emerging concepts are of particular interest. One achieves an increased span for efficiency by incorporating a strut, or truss, to allow for increased wingspan, reduced wing thickness, and reduced sweep without a weight penalty. The second uses a blended wing body (BWB) concept, which reduces the penalty of using a distinct fuselage, resulting in increased efficiency (**Fig. 4**). To decrease

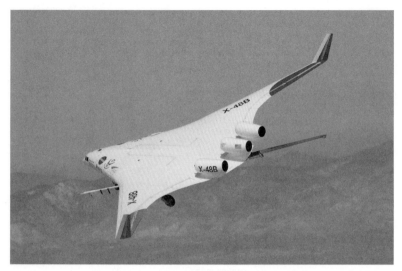

Fig. 4. Boeing X-48, small-scale version of the full-size blended wing body (BWB). (*NASA Dryden Research Center Photo Collection, photograph by Carla Thomas*)

weight, modern aircraft also employ an increasingly large percentage of composite materials in the structure. Finally, as the price of fuel has increased, the possibility of using advanced active flow control has become of interest to reduce the basic parasite drag of the airplane.

For background information *see* AERODYNAMIC FORCE; AERODYNAMIC WAVE DRAG; AERODYNAMICS; AILERON; AIRCRAFT DESIGN; AIRFOIL; COMPOSITE MATERIAL; COMPUTATIONAL FLUID DYNAMICS; FINITE ELEMENT METHOD; MACH NUMBER; SPECIFIC FUEL CONSUMPTION; STRAKE; STREAMLINING; SUBSONIC FLIGHT; SUPERCOMPUTER; SUPERSONIC FLIGHT; TRANSONIC FLIGHT; TURBULENT FLOW; WIND TUNNEL; WING; WING STRUCTURE in the McGraw-Hill Encyclopedia of Science & Technology.

William H. Mason

Bibliography. *Aviation Week & Space Technology*, *Aerospace Source Book*, McGraw-Hill, annually; J. J. Bertin and R. M. Cummings, *Aerodynamics for Engineers*, 5th ed., Pearson/Prentice Hall, Upper Saddle River, NJ, 2009; A. J. Keane and P. B. Nair, *Computational Approaches for Aerospace Design: The Pursuit of Excellence*, John Wiley & Sons, Chichester, England, 2005; D. P. Raymer, *Aircraft Design: A Conceptual Approach*, 4th ed., AIAA, Reston, VA, 2006; R. S. Shevell, *Fundamentals of Flight*, 2d ed., Prentice Hall, Upper Saddle River, NJ, 1989.

Airport wildlife hazard control

Collisions between aircraft and wildlife, principally birds, commonly referred to as birdstrikes, are a relatively new and increasingly hazardous threat to both civil and military aviation worldwide. Although birdstrikes have occurred since people began flying, only in the last 20 years or so has the problem become acute. There are several reasons for this.

Efforts over the last generation to clean up the environment have led to a more benign habitat that has allowed bird populations, especially large flocking

birds, to grow geometrically. Controls on the use of pesticides, hazardous runoff from fields and streams, and improved air quality have all contributed. At the same time, aviation in North America has continued to grow and expand. Conflicts between birds and aviation were inevitable. Birdstrikes in the United States are not declining or leveling off (**Fig. 1**).

When birds or mammals collide with aircraft, serious damage can occur to the aircraft's wings or control surfaces, possibly rendering the aircraft uncontrollable. Likewise, ingestion of birds into a jet or turboprop engine is a serious event. Bird remains can block the airflow through the engine, causing it to overheat or dislodge compressor blades, which can completely destroy the engine (**Fig. 2**).

Previously, federal regulators and aircraft and engine manufacturers agreed on design and construction standards that afforded a certain level of safety should collisions occur. However, these standards have not kept pace with the exponential expansion of large bird populations. For instance, today's engines for mid-size, twin-engine airliners, such as the

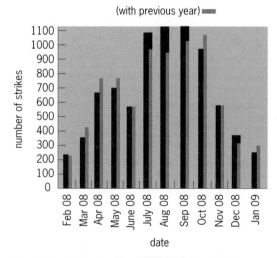

Fig. 1. Report covering 12 months of wildlife strikes. Although strike reports can be in real time, these graphs are presented with a 3-month delay to permit verification of the data. (*Courtesy of the Federal Aviation Administration*)

Fig. 2. An American Airlines MD-80 engine showing fire damage after ingesting a double-crested cormorant on departure from Chicago's O'Hare Airport. (*Courtesy of the Federal Aviation Administration*)

A-320 or B-737, must only be capable of safely shutting down after ingesting one 4-lb bird. The engine does not have to survive. Currently in the United States, there are over 30 species of large flocking birds weighing in excess of 4 lb, with some are as large as 15–20 lb. The fact that these large birds are social and travel in flocks was illustrated most dramatically in January 2009, when a US Airways A-320, during its climb out from New York's LaGuardia Airport, struck a flock of Canada geese. Both engines failed and the aircraft was crash-landed in the Hudson River in downtown New York City. Interestingly, only 3 months before this accident, a similar accident occurred at Rome's Ciampino Airport in Italy. A Ryanair B737-800 struck a large flock of starlings during its approach, lost thrust in both engines, and crash-landed on the runway at Ciampino Airport.

These were not uncommon occurrences. The U.S. Federal Aviation Administration's (FAA) birdstrike database reveals that every year since 1990, multiple turbofan or turboprop aircraft have experienced dual engine damage because of bird ingestions. This FAA database reveals over 7000 reported birdstrikes per year in the United States, but the FAA estimates that only 20% of the birdstrikes are reported to it. Therefore, the true number of collisions between birds and aircraft is closer to 35,000 per year, or over 95 collisions per day, on average, in the United States. The cost of all this damage is astounding. A study by United Airlines and the UK Central Science Laboratory put the dollar cost of birdstrikes to airlines, worldwide, at U.S. $1–1.5 billion annually, due to direct repair cost and loss of service. In a 16-month period between the fall of 2007 and January 2009, there were four catastrophic aircraft accidents caused by birdstrikes in the United States, which left 15 people dead.

Most birdstrikes happen either at the airport or in its immediate vicinity. As a result, the International Civil Aviation Organization (ICAO) amended its Standards and Recommended Practices (SARPs) to mandate wildlife control programs at airports as a standard. As most nations worldwide attempt to comply with ICAO standards, a new emphasis has been placed on this discipline.

Airport wildlife control is generally an attempt to keep birds on and around the airport separated from the aircraft operating at the airport. The strategy includes many different techniques, some passive, such as habitat modification, and others active, such as bird harassment. Obviously, different species of birds require different modification measures.

Airport wildlife control efforts begin with the airport operator conducting a study to ascertain the number and type of birds and other wildlife on and near the airport. Although there is no worldwide standard as to who should conduct this study, it is generally recognized that the party conducting the study should have training and background in aviation wildlife mitigation. In some cases, it is quite easy to detect the problem by simple observation. In other cases, a more thorough study over time may

Fig. 3. Within 2 weeks of completion, starlings and pigeons had started roosting in this canopy constructed over the passenger drop-off area at a major U.S. airport. (*Courtesy of the Federal Aviation Administration; photo by S. Gordon*)

be necessary. The scope and degree of the wildlife problem vary because airport locations vary in their environmental settings, such as near large bodies of water, in areas of farming, or on major migratory flyways.

Passive management. Any plan to control birds and other wildlife around airports should consider that wildlife comes to the airport for only three reasons: eating, drinking, and loafing (which includes nesting). Therefore, not only should bird and wildlife populations be observed, but also the attractants that bring them to the airport must be noted. Both the wildlife and the attractants should be managed. The intent is simply to make the airport and its environs a less desirable place than the surrounding territory. With less food and shelter at the airport, the birds will simply move off to a more desirable habitat.

Passive habitat modification can take place throughout the airport environment. Airport buildings can be designed and constructed, or modified, to take away nesting and perching sites (**Fig. 3**).

The airport operations area (AOA) is the prime focus of passive management. The turf on the field should be managed to an intermediate height of 6–12 in. (15–30 cm). This longer grass discourages many species of bird, as it impedes movement, limits predator detection, reduces interflock communication, and obscures food sources. Longer grass also reduces the amount of mowing the airport operator must do, saving personnel and equipment time.

Landscaping at the airport should be restricted to only those types of plants or trees that provide neither shelter nor food. Ornamental trees such as firs should be avoided, as they provide excellent shelter in inclement weather. Likewise, plants such as hollies, which produce berries, should not be used. Trees that are observed serving as roost or perching sites should be removed or have their interior branches thinned extensively.

Agriculture should not be authorized within the AOA, as virtually every farming crop, whether grain or otherwise, provides some attractant to wildlife. By the same token, manmade food sources should be eliminated. Dumpster lids must remain closed, garbage bags must not be allowed in the open, and personnel must be cautioned regarding wildlife hazards. At one large coastal airport, the prime bird attraction was the taxi queue, where cab drivers fed the birds while waiting for fares.

Waste dumps and landfills must not be located on or near airports. Landfills can be huge bird attractants virtually around the clock. Some nations have established regulations requiring certain distances between landfills and airport runways to ensure safe separation between birds and aircraft operating from those runways.

Water is always a magnet for wildlife, both birds and mammals. After eating, animals migrate to water sources to drink. Drainage ditches, retention ponds, landscaping water features, or simple ground depressions that retain rain water can all be attractants to wildlife. Airport retention ponds should be constructed to drain within 24 h. Retention ponds and other bodies of water can be mitigated by erecting wire grids over the water area. The grids complicate entry into, and exit from, the water by birds, making it less attractive. Smaller water features can be covered with "bird balls," floating plastic balls that completely cover the surface.

Active management. Active wildlife control techniques are applied to those species of wildlife that prove more stubborn or resistant to habitat modification. Active controls can take many forms.

Chemicals can be spread throughout feeding areas to create a taste aversion or gastric distress to the feeding animals, who will avoid the area in the future. Likewise, repellant sprays can used on roosting areas to disturb the roost and harass its occupants.

Startle devices that generate loud sounds, such as propane cannons that fire at random intervals and shell crackers or bangers launched from pistols or shotguns, work effectively when directed toward stubborn wildlife. Pyrotechnics, including shell crackers, are simple but effective and relatively inexpensive to use.

Other auditory devices, such as distress calls, must be used in integration with other dispersal means or birds rapidly habituate to the technique. Likewise, effigies, such as scarecrows, owl effigies, and "scare eyes," are generally ineffective because of habituation.

Lethal control of wildlife is also a part of an integrated active control plan, particularly for mammals. Large mammals, such as deer, at an airport are an obviously threat to aircraft because of their size. Small mammals, such as rodents, are also a problem, as their presence will attract raptors, which feed on rodents. Burrowing mammals can create inadvertent hazards by burrowing next to taxiway or runway surfaces, allowing erosion to undermine those surfaces. Lethal control can consist of traps, poisons, or shooting.

Summary. Airport wildlife control is important, but only part of a safety system to address aviation wildlife hazards. As with any aviation hazard, such as wind shear, volcanic ash, or others, airport wildlife control must be part of a comprehensive and integrated aviation wildlife mitigation plan, which includes technology, training, and regulation.

For background information *see* AIRCRAFT ENGINE; AIRCRAFT PROPULSION; AIRFRAME; AIRPLANE; AIRPORT ENGINEERING; AVIATION; POPULATION ECOLOGY; RISK ASSESSMENT AND MANAGEMENT in the McGraw-Hill Encyclopedia of Science & Technology.

Paul Eschenfelder

Bibliography. Transport Canada, *Airport Wildlife Management Bulletins, Nos. 24–40*, 1999–2008; Transport Canada, *Evaluation of the Efficacy of Products and Techniques for Airport Bird Control, TP13029*, 1998; Transport Canada, *Sharing the Skies*, 2004; Transport Canada, *Wildlife Control Procedures Manual*, 2008; U.S. Federal Aviation Agency, *Wildlife Hazard Management at Airports*, 2004.

Alternative nucleosomal structure

A fundamental innovation that distinguishes eukaryotes from bacteria is the way in which the genome is organized. In bacteria, DNA is a naked circular molecule. In eukaryotes ranging from yeast to humans, DNA is packaged into chromatin, which is composed of beadlike structures called nucleosomes. Each nucleosome contains two copies of the histone proteins, H2A, H2B, H3, and H4, and wraps 147 base pairs (bp) of DNA in a left-handed ramp, resulting in a conserved octameric structure (**Fig. 1**). In addition to packaging DNA, nucleosomes can impede access to regulatory regions and genes. This is referred to as epigenetic control because it provides a layer of information above the genetic code, which itself is embedded within the DNA sequence. The epigenetic status of the genome plays a crucial role in regulating gene expression, DNA repair, DNA accessibility, and DNA segregation. Consequently, it has enormous significance in the biology of normal and diseased cells.

Histone variants, epigenetics, and mitosis. One key example of histones serving an epigenetic function is at cell division during mitosis. At mitosis, a single region on each chromosome is responsible for ensuring accurate segregation of the newly replicated DNA into two daughter cells. At this unique location called the centromere, a centromere-specific histone H3 variant called CenH3 replaces canonical histone H3 within nucleosomes. CenH3 nucleosomes provide the platform on which a large complex of proteins called the kinetochore assembles. The kinetochore, in turn, binds to spindle microtubules, drawing the newly replicated DNA molecules into the two daughter cells. To ensure that chromosomes do not break as a result of multiple microtubule attachments, it is imperative that CenH3 is strictly restricted to a single site per chromosome. Conversely, depleting CenH3 from the cell results in mitotic arrest and cell death, because the kinetochore complex fails to associate with the chromosome. Taken together with the observation that no unique DNA sequence appears necessary for centromere function, it is thought that the primary determinant for centromere specification is the CenH3 nucleosome itself. Therefore, centromeres are considered a prime example of epigenetic specification and inheritance. An example of the consequences of epigenetic dysfunction related to CenH3 is seen in human colorectal cancers. In these diseased cells, CenH3, rather than being restricted to one site, invades noncentromeric regions of the chromosome. These sites attract kinetochore proteins, which can become new centromeres and attach to spindle microtubules, resulting in chromosome breakage during mitosis. Consequently, understanding what features of CenH3 nucleosomes attract kinetochore proteins and how CenH3 location is strictly regulated remain key questions in chromosome biology.

Altered centromeric nucleosomal organization. Observations of the visually distinctive constriction on the chromosome by light microscopy have led to the hypothesis that the molecular structure of the centromere may be unusual. However, the high-resolution structure underlying centromeric chromatin has been elusive. In the past few years, using innovative experimental approaches in the budding yeast *Saccharomyces cerevisiae* and in cultured fruit fly and human cells, new studies have tested this model to reveal that the centromere is indeed unique, even at the level of individual CenH3 nucleosomes.

Normal nucleosomes are rigidly canonical in their organization, invariant across all the diverse eukaryotic organisms. As mentioned previously, they contain two copies of each histone, H2A, H2B, H3, and H4, wrapping 147 bp of DNA around the left-handed octameric protein core (**Fig. 2***a*). However, when the yeast CenH3 is mixed in equal amounts with H2A, H2B, and H4 in a test tube with a kinetochore protein called Scm3, the result is an unprecedented hexameric nucleosome that has displaced one copy of H2A and H2B (Fig. 2*b*). When Scm3 is depleted from living cells, yeast CenH3 fails to associate stably with the native centromeric DNA, suggesting that this unusual CenH3 nucleosome is essential for normal centromere function in yeast.

Topological (configuration) assays are very useful in assessing structural changes in nucleosomes, because the wrapping of DNA around histone complexes is spontaneous and dictated by the normally left-handed, positively charged histone ramp. Such studies have revealed that the curvature of DNA around the yeast CenH3 nucleosome follows a right-handed ramp rather than the canonical left-handed one. In the test tube, fruit fly CenH3 mixed with H2A, H2B, H4, and DNA also yields nucleosomes that wrap DNA in a right-handed ramp rather than the canonical left-handed ramp. The observation of oppositely wrapped DNA in both the yeast and fruit fly CenH3 nucleosomes makes a compelling argument for an unusual structure and suggests that this motif is a unique feature of CenH3 nucleosomes, regardless of which organism they are derived from.

Direct evidence for the alternative CenH3 nucleosome hypothesis comes from cultured cells from

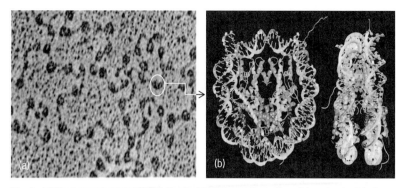

Fig. 1. (*a*) Electron micrograph showing that DNA in a eukaryotic cell is not naked, but organized as "beads on a string" or nucleosomes. (*b*) Nucleosome crystal structure confirms that each bead contains eight molecules of histone proteins (two each of H2A, H2B, H3, and H4), wrapping 147 base pairs of DNA in a left-handed superhelical ramp around the protein core. (*Left panel reproduced with permission from C. L. Woodcock, J. P. Safer, and J. E. Stanchfield, Structural repeating units in chromatin: I. Evidence for their general occurrence, Exp. Cell Res., 97:101–110, 1976; right panel reproduced with permission from K. Luger et al., Crystal structure of the nucleosome core particle at 2.8 Å resolution, Nature, 389:251–260, 1997*)

the fruit fly. In this case, each individual CenH3 nucleosome contains an equal amount of CenH3, H2A, H2B, and H4, cross-links to a tetramer form (rather than a stable octamer), and can be directly visualized as beads measuring one-half of the height of the canonical nucleosomes (Fig. 2*c*).

These studies thus provide the first indication of a histone variant–associated nucleosome that has, in all species studied so far, the centromere with a strikingly alternative nucleosomal structure, referred to as the hemisome, because all its properties are consistent with essentially half of an octameric nucleosome.

Alternative nucleosomal structure and epigenetic inheritance. What purpose could a tetrameric CenH3 nucleosomal structure serve? During DNA replication, when the two parental strands separate, the old nucleosomes are disassembled and must assemble anew on both daughter molecules in equal amounts. New canonical H3 nucleosomes reassemble behind the replication enzymes, providing a full complement of nucleosomes to both daughter DNA molecules. Surprisingly, unlike canonical H3, new CenH3 is added to the chromosome not at the time of DNA replication, but at the end of mitosis. This timing is particularly problematic because, although

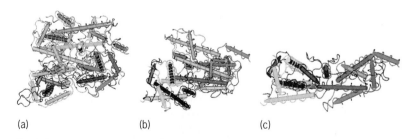

(a)   (b)   (c)

Fig. 2. Proposed organization of CenH3 histone variants into various subnucleosomal structures. (*a*) Traditional octameric organization showing eight molecules (two copies each of H2A, H2B, CenH3, and H4). (*b*) Hexameric organization proposed to occur by displacement of one copy of H2A and H2B by kinetochore protein interaction, resulting in a six-molecule structure. (*c*) Tetrameric organization proposed to occur as stable centromeric nucleosome in vivo containing four molecules: H2A, H2B, CenH3, and H4. [*Structures derived and modified from crystallographic data from the National Center for Biotechnology Information's structure database, MMDB (Molecular Modeling DataBase)*]

the DNA at the centromere is replicated together with the rest of the chromosome, its packaging information is essentially halved. Indeed, no new CenH3 nucleosomes would be available until the end of the next cell cycle to fill up the centromere. During replication, if CenH3 nucleosomes were to segregate randomly, one could envisage a scenario in which one daughter DNA could potentially receive more CenH3 nucleosomes than the other. In addition, CenH3 nucleosomes would be diluted by addition of H3 nucleosomes to gaps in the centromere created by replication, resulting in loss of integrity of centromere domains. Both of these events could then result in loss of chromosomes.

An elegant solution to this problem comes from the novel nucleosomal model for the centromeric nucleosome referred to above (Fig. 2c). During G2 phase, recently assembled CenH3 octamers are envisioned to be intrinsically associated, and would immediately revert into stable tetrameric halves (or hemisomes) during replication, thereby allowing both daughters to get equal numbers of CenH3 nucleosomes (**Fig. 3**). Because no new CenH3 would be assembled until after the next mitosis was completed, these tetramer nucleosomes effectively serve as the sole functional form during mitosis. This solution becomes particularly relevant for organisms, such as the budding yeast, that have only one centromeric nucleosome that must be shared between two daughter DNA molecules after replication, and

to ensure that each daughter has a functional centromere ready for the next mitosis. This model allows for precise and equal segregation of the sole centromere-identifying epigenetic feature to daughter chromosomes after replication, and therefore provides an attractive solution for the epigenetic inheritance of centromeres.

The hemisome model also posits a speculative scenario regarding the evolution of chromatin structure. The Archaea (a group of diverse microscopic prokaryotic organisms, including thermophiles and methanogens) are the earliest organisms known to have developed chromatin structure. The properties associated with eukaryotic CenH3 nucleosomes described above are similar to those seen in the archaeal domain of life. It has been proposed that the Eukaryotae and Archaea share the origins of nucleosomes from a common ancestor predating the split in the evolutionary tree between these two domains. The problem with this scenario has been that (1) archaeal nucleosomes discovered so far are highly flexible *tetramers*, with 80–120 bp of *right-handed* wrapped DNA, whereas all eukaryotic nucleosomes are rigid *octamers* with 147 bp of *left-handed* wrapped DNA; (2) the histone proteins between these two kingdoms are dissimilar; and (3) there are no known organisms that contain an intermediate transition form. However, the finding that eukaryotic CenH3 nucleosomes and archaeal nucleosomes both form tetramers and that both wrap DNA around a right-handed ramp is remarkable. This shared structural identity may derive from a proto-eukaryotic ancestral nucleosome that was co-opted for mitosis in early eukaryotes and evolved separately into the octameric packaging form needed for genome organization.

Future outlook. Observations from recent studies of centromeres have expanded our conceptual framework for understanding how the centromere is organized, how it is assembled, and how it may have evolved. As is the case for all novel concepts, more work must be performed to understand how dynamic biological processes and intrinsic biophysical properties contribute to the alternative CenH3 nucleosomal properties observed thus far, as well as to understand the functional consequences that such nucleosomes have on centromere function. Undoubtedly, ongoing studies and new methodologies will yield exciting and deep insights into these questions aimed at the very heart of eukaryotic biology.

For background information *see* ARCHAEA; CANCER (MEDICINE); CHROMOSOME; DEOXYRIBONUCLEIC ACID (DNA); EUKARYOTAE; GENE; GENETICS; MITOSIS; NUCLEOPROTEIN; NUCLEOSOME in the McGraw-Hill Encyclopedia of Science & Technology.   Yamini Dalal

Bibliography. D. C. Bouck, A. P. Joglekar, and K. S. Bloom, Design features of a mitotic spindle: Balancing tension and compression at a single microtubule kinetochore interface in budding yeast, *Annu. Rev. Genet.*, 42:335–359, 2008; Y. Dalal et al., Structure, dynamics, and evolution of centromeric nucleosomes, *Proc. Natl. Acad. Sci. USA*, 104:15974–15981, 2007; Y. Dalal et al., Tetrameric structure of centromeric nucleosomes in interphase *Drosophila*

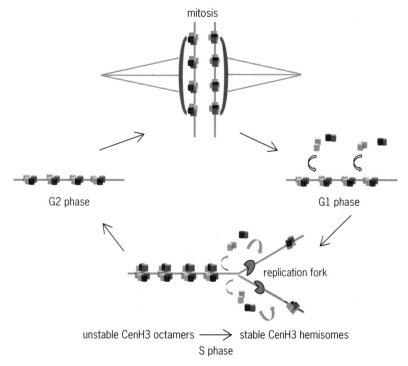

Fig. 3. Model depicting how hemisomal nucleosomes can contribute to epigenetic inheritance of centromeres through replication. As the cell cycle progresses, hemisomes are completed by postmitotic G1 phase assembly by incorporation of CenH3/H4 dimers and H2A/H2B dimers into transiently associated CenH3 octamers. These octamers are intrinsically unstable and dissociate into stable tetramers as the replication fork progresses through the centromere, segregating randomly to each daughter DNA molecule and thus maintaining centromere occupancy and function. Tetramers persist as the dominant CenH3 form through G2 and M phase, where they function to facilitate mitosis via the kinetochore complex and microtubules.

cells, *PLoS Biol.*, 5:e218, 2007; T. Furuyama and S. Henikoff, *Cell*, in press, 2009; L. E. Jansen et al., Propagation of centromeric chromatin requires exit from mitosis, *J. Cell Biol.*, 176:795–805, 2007; G. Mizuguchi et al., Nonhistone Scm3 and histones CenH3-H4 assemble the core of centromere-specific nucleosomes, *Cell*, 129:1153–1164, 2007; M. Schuh, C. F. Lehner, and S. Heidmann, Incorporation of *Drosophila* CID/CENP-A and CENP-C into centromeres during early embryonic anaphase, *Curr. Biol.*, 17:237–243, 2007; H. Wang et al., Single-epitope recognition imaging of native chromatin, *Epigenet. Chromatin*, 1:10, 2008.

Ancient zircons provide a new picture of early Earth

The Hadean Eon, around 4.5–4.0 billion years ago (Ga), is the dark age of Earth history. There is no known rock record from this period. As a result, our knowledge of the growth history of continental crust is equally consistent with the planet then hosting a massive early crust or essentially none at all. Without support from a rock record, our understanding of pre-Archean continental crust comes largely from investigating Hadean detrital zircons (mineral grains from broken-down rock). We know that these ancient zircons yield relatively low crystallization temperatures and that some are enriched in heavy oxygen, contain inclusions similar to modern crustal processes, and show hafnium (Hf) isotope evidence of silicate differentiation by 4.51 Ga. These observations are interpreted to reflect an early terrestrial hydrosphere, early felsic crust in which granitoids were produced and later weathered under high water activity conditions, and even the possible existence of plate boundary interactions, in profound contrast to the traditional view of this period.

The Earth is constantly erasing evidence of its past through a variety of recycling processes. As a result, the geologic record becomes progressively more difficult to read the greater the depth in time one examines. Because there is no known rock record covering the first 500 million years of Earth history, our understanding of the Hadean Eon has come largely from inferences drawn about the state of the early solar system from "dead" planetary objects, such as the Moon, which preserve high-fidelity records of this formative era. The much higher radioactive heat generation expected on Hadean Earth, coupled with the highly visible impact scars left on the lunar surface from this period, contributed to the view that early Earth was an uninhabitable, hellish world (thus "Hadean") dominated by energetic collisions with oceans of magma. The absence of a record of crust older than 4 billion years was widely accepted as evidence that it never existed.

Hadean zircons. About a quarter-century ago, scientists examining ancient Australian quartzites discovered tiny (about twice the width of a human hair) grains of the mineral zircon with ages significantly older than 4 billion years. Following an intensive search involving the dating of over 150,000 zircons, we have now identified and archived thousands of Hadean zircons from localities in Australia, northern Canada, and West Greenland. The oldest grains approach 4.4 billion years in age.

Zircon ($ZrSiO_4$) is close to an ideal mineral for preserving geochemical information. It is inherently resistant to alteration by weathering, dissolution, shock, and chemical exchange, and its tendency to be enriched in radioactive uranium (U) and thorium (Th), relative to daughter-product lead (Pb), makes it a superb chronometer. Although zircon is a chemically simple mineral, it contains a host of trace elements and radioactive isotopes that are valuable probes of the environmental conditions experienced during crystallization. A number of recent studies of these ancient zircons have yielded results that appear to challenge our traditional view of early Earth.

Because zircon is composed of two-thirds oxygen, we are able to measure oxygen isotopic ratios to high precision. Such analyses show that some of the Hadean zircons are enriched in ^{18}O, the heaviest isotope of oxygen, relative to ^{16}O, the lightest. The dominant mechanism for this kind of isotopic enrichment today is the growth of clay minerals, which concentrate the heavy isotope of oxygen, at low (<100°C) temperatures. Because clay minerals contain structurally bound water molecules, the inference is that liquid water must have been present at or near the Earth's surface between 4.0 and 4.4 billion years.

During crystallization, zircons almost invariably overgrow coexisting magmatic phases, incorporating both minerals and melt bubbles as inclusions. Zircons are, in effect, microrock encapsulation systems. By characterizing the nature of these inclusions, it is possible to infer some details about the chemistry of the host magma.

The temperature at which zircon crystallizes from a magma can be determined from knowledge of its titanium concentration. Surprisingly, crystallization temperatures for Hadean zircons calculated in this way cluster at 680°C. This is virtually the lowest temperature at which it is possible for a granite to be molten ("minimum-melting"), and then, only if the melt is essentially saturated in water.

Surveys of large numbers of inclusions in Hadean zircons reveal two kinds of assemblages: a suite of minerals consistent with magma forming in a subduction zone (that is, similar to beneath Mt. St. Helens) environment and another group characteristic of colliding continents (for example, beneath the Himalayas). The chemistry of some of these mineral inclusions can be used as a barometer, and analyses of Hadean zircon indicate that they crystallized at depths of about 25 km. By coupling knowledge of the temperature and depth of zircon crystallization, we have estimated the geothermal gradient under which they formed to be about 30°C/km, or about three times lower than that estimated for Hadean Earth. Because the only magmatic environment on Earth today that is characterized by heat flow of only one-third the global average is where subducting oceanic lithosphere cools the overlying wedge as it descends into the mantle (such as beneath Mt. St. Helens), these results are most simply interpreted as evidence that the zircons formed in a plate boundary environment analogous to the modern subduction zones.

The slow decay of lutetium-176 to hafnium-176 provides us with a long-term tracer of crust-mantle evolution, because parent ^{176}Lu is preferentially retained in the mantle during melting, while daughter ^{176}Hf is exported in the magma that is extracted to create new crust. Thus, the relative enrichment or depletion of ^{176}Hf is a measure of when and under what conditions the silicate portion of the Earth differentiated. Lu–Hf analyses of 4.32–4.01-billion-year-old zircons indicate that the mantle melted very early to form crust that shares similarities with present-day continents. Some grains preserve a memory of such events as early as 4.51 Ga—only 70 million years since the formation of the first solids in the solar system. Thus, all of the energetic, global-scale events broadly believed to have followed from planetary accretion, such as collision with a Mars-sized object to form the Moon, core formation, and magma ocean development, must have occurred within those 70 million years.

Evidence for liquid water at the Earth's surface. The simplest explanation for these and other observations is that they reflect growth of an early (4.51 billion years or older) continental crust in which granites were produced and later weathered under high water activity conditions reflecting the presence of a terrestrial hydrosphere. All this appears to have occurred within a dynamic system similar to modern plate tectonics. This alternative view is profoundly at odds with the traditional paradigm of an uninhabitable, desiccated, continent-free world. While still speculative, the new hypothesis is based on multiple, internally consistent lines of evidence in contrast to the long-standing paradigm, which is essentially mythologic. As virtually all researchers agree that life could not have emerged until there was liquid water at or near the Earth's surface, a significant implication of this new view is that our planet may have been habitable as much as 600 million years earlier than previously thought.

For background information *see* EARTH, AGE OF; HADEAN; ISOTOPE; RADIOACTIVE MINERALS; ROCK AGE DETERMINATION; ZIRCON; ZIRCONIUM in the McGraw-Hill Encyclopedia of Science & Technology.

T. Mark Harrison

Bibliography. T. M. Harrison, The Hadean crust: Evidence from >4 Ga zircons, *Annu. Rev. Earth. Planet. Sci.*, 37:479–505, 2009; T. M. Harrison et al., Early (≥4.5 Ga) formation of terrestrial crust: Lu–Hf, δ^{18}O, and Ti thermometry results for Hadean zircons, *Earth Planet. Sci. Lett.*, 268:476–486, 2008; M. Hopkins, T. M. Harrison, and C. E. Manning, Low heat flow inferred from >4 Gyr zircons suggests Hadean plate boundary interactions, *Nature*, 456:493–496, 2008; G. Turner et al., Pu-Xe, U-Xe, U-Pb chronology and isotope systematics of ancient zircons from Western Australia, *Earth Planet. Sci. Lett.*, 261:491–499, 2007.

Animal navigation

Animals possess a remarkable ability to navigate underwater, on land, and in the air. The distances navigated range from the immediate vicinity of their homes to migration over thousands of kilometers. Animal navigational abilities have been known for centuries, and, in the case of the carrier pigeon, exploited by humans to send messages over long distances. It was not until the late 1800s that scientists began to attempt to understand how animals navigate. Today, the subject of animal navigation continues to draw considerable interest from the scientific community, because the underlying mechanisms involved are still not completely understood.

A fundamental requirement for animal navigation is the ability to sense direction. Animals also use a sense of distance traveled over time and a sense of location to help them navigate. The manner in which animals perceive direction, distance, and location varies greatly across species, time of day, season, and weather conditions. In addition, animals often rely on more than one sensory cue for navigation. This ensures that if one sensory cue is not available, the animal has a backup system. Furthermore, using multiple sensory cues allows for navigation that is more precise than with only a single sensory cue. Because of the multisensory nature of animal navigation, experiments designed to understand how animals navigate are technically challenging. Nevertheless, numerous studies have provided considerable insight into how animals navigate.

Sense of direction. Several animals sense direction by detecting the magnetic field of the Earth. For instance, birds and sea turtles change their orientation in experiments when the orientation of the magnetic field is manipulated around them. The mechanism by which they sense the magnetic field is quite remarkable. The magnetic compass in several bird species is located in the right eye and involves a chemical process that also requires sunlight. Furthermore, it is known that birds do not sense the polarity of the magnetic field, because experiments have shown that altering the polarity of the magnetic field around them does not alter their orientation. Instead, by sensing the inclination angle of the magnetic field, they can determine whether they are moving toward the nearest geomagnetic pole or toward the equator.

The Sun is used by many animals to provide a sense of direction. Clock-shift experiments verify that pigeons and monarch butterflies use their sense of time to determine their orientation based on the position of the Sun in the sky. In these experiments, the animals are displaced from their home and kept in a room without natural sunlight for several days. The cycle of artificial daylight and darkness inside the room is then shifted by several hours. When the animals are released, they attempt to return home, but their departure direction is offset from the true homeward direction based on the error of their internal clock. For example, a 6-h clock shift produces, on average, a homeward orientation error of $90°$. Despite the fact that navigation based on the Sun also requires knowing the time of day, adult pigeons rely more on their sun compass than on their magnetic compass when they are near their home. Furthermore, many animals use their sun compass during the periods just before sunrise and immediately after sunset by observing the polarization pattern that scattered sunlight creates in the sky.

At night, when the Sun is not visible, many animals use the stars to navigate. Experiments conducted with migratory birds placed in planetariums have verified this ability. During their usual period of migration, these birds orient in the same direction in the planetarium that they would fly under the natural sky. If the artificial star pattern is shifted relative to the true night sky, the birds change their orientation accordingly.

Sense of distance traveled. Many animals sense the distance they have traveled using a mechanism that is analogous to the trip odometer of an automobile, which records the distance traveled since it was last reset. When an animal leaves its home to search for food, the odometer and compass work together to tell the animal how far and in what direction the animal has traveled from its home. This allows the animal to return directly home rather than retrace its outbound path.

For land animals, the most direct measure of distance traveled is simply counting steps. This type of odometer is evident in the desert ant. In experiments in which desert ants were intercepted while attempting to return home, if their legs were lengthened by applying miniature stilts, the ants walked past their home. Similarly, the ants stopped short of their home if their legs were surgically shortened.

Honeybees use their eyes to perceive optical flow, that is, the apparent motion of visual cues as they fly. At a given flight speed, the optical flow of nearby visual cues is larger than the optical flow of visual cues that are far away. Thus, optical flow is not a true measure of distance traveled. However, during a return flight to the hive from a foraging site, a bee can gauge the distance traveled in an abstract sense by continually summing the perceived optical flow over time. This bee can then communicate to other bees back in the hive, through a waggle dance, how far they must travel (that is, how much optical flow they must experience) in order to reach a foraging site.

Sense of location. As an animal becomes acquainted with its environment, it can learn to recognize landmarks. Every time it encounters a familiar landmark, it knows where it is in a relative sense. Landmark navigation is certainly important for the desert ant. However, visual landmarks are not a requirement for navigation for all animals. In experiments with pigeons wearing frosted glasses, these birds oriented toward home after being displaced from their home. Because the frosted glasses deprived these animals of the visual acuity required to identify landmarks, they must have used other cues to navigate.

The magnetic field of the Earth also can be used for more than just a directional sense. The logger-

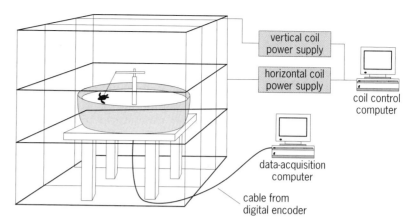

Diagram (not to scale) of the orientation arena, magnetic coil system, and data acquisition system used in studies of sea turtle responses to magnetic fields. (*From K. J. Lohmann and C. M. F. Lohmann, Geomagnetic navigation and magnetic maps in sea turtles, Navigation, 55:115–125, 2008*)

head sea turtle, for instance, can perceive the angle that the magnetic field makes relative to the surface of the Earth (see **illustration**). Since this angle is mostly a function of the turtle's distance from the magnetic poles, the sea turtle has a sense of latitude. The sea turtle can also detect the magnetic field's intensity. Geomagnetic field inclination and intensity vary across the globe, so that different regions are typically marked by different magnetic fields. Thus, the magnetic field can potentially provide animals with navigational landmarks. Turtles can learn the magnetic topography of the areas where they live and acquire "magnetic maps" that are used in navigating to specific geographic areas.

The ability of a pigeon to navigate back to its home after being displaced by more than 100 km (60 mi), even from an unfamiliar location, continues to baffle scientists. This phenomenon has led many to suspect that these animals have a true sense of where they are in the world. This would require their knowing not only the latitude but also the longitude of their location. Several theories have been suggested regarding local variations in the magnetic field, but none of these theories has been sufficiently demonstrated in any animal.

The possibility that animals use odors to infer location has also been considered. There is no doubt that animals track scents to locate food and mates, but the role of odor as a map cue is currently under debate. Experiments with pigeons that are deprived of their sense of smell indicate that they are unable to return home from an unfamiliar location, whereas those that can smell do return home. Furthermore, there is evidence that the ratio of certain odors in the atmosphere is stable enough to allow for navigation based on an olfactory map. However, the odor or combination of odors that pigeons may use to form such a map has yet to be identified.

Application to human navigation. Nature's great navigators use different sensory cues to determine their direction, distance traveled, and location in order to navigate to locations for food and to return home. These navigators continue to fascinate us as we strive to develop new systems to help us navigate accurately by land, air, or sea. The Global Positioning System (GPS) has been a major development in helping humans to navigate, but it does not work well in complex environments such as underwater, under trees, in dense urban environments, or indoors. Therefore, studying the ability of animals to use the various natural phenomena of the Earth to navigate in complex environments is an extremely important research area as we strive to improve our ability to navigate in areas where the GPS does not function properly.

For background information *see* ANIMAL COMMUNICATION; BIOLOGICAL CLOCKS; GEOMAGNETISM; MIGRATORY BEHAVIOR; POLARIZED LIGHT; SATELLITE NAVIGATION in the McGraw-Hill Encyclopedia of Science & Technology. Mikel M. Miller; Adam J. Rutkowski

Bibliography. K. J. Lohmann, C. M. F. Lohmann, and N. F. Putman, Magnetic maps in animals: Nature's GPS, *J. Exp. Biol.*, 210:3697–3705, 2007; T. Merkle and R. Wehner, Landmark guidance and vector navigation in outbound desert ants, *J. Exp. Biol.*, 211:3370–3377, 2008; A. Rozhok, *Orientation and Navigation in Vertebrates*, Springer-Verlag, 2008.

Apolipoprotein C-III

Elevation of triglycerides (hypertriglyceridemia) is an independent risk factor for atherosclerosis-related cardiovascular disease. Similar to low-density lipoproteins (LDLs), triglyceride-rich lipoproteins (TRLs) accumulate in atherosclerotic plaques. Hypertriglyceridemia is associated with an overproduction of very low density lipoprotein (VLDL) particles or delayed catabolism of TRLs, their remnants, and apolipoprotein B (ApoB)–containing lipoproteins. [An apolipoprotein is a protein that combines with a lipid (fat) to form a lipoprotein.] The delayed catabolism of TRLs and their remnants is a direct consequence of reduced lipoprotein lipase activities, hepatic (liver) remnant receptors, and increased apolipoprotein C-III (ApoC-III) levels. ApoC-III resides on a broad distribution of ApoB lipoproteins, including chylomicrons (very large lipoproteins rich in triglycerides, found in blood after the ingestion of fat), VLDLs, LDLs, and high-density lipoproteins (HDLs). It is the major regulator of lipolysis [the enzyme-catalyzed hydrolysis (removal) of fatty acids from triglycerides], noncompetitively inhibiting endothelial-bound lipoprotein lipase, which is the enzyme that hydrolyzes triacylglycerols in TRLs. ApoC-III can also inhibit hepatic lipase as well as the uptake of triacylglycerol-rich lipoprotein remnants by hepatic lipoprotein receptors. In addition, it has direct cellular effects that can contribute to cardiovascular disease.

ApoC-III gene regulation. ApoC-III is synthesized by the liver and to a lesser extent by the intestines. A mature ApoC-III is composed of 79 amino acids with a molecular mass of 8.8 kilodaltons (kDa). The ApoC-III gene is located on human chromosome 11 between the ApoA-I and ApoA-IV genes. It displays strict tissue-specific expression and is controlled by a number of positive and negative regulatory elements.

ApoC-III gene activity is tightly regulated by hormones, hypolipidemic drugs, and cytokines (peptides released by some cells that affect the behavior of other cells, serving as intercellular signals), including interleukin-1 (IL-1) and tumor necrosis factor-α (TNF-α). Inhibition of ApoC-III expression in hepatocytes is caused mostly by transcriptional repression. In contrast to proinflammatory cytokines (such as IL-1 and TNF-α), the anti-inflammatory cytokine transforming growth factor-β activates ApoC-III gene expression. Insulin is also an important regulator of ApoC-III expression, since ApoC-III is upregulated in insulin-deficient diabetic mice. Insulin appears to exert its inhibitory effect on ApoC-III by stimulating the phosphorylation of a certain transcription factor (Foxo1), preventing its entry into the nucleus and interaction with the insulin response element in the ApoC-III promoter. Fibrates, which are drugs used to treat diet-resistant hypertriglyceridemia, also lower ApoC-III, by downregulating gene expression of ApoC-III via transcriptional suppression.

Biochemistry. ApoC-III has been predicted to contain two helices that are amphipathic [having polar and nonpolar groups (or surfaces), thereby allowing interactions with both water and lipids (hydrocarbons)]. It exists in three isoforms, ApoC-III$_0$, ApoC-III$_1$, and ApoC-III$_2$, differing in the number of sialic acid molecules that terminate the carbohydrate portion of the proteins. The three forms contribute approximately 10%, 55%, and 35%, respectively, of the total ApoC-III in the circulation. All have the same plasma half-life, suggesting similar synthesis and degradation processes and probably similar physiological functions. Lipid interactions are required for ApoC-III's solubility, correct folding, and function. The three-dimensional structure of ApoC-III in complex with micelles has recently been determined. It consists of six ~10-residue helices that are connected via semiflexible hinges and are curved so that they can wrap tightly around the micelle. The semiflexible hinges allow ApoC-III to adapt to the different diameter particles with which it associates: chylomicrons [75–2000 nanometers (nm)], VLDLs (25–75 nm), LDLs (18–25 nm), and HDLs (5–12 nm). The lipid-interacting region of ApoC-III extends over all six helices and originates mainly from hydrophobic residues that point with their side chains to the interior of the micelle. Seven negatively charged residues line the polar surface of the ApoC-III C-terminus, whereas the N-terminus is negatively charged.

Physiology. ApoC-III has a number of physiological effects that are especially important for metabolism and cell signaling (see **illustration**).

Effects on triglyceride and lipoprotein metabolism. ApoC-III is associated with ApoB-containing lipoproteins and HDLs, and is exchanged rapidly between different lipoprotein particles. This exchange appears to involve a simple aqueous diffusion mechanism and is not dependent on collisions between lipoprotein particles. During hydrolysis of VLDL triglycerides by lipoprotein lipase, ApoC-III redistributes from VLDLs to HDLs. This distribution of ApoC-III between HDL particles and non-HDL particles is a strong predictor of coronary atherosclerosis. In normolipidemia, the majority of the ApoC-III is associated with HDL particles. However, in hypertriglyceridemia, the majority is bound to other TRL particles. As in other apolipoproteins, ApoC-III contains amphipathic helices that promote its binding to the lipid interface in lipoprotein particles, and excess ApoC-III can lead to displacement of other apolipoproteins from lipid-rich particles. The ability of ApoC-III to displace ApoC-II from lipoprotein particles is thought to reduce ApoC-II activation of lipoprotein lipase. ApoC-III may also act as a direct noncompetitive inhibitor of lipoprotein lipase, although there is controversy as to the identity of the inhibitory sites on ApoC-III for lipoprotein lipase.

ApoC-III can also displace ApoE from lipoproteins. As a consequence, raised plasma ApoC-III reduces ApoE-mediated clearance of triglyceride-rich remnant lipoproteins. In ApoC-III transgenic mice, the number of VLDL particles is increased as a result of diminished ApoE-mediated lipoprotein uptake. Definitive evidence for ApoC-III regulating triglyceride metabolism has been obtained using ApoC-III-deficient mice. In these mice, fasting plasma triglyceride levels are reduced by about 70% and postprandial (after a meal) hypertriglyceridemia is abolished. Also, ApoC-III transgenic mice are severely hypertriglyceridemic.

Effects on endothelial and monocyte/macrophage cell signaling. ApoC-III is an important mediator of cell signaling in endothelial cells and monocytes/macrophages (white blood cells). ApoC-III-rich VLDLs augment the adhesion of monocytes to endothelial cells by increasing the expression on endothelial cells of vascular cell adhesion molecule-1 (VCAM-1) and intercellular cell adhesion molecule-1 (ICAM-1). This can contribute to atherosclerosis by increasing monocyte and lymphocyte recruitment. This effect on adhesion molecules is the consequence of ApoC-III activating protein kinase Cβ and nuclear factor-κB (NF-κB) in the endothelial cells. Repetitive exposure of endothelial cells to TRLs can exacerbate inflammatory responses by lowering the threshold at which, for example, TNF-α elevates VCAM-1 expression and by increasing monocyte recruitment. This effect appears to be mediated via NF-κB.

ApoC-III increases phosphatidylcholine-specific phospholipase C activity in monocytes, resulting in protein kinase Cα activation. As in endothelial cells, ApoC-III activates NF-κB in monocytes via protein kinase Cα to augment expression of β_1-integrins (which are receptors for VCAM-1). ApoC-III also activates Rho, which is important for activation and clustering of integrins and downstream monocyte-endothelial interactions. This ability to augment adhesion of monocytic cells to endothelial cells is dependent on the nature of the lipoprotein particles with which the ApoC-III is associated. ApoC-III associated with VLDL or LDL particles or ApoC-III alone stimulates this interaction, whereas ApoC-III associated with HDL does not. Activation of toll-like receptor 2 (TLR2, a signal transducer) accounts for these effects, since ApoC-III binds to TLR2 on monocytes with high affinity. Activation of NF-κB can also increase expression of proinflammatory cytokines. ApoC-III increases the expression of TNF-α, IL-1, and IL-6, suggesting that it may directly contribute to inflammation in atherosclerotic lesions.

ApoC-III can also induce endothelial dysfunction by inhibiting insulin signaling, specifically by preventing insulin receptor substrate-1 (IRS-1) phosphorylation at Tyr989 and downstream phosphatidylinositol-3 kinase/Akt activation. Endothelial nitric oxide synthase (eNOS) is activated by phosphorylation via insulin-activated Akt, and ApoC-III dose-dependently attenuates insulin-stimulated eNOS without affecting its expression. ApoC-III stimulates extracellular signal–regulated kinase 1/2 (ERK1/2) and protein kinase Cβ in endothelial

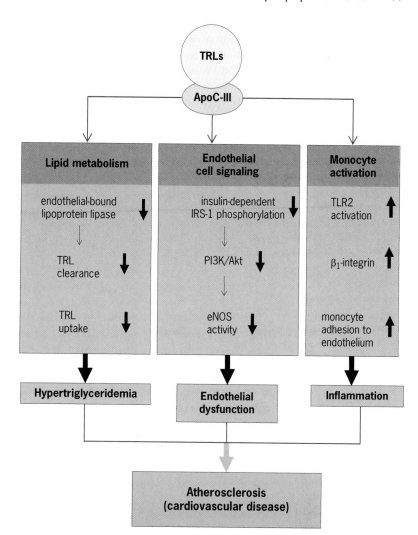

Critical effects of apolipoprotein C-III (ApoC-III) that may contribute to atherosclerosis-related cardiovascular disease. ApoC-III in triglyceride-rich lipoproteins (TRLs) inhibits lipoprotein lipase, thereby delaying clearance of TRLs and attenuating uptake by the liver. This may lead to increased accumulation of lipids within developing atherosclerotic lesions. It also attenuates endothelial nitric oxide synthase (eNOS) activity to induce endothelial dysfunction, and it activates monocytes/macrophages. This can lead to increased adhesion of monocytes to the endothelium, their recruitment to developing atherosclerotic lesions, and also increased inflammation within lesions. Terms: IRS-1, insulin receptor substrate-1; PI3K/Akt, phosphatidylinositol-3 kinase/Akt; TLR2, toll-like receptor 2.

cells, inducing phosphorylation of IRS-1 on Ser616 and thereby preventing insulin-stimulated Tyr989 phosphorylation. eNOS is important for nitric oxide generation in blood vessels, not only acting as a vasodilator but also exerting atheroprotective effects by limiting endothelial cell activation. Nitric oxide also attenuates smooth muscle cell migration and proliferation, platelet aggregation, and NF-κB. ApoC-III thus appears to induce insulin resistance at the level of the endothelium.

ApoC-III and cardiovascular disease. A body of evidence is now accumulating that indicates that ApoC-III may directly contribute to cardiovascular disease. For example, overexpression of ApoC-III in mice that are deficient in the LDL receptor—which reflects a commonly inherited human lipid disorder, familial combined hyperlipidemia—results in augmented atherosclerosis. Also, individuals with familial combined hyperlipidemia have large quantities of VLDLs and LDLs and develop premature coronary artery disease. The recent discovery that a null mutation in human ApoC-III that results in reduced expression of ApoC-III confers a favorable plasma lipid profile and apparent cardioprotection also supports a proatherogenic role for ApoC-III. In a prospective study involving more than 2200 subjects over a 15-year period, those that died had higher plasma ApoC-III concentrations. Moreover, in a randomized placebo-controlled trial of pravastatin (a cholesterol-lowering drug) that followed over 4100 subjects with myocardial infarction, plasma concentrations of VLDLs and ApoC-III in VLDLs and LDLs were found to be more specific measures of coronary heart disease risk than plasma triglyceride concentrations. Other studies have also causally related triglyceride-rich ApoC-III-containing particles with severity of coronary artery disease.

Conclusions. ApoC-III has significant effects on lipoprotein metabolism and probably on the accumulation of lipoproteins within developing atherosclerotic lesions. Along with findings that indicate that ApoC-III can also promote endothelial cell dysfunction and activate monocytes, this strongly suggests that ApoC-III is involved in atherosclerosis-related cardiovascular disease. Although these effects alone are unlikely to initiate atherosclerosis, the effects are likely to be synergistic when acting in cooperation with other proatherogenic stimuli in patients with developing atherosclerosis.

For background information *see* ARTERIOSCLEROSIS; CARDIOVASCULAR SYSTEM; CHOLESTEROL; INFLAMMATION; INSULIN; LIPID; LIPID METABOLISM; LIPOPROTEIN; LIVER; METABOLIC DISORDERS; TRIGLYCERIDE (TRIACYLGLYCEROL) in the McGraw-Hill Encyclopedia of Science & Technology.

Alexander Bobik

Bibliography. M. M. Esther et al., Apolipoprotein C-III: Understanding an emerging cardiovascular risk factor, *Clin. Sci.*, 114:611–624, 2008; A. Kawakami and M. Yoshida, Apolipoprotein C-III links dyslipidemia with atherosclerosis, *J. Atherosclerosis Thromb.*, 16:6–11, 2009; A. Kawakami et al., Apolipoprotein CIII links hyperlipidemia with vascular endothelial cell dysfunction, *Circulation*, 118:731–742, 2008; T. I. Pollin et al., A null mutation in human ApoC3 confers a favorable plasma lipid profile and apparent cardioprotection, *Science*, 322:1702–1705, 2008; P. G. Scheffer et al., Increased plasma apolipoprotein C-III concentration independently predicts cardiovascular mortality: The Hoorn Study, *Clin. Chem.*, 54:1325–1330, 2008; N. S. Shachter, Apolipoprotein C-I and C-III as important modulators of lipoprotein metabolism, *Curr. Opin. Lipidol.*, 12:297–304, 2001.

Applications of Bayes' theorem for predicting environmental damage

Ecosystems are inherently complex, and despite efforts to identify and model causal chains linking ecosystem disturbances with ecosystem response, there are inevitable discrepancies between observed and predicted conditions in the natural environment. Uncertainty, variability, and change all contribute to these differences, yet they are often ignored in predicting environmental problems. Statistical modeling techniques represent a general classification of tools that can help address discrepancies between predictions and observations, and Bayesian statistics in particular has recently been demonstrated to be a novel and effective tool for forecasting environmental pollutant problems because of its unique approach to quantifying uncertainty and variability.

Bayesian statistics. In 1763, an essay by Reverend Thomas Bayes, "Essay Towards Solving a Problem in the Doctrine of Chances," was published in *Philosophical Transactions of the Royal Society of London*. More than 200 years later, the fundamental elements of this essay, including the introduction of a probabilistic relationship commonly referred to as Bayes' theorem (described in detail later in this article), form the foundation of Bayesian statistical analysis, a class of robust mathematical approaches to solving inverse probability problems.

Common strategies for statistical problem solving can be divided into three categories, each involving a different approach to quantifying the likelihood of an event relative to a set of all possible events. The first approach can be thought of as using a priori beliefs, which, in the case of a single roll of a six-sided die, might reflect an expectation that the die is fair, and therefore that the probability of each of the six possible outcomes (that is, 1, 2, . . . , 6) is exactly 1/6. A second approach is based on empirical evidence, in which our understanding of the underlying probability of events is based entirely on data. In the case of the six-sided die, this approach might involve rolling the die repeatedly and estimating the probability of each outcome as its observed relative frequency. In environmental problem solving, of course, this approach is often hindered by limited data and other complicating factors. Bayesian statistics, the third approach, provides a mechanism for combining a priori beliefs with potentially sparse empirical evidence to derive a posterior probability distribution. We describe this approach within the context of Bayes' theorem in the following section.

Bayes' theorem. Bayes' theorem can be written as Eq. (1), where $P(A)$ and $P(B)$ represent the marginal

$$P(A|B) = \frac{P(B|A)P(A)}{P(B)} \qquad (1)$$

probabilities of events A and B, respectively, while $P(A|B)$ and $P(B|A)$ represent the conditional probabilities of event A given that event B has occurred, and of event B given that event A has occurred, respectively. The probability $P(A|B)$, in a Bayesian framework, is referred to as the posterior probability of event A, given that event B has occurred. In this context, Bayes' theorem states that the posterior probability of event A (that is, the probability of event A given that event B has occurred) is equal to the likelihood [written $P(B|A)$] times the prior probability distribution of event A [that is, $P(A)$], divided by the marginal distribution of event B. In this way, the prior probability distribution, the likelihood, and the posterior probability distribution provide the framework for and serve as the necessary elements of a Bayesian statistical problem.

Applications of Bayes' theorem. In more practical terms, Bayes' theorem allows scientists to combine a priori beliefs about the probability of an event (or an environmental condition, or another metric) with empirical (that is, observation-based) evidence, resulting in a new and more robust posterior probability distribution.

Understanding pollutant removal infrastructure performance. **Figure 1** presents an example of how Bayes' theorem can be applied to solve environmental problems. In this hypothetical example, we are trying to improve our understanding of how effective stormwater management infrastructure systems are at removing sediment from stormwater runoff. While sediment often carries nutrients, metals, and other contaminants, sediment itself is also a pollutant in many environmental systems. In this problem, we represent the fraction of sediment removed by a stormwater management system as θ. Figure 1 presents the evolution of this understanding in a Bayesian framework, beginning with the development of a prior probability distribution. The prior probability distribution for θ is based on pollutant removal rate values in a published database documenting hundreds of studies, and is expressed in Fig. 1 first as a histogram of historic values (Fig. 1a), and then as a dashed line approximating the pollutant removal rate prior probability distribution (Fig. 1b). Hypothetical sediment removal rates from a new study site are then introduced through a

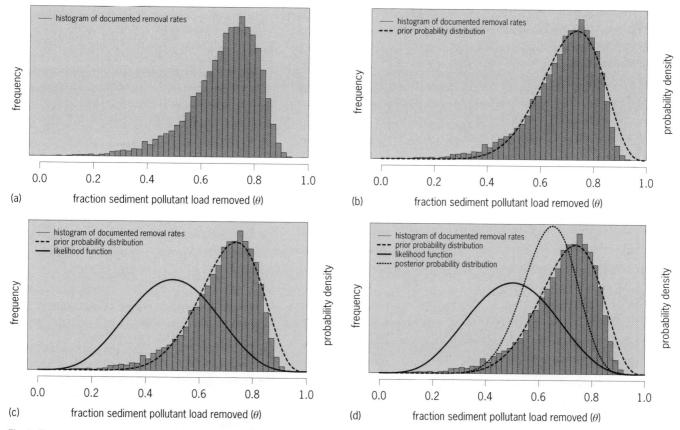

Fig. 1. Example of applying Bayes' theorem to a simple environmental pollution problem. (a) Histogram summarizing historical stormwater management system sediment pollutant load removal rate information, (b) approximation (dashed line) of the associated prior pollutant load removal rate probability distribution, (c) likelihood function (solid line) reflecting evidence from hypothetical site-specific monitoring data, and (d) posterior sediment removal rate probability distribution (dotted line). The left-hand y axis in each panel (labeled "frequency") applies to the corresponding histogram, and indicates the number of times each value of θ was observed in the historical information. Actual numbers were removed from the axis to simplify the figure.

likelihood function (solid line in Fig. 1*c*), and finally the posterior probability distribution is calculated using Bayes' theorem (and represented by a dotted line in Fig. 1*d*).

Mathematically, Fig. 1 approximates the underlying histogram as a beta Be($\theta | \alpha, \beta$) probability distribution with mean $\alpha/(\alpha + \beta)$ and variance $\alpha\beta/(\alpha + \beta)^2(\alpha + \beta + 1)$, with parameters α and β set to 11 and 4.6, respectively. The likelihood is derived by modeling the hypothetical sediment removal rates from a new study site using a binomial probability distribution Bi($x | n, \theta$) with mean $n\theta$ and variance $n\theta(1 - \theta)$, where x, in general, represents the number of positive outcomes out of n trials, and θ is the probability of a positive outcome in each trial. In this example, x represents the total mass of pollutant removed by the stormwater management infrastructure at a new study site, and n represents the total mass of pollutant entering the site. When expressed as a function of the unknown parameter θ, however, the likelihood [Eq. (2)] is a beta Be($\theta | x+1, n-x+1$) probability distribution with parameters n and x set to 8 and 4, respectively. Using Bayes' theorem, we combine the prior distribution and the likelihood to derive the posterior distribution for θ as follows in Eqs. (2) and (3), where Eq. (3) is a beta Be(α', β')

$$\overbrace{p(\theta | x)}^{Posterior} \propto \overbrace{\theta^{\alpha-1}(1-\theta)^{\beta-1}}^{Prior} \times \overbrace{\theta^x(1-\theta)^{n-x}}^{Likelihood} \quad (2)$$

$$\theta^{\alpha+x-1}(1-\theta)^{\beta+(n-x)-1} \quad (3)$$

probability distribution with $\alpha' = \alpha + x$ and $\beta' = \beta + (n - x)$. Note that the right-hand side of Eq. (2) does not include a denominator, which we might expect based on Bayes' theorem, Eq. ((1)), because it is simply a proportionality constant and does not affect our calculation of the posterior distribution. Put differently, once we recognize that Eq. (3) is a beta distribution, the values of α' and β' are the only information we need to formulate the posterior distribution for θ.

Predicting water quality conditions. Water quality is often measured by the concentration of one or more in situ pollutants (such as nutrients, bacteria, and organic compounds), and the suitability of a particular water body for its intended use (such as drinking water, recreation, or agricultural use) depends on whether or not the measured pollutant concentrations exceed water quality standard numeric limits. Because these pollutants often cannot be measured directly, scientists typically measure indicators that serve as potential surrogates for the pollutant of concern. The strength of the relationship between an indicator concentration and the concentration of the pollutant it supposedly represents varies widely depending on the type of pollutant. For example, in recreational and shellfish-harvesting waters throughout the United States, water quality is based on the concentration of nonpathogenic fecal indicator bacteria (FIB) such as fecal coliforms and *Escherichia coli*. These bacteria are used as a conservative indicator of fecal contamination and of the potential presence of harmful waterborne pathogens, which, while more directly linked to human and environmental health, are also much more difficult and costly to measure. Regardless of the specific pollutant and associated indicator, it is clear that not only the pollutant-indicator relationship, but also the spatial and temporal frequency of sampling and other factors might collectively contribute to uncertainty and variability in environmental condition forecasts. Here, we present a Bayesian approach to assessing water quality conditions using fecal coliform concentration measurements (reported in organisms per 100 ml) in a shellfish harvesting area as an example.

Like many other pollutants, FIB concentrations are commonly assumed to follow a lognormal LN (μ, σ) probability distribution with log-concentration mean (μ) and log-concentration standard deviation (σ). While this common probability model acknowledges natural spatial and temporal variability in FIB dispersion patterns, it (like other simple probability models) often fails to explicitly acknowledge other, more subtle sources of variability, including intrinsic sources arising from FIB concentration measurements and how FIB concentrations are calculated, all

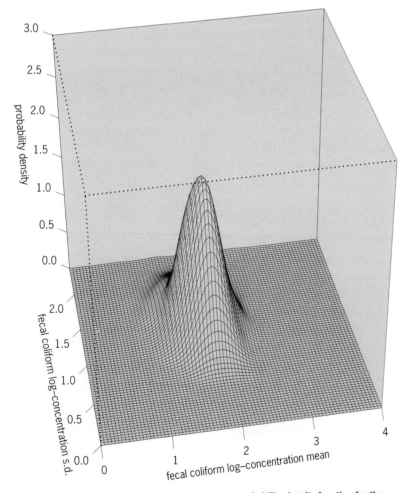

Fig. 2. Smoothed contour plot of the joint posterior probability density function for the fecal coliform log-concentration mean μ and log-concentration standard deviation σ at a shellfish-harvesting area water quality monitoring site in eastern North Carolina. Using Bayes' theorem to develop posterior probability density functions and presenting the results graphically (as shown here) provides regulatory and planning agencies with valuable information regarding the relationship between variability in experimental information and uncertainty in future water quality conditions.

of which can lead not only to uncertainty in FIB concentration predictions, but to uncertainty in probability distribution parameters (that is, μ and σ) as well. In a Bayesian framework, we can explicitly acknowledge these uncertainties by first placing a prior probability distribution on the population parameters μ and σ (which may account for a priori beliefs about their potential values), then developing a likelihood function for μ and σ based on empirical evidence (in this case, using water quality samples), and, finally, deriving a joint posterior probability distribution for both. Results of this procedure are presented in **Fig. 2**, which includes a smoothed contour plot of the joint posterior probability density for the fecal coliform log-concentration mean (μ) and standard deviation (σ) for a sample site in eastern North Carolina.

Guiding environmental management decisions. Perhaps equally important as reflecting uncertainty in water quality predictions is understanding how that uncertainty might propagate into water quality–based management decisions. In a management context, the predicted conditions presented in Fig. 2 might be used to guide beliefs about the likelihood that future samples might indicate both a violation of the appropriate standards and a potential threat to human and environmental health. For example, water quality standards for shellfish-harvesting waters indicate it is unsafe to harvest shellfish when either the fecal coliform concentration median, geometric mean, or 90th percentile of a minimum of 30 water quality samples exceeds 14, 14, and 43 (all in organisms per 100 ml), respectively. When water quality sample concentrations exceed these numeric limits, the corresponding shellfish-harvesting area is closed, and signs are often posted warning the public of potential health risks (**Fig. 3**).

To better understand the uncertainty in fecal coliform concentration predictions, these numeric limits are translated into corresponding maximum allowable combinations of the fecal coliform log-concentration mean (μ) and log-concentration standard deviation (σ). These maximum allowable μ, σ

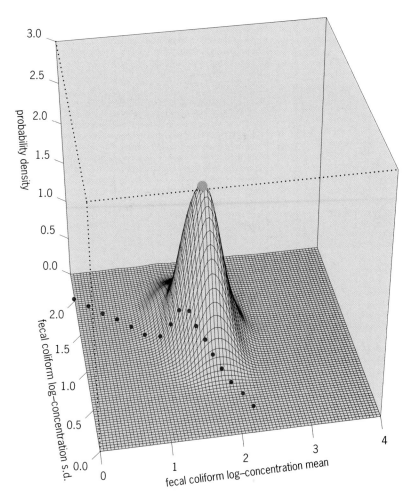

Fig. 4. Smoothed contour plot of the joint posterior probability density function for the fecal coliform log-concentration mean μ and log-concentration standard deviation σ at a water quality monitoring station in eastern North Carolina. The dotted line indicates maximum allowable combinations of μ and σ expected to achieve an acceptably low risk to human health. Combinations of μ and σ below and to the left of this line represent compliance, and combinations of μ and σ above and to the right of this line represent violations. The relative volume of the three-dimensional joint probability space "sliced off" by the dotted line represents the confidence of compliance with the proposed water quality standard. In this example, the confidence of compliance is about 0.03 (or 3%). As with Fig. 2, this figure demonstrates graphically the relationship between experimental information and uncertainty in future water quality conditions. However, this figure also demonstrates the implicit (and often ignored) relationship between uncertainty in future water quality conditions and water quality–based management decisions (as indicated by the dotted line).

Fig. 3. Posted signs, such as the one shown here, indicate that coastal waters might be unsafe for harvesting and eating shellfish. The decision to close a shellfish-harvesting area, however, can change depending on how uncertainty in water quality measurements is addressed. Bayesian statistics represents a novel approach for addressing this type of uncertainty in environmental systems.

pairs, when projected onto the three-dimensional joint (μ, σ) posterior probability space (dotted line in **Fig. 4**), provide an indication of how likely the water quality conditions are to yield a water quality sample in violation of the given standards. Put differently, we can imagine the dotted line in Fig. 4 "slicing off" a portion of the three-dimensional joint probability space to the bottom left of the figure, and the relative volume of this portion, sometimes called the confidence of compliance, can be thought of as the degree of confidence one can have that the water body will comply with water quality standards. In this example, the confidence of compliance is about 0.03 (or 3%).

To contrast the Bayesian-based confidence of compliance result with more common non-Bayesian strategies, a dot is plotted in Fig. 4, representing a potential point estimate of the most likely combination

Estimated confidence of compliance and deterministic assessment of future violations for monitoring stations in an Eastern North Carolina estuary. Assessment results from station 25 are presented graphically in Figs. 2 and 4. By propagating experimental information variability into uncertainty in future water quality conditions, the Bayesian assessment allows regulatory and planning agencies to understand the relative risks associated with restricting (or allowing) public access to a water-resource area (see, for example, Fig. 3). In contrast, the deterministic assessment indicates only whether the water quality standard is violated without any indication of how severe the violation is and, consequently, the magnitude of potential risks to human health.

Station	Bayesian assessment (confidence of compliance, %)	Deterministic assessment (will standard be violated?)
3	52	no
4	44	yes
7	<1	yes
8	14	yes
9	93	no
25	3	yes
28	96	no
29	<1	yes
35	80	no
41	<1	yes
84	13	yes

of μ and σ. A deterministic prediction of water quality conditions would probably be based solely on these point estimates, an approach that clearly ignores much of the potential variability in the future fecal coliform concentrations, and might lead to an oversimplified management assessment based not on a confidence of compliance, but on a simple statement of whether or not the water body violates the standard. In the case of the assessment results presented in Fig. 4, the deterministic approach would lead us to believe that future conditions will violate the given standard. A summary of monitoring assessment results for the station presented in Figs. 2 and 4, along with other neighboring water quality monitoring stations, is presented in the **table**. These results demonstrate how a Bayesian approach to predicting environmental conditions and to guiding management decisions provides a relatively robust approach to quantifying risk and protecting human and environmental health.

[Disclaimer: The U.S. Environmental Protection Agency through the Office of Research and Development funded and managed some of the research described here. The present article has been subjected to the agency's administrative review and has been approved for publication.]

For background information *see* BAYESIAN STATISTICS; ENVIRONMENTAL ENGINEERING; ENVIRONMENTAL MANAGEMENT; WATER POLLUTION in the McGraw-Hill Encyclopedia of Science & Technology.   Andrew Gronewold, Daniel Vallero

Bibliography. T. Bayes, An essay towards solving a problem in the doctrine of chances, *Phil. Trans. Roy. Soc. Lond.*, 53:370–418, 1763; D. A. Berry, *Statistics: A Bayesian Perspective*, Duxbury Press, Belmont, CA, 1996; Food and Drug Administration and Interstate Shellfish Sanitation Conference, National Shellfish Sanitation Program: Guide for the Control of Molluscan Shellfish, 2005; A. D. Gronewold and M. Borsuk, A software tool for translating deterministic model results into probabilistic assessments of water quality standard compliance, *Environ. Model. Software*, 24(10):1257–1262, 2009; A. D. Gronewold et al., An assessment of fecal indicator bacteria-based water quality standards, *Environ. Sci. Tech.*, 42(13):4676–4682, 2008; P. C. Milly et al., Stationarity is dead: Whither water management? *Science*, 319(5863):573–574, 2008.

Arctic engineering

Arctic engineering deals with the specialized technical problems of designing structures, facilities, and vehicles to work in the rigors of Arctic conditions. These include low temperatures, snow and ice, floating ice, permafrost, and seasonal frozen ground.

Attributes of the environment. The low temperatures on the ice-covered polar seas are moderated by the relatively warm ocean water, and the extremes of the Antarctic are not observed. On Arctic sea ice, a typical low is around $-51°C$ ($-60°F$); on land, especially in upper Siberia, temperatures may be as low as $-68°C$ ($-90°F$).

Permafrost is earth that has a temperature less than $0°C$ ($32°F$) and that persists over at least two full winters and the intervening summer. The term permafrost can also be used to describe the areal extent of the below-$0°C$ ($32°F$) condition. In the continuous permafrost zone, permafrost is found everywhere under the ground surface to considerable depth, except for layers of ground below large water bodies. In the discontinuous permafrost zone, permafrost is not as thick and exists in combination with areas with materials that are softened by thawing.

Seasonally frozen ground, known as the active layer, is the top layer of ground in which the temperature fluctuates above and below $0°C$ ($32°F$) during the year. The thickness of this layer is of order 1 m (3 ft) in the Arctic southern reaches, decreasing by a factor of 7 in the far north.

Snow and ice. In the Arctic, the record annual snowfall is 31 m (102 ft), in the mountains of Siberia. At sea level, snowfall is typically much less. As a result, ice is of greater importance.

Freshwater ice. In the Arctic, the presence of river and lake ice is of great economic importance in terms of transportation (ice roads, shipping) and damage to infrastructure and property (ice jams). Icing (rime ice) on roads and power lines, as well as atmospheric icing on aircraft, also pose transportation and infrastructural hazards. Everyday freshwater ice has a hexagonal crystal structure; rime ice contains a high proportion of trapped air. Lake ice formed under calm weather conditions forms a macrocrystalline structure, the optical c axis of each grain being predominantly vertical; should there be wind and/or snow falling, a columnar structure forms, with the c axes predominantly (randomly) horizontal. The more turbulent conditions in rivers leads to the formation of frazil ice, a collection of loose, randomly oriented, needle-shaped ice crystals in water. Frazil ice is also the first stage in the formation of sea ice.

First-year sea ice. First-year sea ice is ice that has not yet survived a summer. First-year sea ice is only partially solidified, incorporating a structured mixture of freshwater ice, brine, gas inclusions, and, often, particulate and biological matter. This thin, weak, and restless solid layer separates two much larger entities, the ocean and the atmosphere. The extent and thickness of sea ice is crucial to global climate models and climate predictions. In the polar oceans, an ice type called frazil or grease ice forms first. In quiet conditions the frazil crystals soon freeze together to form a continuous thin sheet of young ice. In its early stages, when it is still transparent, it is called nilas. Once nilas has formed, a quite different growth process occurs, in which water molecules freeze on to the bottom of the existing ice sheet, a process called congelation growth. This growth process yields first-year ice, which in a single season may reach a thickness of 1.5–2 m (4.9–6.6 ft).

Multiyear sea ice. Multiyear sea ice is ice that has survived one melt season or more. Warm temperature cycling has drained much of the brine and made the ice stronger. Often several meters thick, multiyear ice masses pose a major threat to offshore structures and shipping. Rafting and ridges formed when ice sheets collide can reach a thickness of 24 m (80 ft).

Engineering properties of sea ice. Sea ice is characterized by a top granular layer (formed by snow and frazil ice) underlain by a columnar layer. In the columnar zone, sea-ice grains each harbor a cellular substructure consisting of evenly spaced parallel ice platelets. The salt is stored between these platelets as liquid–solid inclusions. In cold sea ice, these inclusions are not connected. As the ice temperature rises, however, the pores become interconnected, strongly influencing thermal conductivity, permeability, ductility, and cracking behavior. The *c* axes are all aligned within a grain; under certain conditions the *c* axes of all of the grains can be aligned. The temperature, grain size, and salinity each vary through the depth of the floating sea-ice sheet.

The design of offshore structures and marine transportation and drilling systems, principally for petroleum exploration and production, requires knowing the design strength of ice under compression. The effective pressure versus contact area has been measured for a variety of ice–structure, ice–ship, ice floe–ice island interaction scenarios, and a significant dependence on contact area is evident. A pressure-area relationship of the form $p = kA^{-q}$ is much discussed in the different design codes and practices, [where p = ice pressure, A = loaded area, k and q are constants ($0 < q < 1$)].

Freshwater ice is far more brittle than sea ice. Warm sea ice is far more ductile than cold sea ice. The concept of brittleness is best phrased in terms of the magnitude of L/δ, where L is the relevant structural size and δ is a fracture-related length scale with $\delta = EG_F/\sigma_t^2$, in which E is the elastic modulus, G_F is the size-independent fracture energy, and σ_t is the local tensile strength of the ice (**Fig. 1**).

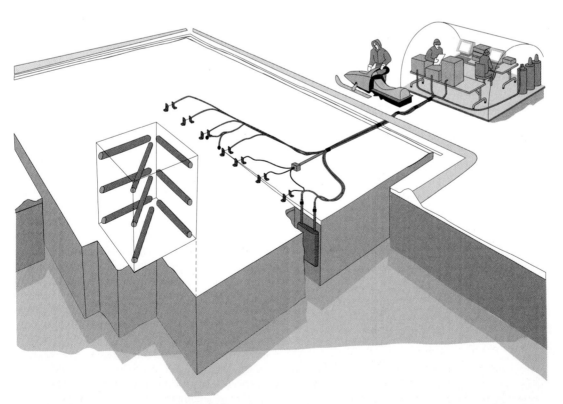

Fig. 1. Large-scale in situ fracture test of free-floating sea ice. Pillow-like flat jacks apply a crack-opening force. Hydraulic pressure source and data recorders are inside the shelter. Sensors measure displacements across and ahead of the crack. During post-testing back in the laboratory, the stress-versus-strain behavior is measured on cores oriented along three axes and from three levels in the sea-ice sheet.

Compressive strengths are highly scale-dependent. At structural scale, the ice strength is limited by crushing, buckling, radial cracking, spalling, and flaking. The geophysical strength is limited by the formation of leads, rafting, and ridging. There is a dependence on rate of loading (ship impact velocity or ice sheet velocity), temperature, and creep.

The above pressure–area curve, $p = kA^{-q}$, is a scaling law. All scaling laws have limitations. This law is relevant at a structural scale of up to 100 m (328 ft), for specific ranges of the ice sheet velocities, ice sheet thicknesses, and aspect ratios (structure width divided by ice thickness). The bearing capacity P of floating ice sheets has been predicted to scale as $P = ch^2$, whereas (with sufficient ice extent) the average ridging force F is predicted to scale as $F = bh^{3/2}$, where b and c are constants and h is the ice thickness.

Engineering properties of frozen ground. Frozen ground is composed of solid grains (mineral and organic), ice, unfrozen water, and gases. This composition changes continuously with varying temperature and applied stress. When ice fills most of the pore space, even though unfrozen water is still present, the mechanical behavior of frozen ground is similar to that of the ice. When frozen ground thaws, it is typically weaker than the soil that existed before freezing.

Structures. Buildings for Arctic use are often prefabricated at a warmer location and barged or trucked to their site. Load-out and placement may be by lifting, skidding, or movement on crawler transporters. Metal and fiber-composite construction is preferred for lightness and to withstand the rigors of placement. Steel exposed to the cold requires attention to its low-temperature properties. Interiors are usually insulated and heated. For buildings in which permanency and levelness is required, deep pile foundations may be used, extending below the zone of seasonal freeze–thaw cycles and insulated to prevent the downward transmission of heat into the permafrost. For temporary structures, such as drilling rigs, a thick gravel pad may suffice.

Bridges may be built on pilings where elevation is the main function. Bridges crossing bodies of water must deal with the forces imposed by ice. Large-diameter, heavy-wall steel or prestressed concrete piling may suffice in shallow water with seasonal ice, but deeper water or more severe ice conditions requires the use of specialized substructures as described below. Welded steel in bridges is subject to fatigue cracking as well as brittleness at low temperature, and notch (fracture) toughness is always a requirement.

Piers and jetties are similar to bridges or artificial islands. Piers subjected to tides may accumulate large volumes of layered ice from wetting–freezing cycles.

Artificial islands are built by barging or ice-road trucking of sand or gravel landfill to the site. Their perimeter must be armored to withstand attack by storm waves in summer and pack ice in winter, using large rocks, concrete-filled jumbo sandbags, or prefabricated caissons. Sufficient height must be provided to avoid encroachment by wave crests and ice rubble buildup. A fence to keep out polar bears may also be desirable.

Rafts and caissons are floated to their offshore site and sunk to rest on a prepared pad at the seafloor, or on a preinstalled foundation mat structure. In water deeper than the winter ice thickness, the seafloor soils are not frozen, although there may be relict permafrost at some depth below the mudline. Skirt walls may be punched into the seafloor to provide resistance to sliding from lateral ice leads. The perimeter walls may be vertical or sloped, and are reinforced to resist sea-ice contact forces as high as 1400 metric tons per square meter. Lower average pressures apply to larger areas.

Arctic cones are designed to "break the back" of multiyear ice ridges, by forcing them upward as they encroach on the structure (**Fig. 2**). Otherwise, their design and installation considerations are similar to that of other caissons.

Tower-type structures have a few large-diameter legs at the waterline, which resist first-year ice by crushing into the ice sheet (**Fig. 3**). Reinforcing structure is well below the waterline, to avoid ice contact, and the working deck superstructure is elevated far above the waterline. Concrete gravity structures rest on a mat at the seafloor. Steel tower structures are supported by multiple pilings driven through each leg. For both types of structures, oil wells are protected inside the legs.

Floating structures have been used to drill in summer and in seasonal ice. The circular drillship *Kulluk* operated successfully with a heavy omnidirectional mooring in the Canadian Beaufort Sea until oil prices collapsed in the 1990s. Its inverted-slope perimeter broke sheet ice by forcing it downward. If a large ice ridge approached, the strategy was to plug the oil well temporarily, disconnect the mooring, and drift with the ice pack. There were several stages of safety alerts for such an event—green, yellow, red, and black. Black meant the crew could not be evacuated before going adrift. Recently, a dynamically positioned drillship was used to take geotechnical cores near the North Pole, with icebreakers breaking up heavy pack ice if it approached the ship.

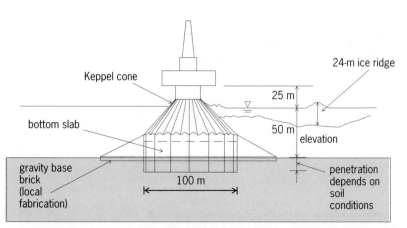

Fig. 2. Concept for an Arctic cone structure in the Chukchi Sea.

Submerged structures, on the sea floor and out of harm's way from ice, have been used in a large gas field in the Barents Sea, offshore Norway, within sub-sea pipelining distance to shore. Drilling was done from surface vessels during the summer.

Facilities. In winter, vehicles can travel directly on frozen ground, with grading and snow removal for wheeled vehicles. Thick gravel roads are commonly used for year-round access. The thickness helps to in-sulate frozen ground and spreads the load over soft spots. Periodic regrading keeps the roads level. Work pads are made thicker, as regrading is less conve-nient, and the work generates more heat.

Pipelines must often be insulated to keep the trans-ported product mobile. Water is an obvious problem, which may require heat tracing in addition to insu-lation. Crude oil can turn to wax or asphalt when chilled. Natural gas can build up hydrate plugs unless it is thoroughly desiccated. Onshore pipelines may be elevated on pilings to keep the warm product from melting the permafrost. The pilings are fitted with heat exchangers and filled with a heat-transfer medium to refrigerate the supporting permafrost in summer. The line may follow a zig-zag path to ac-commodate thermal expansion and contraction. It is anticipated that offshore pipelines will be laid in trenches to avoid damage from iceberg keels, and will be robust enough to withstand plastic strains from thermal loading and ground movement.

Antifreeze works only down to $-40°C$ ($-40°F$). Where it gets colder, internal-combustion engines must either be kept running, or drained upon shut-down. Special lubricants have been developed to maintain the proper viscosity at low temperatures. The wrong rubber in tires and seals can lose its flex-ibility. The wrong steel can lose its ductility, and be subject to brittle fracture. Either special materials or heated enclosures must be used.

In addition to the structural requirements for buildings, habitations require insulation, heating, fresh air, electrical power, lighting, fresh water, san-itary facilities, and wastewater treatment. Providing these in the harsh but fragile Arctic environment can present special challenges. Quartering for Arc-tic workers must be a self-sufficient community, and also needs amenities to help with problems arising from long deployments, isolation, darkness, and con-finement. Ample storage is needed for food and fuel to cover periods when resupply is not possible.

Vehicles. Dogsleds and snowmobiles are tradi-tional vehicles that can operate over rough terrain, on snow, ice, and short stretches of frozen ground.

In the Arctic winter, heavy trucks can operate on river ice and shore-fast sea ice for construction, rig moves, and seasonal resupply of locations far from the main base. Offshore, seawater is pumped up through the ice sheet and allowed to spread and freeze, thickening the natural sea ice over a width similar to an Interstate highway right-of-way. These roads can accommodate trucks heavier than the legal limit for highways, traveling at top speed. River ice is usually allowed to grow naturally, and measured to see when it is thick enough for traffic. The ice sheet

Fig. 3. Steel tower-type offshore drilling platform, Cook Inlet, Alaska, in first-year sea ice.

deflects into a broad bowl, picking up extra buoy-ancy to support the traffic load, similar to the behav-ior of a beam on an elastic foundation. Because river ice is stronger and often thinner than sea-ice roads, the moving bowl under the trucks presents an inter-esting problem in dynamics, with speed and spacing limits being imposed, for example, for trucks passing in opposite directions.

Ice-class ships. Several designated classes of ships are designed to operate in ice conditions of various severity, ranging from thin first-year ice to thicker broken ice with the assistance of icebreakers, to solo operation in multiyear ice. The ice-breaking tanker *Manhattan* was an example of the latter. Scantlings—side and bottom shell plating, and asso-ciated stiffeners—are heavier than for other ships of comparable size, with the design rules based on experience. Subdivision of the hull into many com-partments accommodates the occasional holing.

Icebreakers clear the way for lesser ships, operate to resupply remote bases, and occasionally rescue stranded whales. Thin ice is broken by crushing as the ship advances into the ice sheet. Thicker sheet ice is broken by beaching the ship on top of the ice and letting deadweight break the ice in bending. Power going astern is important for ships doing this, in case they have to back off and try again. Some ice ridges can be broken by repeated ramming, or the icebreaker finds another route. Structural steel used for the hull must withstand repeated plastic strains at low temperature. API Spec 2Y grade 50 is an adaptation of icebreaker steel for use in offshore structures.

For background information *see* ARCTIC AND SUBARCTIC ISLANDS; ARCTIC OCEAN; CAISSON FOUNDATION; ENGINEERING GEOLOGY; ICEBERG; ICE-BREAKER; PERMAFROST; PETROLEUM ENGINEERING; PETROLEUM GEOLOGY; PILE FOUNDATION; PIPELINE;

PRESTRESSED CONCRETE; SEA ICE; STEEL in the McGraw-Hill Encyclopedia of Science & Technology.
      Peter W. Marshall; John P. Dempsey; Andrew C. Palmer
      Bibliography. A. B. Cammaert and D. B. Muggeridge, *Ice Interaction with Offshore Structures*, Van Nostrand Reinhold, 1988; ISO 19906, *Petroleum and natural gas industries: Arctic offshore structures*, Draft International Standard; T. J. O. Sanderson, *Ice Mechanics: Risks to Offshore Structures,* Graham & Trotman, London, 1988; P. Wadhams, *Ice in the Ocean*, Gordon & Breach, 2000.

Atapuerca fossil hominins

Since the discovery of human fossils at the Sierra de Atapuerca in northern Spain in 1979, there has been an extensive series of finds at several sites within this complex of caves. Overall, these human fossils are important finds that have contributed to clarifying the patterns and processes of human evolution in Europe and, by extension, to understanding the evolution of the genus *Homo*. Many questions have been posed and partially solved by the study of the Atapuerca finds: When did the first humans arrive in Europe? What was their genetic origin? How many waves of immigrants reached this continent? What was the origin of the Neanderthals? Thus, the Atapuerca finds have proven to be of key importance for the study of human evolution.

Geographic context of Atapuerca. The Atapuerca hill is located near the town of Burgos, in northern Spain, and contains a complex cave system and underground galleries that are partially interconnected. During the past 1.2 million years, some of these galleries have occasionally become exposed or connected to the exterior, thereby allowing sediments and animal remains to accumulate inside the caves. Hence, the ages of the sediments and fossils vary a great deal, depending on when outside sediments began to infill the cavities. Two sets of sites can be distinguished within the Atapuerca hill (**Fig. 1**). The first is the so-called railway trench (denoted by the letter T), where the Sima del Elefante (TE), Gran Dolina (TD), and Galería (TG) are located, listed in chronological order. The second is located in a nearby network of karstic galleries named Cueva Mayor–Cueva del Silo. (Karst is a topography that forms over limestone, dolomite, or gypsum and is characterized by sinkholes, caves, and underground drainage.) The important site of Sima de los Huesos (SH) is located inside this system. Although other sites at Atapuerca (for example, Portalón de Cueva Mayor and Cueva del Mirador) have contributed significantly to the archeological record of Holocene age (younger than 10,000 years), only sites that have provided human remains of Pleistocene age (lasting from about 1.8 million to 10,000 years ago) will be treated here.

Homo antecessor. The oldest sediments containing Pleistocene human fossils come from the base of the Sima del Elefante site, which is a large cave comprising a rich Quaternary record. (The Quaternary is the second period of the Cenozoic era, encompassing the last 2 million years.) A mandible fragment and a lower premolar tooth of the same individual were found in 2007 at the TE9 level, associated with mode I flint stone tools (Oldowan technology) and evidence of animal butchery dating back to approximately 1.2–1.1 million years ago (Mya). The specimen preserves some primitive features within the genus *Homo*, whereas other features are clearly derived, such as a vertically oriented symphysis (the anterior region of the lower jaw) with a rudimentary chin. Also, the internal side of the symphysis is very smooth, unlike those of other more primitive African and Asian contemporaneous specimens. What, then, does this mean about the taxonomic status of these new finds? The hominins represented by the TE9 remains have been preliminarily classified as belonging to the species *Homo antecessor* and presently are considered to be the first hominins in Europe.

The species *H. antecessor* was first proposed by the Atapuerca Research Team (ART) in 1997 to accommodate a hominin with a unique combination of primitive and evolved features coming from the so-called Aurora stratum, a level within the TD6 stratigraphic unit at the Gran Dolina site, dated to 0.8 Mya. Today, a collection of more than 110 remains belonging to a minimum of 10 individuals are under investigation; these are essential for understanding the origin of modern humans and Neanderthals. The anatomy of *H. antecessor* has been interpreted as being congruent with the hypothetical last common ancestor of Neanderthals and modern humans (**Fig. 2**). Dental morphology fits well with the primitive pattern, whereas the mid-face (the maxilla and cheek bones) shows the earliest evidence of what has been previously considered as a modern human face: that is, a shortened face with a characteristic

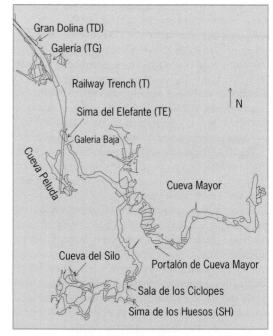

Fig. 1. Map of the major Atapuerca sites.

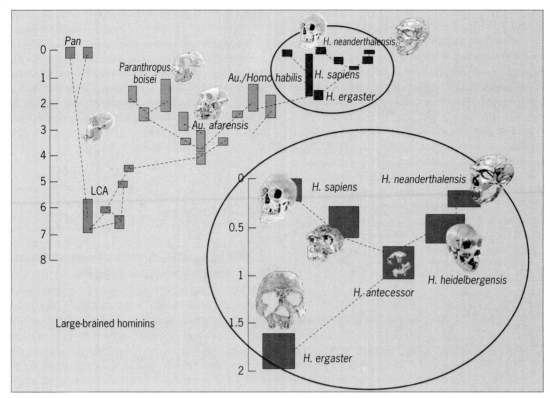

Fig. 2. Phylogenetic model for the hominins in which the various species identified at the Atapuerca sites can be located (background). An enlarged and more detailed picture of potential evolutionary relationships of the Atapuerca hominins (for example, *Homo antecessor* and *H. heidelbergensis*) can be seen encircled in the foreground. Both scale bars indicate time scales in millions of years. LCA = last common ancestor.

depression (the canine fossa) just below the cheek-bones. This combination of features had not been previously detected in any other hominin species, and the new name *H. antecessor* was proposed (in Latin, *antecessor* means pioneer, or the one arriving first).

There is an ongoing debate about whether the fossils from TD6 belong to a new species and, if so, which other fossils should be included in it. Attempts to incorporate the TD6 fossils into other previously named taxa (for example, *H. mauritanicus*) have been discussed recently, but the most reasonable conclusion is that the proposed new species, *H. antecessor*, should be retained.

Point of origin and relation to Neanderthals. The Atapuerca Early Pleistocene fossils raise several other questions. For example, where did the earliest populations colonizing Europe come from? Two models are under investigation. One model proposes that the species *H. antecessor* originated in Africa at some time during the Early Pleistocene and later expanded across Europe, and possibly also into Asia. A recent alternative, though, proposes that *H. antecessor* is a species of Eurasian origin, possibly having a pan-Eurasian distribution from China to Europe. In this latter interpretation, Asian hominin populations would have had a major impact on the early colonization of Europe. More fossil remains and analyses are needed before this question can be settled.

An African origin has been proposed for the populations represented in the European fossil record of the Middle Pleistocene age (0.78–0.125 Mya). These

fossils are very robust in appearance, with large brain sizes, and most display features, in variable combinations, typical of classic Neanderthals (those hominins living in Europe between 80,000 and 35,000 years ago). Middle Pleistocene European hominins are classified as a distinct species named *H. heidelbergensis*. Fossils of that age have also been found at the Galería and Sima de los Huesos sites. The Galería site has provided two fragmentary specimens (mandible and parietal bone), but a remarkable fossil sample is being recovered from Sima de los Huesos ("pit of the bones"). Beautifully preserved skulls and postcranial bones of at least 28 individuals are present, and all the skeletal parts are represented, including very small bones, such as those of the middle ear and pedal phalanges.

Study of the SH sample has clarified the complex pattern of variation in the European Middle Pleistocene populations and the origin of the Neanderthals. Fossil hominins from the SH site present clearly derived features previously considered exclusive to Neanderthals that are distributed throughout the skeleton, but significantly are concentrated in the facial skeleton. Thus, the morphologies of various skull features, such as the supraorbital torus, the zygomatic region, and the mandible, provide evidence for considering the European Middle Pleistocene populations as the putative direct ancestors of the Neanderthals.

There have been several attempts to date the SH sample, with the latest study giving an age of 530,000 years. If the dating is correct, this would mean that populations with clearly derived Neanderthal features existed in Europe along with other populations that have much less evidently derived Neanderthal features (for example, specimens from Aragó, France, and, more significantly, Mauer, Germany). If this were the case, there would have been a very large diversity of human populations in Europe.

Further evidence and hypotheses. Besides the appearance of Neanderthal features in Middle Pleistocene hominin populations, a major change in the archeological record is detected at around 0.6–0.5 Mya. Prior to this date, lithic (stone) industries appearing in the sites are of mode I or Oldowan technology (simple flaked cobbles or blocks of stone). However, after this date, mode II lithic tools or Acheulean technology (using bifacial knapping, that is, breaking off pieces to shape a tool so that both sides are flattened to form a V-shaped cutting edge), which first originated in Africa, start to appear in the European archeological record.

The cultural changes observed in the Paleolithic record of Europe can be interpreted as the result of cultural diffusion from Africa, but with genetic continuity of local populations. An alternative hypothesis holds that *H. antecessor* was genetically replaced (or absorbed) by a new wave of immigrants coming from Africa. These new colonizers were the bearers of the new mode II technology, and they were also the direct ancestors of the Neanderthals. Currently, both possibilities, continuity and replacement, can be defended. A major argument against local continuity is the lack of derived Neanderthal features in the TD6 hominins, although some dental features may support the hypothesis of genetic continuity. In addition, a common morphological background could be put forward as evidence for the possible genetic links between Early and Middle Pleistocene populations.

Thus, a major question is whether *H. antecessor* populations evolved locally in Europe to give rise to the ancestors of the Neanderthals, for example, *H. heidelbergensis*. The issue of continuity could also be applied to the link between TE9 hominins (1.2–1.1 Mya) and TD6 hominins (0.8 Mya). Did they belong to the same wave of colonizers to Europe? The possibility of different waves/species arriving in Europe is theoretically possible. Given the present evidence, the most parsimonious hypothesis is that both samples came from the same ancestral species.

These various questions are linked by the issue of genetic relationships between human populations throughout the Pleistocene of Europe. The null hypothesis to be tested in the coming years, which is directly derived from the Atapuerca finds, is whether or not there is continuity of the lineages represented by *H. antecessor*, *H. heidelbergensis*, and *H. neanderthalensis*.

For background information *see* ANTHROPOLOGY; ARCHEOLOGY; DATING METHODS; EARLY MODERN HUMANS; FOSSIL HUMANS; NEANDERTALS; PALEOLITHIC; PALEONTOLOGY; PHYSICAL ANTHROPOLOGY; PLEISTOCENE; PREHISTORIC TECHNOLOGY in the McGraw-Hill Encyclopedia of Science & Technology.
Antonio Rosas

Bibliography. E. Aguirre et al., The Atapuerca sites and the Ibeas hominids, *Hum. Evol.*, 5:55–57, 1990; J. L. Arsuaga, *The Neanderthal's Necklace: In Search of the First Thinkers*, John Wiley, Chichester, U.K., 2003; J. M. Bermúdez de Castro et al., The Atapuerca sites and their contribution to the knowledge of human evolution in Europe, *Evol. Anthropol.*, 13:25–41, 2004; E. Carbonell et al., Out of Africa: The dispersal of the earliest technical systems reconsidered, *J. Anthropol. Archaeol.*, 18:119–136, 1999; A. Rosas, Human evolution in the last million years: The Atapuerca evidence, *Acta Anthropolog. Sinica Suppl.*, 19:47–56, 2000.

Autonomous passenger vehicles

Today, most industrial robots operate while bolted to factory floors, restricted to operating within steel cages in order to protect people from injury. Conveyor belts bring raw materials within reach of each robot for assembly, welding, inspection, and so forth, and take the finished products away. Such settings are called structured environments. Extraordinary human effort is required to prepare the environment, and design the manufacturing processes within that environment, so that robots can function effectively and efficiently there.

A new stage has begun in the development of robotics, in which mobile robots are beginning to leave the factory to move and work alongside people in environments designed for humans rather than for robots. Modern robots have begun to walk, roll, fly, and swim anywhere people can go—even in places where it is too dangerous for people to go, such as inside active volcanoes or near explosives. As robots migrate into environments historically occupied by people, they must become capable of operating without the crutch of a tailor-made world.

One promising research area has been the development of autonomous, or self-driving, vehicles. The goal of autonomous vehicle research is the realization of full-size passenger vehicles that can drive with no human involvement beyond the specification of a destination. One can think of such vehicles as providing the same service as a taxi, but without the human taxi driver. The key technical challenges inherent in robotic driving include handling roadways that have not been encountered before, and reacting appropriately to unpredictable vehicles, pedestrians, and other aspects of the environment.

Why develop self-driving cars? Widely available self-driving cars would accrue a number of immediately apparent societal benefits, including increased safety, productivity, and energy efficiency.

Car accidents cause more than 40,000 fatalities annually in the United States alone and about 1.3 million worldwide. Self-driving cars would reduce the frequency and severity of accidents, thereby reducing the attendant human misery and societal costs of these events.

Commuters devote hours of intense attention to the task of surviving the drive from home to work and back each day. Self-driving cars would free former drivers to use their brains for other purposes, hugely increasing productivity.

Self-driving cars, even those acting independently of other vehicles and with no direct influence on the traffic infrastructure, can drive smoothly, saving gasoline and decreasing the amount of energy dissipated as useless heat by braking. Self-driving cars capable of cooperating with other vehicles, or with the signaling infrastructure, could save even more energy, for example, by eliminating stop-and-start driving and reducing waiting at traffic lights. Another potential benefit of self-driving cars would be the elimination of the asymmetry in the burdens borne by between carpool drivers and passengers.

Finally, for researchers, the development of self-driving cars promises to provide decades of intellectual excitement and engineering challenges.

What is required for autonomous driving? Seven principal technical challenges must be met in order to construct a safe, self-driving car capable of traveling along existing road networks and interoperating with existing human-driven traffic: drive-by-wire operation, representation, perception, planning, control, the human–robot interface, and the creation of a suitable platform.

Drive-by-wire operation. Any robot vehicle is based on a "drive-by-wire" mobility platform, in which the car's physically controllable "degrees of freedom"—including its transmission, gas, brake, and steering—must be placed under the control of a computer, typically through the use of electric motors and suitable hardware, firmware, and software to control them.

Representation. A robot car must have a useful internal representation of roadways (how they are delineated, by curbs, grass, painted markings, and so forth; and how they start, stop, split, and merge) and of traffic (how to safely start, follow, merge, turn into or across traffic, and stop). The robot must also have valid predictive models of how other entities in the world—trucks, cars, motorcycles, and pedestrians—are likely to behave at any moment. This behavior is of course highly dependent on the entity's location and surroundings: travel lane, exit or entrance lane, sidewalk, crosswalk, and so forth.

Perception. Of course, no vehicle can make useful progress in the world without an accurate, timely model of its surroundings. A robot car must continuously answer, and update its answers to, questions such as: Where is the drivable road surface? Are there hazards (such as potholes or road debris), obstacles (such as traffic islands, curbs, bollards, or other vehicles), or vulnerable entities (such as motorcyclists, bicyclists, or pedestrians) in or near the space likely to be occupied soon by the car? If so, how are these entities likely to move or change as the car approaches?

In practice, answering such questions involves selecting suitable sensors, positioning them in or on the vehicle so as to provide sufficient observations of the vehicle's surroundings, and integrating software

Fig. 1. Cargo compartment of Team MIT's vehicle at the 2007 DARPA Urban Challenge, which carried more than 40 central processing units (CPUs) in a built-in air-cooled mobile machine room, illustrating the enormous computational resources (by current standards) required for autonomous driving. About half of these CPUs were dedicated to machine vision processing used for finding painted lane markings on the pavement. (*Jason Dorfman, CSAIL/MIT*)

to process the raw sensor data gathered as the vehicle moves into a useful interpretation of what entities are close to the vehicle or likely to come close in the near future. Sensor integration requires spatiotemporal calibration of the sensors, that is, knowledge of how data from different sensors is related across space and time. Detection and classification of objects near the vehicle is highly challenging, in part because the sensors and interpretation algorithms available today are far less sophisticated than the human visual and cognitive systems.

Planning. Once the car has an underlying representation of the world, and can perceive its surroundings, it must be able to make progress toward its "goal," that is, the destination specified by its user. (The user may or may not be a passenger; for example, a car could be summoned to its user from a parking area, or dispatched by its user to deliver a package.) In robotics terminology, the car must "plan" actions that advance it toward its goal, if possible.

Fig. 2. Team MIT's vehicle, "Talos," driving unoccupied at the 2007 DARPA Urban Challenge. (*Jason Dorfman, CSAIL/MIT*)

Fig. 3. Intersection on the 2007 DARPA Urban Challenge course, with human observers behind a concrete barrier. Because of the unpredictability of the current generation of robotic vehicles, DARPA removed all pedestrians from the Urban Challenge course and arranged for observers to remain behind barriers for safety. (*Jason Dorfman, CSAIL/MIT*)

Such planning involves high-level tasks, such as localizing and orienting within the world (for example, choosing to head north); medium-level tasks, such as choosing a particular north-bound road or lane; and low-level tasks, such as deciding to divert briefly around a pothole, parked car, or other vehicle. The planning task has a static aspect, in which prior information about road topology is used, and a dynamic aspect, in which live sensor information (for example, about debris on the road) or infrastructural information (for example, about traffic conditions ahead) is used. Finally, in case of unexpected occurrences such as road blockages, the vehicle must be able to replan, that is, abandon its current low-level, medium-level, or high-level plan and generate a new one. And, of course, all robots must be able to gracefully handle the case in which no feasible plan exists, that is, in which no progress may be made toward

Fig. 4. One of the world's first accidents between robotic vehicles, involving vehicles from Cornell and MIT. Each vehicle exhibited nonhuman driving behavior that was later found to have contributed to the accident. (*Jason Dorfman, CSAIL/MIT*)

the goal. In the case of a robot vehicle, the proper behavior in this circumstance might be to wait patiently, or request for help.

Control. At any moment, an autonomous vehicle must maintain a low-level plan consisting of a spatiotemporal trajectory through the world that will advance it toward its medium-level goal. For instance, the low-level plan might include a curving path through an intersection, chosen so as to exit the intersection roughly centered and aligned with the outgoing lane. Such plans also include desired bounds on speed, in addition to those for position and orientation.

The "control" problem in robotics consists of exercising the vehicle's drive-by-wire degrees of freedom so as to achieve a desired trajectory. To achieve robust control, the vehicle's "dynamics"—its response to longitudinal and lateral forces, and to gravity and road conditions—must be characterized, typically through a calibration procedure undertaken before mission-critical operation is attempted. Analogous to replanning, the vehicle's control objectives may have to be reconsidered on very short time scales in the event of unexpected occurrences such as encountering wet pavement or having a tire blow out.

Interface. The sixth essential element of a useful robotic vehicle capability is the robot's interface, that is, the means by which its human user tells the robot what to do (for example, through text or speech).

Platform. Designers of autonomous driving systems must also keep in mind the critical engineering constraint that any proposed implementation of the elements listed above will require significant resources, including computation, memory, storage, and networking, as well as internal and external sensors. Any plausible solution must provide sufficient resources for correct operation, while observing operational limits on power, volume, weight, temperature, and so forth (**Fig. 1**).

DARPA Urban Challenge. The state of the art in robotic vehicle development is well represented by the group of vehicles competing in the final round of the Defense Advanced Research Projects Agency (DARPA) Urban Challenge in November 2007 (**Fig. 2**). DARPA provided a challenge course somewhat more structured than a car would face in the fully general driving task, by clearing pedestrians from the course (**Fig. 3**), and employing only intersections involving stop signs. Moreover, DARPA supplied detailed prior information about the course to each competing vehicle, in the form of a USB stick with a plaintext Road Network Description File (RNDF) containing highly accurate Global Positioning System (GPS) coordinates for "waypoints" along each roadway, the topological and geometric properties of each intersection, and the location of all stop signs. There were no traffic signals, and all informational signs such as speed limit signs were encoded in the same plaintext file, so that the robots did not need to perceive signage in the environment.

All this information was provided to each robot 48 hours before the start of the competition, and each team was allowed to manually edit and annotate the roadmap in order to provide denser waypoint information to its vehicle. Finally, after one vehicle was immobilized by its failure to achieve high-quality GPS reception, DARPA took steps to remove sources of GPS interference from the course.

In short, DARPA took many steps to make things easier for the robots, but even under these conditions the competition was dauntingly difficult. DARPA arranged for dozens of human-driven cars to move through the course. In addition, the fact that all robots would be operating simultaneously, yet independently, introduced significant randomness into the competition: A robot might predict the behavior of a "polite" human driver fairly accurately, but it proved impossible for even the human observers to predict the behavior of the robots (**Fig. 4**).

Throughout the qualifying rounds and final competition, the human–robot interface was limited to two channels: a Mission Description File (MDF) and a "remote E-stop" switch. Each MDF consisted of a list of waypoints from the RNDF, to be visited in order.

The remote E-stop switch, designed to bring a robot to a rapid halt if necessary, was controlled by the human driver of a chase car assigned to each competing robot, and by other DARPA personnel observing the course. Of 89 original entering teams, 35 were judged by DARPA as sufficiently capable to participate in the National Qualifying Event (NQE), in October 2007. Those 35 vehicles were subjected to a number of individual performance tests by DARPA, which judged 11 vehicles capable enough to participate in the final competition. Of those 11 vehicles, 5 were removed during competition, either because they drove in an unsafe manner, or simply got stuck. Of the original 89 entrants, six vehicles managed to cross the finish line.

Prospects. Over the coming decades, every aspect of robotic vehicles will improve, enabling self-driving vehicles to match, and eventually exceed, the capability of human drivers. Sensors will require less power, observe wider fields of view and longer ranges, and provide faster refresh rates. Improved radars and signal processing algorithms will enable future vehicles to "see" through fog, rain, snow, and dust. Algorithms for interpreting sensor data will become more capable, achieving (for example) more accurate classification of roads, buildings, static obstacles, vehicles of multiple types, and pedestrians. Computational resources and network bandwidth will continue to increase, and storage systems will become more capacious. Predictive models of vehicles and pedestrians will improve. Vehicles will communicate with one another to detect and avoid imminent collisions, and with the roadway infrastructure to improve traffic flow and decrease energy usage. *See* INTERVEHICLE COMMUNICATIONS.

Human–robot interfaces will improve, getting nearer to the "holy grail" of entirely natural interaction such as that between humans. Robots will understand spoken commands, and will ask questions to clarify user intent when needed. Robot vehicles will also interact with humans other than their user. For example, a robot vehicle may indicate to pedestri-

ans that it is aware of their presence, use directional sound to warn pedestrians or other vehicles of imminent dangers, or even use visible light to "paint" areas of the road surface that will be, or may be, occupied by the vehicle in the near future.

For background information *See* COMPUTER VISION; CONTROL SYSTEMS; DRONE; GUIDANCE SYSTEMS; ROBOTICS; UNDERWATER VEHICLES in the McGraw-Hill Encyclopedia of Science & Technology.

Seth Teller

Bibliography. J. Bares and D. Wettergreen, Dante II: Technical description, results, and lessons learned, *Int. J. Robot. Res.*, 18(7):621–649, 1999; L. Fletcher et al., The MIT-Cornell collision and why it happened, Special issue on the 2007 DARPA Urban Challenge, part 3, *J. Field Robot.*, 25(10):775–807, 2008; J. Leonard et al., A perception-driven autonomous urban vehicle, Special issue on the 2007 DARPA Urban Challenge, part 3, *J. Field Robot.*, 25(10):727–774, 2008.

Avalanches and phase transitions

Many systems change from one phase to another as a function of some external parameter. For example, in boiling, liquid changes to gas as a function of temperature or pressure. The transition can occur continuously or abruptly from a starting phase, through an intermediate or a mixed phase (both phases coexisting), until reaching the final phase. In the boiling example, the transition as a function of temperature will be abrupt, with more and more liquid turning to gas at the transition temperature until there is only gas and the transition ends. In recent decades, a growing number of examples have been found of systems in which the transition behaves differently. In such systems, while the relevant parameter is changing, there are many occurrences in which a discrete "amount" of one phase changes suddenly to the other in what is called an avalanche event. One can find characteristics that are universal across different systems that have a phase transition through avalanches and do not depend on the detailed mechanism of the phase transition. The avalanches span a broad range of sizes and show a power-law distribution of avalanche magnitude; that is, there are many small avalanches, fewer medium-size ones, and only a few big avalanches. This behavior appears also in complex systems that do not possess a phase transition at all, pointing toward an even more general behavior of complex systems. Transitions through avalanches are also referred to as crackling phenomena.

Avalanches. The term avalanche is borrowed from the phenomena of snow or land avalanches. An avalanche can occur in a system that is not in equilibrium. A small perturbation to the system can have an effect that does not scale with the perturbation. For example, a skier (the perturbation) could trigger a massive amount of snow to slide down a mountain. The state before the avalanche is called a metastable state. Systems that are out of equilibrium can advance

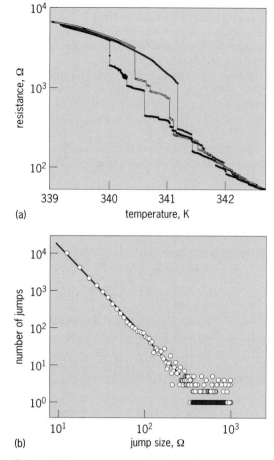

(a)

(b)

Fig. 1. Resistance avalanches in the metal-insulator transition of VO_2. (*a*) Three resistance-versus-temperature scans. Avalanches are seen as jumps in resistance. (*b*) Histogram of avalanches, showing power-law behavior.

toward equilibrium through a series of avalanches connecting metastable states.

Power laws, critical phenomena, and universality. Many of the systems that have a phase transition through avalanches show emergent time and length scales. The avalanches follow a power-law distribution. This means that the probability, p, of having an avalanche of magnitude A is $p(A) \sim A^{-\alpha}$, where α is the characterizing exponent, called the critical exponent. Additional power laws can be found, such as for the length of time an avalanche spans or the frequency of avalanches.

Similar behavior is found near the critical points of systems that go through a phase transition. Many system properties, such as heat capacity and correlation lengths, have power-law behavior with coefficients known as critical exponents. Thus, near a critical point, one would expect that, in systems where the phase transition is through avalanches, the distribution of avalanche size would follow a power law. For this reason, phase transitions with avalanches are said to possess critical behavior, even though it is not simple to identify the critical point. Two important questions scientists are trying to answer are:

1. Do these systems really have critical behavior, and if so, how do they naturally evolve into a critical state without the need for fine-tuning parameters (as is generally required in order to reach critical points in thermodynamics)?

2. Can we identify different classes of systems that follow similar behavior? These are known as universality classes. They may depend on the dimensionality of the system or the types of dynamics and interaction that a system possesses (such as long-range versus short-range interaction).

Examples of phase transitions with avalanches. A growing number of systems have been found to exhibit phase transitions through avalanches, and there are as many different ways in which the avalanches are detected, making it difficult to resolve universal classes. Four such examples will be given.

Resistance avalanches in the VO$_2$ phase transition. The resistance of vanadium oxide (VO_2) increases by a factor of 10,000 when cooled below 68°C (154°F or 341 K). This behavior is called a metal-insulator transition. Avalanches are seen as sharp jumps in resistance when the temperature is decreasing (or increasing) across the transition (**Fig. 1***a*). The magnitude of the avalanche is proportional to the amount of material switching from the metallic phase to the insulating one. It is evident from the figure that there are more small avalanches than large. The histogram of resistance-jump sizes plotted on a log-log scale shows the power-law dependence of the jump-size distribution (Fig. 1*b*). The critical exponent is $\alpha \sim 2.5 \pm 0.2$.

Barkhausen effect. In measurements of the magnetization of some ferromagnets (usually, but not exclusively, granular materials) as a function of an external magnetic field, the magnetization is observed to change in avalanches. These changes are detected by measuring the voltage on a wire loop close to the magnetic sample. The voltage measured during an avalanche is related to the change in magnetization. Both the magnitude and the duration of the avalanches are found to follow a power law. In different experiments, critical exponents in the range of 1.3–1.5 were found for the magnitude.

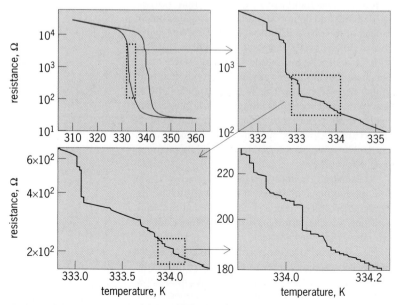

Fig. 2. Self-similarity for the VO$_2$ system. Each consecutive zoom-in has similar features.

Martensitic transitions and acoustic emission. The lattice structure of a martensite shape memory alloy changes when it is heated above a critical temperature. During the spatial transition, parts of the alloy switch structure in an avalanche, causing a sound wave to propagate. By measuring the acoustic emission, one can detect that the transition indeed takes place through avalanches. The amplitude of the acoustic spectrum shows critical behavior with $2.2 < \alpha < 3$.

Capillary condensation in porous media. Superfluid helium-4 (^{4}He) will condense inside a porous material above some critical pressure. When one pumps on the ^{4}He, it starts to drain from the pores in avalanches. These were detected by measuring the capacitance of the porous material, which decreases when the amount of condensed ^{4}He in the pores changes as a result of a change in the effective dielectric constant. These experiments find $1.5 < \alpha < 2$.

Self-similarity. Systems with power-law distributions usually possess self-similarity. A self-similar object will look similar to parts of itself on various scales. This is the case in a fractal structure. The phenomenon is also referred to as scale invariance. Self-similarity is evident in the measurements of the metal-insulator transition in VO$_2$ (**Fig. 2**). For all resistance scales, the general features look similar. There are a few large jumps and more small jumps in all the graphs.

Self-organized criticality. Power-law distributions are found in many different fields in and out of physics, which at a first glance do not have much in common with each other. One notable example is the Gutenberg-Richter law, where the distribution of earthquake intensity follows a power law. Additional systems exhibiting power laws include snow and sand avalanches, the intensity and frequency of solar corona bursts, the noise emitted by crumpling paper, the way milk propagates into cereal, the connectivity of nodes in the Internet, and even fluctuations in the stock market. All these systems are considered to show critical behavior.

Such universal behavior has encouraged the search for explanations as to why complex systems with very different types of interactions and microscopic details behave in a similar way at the macroscopic level. It does not seem probable that in all the different cases there is an experimenter who fine-tunes the parameters to bring the system to the critical state. The theory of self-organized criticality (which became popular following a paper written by Per Bak, Chao Tang, and Kurt Wiesenfeld in 1987) claims that many systems evolve naturally toward a critical point; the system "self-organizes" into a critical state. There is still debate in the scientific community as to whether this theory truly applies to the different systems in which power laws are observed. Still, the idea of self-organized criticality has increased interest in this field. Theoretical methods such as the renormalization group have been able to classify universality classes. For example, ferromagnets and breakfast cereal are in the same class.

Theoretical models and simulations. Theoretical models should have a set of rules that will capture the important features of systems belonging to a universality class, since all such systems are expected to have similar behavior that will not depend on the microscopic details.

Sand piles. One family of models that show avalanches and self-organized criticality is the sand piles, which belong to the cellular automaton models. We consider a square grid with four nearest neighbors per site. Sand is added randomly to the different sites. If a site has more than three grains, one grain will spill to each of the neighboring sites, or out of the system. This in turn can cause other sites to "spill," causing an avalanche. The system naturally self-organizes into the critical state, in which adding one grain can cause a small avalanche, cause a large avalanche, or do nothing at all (**Fig. 3a**). The resulting power-law distribution has a slope of -1 (**Fig. 3b**). Small modifications of this model are used to simulate earthquakes. One drawback is that this model does not capture behavior in

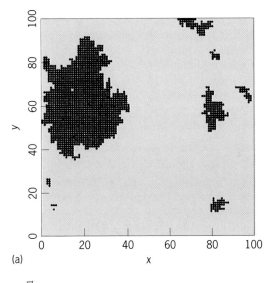

(a)

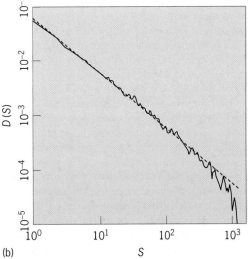

(b)

Fig. 3. Simulations of sand pile avalanches in two dimensions. (*a*) Results, at some particular time, of avalanche events with different magnitudes. (*b*) Histogram of avalanche size with power-law behavior. *S* is the size of the avalanche in pixels, and *D*(*S*) is its probability of occurring. (*From P. Bak, C. Tang, and K. Wiesenfeld, Self-organized criticality, Phys. Rev. A, 38:364–373, 1988*)

finite-size systems, like those presented in the earlier examples.

Random-field and random-bond Ising models. The Ising model is used to describe the ferromagnetic phase transition. The random-field and random-bond Ising models add frozen disorder as fluctuations in the local magnetic fields or in the interaction strengths between the different sites. The transition occurs as a function of an external magnetic field, where switching of one site can cause an avalanche of switching sites. While this model gives a good account of the transition between two different phases, the amount of disorder must be tuned in order to have power-law distribution, and the system does not self-organize.

For background information *see* ACOUSTIC EMISSION; AVALANCHE; BARKHAUSEN EFFECT; BOILING; CELLULAR AUTOMATA; COMPLEXITY THEORY; CONDENSATION; CRITICAL PHENOMENA; EARTHQUAKE; FRACTALS; ISING MODEL; METASTABLE STATE; PHASE TRANSITIONS; RENORMALIZATION; SHAPE MEMORY ALLOYS in the McGraw-Hill Encyclopedia of Science & Technology.   Amos Sharoni

Bibliography. P. Bak, *How Nature Works: The Science of Self-Organized Criticality*, Springer-Verlag Telos, New York, 1999; G. Ódor, Universality classes in nonequilibrium lattice systems, *Rev. Mod. Phys.*, 76(3):663–724, 2004; J. P. Sethna, K. A. Dahmen, and C. R. Myers, Crackling noise, *Nature*, 410:242–250, 2001.

Bat guano: record of climate change

Bats are a ubiquitous group of flying mammals found on every continent except Antarctica, with highest abundance and diversity in the tropics and subtropics. Some species are very gregarious and may roost together in caves in substantial numbers. For example, the Mexican free-tailed bat may roost in maternity colonies reaching in excess of 20 million individuals in the semiarid parts of northern Mexico and the southwestern United States. These high population densities can result in bleached fur on the bats as a result of the high concentration of ammonia given off by microbial processing of bat urine and excrement (guano). Both big and smaller populations of bats produce considerable quantities of guano, which, over thousands of years, can lead to deposits many meters thick on the cave floor. Sizable guano deposits have been mined for fertilizer, with the remaining deposits now serving as valuable archives of past environmental change.

Any sediment that accumulates over time has the potential to unlock secrets of the past, and bat guano is no exception. Guano contains several environmental proxy indicators and can also be reliably dated—two key requirements for deciphering a record of past climate change. Bat guano, like all organic matter, is composed mostly of the elements carbon, nitrogen, hydrogen, and oxygen. The relative abundance (ratio) of each element's stable isotopes in a sample of guano is related ultimately to the local climate at the time the guano was produced. The stable isotopic composition of bat guano can be precisely measured by isotope-ratio mass spectrometry, and from these data the environment at the time of guano production can be deduced. In stable-isotope ecology, the general rule is that "you are what you eat." For insectivorous bats, this means the isotope ratios of the guano are approximately the same as the average stable isotope composition of the insects the bats consumed. Most gregarious caverniculous bats are nonselective in their feeding behavior, sampling insects around their roost. The insects are then processed in the gut of the bats, and the exoskeletons, along with metabolic wastes, are dropped onto the cave floor together with the fecal material of thousands of other bats. By sampling material down through a bat guano deposit, we can measure the changing isotope ratios of carbon, nitrogen, hydrogen, and oxygen of insects in the past. But what do these isotopic ratios mean?

Stable isotopes of guano. Stable carbon isotope ratios of plants ($\delta^{13}C$) are related primarily to the photosynthetic pathway used to convert carbon dioxide to organic carbon. There are three major pathways: C_3 (Calvin), C_4 (Hatch-Slack), and CAM (Crassulacean acid metabolism). Because each pathway uses a different set of chemical reactions to transport carbon, large differences in carbon isotope ratios occur between C_3 plants and C_4 plants. Because C_3 plants (trees and cool-climate grasses) dominate in forests and temperate regions and C_4 plants (tropical grasses) dominate in hot arid and semiarid environments, the carbon isotope composition of plant material can be related to the environment in which the plant grew. In more arid climates, where CAM plants such as cacti and agaves may be locally abundant, stable carbon isotope ratios are similar to C_4 plant ratios. Thus, the stable carbon isotopic composition of insects feeding on the plants, and ultimately the guano of bats feeding on the insects, represents an integrated measure of the carbon isotope composition of the vegetation. The carbon isotope composition of bat guano is lower in C_3-dominated ecosystems and higher in C_4- and CAM-dominated ecosystems, and the distribution of these plant types is strongly correlated with climate.

Interpreting variations in nitrogen isotope ratios ($\delta^{15}N$) in bat guano is more problematic. In the southern United States, there was no relationship between fresh guano $\delta^{15}N$ values and obvious climate indices, suggesting that local soil $\delta^{15}N$ values may be most important in determining guano $\delta^{15}N$ value. As guano degrades, ammonia is given off and the residual nitrogen is locked in unique guano minerals that have relatively high nitrogen isotope values compared with the original guano, obscuring the original isotope signal.

Stable isotope values of hydrogen (δD) and oxygen ($\delta^{18}O$) in bat guano are ultimately related to local precipitation, and the processes controlling isotope ratios of water are well known. Seasonal changes and latitudinal variation in insolation patterns drive precipitation and evaporation. As energy and moisture is distributed from the tropics to the poles, changes in the isotopic composition of both hydrogen and oxygen occur. Cooler temperatures at higher latitudes and higher altitudes cause rain to condense with higher $\delta^{18}O$ and δD values compared with the isotopic composition of the remaining water vapor. As water vapor moves to higher latitudes and higher altitudes, $\delta^{18}O$ and δD values decrease. This means that as one moves to higher latitudes or higher altitudes with lower average temperatures, $\delta^{18}O$ and δD values of precipitation decrease. This also means that winter precipitation has a more negative isotope composition than summer precipitation, and that tropical storms and cyclones are also depleted in the heavy isotopes of water. (The more it rains in the tropics, the lower are the δD and $\delta^{18}O$ values.) These patterns are further modified by variations in moisture source, humidity, seasonality of precipitation, and storm-track patterns.

Paleorecords. We have so far discussed the relationship between the isotope composition of contemporary guano and climate. However, it is possible that the isotope ratios of guano can change after deposition as a result of decomposition by fungi and bacteria, and mineral material in the deposits can also contain hydrogen and oxygen. These problems can be overcome by extracting the originally deposited, intact insect cuticles (exoskeletons) from the bulk guano sediment (**Fig. 1**). Comparisons of carbon isotope values of these extracted remains with bulk guano sediment values indicate that similar profiles are obtained, suggesting little postmodification of carbon. Because of contamination and diagenesis, the nitrogen, oxygen, and hydrogen isotope composition of the bulk guano is very different from the composition of the insect cuticles. This suggests that although carbon isotopes in bulk guano can be used reliably to infer past climate, it is preferable to extract and analyze the original insect cuticles for other elements. Moreover, insect cuticles provide a more robust material for accurate radiocarbon dating.

In semiarid regions, time transgressive records are limited to high elevations, where lacustrine sediments are more reliably preserved over time. Guano deposits provide an attractive alternative for the development of palaeoenvironmental records at lower elevation sites. For example, in the Grand Canyon, variations in carbon and hydrogen isotope profiles of a bat guano profile indicate a change in monsoonal strength that closely follows climate in the North Atlantic (**Fig. 2**). C_3 vegetation dominated during the globally cooler climates of the last glacial period, the Younger Dryas, and the 8.2-kyr (thousand-year) event, indicating more winter precipitation and/or cooler temperatures (Fig. 2). There was a slow increase after the Younger Dryas toward a stronger summer monsoon and a generally more arid and warmer climate.

Good depositional records are also rare in the tropics and are often biased toward perpetually wet swamps or areas close to river channels. In the now-wet tropical environment of Peninsular Malaysia, a bat guano deposit $\delta^{13}C$ profile provides evidence of

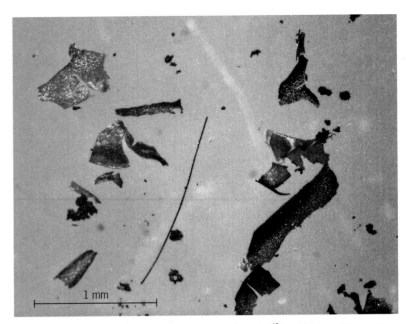

Fig. 1. Insect cuticles recovered from bat guano sediment. $\delta^{13}C$ values from this sample indicate that there was an open savannah environment in Peninsular Malaysia about 25,000 years ago.

a much drier past, with open savannah present during the last glacial period (**Fig. 3**). Similarly, carbon isotopes in guano have been used to infer that parts of the Philippines that are now covered by tropical forest were covered by open savanna vegetation during the last ice age, implying lower and more seasonally distributed rainfall in the past. Such findings are unique and help to explain the trajectory of past environmental change in this globally significant "biodiversity hotspot" region. Moreover, understanding environments of insular Southeast Asia have significant implications for understanding early human disper-

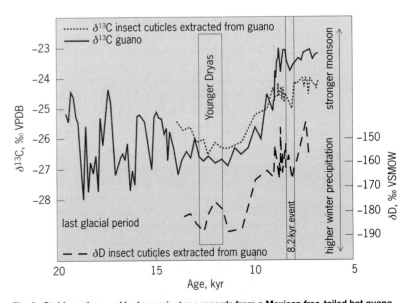

Fig. 2. Stable carbon and hydrogen isotope records from a Mexican free-tailed bat guano deposit from the Grand Canyon. Both records document a drastic change from a winter precipitation–driven climate to one characterized by the North American monsoon. Isotope standards: Vienna Pee Dee Belemnite (VPDB) for reporting carbon stable isotope ratios and Vienna Standard Mean Ocean Water (VSMOW) for reporting stable hydrogen isotope ratios, both in ‰ (parts per thousand).

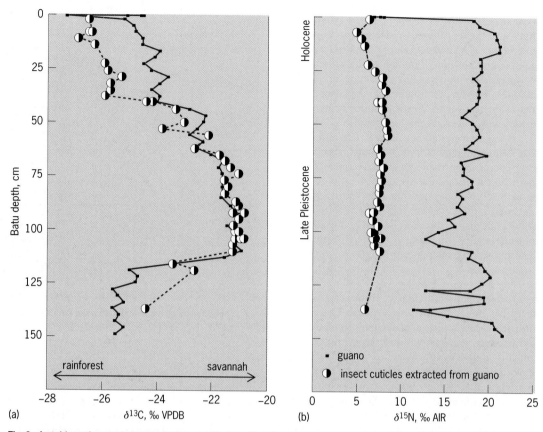

Fig. 3. A stable carbon and nitrogen isotope profile from Batu Caves near Kuala Lumpur, Malaysia. (*a*) High carbon isotope values during the last glacial period archive a savannah ecosystem during the Pleistocene that changed to the tropical rainforest present today. (*b*) Nitrogen isotope values of extracted insect cuticles are much lower than nitrogen-bearing minerals in the bulk guano sediment and similar to those in fresh guano at the surface. Isotope standards: Vienna Pee Dee Belemnite (VPDB) for reporting carbon isotopes and atmospheric air (AIR) for reporting nitrogen isotopes, both in ‰ (parts per thousand).

sal in the region. Early humans arrived in the region 45,000–60,000 years ago, and may have encountered an inland coastal route similar to what they had previously experienced rather than dense tropical rainforest, as suggested by some models. Guano-derived records are now helping better define the environments of the past in this region and assist in developing a deeper understanding of the environmental drivers of human migration in prehistory.

Summary. This review has been concerned with variations of the stable isotope composition of guano as a paleoclimate record. We have briefly discussed how stable isotopes provide a robust archive of past climate and environment. However, it should be remembered that guano also contains several other proxy materials, such as pollen, charcoal, geochemical, and organic chemical proxies, which also archive environmental information. Reliable continuous records of past climate change from continental regions are harder to locate than in the marine environment. Particularly rare are well-dated tropical and semiarid records, with the majority of those that exist coming from lacustrine environments and thus biased toward inherently wetter regions that may have had locally unrepresentative vegetation. Undoubtedly, guano records have their own biases, but it is only through comparing many proxies that we can accurately determine climates of the past.

For background information *see* CAVE; CHIROPTERA; CLIMATE HISTORY; CLIMATOLOGY; INSOLATION; ISOTOPE; MASS SPECTROMETRY; PALEOCLIMATOLOGY; PHOTOSYNTHESIS; RADIOCARBON DATING in the McGraw-Hill Encyclopedia of Science & Technology.   Christopher Wurster;
Michael Bird; Donald McFarlane

Bibliography. D. G. Constantine, Bats in relation to the health, welfare, and economy of man, in *Biology of Bats*, vol. 2, pp. 319–449, edited by W. A. Wimsatt, Academic Press, 1970; J. R. Ehleringer, T. E. Cerling, and B. R. Helliker, C4 photosynthesis, atmospheric CO_2, and climate, *Oecologia*, 112(3):285–299, 1997; D. R. Gröcke et al., Stable hydrogen-isotope ratios in beetle chitin: Preliminary European data and re-interpretation of North American data, *Quaternary Sci. Rev.*, 25(15–16):1850–1864, 2006; H. Mizutani, D. A. McFarlane, and Y. Kabaya, Carbon and nitrogen isotopic signatures of bat guanos as a record of past environments, *J. Mass Spectrom. Soc. Jpn.*, 40(1):67–82, 1992; C. M. Wurster et al., Stable carbon and hydrogen isotopes from bat guano in the Grand Canyon, USA, reveals Younger Dryas and 8.2 ka events, *Geology*, 36(9):683–686, 2008.

Bdelloid rotifers

Bdelloid rotifers are microscopic freshwater inverte-brate animals less than 1 mm in length. They are found in nearly every possible freshwater aquatic habitat, including those that remain wet for only a short period of time (**Fig. 1**). Originally discov-ered by Anton van Leeuwenhoek in the late 1600s, bdelloids were later noticed to possess no males. Four hundred described bdelloid species consti-tute the class Bdelloidea, which is a sister taxon to the facultatively asexual rotifers (that is, rotifers in which reproduction can occur asexually, but is not obligatory, being capable of adapting to different

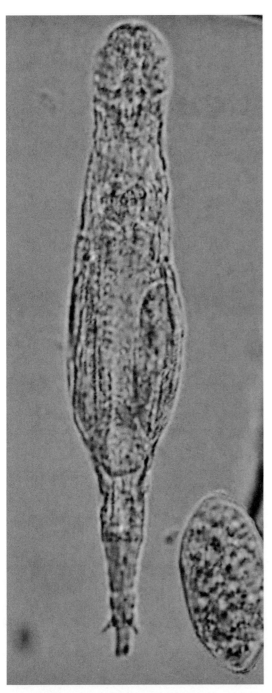

Fig. 1. Bdelloid rotifer *Adineta vaga* and its egg.

conditions to reproduce sexually) of the class Mono-gononta, in the phylum Rotifera. Bdelloids reproduce exclusively asexually, by parthenogenesis, whereby females lay eggs produced from resting oocyte nu-clei after two mitotic divisions, not accompanied by chromosome pairing or reduction in chromosome number. Recently, new information on the molecu-lar structure of bdelloid genomes has revealed their unique features, possibly conferred by their asexual-ity and unusual lifestyle involving repeated cycles of desiccation and rehydration.

Genome structure. Bdelloid genomes contain pairs of chromosomes with overall sequence divergence of a few percent, which is a value comparable to the average divergence between allelic gene pairs (al-leles being the alternate forms of a gene at a locus on a chromosome) in species with very large popu-lation sizes, such as ascidians (a class of tunicates). While monogonont rotifers are sexual diploids (hav-ing two complete chromosome pairs in a nucleus), bdelloids appear to be degenerate tetraploids, with chromosomes present in quartets, each comprising two colinear pairs, and with only a minority of genes common to both pairs, in the same order and ori-entation. (The term *degenerate* as used here refers to the fact that some genes have been lost, but the order has stayed the same in the genome over subse-quent generations.) The quartet structure arose prior to separation of the major bdelloid families. Gene copies from different colinear pairs are extremely divergent, which was initially interpreted as diver-gence between alleles accumulated following the an-cient loss of sex. Variable levels of divergence within each colinear pair presumably reflect the operation of homogenizing processes such as gene conversion and mitotic crossing-over [a process whereby one or more gene alleles present on one chromosome may be exchanged with their alternative alleles on a ho-mologous chromosome to produce a recombinant (crossover) chromosome].

Desiccation and radiation resistance. The ability to survive desiccation without assuming a special-ized developmental form, such as the desiccation-resistant cysts of certain nematodes, was acquired by bdelloids after their separation from monogononts. While dried resting eggs of monogonont rotifers re-main capable of hatching for months, adult mono-gononts, unlike bdelloids, cannot undergo anhydro-biosis (life without water). When humidity starts to decrease, bdelloid rotifers undergo physiological ad-justments and contract into a compact body shape called a tun. Desiccated animals retain very little water, and they respond to drastic water loss by producing protective hydrophilic molecules, such as late embryogenesis abundant (LEA) proteins, which may substitute for water by forming an extended net-work of hydrogen bonds.

Suspended animation is linked to exceptionally ef-ficient deoxyribonucleic acid (DNA) repair in the bacterium *Deinococcus radiodurans*, which is a species that is capable of reassembling its chromo-somes after they have been shattered by several thou-sand grays (Gy) of gamma irradiation. Because such strong sources of radioactivity are naturally unknown on Earth, resistance to ionizing radiation can hardly be a trait under selection per se. However, prolonged desiccation, akin to ionizing radiation, is known to produce double-strand breaks (DSBs) in *D. radiodurans*, and therefore the ability to repair these potentially lethal DNA lesions efficiently may provide a crucial evolutionary advantage to organisms that dwell in drying habitats. Not surprisingly, bdelloid rotifers were also found to be extraordinarily resistant to ionizing radiation, being able to repair many hundreds of DSBs and resume reproduction after receiving nearly 1000 Gy of gamma irradiation from a cesium-137 (^{137}Cs) source. In contrast, desiccation-sensitive monogonont rotifers exhibit no special resistance and are as sensitive to gamma radiation as representatives of most other invertebrate phyla.

Two principal pathways of DSB repair operate in all eukaryotes: the first copies genetic information from a homologous copy of the damaged DNA sequence, whereas the second simply joins broken ends together. Homologous repair can mend the broken sequence without adding or omitting nucleotides. In contrast, nonhomologous end joining usually removes a few bases from each side of the break before sealing it, and in doing so leaves characteristic lesions in the form of small deletions in DNA. Homologous repair is essential for complete reassembly of the *D. radiodurans* genome disintegrated by ionizing radiation, and it is likely that bdelloids also take advantage of this repair pathway.

Transposable elements. Homologous DSB repair is a logical choice when a break is made within a unique sequence. However, severe chromosomal abnormalities may appear if the break occurs within a dispersed repetitive sequence, such as a transposable element (TE; a mobile segment of DNA that can move from one chromosomal site to another). In asexual populations, depending on the effective population size and other parameters, and in the absence of horizontal (lateral) transmission [also known as horizontal gene transfer (HGT), in which genetic material is exchanged among organisms that are very distantly related], deleterious (harmful) TEs are expected either to be lost or domesticated (adapted to a form that may be beneficial), or to increase indefinitely, thereby eventually driving the host population to extinction. Loss of TEs may occur via stochastic (random) excision, and their limitation and mutational decay could be facilitated by silencing mechanisms. In sexual species, it is thought that an essential factor limiting TE proliferation is synergistic selection against chromosomal abnormalities resulting from ectopic crossing-over (that is, crossing-over in an abnormal position) between dispersed homologous elements during meiosis. In bdelloids, DNA breakage and repair associated with repeated cycles of desiccation and rehydration may provide a nonmeiotic route for the elimination of TEs, imposing synergistic selection against TEs. Frequent rounds of DNA breakage and repair would be constantly selecting for genotypes with the lowest number of TEs, minimizing the chance of producing deleterious chromosomal rearrangements in bdelloids recovering from desiccation and sufficient to neutralize the propensity of TEs to multiply exponentially. Thus, DNA repair following desiccation would essentially act to limit the TE load.

Bdelloid genomes contain a variety of TEs, both intact and decayed, as well as domesticated, all of which are, however, highly compartmentalized in the regions close to chromosome ends (telomeres) and are virtually absent from gene-rich regions. These TEs include low-copy-number DNA transposons, retrovirus-like elements, and specialized telomere-associated retroelements capable of transposing specifically to chromosome ends. In the absence of sexual reproduction, the presence of parasitic transposons in an anciently asexual population implies their entry by horizontal transmission, albeit kept at a low copy number by mechanisms of silencing and selection. Indeed, bdelloid telomeric regions contain numerous TEs and foreign genes that are both eukaryotic and prokaryotic in origin, providing evidence for unprecedented levels of HGT in metazoans (the multicellular animals that make up the major portion of the animal kingdom).

Horizontal gene transfer. Subtelomeric regions of bdelloid genomes contain large numbers of protein-coding sequences that appear more similar to nonmetazoan homologs than to metazoan ones, or for which no metazoan counterparts have been identified at all (**Fig. 2**). Several foreign genes of apparently bacterial origin are interrupted by canonical spliceosomal introns, which are absent in bacteria. [Introns are intervening genomic sequences that are removed from the corresponding ribonucleic acid (RNA) transcripts of genes; spliceosomal introns are removed from RNA transcripts through a series of transesterifications mediated by a large ribonucleoprotein complex called the spliceosome.] While the majority of nonmetazoan genes encode full-length proteins, others contain defects in the coding sequence and thus can be classified as pseudogenes. A few foreign

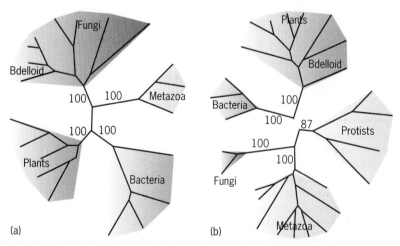

(a)   (b)

Fig. 2. Evidence for HGT in bdelloid rotifers. The phylogenetic trees are shown for two genes coding for the enzymes galacturonidase (*a*) and UDP-glycosyltransferase (*b*). The corresponding genes from the bdelloid rotifer *Adineta vaga* do not cluster with genes from other metazoan animals, but instead are grouped with fungal genes *a* or plant genes *b*. Numbers indicate the percentage of phylogenetic support for each group.

genes are shared between very restricted groups of bacteria and fungi and bdelloid rotifers, but not other metazoans, making the probability of their vertical descent from the common ancestor vanishingly low. Although no direct evidence of a contemporary HGT event has been detected, the presence of nonmetazoan pseudogenes at eroding chromosome ends implies relatively recent acquisition, as such decaying inserts are rapidly removed in the absence of purifying selection. Ectopic DNA repair, via selection against aneuploidies (deviations from the normal number of chromosomes), would act to concentrate TEs and alien DNA in telomeric regions. Addition of foreign DNA to deprotected chromosome ends may also occur directly.

While HGT is the norm among prokaryotes and lower eukaryotes, plants and animals typically depend on internal sources of genetic novelty, such as gene duplication or diversification of gene expression, and on recombination of preexisting polymorphisms for successful adaptation to the ever-changing environment. HGT in Metazoa usually results from symbiotic or parasitic relationships between the nonmetazoan donor and the metazoan host that have exhibited a long-term close association. For example, a large portion of the genome of *Wolbachia*, which is a common bacterial endoparasite of insects and worms, has been detected in the fruit fly genome, and certain *Wolbachia* genes have also been found in worms. Yet another case of metazoan HGT has been described in the sea slug, which preys on algae, but can retain photosynthesizing algal plastids within its digestive epithelium, and has captured an algal nuclear gene, *psbO*, which is required for plastid viability.

The dominance of vertical inheritance in Metazoa rests on anatomical barriers between germline nuclei and somatic tissues that may often come in contact with exogenous DNA. If the barrier becomes compromised, though, one should not be surprised to find substantial levels of HGT even in Metazoa. As bdelloid germline nuclei are already formed at the moment of birth and reside in the immediate vicinity of the gut, they may become a target for ingested foreign DNA should the integrity of the gut lining be jeopardized (**Fig. 3**). Bdelloid rotifers do not discriminate among food particles, and at any given time their guts may contain semidigested DNA of diverse phylogenetic origin. The phylogenetic spectrum of ingested DNA may be even wider than their immediate food sources, as free DNA in aquatic and soil environments is absorbed on mineral surfaces or bound by complex organic molecules, and may easily find its way into the bdelloid gut. A single desiccation event may suffice to induce transient damage to the gut and the ovary, allowing the passage of foreign DNA. Once exogenous DNA reaches the resting oocytes, it may be integrated into the germline DNA, either homologously or ectopically. If conspecific DNA (that is, DNA of the same species), possibly released from nearby dead animals, is absorbed together with other exogenous DNA, it could potentially replace endogenous sequences by homolo-

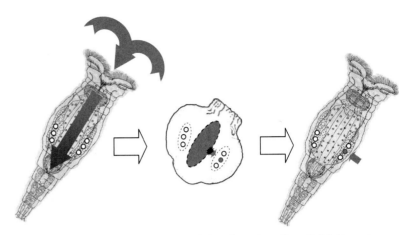

Fig. 3. A possible pathway for desiccation-induced transformation of bdelloid oocyte nuclei. Semidegraded foreign DNA (in dark gray) from ingested food particles may enter the bdelloid body cavity during a period of desiccation and, upon recovery, become incorporated into resting oocyte nuclei (depicted schematically as filled circles).

gous recombination. Such rare events may provide the physical foundation for genetic exchange among conspecific individuals in the absence of sexual reproduction.

For background information *see* ANIMAL REPRODUCTION; BDELLOIDEA; CHROMOSOME; CROSSING-OVER (GENETICS); DEOXYRIBONUCLEIC ACID (DNA); GENE; GENETICS; METAZOA; MONOGONONTA; RECOMBINATION (GENETICS); ROTIFERA in the McGraw-Hill Encyclopedia of Science & Technology.

Eugene A. Gladyshev; Irina R. Arkhipova

Bibliography. I. R. Arkhipova and M. Meselson, Deleterious transposable elements and the extinction of asexuals, *BioEssays*, 27:76–85, 2005; E. A. Gladyshev and M. Meselson, Extreme resistance of bdelloid rotifers to ionizing radiation, *Proc. Natl. Acad. Sci. USA*, 105:5139–5144, 2008; E. A. Gladyshev, M. Meselson, and I. R. Arkhipova, Massive horizontal gene transfer in bdelloid rotifers, *Science*, 320:1210–1213, 2008; P. J. Keeling and J. D. Palmer, Horizontal gene transfer in eukaryotic evolution, *Nat. Rev. Genet.*, 9:605–618, 2008.

Beta-delayed neutron emission

A number of different processes, including alpha-, beta-, and gamma-ray emission, exist by which radioactive nuclei are transformed into stable nuclei. Because these processes are of importance in any situation involving radioactive nuclei, understanding them is important to subjects as disparate as the production of nuclei in stellar atmospheres and the operational characteristics of nuclear fission reactors.

Nuclear landscape. The nucleus of any atom is composed of some combination of neutrons and protons, which is responsible for nearly all the mass of the atom. However, not all conceivable combinations of neutrons and protons can exist, and among those that do, only certain combinations lead to stable nuclei. The basic reason for this lies in the mass–energy relationship as first proposed by Albert Einstein. Because neutrons and protons, the nucleons, have

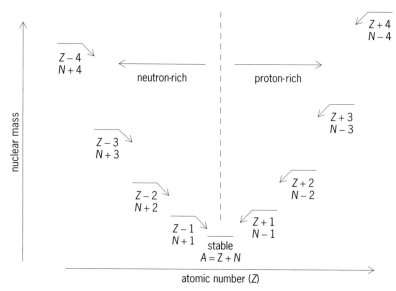

Fig. 1. The valley of beta stability. Radioactive nuclei decay in order to reach a lower total mass energy. Shown here is a hypothetical plot of the nuclear mass as a function of the element number Z with a constant atomic mass number A. The arrows indicate the beta decay path along the mass chain. Z = number of protons; N = number of neutrons; A = number of nucleons ($N + Z$).

approximately the same mass, any atom containing a combination of N neutrons and Z protons that adds to the same atomic mass number ($A = Z + N$) should have basically the same mass. However, slight differences in atomic mass exist such that only one or two stable nuclei are observed for each value of A. If a plot is made of the nuclear mass for a particular atomic mass number as a function of the atomic number Z, one observes a minimum for the stable nuclei with an increase in mass as one proceeds in either direction from the minimum (**Fig. 1**). Furthermore, if a three-dimensional surface is produced showing this curve for all atomic mass numbers, then a valley with hills on each side is observed. This is referred to as the valley of stability. Nuclei with excess neutrons are called neutron-rich, whereas those with excessive protons are called proton-rich.

Beta decay. To understand beta decay, one just needs to view the valley of stability as a potential-energy path. Just as a marble placed on a slope will roll down as it converts gravitational potential energy into kinetic energy, sound, and heat, nuclei will convert mass energy into other forms as they decay down to stability. In beta decay, the energy usually goes into a beta particle and neutrino and their kinetic energies, and gamma rays emitted by the excited nuclei produced in the decay. Of course, the process is more complicated than for a rolling marble, because there are barriers that slow the process, resulting in the observed radioactive half-lives. These half-lives become shorter as the energy available to the decay increases, but details of the decay are also important. The beta decay of a typical neutron-rich nuclide is shown in **Fig. 2**. The beta decay parent has a mass energy higher than that of its daughter. The Q value represents the maximum energy available to this decay. In any nucleus, it is possible to excite the nucleons into higher-energy quantum me-

chanical states that will emit gamma rays in order to reach the lowest energy state, the ground state, of the nucleus. If the Q value for the beta decay is greater than these excitation energies, then the decay can proceed through these states, producing a spectrum of gamma rays that is specific to that decay. In addition, if the Q value exceeds the neutron separation energy, that is, the energy at which a neutron will be ejected from the daughter nucleus, then unbound states will be populated and beta-delayed neutron emission may occur. Beta decay to bound states will lead to a daughter with the same atomic mass number, but beta-delayed neutron emission will result in a daughter with one less nucleon. The probability for this decay process has implications for a number of physical situations, two of which are highlighted here.

Nucleosynthesis. Although most of the material universe is composed of hydrogen and helium, elements up to uranium are observed in nature. The process by which the heavier nuclei are produced, termed nucleosynthesis, can proceed along a path near stability over long periods of time, but the observed elemental abundances cannot be explained solely by these processes. Instead, processes have been proposed that follow paths far from stability and involve very short-lived radioactive nuclei. These processes must occur very quickly with very high temperatures and material densities in explosive environments. One such process is the rapid neutron capture process that is believed to occur in core-collapse supernovae. The nuclei involved in the process are determined by a balance between the ability of a nucleus to capture a neutron and the possibility that the nucleus will absorb energy by way of a gamma ray and emit a neutron. A short time after the supernova event begins, the temperature and density drop to a point where the process can no longer proceed, and all the nuclei present begin to beta decay back to stability. Because beta-delayed neutron emission results in a change in the atomic mass number of the nuclei decaying, the final observed elemental abundances are affected.

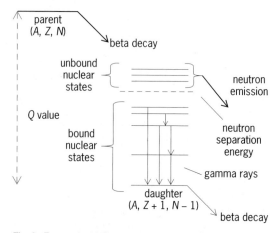

Fig. 2. Energy-level diagram representing the beta decay of a neutron-rich parent nucleus. If unbound states can be populated in the decay, then neutrons will be emitted, resulting in a change in the atomic mass number.

Nuclear power reactors. All current nuclear power reactors operate under the same basic principle. A nuclear fuel, typically uranium, absorbs neutrons, causing it to split into two fragments with the release of large amounts of energy and additional neutrons. In a properly designed nuclear reactor core, the released neutrons from one fission event result in additional fissions in a controlled, self-sustaining chain reaction. The two fission fragments produced are neutron-rich and must decay back to stability, and in a number of cases produce beta-delayed neutrons that can also result in fission. Maintaining a chain reaction requires a delicate balance in the number of neutrons in the reactor core. As a consequence, the presence of beta-delayed neutrons, even though they are a small fraction of the total neutrons, can have a significant effect on the operation of the reactor. In particular, the beta-delayed neutrons provide time for inserting control rods to absorb neutrons to regulate flux and reaction rates. Therefore, understanding the operational characteristics of a nuclear power reactor requires an understanding of the beta-delayed neutron emission from the fission fragments. Furthermore, the properties of the radioactive waste produced by nuclear reactors depend on the beta-delayed neutrons. This nuclear waste represents a major biological hazard (radiotoxicity), which must be addressed. Scientists are working on new designs for nuclear power reactors that will be more efficient in energy production and result in nuclear waste that is made less radiotoxic by the process of transmutation, either within the reactor or during processing of the fuel after it is removed from the reactor. These designs require complex modeling of the reactor cores, which must include many parameters including the production of beta-delayed neutrons. Unfortunately, imprecise experimental data, which leads to inaccuracies in theoretical modeling and extrapolations of the beta-delayed neutron emission process, can lead to unnecessary and erroneous requirements in the operation of fission reactors as well as waste handling and transmutation.

Measuring beta-delayed neutron probabilities. Absolute beta-delayed neutron emission probabilities are very difficult to measure because of problems with contamination of the samples and detection of the neutrons. Although production of the neutron-rich radioactive nuclei is a simple process, it is very difficult to separate out the specific parent nuclide to be studied. The standard technique involves separating the fission fragments according to their atomic mass number and observing the decay of this sample. Unfortunately, the very neutron-rich nuclei of interest are produced at a much lower rate than those closer to stability. The process thus becomes one of separating the needle from the haystack. In the past, most measurements of beta-delayed neutron probabilities used a neutron detector to obtain the selectivity needed to study the nuclide of interest along with a beta detector to estimate the number of nuclei that decay. However, this technique requires determining the absolute efficiency of the neutron detector,

which is a difficult task and can lead to significant uncertainties.

Using the Holifield Radioactive Ion Beam Facility located at Oak Ridge National Laboratory, a new set of measurements has been performed using a different technique. First, a high-resolution separator is used to provide better isolation of the nuclide of interest so that it becomes the dominant component of the radioactive sample being studied. Second, the radioactive ions are accelerated so that they can be passed through a detector, allowing the absolute number of the nuclide of interest to be measured before collecting the sample for study. Finally, instead of observing the neutrons emitted by the decays, the characteristic gamma rays associated with the decays are used to determine the probability that the beta-decaying parent results in nuclei in the "A" and "$A - 1$" mass chains. Since performing an absolute calibration of a gamma-ray detector is a much easier and more precise task than the former technique, the results obtained are more accurate. Measurements were made for isotopes of copper with atomic mass numbers 76, 77, and 78, and for the gallium isotope with atomic mass number 83. It was found that the measured beta-delayed neutron emission probabilities were two to four times larger than the previously accepted "best" values reported in databases of nuclear decay properties. The information in these databases is used in the theoretical extrapolations and complex modeling calculations by which scientists study both nucleosynthesis and nuclear reactor cores. Although these specific results do not have a major effect on model calculations because they are just a small part of a large calculation, they do call into question the accuracy of the information in the databases, which were based on measurements made with low-resolution separators. Additional experiments are needed to determine if the problem is localized or more systemic within the databases.

Implications for nuclear structure. The probability for beta-delayed neutron emission depends primarily on the window of opportunity available for this decay mode, which is specified by the difference between the Q value and the neutron separation energy. In general, the larger this energy difference, which is determined from the masses of the parent and daughter, the greater is the probability of beta-delayed neutron emission. However, an important component of this process involves the quantum structure of the states in the parent and the daughter connected by the decay. In the shell model of nuclear structure, we visualize the filling of nuclear states for both protons and neutrons in a specific order along with a nucleon–nucleon interaction that dictates the structure of the quantum states. The expected ordering of the nuclear states is based mainly on observations near stability. A comparison of the experimental results to theoretical calculations showed that including a reordering of the nuclear states for protons as neutrons are added to the isotopes was required to obtain good agreement. Because many of the nuclei needed to understand nuclear astrophysics and the nuclear fuel cycle have only theoretical calculations, it is important to understand how the changes in nuclear structure away from the valley of stability will modify the predicted results.

For background information, *see* DELAYED NEUTRON; ENERGY; NUCLEAR FUEL CYCLE; NUCLEAR REACTOR; NUCLEAR STRUCTURE; NUCLEOSYNTHESIS; RADIOACTIVE WASTE MANAGEMENT; RADIOACTIVITY; REACTOR PHYSICS; SUPERNOVA; TRANSMUTATION in the McGraw-Hill Encyclopedia of Science & Technology.   Jeff Allen Winger

Bibliography. D. D. Clayton, *Handbook of Isotopes in the Cosmos*, Cambridge University Press, 2003; B. E. J. Pagel, *Nucleosynthesis and Chemical Evolution of Galaxies*, Cambridge University Press, 2009; P. D. Wilson, *The Nuclear Fuel Cycle: From Ore to Waste*, Oxford University Press, 1996; J. A. Winger et al., Large β-delayed neutron emission probabilities in the ^{78}Ni region, *Phys. Rev. Lett.*, 102:142502, 2009.

Beyond CMOS technology

In the past several decades, the scaling of silicon metal-oxide-semiconductor field-effect-transistor (MOSFET) technology has been the primary driving force behind the prosperity of the global microelectronics industry. By reducing the size of transistors in an integrated circuit, almost every measure of the integrated circuit's capabilities is improved, such as higher transistor density, lower cost per transistor, and faster transistor speed. Over the past 40 years, transistor feature size has been reduced from 10 micrometers to approximately 30 nanometers. Although most of the time scaling is simply reducing the feature size, during certain periods the industry has made major changes by moving from bipolar transistor to *p*-channel and *n*-channel MOS, and finally to complementary MOS (CMOS) planar transistors in the 1980s. Since then CMOS has remained the dominant technology. As the MOSFET size approaches tens of nanometers, this scaling trend is facing an inevitable end, which calls for beyond-CMOS technologies to continue the technological advancement for the microelectronics industry.

CMOS scaling limits. It is important to first understand what factors limit CMOS scaling. Although an obvious challenge is the fabrication and accurate control of extremely small transistors, engineering innovations have always found solutions to these obstacles. In current and future technology, one of the most important scaling limits is power density and heat dissipation. As CMOS transistor density increases with scaling, power density and heat generated per unit area increase. The power density of CMOS is approaching a level beyond the capability of practical cooling techniques, which imposes a limit on device density.

Another critical limit is the degradation of transistor performance caused by parasitic components. While ideal MOSFET performance should improve

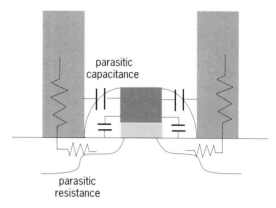

Fig. 1. Schematic diagram of planar CMOS transistor, showing parasitic resistances and capacitances.

with scaling, in reality some undesirable and unintended resistance and capacitances (known as parasitic components) always exist, which degrade transistor performance (**Fig. 1**). CMOS scaling increases the dominance of parasitic components in overall transistor performance. Beyond about 32 nm, performance gain from scaling may even disappear due to parasitic effects, leaving cost reduction as the primary motivation of further CMOS scaling.

Overall CMOS system performance is determined not only by transistors, but also by interconnects. In current CMOS technology, system speed is constrained more by communication delay over interconnects than by transistor delay. Energy is consumed by both active and nonactive (also known as "standby") devices. Standby devices have surpassed active devices to dominate the overall CMOS power consumption. These performance metrics deteriorate with further reduction of transistor size and impose limits on CMOS scaling.

Beyond-CMOS devices can be divided into two main categories: (1) nonclassical CMOS, where new channel materials or new transistor structure is employed; and (2) emerging device concepts, where alternative state variables are utilized to construct novel logic devices superior to CMOS in certain areas. **Figure 2** shows the hierarchy of memory and logic devices, with beyond-CMOS technologies circled by broken lines. Nonclassical CMOS devices are based on similar device physics as conventional CMOS and are compatible with silicon CMOS architectures; however, their performance gain may be limited. Emerging devices are usually based on significantly different physics, with potentially orders-of-magnitude performance gain and also high risk.

Nonclassical CMOS. Nonclassical CMOS structures include ultra-thin-body (UTB) silicon-on-oxide (SOI), multigate transistors, band-engineered transistors, and channel replacement materials. UTB structure provides the extremely thin channel dimensions required for CMOS scaling. While scaling of planar CMOS structures is limited by factors such as the short channel effect (SCE), multigate transistors with improved electrostatics have been developed to extend scaling further. In band-engineered transistors, strain is introduced in the channel to improve the mobility of electrons or holes. Higher-mobility materials (such as narrow band gap III–V compound semiconductors) may replace silicon as the MOSFET channel material to extend transistor performance beyond the level attainable by silicon. Carbon nanotubes, semiconductor nanowires, and graphene (a single-layer sheet of carbon atoms) are also expected to have properties superior to those of silicon. For each of these channel replacement materials, various challenges exist: Compound semiconductors lack a high-quality oxide-semiconductor interface for effective field-effect operation; carbon nanotubes have mixed semiconducting and metallic components, in addition to the alignment challenges; and graphene is a zero-band-gap semiconductor that requires extreme patterning and edge control to open a limited band gap. These technologies embody similar logic implementation as silicon CMOS and hence are compatible with CMOS technology and architecture.

Emerging memory and logic devices. Silicon CMOS has provided both memory and logic devices, which is a great advantage for system integration. On the other hand, beyond-CMOS technologies for memory and logic are likely to arise from different origins. Beyond-CMOS memories are generally expected to be nonvolatile, fast, energy-efficient, and extremely dense. Major types of emerging memory devices include phase-change memory (PCM), magnetic random-access-memory (MRAM), ferroelectric RAM (FeRAM), charge-trapping memory, polymer-based memory, molecular memory, and resistive random-access-memory (RRAM). Although memory technologies are considered universal in application, beyond-CMOS logic devices are likely to be more application-specific. Some important examples of emerging logic devices and technologies include resonant tunneling diode/transistor (RTD/RTT), single-electron transistor (SET), rapid single flux quantum (RSFQ) logic, quantum cellular automata (QCA), spintronic devices (including ferromagnetic logic and spin transistors), and molecular logic devices. Both beyond-CMOS memory and logic devices must possess certain compelling attributes to justify the substantial investments needed to build a new infrastructure.

CMOS technology uses electron charges as a state variable to encode binary logic information ("0" and "1"). Analysis has shown that MOSFET represents a nearly ideal implementation of a charge-based logic switch. Therefore alternative state variables have been explored for beyond-CMOS technologies. For example, particles such as electrons and nuclei have a spin degree of freedom, a quantum mechanics concept. Their spins can be either up or down, a natural candidate for binary logic representation. The **table** shows a list of state variables studied in emerging-device research. Emerging devices based on some unique features of these novel state variables may enable unprecedented applications. For example, while silicon CMOS circuits are composed of two-input logic gates, multiple-input majority logic gates can be easily built from some novel devices (such as spintronic devices). Majority logic gates output the

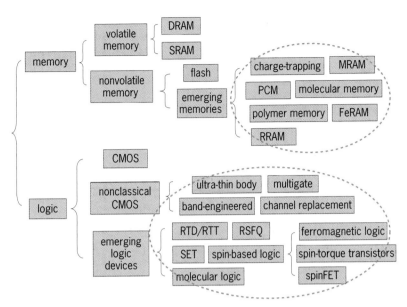

Fig. 2. **Hierarchy of memory and logic devices. Broken lines circle beyond-CMOS technologies.**

majority decisions of the inputs and are more efficient in implementing logic functions than CMOS. Some novel logic devices also have built-in memory capabilities; for example, their logic states remain unchanged after switching. This memory-in-logic capability may enable more efficient logic architectures, and would be very useful in reprogrammable logic where current design based on silicon CMOS is inefficient because of large areas of dedicated memories. *See* PURE SPIN CURRENTS.

Novel architectures. While transistor performance gain from CMOS scaling is diminishing, beyond-CMOS solutions may arise at the architecture level. Homogeneous multicore architectures have become an industry trend, where multiple general-purpose CMOS processors are homogeneously integrated on

Alternative state variables for beyond-CMOS logic devices	
State variables	Devices and explanations
Charge	Silicon CMOS; channel replacement (such as III–V compound semiconductors, nanowires, carbon nanotubes, graphene) Tunneling devices Single-electron transistor (SET)
Electrical dipole	Ferroelectric devices
Exciton	Charge-neutral electron-hole pair
Spin	SpinFET, spin-torque-transfer (STT) devices Nanomagnetic logic [magnetic quantum cellular automata (MQCA)] Spin-wave logic
Pseudospin	Pseudospin devices in bilayer graphene
Phonon	Thermal management and logic implementation using phonons
Photon	Optical computing, plasmonic devices
Mechanical position	Nanoelectromechanical systems (NEMS)
Phase	Use property change associated with phase transition for logic information

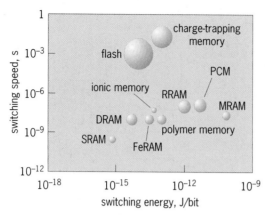

Fig. 3. Comparison of beyond-CMOS memory technologies with CMOS-based memories (flash, SRAM, and DRAM) with regard to switching speed, switching energy, and operation voltage (proportional to symbol size).

the same chip to provide a performance gain that is possibly linear with the number of processors. Heterogeneous architectures integrate specialized processor or on-chip memories with general-purpose CMOS processors. These heterogeneous cores may execute specialized functions with better performance or cost than CMOS-only cores. For example, a so-called CMOL structure fabricates nanogrids atop traditional CMOS, where CMOS provides external communication, current drive, and signal restoration, and the nanogrids provide an ultradense programmable interconnect capability to enhance CMOS functions. Morphic architectures are adapted to effectively address a particular problem set, often with their inspiration drawn from biological or scientific computational paradigms. Bio-inspired hybrid computation may be well suited for the processing of visual and auditory information. In a cellular nonlinear network (CNN), elementary processing elements are organized as a connected array with problem-dependent layout geometry and connection net-

works. Heterogeneous architectures and morphic architectures may provide orders-of-magnitude performance gains.

Evaluation criteria. With a large variety of novel device concepts proposed based on different physical principles, it is important to evaluate these concepts using consistent criteria and benchmark them against silicon CMOS. The following criteria are considered important. Scalability is the major incentive to explore these novel devices. The devices need to provide performance metrics of cost and speed superior to ultimate CMOS. Energy efficiency appears to be a dominant criterion for these devices, especially alternative state variable devices, as potential solutions to thermal-constrained scaling of CMOS. A high ON/OFF ratio is favorable for operation and design, especially for memory devices. Gain is needed in these novel logic devices for gate fan-out; otherwise signal restoration has to be designed separately. These devices have to demonstrate reasonable operation reliability and tolerable error rate. Preferably their operation temperature is at room temperature or higher. If these devices are compatible with CMOS technologies and architectures, they may benefit from the tremendous investment in the silicon CMOS infrastructure. **Figure 3** compares the switching speed and switching energy of some beyond-CMOS memory technologies with CMOS-based memories (flash, SRAM, and DRAM). The size of the symbols in Fig. 3 is proportional to the operation voltage of these memories. Faster switching speed, smaller switching energy, and smaller operation voltage are desirable for memory technologies. Although SRAM and DRAM have almost the best performance in terms of these three parameters, they are not nonvolatile (being able to retain data when power is off) memories. Flash memory dominates today's nonvolatile memory industry because of its maturity, but most beyond-CMOS memory technologies are nonvolatile and may provide better overall performance than flash memory. **Figure 4** compares the energy-delay product (switching energy × logic gate delay) and density of some beyond-CMOS logic technologies with CMOS by normalizing both parameters to the values of silicon CMOS. Smaller energy-delay product and higher density are desirable for logic technologies. Unlike beyond-CMOS memory technologies, most beyond-CMOS logic technologies do not appear to be superior to CMOS in terms of these two important logic parameters. This is partially because their development is still in an early stage. Although some of these emerging logic devices may not surpass CMOS in overall performance, they may provide novel functions to augment CMOS.

Similar to CMOS, performance of beyond-CMOS devices is also degraded by parasitic components. Therefore, the validity of their evaluation depends on the understanding of parasitics, which is often incomplete or not yet optimized in the emerging devices. In addition, because beyond-CMOS devices are often better suited for specialized functions and novel architectures, function-level evaluation may be more appropriate than device-level evaluation.

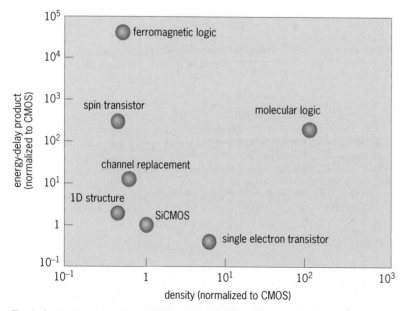

Fig. 4. Comparison of beyond-CMOS logic technologies with silicon CMOS by normalizing the parameters of density and energy-delay product to those of silicon CMOS.

In summary, various beyond-CMOS memory and logic technologies have emerged. A truly competitive beyond-CMOS technology may arise from systematic approaches that combine innovations in devices, fabrication, algorithms, and architecture. CMOS will probably remain a critical part of these novel technologies to provide functions such as signal restoration and communication interface. More information about CMOS scaling and emerging research devices can be found in the International Technology Roadmap for Semiconductors (ITRS).

For background information *see* EXCITON; INTEGRATED CIRCUITS; MAGNETORESISTANCE; MICRO-ELECTRO-MECHANICAL SYSTEMS (MEMS); MOLECULAR ELECTONICS; OPTICAL INFORMATION SYSTEMS; PHONON; SEMICONDUCTOR MEMORIES; SPIN (QUANTUM MECHANICS); TRANSISTOR in the McGraw-Hill Encyclopedia of Science & Technology.   An Chen

Bibliography. R. Cavin et al., Emerging research architectures, *Computer*, 5:33–37, 2008; R. Chau et al., Benchmarking nanotechnology for high-performance and low-power logic transistor applications, *IEEE Trans. Nanotech.*, 4(2):153–158, 2005; J. A. Hutchby et al., Extending the road beyond CMOS, *IEEE Circuits Dev. Mag.*, 3:28–41, 2002; S. E. Thompson and S. Parthasarathy, Moore's law: The future of Si microelectronics, *Mater. Today*, 9(6):20–25, 2006.

Biotechnology and the environment

Scientists and engineers who engage in environmental practices must have a common understanding and application of living systems, that is, biosystems. Biotechnologies present both challenges and opportunities for environmental science and engineering.

Biotechnologies can be visualized as sets of biological reactions occurring at various scales in the environment. The reactions may lead to desirable results, such as the degradation of toxic substances into harmless compounds. Biological reactions may also lead to undesirable results, such as the introduction of genetically modified organisms to an ecosystem or the generation of toxic chemicals.

In environmental systems, thermodynamic processes occur over a broad domain, having scales ranging from just a few angstroms to global. For example, the processes that lead to a contaminant moving and changing in a bacterium may be very different from those processes at the lake or river scale, which in turn are different from those processes that determine the contaminant's fate as it crosses an ocean. This is simply a manifestation of the first law of thermodynamics: Energy or mass is neither created nor destroyed, only altered in form. This also means that energy and mass within a system must be in balance—what comes in must equal what goes out. These fluxes can be measured to yield energy balances within a region in space through which a fluid travels. This region, the control volume where balances occur, can take many forms. For any

control volume, the calculated mass balance is

$$\left[\begin{array}{c}\text{Quantity of}\\\text{mass per unit volume}\\\text{in a medium}\end{array}\right] = [\text{Total flux of mass}]$$

$$+ \left[\begin{array}{c}\text{Rate of production or loss}\\\text{of mass per unit volume}\\\text{in a medium}\end{array}\right] \quad (1)$$

or, stated mathematically,

$$\frac{dM}{dt} = M_{\text{in}} - M_{\text{out}} \quad (2)$$

where M = mass and t = specified time interval.

If we are concerned about a specific chemical (for example, environmental engineers worry about losing good ones, such as oxygen, or forming bad ones, such as toxic dioxins), Eq. (2) needs a reaction term (R) [Eq. (3)].

$$\frac{dM}{dt} = M_{\text{in}} - M_{\text{out}} \pm R \quad (3)$$

Biosystems include the interrelationships of the abiotic (nonliving) and biotic (living) environments. Biotechnology has been an application of the concept of the "trophic state" for much of human history. Organisms, including humans, live within an interconnected network or web of life (**Fig. 1**).

Ecologists attempt to understand the complex interrelationships shown in **Fig. 2**, and consider humans, too, among the consumers. Food chains illustrate the complexity of biosystems. Species at a higher tropic level are predators of lower-level species, so materials and energy flow downward. The types and abundance of species and interaction rates vary in time and space (Fig. 2). From a biotechnological standpoint, the introduction of

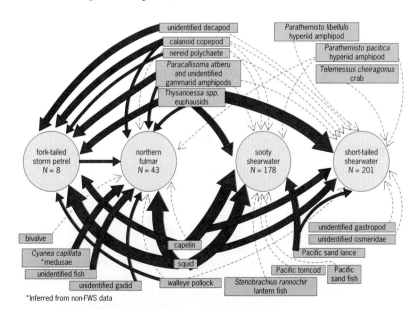

Fig. 1. Flow of energy and mass among invertebrates, fish, and seabirds (Procellariform) in the Gulf of Alaska. The larger the width of the arrow, the greater is the relative flow. Note how some species prefer crustaceans, such as copepods and euphausiids, but other species consume larger forage species, such as squid. (*G. A. Sanger, Diets and food web relationships of seabirds in the Gulf of Alaska and adjacent marine areas, U.S. Department of Commerce, National Oceanic and Atmospheric Administration, OCSEAP Final Rep. no. 45, pp. 631–771, 1983*)

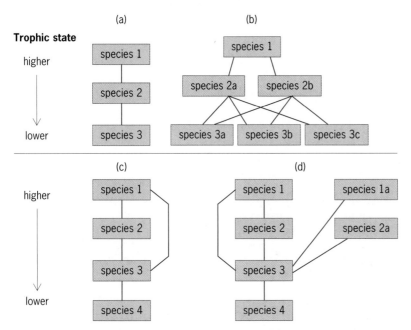

Fig. 2. Flow of energy and matter in biosystems moves from higher trophic levels to lower trophic levels. Lines represent interrelationships among species. (*a*) Linear biosystem, (*b*) multilevel trophic biosystem, (*c*) omnivorous biosystem, and (*d*) multilevel biosystem with predation and omnivorous behaviors. (*Based on information from T. E. Grandel, Annu. Rev. Energy Environ., 21:69–98, 1996*)

modified species or changes in environmental conditions (such as the introduction of nutrients and toxic by-products) can change these trophic interrelationships.

All species living in biosystems consist of molecular arrangements of the elements carbon, oxygen, and hydrogen, and most contain nitrogen. These four elements, known as the "biophile" elements, have an affinity for each other and form complex organic compounds. In fact, less than 200 years ago, such organic compounds were thought to be able to be produced only within natural biological systems. The smallest biosystems, the viruses, bacteria, and other microbes, can be seen as biochemical factories. For much of human history, the systems within microbes have been considered to be black boxes.

Biosystematic processes provide microbes with remarkable proficiencies to adapt to various hostile environments. Some produce spores, and many have durable latency periods. All have the ability to reproduce in large numbers when environmental conditions become more favorable. The various systems that allow for this efficient survival have become better understood in recent decades, to the point that cellular and subcellular processes of uptake and absorption, nutrient distribution, metabolism, and product elimination have been characterized, at least empirically. More recently, the genetic materials of DNA and the various forms of RNA have been mapped. As genes have become better understood, so has the likelihood of their being manipulated. Such manipulation is the stuff of biotechnology.

Systematic view of biotechnological risks. Estimating and predicting the risks associated with the manufacture and use of biotechnologies is complicated by the diversity and complexity of the types of technologies available and being developed, as well as the seemingly limitless potential uses of these processes. A risk assessment is the evaluation of scientific information on the hazardous properties of environmental agents, the dose–response relationship, and the extent of exposure of humans or environmental receptors to those agents. The product of the risk assessment is a statement regarding the probability that humans (populations or individuals) or other environmental receptors, so exposed, will be harmed and to what degree (risk characterization). As more products and by-products are developed using biotechnologies, the potential for environmental exposure has increased. Potential release sources include direct and/or indirect releases to the environment from the manufacture and processing of biochemicals generated (for example, proteins) or the release of the modified organisms themselves.

The transfer of genetic material between separate populations (gene flow) is an example of the downstream risks from biotechnologies. To date, agricultural biotechnological products have come from microorganisms. The exact data requirements for each product have been developed on a case-by-case basis. All of the products have been proteins, either related to plant viruses or based on proteins from the common soil bacteria *Bacillus thuringiensis* (Bt). The general data requirements include product characterization, mammalian toxicity, allergenicity potential, effects on nontarget organisms, environmental fate, and, for the Bt products, insect resistance management to protect from losing use of both the microbial sprays and the Bt plant-incorporated protectants (PIPs).

A transgene is an exogenous gene that has been introduced into the genome of another organism, and a transgenic species is one whose genome has been genetically altered. For instance, if a biotechnology is used as a PIP, the movement of transgenes from a host plant into weeds and other crops presents a concern that new types of exposures will occur. Bt corn and potato PIPs that have been registered to date have been expressed in agronomic plant species that, for the most part, do not have a reasonable possibility of passing their traits to wild native plants. Most of the wild species in the United States cannot be pollinated by these crops (corn and potato) because of differences in chromosome number, phenology, and habitat. There is a possibility of gene transfer from Bt cotton to wild or feral cotton relatives in Hawaii, Florida, Puerto Rico, and the U.S. Virgin Islands. Regulators have prohibited the sale or distribution of Bt cotton in these areas, where feral populations of cotton species similar to cultivated cotton exist. These containment measures prevent the movement of the registered Bt endotoxin from Bt cotton to wild or feral cotton relatives.

Researchers have reviewed the potential for gene capture and expression of Bt plant-incorporated protectants by wild or weedy relatives of cultivated potato in the United States, its possessions, and territories. Based on data submitted by the registrant and a review of the scientific literature, regulators have concluded that there is no foreseeable risk of unplanned pesticide production through gene capture and expression of the Colorado potato beetle control protein, Cry3A (the only one introduced into potato), in wild potato relatives in the U.S. tuber-bearing *Solanum* species, including *S. tuberosum*. *S. tuberosum* cannot hybridize naturally with the non-tuber-bearing *Solanum* species in the United States. Three tuber-bearing wild species of *Solanum* occur in the United States: *Solanum fendleri*, *Solanum jamesii*, and *Solanum pinnatisectum*. However, successful gene introgression into these tuber-bearing *Solanum* species is virtually excluded because of constraints of geographical isolation and other biological barriers to natural hybridization. These barriers include incompatible (unequal) endosperm balance numbers that lead to endosperm failure and embryo abortion, multiple ploidy levels, and incompatibility mechanisms that do not express reciprocal genes to allow fertilization to proceed. No natural hybrids have been observed between these species and cultivated potatoes in the United States. The extent to which these findings will continue for the potato or will be similar to those of other species depends on the unique genomic characteristics of those species.

The constraints, drivers, and boundary conditions of the control volume wherein gene flow may occur must be understood to predict possible risks of genetically modifying these plant species. A systematic question is the extent to which such transfers present problems in microbial populations, such as genetically modified bacteria introduced in oil spills or hazardous waste sites to degrade toxic compounds, industrial applications of microbial reactors to produce pharmaceuticals and other chemicals that are released into the environment, and animal populations (for example, genetically modified fish released into the surface waters) or microbes that undergo transformation (for example, similar to the recent H1N1 pandemic, but from a genetically modified microbe).

Applied thermodynamics. Biotechnology began as a passive approach. For example, sanitary engineers noted that natural systems, such as surface waters and soil, were able to break down organic materials. As they studied these processes, they realized that various genera of microbes had the ability to use detritus on forest floors, suspended organic material in water, and organic material adsorbed onto soil particles as sources of energy for growth, metabolism, and reproduction. The engineers correctly hypothesized that a more concentrated system could be fabricated to do the same thing with society's organic wastes. Thus, trickling filters, oxidation ponds, and other wastewater treatment systems are merely supercharged versions of natural systems.

In such a passive system, the same microbes are used as those in nature, but they are allowed to acclimate to the organic material that needs to be broken down. The microbes' preference for more easily and directly derived electron transfer (that is, energy sources) is overcome by permitting them to come into contact only with the chemicals in the waste. Thus, the microbes adapt their biological processes to use these formerly unfamiliar compounds as their energy sources and, in the process, break them down into less toxic substances. Ultimately, the microbes degrade complex organic wastes to carbon dioxide and water in the presence of molecular oxygen, and methane and water when molecular oxygen is absent. These processes are known as aerobic and anaerobic digestion, respectively.

Numerous examples of passive systems have been put to use with the evolution of complex societies. For example, passive biotechnologies were needed to allow for large-scale agriculture, including hybrid crops and nutrient cycling in agriculture and vaccines in medicine.

Very recently, more active systems have been used increasingly to achieve such societal gains, but at an exponentially faster pace. In addition, scientists have developed biotechnologies that produce products that simply would not exist in passive systems.

The relationships between organisms and their environments are revealed in cycles of matter and energy into and out of the organism. This means that the organism itself is a thermodynamic control volume. In turn, the population of organisms is part of the larger control volumes (such as microbes in the intestine, the intestine in the animal, the herd as prey in a habitat, and the habitat as part of an ecosystem's structure).

Smaller control volumes assimilate into larger ones. Within reactors are smaller-scale reactors, such as within the fish liver, on a soil particle, or in the pollutant plume or a forest, as shown in **Fig. 3**. Thus, scale and complexity can vary by orders of magnitude in environmental systems. For example, the human body is a system, but so is the liver, and so are the collections of tissues through which mass and energy flow as the liver performs its function. Each hepatic cell in the liver is a system. At the other extreme, large biomes that make up large parts of the Earth's continents and oceans are systems from the standpoint of biology and thermodynamics.

The interconnectedness of these systems is crucial to understanding biotechnological implications, because mass and energy relationships between and among systems determine the efficiencies of all living systems. For example, if a toxin adversely affects a cell's energy and mass transfer rates, it could have a cumulative effect on the tissues and organs of the organism. And if the organisms that make up a population are less efficient at surviving, then the balances needed in the larger systems, such as ecosystems and biomes, may be changed, causing problems at the global scale. Viewing this from the other direction, a larger system can be stressed, such as by changes in ambient temperature levels or increased concentrations of contaminants in water bodies and the atmosphere. This results in changes all the way down to the subcellular levels; for example, higher temperatures or the presence of foreign chemicals at a cell's membrane will change the efficiencies of uptake, metabolism, replication, and survival. Thus, the changes at these submicroscopic scales determine the value of any biotechnology.

Predicting environmental implications. A particular measure of the success of emerging technologies is whether they induce human and ecological effects. That is, the system must be sustainable. This means that successful new technologies must not lead to unacceptable risk, including risks that are disproportionately distributed in certain subpopulations.

The "feedbacks" in **Fig. 4** are crucial to environmental biotechnology, wherein bioengineers optimize the intended products and preserve (limit the effects) on the energy and mass balances. Sometimes, the bioengineer must conclude that there is no way to optimize both. In this instance, the ethical bioengineer must recommend the "no go" option. That is, the potential downstream costs are either unacceptable or the uncertainties of possible unintended, unacceptable outcomes are too high. Usually, though, scientists can model a number of permutations and optimize solutions from more than two variables (such as species diversity, productivity and sustainability, costs and feasibility, and bioengineered product efficiencies). The challenge is to know the extent to which the model represents the realities, as they vary in time and space.

Biotechnologies can introduce hazards in time and space. The benefits of medical, industrial, agricul-

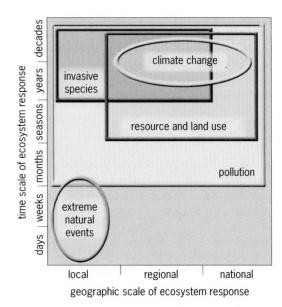

Fig. 4. The response to stressors has temporal and spatial dependencies. Near-field stressors can result from a spill or emergency situation. At the other extreme, global climate change can result from chronic releases of greenhouse gases with expansive (planetary) impacts if global temperatures rise significantly. (*R. Araujo, U.S. EPA, personal communication with author*)

tural, and environmental biotechnologies must be weighed against these possible hazards. Scientists continue to develop models to characterize hazards. However, risk–benefit analyses are difficult, because the science underpinning biotechnologies is still emerging and is fraught with uncertainties.

For background information *see* BIOTECHNOLOGY; ECOLOGY; ENVIRONMENT; ENVIRONMENTAL MANAGEMENT; FOOD WEB; GENETIC ENGINEERING; GENOMICS; MOLECULAR BIOLOGY; PESTICIDE; RISK ASSESSMENT AND MANAGEMENT; THERMODYNAMIC PRINCIPLES; TROPHIC ECOLOGY in the McGraw-Hill Encyclopedia of Science & Technology.
                                                            Daniel Vallero

Bibliography. R. E. Evenson and V. Santaniello (eds.), *The Regulation of Agricultural Biotechnology*, CABI, Cambridge, MA, 2004; Food and Agriculture Organization of the United Nations, *The State of Food and Agriculture, 2003–2004, Agricultural Biotechnology: Meeting the Needs of the Poor?* 2004; T. E. Grandel, On the concept of industrial ecology, *Annu. Rev. Energy Environ.*, 21:69–98, 1996; U.S. Environmental Protection Agency, Nanotechnology White Paper, EPA 100/B-07/001, February 2007; U.S. General Accountability Office, Genetically modified foods: Experts view regimen of safety tests as adequate, but FDA's evaluation process could be enhanced, GAO-02-566, 2002.

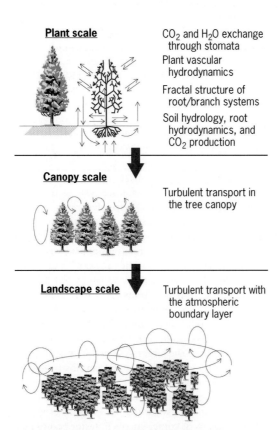

Plant scale

CO_2 and H_2O exchange through stomata

Plant vascular hydrodynamics

Fractal structure of root/branch systems

Soil hydrology, root hydrodynamics, and CO_2 production

Canopy scale

Turbulent transport in the tree canopy

Landscape scale

Turbulent transport with the atmospheric boundary layer

Fig. 3. Hierarchical scales applied to trees. Although the flow and transport equations do not change, the application of variables, assumptions, boundary conditions, and other factors are scale- and time-dependent. (*G. Katul, Modeling heat, water vapor, and CO_2 transfer across the biosphere-atmosphere interface, Pratt School of Engineering, December 1, 2001*)

Bipolar disorder

Bipolar disorder, also known as manic-depressive illness, is an episodic but lifelong mental disorder. The first recognizable descriptions of mania and depression date back to the writings of Aretaeus of

Cappadocia (a Greek physician who lived around 150–200 C.E.). The modern history of bipolar disorder begins in the midnineteenth century with the concept of *folie circulaire* ("circular insanity") proposed by the French psychiatrist Jean-Pierre Falret. Later, it was defined by the work of the German psychiatrist Emil Kraepelin, around the beginning of the twentieth century.

Bipolar disorder is characterized by sudden and often unexplained mood swings, ranging from delirious mania to severe depression. These mood changes are regularly accompanied by other mental and behavioral symptoms, such as fluctuations of volition, activity level, and cognitive functioning.

Symptomatic criteria for bipolar disorder have been conceptualized in diagnostic manuals [the two most important being the *Diagnostic and Statistical Manual of Mental Disorders (DSM-IV)* by the American Psychiatric Association and the *International Classification of Diseases (ICD-10)* by the World Health Organization], with only minor differences between these manuals. Bipolar disorder may manifest itself with different grades of severity. The best-known form is bipolar I disorder, characterized by at least one episode of mania, and, in the overwhelming majority of patients, also by depressive episodes. Bipolar II disorder is defined by at least one (but usually several) depressive episode and at least one episode of elated mood (hypomania), which does not yet fulfill diagnostic criteria for mania. Cyclothymia (or cyclothymic disorder) describes a constant mood instability ranging from hypomania to mild depression, lasting for at least 2 years. Other rare bipolar spectrum disorders include recurrent mania or hypomania with no mood dips.

Traditionally, mania has been considered as the polar opposite of depression. Subjects with mania are said to be cheerful and overly optimistic, as well as possessing an inflated self-esteem. However, it now has become clear from several descriptive studies that a substantial proportion of manic individuals also exhibit dysphoric (distressed) or even depressive features. Vice versa, there are a reasonable number of individuals who, while depressed, also display symptoms commonly attributed to mania. Thus, an episodic disturbance of mood including both manic and depressive features, so-called mixed states, may be more characteristic of bipolar disorder in a fair proportion of subjects.

The diagnosis of bipolar disorder is usually made by a trained psychiatrist or psychologist. Self-rating questionnaires such as the Mood Disorders Questionnaire (MDQ) or the Hypomania Rating Scale (HCL-32) can be useful screening instruments, but their outcome needs to be verified by an in-depth interview. A medical condition or the use of medications or substances with a probability to induce mood aberrations ("secondary mania") also needs to be excluded.

Course of bipolar disorder. The long-term course of bipolar disorders can be quite variable. Data from large cohorts of bipolar patients suggest that patients are on average symptomatic—that is, in a manic, hy-pomanic, mixed, or depressive state—for about 50% of the time once the diagnosis has been made. The vast majority of time spent unwell is in depression, and the length of a depressive episode exceeds that of a manic episode by two to three times on average. This clearly underlines the importance of optimized treatment and prophylaxis of depression.

Women tend to have more depressive episodes than manic episodes, whereas men show a more even pattern of distribution. In addition, female gender appears to be a predictor for more hypomanic mixed states, whereas manic mixed states seem to be evenly distributed between genders. The frequency of new episodes can also vary considerably; patients with four or more episodes per year are usually characterized as "rapid cyclers," although the latest research does not confirm that rapid cycling specifies a distinct and homogeneous subgroup.

How frequent is bipolar disorder? Bipolar disorder is not a rare condition; recent epidemiological studies indicate a lifetime prevalence of bipolar I disorder between 0.5 and 1%. Similar, sometimes slightly higher rates have been estimated for bipolar II disorder. Men and women appear to be equally affected by bipolar I disorder; for bipolar II disorder, a predominance of females has been described.

Bipolar disorder impacts not only mental health but also many other areas: employment, relationships, general quality of life, and even physical health. In addition, other mental disorders such as anxiety disorders or substance abuse are frequent comorbidities in bipolar disorder, complicating the treatment, course, and outcome. Difficulties in education and professional career are especially evident as bipolar disorder typically manifests itself in adolescence and early adulthood, at times where the familiar and professional basis is laid for the rest of life. The true incidence of bipolar disorder in children and adolescents remains somewhat nebulous as diagnostic criteria overlap with other conditions [for example, attention deficit hyperactivity disorder (ADHD)]. In addition, both periods of mania and depression may not be as obvious; symptoms do differ from adults and psychotic features are not rare. Finally, early substance abuse may obscure the diagnosis, but may also lead to an earlier exacerbation of the disorder. Different rates of early substance abuse and diagnostic habits may thus explain diverging figures for early-onset bipolar disorder across countries.

What is the neurobiology behind bipolar disorder? The etiology and pathophysiology of bipolar disorder is complex. Although a variety of biological, psychological, and social factors may contribute, a single paradigm cannot explain all aspects of the occurrence, course, and severity of bipolar disorder. A genetic predisposition is likely. Several studies have shown that rates of bipolar disorder in first-degree relatives of affected individuals are elevated up to ten times over rates found in the general population. Although most research has been done in bipolar I disorder, a similar heredity seems to be true in bipolar II disorder. Genetic studies so far support a hereditary factor, although it is obviously a multichromosomal disposition, with no single gene locus, that could explain the disorder. Some genes that are involved in the metabolism of the biogenic amines, serotonin and dopamine, as well as some genes encoding for intracellular signal transduction, may possibly show aberrations in bipolar disorder. There is also accumulating evidence that intracellular alterations that impact neuron durability may play a crucial role in the long-term prognosis of bipolar disorder; some commonly used medications such as lithium or anticonvulsants have demonstrated positive effects on cellular survival.

Psychological explanations for the basis of mood disorders include cognitive, behavioral, and psychoanalytic theories; however, their main focus is depression, whereas mania remains largely unexplained. Only a few promoters of deep psychology (that is, psychological approaches to therapy and research that take the unconscious into account) have developed theories on mania; they mainly suggest that the role of mania is to fight back and suppress depression.

Social and environmental factors are also recognized as important contributors to the actual manifestation of bipolar disorder. First episodes of both depression and mania frequently manifest themselves in times of increased stress (positive or negative). The importance of a stressor for consecutive episodes, however, seems to decrease to a point where the timing of a new episode appears to become unpredictable.

Treatment of bipolar disorder. Current treatment of bipolar disorder is based on two areas of expertise: (1) biological treatment, which includes both medication and, if indicated, physical treatments (for example, electroconvulsive therapy); and (2) psychotherapy, including psychoeducation (wherein the patient is provided with knowledge about the psychological condition, the causes of that condition, and the reasons why a particular treatment might be effective for reducing symptoms). Additionally, any necessary social support should be arranged to attenuate the level of stress that may otherwise compromise treatment success.

Biological treatment can be roughly divided into acute treatment of mania, mixed states, or depression, and long-term prophylactic treatment that aims to prevent a recurrence. Expert opinion and research evidence both support the need for a mood stabilizer (a substance with both acute and prophylactic efficacy) to be used throughout the course of the disorder. Depending on the prevalent polarity (either more manic or depressive episodes), the choice includes lithium or various anticonvulsants (for example, valproate, carbamazepine, or lamotrigine) as commonly accepted mood stabilizers. Some so-called atypical antipsychotics have also demonstrated efficacy both against acute episodes and for maintaining a stable mood (these are designated as "atypical" because they are usually more recent, second-generation antipsychotics with fewer side effects than "typical," or first-generation, antipsychotics). Two of them (namely, quetiapine and olan-

zapine) have demonstrated bimodal efficacy in controlled studies, meaning that they treat and prevent both manic and depressive episodes. The use of typical antipsychotics and of antidepressants, however, remains controversial. Typical antipsychotics (for example, haloperidol) may be highly effective in mania, but at the expense of extrapyramidal (motor system) side effects such as stiffness and tremor, and may lack prophylactic efficacy. On the other hand, some antidepressants may increase the risk of a switch into a manic episode without necessarily providing additional benefit for the treatment of acute depression.

Biological treatment has a reasonable evidence base for adults with bipolar I disorder. However, there is little evidence investigating specific treatments in bipolar II patients. There is also a paucity of controlled studies on the treatment of this condition in the elderly and in children and adolescents; these age groups differ from adults in their rate of metabolism and thus may show differences in efficacy and tolerability of a given medication. It is only recently that controlled studies with some mood stabilizers and atypical antipsychotics have been conducted in adolescents.

Psychological treatments in bipolar disorder mainly follow a cognitive-behavioral approach. In combination with medication, they have demonstrated additional benefit in treating depressive episodes and maintaining mood stability. A technique with proven prophylactic effects, psychoeducation, can be administered in groups, making it also more cost-effective than single face-to-face psychotherapy.

For background information *see* AFFECTIVE DISORDERS; ANXIETY DISORDERS; ATTENTION DEFICIT HYPERACTIVITY DISORDER; BEHAVIOR GENETICS; BRAIN; ELECTROCONVULSIVE THERAPY; NEUROBIOLOGY; PSYCHOPHARMACOLOGY; PSYCHOSIS; PSYCHOTHERAPY in the McGraw-Hill Encyclopedia of Science & Technology.   Heinz Grunze

Bibliography. K. N. Fountoulakis, The contemporary face of bipolar illness: Complex diagnostic and therapeutic challenges, *CNS Spectr.*, 13:763–769, 2008; F. K. Goodwin and K. R. Jamison, *Manic-Depressive Illness*, 2d ed., Oxford University Press, New York, 2007; D. J. Miklowitz, Adjunctive psychotherapy for bipolar disorder: State of the evidence, *Am. J. Psychiatry*, 165:1408–1419, 2008; A. R. Newberg et al., Neurobiology of bipolar disorder, *Expert Rev. Neurother.*, 8:93–110, 2008; G. Sanacora, Reviewing medications for bipolar disorder: Understanding the mechanisms of action, *J. Clin. Psychiat.*, 70(1):e02, 2009.

Body self-perception

When we look at ourselves, we immediately recognize our body as our own. The question of how this comes to be has been discussed by philosophers and psychologists for centuries. Recently, cognitive neuroscience studies have begun to identify the perceptual processes and brain mechanisms involved in body self-perception. This includes experiments investigating how we feel ownership of our limbs and our entire bodies, why we experience that we are "inside" our physical body, and how the brain distinguishes between sensory signals from objects in the external world and from parts of the body. This research is important because the understanding of how we recognize our own bodies is a significant first step for understanding self-awareness more generally. Furthermore, it can also lead to important new medical and industrial applications. For example, in building prosthetic limbs that feel more like real limbs, and simulated bodies in virtual reality (the computer-generated simulation of an environment) that feel just like real bodies.

Investigations of mechanisms and processes. The first evidence that specific mechanisms are involved in body self-perception came from the clinical literature. Patients who have suffered a stroke affecting the frontal and parietal regions, mainly in the right hemisphere of the brain, can develop conditions with disturbed perception of their own body. Some of these patients perceive parts of their bodies as belonging to someone else (a condition known as somatoparaphrenia) or develop asomatognosia (a condition in which the patient develops a deficit in body awareness that can take the form of denying, ignoring, forgetting, disowning, or misperceiving the body). Although these cases indicate that the frontal and parietal association cortices are associated with body perception, they do not pinpoint the specific brain mechanisms involved because typically the lesions are large and affect multiple areas, including the underlying white matter tissue (the axonal compartment of myelinated nerve fibers).

Behavioral and brain imaging studies in healthy individuals can directly aid in investigations of the perceptual and neuronal processes underlying body self-perception. However, experimentally manipulating body self-perception is a challenge because "the body is always there," as the American psychologist and philosopher William James remarked over a century ago. One method of tackling this issue is to use perceptual illusions. The study of illusions is a classical approach adopted in psychology to learn more about the basic processes that underlie normal perception. One particularly informative illusion is the "rubber hand illusion," where people experience a prosthetic hand as their own hand. When synchronous touches are applied to a rubber hand that is in full view and the real hand, which is hidden behind a screen, most individuals will sense the touches on the rubber hand and experience the artificial limb as their own. There are two commonly used objective tests for this effect. First, when people who are experiencing the illusion are asked to close their eyes and point toward their stimulated hand using their other hand, they tend to point toward the rubber hand rather than the real hand. Second, physical threats to the rubber hand are scary to the participants and result in increased sweating of the palms, which can be registered with "skin conductance responses" (SCRs). Importantly, the illusion

breaks down if the touches applied to the two hands are asynchronous, if the rubber hand is not aligned in parallel with the person's real hidden arm, if the rubber hand is replaced by an object that does not resemble a limb, or if the direction of the strokes applied to the two hands is not the same. These observations show that spatially and temporally congruent visual and tactile signals in arm-centered reference frames are crucial for the feeling of ownership of a limb.

Recent experiments. The self-attribution of entire bodies seems to depend on similar processes, as demonstrated by recent experiments. In one experiment, people experienced an illusion that they were outside their real body ("out-of-body illusion"). The participants wore head-mounted displays (HMDs, display devices that are worn on the head in front of their eyes) that were connected to two closed-circuit television cameras placed about 1.5 m behind them. The two cameras provided a stereoscopic image, and the participants could thus see themselves from the point of view of the cameras, that is, from the back. The experimenter then jabbed a rod toward a location just below the cameras while simultaneously touching the participant's chest, which was out of view. The visual impressions of a hand approaching a point below the cameras and the felt touches on the chest led the participants to experience the illusion of being located 1.5 m behind their real body. Interestingly, many individuals reported the feeling that their real body, which they observed from the back, belonged to someone else; that is, they seemed to experience a partial loss of self-identification with that body (**Fig. 1**). Similar to the rubber hand illusion, physical threats to the "illusory body" below the cameras produced enhanced SCRs. This study illustrates how the perception of where one is located in space is determined by the visual first-person perspective in combination with

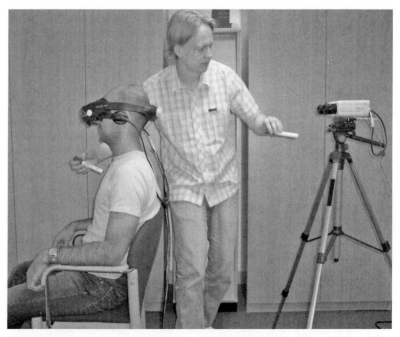

Fig. 1. Experimental setup to induce an out-of-body experience.

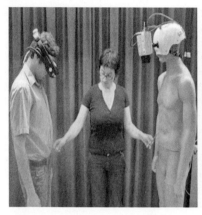 

Fig. 2. Experimental setup to induce the perceptual illusion of having a new artificial body (left panel); the view from the cameras (right panel).

correlated visual and tactile signals in body-centered reference frames.

A subsequent study demonstrated more directly that people can perceive a new body as their own. In these experiments, the two cameras were attached to a helmet worn by a life-size mannequin and positioned so that they were looking down on the mannequin's body. Thus, when the participants wore the HMDs connected to these cameras and looked down, they saw the mannequin's body where they would expect to see their own real body (**Fig. 2**). When the experimenter used a couple of pens to simultaneously touch the mannequin's belly and the person's belly at corresponding sites for a minute, the majority of the participants began to experience the artificial body as their own. Impor-

tantly, this "body-swap illusion" works only when a humanoid body is used; when the mannequin is replaced with a rectangular block of wood, the illusion breaks down immediately.

The body-swap illusion can easily be produced with another human individual simply by attaching the cameras to a helmet worn by another person. In one dramatic example of this, the test person experienced "owning" the scientist's body, which was facing his or her real body, while shaking hands with it. The cameras were mounted on the scientist's head and connected to the HMDs worn by the participants, who then looked at themselves from the scientist's perspective. When the scientist and the participant repeatedly squeezed their hands in a synchronized fashion, most participants experienced an illusion of being "inside" the scientist's body and owning the scientist's hand (**Fig. 3**). Strikingly, people were more scared when they saw a knife close to the scientist's arm than when the knife approached their own real arm during the illusion, as indexed by the SCRs.

Taken together, these observations demonstrate that there are a number of factors that contribute to the perception of an object as one's own body. First, the object has to look sufficiently similar to a human body. Second, visual, tactile, proprioceptive, and other sensory signals from the body must be temporally and spatially correlated in coordinate systems centered on the body. Third, when executing voluntary movements, the sensory feedback must match the expected sensory feedback from the intended movements. Fourth and last, the visual information from the first-person perspective plays an important role in establishing the location of the perceived body relative to environmental landmarks and in defining the "origin" of the body-centered reference frame. The perception of one's body is thus continuously constructed on the basis of the available sensory evidence guided by these principles, demonstrating a remarkable dynamic nature of the body representation.

Brain and neuronal involvement. Brain imaging studies in humans and neurophysiological studies in nonhuman primates suggest that the neuronal substrates of body self-perception involve multisensory areas in the frontal and parietal lobes that receive convergent visual, tactile, and proprioceptive afferent inputs. Of particular interest are neurons in the ventral premotor cortex and areas in the intraparietal cortex that integrate multisensory information in limb-centered reference frames from the space near the body. These neurons are strong candidates for mediating the perception of a limb as one's own because human functional magnetic resonance imaging experiments have found significantly increased activation in these areas when people experience the rubber hand illusion. Furthermore, the stronger the activity in these areas, the stronger the participants report that they are experiencing the illusion, and the stronger the neuronal responses in areas related to pain anticipation when the rubber hand has been injured. It is likely that ownership of entire bodies

Fig. 3. Experimental setup to induce the illusion of swapping bodies with another individual.

involves similar multisensory mechanisms, perhaps with the addition of spatial processing in the right inferior parietal cortex related to the identification of where the body is located in the environment.

Outlook. Understanding the perceptual and brain basis of body self-perception represents a major advance in the study of body awareness and self-consciousness. Moreover, the clarification of the principles that determine whether or not an object is perceived as oneself can contribute to the development of new clinical and industrial applications where the self-perception of the body is deliberately manipulated. For example, some data indicate that one could use the rubber hand illusion to enhance the feeling of ownership of artificial limbs used by amputees. Furthermore, the projection of ownership onto simulated bodies represents a new direction in virtual reality research, which could enhance user control, realism, and the feeling of "presence" in industrial, educational, and entertainment applications.

For background information *see* BRAIN; COGNITION; COMPUTER VISION; CONSCIOUSNESS; INFORMATION PROCESSING (PSYCHOLOGY); NERVOUS SYSTEM (VERTEBRATE); PERCEPTION; PSYCHOLOGY; PSYCHOPHYSICAL METHODS; SENSATION; VIRTUAL REALITY in the McGraw-Hill Encyclopedia of Science & Technology.   Valeria I. Petkova; Henrik Ehrsson

Bibliography. M. Botvinick and J. Cohen, Rubber hands "feel" touch that eyes see, *Nature*, 391:756, 1998; H. H. Ehrsson, Rubber hand illusion, in T. Bayne, A. Cleeremans, and P. Wilken (eds.), *The Oxford Companion to Consciousness*, Oxford University Press, Oxford, 2009; H. H. Ehrsson, The experimental induction of out-of-body experiences, *Science*, 317:1048, 2007; V. Petkova and H. Ehrsson, If I were you: Perceptual illusion of body swapping, *PLoS One*, 3(12):e3832, 2008.

Brand security in packaging

Brands are developed and supported to provide consumers with a trusted experience or expected quality level, which is expected to result in repeat sales and a premium price over unknown products. The protection of brands from attacks, illegal infringement, or risks is called brand security and is also sometimes referred to as brand protection, brand integrity, or product protection.

Packaged and bulk consumer products include food and beverages, consumer packaged goods (CPGs), consumer electronics, automotive parts, pharmaceuticals and healthcare products, clothing and accessories, toys, and tobacco. Brand security covers a wide range of threats including counterfeiting, diversion/parallel trade, tampering, child resistance, wholesale theft, shoplifting, warranty fraud, and return fraud. Counterfeit threats do not include currency counterfeiting, document forgery, copyright piracy, or artwork forgery. In a corporation, the packaging and brand integrity functions work together to implement solutions that leverage current strengths and directly address specific risks. With a strategic focus and an understanding of the root causes of the threat, usually a packaging component or supply chain process can provide security for multiple types of threats. A major focus of brand security has been on anticounterfeit and antidiversion strategy, and in authentication of product.

Definitions. When considering brand security and anticounterfeiting, it is important to consider the risks to the brand and the motivations of the fraudsters.

Counterfeiting. Counterfeiting is the copying of a product or package to deceive others into believing the product or package is genuine. Counterfeit products are also called knockoffs or fakes. Michigan State University research defined counterfeiting as both a macro term for this whole category of actions as well as a micro term, where everything about the product, package, documentation, and supply chain are fraudulent. The types of counterfeiting include adulteration, tampering, theft, unauthorized production (including licensee fraud, used-product remanufacturing, illegal repackaging, and unauthorized refilling), diversion (including illegal parallel trade, smuggling, and origin laundering), simulation, and full counterfeiting. An example of adulteration-type counterfeiting is adding melamine, a counterfeit additive, to pet food. Counterfeiting is illegal because it is an intellectual property rights infringement consisting of the unauthorized use of the trademark (such as a logo or brand name) or a patent (such as a recipe or design).

Diversion. Diversion is the distribution of a genuine product outside of its intended market. It is also referred to as parallel trade, gray market, secondary market, product arbitrage, or smuggling. Depending on local laws and the specific activity, diversion is not usually illegal. Nevertheless, it creates a transfer of the ownership of product, which is a major opportunity for counterfeit product to be introduced into the supply chain.

Piracy. A term often incorrectly used and interchanged with counterfeiting, piracy is the unauthorized use of a copyrighted work such as a song, a movie, a book, or computer software.

Counterfeiting. The root motivation for counterfeiting is economic gain, and vulnerabilities include profit potential, a product that is cheap and easy to copy, unsatisfied market demand, difficulties in detection or proof, and lack of deterrent laws or enforcement. Reasons for the growth in counterfeiting include:

1. Availability and growth of technology.

2. Increased globalization.

3. Low legal penalties.

4. The influence and prevalence of organized crime.

Scope and scale. Because of the clandestine nature of counterfeiting, a precise quantitative measurement of product counterfeiting is not only unknown but also unknowable. Even though we have data on the value of counterfeit products seized at U.S. borders, with the increased stealth and production

TABLE 1. Examples of counterfeit products

Product	Counterfeit attribute
Toothpaste	Contains diethylene glycol
Pet food	Contains melamine
Infant formula	Expired and relabeled
Electrical extension cord	Catches fire
Cell-phone battery	Explodes
Car parts	Fake
Aircraft parts	Illegally remanufactured
Vaccines	Expired
Blood thinner	Contains undeclared allergens
Cholesterol drugs	Wrong active ingredient
Birth control pills	No active ingredient

Fig. 1. Hand-held label authentication device. (*Courtesy of Reconnaissance International***)**

quality of the counterfeiters, what percent is caught is only a guess. In addition, much of the data is considered classified by governments or confidential by companies. From numerous examples, we know that the threat to public health is tangible and the impact on the global economy is in the $500 billion range, frequently estimated at 5–7% of world trade. The annual tax loss for New York City is about $1 billion, and for the City of Los Angeles it is about $400 million. The U.S. Federal Bureau of Investigation has called counterfeiting "possibly the crime of the 21st century," and the Council of Europe has called medical drug counterfeiting "a silent pandemic." What is surprising is that probably no more than 10% of all counterfeits are luxury goods.

Table 1 shows some examples of counterfeit consumer products, vehicle parts, and pharmaceuticals. The range of criminals and the range of actions will continue to be more aggressive, bolder, and more effective at infiltrating the legitimate supply chain.

Strategic focus. The strategic focus for brand security should be on the specific risks, such as lost sales, lost brand premium or reputation, lost distribution network control or exclusivity, consumer liability, and business liability of returned product or recall. Consideration must be made for where the product or supply chain is being compromised, where the product will be authenticated, who will authenticate and by what methods, and how the gathered intelligence will be used (**Fig. 1**).

Countermeasures. The brand security or anticounterfeiting industry is estimated at between $2 and $40 billion annually, depending on how the boundaries are drawn. For example, a hologram could be used for decoration or for authentication, or both (**Fig. 2**). It must be emphasized that there is no single solution that will provide complete brand security. An interdisciplinary and systems anticounterfeit strategy approach is necessary.

Brand security countermeasures fall into specific categories, based on the specific risk. Security is implemented based on specific objectives, including supply chain optimization, traceability, and authentication. The goal of authentication is deterrence, by making it more difficult to counterfeit, and detection, by including some way to uniquely identify the product. However, all countermeasures must consider the probable response by counterfeiters. Authentication can either prove a product genuine or fake. The goal of the action must be considered, such as to detect and remove product from the marketplace, market surveys and intelligence gathering, or for prosecution in court.

The 2004 U.S. Food and Drug Administration report, *Combating Counterfeit Drugs*, outlined several actions needed for a multilayered, public and private anticounterfeiting initiative:

1. Secure the product and packaging.
2. Secure the movement of drugs through the supply chain.
3. Secure business transactions.
4. Ensure appropriate regulatory oversight and enforcement.
5. Increase penalties.
6. Heighten vigilance and awareness.
7. Secure international cooperation.

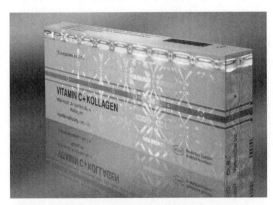

Fig. 2. Hologram (produced by Pura Barutuma PT, Indonesia) on a medicine package. (*Courtesy of Reconnaissance International***)**

TABLE 2. Review of anticounterfeiting technologies

Anticounterfeiting technologies	Description	Examples
Overt	Visible components, features, or markings.	Unique package design, complex printing, holograms, special inks, special printing, etc.
Covert	Not readily apparent or not visible without a special tool or process.	Microprint, special inks or printing, embedded printing, microtags, etc.
Forensic or reserved	Not usually discernable without special equipment, usually in a laboratory. This is often considered extremely confidential.	The previous examples but in smaller or more complex systems.
Traceability	System to monitor where a product has been, where it is going, possibly including a unique serial number for identification.	Serial numbers, bar codes, invoice tracking, automatic identification such as RFID

All features are usually considered extremely confidential and revealed only in a court action. The technologies are frequently changed or updated to add complexity to the system. Some components are commercially available, though many systems or components are proprietary. A multilayered approach is taken to increase the complexity of technologies a counterfeiter would need to master the operation. In some situations, upgrading to a new printing or packaging technology, or to collateral materials, may raise the time or cost to produce product, and thus deter counterfeiters (**Fig. 3** and **Table 2**).

Implementation. The entire marketplace of brand owners, distributors and shippers, academics, nongovernmental organizations, and governments must work together to understand and control counterfeiting and diversion. All parties must improve procurement and handling procedures to reduce the opportunity for introduction or distribution of counterfeits. All parties must work together, first to make sure that strong intellectual property rights laws are in place, and then to work together to assure efficient enforcement. In some cases, even in the United States, some counterfeiting was not explicitly illegal until recently. All parties must work together to incorporate and coordinate components and systems to increase the transparency of the product and distribution.

The steps to implementing an anticounterfeit strategic solution are to (1) gather information on company and industry threats, (2) gather intelligence on the counterfeiters and their technical and supply chain capabilities, (3) review current systems and features, (4) agree on action to either deter or detect, (5) decide on the level of complexity and system integration, (6) decide on the component and system, and then (7) constantly review the marketplace for new, emerging, or evolving threats.

For background information *see* FOOD MANUFACTURING; FOOD SCIENCE; HOLOGRAPHY; INK; PHARMACEUTICALS TESTING; PRINTING; SUPPLY CHAIN MANAGEMENT in the McGraw-Hill Encyclopedia of Science & Technology.   John Spink

Bibliography. Coalition Against Counterfeit and Piracy (CACP), *No Trade in Fakes, Brand Integrity Tool Kit*, U.S. Chamber of Commerce, Coalition Against Counterfeit and Piracy (CACP), 2006; K. Eban, *Dangerous Doses: How Counterfeiters Are Contaminating America's Drug Supply*, Harcourt, New York, 2005; Food and Drug Administration, *Combating Counterfeit Drugs*, U.S. Food and Drug Administration, 2004; Grocery Manufacturer's Association, *Global Product Counterfeiting*, 2007; D. M. Hopkins, L. T. Kontnik, and M. T. Turnage, *Counterfeiting Exposed: Protecting Your Brand and Customers*, John Wiley & Sons, New Jersey, 2003; Organization for Economic Co-operation and Development, *The Economic Impact of Counterfeiting and Piracy, Executive Summary*, 2007.

Fig. 3. Meditag® polarizing hologram label, including serial number, used to protect Malaysian pharmaceutical products. (*Courtesy of Reconnaissance International*)

Carbon capture and storage

International concern about climate change is increasing, and this major world issue is not without its controversy. At the heart of debate is whether human-made carbon dioxide (CO_2) emissions affect the Earth's climate; most in the scientific community agree that such anthropogenic greenhouse gases are the primary cause of global warming. Recent studies also suggest other reasons to consider reducing CO_2 emissions to the atmosphere, including an across-the-globe increase in the ocean's acidity, thought to be responsible for major effects on coral reefs and other significant oceanic ecosystems.

It is clear that a single approach to solve the global carbon emissions problem does not exist. Rather, a portfolio of options must be deployed, including increasing use of renewable energy sources such as

Fig. 1. Schematic of geologic carbon sequestration, showing CO₂ piped from source to injection wells for storage in different subsurface reservoir formation options. (*Genevieve Young*)

alluvium

freshwater

confining layer

oil reservoir

confining layer

gas reservoir

confining layer

unmineable coal

saline water

capture the CO_2 during or after the fossil-fuel combustion process and sequester it below the Earth's surface in deep rock formations. Given that the CO_2 was originally in place underground but in the form of oil, natural gas, or coal, it is perhaps logical to "put it back" in the form of pure CO_2 and store it permanently. This approach is commonly called geologic carbon sequestration.

Geologic carbon sequestration. The three primary types of geologic carbon sequestration (**Fig. 1**) are (1) injection and storage in oil and gas reservoirs, (2) injection and storage in coal beds, and (3) injection and storage in deep brine aquifers. Sequestration in oil and gas reservoirs is usually considered only if the reservoirs are depleted or otherwise abandoned. Active fields are typically not considered as options because of the possibility that injection and storage operations might interfere with oil and gas production. Regardless of field status in this context, abandoned wells are a major concern in oil and gas field sequestration, because these may serve as leakage pathways to the surface. Similarly, for storage in coal beds, the most viable candidates are not mineable (Fig. 1), because such beds are not valuable for coal production. Deep saline formations are extremely attractive because they offer the greatest storage capacity and the brines within them are not considered a useful commodity. More important, the deeper the formation, the less is the risk of CO_2 making its way back to the surface through surface-exposed faults or abandoned wells, because such features do not typically penetrate to deeper formations.

CO₂ storage. Injected CO_2 will only stay in its intended storage reservoir if it is effectively "trapped" by one or more trapping mechanisms, processes that inhibit fluid migration. The four primary trapping mechanisms are hydrostratigraphic, residual gas, solubility, and mineral trapping.

Hydrostratigraphic trapping. Hydrostratigraphic trapping refers to trapping of CO_2 by low-permeability confining layers (Fig. 1). In general, CO_2 is trapped in permeable rock units in which the fluid flow is constrained by upper and lower, less permeable "barrier" lithologies. Such top and bottom seals are often formed by shale or salt units; lateral flow barriers may be due to rock composition changes or to faults. Faults and fractures may affect fluid flow. In some cases, faults/fractures may be sites for preferential fluid flow, whereas in other cases they may inhibit fluid flow. Deep saline units typically have large lateral extents, whereas oil and gas reservoirs are typically much smaller. Although reservoirs may be classified by the nature of trapping mechanism, the geologic community tends to distinguish them on the basis of lithology—that is, clastics such as sandstone versus carbonates such as limestone.

Residual gas trapping. At the interface between two different liquid phases (such as CO_2 and water), the cohesive forces acting on the molecules in either phase are unbalanced. This imbalance exerts tension on the interface, causing the interface to contract to as small an area as possible. The importance of this

wind, solar, and geothermal, increasing non–fossil-fuel energy sources such as nuclear power, and increasing efficiency in general. One limitation of the portfolio approach is the time scales of deployment. To replace all CO_2-emitting power plants with new forms of energy will require decades. Therefore, many strategies for immediate CO_2 emissions reduction are being considered. One such strategy is to

interfacial tension in multiphase flow is paramount; physical interactions in CO_2–brine–oil–gas flow are more sensitive to interfacial tension than many other fluid properties. Interfacial tension may trap CO_2 in pores if fluid saturations are low. The threshold at which this occurs is called the irreducible saturation of CO_2, and is a key concept for defining residual gas trapping. The magnitude of residual CO_2 saturation within rock, and thus the amount of CO_2 that can be trapped by this mechanism, is a function of the rock's pore network geometry as well as fluid properties. Geologic conditions that affect the amount of CO_2 trapped as a residual phase include rock density and porosity, burial effects, temperature and pressure gradients, CO_2 density and viscosity under different pressure and temperature conditions, and engineering parameters such as injection pressure, induced flow rates, and well orientation.

Solubility trapping. Perhaps the most fundamental type of trapping is dissolution, or solubility trapping. First, CO_2 dissolves to an aqueous species [Eq. (1)], followed by rapid dissociation of carbonic

$$CO_2(g) + H_2O \rightarrow H_2CO_3 \qquad (1)$$

(relatively slow rate)

acid to produce bicarbonate [Eq. (2a)] and carbonate [Eq. (2b)] ions, while lowering pH. This leads to

$$H_2CO_3 \rightarrow H^+ + HCO_3^- \qquad (2a)$$

(relatively fast rate)

$$HCO_3^- \rightarrow H^+ + CO_3^{2-} \qquad (2b)$$

(relatively fast rate)

a series of additional reactions and mineral trapping, discussed in the next section. The amount of sequestration that is possible through solubility trapping is very limited per unit mass of water, because groundwater (brine) can only dissolve up to a few percent or less, depending on pressure (P), temperature (T), and salinity. Over large volumes of reservoir, however, solubility trapping may provide a significant amount of storage.

Mineral trapping. Mineral trapping refers to the process in which CO_2 reacts with cations to form mineral precipitates in the subsurface. The reactions, especially reaction rates and associated processes that affect rates (such as complexation, pH buffering, and so on), are complicated and make estimates of CO_2 storage capacity difficult. However, mineral trapping is assumed to be a relatively safe mechanism that may sequester CO_2 for millions of years. Although mineral trapping may not be permanent, it can certainly render CO_2 immobile for very long time scales. The main sources of uncertainty associated with mineral trapping are associated with the kinetic rate coefficients and reaction-specific surface areas of minerals for the many homogeneous and heterogeneous reactions.

Potential subsurface storage capacity. In 2008, the U.S. Department of Energy assembled a summary of the total emissions of CO_2 per year in North America,

and also estimated the total potential CO_2 subsurface storage capacity for North America. According to their data, the minimum total emissions of stationary CO_2 sources in North America are 3.6 billion tons of CO_2 per year. The potential CO_2 storage capacity of North American oil and gas reservoirs is 143 billion tons, the minimum storage capacity of unmineable coal seams is 188 billion tons, and the minimum capacity of deep saline rock formations is 3620 billion metric tons.

Rather than expressing the potential storage capacity in tons, it may be more meaningful to express storage capacity as the number of years of emissions equivalent. Specifically, dividing the minimum total potential CO_2 storage capacity in billions of tons (3951) by the stationary CO_2 emissions source rate (3.6) suggests that almost 1100 years' equivalent of current CO_2 emissions could be injected and stored in the subsurface of North America. Of course, this figure applies only if emissions rates do not change and if associated wells and infrastructure are installed for all sources. More important, a number of limitations must be overcome.

Limitations. All CO_2 trapping mechanisms may fail. Critical objectives are to ascertain the physical and chemical processes of each failure mode and to minimize uncertainties in the characterization, and potential range of response, of those processes under sequestration conditions. Major failure modes and risks include (1) unintended migration by preexisting but unidentified faults, fractures, or other fast-flow paths, (2) unintended migration by stress-induced or reactivated fractures or faults, (3) induced seismicity by stress-induced or reactivated fractures or faults, (4) unintended migration by reaction-induced breaching of a seal layer, (5) unintended lateral flow to unintended areas, (6) catastrophic events (such as unexpected earthquakes), and (7) wellbore failure events.

Risk assessment has different functions depending on the stage in the process at it is implemented. The success of risk assessment depends directly on characterization and monitoring plans for a given sequestration site. Risk minimization plans generally include two primary aspects: (1) programmatic risks, including resource and management risks, which may impede project progress or costs; and (2) sequestration (technical) risks inherent to the scientific and engineering objectives of a sequestration project. **Figure 2** illustrates that technical risks are highest during injection, but programmatic risks may also follow this trend.

Perhaps the most fundamental approach to minimizing risks is to select a storage site with multiple alternating seals and reservoirs above the primary (intended) reservoir, sometimes described as stacked reservoirs. Even when stacked reservoirs are present, other measures must be taken to minimize risk of failure. Effective surface and subsurface monitoring must be in place, especially during injection and for perhaps decades after injection ceases (Fig. 2). The injection field must include mitigation measures such as quick capability to cease injection and

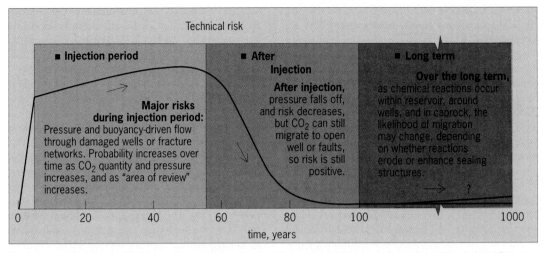

Fig. 2. Schematic of potential risk over time during geologic carbon sequestration operations. (*Grant Bromhal and Doug Jensen*)

perhaps even ability to reverse the flow on wells to reduce reservoir pressure on demand.

The other critical limitation of geologic carbon sequestration is cost. In 2008, the cost of sequestering a ton of CO_2 was thought to approach \$10 (U.S.). Separation of CO_2 from power-plant flue gases is thought to cost three or four times as much (\$30 to \$40 per ton). Given that a typical power plant emits on the order of millions of tons of CO_2 per year, this would add significantly to the cost of electricity.

Practical aspects. Cost limitations will always play a role, and they cannot be completely eliminated. However, to minimize risks, the practical aspects of engineering and operations become critical.

Perhaps the most important practical issue is monitoring. Monitoring design, deployment, and associated subsurface-to-surface modeling are the primary tools for determining how well a specific sequestration project will meet or exceed performance targets stipulated by national and international guidelines. Monitoring and modeling are the means of assessing and confirming storage capacity, verifying indirect sequestration and associated costs, and verifying monitoring technology efficacy. Monitoring technologies vary in spatial coverage (area, depth, and volume) as well as spatial and temporal resolution, depending on whether they are "direct" or "indirect" methods. The most common direct monitoring technologies include CO_2 flux measurements at the surface, trace-gas and isotopic tracers in groundwater and injected CO_2, water chemistry sampling and analyses, and mineralogic analyses of rock surfaces exposed to the injection stream. The most common indirect monitoring technologies are geophysical methods such as two-dimensional (2D) and three-dimensional (3D) seismic imaging, electrical conductivity, electromagnetic methods, and gravimetric imaging.

A major limitation of monitoring design is optimization of the combination of technologies to ensure that the full range of spatial/temporal scales is covered. One approach to this problem is to estimate the maximum resolution/scale of various monitoring technologies, implement these estimates in simulation models of the field site, and then use the models to guide design. Then, as injection proceeds, the actual resolution and effective scales of technologies are determined and used to adjust the simulation models accordingly. Iteratively, the model results are used to redesign and redeploy monitoring arrays, as appropriate. A detailed summary and discussion of monitoring methods is found in a 2009 report published by the National Energy Technology Laboratory. A combination of technologies is typically applied at an injection site.

Another critical aspect of geologic sequestration is indemnification of sequestration sites. Specifically, several insurance companies offer short-term, year-to-year coverage of sequestration sites, but long-term liability is still an outstanding problem. Quantitative risk analysis is necessary for short-term insurance policies and long-term indemnification, and combined monitoring and modeling provide the fundamental data required to assess risk and develop the associated risk-mitigation plans. Injection performance and monitoring data are simultaneously integrated in reservoir simulation models and model forecasts are updated accordingly. These model results may be used to update risk criteria, including any new and unanticipated impacts and processes. In addition, these integrated model and monitoring results may be used to evaluate how the system may evolve after injection ceases. Such an adaptive risk assessment framework can then in turn be used to guide subsequent monitoring and data collection activities, as well as injection engineering design. In summary, combined monitoring, modeling, and risk assessment data may be used to guide engineering operations and vice versa, in the form of a continuous feedback loop.

Perhaps the other major practical issue that determines the size, scope, and limiting goals of a geologic sequestration project is the regulatory regime. A major regulatory component is the U.S. Safe Drinking Water Act, which stipulates injection-well design and use, containment verification requirements, 50 years of postinjection site care, financial assurance requirements, and continual monitoring. This and other federal legislation activities are evolving. Regulations vary by state but stipulate critical aspects of sequestration projects, including site selection, operation and closure requirements, subsurface pore-space or CO_2 ownership, assurance (proof) of long-term financial support of operations, liability, and general jurisdictional issues.

Outlook. In 2009, over 25 geologic sequestration field tests were at various stages of design and deployment in the United States. An additional 25 or so were ongoing or slated for deployment soon in other countries. Most of these tests use different technologies, including different engineering designs, different monitoring approaches, different risk assessment protocols, and different risk minimization strategies. And many of these tests are relatively small in scale—small injection rates compared to typical power-plant emissions output. The uncertainties or error associated with the evaluation and design of large-scale sequestration operations are significant. For large-scale geologic sequestration to be deployed and sustainable over the long term, a realistic (field-based) evaluation of uncertainties, and how these uncertainties affect risk assessment and minimization strategies, must be done. Additionally, the community needs a detailed and meaningful field assessment of CO_2 trapping mechanisms, as well as the physical and chemical factors that may cause the mechanisms to lose efficacy under realistic (field) conditions.

For background information *see* CARBON; CARBON DIOXIDE; CARBONATE MINERALS; COALBED METHANE; OIL FIELD WATERS; PETROLEUM GEOLOGY; RISK ASSESSMENT AND MANAGEMENT; SEISMIC STRATIGRAPHY in the McGraw-Hill Encyclopedia of Science & Technology.   Brian J. McPherson

Bibliography. S. Bachu, W. D. Gunter, and E. H. Perkins, Aquifer disposal of CO_2: Hydrodynamic and mineral trapping, *Energy Conserv. Manag.*, 35:269–279, 1994; C. Doughty and K. Pruess, Modeling supercritical carbon dioxide injection in heterogeneous porous media, *Vadose Zone J.*, 3:837–847, 2004; K. S. Lackner, Climate change: A guide to CO_2 sequestration, *Science*, 300(5626):1677–1678, 2003; F. M. Orr, Jr, Storage of carbon dioxide in geologic formations, *J. Petrol. Tech.*, 56(9):90–97, 2004; S. Pacala and R. Socolow, Stabilization wedges: Solving the climate problem for the next 50 years with current technologies, *Science*, 305 (5686):968–972, 2004; U.S. Department of Energy, *Carbon Sequestration Atlas of the United States and Canada*, 2d ed., Washington, DC, 2008; G. J. Weir, S. P. White, and W. M. Kissling, Reservoir storage and containment of greenhouse gases, *Energy Convers. Manag.*, 36(6–9):531–534, 1995.

Catalytic aftertreatment of NOx from diesel exhaust

Diesel fuel is used to run vehicles and machinery in a variety of applications, including transportation, construction, and portable electricity generation. Pollutants from diesel exhaust—in particular, nitrogen oxides (NO and NO_2), commonly referred to as "NOx," and particulate matter (PM) in the form of soot—represent a significant fraction of the totals emitted in urban areas. Until the 1990s, emissions from diesel engines were not restricted to the same extent as those from gasoline vehicles. However, during the past 15 years, there has been an aggressive policy to reduce the emissions of NOx and PM from diesel exhaust for heavy-duty diesel vehicles (**Fig. 1**) and nonroad diesel engines (see **table**). A set of photochemically catalyzed reactions involving NOx and hydrocarbons that occur in the atmosphere produces ground-level ozone, a respiratory irritant and a constituent of smog. In addition, particulate soot contains known carcinogens. Thus, the elimination of NOx and PM is essential for improving the air quality of major urban areas.

The combustion of diesel fuel is more efficient than that of other fuels, such as gasoline, elevating its stature in terms of reducing carbon dioxide emissions. One reason for its higher efficiency is that diesel combustion operates in a fuel-lean mode. This produces an exhaust gas that contains a considerable amount of unconverted oxygen (5–10% by volume), whereas the exhaust of a gasoline-powered vehicle contains only trace amounts (0.5–2%). Both engine types produce undesired pollutants, including unconverted volatile organic hydrocarbons (HC), carbon monoxide (CO), and NOx, which is produced during high-temperature fuel combustion. The "three-way catalytic converter" (or TWC) is the standard aftertreatment technology for minimizing the three pollutants HC, CO, and NOx in gasoline vehicle exhaust. Several chemical reactions are

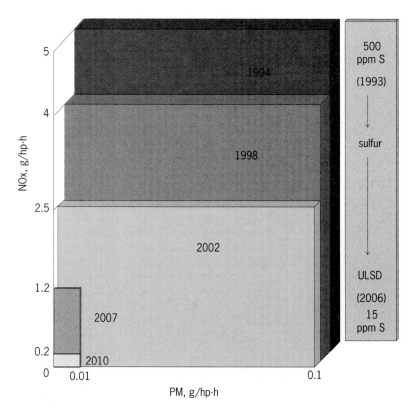

Fig. 1. NOx and particulate matter emission limits for heavy-duty diesel vehicles.

U.S. EPA Tier 4 emission standards in grams per brake horsepower-hour g/(kWh) for nonroad diesel engines			
Rated power, P	Year (phase-in)	PM, g/kWh	NOx, g/kWh
$P < 19$ kW	2008	0.3	5.6*
$19 \leq P < 56$ kW	2013	0.022	3.5*
$56 \leq P < 130$ kW	2012–2014	0.015	0.3
$130 \leq P < 560$ kW	2011–2014	0.015	0.3
$P \geq 560$ kW (generators)	2011–2014	0.022	0.5
all other		0.03	2.6

*NOx and NMHC (nonmethane hydrocarbons).

carried out on the TWC catalyst, which consists of a high-surface-area alumina washcoat impregnated with the precious metals platinum, palladium, and rhodium (Pt, Pd, and Rh) and ceria (CeO_2). The HC and CO are converted to CO_2 and H_2O, while the NOx is converted to molecular nitrogen (N_2). There is enough O_2 present in gasoline engine exhaust to accomplish the oxidations, but not enough to inhibit the reduction of NOx to N_2. On the other hand, while the excess O_2 in diesel exhaust enables the efficient oxidation of HC, CO, and particulate soot, the reduction of NOx using conventional TWC technology is ineffective.

As a result of the more stringent emissions rules and the emission aftertreatment challenges, research is underway to develop cost-effective and reliable technologies to eliminate NOx and particulate soot from diesel exhaust. Indeed, the modern diesel vehicle is becoming a chemical plant on wheels. Here we consider NOx emissions; the elimination of PM involves an entirely different set of challenges.

The NOx challenge is to create favorable conditions in the exhaust for reducing NOx. This means either using a reductant and catalyst that are effective in a lean environment or creating a reducing environment, at least intermittently, to allow conventional NOx reduction chemistry to occur. Either approach comes at a cost in terms of additional fuel consumption and investment. These costs must be minimal; otherwise, the clean diesel vehicle would not be competitive in the marketplace. Two competing technologies are emerging that meet these demands. The first is selective catalytic reduction of NOx with urea as the reductant source, and the second is NOx storage and reduction with diesel fuel as the reductant source.

Selective catalytic reduction. Selective catalytic reduction, commonly abbreviated as SCR, was developed from the successful large-scale technology for eliminating NOx from the exhaust of power plants and other large-scale combustion processes. Developed in the 1970s and 1980s, SCR uses anhydrous ammonia (NH_3) as the reductant. The main chemical reactions [(1) and (2)] are successfully carried

$$2NH_3 + 2NO + 0.5O_2 \rightarrow 2N_2 + 3H_2O \quad (1)$$

$$2NH_3 + NO_2 + 0.5O_2 \rightarrow 1.5N_2 + 3H_2O \quad (2)$$

out on a catalyst containing a mixture of vanadia and titania (V_2O_5/TiO_2), supported on a honeycomb support structure. The primary success of this process is the high selectivity achieved in promoting

the NOx reduction in a large excess of oxygen. In addition, the catalyst maintains high activity in the presence of combustion products, H_2O and CO_2, together with tolerance for commonly present sulfur-containing compounds, such as sulfur dioxide (SO_2), since many fuels contain small amounts of sulfur.

Researchers recognized that NOx reduction in diesel exhaust must meet similar performance requirements, but with a couple of new challenges. First, an alternative, portable source of ammonia is needed because using pressurized anhydrous NH_3 on a vehicle is not acceptable from a safety standpoint. Second, an alternative catalyst is needed because of the toxicity of the vanadia. This led to the discovery and development of aqueous urea (H_2NCONH_2) as the ammonia source and a copper- or iron-promoted zeolite as the catalyst. A concentrated solution of aqueous urea, when injected into a hot exhaust stream, leads to the vaporization of the water and decomposition of the urea into a mixture of NH_3 and CO_2. The exhaust containing the NH_3 and NOx flows through a monolith structure whose walls are coated with the Cu- or Fe-promoted zeolite catalyst. The injection of the urea solution is controlled by an onboard system that relies on measurements of the engine load and/or exhaust NOx concentration. This urea SCR technology is currently recognized as the best option for heavy-duty diesel vehicles for which the cost of the urea system can be justified.

NOx storage and reduction. Notwithstanding the successful development of urea SCR technology, research efforts are underway to develop alternatives that avoid the need for the onboard urea system. The incentive is to avoid this fixed cost by exploiting the reductive capability of diesel fuel itself. To this end, NOx storage and reduction, commonly abbreviated as NSR, and variants are being researched. NSR is carried out in a lean NOx trap (LNT), which is essentially an advanced version of the classical three-way catalytic converter mentioned earlier. The main difference is that the NOx reduction must be accomplished through periodic operation on a multifunctional monolithic catalyst.

In the first of two sequential steps, NOx contained in the lean exhaust from the diesel engine is trapped by the LNT. This is accomplished by the catalytic oxidation of the NO to NO_2 on the precious-metal function (typically Pt) of the LNT catalyst. The NO_2 is then trapped by reaction with an alkaline earth metal oxide, such as barium oxide (BaO), forming a mixture of barium nitrite [$Ba(NO_2)_2$] and barium nitrate [$Ba(NO_3)_2$] (**Fig. 2**). The duration of the

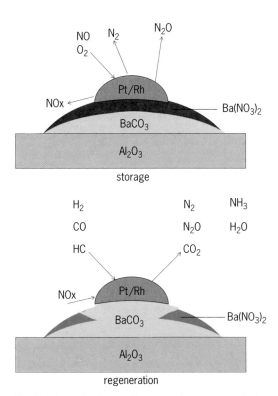

Fig. 2. Schematic diagram illustrating the processes during NOx storage and reduction.

storage is typically 1–3 min, depending on a number of factors, such as the exhaust temperature and flow rate. Before any NOx breaks through, because any NOx that is not trapped leaves the LNT unconverted, the second step, called the LNT regeneration, is carried out. In this step, the engine is briefly operated in a fuel-rich mode to generate an exhaust that is devoid of oxygen and contains a higher concentration of incomplete combustion products such as CO, H_2, and low-molecular-weight hydrocarbons. These fuel by-products are effective reductants, and they react readily with the stored NOx components on the precious-metal crystallites (Pt, Rh). The duration of this step is very short, lasting only a few seconds. Unlike the rather simple chemistry during the storage step, the regeneration is very complex, with a large number of competing reactions producing several nitrogen-containing products. During the initial stage of the regeneration, the role of the reductant is to react with any oxygen that is trapped on the catalyst, usually in the form of oxygen atoms adsorbed on the precious metal. This serves to clean the metal surface, which enables the desired NOx reduction chemistry to occur. Recent studies have revealed that the nitrates and nitrites decompose near the interface with the Pt crystallites, leading to NO adsorbed on the Pt surface. The NO reacts with adsorbed hydrogen and CO [reactions (3)–(6)].

$$NO + 2.5H_2 \rightarrow NH_3 + H_2O \qquad (3)$$

$$NO + H_2 \rightarrow 0.5N_2 + H_2O \qquad (4)$$

$$NO + 0.5H_2 \rightarrow 0.5N_2O + 0.5H_2O \qquad (5)$$

$$NO + CO \rightarrow 0.5N_2 + CO_2 \qquad (6)$$

Other reactions involve water reacting with CO and hydrocarbons, producing more H_2. These reactions are reversible, so the reverse reaction can inhibit the reaction extent [reactions (7) and (8)].

$$CO + H_2O \leftrightarrow H_2 + CO_2 \qquad (7)$$

$$C_2H_4 + 4H_2O \leftrightarrow 6H_2 + 2CO_2 \qquad (8)$$

Reactions involving other hydrocarbons that are similar to the example involving ethylene in reaction (8) also occur. As noted from these reactions, N_2 is not the only N-containing product during LNT regeneration. The formation of the greenhouse gas N_2O is to be minimized, especially at low temperatures. Ammonia is a particularly intriguing by-product that should not be allowed to exit the reactor. However, we know from SCR that NH_3 can itself be an effective reductant, and the same is true here. This property has led to some interesting concepts that are highlighted below.

Recent research. Recent research has shed light on the very interesting and complex spatiotemporal phenomena encountered in the lean NOx trap. One phenomenon that occurs if the regeneration gas feed contains oxygen, and if the regeneration is accomplished by injecting diesel fuel into the exhaust stream, is the rapid oxidation of the reductants and the concomitant release of heat. A "hot spot" is formed at the front end of the monolith and propagates downstream. Generally, the rate of this heat propagation is slower than the duration of the regeneration, so the LNT will become hotter during the subsequent NOx storage step. This will have an adverse effect on the performance of the LNT because nitrates can decompose if the temperature is too high. Recent research has revealed another interesting effect that occurs. During the regeneration, hydrogen reacts readily with the NOx released from the stored nitrites and nitrates, forming ammonia. Distinct fronts of hydrogen and ammonia propagate through the reactor. The H_2 reacts to form NH_3, which itself reacts with stored NOx to form N_2. The H_2 and NH_3 fronts travel at a rate determined by the feed rate of the reductant, the amount of stored NOx, the temperature, and the properties of the catalyst. Eventually, the ammonia and hydrogen fronts exit the reactor, having regenerated most of the trap. The formation of NH_3 is both problematic and enticing, as we discuss below.

The design and control of the urea SCR and NSR units are extremely challenging. In each case, the performance measures of the NSR process are the cycle-averaged NOx conversion and N_2 selectivity. The conversion is the percentage of exhaust NOx that reacts, and the N_2 selectivity is the percentage of reacted NOx that is converted to N_2. In practice, the NOx conversion should exceed 95%, while the N_2 selectivity should exceed 90%. The reactors must have sufficient capacity to convert the exhaust NOx. Since the exhaust itself is time dependent, a sophisticated operating strategy is needed to minimize the breakthrough of unconverted NOx and the amount of additional fuel needed to regenerate the trap. The latter is the so-called fuel penalty, which must be kept below 2% of the total fuel usage. The timing of the urea injection for the SCR process and the engine modulation for the LNT process require an intelligent and sophisticated control system that uses real-time measurement of the engine and exhaust parameters. NSR is especially challenging because it requires periodic operation that is superimposed on the intrinsic transient nature of the vehicle operation.

In the last few years, research has focused on expanding the operability and durability as well as on reducing the cost and complexity of both SCR and NSR. To date, SCR has the upper hand, especially for heavy-duty vehicle applications. SCR catalysts are less costly and more durable than NSR catalysts, although NSR is a simpler technology to operate. In Europe, there has been extensive development of a urea infrastructure concurrent with the production of vehicles with installed SCR units. The United States lags behind Europe in the development of SCR, due in part to slower acceptance by the U.S. EPA. This delay has provided some time for the discovery and development of alternative technologies.

One, in particular, exploits the interesting capability of NSR to generate "undesired" ammonia. No urea solution is needed, avoiding the major handicap of the SCR technology. The technology involves a series configuration in which the LNT is positioned upstream of an SCR unit. The concept is as follows. The LNT performs its conventional function of trapping NOx as lean vehicle exhaust passes through it, and the trapped NOx is reduced during the intermittent regeneration with a rich gas, typically created by running the engine rich. During the regeneration, some of the stored NOx is converted to ammonia. While NH_3 would be considered a by-product during conventional LNT operation, it is exploited in the NSR-SCR technology. Rather than escaping, unconverted NOx that breaks through the LNT reacts downstream on the SCR reactor where NH_3 from a previous LNT regeneration has been trapped. Thus, in the NSR-SCR technology, NH_3 is generated in situ to carry out some of the NOx reduction by catalytic reaction on the SCR catalyst. Moreover, the cost of the precious metal in the LNT can be reduced by shifting most of the NOx reduction downstream to the SCR unit.

The development of clean diesel technology as described here is a critical element of more fuel-efficient transportation. Clean diesel vehicles are a bridge technology between today's vehicles and the eventual emergence of electric cars that run with batteries or fuel cells.

For background information *see* AIR POLLUTION; ALTERNATIVE FUELS FOR VEHICLES; CATALYSIS; CATALYTIC CONVERTER; DISESEL ENGINE; DIESEL FUEL; NITROGEN OXIDES; OXIDATION-REDUCTION; SMOG; UREA; ZEOLITE in the McGraw-Hill Encyclopedia of Science & Technology.

Michael P. Harold; Vemuri Balakotaiah

Bibliography. R. D. Clayton, M. P. Harold, and V. Balakotaiah, Performance features of Pt/BaO lean NOx trap with hydrogen as reductant, *AIChE J.*, 55:687–700, 2009 (doi: 10.1002/aic.11710); W. S. Epling et al., Further evidence of multiple NOx sorption sites on NOx storage/reduction catalysts, *Catal. Today*, 96:21–30, 2004 (doi: 10.1016/j.cattod.2004.05.004); A. Guthenke et al., in G. B. Marin (ed.), *Advances in Chemical Engineering*, vol. 33, p. 103, Academic Press, 2008.

Cell phone cameras

Modern cell phone cameras are enabled by solid-state imaging devices and electronics that instantly capture and store high-resolution pictures. The tiny light-sensing chip, or "camera on a chip," is based on semiconductor technology similar to that in computer and memory chips. Photons strike the surface of the photosensor chip to produce electrical signals corresponding to the actual image. This chip is comprised of a pixel (picture element) grid containing millions of individual light-sensing cells. The newer 10-megapixel (MP) camera chip has 30 million individual sensors ($3\times$ for color), each of which produces an electrical signal for instant display, storage, and transmission to others. The chips also may have amplifiers built into each pixel. Because picture information is stored electronically in a standard format, such as JPEG for still images, data can be sent to a phone, computer, or website in seconds. Each display pixel is "lit" to reproduce the same color and light intensity that was captured by the corresponding sensor in the camera.

Complementary metal oxide system, or CMOS, is now the preferred camera chip technology. This is the same technology used to make computer and memory chips. CMOS technology is highly developed and is used as the world standard for most solid-state electronics. The replacement of CCDs (charged-coupled devices) with CMOS for many camera chips continues to help reduce cost, increase performance, and enable miniaturization. However, a major change in device packaging technology, the enclosing and protecting of camera chips, is enabling smaller and higher-quality cameras at bargain prices. The camera is now a standard feature of the phone, and the world-installed base has probably exceeded 2 billion.

Color is a part of phone cameras, but unlike photographic film, these chips are intrinsically "color-blind"—pixels sense light intensity to produce a grayscale picture. Innovation provides an elegant solution for yielding color capability. Color CMOS chips use red, green, and blue pixel primary color filters in the chip sensor array. Photosensors are color-coated so that each pixel zone measures one of the three primary colors. Color systems can use planar or stacked arrays for color measurement (**Fig. 1**). This is why there are three times more pixels than the camera pixel rating. Combining the electronic data produces a palette of millions of colors encompassing the entire visible spectrum and even beyond. The result is quality color pictures from the cell phone camera stored as digital data that will not fade with time.

Device package. Electronic components require an enclosure, or "package," that serves two requirements: protection of the delicate device within and enablement of electrical connections to a printed circuit board (PCB). Designing and building the tiny camera chip package has been a major challenge compared to more conventional electronics packaging for memory chips and other conventional electronics. The camera package must allow light to enter without distortion to project a sharp image onto the photosensor array. Standard electronic packaging uses opaque materials in contact with electronic chips, so conventional packaging cannot be used. Equally challenging is the relentless demand for extreme miniaturization in today's feature-packed phones. The camera chip package must therefore be compact and provide an optical window. Advances in cell phone cameras are as much the result of packaging technology as chip developments. Recently, a packaging breakthrough incorporated the optical system into the package containing the camera chip. This integration will boost performance, reduce size, simplify assembly, and lower cost.

The newest camera chip packaging occurs at the wafer level. Camera chips are produced on a thin silicon semiconductor wafer that can contain tens of thousands of chips. In conventional packaging, chips are typically diced, or singulated, into the individual units before placement into the package. The new approach, known as wafer-level packaging (WLP), changes everything. The package can now be fabricated while the chips are still in wafer form. When the last packaging step is complete, the entire wafer, now an array of cameras, is diced into finished modules. Tessera Technologies, Inc., recently introduced a WLP, or WLO (wafer-level optics), process in which the optical components are fabricated on etched glass or replicated (forms lens and body) plastic wafers. Next, the entire camera structure, consisting of two or more joined wafers, is aligned and assembled at the wafer level and finally diced into individual camera modules; lenses may also be attached individually. **Figure 2** compares the tiny WLP camera to a conventional assembly still used in many cameras.

Other components. The camera chip is just one element in the total system. Supporting hardware and electronics include a power controller, signal processor, optics, image stabilization, and flash. Software features continue to evolve, with even a "smile detector" available. Digital photoimaging requires analog-to-digital data conversion and multiplexing (combining multiple signals for transmission over a single line) to deal with the millions of signals from the camera chip. These functions, and several others, can be built into the imaging chip to greatly simplify camera assembly. The camera system may include dynamic lens elements for zooming, a high-speed shutter, lighting, memory storage, and sophisticated software. Optic elements must project a real-world macro image onto the camera chip surface with the proper scaling. The scene might be a distant landscape several miles wide, and the light image must be

(a)

(b)

blue ——
green ——
red ——

Fig. 1. Color mechanisms by which primary color filters are affixed to photosensing elements to provide electronic output signals that can combine to display millions of colors. (R = red; G = green; B = blue.) Configurations can be (*a*) in a single plane as in the Bayer pattern or (*b*) stacked as in the Foveon design.

faithfully reduced to the tiny size of the chip's pixel array. Cell phones have challenged lens technology with the demand for extreme miniaturization, inflexible cost constraints, and the necessity of efficient high-volume manufacturability. Consumers expect ever-increasing performance from new phone models, approaching that of stand-alone cameras. Camera phone optics has used lower-cost, easily shaped plastic materials, and a simple fixed lens with electronic

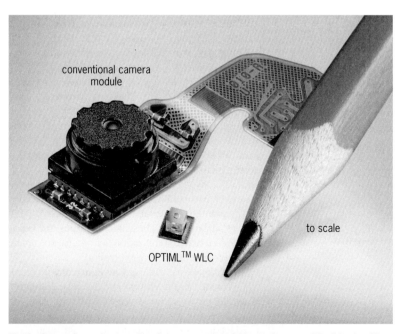

conventional camera module

OPTIML™ WLC

to scale

Fig. 2. Comparison of conventional camera module (chip, package, and flexible circuit) and new wafer-level camera system. (*Reproduced with permission from Tessera Technologies, Inc.*)

zooming, but this is changing as well. Many cell phones also have a tiny "flash" unit based on solid-state high-intensity light-emitting diodes (LEDs) or traditional xenon bulbs. Software for optimizing and assisting in taking pictures is now an important feature.

Trends and future developments. Until recently, cell phone designers had focused on continual miniaturization. The advent of the versatile and function-rich "smart phone" interrupted this trend, with users preferring more functionality, a larger display screen, and a full keyboard over smaller-size and talk-only capability. The larger size of smart phones and other multipurpose products has increased the challenge for camera technology. Better cameras must do more and do it better, while still fitting into space-constrained and thinner housings. The convergence of the "do-all" smart phones and other high-performance mobile products means that more components are being compressed into shrinking volumes. This "thin" trend has also required lower package height, or profile while boosting megapixel count.

Video is rapidly replacing still photos as the industry standard, and this trend is pushing camera chip makers to boost the "speed" or refresh time while accentuating video capability. A video is essentially a continuum of still images that appear to move when played back more rapidly than the human vision "refresh time." Video thus requires faster image capture, better picture stabilization, advanced automatic adjustment, and the ability to handle much larger amounts of data very rapidly. Smart phones are now featuring 3.2-MP cameras for video. Video publishing will be the trend during 2010 and beyond, and there will be a continuing increase in camera performance far into the future. Expect new phones to have camera systems that boast 8–12 MP, double-digit digital zooming, face and smile detection, effective image stabilizing, autofocus, and location tagging provided by built-in Global Positioning System (GPS).

Phone data transmission rates will match the new high-definition (HD) video phone cameras as 3G, 4G, WiMAX, and other very-high-speed wireless schemes are deployed. The entire phone infrastructure—cameras, phone electronics, data processing, software, and signal transmission—will emphasize video and effortless posting to the web as our lives become increasingly net-centric. Advances in lens technology will involve paradigm changes such as deformable shapes and electronically alterable smart materials. Mechanical zooming using tiny motors, and even micro-electro-mechanical systems (MEMS), will be alternatives. These technologies are being adopted for image stabilization, autofocus, automatic dynamic zooming, and optimization.

Special features, such as low light enhancement, will provide "night vision" and increased spectral range, including infrared vision. Although such extrasensory features might seem to be geared toward clandestine operations, they will also have value for healthcare, in that the phone will enable telemedicine advances, including visualization of the arterial system. The cell phone will become increasingly important as a healthcare/wellness device, and the camera will be central. Soon, "remote medical care" advances will make the cell phone a communications center for sensors, monitors, and specialized external cameras, mostly linked by wireless protocol. The doctor, and even specialists, will make house calls again, but by cell phone.

For background information *see* CAMERA; COLOR; ELECTRONIC PACKAGING; MICRO-ELECTRO-MECHANICAL SYSTEMS (MEMS); MOBILE COMMUNICATIONS; PHOTOGRAPHY in the McGraw-Hill Encyclopedia of Science & Technology.   Ken Gilleo

Bibliography. M. Chen, OmniVision has received 3.2-megapixel CMOS image sensor (CIS) orders for Apple's next-generation iPhone, *Digi-Times*, April 3, 2009; M. Feldman, Wafer-level camera technologies shrink camera phone handsets, *Photonics Spectra*, 41(8):58–60, 2007; R. Voelkel and R. Zoberbier, Inside wafer-level cameras, *Semiconductor Int.*, 32(2):28–32, 2009.

Cluster ion mass spectrometry

The chemical analysis of multicomponent molecular solids poses a particularly challenging problem in materials science, biology, and nanotechnology. Recent developments in cluster ion mass spectrometry research are opening new directions for this field. With this tool, an energetic primary ion beam is created using various types of molecular ions to create a projectile consisting of many different atomic components. Next, this beam is accelerated to 10–20 keV of kinetic energy, tightly focused to a submicrometer probe size and directed to the surface of the sample. As this energetic cluster projectile interacts with the target, positive and negative secondary ions are emitted from near the impact point as a result of a complex series of energy transfer events. These secondary ions are extracted into a mass spectrometer to reveal the chemical composition of the desorbed species. Determination of this mass distribution is referred to as secondary ion mass spectrometry or SIMS. The chemical nature of these ions is strongly dependent on the type and energy of the projectile. For purely atomic ion projectiles, the secondary ion intensity is low, and the chemical composition is often different from that on the surface itself, limiting the analytical utility. As the number of atoms in the cluster projectile is increased, however, the number of secondary ions and the amount of reliable chemical information associated with the mass spectrometry increases greatly. With a microfocused cluster ion beam probe, imaging of the two- and three-dimensional (2D and 3D) chemical composition is possible for a wide variety of materials, ranging from inorganic and organic multilayer structures to biomaterials, sensors, and single biological cells.

SIMS imaging. Stimulated desorption can be achieved using many different strategies, including an intense laser pulse, a strong electric field, or an energetic ion beam. Matrix assisted laser desorption ionization (MALDI) mass spectrometry is perhaps the most widely used method, because the resulting mass spectra generally consist of only the molecular ion and lighter fragment ions, and the molecular weight of desorbed molecules covers a range of up to several hundred thousand atomic mass units. Ion beams have two special properties for desorption that are being exploited by mass spectrometry researchers and that are not possible using MALDI. First, because the energy is deposited in the first few atomic layers of the material, the secondary ions originate from this region, allowing surface-specific information to be obtained. Second, ion beams can be focused to a submicrometer size, defining the x, y position of these ions. By rastering (scanning) the incident ion beam over an area and acquiring mass spectra on the fly, it is feasible to acquire mass spectral image information with spatial resolution exceeding that of normal optical methods. These measurements have been possible with the development of time-of-flight (TOF) mass analyzers, where as many as 10^4 mass spectra per second may be acquired. Hence, a typical image consisting of 256×256 pixels, or 65,536 individual mass spectra—each with its own x, y coordinate—can be recorded in just a few seconds. There is obviously a lot of information in this sort of data set, accounting for the fact that molecule-specific imaging is a rapidly growing segment of the mass spectrometry field.

Cluster impact dynamics. Cluster ion projectiles play a key role in contributing to the success of these experiments. Early SIMS investigations were forced to rely on atomic projectiles, which could be focused to a very small spot size but which suffered a number of detrimental characteristics. For example, the energy of the atomic projectile was deposited much deeper into the sample, leading to inefficient desorption of surface molecules. In addition, the cascade of colliding atoms after the impact event led to considerable chemical damage buildup, meaning that the dose of incident ions had to be kept very small to avoid changing the sample chemistry. Interestingly, it has been known since at least 1989 that cluster bombardment greatly enhances the yield of molecular ions. The idea is that each atom in the cluster carries only a fraction of the total kinetic energy. When the projectile collides with the sample, it immediately breaks apart into its atomic components. Each atom initiates its own cascade of moving particles, but with lower kinetic energy. Hence, the energy is deposited closer to the surface and in a more concentrated fashion.

During the 1990s, interest in these sources increased, but technical difficulties and the lack of commercial instrumentation prevented widespread adaptation. About 10 years ago, however, a liquid-metal ion source, made first with gold or bismuth, became available, allowing the routine generation of Au_3^+ or Bi_3^+ projectiles. At about the same time, a reliable C_{60}^+ (fullerene) gas ion source was introduced to the community. These systems were rapidly deployed throughout the world. The C_{60}^+ source is of particular interest because of the larger number of

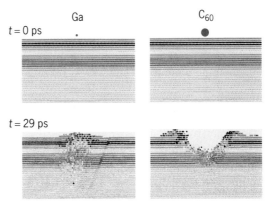

Fig. 1. Cross-sectional view (14.3 nm in width and 7.3 nm in depth) of the temporal evolution of a typical collision event leading to ejection of atoms due to 15-keV Ga and C_{60} bombardment of a Ag(111) surface at normal incidence. The atoms are colored by original layers in the substrate. The projectile atoms are black.

atoms in the cluster. Each C atom in a 20-keV beam has an energy of just 333 eV. The difference in the dynamics between a single atomic impact and 60 simultaneous low-energy impacts is illustrated in **Fig. 1**. These pictures are generated using molecular dynamics computer simulations of the bombardment event, and clearly reveal the mesoscopic scale of the disruption. The formation of a craterlike feature on the sample is characteristic of cluster impact. Perhaps it is the collective motion associated with the formation of this crater that leads to such effective molecular desorption.

Molecular depth profiling. In addition to improved mass spectra, surface damage accumulation is greatly reduced during cluster bombardment. This observation raises the possibility that complex materials, such as multilayer polymers, simple organic thin films, biofilms, or even single biological cells can be examined in depth by using the cluster beam probe to peel away molecular layers without insult. The erosion rates are typically several tens of nanometers per second using commercially available sources, and depth profiles a few micrometers into the sample are feasible to acquire in less than an hour. The cluster bombardment process keeps the surface quite smooth, allowing the depth scale to be maintained without significant interlayer mixing. This added capability of acquiring information in the z direction, when combined with the imaging capability mentioned above, suggests that 3D analysis of the sample is indeed possible. Several groups are working on developing protocols for these measurements, and results have been reported for drug distributions in polymer-based carriers, polymer composites, multilayer organic thin films, and single biological cells.

An example of a depth profile using a model system comprised of multilayer lipid thin films prepared with a Langmuir-Blodgett trough is shown in **Fig. 2**. This model system has provided a great deal of insight into the type of data that can be obtained, because the interface is atomically abrupt and little topography is observed during erosion through six, approximately 50-nm lipid layers. The results, shown

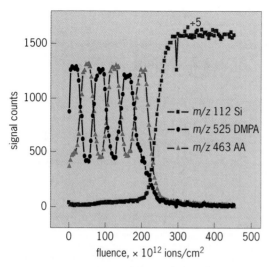

Fig. 2. Molecular depth profile of a multilayer film of dimyristoyl phosphatidic acid (DMPA) and arachidic acid (AA) using 40-keV C_{60} bombardment at a 73° angle of incidence with the sample held at 100 K. Each of the six layers is about 50 nm thick (*Zheng et al., 2008*). The Si^{4+} ion at *m/z* 144 is used to monitor the presence of the Si/lipid interface.

in Fig. 2, reveal that the best depth resolution of 8 nm is achieved when the sample is kept at cryogenic temperatures and bombarded at glancing angles of incidence. Because the erosion rate is uniform across the film, it is feasible to record images during the depth profile, stack them up, and color code them with respect to mass. A representation resulting from this procedure is shown in **Fig. 3**.

Outlook. The cluster mass spectrometry field is in a developing stage at the present time and there are still hotly debated issues. For example, it is not yet clear if there is a best projectile that provides the highest x, y, and z resolution as well as the best mass spectra, or whether multiple beams will need to be employed. Only limited classes of materials have been investigated, and it is not clear how general the molecular depth profiling experiments will prove to be. The interest level is high, however, because there

are not many ways to acquire 3D chemical information with submicrometer lateral resolution and depth resolution on the nanometer scale.

For background information *see* ANALYTICAL CHEMISTRY; DESORPTION; FULLERENE; ION SOURCES; MASS SPECTROMETRY; MASS SPECTROSCOPE; SECONDARY ION MASS SPECTROMETRY (SIMS); TIME-OF-FLIGHT SPECTROMETERS in the McGraw-Hill Encyclopedia of Science & Technology.   Nicholas Winograd

Bibliography. D. G. Castner, Surface science: View from the edge, *Nature*, 422:129–130, 2003; J. C. Vickerman, Molecular SIMS: A journey from single crystal to biological surface studies, 603:1926–1936, 2009; N. Winograd, The magic of Cluster SIMS, *Anal. Chem.*, 77:142A–149A, 2005; S. G. Ostrowski, Mass spectrometric imaging of highly curved membranes during single cell mating, *Science*, 305:71–73, 2004; Zheng et al., Chemically alternating Langmuir-Blodgett thin films as a model for molecular depth profiling by mass spectrometry, *J. Am. Soc. Mass Spectrom.*, 19:96–102, 2008.

Cochlear wave propagation

The auditory peripheral organ consists of the outer, middle, and inner ears. The cochlea (from the Greek word for snail), in the inner ear, is the auditory sensory organ. The bony, spiral cochlear shell is filled with fluid and separated by Reissner's membrane and the basilar membrane into three ducts: the scala tympani, scala media, and scala vestibuli (**Fig. 1a**). Because of the flexibility of Reissner's membrane, the cochlea is often simplified to a two-duct mechanical system, in which the two ducts are open to each other at the apex and sealed at the cochlear base by the stapes on the scala-vestibuli side and the round window membrane on the scala-tympani side (Fig. 1b).

In the normal ear, incoming sounds vibrate the eardrum, propagate along the middle ear bony chain, and enter the cochlea at the oval window. The vibration of the rigid stapes footplates disturbs the cochlear fluid and initiates the vibration of the basilar membrane. Mechanically sensitive auditory sensory cells on the basilar membrane can sense the vibration and convert mechanical stimuli into electrical signals. The auditory nerve encodes the acoustic information in the form of electrical pulses and sends it to the brain. Such acoustic information, for example, environmental sounds and sounds for daily communication, may vary a millionfold in magnitude and a thousandfold in frequency, and arrival times of critical information in the cochlea may be segregated by less than a millisecond. Remarkably, this real-time signal processing is accomplished entirely by the cochlea. The question of how the cochlea obtains its remarkable sensitivity, wide dynamic range, and frequency selectivity has been fascinating scientists for years. The most direct method of studying the hearing mechanism is to measure sound-induced subnanometer vibration in the cochlea. In the living cochlea, the vibration at a single location of the

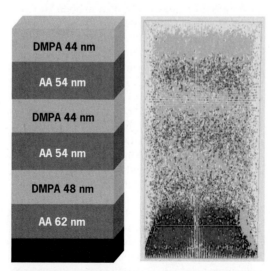

Fig. 3. 3D representation of the six-layer film, whose depth profile is shown in Fig. 2.

basilar membrane shows high sensitivity, nonlinear compressive growth (that is, the growth rate decreases with the stimulus level), and sharp tuning. In recent years, with improved sensitivity of instruments, the spatial pattern of the basilar membrane vibration has been measured in sensitive living cochleae. In addition, increasing experimental data indicate that inner ear–generated sound exits the cochlea mainly through the cochlear fluid as a longitudinal wave rather than as the widely expected transverse wave along the basilar membrane.

Forward propagation of cochlear wave. Sound-induced pistonlike motion of the stapes in the oval window displaces the cochlear fluid, producing fluid pressure changes behind the stapes footplate. It has been thought that the stapes vibration results in two waves. One is a pressure wave in both cochlear ducts, traveling with the speed of sound in water. This wave produces no basilar membrane vibration because of the equal pressure in both cochlear ducts. The other is the cochlear displacement wave or transverse wave along the basilar membrane (often called the cochlear traveling wave in auditory literature). Owing to the impedance difference between the stiff stapes footplate in the oval window and the flexible round window membrane, the sound wave results in a pressure difference across the basilar membrane, which vibrates the membrane. This cochlear wave is commonly described as a transverse wave. It starts at the base and propagates toward the apex. As the wave propagates, its magnitude increases and reaches the maximum at the characteristic frequency (CF) location (**Fig. 2***a*). Magnitude and phase of the basilar membrane vibration are presented with respect to the stapes vibration in Fig. 2 and **Fig. 3**. The characteristic frequency location is frequency dependent. Cochlear responses to high-frequency stimuli peak near the base, and low-frequency responses are near the apex. This frequency-location relationship is the basis of cochlear frequency selectivity. The phase of the basilar membrane vibration is a function of the distance from the cochlear base. It progressively decreases as the distance from the base increases (Fig. 2*b*), indicating that the speed of the wave varies with the longitudinal location. At the base the wave propagates fast and, therefore, for a sinusoidal stimulus, the wavelength is relatively long. As the wave travels toward the apex, it progressively becomes slower and the wavelength shortens (Fig. 2*c*).

In contrast to responses in dead cochleae or under insensitive conditions, basilar membrane vibrations in living cochleae are sensitive, nonlinear, and sharply tuned. The basilar membrane response in a sensitive cochlea to a stimulus at low levels at frequencies near the characteristic frequency is much larger than that in an insensitive cochlea. As the sound level increases, the increase rate of the responses decreases, resulting in nonlinear compressive growth. This nonlinear feature provides the cochlea a wide dynamic range, which is essential for processing sounds with millionfold variation in magnitude. Experiments conducted in different species

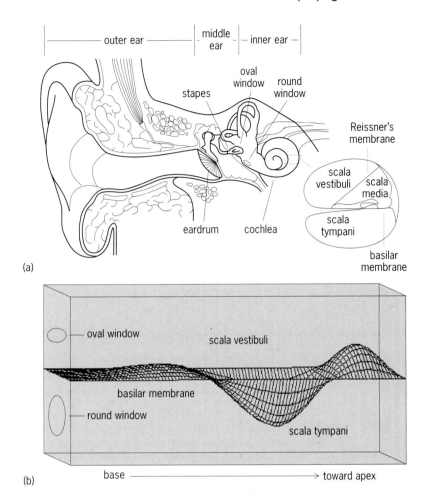

Fig. 1. **Auditory peripheral organ. (*a*) Diagram showing the outer, middle, and inner ears, and a cross section of the cochlea. Sound vibrates the eardrum and stapes and enters the cochlea through the oval window. (*b*) Sound-induced pressure difference across the basilar membrane launches a transverse wave propagating from base toward apex.**

demonstrate that the magnitude of the basilar membrane vibration at a given longitudinal location is a function of frequency. Vibration peaks at the characteristic frequency, and its magnitude then decreases quickly as the stimulus frequency moves away from the characteristic frequency. This sharply tuned response in the sensitive cochlea is critical for distinguishing one tone from the others with slightly different frequency. The spatial patterns of the basilar membrane vibration in living cochleae (solid lines in Fig. 2) are different from those in postmortem ones (broken lines in Fig. 2). The most striking difference is that at low sound levels the vibration occurs at a very restricted area centered at the characteristic frequency location, while there is no detectable low-level response under insensitive conditions. Longitudinal data from sensitive gerbil cochleae show that, for 16-kHz tones, the phase delay from the basal to apical end of the observed location is up to 6π radians (three full cycles or $1080°$) over only approximately 1 mm distance. The detectable basilar membrane response to a low-level 16-kHz tone distributes over only a 0.6-mm range along the basilar membrane. This localized response is essential for the cochlear frequency selectivity. Data obtained at different frequencies indicate that cochlear sharp

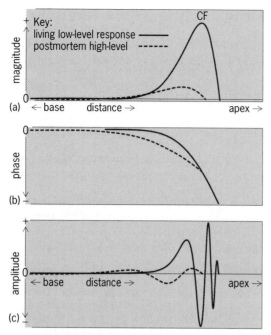

Fig. 2. Basilar membrane response to a low-level tone in living sensitive cochlea (solid lines) and responses to a high-level tone in postmortem cochlea (dotted lines). (*a*) Magnitude. (*b*) Phase. (*c*) Instantaneous waveform. Basilar membrane vibration in living cochlea is more sensitive and localized than that in postmortem cochlea.

tuning results from the localized basilar membrane vibration. Spatial patterns also show compressive nonlinear growth, a shorter wavelength, and a slower propagation velocity along the cochlear length than those in insensitive cochleae.

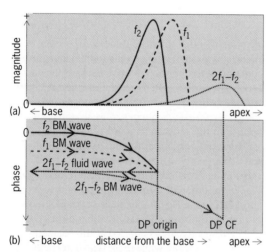

Fig. 3. The compression-wave mechanism for the backward transmission of the otoacoustic emission in the cochlea. (*a*) Magnitudes and (*b*) phases of basilar membrane (BM) vibrations are shown at frequencies f_1, f_2, and $2f_1 - f_2$, as functions of the distance from the cochlear base. Distortion product at frequency $2f_1 - f_2$ is generated at location labeled "DP origin," where BM waves f_1 and f_2 overlap. Distortion product propagates to the stapes through the fluid as a compression wave, and the stapes vibration results in otoacoustic emission in the ear canal. The stapes vibration also causes a forward transverse wave, which peaks at location labeled "DP CF." Because sound in water is much faster than BM transverse wave, the backward propagation does not result in significant phase or time delay.

Backward propagation of cochlear wave. The normal cochlea not only senses but also generates sounds. Ear-generated sounds, called otoacoustic emissions (OAEs), can be measured in the ear canal using a tiny sensitive microphone. OAE has been widely used in clinics for hearing screening in infants and for diagnosing hearing loss. Knowledge of how the OAE exits the cochlea and whether the incoming sound can propagate backward along the basilar membrane is important for OAE applications and for understanding the fundamental hearing mechanism. There are two competing theories: the backward-traveling-wave theory and the compression-wave theory. According to the backward-traveling-wave theory, the OAE propagates from its generation site to the cochlear base as a transverse vibration along the basilar membrane at the same speed as a forward transverse wave. The compression-wave theory posits that OAE propagates to the cochlear base via longitudinal waves in the cochlear fluids at the speed of sound in water.

A number of experiments have been conducted in recent years to test these hypotheses. When two tones at nearby frequencies f_1 and f_2 (f_2 greater than f_1) are presented to a sensitive cochlea, distortion products at frequencies different from f_1 or f_2 are generated due to the nonlinear responses of the basilar membrane. Among them, the distortion product at frequency $2f_1 - f_2$ is often the strongest and most studied. The longitudinal pattern of the basilar membrane vibration, measured using a scanning laser interferometer, shows that the maximum vibration at the emission frequency occurs at its characteristic frequency place rather than at its generation site, where f_1 and f_2 overlap (Fig. 3*a*). The vibration phase (the dotted curve in Fig. 3*b* labeled "$2f_1 - f_2$ BM wave") decreases with distance from the cochlear base. The phase-decrease rate increases as the vibration approaches the emission characteristic frequency location. In contrast to the prediction from the backward-traveling-wave theory, these data showed a forward transverse wave and no detectable backward wave on the basilar membrane. The phase difference between the stapes and the basilar membrane vibration showed that the distortion product arrives at the stapes earlier than at the basilar membrane, which is also inconsistent with the backward-traveling-wave theory. To confirm these experimental findings, vibrations of the basilar membrane at different emission frequencies were measured at two locations along the cochlear length. Data show that the vibration peaks at the characteristic frequency location, and waves at emission frequencies always arrive at a basal location earlier than a more apical location.

Since the OAE can be measured in the ear canal, the above results indicate that the OAE propagates from its generation place to the stapes through the cochlear fluid compression wave at the speed of sound in water, and consequently results in a detectable OAE in the ear canal. As with an external sound-induced vibration, the stapes vibration also causes a forward transverse wave along the basilar membrane. These studies have elicited vigorous debate on the cochlear mechanism, because classical cochlear models predict the existence of the backward propagation of the cochlear transverse wave along the basilar membrane.

Though there are experiments apparently supporting the backward-traveling-wave theory, increasing experimental data are consistent with the cochlear-compression-wave mechanism. As new experimental findings emerge and are confirmed, the classical theory of hearing is challenged to adapt to the new data. Resolving the current conflict between theory and data will advance understanding of the basic mechanism of hearing in a fundamental way.

For background information *see* ACOUSTIC IMPEDANCE; EAR (VERTEBRATE); HEARING (HUMAN); HEARING (VERTEBRATE); LOUDNESS; PITCH; SOUND in the McGraw-Hill Encyclopedia of Science & Technology.   Tianying Ren

Bibliography. E. de Boer et al., Inverted direction of wave propagation (IDWP) in the cochlea, *J. Acoust. Soc. Am.*, 123:1513–1521, 2008 (doi:10.1121/1.2828064); W. Dong and E. S. Olson, Supporting evidence for reverse cochlear traveling waves, *J. Acoust. Soc. Am.*, 123:222–240, 2008 (doi:10.1121/1.2816566); W. He et al., Reverse wave propagation in the cochlea, *Proc. Natl. Acad. Sci. U.S.A.*, 105:2729–2733, 2008 (doi:10.1073/pnas.0708103105); T. Ren, Longitudinal pattern of basilar membrane vibration in the sensitive cochlea, *Proc. Natl. Acad. Sci. U.S.A.*, 99:17101–17106, 2002 (doi:10.1073/pnas.262663699); T. Ren, Reverse propagation of sound in the gerbil cochlea, *Nat. Neurosci.*, 7:333–334, 2004 (doi:10.1038/nn1216); W. S. Rhode, Distortion product otoacoustic emissions and basilar membrane vibration in the 6–9 kHz region of sensitive chinchilla cochleae, *J. Acoust. Soc. Am.*, 122:2725–2737, 2007 (doi:10.1121/1.2785034).

Coevolution between flowering plants and insect pollinators

Coevolution is a term used to describe the mutual changes in two or more species, usually one following the other, that affect their interactions. Flowering plants (angiosperms) and their pollinators are often used as the classic example of this evolutionary phenomenon. The plant and the pollinator place evolutionary pressure on each other for changes in morphology, physiology, or habits that benefit both.

Early history. The flowering plants evolved about 130–140 million years ago (mya). Insects are considered to be the earliest pollinators of flowering plants, and their interaction with the reproductive organs of the earliest angiosperms may be responsible for the evolution of flowers. Beetles (Coleoptera) and flies (Diptera) were the first groups of insects that were already present when the first angiosperms appeared, and they also are important pollinators of present-day basal angiosperms (that is, early forms of angiosperms that are still living today). Flowering plants are pollinated through a variety of mechanisms. These mechanisms include pollen being transported via wind, water, and numerous groups of insects, reptiles, birds, and mammals, including humans. The particular pollinator and the particular methods by which pollination is achieved are referred to as the pollination syndrome. Various related or unrelated plant species that share pollinators, and methods to attract pollinators, with other plant species are considered a pollination guild. Thus, a particular plant will share pollinators with several other plants in the same guild. Likewise, a specific plant species may be visited by a variety of animal pollinators. The unique, sophisticated pollination mechanisms developed by the plants and the pollinators are the subject of much research, as the results very often affect the well-being of humans.

As the flowering plants evolved into the approximately 300,000 species known today, many species developed specialized floral organs and methods, including color and scents, to ensure pollination by only one or a few pollinators. In this way, these plants would not "waste" the nectar reward presented and could ensure that the pollen would be delivered to another flower of the same species, thus completing the process of pollination and the start of fertilization. The random, often haphazard process of wind or water pollination could be eliminated through the relatively complex pollination syndromes developed mutually between plant and pollinator. In part, these pollination syndromes are responsible for the success of the flowering plants today. Through natural selection, flowering plants and their pollinators have mutually benefited from their shared ecological services. The plant is pollinated while the pollinator is rewarded with a high-energy food of pollen or nectar.

Pollen grain clumping. The origin and development of this mutual relationship between flowering plants and pollinators has recently been the topic of research for scientists studying the kinds of pollen grains recovered from ancient river and lake sediments dated about 100 million years old. The river and lake sediments from the Dakota Formation (100 mya, southern Minnesota) yield an unusual arrangement of pollen: clumped pollen grains (**Fig. 1**). Clumped pollen grains, consisting of over 100 pollen grains glued together in a sticky mass, are much easier to transfer to an insect pollinator than an individual grain. Of course, the transfer of a large number of pollen grains to the female part of a flower (the stigma) is an advantage for the plant because it provides a greater number of possible pollen grains to perform fertilization, and this large number of grains with genetic differences is believed to increase fitness through competition. The evidence from the pollen record of the Dakota Formation, including the clumping of pollen, recorded as early as 100 million years ago, supports the hypothesis that insects, perhaps including bees, were the initial pollinators of these early flowering plants, and that more specialized pollination methods, involving elaborate or unique flower-animal interactions, are derived in nature.

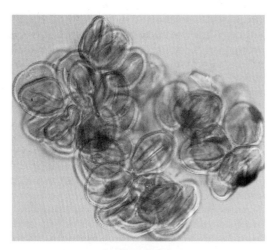

Fig. 1. Fossil pollen (90 million years old) showing the clumping common in the Cretaceous sediments of Minnesota. (*Photo by Shusheng Hu*)

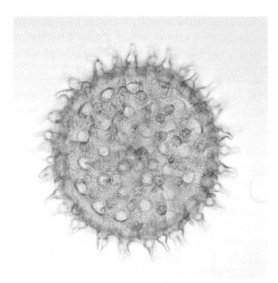

Fig. 2. Pollen grain of *Ipomoea squamosa* (a member of the Morning Glory family) showing the large spines that attach to the body hairs of insects for transport to another *Ipomoea* flower. (*Photo by D. M. Jarzen*)

The individual pollen grains within these clumps were identified based on their morphology, their shape, the number and structure of apertures (pores or slits in the pollen wall where the pollen tube emerges), and their surface sculpture. Typical wind-pollinated pollen is often small, smooth, spherical, produced in great quantities, and easily carried by even the slightest currents of wind. Insect-pollinated flowers, on the other hand, produce pollen that is larger, ornamented with processes such as spines or netlike structures (**Fig. 2**) that would be easily attached to the hairy bodies of insects, and may be sticky. Pollen clumping was an early step by flowering plants to ensure that a large number of pollen grains would be carried by insect pollinators, thus using them to transfer the plant's male genetic material. The clumping of the pollen was a mechanism by which the plant could overcome the limitations of smaller, lightly ornamented pollen grains, by presenting a larger mass of pollen to its pollinator. It

was an advantage to the plant that produced these clumps because the clumps carried scores of pollen grains to the female portion of other flowers.

The pollination syndromes of basal angiosperms living today have been tabulated to determine their pollinators. In the most basal of the angiosperms, that is, the families including the Amborellaceae, Nymphaeaceae, Illiciaceae, and others that are considered to be ancient lineages, the pollinators were found to include beetles, flies, and Hymenoptera (mostly bees). Recent evidence suggests that the bees originated as early as the mid-Cretaceous, between 110 and 90 million years ago (**Fig. 3**). This same period was a time of rapid and major radiation for the flowering plants, further suggesting that the increase in specialized pollination techniques such as clumping was linked with insect pollination.

Meganosed fly. One rather bizarre example of the potential of coevolutionary forces is displayed by the meganosed fly (*Moegistorhynchus longirostris*) [**Fig. 4**] of southern Africa and the flowers that it pollinates. All of the flower species that the fly visits have very long floral tubes that can be up to several inches long. The "nose" of the fly is really a very long proboscis with which the fly probes the floral tube in search of nectar. This is rather unique since a fly toting a 4-in.-long (10 cm) proboscis is indeed a strange sight, but the strangeness does not end there. The flowers visited by the meganosed fly deposit their respective pollen on a different part of the fly's body. As the fly visits another flower, it may receive pollen of another species on a different part of its body. When eventually it visits flowers of the species from which it received pollen, the female parts of those flowers are exactly situated to receive the pollen only from that particular part of the fly's body. In this way, the pollen received is delivered to the same species, guaranteeing pollination.

Conclusions. Paleobotanists, palynologists (scientists who study extant and extinct pollen grains and spores), and floral biologists are looking at the

Fig. 3. A bee visits a rose flower in Indiana. (*Photo by D. L. Dilcher*)

Fig. 4. Flowers of *Pelargonium suburbanum* are probed by the South African meganosed fly (*Moegistorhynchus longirostris*). Pollinaria (sacs of pollen) from an earlier visit to a nearby orchid dangle from its proboscis. The proboscis length of this fly is closely matched to the length of the flower tubes of a guild of plants that have coevolved with this specialist pollinator. (*Photo by Steven D. Johnson*)

so-called basal angiosperms of today and learning much about the nature of pollination syndromes, pollen and nectar as food sources, and specialized pollination techniques. By investigating the pollination mechanisms for these plants, clues may be learned regarding the pollination systems employed in the earliest flowering plants. Without insect pollination, the flowering plants would not have evolved into the diversity that is seen and experienced today. It is the coevolution of flowers and insect pollinators that has led to the wonderful world of flowers.

For background information *see* COLEOPTERA; DIPTERA; FLOWER; HYMENOPTERA; INSECTA; MAGNOLIOPHYTA; PALEOBOTANY; PALYNOLOGY; PLANT-ANIMAL INTERACTIONS; PLANT EVOLUTION; PLANT REPRODUCTION; POLLEN; POLLINATION in the McGraw-Hill Encyclopedia of Science & Technology.

David M. Jarzen; David L. Dilcher

Bibliography. J. Brackenbury, *Insects and Flowers: A Biological Perspective*, Blandford, London, 1995; S. Hu et al., Early steps of angiosperm-pollinator coevolution, *Proc. Natl. Acad. Sci. (USA)*, 105(1): 240–245, 2008; B. Meeuse and S. Morris, *The Sex Lives of Flowers*, Facts on File, New York, 1984; L. A. Sessions and S. D. Johnson, The flower and the fly, *Natural History*, 114:58–63, 2005; L. B. Thien et al., Pollination biology of basal angiosperms (ANITA Grade), *Am. J. Bot.*, 96(1):166–182, 2009.

Colony collapse disorder

Colony collapse disorder (CCD) is a condition in honey bee colonies characterized by a rapid loss of the adult bee population with an absence of dead bees in and around affected colonies (see **illustration**). CCD first came to light in the fall of 2006 and is the most recent in a growing list of threats to managed honey bee colonies in the United States.

Decline of bee populations. The number of managed bee colonies has declined 61% from a high of 5.9 million colonies in 1945 to 2.3 million colonies

in 2008. At least 90 different agricultural crops rely on insect pollination. Maximum production and the quality of these crops depend on this insect-mediated pollination. Eighty percent of the pollination services provided by insects is provided by honey bees. Further reductions in colony numbers may have impacts on American fruit and vegetable producers that are dependent on honey bee pollination. The value of honey bee pollination in the United States is estimated to be more than $14 billion per year.

Descriptions of colony losses similar to CCD have been documented in the United States at least 18 times since the early 1900s. Past incidences, designated by names such as *May disease*, *disappearing disease*, and *fall dwindle disease*, affected only discrete areas of the country. CCD, however, distinguishes itself from these previous events both in the geographic scope that it encompasses and in the extent of the losses suffered by U.S. beekeepers.

In the winter of 2006–2007, 31% of all U.S. overwintered colonies were lost. During the winter of 2007–2008, another 36% of colonies vanished. Although most operations did not suffer symptoms indicative of CCD, those beekeepers that reported symptoms indicative of CCD lost more than twice as many colonies as those that did not report CCD-like symptoms.

Concurrent with reports of large losses by U.S. beekeepers since 2006, Canada, France, Germany, Spain, Belgium, the Netherlands, and Japan have submitted reports of equal or greater losses. Colony losses in these countries, though, do not all share the symptoms that define CCD and are probably not attributable to this particular disorder. It is likely that these losses are caused by a host of other threats to honey bee health, including Varroa mite parasitism, infection with disease agents such as the fungus *Nosema*, sublethal effects of pesticides, and poor nutrition. Most beekeepers in the United States do

A colony that died from colony collapse disorder (CCD). Note the large number of frames of brood that have been completely abandoned by the adult bee population, indicating that, shortly before dying, this colony had a large, healthy population of adult worker bees.

not ascribe colony mortality to CCD, but to the poor quality of queen bees, colony starvation, and Varroa mite infestation.

Causes of colony collapse disorder. No single cause has been found that explains all cases of CCD. It is clear that bees exhibiting this condition suffer from some form of illness, and they tend to be coinfected with many common bee pathogens, including simultaneous infection with multiple viruses and fungal pathogens. This suggests that a single factor or a combination of factors may compromise the immune systems of the bees, thereby rendering them susceptible to pathogens that healthy colonies would be able to fend off.

One widely reported hypothesis regarding the underlying cause for the weakening immune systems of bees ascribes blame to the increase in cell phone usage. This theory has largely been dismissed for lack of evidence. Another often-suggested cause attempts to connect immunosuppression in bees and genetically modified (GM) crops such as Bt corn [that is, corn that has been altered genetically with a gene from an insect pathogen, *Bacillus thuringiensis* (Bt), that encodes a protein that is toxic to insect pests that eat and destroy corn stems]. However, experiments in which Bt corn pollen was fed to bees failed to document any negative effect on bee health. The transportation of bees for thousands of miles across the country to pollinate crops has also been considered as a cause. Although the trucking of colonies may result in some stress to the bees, exploration in this area has failed to report a difference in mortality rates between beekeepers that move colonies and those that do not, suggesting that the impact of transporting bees is minimal and is unlikely to be a leading factor in CCD.

Concerted efforts to identify a factor or combination of factors that permit pathogen infections to take hold in CCD colonies have focused on three broad areas: (1) a newly introduced or newly mutated bee pathogen, (2) exposure to pesticides, and (3) poor nutrition.

Initial studies of CCD identified the presence of one virus, Israeli acute paralysis virus (IAPV), in collapsing colonies as highly predictive of the condition. Experiments in which colonies were exposed to IAPV have shown that infected colonies died with symptoms characteristic of CCD. However, IAPV is only one of more than 18 viruses that have been found in honey bees, and infection with many of these other viruses is also likely to cause the main symptom of the CCD condition.

The phenomenon of "missing bees" or "dead colonies without dead bees" is not as mysterious as it may sound. In other insect societies, such as termites, sick insects demonstrate a behavior known as "altruistic suicide." For instance, termites that are infected with some fungal pathogens walk away from their colony to die. This behavior is presumably an attempt to prevent the passing on of infection from the sick individual to its sisters. Since bees are also social insects, it seems likely that sick individuals leaving the colony to die may be a common evolved response to disease prevention in insect societies.

While IAPV (and other bee viruses) can cause the symptoms of CCD, field surveys illustrate that many colonies with virus infections survive, suggesting that IAPV and other viral infections are secondary and not causal. Many different known threats, acting alone or in combination, are thought to be responsible for an immunosuppression in bees that allows viral infections to take hold. These factors include Varroa mite parasitism, in-hive pesticides, farmer-applied pesticides, and poor nutrition.

Varroa mites. Varroa mites are large parasitic mites that live on adult and developing bees. Varroa mites reproduce on developing larval and pupal bees. When feeding, they pierce the exoskeleton of the bees to feed on their hemolymph (blood). In the process of feeding, they can transfer bee viruses to uninfected individuals. Furthermore, when feeding, they "spit" into their host, and a protein in that "spit" attacks the immune system, making the bee more susceptible to a host of honey bee pathogens.

In-hive pesticides. To control Varroa populations, beekeepers often treat colonies with synthetic pesticides, including coumaphos and fluvalinate. These miticides (pesticides that kill mites) are lipophilic (lipid- or fat-loving), which is a property that permits them to be absorbed into wax at increasing levels, thereby leading to a buildup of these products in the comb of honey bee colonies. High levels of these products are known to have a negative impact on honey bee health and the development of larval bees.

Farmer-applied pesticides. When bees forage on flowers, they readily pick up pesticides that have been sprayed on crops. If the level of pesticide to which the bees are exposed is not enough to kill them, the foraging bees can bring these pesticides back to the colony, storing them in the pollen reserves of the colonies and accumulating them in comb wax. Sublethal exposure to pesticides remains one of the possible factors contributing to CCD, as these pesticides are likely to have a negative effect on bee health.

Poor nutrition. Honey bees are vegetarians, collecting and consuming pollen to meet all their protein needs. Poorly nourished bees are more likely to become ill. The lack of adequate pollen for colonies is a factor that may contribute to CCD. Pollen scarcity can result from environmental factors (for example, drought) that restrict the amount of forage available to bees. Different pollens also contain different proportions of amino acids, so a diet of pollen from different floral sources may be an important factor in bee health. Honey bees feeding exclusively on one crop for an extended period of time, which is the case in some large monocrop areas, has been suggested as a contributing cause of CCD.

Potential effect on agriculture. In general, the potential effect of CCD as it relates to losses of managed honey bee colonies is of particular concern. An estimated one-third of all food is directly or indirectly the result of honey bee pollination. Most fruit, nut, and vegetable production in the United States is reliant on a movable pollination force. Comparatively few (an estimated 900) commercial beekeepers manage the 1.5 million colonies necessary to meet the pollination needs of the United States. In a single year, a typical beekeeper operation may move bee colonies from the citrus groves of Florida to the apple orchards of Pennsylvania, and then on to the blueberry fields of Maine, the cranberry bogs of Massachusetts, the pumpkin patches of Pennsylvania, and finally the almond groves of California. It is the devastating effect that CCD has on these operations that is of greatest concern, as repeated years of losses in excess of 30% is likely to lead many beekeepers to go bankrupt, thereby potentially leading to reduced production in pollinator-dependent crops.

For background information *see* ACARI; AGRICULTURE; BEEKEEPING; HYMENOPTERA; IMMUNOSUPPRESSION; INSECT DISEASES; PATHOLOGY; PESTICIDE; POLLEN; POLLINATION; POPULATION ECOLOGY; SOCIAL INSECTS; VIRUS in the McGraw-Hill Encyclopedia of Science & Technology.   Dennis vanEngelsdorp

Bibliography. D. Cox-Foster and D. vanEngelsdorp, Solving the mystery of the vanishing bees, *Sci. Am.*, April:40–55, 2009; D. L. Cox-Foster et al., A metagenomic survey of microbes in honey bee colony collapse disorder, *Science*, 318:283–286, 2007; D. vanEngelsdorp et al., A survey of honey bee colony losses in the U.S., Fall 2007 to Spring 2008, *PLoS ONE*, 3:e4071, 2008.

Color vision in mantis shrimps

Color vision is the ability to distinguish among light stimuli based solely on differences in their spectra, independent of brightness, polarization, or any other stimulus feature. To make such discriminations, at least two receptor classes with different sensitivity spectra must exist in the retina. In vertebrates, this generally means that two cone spectral classes are present. While invertebrate photoreceptors vary in their cellular structure, two or more types must nevertheless be present to provide color vision. Because color vision involves internal processing to distinguish among stimuli, it cannot be demonstrated by physiological means alone. The definitive proof that an animal possesses this ability requires rigorous behavioral testing to show that it can make the required discriminations. Often, this is difficult to do, so it is not always a simple matter to show that color vision is present in a given species.

Humans tend to think that color vision is rare and special, limited to the most advanced vertebrates, but this visual modality is widespread among animals. Moreover, in many species, ranging from insects to birds, color vision is based on more receptor types than in humans, and thus can potentially enable more subtle discriminations. Humans have three color receptor classes, provided by three cone types: blue-sensitive, green-sensitive, and red-sensitive. Most birds and many fishes have cones like these plus a fourth class that is sensitive to shorter

wavelengths, violet or ultraviolet. Butterflies, and perhaps other insects, have up to five primary color receptor types, ranging from the ultraviolet to the red. Surprisingly, many species of an obscure group of marine crustaceans known as mantis shrimps, or stomatopod crustaceans, possess at least eight color receptor classes and possibly more. No other animal approaches this level of complexity.

Structure of stomatopod crustacean compound eyes. Stomatopod color vision is made possible by the unusual organization of their compound eyes. All compound eyes are based on multiple units called ommatidia, each with a set of optical structures and other cells near the surface of the eye (forming the facets) overlying a specialized photoreceptive region. Crustacean photoreceptors are composed of a set of eight receptor cells (called "retinular cells"), numbered 1 to 8. Retinular cell number 8, abbreviated R8, is situated at the top of this assembly, whereas cells R1 to R7 form a ring of elongated cells beneath, so that the entire receptor group forms a column of cells extending through most of the length of the ommatidium.

In mantis shrimp eyes, the arrangement of ommatidia is unique among crustaceans. The eye is divided into three regions: dorsal and ventral halves, similar in size and organization, and an equatorial midband separating them, constructed from six parallel rows of ommatidia (**Figs. 1** and **2**). Ommatidia of

(a)

(b)

Fig. 1. Eyes and color signals in the stomatopod crustacean *Odontodactylus scyllarus*. (*a*) Frontal view of the compound eyes. Note that each eye is divided into two roughly equal halves by an equatorial midband, which consists of six parallel rows of ommatidia. (*b*) Posterior view of the animal, showing the colors and patterns displayed on the animal's body; these are potentially used in mating, signaling, and aggressive displays. (*Photographs by R. L. Caldwell*)

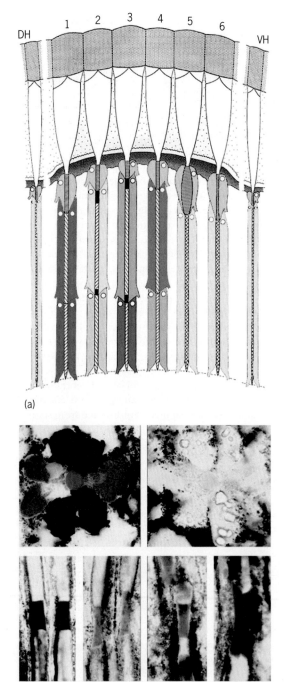

(a)

(b)

Fig. 2. Eye structure typical of stomatopod crustaceans.
(a) Schematic diagram of a vertical section through the eye
perpendicular to the plane of the midband. The relatively
simple ommatidia of the dorsal and ventral hemispheres of
the eye are labeled DH and VH, respectively. These consist
of two tiers of photoreceptors, the distal R8 and the
proximal R1–R7. Ommatidia of the two most ventral rows of
the midband, rows 5 and 6, have a similar organization. The
four dorsal rows, rows 1 through 4, have three receptor
layers, as R1–R7 are split into two tiers. All R8 cells
throughout the retina are sensitive to ultraviolet light.
R1–R7 cells in the hemispheres and rows 5 and 6 are used
for spatial vision or polarization vision. The tiered receptors
of rows 1 to 4 are responsible for color vision. The tiers of
color receptors in rows 2 and 3 are usually separated by
photostable filters, indicated by the black sections within the
photoreceptors. (b) Photographs of these filters as they
appear in fresh-frozen sections. The top row shows cross
sections, and the bottom row shows long sections of these
filters. (*Photographs by N. J. Marshall*)

the dorsal and ventral hemispheres resemble those of
other crustaceans (Fig. 2a), with a small, ultraviolet-
sensitive R8 cell at the top of the receptor group
and an extended receptor region below formed by
R1–R7. These ommatidia are used for spatial and
motion vision. In contrast, the midband ommatidia
house the receptors responsible for color vision (as
well as for specialized visual analysis of polarized
light). The six parallel rows of ommatidia, designated
rows 1 through 6 from dorsal to ventral, are individually specialized and have several structural features
that are not seen in other crustacean eyes (Fig. 2a).
The two most ventral rows, 5 and 6, are most similar to the ommatidia of the hemispheres, but are
larger and have significantly enlarged R8 cells. Their
receptor sets are highly specialized to analyze light
polarization. Ommatidia of the dorsal four rows of
the midband are among the most unusual found in
any compound eye, and their specializations foster a
unique, highly developed color sense.

Unlike other ommatidia, receptors in the four dorsal midband rows have three tiers. The ultraviolet-
sensitive R8 receptor is, as always, placed at the top
(Fig. 2a). The R1–R7 group, however, is subdivided
into two additional tiers: one of three retinular cells
(R1, R4, and R5) and one of four (R2, R3, R6, and
R7). Furthermore, the receptor cells of the two tiers
contain different visual pigments and are thus sensitive to different wavelengths of light. Cells in the
upper tier invariably are sensitive to light in a shorter-
wavelength range than those of the lower tier. Because light is absorbed as it travels down the tiered
receptor group, it first has the ultraviolet component
largely removed by the R8 cells. It then enters the
upper tier of the main receptor, where shorter wavelengths are absorbed by the visual pigment present
here. Finally, the remaining transmitted light travels
to the lowest receptor group and is absorbed by the
longer-wavelength-absorbing visual pigment of this
tier. This serial filtering successively trims the spectrum of light within the photoreceptor, sharpening
color perception.

The ommatidia of midband rows 2 and 3 take this
filtering theme further by adding strongly absorbing,
photostable pigments to modify light. The locations
of the filters are indicated by the black segments between tiers in Fig. 2a. In *Haptosquilla trispinosa*,
the species illustrated here, there are four types of
these filters, with two each in rows 2 and 3. While
having four types is most common, some other stomatopod species have fewer filter classes. Examples
of filters from fresh, unfixed material from various
species are illustrated in Fig. 2b.

Receptor tuning by serial filtering. As just described,
light becomes modified at several stages as it moves
through the successive tiers of each photoreceptor.
This process is illustrated in **Fig. 3**, which represents the situation in the photoreceptors of row 2 in
Haptosquilla trispinosa (refer also to Fig. 2). The top
pair of panels in Fig. 3 shows the absorption spectra of the filter pigments (left) and the visual pigments (right). The filter nearest the top of the receptor group ("distal") appears similar in color to the

filter seen in cross section in the top-right panel of Fig. 2*b*, whereas the deeper filter ("proximal") is similar in color to that seen in the second panel in the bottom row of Fig. 2*b*. The visual pigments (Fig. 3, top right) have peak absorbances near 505 nm in the top (distal) receptor group and 535 nm in the deeper (proximal) group.

The three lower panels in Fig. 3 illustrate how filtering produces sharply tuned spectral sensitivity functions. At the top is the predicted sensitivity of the two receptors in the absence of any filtering, as in single, isolated receptors (that is, not as part of the complex serial design of mantis shrimps). The broad sensitivity spectra overlap over most of their spectral coverage. The middle panel illustrates the tiered retina in the absence of the photostable filters. Here, the distal tier filters the deeper proximal tier, which consequently has a narrowed sensitivity spectrum. The photostable filters are not required for the spectral tuning, so even in rows 1 and 4, which lack filters, the tiered structure is still useful. The effect of adding the filters is seen in the bottom panel of Figure 3. The two receptor tiers become very sharply tuned and are spectrally well placed to analyze small shifts in color in the spectral region where the curves overlap, from about 550 to 625 nm. The filtering also shifts the two sensitivity spectra to longer wavelengths compared to the absorption spectra of the visual pigments.

This theme of tuning and sharpening, repeated in each of the dorsal four rows of the midband, produces eight spectral functions that cover a range exceeding 300 nm (**Fig. 4**). Such a color vision system can exist because each receptor class is devoted only to a restricted segment of the entire range. In comparison, in humans, the three cone types, being unfiltered, overlap in their sensitivities through most of the visible spectrum, and adding more classes of color receptors would provide little, if any, improvement in color vision. With their eight classes of specialized receptors, stomatopod crustaceans have achieved the most complex and competent color vision system known.

Functions of color vision in mantis shrimps. Do these features of mantis shrimp retinas provide these animals with color vision? Recent behavioral experiments establish that they can learn to discriminate among objects with different colors. Their color vision probably evolved to enhance visual contrast in the hazy, light-scattering environments of the shallow waters in which most species live. Some species with excellent color vision are fairly plain in their appearance, but many others have beautiful colored markings that are used in communication during mating or aggressive encounters (Fig. 1*b*). As in colorful fish, birds, and butterflies, color vision in these animals has favored the evolution of color signals that are useful for intraspecific signaling.

Perception of polarized light and its relationship to color vision. Since the specializations for polarization vision are anatomically (and possibly functionally) related to those for color vision, it is interesting to compare polarization vision to color vision in

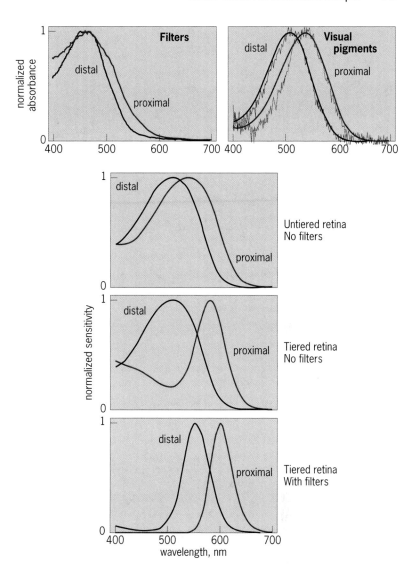

Fig. 3. An illustration of how retinal tiering and filtering can produce very narrow spectral sensitivity functions, using the second midband row of the stomatopod crustacean *Haptosquilla trispinosa* as an example. The top panels show the absorption spectra of the filter pigments (left) and the visual pigments (right), measured using a microspectro-photometer in freshly cut sections of unfixed eyes. The jagged curves in the visual pigment panel show averaged data, and the smooth curves are ideal visual pigment spectra fitted to the data. The lower three panels show the successive effects of retinal tiering and filtering. See the text for further explanation. [*Modified version reproduced with permission from F. Prete (ed.), Complex Worlds from Simpler Nervous Systems, MIT Press, Cambridge, Mass., 2004*]

stomatopods. Receptors in rows 5 and 6 of the midband show a number of features that are compatible with an ability to discriminate polarized light, and stomatopods have been trained to select objects that differ only in polarized light patterns. Recent work with these animals has revealed that some species of mantis shrimps detect and respond not only to linearly polarized light, which is common in nature, but also to the much rarer circularly polarized light. The specialized midband receptor groups are physiologically capable of discriminating right from left circularly polarized light, and some mantis shrimp species display patterns of either (or both) linearly or circularly polarized light. It is likely that polarization vision is simply another aspect of the perceptual specializations that mantis shrimps have evolved to

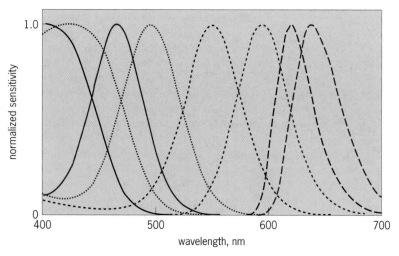

normalized sensitivity

wavelength, nm

Fig. 4. The eight narrowly tuned receptor classes in midband rows 1–4 of the stomatopod species *Haptosquilla trispinosa* (see Figs. 2*a* and 3 as well). Note how the combination of tiering and filtering produces a series of sharply tuned sensitivity functions that cover the spectrum effectively from below 400 to beyond 700 nm.

enhance their ability to detect low-contrast objects and animals in the light-scattering environments of shallow marine waters.

For background information *see* ANIMAL COMMUNICATION; COLOR; COLOR VISION; CRUSTACEA; EYE (INVERTEBRATE); EYE (VERTEBRATE); LIGHT; PHOTORECEPTION; PIGMENT (MATERIAL); POLARIZED LIGHT; STOMATOPODA; VISION in the McGraw-Hill Encyclopedia of Science & Technology.   Thomas W. Cronin

Bibliography. T. W. Cronin, R. L. Caldwell, and J. Marshall, Tunable colour vision in a mantis shrimp, *Nature*, 411:547–548, 2001; T. W. Cronin and J. Marshall, The unique visual world of mantis shrimps, pp. 239–268, in F. Prete (ed.), *Complex Worlds from Simpler Nervous Systems*, MIT Press, Cambridge, Mass., 2004; T. W. Cronin, N. J. Marshall, and M. F. Land, Vision in mantis shrimps, *Am. Sci.*, 82:356–365, 1994; J. Marshall, T. W. Cronin, and S. Kleinlogel, A review of stomatopod eye structure and function, *Arthropod Structure Development*, 36:420–448, 2007; N. J. Marshall, J. P. Jones, and T. W. Cronin, Behavioural evidence for color vision in stomatopod crustaceans, *J. Comp. Physiol.*, A179:473–481, 1996.

Community genetics

Community genetics looks at how genetic variation within a species, or group of species, changes other species in an ecosystem or the ecosystem as a whole. These changes can involve species presence or abundance; the behavior, physiology, or evolution of a species; or the rates of ecosystem processes. The question at the core of community genetics is "How do genotypes differentially influence associated species and ecosystems?" Community genetics takes ideas from population genetics and community ecology, and in so doing provides a way of integrating the disciplines of ecology and evolution.

Background. The phrase "community genetics" was coined by Janis Antonovics in 1992 to describe research that analyzed "evolutionary genetic processes" occurring "among interacting populations in communities." Antonovics recognized that a single species comprises a number of genetically distinct genotypes and that the different genotypes might influence in different ways the environment in which they lived and other individuals with which they interacted. These effects could be small, but they could also be large—in some cases, large enough to alter, or even reverse, the eventual outcome of the interaction. The discipline of community genetics also incorporates the concept of an "extended phenotype." The phenotype is the form or character of an individual, be it animal or plant—for example, the length of a leg or the number of leaves. It is made up of a genetic component, determined by the genes expressed by that individual, and an environmental component, determined by the environment experienced by the individual. The idea of extended phenotype is that the influence of an individual can be demonstrated to extend beyond itself and its immediate associates to the wider ecological community. In effect, the individual becomes the environment for something else; for example, a single plant may represent most of the environment experienced by an insect herbivore. A number of methods have been used to study community genetics effects, and these can be broadly divided into quantitative genetics or community-focused approaches.

Quantitative genetics approach. A quantitative genetics perspective looks for reciprocal effects of genetic variation on interacting partners in a trophic system or among members of an ecological guild. An example of a trophic system is the hierarchy from a plant, through an aphid herbivore, to an aphid predator and beyond. An ecological guild is a group of species with similar requirements or that play a similar role within an ecological community—

Fig. 1. An example model system used to determine community genetics effects taking a quantitative genetics approach: a single species of aphid living and feeding on a single species of plant and being preyed upon by ladybird beetles. Different plant genotypes are represented by different plants, as are different aphid genotypes (different aphids). (*Copyright by V. Ogilvy*)

for example, species occupying similar niches. The quantitative genetics approach uses model systems such as the one shown in **Fig. 1**. Here, a number of genotypes of a single aphid species (black, gray, and white) are living on, and eating from, a number of genotypes of a single plant species (different shades of plant). Ladybird beetles are then added to the system to feed on the aphids. After some time, the number of aphids on each plant is counted, and the sizes of the plants and ladybird beetles are measured. It may be that there are different numbers of aphids on each genotype of plant, in which case plant genotype has had an effect on the population growth of the aphids. Aphid genotype may also affect plant size, in which case plant size varies according to the genotype of aphid feeding on it. It could also be that different combinations of plant and aphid genotypes show different responses in plant size and aphid number. If this is the case, then the effect of plant and aphid on each other depends on which specific combination of plant and aphid genotype is present. If we then look at the ladybird beetles, we might see an effect of aphid genotype, in which case the different genotypes of aphids might result in different sizes of ladybird beetles. We might also see an effect of plant genotype, where we will find that the different genotypes of plant result in different sizes of ladybird beetles. Again, we may see an effect of both plant and aphid genotype, where the size of the ladybird beetle depends on the specific combination of plant and aphid genotype. Therefore, genetic differences in the plants may be detected in both aphids and ladybird beetles, and the effect of the plant genotype has extended beyond its direct interaction with the aphid to the higher trophic level of the ladybird beetle. Genetic differences in aphids may also be detected in plants, showing that a reciprocal genetic effect is also present. These are community genetic effects. Quantitative genetics experiments can be large and complicated, but they are a powerful way of determining whether the genotype of one or a number of species affects the performance of other species at many tropic levels in the community in which they live.

Community-focused approach. The community-focused approach looks at the effects of genetic variation within a dominant or keystone species on other species or functions of that community. A dominant species is one that makes up a large proportion of the biomass present in the community. An example is a common tree species of a forest. A keystone species has a proportionally larger effect on the community in which it resides than its biomass alone might suggest. The classic example of a keystone species is the sea star, *Pisaster ochraceous*, in the intertidal communities of western North America, whose removal causes dramatic changes to the whole community. When the *Pisaster* is present, it preys on the mussel, *Mytilus californianus*, leaving rocks free for colonization by other invertebrates. With removal of the sea star, the community changes to one dominated by the mussel,

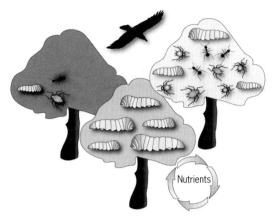

Fig. 2. An example of a system used to determine community genetics effects taking a community-focused approach: a dominant tree species supporting a community of herbivore species, which in turn feeds a community of predator species. Different tree genotypes are represented by different shades of tree, and herbivore communities are shown by the different species (caterpillar, ant, and aphid). The predator community is represented by a bird, and nutrient cycling within the system is represented by the circle of arrows. (*Copyright by V. Ogilvy*)

which eventually covers all surfaces of the rocks and excludes other invertebrates and algae, resulting in a community that is much less species-diverse. An example of the community-focused approach is illustrated in **Fig. 2**. Here, we are looking at a dominant tree species with genetic variation (shown by the different shades: light, medium, and dark) growing in a natural woodland. Other species interact directly with this tree, for example, herbivores eating the leaves (shown by the caterpillar, ant, and aphid). It may be that the abundance and type, or community structure, of herbivores on each tree depends on the specific genotype of that tree. When this occurs, then trees with similar genotypes will have similar communities of herbivores on them. If this is so, then the genotype of the tree, or genetic variation within the tree species, is affecting the community structure of the herbivores and thus having an effect on the wider ecological community. As an extension to this idea, tree genotype may also influence predator community structure feeding on the herbivores (represented by the bird). In addition, different genotypes of tree could have different degradation rates for fallen leaves, which may in turn affect the cycling of nutrients within the system (represented by the circle of arrows). Such effects have already been shown to occur in various natural systems and demonstrate the extended phenotype.

Applications. Community genetics approaches are being used to try and answer some important and timely ecological questions. For example, just how important is genetic variation in determining the outcome of ecological interactions between species? What role do microevolutionary processes play in ecological systems? Does greater within-species genetic variation mean that there is greater species diversity in the community? Are more genetically diverse and species-diverse communities better able to adapt to or resist global environmental change? If we are looking to conserve species, should we also be looking to conserve their genetic diversity, and how do we achieve this best? Can we manage agricultural pests with a genetically diverse approach to crop planting? What effect will genetically modified crops have on beneficial insects as well as pests? Is resistance to disease dependent on the genotype of the host, the genotype of the disease, or a combination of both? Does this change the way that we should be treating infected individuals?

A community genetics approach can be taken with a vast range of biological questions, encompassing areas from conservation biology to agriculture and disease management. It is also one of the few ways in which we can begin to understand interactions within complex biological systems such as the one in which we all live.

For background information *see* BIODIVERSITY; ECOLOGICAL COMMUNITIES; ECOLOGY; ECOSYSTEM; GENETICS; GENOMICS; POPULATION ECOLOGY; POPULATION GENETICS; POPULATION VIABILITY; TROPHIC ECOLOGY in the McGraw-Hill Encyclopedia of Science & Technology.   Jennifer Rowntree;
Richard Preziosi

Bibliography. A. A. Agrawal (ed.), Special feature: Community genetics, *Ecology*, 84(3):543–601, 2003; R. Dawkins, *The Extended Phenotype*, Oxford University Press, Oxford, 1982; R. S. Fritz and E. L. Simms (eds.), *Plant Resistance to Herbivores and Pathogens: Ecology, Evolution, and Genetics*, University of Chicago Press, Chicago, 1992; M. T. J. Johnson and J. R. Stinchcombe, An emerging synthesis between community ecology and evolutionary biology, *Trends Ecol. Evol.*, 22(5):250–257, 2007.

Computational photography

Photography, literally, drawing with light, is the process of making pictures by recording the visually meaningful changes in the light reflected by a scene. This goal was envisioned and realized for plate and film photography over 150 years ago by pioneers Joseph Nicéphore Niépce, Louis-Jacques-Mandé Daguerre, and William Fox Talbot, whose invention of the negative led to reproducible photography. Niépce created the first permanent photograph in 1826, titled *View from the Window at Gras*.

Digital photography is "filmlike," in the sense that the film or plate is replaced by an electronic sensor. The goals of the classic film camera, which are at once enabled and limited by chemistry, optics, and mechanical shutters, are pretty much the same as the goals of the digital camera. Both work to copy an image formed by a lens on a sensor, without imposing judgment, understanding, or interpretive manipulations; that is, both film and digital cameras are faithful but mindless copiers. As with conventional film and plate photography, filmlike photography presumes (and often requires) artful human judgment, inter-

vention, and interpretation at every stage to choose viewpoint, framing, timing, lenses, film properties, lighting, developing, printing, display, search, index, and labeling.

This article will explore a progression away from film and filmlike methods to a more comprehensive technology that exploits plentiful low-cost computing and memory with sensors, optics, probes, smart lighting, and communication.

Nature of computational photography. Computational photography (CP) is an emerging field that attempts to record a richer, multilayered visual experience. It captures information beyond a simple set of pixels and renders the recorded information as a machine-readable representation of the scene.

Computational photography exploits computing, memory, interaction, and communications to overcome the inherent limitations of photographic film and camera mechanics that have persisted in digital photography, such as constraints on dynamic range, limitations of depth of field, field of view, resolution, and the extent of subject motion during exposure.

Computational photography enables new classes of recording the visual signal, such as the "moment," shape boundaries for nonphotorealistic depiction, foreground versus background mattes, estimates of three-dimensional (3D) structure, "relightable photos," and interactive displays that permit users to change lighting, viewpoint, and focus for capturing some useful, meaningful fraction of the "light field" of a scene as a four-dimensional (4D) set of viewing rays.

Computational photography enables synthesis of "impossible" photos that could not have been captured with a single exposure in a single camera, such as wraparound views (multiple-center-of-projection images), fusion of time-lapsed events, the motion microscope (motion magnification), video textures, and panoramas. It supports seemly impossible camera movements such as the "bullet time" sequences made with multiple cameras using staggered exposure times and free-viewpoint television (FTV) recordings.

It encompasses previously exotic forms of imaging and data-gathering techniques in astronomy, microscopy, tomography, and other scientific fields.

Elements of computational photography. Digital photography involves a lens, a two-dimensional (2D) planar sensor, and a processor that converts sensed values into an image. In addition, it may entail external illumination from point sources (for example, flash units) and area sources (for example, studio lights).

Computational photography can be categorized into the following four elements and refined by considering the external illumination and the geometric dimensionality of the involved quantities (see **illustration**).

Generalized optics. Each optical element is treated as a 4D ray bender that modifies a light field. The incident 4D light field for a given wavelength is transformed into a new 4D light field. 4D refers here to the parameters (in this case 4) necessary to select one light ray. The light field is a function that describes the light traveling in every direction through every point in a 3D space. This function is alternately called the photic field, 4D light field, or lumigraph.

The optics may involve more than one optical axis. In some cases, perspective foreshortening of objects based on distance may be modified, or depth of field may be extended computationally by wavefront-coded optics. In some imaging methods and in coded-aperture imaging used for gamma-ray and x-ray astronomy, the traditional lens is absent entirely. In other cases, optical elements, such as mirrors outside the camera, adjust the linear combinations of ray bundles reaching the sensor pixel to adapt the sensor to the imaged scene.

Generalized sensors. All light sensors measure some combined fraction of the 4D light field impinging on it, but traditional sensors capture only a 2D projection of this light field. Computational photography attempts to capture more; that is, a 3D or 4D ray representation using planar, nonplanar, or volumetric sensor assemblies. For example, a traditional out-of-focus 2D image is the result of a capture-time decision, whereby each detector pixel gathers light from its own bundle of rays that do not converge on the focused object. A plenoptic camera, however, subdivides these bundles into separate measurements. Computing a weighted sum of rays that converge on the objects in the target scene creates a digitally refocused image and permits multiple focusing distances within a single computed image. Generalizing sensors can extend both their dynamic range and their wavelength selectivity. While traditional sensors trade spatial resolution for color measurement (wavelengths) using red (R), green (G), or blue (B) filters or a Bayer grid (RGB array) on individual pixels, some modern sensor designs determine the photon wavelength by sensor penetration, permitting several spectral estimates at a single pixel location.

Generalized reconstruction. Conversion of raw sensor outputs into picture values can be more sophisticated. While digital cameras demosaic (interpolate the Bayer grid), remove fixed-pattern noise, and hide "dead" pixel sensors, recent work in computational photography leads further. Reconstruction might combine disparate measurements in novel ways by considering the camera's intrinsic parameters used during capture. For example, the processing could construct a high-dynamic-range image out of multiple photographs from coaxial lenses or sensed gradients, or compute sharp images of a fast-moving object from a single image taken by a camera with a "fluttering" shutter. Closed-loop control during photographic capture itself can be extended, exploiting the exposure control, image stabilization, and focus of traditional cameras as opportunities for modulating the scene's optical signal for later decoding.

Computational illumination. Photographic lighting has changed very little since the 1950s. With digital video

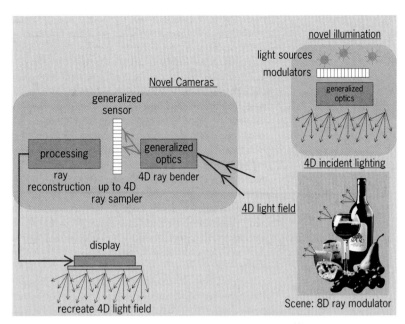

Elements of computational photography. (*From R. Raskar, Computational photography: epsilon to coded photography, pp. 238–253, in F. Nielsen, ed., Emerging Trends in Visual Computing: LIX Fall Colloquium, ETVC 2008, Pallaiseau, France, November 2008, Revised Invited Papers, Springer, Berlin/Heidelberg, 2008*)

projectors, servos, and device-to-device communication, we have new opportunities for controlling light sources with as much sophistication as we have for controlling digital sensors. Spatiotemporal modulations of lighting might better reveal the visually important contents of a scene, for example. Harold Edgerton showed that high-speed strobes offered tremendous appearance-capturing capabilities. New advantages might be realized by replacing "dumb" flash units, static spot lights, and reflectors with actively controlled spatiotemporal modulators and optics. We are already able to capture occluding edges with multiple flashes, exchange cameras and projectors by Helmholz reciprocity, gather relightable actors' performances with light stages, and see through muddy water with coded-mask illumination. In every case, better lighting control during capture allows for richer representations of the photographed scenes.

Present: epsilon photography. At their best, cameras define a "box" in the multidimensional space of imaging parameters. The first and most obvious thing that can be done to improve digital cameras is to expand this box in every conceivable dimension. The goal is to build a "supercamera" that has enhanced performance in terms of the traditional parameters, such as dynamic range, field of view, or depth of field. In this project, computational photography becomes epsilon photography, in which the scene is recorded via multiple images that vary at least one camera parameter by some small amount, or epsilon. For example, successive images (or neighboring pixels) may have different settings for parameters, such as exposure, focus, aperture, view, illumination, or timing of the instant of capture. Each setting allows recording of partial information about the scene and the final image is reconstructed by combining all the useful parts of these multiple observations. Epsilon photography is the concatenation of many such boxes in parameter space; that is, multiple film-style photos computationally merged to make a more complete photo or scene description. While the merged photo is superior, each individual photo is still useful and comprehensible. The merged photo contains the best features from the group. Thus, epsilon photography corresponds to the low-level vision: estimating pixels and pixel features with the best signal-to-noise ratio.

Field of view. A wide field-of-view panorama is achieved by stitching and mosaicing pictures taken by panning a camera around a common center of projection or by translating a camera over a near-planar scene.

Dynamic range. A high dynamic-range image is captured by merging photos at a series of exposure values.

Depth of field. An image entirely in focus, foreground to background, is reconstructed from images taken by successively changing the plane of focus.

Spatial resolution. Higher resolution is achieved by tiling multiple cameras (and mosaicing individual images) or by jittering a single camera.

Wavelength resolution. Conventional cameras sample only three basis colors. Multispectral imaging (from multiple colors in the visible spectrum) or hyperspectral imaging (from wavelengths beyond the visible spectrum) is accomplished by successively changing color filters in front of the camera during exposure by using tunable wavelength filters or diffraction gratings.

Temporal resolution. High-speed imaging is achieved by staggering the exposure time of multiple, low-frame-rate cameras. The exposure durations of individual cameras can be nonoverlapping or overlapping.

Photographing multiple images under varying camera parameters can be done in several ways. Images can be taken with a single camera over time. Images also can be captured simultaneously using "assorted" pixels, where each pixel is tuned to a different value for a given parameter. Just as some early digital cameras captured scanlines sequentially, including those that scanned a single, one-dimensional (1D) detector array across the image plane, detectors are conceivable that intentionally randomize each pixel's exposure time to trade off motion blur and resolution, as was previously explored for interactive computer graphics rendering. Simultaneous capture of multiple samples can also be recorded using multiple cameras, with each camera having different values for a given parameter. Two designs are currently being used for multicamera solutions: a camera array and single-axis multiple-parameter (coaxial) cameras.

Future: coded photography. To go far beyond the limits of the best possible film camera, instead of high-quality pixels, the goal will be to capture and convey the midlevel cues, such as shapes, boundaries, materials, and organization. Coded photography reversibly encodes information about the scene in a single photograph (or a very few photographs) so that the corresponding decoding allows powerful decomposition of the image into light fields, motion-resolved images, global/direct illumination components, or distinction between geometric versus material discontinuities.

Instead of increasing the field of view just by panning a camera, we might want to create a wraparound view of an object. Panning a camera allows us to concatenate and expand the box in the camera parameter space in the dimension of field of view. A wraparound view spans multiple disjoint pieces along this dimension. We can virtualize the notion of the camera itself if we consider it as a device for collecting bundles of rays, leaving a viewed object in many directions (not just toward a single lens), and virtualize it further by gathering each ray with its own wavelength spectrum.

Coded photography is the notion of a photographic method in which individual (ray) samples or data sets may not be comprehensible as images without further decoding, rebinning, or reconstruction. For example, a wraparound view might be built from multiple images taken from a ring or a sphere of camera positions around the object, but take only a few pixels from each input image for the final result. Coded-aperture techniques, inspired by work in astronomical imaging, try to preserve the high spatial frequencies of light that passes through the lens so that out-of-focus, blurred images can be digitally refocused or resolved in depth. By coding illumination, it is possible to decompose radiance in a scene into direct and global components. Using a coded-exposure technique, the shutter of a camera can be rapidly fluttered open and closed in a carefully chosen binary sequence as it captures a single photo. The fluttered shutter encodes the motion that conventionally appears blurred in a reversible way, such that we can compute a moving but unblurred image. Other examples include confocal synthetic aperture imaging that lets us see through murky water, and techniques to recover glare by capturing selected rays through a calibrated grid. Additional novel abilities might be possible by combining computation with sensing novel combinations of scene appearance.

The next phase of computational photography will go beyond the radiometric quantities and challenge the notion that a synthesized photo should appear to come from a device that mimics a single-chambered human eye. Instead of recovering physical parameters, the goal will be to capture the visual essence of the scene and scrutinize the perceptually critical components. This "essence photography" may loosely resemble depiction of the world after high-level vision processing. In addition to photons, additional elements will sense location coordinates, identities, and gestures via novel probes and actuators. Sophisticated algorithms will exploit priors based on natural image statistics and online community photo collections. Essence photography will spawn new forms of visual artistic expression and communication.

We may be converging on new, more capable parameters in computational photography that we cannot fully recognize, with quite a bit of innovation to come.

For background information *see* ADAPTIVE OPTICS; CAMERA; COMPUTER GRAPHICS; COMPUTER VISION; GAMMA-RAY ASTRONOMY; IMAGE PROCESSING; OPTICS; PHOTOGRAPHY; PHOTON; RADIOMETRY in the McGraw-Hill Encyclopedia of Science & Technology.   Ramesh Raskar

Bibliography. J. Hays and A. A. Efros, Scene completion using millions of photographs, Article No. 4, in *ACM Transactions on Graphics*, vol. 26, no. 3 (*Proceedings of ACM SIGGRAPH 2007*), 2007; M. McGuire et al., Defocus video matting, *ACM Transactions on Graphics*, 24(3):567–576 (*Proceedings of ACM SIGGRAPH 2005*), 2005; S. Nayar, Computational Camera and Programmable Imaging, *Symposium on Computational Photography and Video*, MIT Computer Science and Artificial Intelligence Laboratory, 2005; R. Raskar, Computational photography: Epsilon to coded photography, pp. 238–253, in F. Nielsen (ed.), *Emerging Trends in Visual Computing: LIX Fall Colloquium, ETVC 2008, Pallaiseau, France, November 2008, Revised Invited Papers*, Springer, Berlin/Heidelberg, 2008; N. Snavely, S. M. Seitz, and R. Szeliski, Photo tourism: Exploring photo collections in 3D, *SIGGRAPH 2006 Conference Proceedings*, pp. 835–846, ACM Press, New York, 2006.

Conical bearingless motor-generators

A motor-generator consists of a component that spins, called the rotor, and a stationary component, called the stator. The rotor is usually suspended from the stator by a mechanical bearing, such as a ball bearing, roller bearing, or journal bearing. This method of suspension makes sense for most applications, as it is very economical. However, certain high-performance applications either require or can be greatly enhanced by a noncontacting suspension system. In applications where the rotor moves at an extremely high speed (and the friction losses of a mechanical system would be prohibitively large), where access for bearing maintenance (for example, lubrication) is limited, or where long life is required (mechanical bearings are a wear item, and have a limited lifetime), a noncontacting suspension might be preferred. This article describes a noncontacting system in which the rotor is suspended on a magnetic field, allowing five-axis levitation with a new type of bearingless motor-generator.

One example of a system that operates at extremely high speeds and must have a long life and minimal frictional losses is the flywheel system used for energy storage. At NASA Glenn Research

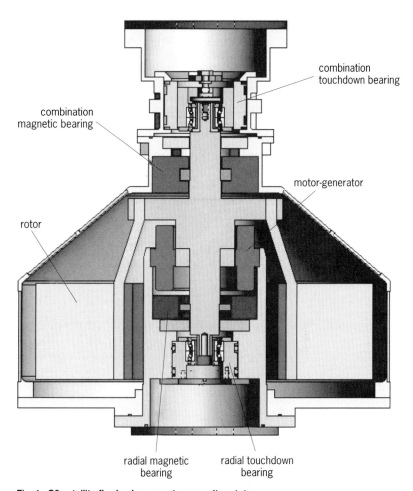

Fig. 1. G3 satellite flywheel energy storage unit prototype.

Center, work has been done to develop flywheel energy-storage systems to replace batteries on spacecraft (**Fig. 1**). Whereas batteries store energy chemically, flywheel systems store energy kinetically in a

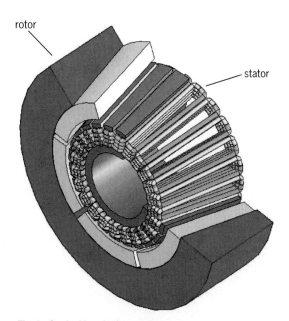

Fig. 2. Conical bearingless motor.

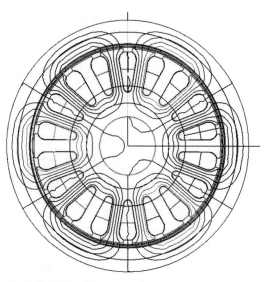

Fig. 3. Motor flux lines.

rotating mass called the flywheel rotor. In the flywheel system, the rotor is used to store mechanical energy by spinning at a high speed, the magnetic bearings levitate the rotor, and the motor-generator is used to convert energy back and forth between the electrical form and the mechanical form.

In a spacecraft application, excess energy generated by solar cells must be stored onboard. With the flywheel system, excess energy would be delivered to the flywheel rotor by means of the motor-generator, creating a torque to increase rotational speed. When the spacecraft is no longer exposed to direct sunlight, energy would be removed from the flywheel by means of the motor-generator, creating a torque to decrease rotational speed. In addition to replacing the functionality of batteries, flywheels can also be used in groups to control the attitude of

the spacecraft (its roll, pitch, and yaw). As a result, flywheels could also be able to replace the momentum wheels that would normally control spacecraft attitude. Flywheels have many advantages over traditional chemical batteries, including better system energy density, longer life, increased power capability, and the ability to perform attitude control.

Other motor-generator applications that would benefit from a magnetically levitated rotor are those that require operation in extreme environments, such as systems running in very high or low pressure or temperature environments (for example, in a gas turbine engine). These environments would not allow mechanical bearings to survive. To show the feasibility of a high-temperature magnetic bearing operation, a research program at NASA Glenn Research Center has successfully demonstrated magnetic levitation and rotation of a rotor in a 1000°F (540°C) environment.

Dedicated magnetic bearings. Early implementation of magnetic suspension was achieved using three separate magnetic bearings, two radial and one axial, to levitate a rotor. In this system, the motor was attached at one end of the shaft to facilitate rotation. This type of system allows full rotor suspension and the ability to provide torque to the shaft with the motor-generator without any mechanical wearing surfaces.

Motors and magnetic bearings both have finite iron and copper dedicated to rotation and levitation, and every complete system needs to be sized to meet maximum application requirements. In the system described above, the motor and the magnetic bearings are sized separately, and resources (such as copper or iron) from one function cannot be used for the other. This is a disadvantage in systems where peak force or torque requirements drive a larger system size. Developing a system in which both iron and copper could be shared for both motoring and levitation would be advantageous.

Bearingless motor-generators. Because of the above deficiencies, work has been done to develop bearingless motor-generators. The initial systems basically combined one of the magnetic bearings with a motoring function. Although these systems used common iron for motoring and levitating, they used separate copper windings, so the copper resources could not be shared. This scheme was applied to the major motor types, including synchronous reluctance motors, induction motors, permanent magnet motors, and switched reluctance motors.

Conical bearingless motor-generator. While the bearingless motor-generators are a great improvement over the separate motor-magnetic bearing systems, they have several disadvantages in that they are capable of levitating in only one radial plane or axial axis, and they use separate bearing and motoring windings, which does not allow all of the copper to be used for both functions. This deficiency was eliminated by the conical bearingless motor-generator (CBMG), which uses the same iron and the same copper windings for both levitation and motoring.

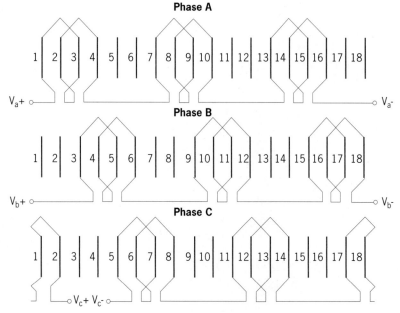

Fig. 4. Standard motor windings.

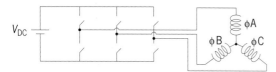

Fig. 5. Standard motor drive.

A conical bearingless motor is shown in **Fig. 2**. Note the unconventional inside-out configuration of the motor, where the rotor is on the outside and the stator is on the inside. There is always an attractive force between the stator and the rotor in motors. This force is exactly like the force that causes magnetic materials, such as iron, to be attracted to magnets. Motors always have an even number of poles, which implies that the lines of flux always form complete loops, as seen in **Fig. 3**.

While it is possible to increase or decrease the flux in a motor, the windings in a standard motor are balanced by design, so that whatever is done to one pole pair is done to all of them. The series-connected pole-pair windings of a standard motor are shown in **Fig. 4**, where each phase has three associated pole pairs. An increase in the flux in a pole pair will create a stronger attractive force in a particular pole pair. However, the attractive force is increased in all the other pole pairs as well, and because of motor symmetry, all of these forces will add to zero. This is desirable in a motor that is supported by mechanical bearings, because it minimizes vibrations due to the lateral force. All three phases are connected in a "Y" configuration, with two switches controlling each phase, resulting in a six-switch system (**Fig. 5**).

In the new conical bearingless motor-generator, the pole-pair windings are left unconnected, and each pole pair is broken out into independent windings (**Fig. 6**). This motor has three pole pairs, each of which has three phases, resulting in the need for 18 switches in the drive electronics (**Fig. 7**). Although more switches are required for this new configuration, these switches have lower power ratings than switches made for the standard motor, because each set of six is now only carrying one-third of the motor's total power.

CBMG control system. The CBMG has three stator windings in every pole pair. Physically, these three phases remain stationary, while the rotor rotates around them. It is helpful from a control standpoint to mathematically transform these three stationary phases into two phases, the *d* axis and the *q* axis, which rotate at the same speed as the rotor. This control technique is called field-oriented motor control, where the *d*-axis winding acts to either increase or decrease the permanent magnet flux, while the *q*-axis winding acts to create either a positive or negative torque. These theoretical windings are shown in **Fig. 8**.

In the CBMG control scheme, the *q*-axis current in each pole pair is set to the same value, so that the torque generation is divided equally between the

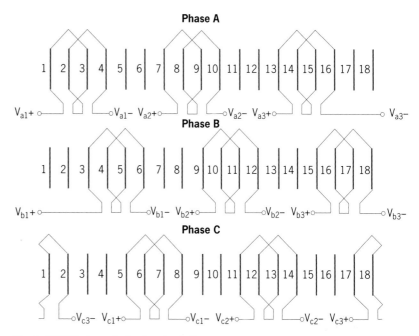

Fig. 6. CBMG independent pole-pair windings.

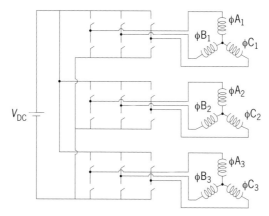

Fig. 7. CBMG independent pole-pair drive.

pole pairs. Unlike traditional motor control, where the *d*-axis current is set to zero (since by design, there is no way to unbalance the flux), the CBMG controller varies the *d*-axis current in the different pole pairs to create a flux differential on the rotor. This flux differential is what creates the levitating force.

The three *d*-axis currents, controlled in the positive and negative directions, can generate six possible force vectors (**Fig. 9**). The controller uses vector math to combine these vectors to create an arbitrary force in any direction.

Standard motors typically have cylindrically shaped rotors and stators. If the CBMG were cylindrical in shape, the forces from the flux imbalance would be exerted only in the radial direction. However, the CBMG was designed with a conical air gap, allowing a portion of the force to be exerted in the axial direction. When these motors are used in a pair, with the cones in opposing directions, a rotor may be fully suspended, radially and axially. In this

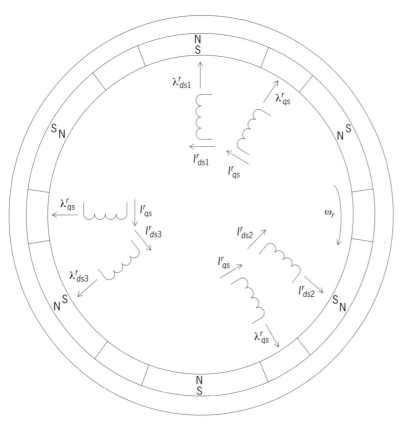

Fig. 8. Theoretical *d*- and *q*-axis windings.

Fig. 10. CBMG prototype with five-axis levitation and rotation.

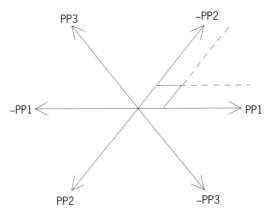

Fig. 9. Available force vectors.

configuration, the *z*-axis force is proportional to the difference of the sum of the *d*-axis currents in the top motor and the sum of the *d*-axis currents in the bottom motor.

A prototype CBMG and control system was designed and built at NASA Glenn Research Center. The rotor was successfully levitated in all axes, while simultaneously rotating. This represents the first five-axis levitation and motoring of this type. The prototype CBMG is shown in **Fig. 10**.

For background information *see* ANTIFRICTION BEARING; FLYWHEEL; GENERATOR; MAGNETIC LEVITATION; MOTOR; MOTOR-GENERATOR SET in the

McGraw-Hill Encyclopedia of Science & Technology.

Peter Kascak; Timothy Dever; Ralph Jansen

Bibliography.   A. Chiba, D. T. Power, and M. A. Rahman, No load characteristics of a bearingless induction motor, *IEEE Trans. Magn.*, 27(6):5199–5201, 1991; A. Chiba, K. Chida, and T. Fukao, Principles and characteristics of a reluctance motor with windings of magnetic bearing, *Proc. International Power Electronic Conference*, Tokyo, pp. 919–926, 1990; K. Dejima, T. Ohishi, and Y. Okada, Analysis and control of a permanent magnet type levitated rotating motor, *IEEJ Proc. Symp. Dynamics of Electro Magnetic Force*, pp. 251–256, June 1992; P. Kascak, R. Jansen, and T. Dever, U.S. Patent 7,456,537 B1, November 25, 2008; P. Kascak et al., Demonstration of attitude control and bus regulation with flywheels, *Industry Applications Conference, 2004, 39th IAS Annual Meeting*, Seattle, October 3–7, 2004 (*Conference Record of the 2004 IEEE*), vol. 3, pp. 2018–2029, 2004; P. Kascak et al., Bearingless five-axis rotor levitation with two pole pair separated conical motors, *Industrial Applications Society Annual Meeting*, Houston, October 4–8, 2009; A. Provenza, G. Montague, and M. Jansen, High temperature characterization of a radial magnetic bearing for turbomachinery, *J. Eng. Gas Turbines Power*, 127(2):437–445, 2005; K. Shimada et al., A stable rotation in switched reluctance type bearingless motors, *IEEJ Technical Meeting on Linear Drives*, 1997, LD-97-116, 1998; L. Teschler, New spin for flywheel technology, *Mach. Des.*, September 16, pp. 72–80, 2004.

Control of shoot branching in plants

Much of plant shape is defined by the number and location of branches. A combination of genetics and environment controls plant branching. Plants growing beneath the canopy of a dense forest will not produce many side branches. As a consequence, the available energy can be channeled into the main shoot so it can reach upward into sunlight. On the other hand, species growing on an open plain will be highly branched to extend outward and optimize sun exposure. Breeding desired branching traits has been very important in the course of human agriculture. For example, modern maize (*Zea mays*, corn) has fewer branches (tillers) than the wild teosinte progenitor. This reduction in tillers seems to be due to enhanced expression of a single gene, *TEOSINTE BRANCHED1* (*TB1*), which inhibits bud outgrowth in plants.

Plant hormones stop buds from growing. Side branches start out as tiny buds produced in the axils of leaves. (The axil is the angle between the base of a leaf and a stem.) The buds then either grow out unhindered to form a branch or enter a state of dormancy, where they are metabolically active but nongrowing. The interaction of plant hormones is thought to control whether a bud will stop growing, and if and when it will start growing again. Auxin is the most well known of these hormones. The predominant auxin, indole-3-acetic acid (IAA), is produced in the young leaves of actively growing shoot tips. It is transported down the plant stem in the vascular cambium and blocks bud outgrowth along the stem. Auxin is known to be involved because, when the shoot tip is removed, side branches below the decapitation site will start to grow out, and the application of IAA to the stump will normally reinhibit these buds. However, it has been a mystery how auxin acts to stop buds from growing, because it never actually goes into the buds. Various theories involving second messengers and correlative dominance (in which correlative signals are sent from a dominant to a dominated organ) have been postulated to explain this. For instance, cytokinins are another class of plant hormones, which promote bud release when applied directly to buds. They have been proposed as a possible secondary signal, which might be restricted from entering buds by auxin in the stem.

Branching mutants and novel graft-transmissible signal. Major breakthroughs were made in this area when various branching mutants were progressively isolated in laboratory plant species. Not surprisingly, many branching mutants were defective in known plant hormones such as auxin. However, plant height, leaf and root growth, and other developmental traits were also greatly affected in these mutants. A very distinctive class of mutants with excessive branching was found that did not correlate well with the negative side effects and levels of known plant hormones. These mutants were *ramosus* (*rms*) in pea, *more axillary growth* (*max*) in Ara-

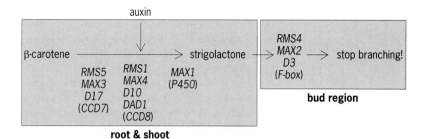

Fig. 1. Model of strigolactone production. Auxin promotes genes (listed below arrow) required for strigolactone biosynthesis. This can occur in roots and shoots. Strigolactones move into the bud region and are perceived by the F-box response protein, which signals to stop branching.

bidopsis thaliana, *dwarf* (*d*) in rice, and *decreased apical dominance* (*dad*) in petunia (**Fig. 1**). Elegant grafting studies in pea showed that particular mutant shoots grafted to nonmutant (wild-type) rootstocks restored branch repression, indicating that a novel inhibition signal could move up from wild-type roots to restore branch inhibition in mutant shoots. Interestingly, IAA could not inhibit branching in mutant plants. However, IAA was found to stop branching after decapitation in mutants grafted to wild-type rootstocks. This auxin/grafting experiment demonstrates that applied IAA, or some signal induced by it, moved down to the wild-type roots and enhanced production of a signal, which then moved back up into mutant shoots to inhibit buds. This auxin-induced hormonal signal that moves upward in the plant was called SMS (shoot multiplication signal). Local production of SMS in the stem is also sufficient to inhibit branching. In addition, whereas some SMS mutants were evidently defective in synthesis of SMS, others could not be rescued by grafting and were classed as defective in response to SMS. One of the genes responsible for this response has been identified as encoding an F-box protein (an F-box is a protein motif of approximately 50 amino acids functioning as a site of protein–protein interaction). Certain F-box proteins are hormone receptors in signaling pathways in plants. Interestingly, a possible outcome of SMS signaling is to enhance expression of the *TB1* gene.

Carotenoid connection. SMS synthesis mutant genes were cloned and, based on homology with a characterized gene family, were found to encode two CAROTENOID CLEAVAGE DIOXYGENASE (CCD) enzymes (Fig. 1). As mentioned above, cloning also revealed that an F-box protein controls SMS response. A fourth gene, *MAX1*, encodes a cytochrome P450 enzyme (Fig. 1). (Cytochrome P450 enzymes mediate a wide range of oxidative reactions involved in the biosynthesis of plant secondary metabolites.) Heterologous expression of the *CCD* genes from *Arabidopsis* or rice in *Escherichia coli* showed that they could act to cleave β-carotene and produce apocarotenoid secondary metabolites. The breakthrough idea that SMS might be a strigolactone was suggested recently when strigolactones were proposed to be derived from apocarotenoids. Strigolactones are a class of carotenoid-derived signal molecules with characteristic lactone rings. They are exuded from the roots and in most plants promote symbiosis with fungi (arbuscular mycorrhizae), an ancient and important interaction that greatly enhances nutrient and water uptake in plants. In contrast, the seeds of the *Striga* and *Orobanche* species of parasitic plants respond to this signal as a cue to germinate, with devastating consequences to crops. Chemicals with similar structure to strigolactones can occur in smoke, which may trigger seed germination after bushfires.

Novel plant hormone. Recent investigations have revealed that strigolactones may be the novel branching hormones and have demonstrated that the SMS genes act in a strigolactone synthesis and response pathway. Testing of SMS synthesis mutants showed them to be deficient in strigolactones, and applying or feeding strigolactones back to the mutants reduced their branching. Interestingly, only minute concentrations of strigolactone (for example, 10 nM or 10^{-8} M) were necessary to obtain an effect and feeding showed that strigolactone, or a downstream product, could move upward to act in the shoot. As expected, the SMS response mutants were not deficient in strigolactones and were not rescued by strigolactone treatments. The precise mobile and active strigolactones or products involved in shoot branching have not yet been elucidated.

Removal, replacement, low active concentration, and mobility together suggest that strigolactones act as hormones in the classic sense. Strigolactones also block branching after decapitation, fulfilling the requirements for the second messenger for auxin action. Strigolactones and cytokinins may act antagonistically to control bud outgrowth, whereas auxin acts to increase strigolactone levels and decrease cytokinin levels (**Fig. 2**).

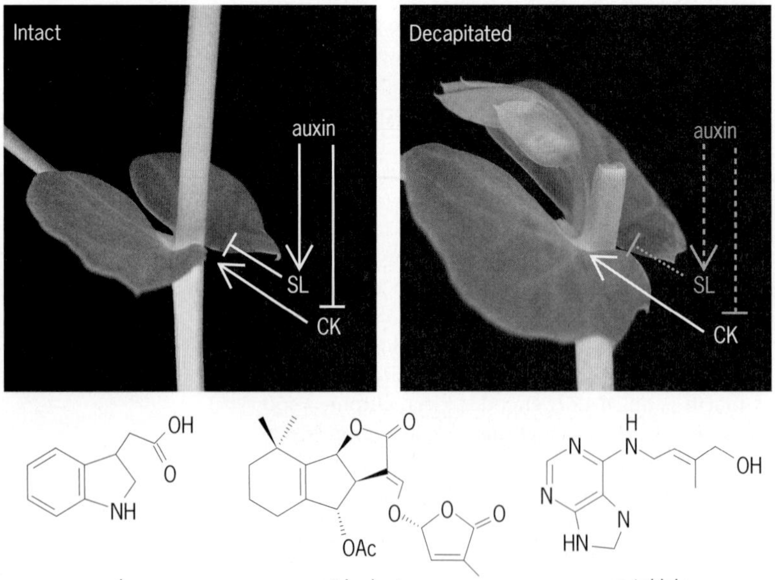

Do strigolactones move into buds? *CCD* genes are expressed in vascular tissues, and data suggest that the SMS signal moves upward in the transpiration stream. Also, the F-box response gene is expressed in buds and vasculature. It is tempting to think that strigolactones are made in the stem vasculature, are excreted into the transpiration stream in xylem, and then move upward into buds, where they act directly to block growth. However, as mentioned above, it is not absolutely known yet if strigolactones move in the plant and where they might go, or whether it is some other downstream product that moves. Measuring strigolactones in xylem sap and shoots of grafted plants will help to determine this, along with testing possible molecular interactions between strigolactones and the F-box response protein. Up to this point, the quantification of strigolactones has been difficult because of their extremely low concentrations in plant stems. Hopefully, technology will improve, and tracing radiolabeled strigolactone movement will also help to verify mobility.

Crosstalk with cytokinins. As shown in Fig. 2, auxin is thought to repress synthesis of cytokinins locally in the shoot, which may limit their availability for buds. Cytokinins can act to increase cell division and will trigger bud release when applied to buds. It is not known, though, how they relate to strigolactones. Increased cytokinin levels might be expected in SMS mutants. However, they actually have low levels of cytokinins in the xylem. In contrast to the outdated dogma that the roots are the only source of cytokinins, the cytokinin content is not greatly affected in the mutant shoots. Perhaps important differences in cytokinin content are effectively hidden or partitioned away in particular cells or compartments in the shoot. Also, it is possible that auxin acts to produce strigolactones that then block cytokinin movement to buds, and that this occurs at the same time that auxin independently reduces cytokinin levels. This would enable a fine-tuning of branch numbers in intact plants without relying on changes in auxin supply from the shoot.

Role of auxin transport. Auxin transport has been proposed to be important for axillary bud outgrowth. Recent data suggest that strigolactones do not inhibit the initial stages of bud outgrowth by affecting auxin transport in the stem or axillary shoot. However, prevention of auxin transport in growing axillary shoots is likely to prevent their continued growth.

Auxin is too slow. Large plants have a distinct problem in relying on auxin depletion after decapitation to rapidly release buds below the decapitation point. This is because auxin moves at only 1 cm (0.4 in.) per hour. Therefore, it should theoretically take 8 days for lower buds on a 2-m-tall (6.6-ft-tall) plant to be released to grow after decapitation, which is simply not the case in reality. Carefully timed measurements have shown that pea buds that are 20 cm (8 in.) away from the decapitation site start growing within 6 h, and with no loss of auxin at that node. The

Fig. 2. Simplified view of the interaction of three branching hormones. Auxin promotes production of strigolactones (SL) while limiting that of cytokinins (CK) at bud nodes. This blocks bud outgrowth in intact plants (*left*). After decapitation, auxin and strigolactone supply is lost, whereas cytokinins increase to promote bud growth (*right*). The chemical structures shown at the bottom represent an endogenous auxin (indole-3-acetic acid), strigolactone (orobanchyl acetate), and cytokinin (zeatin) from garden pea (the plant shown). (*Plants supplied by Kerry Condon; photographs by Philip Brewer*)

mechanism and signal for this faster decapitation response is completely unknown. Finding genes that respond rapidly to decapitation may be a way to unlock this puzzle.

Novel feedback signals. The system for producing strigolactone appears to be strongly balanced by feedback signaling. When there is a tendency for excessive branching or depleted strigolactone, it seems that extra strigolactone can be produced to counteract the branching. This is likely to occur through increased auxin production, but there is evidence for a second feedback signal. It will be necessary in the future to identify this other unknown signal. So far, a large number of hormones have been found in animal studies, so it will not be surprising if many novel plant hormonal signals come to light in the future.

How to apply this knowledge? In recent years, knowledge of how branching is controlled by hormones has rapidly improved. However, applying auxin and cytokinins to plants to change branching has not been useful because of their adverse effects on other aspects of plant growth. Now, strigolactones, suggested to be a novel group of branching-specific hormones, might be used to optimize plant architecture with minimal side effects. Combined use of strigolactones and strigolactone inhibitors may even lead to the ability to switch branching on and off at will. Knowledge of the biosynthesis of strigolactones and of the identity of the active compounds for each process involved may also help to develop plant lines with enhanced branching, along with improved mycorrhizal symbiosis, reduced infection by parasitic plants, and potentially increased β-carotene levels. Understanding how branching is regulated will continue to contribute to increasing plant productivity in the future.

For background information *see* APICAL DOMINANCE; AUXIN; BUD; CAROTENOID; CYTOKININS; GENE; PLANT ANATOMY; PLANT GROWTH; PLANT HORMONES; PLANT ORGANS; PLANT PHYSIOLOGY; STEM in the McGraw-Hill Encyclopedia of Science & Technology.   Philip B. Brewer; Christine A. Beveridge

Bibliography. C. A. Beveridge, Axillary bud outgrowth: Sending a message, *Curr. Opin. Plant Biol.*, 9:35–40, 2006; E. A. Dun et al., Strigolactones: Discovery of the elusive shoot branching hormone, *Trends Plant Sci.*, in press, 2009; V. Gomez-Roldan et al., Strigolactone inhibition of shoot branching, *Nature*, 455:189–194, 2008; P. McSteen, Hormonal regulation of branching in grasses, *Plant Physiol.*, 149:46–55, 2009; M. Umehara et al., Inhibition of shoot branching by new terpenoid plant hormones, *Nature*, 455:195–200, 2008.

Cosmic acceleration and galaxy cluster growth

Galaxy clusters are the largest and most massive gravitationally bound objects in the universe. Even though they are made up of thousands of galaxies, each with tens of billions of stars, the bulk of the visible mass in clusters (that is, mass in atoms and

Fig. 1. Superposed optical and x-ray images of A85, a typical galaxy cluster (distance 740 million light-years) from the nearby sample observed with the *Chandra X-ray Observatory*. The diffuse glow is thermal bremsstrahlung radiation from gas at a temperature of 7.5×10^7 K. The optical image (black sky with galaxies and stars) is from the Sloan Digital Sky Survey. Most of the cluster mass is dark matter and only about 20% of the baryons are in stars, the remainder being hot, x-ray–emitting gas. (*NASA, Chandra X-ray Observatory, Smithsonian Astrophysical Observatory, A. Vikhlinin et al.*)

nuclei) resides in a hot (10^7–10^8 K) gas that radiates predominantly in the 1–10-keV (equivalent to photons of wavelength 1.24–0.124 nm) x-ray band (**Fig. 1**). Clusters are unmistakable lighthouses in the x-ray universe—they are x-ray luminous (10^{36}–10^{38} J/s) and large, with typical sizes of about 1 megaparsec (3×10^{22} m = 3 million light-years). They are detectable to large distances (redshift $z \sim 1$, when the universe was less than half its present age) and appear as extended sources (they are resolvable) at all distances with the current generation of x-ray telescopes. While the cluster gas is hot, it also has very low density, typically 0.1–1 atom/liter, and yet even this low density is sufficient to illuminate the intracluster medium with thermal bremsstrahlung radiation, which is emitted by electrons that are accelerated in collisions with ions. In addition, extensive studies show that the x-ray properties are tightly coupled to the total cluster mass. Therefore, clusters are excellent sources for studying the evolution of the universe.

New test for an accelerating universe. The most remarkable recent discovery about the evolution of the universe was made using type Ia supernovae. Such supernovae are seen with the collapse of a white dwarf star in a binary system and its subsequent explosion. Type Ia supernovae are believed to behave as standard candles so their absolute distances can be determined by comparing their apparent brightness to their known absolute brightness. At the same time, the velocity of recession of the supernova's host galaxy can be measured. In the standard picture of the universe, there is only the gravity of matter, and it has always, since the big bang, retarded the expansion of the universe. Hence, the expansion velocities that we observe with type Ia supernovae and their host galaxies should vary as a predictable function of distance. Instead, the predicted retarding effect of gravity was found to be less than expected. In the context of Albert Einstein's general theory of relativity, the effect of gravity was being partially countered by an unknown accelerating agent.

Until 2008, all observational tests for an accelerating universe, like those from supernovae, studied the expansion of the universe as a function of distance. X-ray observations of galaxy clusters also provide such a test by using a constant cluster baryon fraction as a standard to measure the distance to clusters, because the observationally derived baryon fraction depends directly on distance to the cluster. A different approach to determining the acceleration of the universe is to measure the growth of clusters, the increase in their mass over cosmologically long times.

This new test measures the effect of competition between gravity and dark energy on the growth of galaxy clusters. While gravity produces growth of these clusters with time, a competing acceleration, driven by dark energy, retards that growth.

Measuring cluster growth. The new x-ray study of cluster growth was carried out by measuring the masses of the cluster population when the universe was 50–70% of its present age and comparing these masses to those of the cluster population today. These measurements required two well-defined samples, which were selected using surveys from the *ROSAT* (*ROentgen SATellite*), a joint German–United States mission with a large field of view that was designed with the express purpose of carrying out large-area surveys of the x-ray sky. From a shallow, nearly all-sky survey, 49 of the brightest nearby clusters (in the redshift range $z = 0.05$–0.25, corresponding to 0.7–3 billion years ago) were studied in detail using the National Aeronautics and Space Administration's *Chandra X-ray Observatory* to measure their mass distributions. Using deeper *ROSAT* observations covering only 400 square degrees, a second, more distant set of clusters was discovered. This sample of 37 distant clusters provides a measure of cluster masses 4–7 billion years ago, when the universe was about 60% of its present age. These distant clusters also were studied in detail with *Chandra*.

For the sample of 49 nearby clusters, the detailed Chandra studies were used to derive their masses directly from the x-ray observations under the assumption that the hot gas maps the gravitational potential of all the matter. From the equation of hydrostatic equilibrium, the gas density and gas temperature can be used to measure the total gravitating mass of the cluster. These same nearby clusters were then used to derive relations between this total mass and "easy"-to-measure observables such as the average gas temperature. The gas temperature is readily measured from the x-ray observations, and temperature provides a classic measure of the gravitational potential or mass of a system. The need for these "easy"-to-measure observables arises because distant are faint and cannot be observed in as much detail as nearby systems. The masses of the distant clusters were estimated using a mass proxy, for example, the gas temperature.

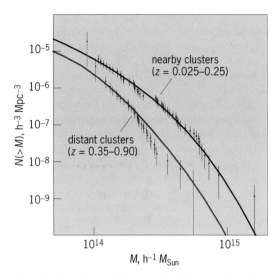

Fig. 2. Estimated cumulative mass functions (number of clusters N with a mass exceeding the mass M on the horizontal axis per unit survey volume) for the nearby (black) and distant (color) cluster samples derived for a standard cosmology with dark energy. The solid lines show the mass function models; the vertical lines show the data and their uncertainties. Here, h is the Hubble constant in units of 100 km s^{-1} Mpc^{-1}, and M_{Sun} is the solar mass. The deficit in the distant sample (color) compared to its model at a mass of about 3–4 $\times$ 10^{14} M_{Sun} is not significant. The growth from earlier times (color curve and data) to the present (black curve and data) is four times less than would be expected in a universe with no acceleration.

The key to the measurement of the effect of dark energy is to compare the distribution of masses of the nearby cluster sample (technically, their mass function, the number of clusters per unit volume as a function of mass; **Fig. 2**), measured directly, to the distribution of masses of the distant sample, using a mass proxy such as the average gas temperature. The comparison of nearby and distant samples relies on the theory of the growth of structure and the formation of clusters, carefully calibrated with state-of-the-art computer simulations.

The comparison of the calculated distribution of masses in the nearby universe with that when the universe was 60% of its present age shows unmistakable signs of acceleration. The x-ray studies show that there are more clusters today than at earlier times, and they are more massive. However, there are not as many, nor are they as massive, as would have been expected in a universe without acceleration. Without acceleration, we would expect to find 50 times more clusters, whereas the actual growth of the cluster population is a factor of only about 12. In conclusion, there are more clusters today, but about four times fewer than would have been expected if there were no acceleration to counter the effects of gravity.

Constraints on dark-energy content and equation of state. The two cluster data sets—nearby and distant—are used to constrain simultaneously the cosmological parameters that govern the growth of structure and evolution of the universe. Two key parameters characterize the dark energy. The first parameter is the fractional contribution that the dark energy makes to the total mass-energy content of the

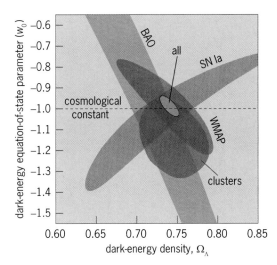

Fig. 3. The allowed regions (shaded) derived from different cosmological measurements of w_0, the dark-energy equation-of-state parameter, and Ω_Λ, the dark-energy density as a fraction of its contribution to the total mass-energy density of the universe. Comparison of the dark-energy constraints from x-ray clusters (the region labeled "clusters") and from other individual methods—supernovae (SN Ia); baryon acoustic oscillations, detected by measuring the distribution of galaxies over very large volumes (BAO); and *WMAP* measurements of the cosmic microwave background (WMAP)—are in agreement with each other and with prediction from the theory of general relativity (the horizontal line at $w_0 = -1$). The small dark region labeled "all" shows the constraints derived by combining all the separate measurements. The great reduction in the uncertainty (to 5% in the equation-of-state parameter) illustrates the complementarity of the different methods and the power of extracting all the information including the hidden ties between the cosmological data sets. Einstein's prediction agrees with the combined measurements from all the observations.

universe. This parameter is called Ω_Λ and is about 74% of the total mass-energy content. (Dark matter makes up about 22%, and normal matter about 4%.) The second parameter characterizes the nature of the dark energy. This second parameter, the equation-of-state parameter w_0, is the ratio of the dark-energy pressure to its density—mathematically, $w_0 = p_\Lambda/\rho_\Lambda$, where p_Λ is the pressure and ρ_Λ is the density attributed to the cosmological constant. A qualitative way to think of the dark energy and its equation-of-state parameter is to think of it as a measure of the "springiness" of space and, in the simplest interpretation, it is assumed to be a constant during the evolution of the universe.

Experiments to measure Ω_Λ and w_0 constrain their allowed ranges by simultaneous fits; hence, the allowed ranges are coupled together. **Figure 3** shows the regions of the Ω_Λ–w_0 plane that are allowed by current observations. Observations of the microwave background from NASA's *Wilkinson Microwave Anisotropy Probe* (*WMAP*), studies of supernovae, measures of the distribution of galaxies in the sky (technically referred to as baryon acoustic oscillations, because they are the imprint left by sound waves in the early universe on the baryons that subsequently formed galaxies), and now the x-ray measured growth of galaxy clusters all find consistent values for the dark-energy properties; that is, they

all have overlapping regions. From the x-ray cluster studies of the growth of clusters from 6 billion years ago to the present, the value of the dark-energy equation-of-state parameter was found to be $w_0 = -1.14 \pm 0.21$.

General relativity and the cosmological constant. Dark energy is a peculiar property attributed to empty space to explain the unexpected acceleration of the universe. Although it was first introduced by Einstein as a cosmological constant to counteract the effects of gravity by matter and thereby allow a static solution to his general relativistic equations, modern physics attributes the same effect to the energy associated with the vacuum of empty space predicted by quantum field theory. In the standard model, dark energy accounts for about 74% of the mass-energy of the universe. (One cubic meter of dark energy is equivalent to the mass-energy of one hydrogen atom.)

In Einstein's theory of general relativity with a cosmological constant, the equation-of-state parameter of the vacuum energy density is a true constant and is exactly equal to -1. As shown in Fig. 3, all the current measurements are consistent with $w_0 = -1$, the value expected if the cosmological acceleration arises from a cosmological constant–like term. Even more striking is the observation that, if we combine all the measurements to yield a single estimate of the parameter w_0, we find a best estimate within a few percent of Einstein's prediction with an uncertainty of 2–5% depending on the details of other assumptions about the universe. Thus, to an accuracy of a few percent, Einstein's general relativity is in agreement with the observations, a remarkable success.

If Einstein's general relativity is the correct explanation, what does it mean to have 74% of the mass-energy of the universe in the form of dark energy? First, the universe has only "recently" (in the past ~5 billion years) entered a period of accelerated expansion. In the future, the dark energy will cause the acceleration to continue so that distant galaxies will recede faster and faster. In our local neighborhood, the mass density is sufficiently high that our Milky Way Galaxy will merge with our nearest neighbor, the Andromeda Galaxy (Messier 31). However, all other galaxies will eventually (in 50–100 billion years, 4–8 times the present age of the universe) recede from sight as the universe continues its accelerated expansion.

For background information *see* ACCELERATING UNIVERSE; CHANDRA X-RAY OBSERVATORY; COSMOLOGICAL CONSTANT; COSMOLOGY; DARK ENERGY; GALAXY, EXTERNAL; HYDROSTATICS; RELATIVITY; SUPERNOVA; WILKINSON MICROWAVE ANISOTROPY PROBE; X-RAY ASTRONOMY; X-RAY TELESCOPE in the McGraw-Hill Encyclopedia of Science & Technology.
William R. Forman; Alexey Vikhlinin

Bibliography. C. Day, Galaxy clusters tighten constraints on the cosmic accelerator, *Phys. Today*, 62(3):14, 2009; J. A. Frieman, M. S. Turner, and D. Huterer, Dark energy and the accelerating universe, *Annu. Rev. Astron. Astrophys.*, 46:385–432, 2008; D. Overbye, Dark energy stunts galaxies' growth, *The*

New York Times, December 16, 2008; A. Vikhlinin et al., *Chandra* cluster cosmology project. II. Samples and x-ray data reduction, *Astrophys. J.*, 692:1033–1059, 2008; A. Vikhlinin et al., *Chandra* cluster cosmology project III: Cosmological parameter constraints, *Astrophys. J.*, 692:1060–1074, 2008.

Dawn ion propulsion system

The Dawn mission is designed to perform a scientific investigation of the main-belt asteroid (4) Vesta and the dwarf planet (1) Ceres in order to obtain new information to help answer questions about the formation and evolution of the early solar system. It will be the first mission to orbit a main-belt asteroid and the first to orbit two target bodies. The mission is enabled by an ion propulsion system, which the spacecraft will use to rendezvous with and go into orbit about each of these bodies. Dawn is the ninth project in the Discovery Program of the National Aeronautics and Space Administration (NASA), and was launched on September 27, 2007.

Dawn's ion propulsion system is the most advanced propulsion system ever built for a deep-space mission. Aside from the Mars gravity assist, it provides all of the postlaunch primary propulsion required to perform the mission. This includes the transfer to Vesta, getting into orbit around Vesta, transfer between various science orbits at Vesta, escape from Vesta, the transfer to Ceres, orbit capture at Ceres, and transfer between different science orbits at Ceres. The trajectory of the *Dawn* spacecraft is given in **Fig. 1**, which indicates periods when the

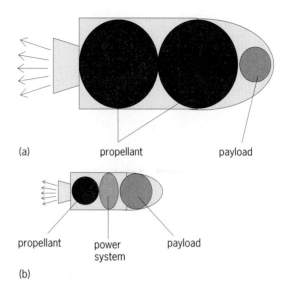

(a)   propellant   payload

(b)

propellant   power system   payload

Fig. 2. Comparison of spacecraft carrying the same payload with (*a*) chemical propulsion and (*b*) ion propulsion. Ion propulsion systems enable the use of much less propellant for the same size payload, but require the addition of an on-board power system.

ion propulsion system is thrusting and coast periods. It is clear from this figure that the ion propulsion system is operating for a significant fraction of the total mission duration. In fact, to complete the Dawn mission the ion propulsion system must operate for just over 50,000 h.

Reduction in propellant mass. The ion propulsion system uses xenon as the propellant, and to accomplish all of the required propulsive maneuvers the *Dawn* spacecraft carried an initial propellant load of 425 kg (937 lb) of xenon. To provide the same propulsive maneuvers with a conventional chemical bipropellant propulsion system would require more than 6000 kg (13,000 lb) of propellant. Thus, the ion propulsion system has reduced the amount of propellant required by more than a factor of ten. This difference is illustrated in **Fig. 2**, where for the same-sized payload the spacecraft using chemical propulsion is dominated by the mass and volume of the chemical propellants, while the spacecraft using ion propulsion requires much less propellant but requires the addition of a power source to operate the ion propulsion system.

The substantial reduction in propellant mass enabled by ion propulsion results directly from the ion propulsion system's ability to use each kilogram of propellant much more effectively than the best chemical rockets. How effectively the propellant is used depends almost entirely on the exhaust velocity, that is, the speed at which the propellant is expelled from the spacecraft. In chemical propulsion systems the energy used to accelerate the propellant is carried along with the propellant in the form of its chemical potential energy. There is only so much chemical energy stored in the unreacted chemical propellant, and this puts a fundamental limit on how much kinetic energy can be added to each kilogram of propellant. This, in turn, represents a

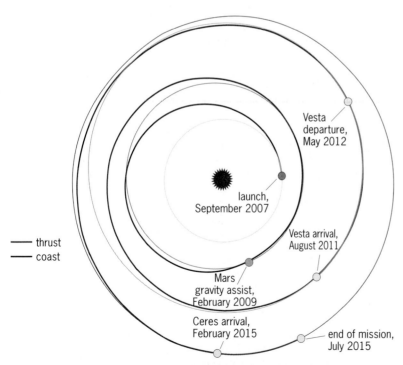

— thrust
— coast

Vesta departure, May 2012

launch, September 2007

Vesta arrival, August 2011

Mars gravity assist, February 2009

Ceres arrival, February 2015

end of mission, July 2015

Fig. 1. *Dawn* spacecraft trajectory over the 8-year mission. The ion propulsion system, which is used for all of the postlaunch propulsive maneuvers (except for the Mars gravity assist), must operate for a large fraction of the total mission duration.

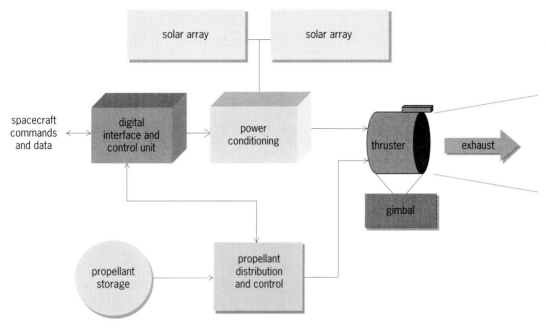

Fig. 3. Basic components of an ion propulsion system. Structural and thermal hardware are not shown.

fundamental limit on the maximum exhaust velocity achievable with a chemical-propellant rocket. The performance of chemical propulsion systems, therefore, is said to be energy-limited.

Ion propulsion systems achieve their substantial propellant mass reductions by decoupling the propellant from the energy used to accelerate it. In the *Dawn* ion propulsion system the propellant is the chemically inert, noble gas xenon. A separate power source is used to add kinetic energy to the propellant. This enables much more energy to be added to each kilogram of propellant, resulting in much higher exhaust velocities and therefore much more effective use of the propellant mass. The amount of thrust produced by an ion thruster depends on how much electrical power is available. The performance of ion propulsion systems, therefore, is said to be power-limited.

Components. The basic components of an ion propulsion system are shown in **Fig. 3**. The power source in this case is the solar array, which converts sunlight into electrical power. The "power conditioning" box in this figure converts the solar array power into the currents and voltages required to start and operate the ion thruster. The "digital interface and control unit" accepts commands from the spacecraft regarding what thrust level to operate the ion thruster. It then controls both the power conditioning unit and the propellant distribution and control system to accomplish this. It also collects data on the operation of the ion propulsion system and reports this to the spacecraft. The "propellant storage" represents the propellant tanks containing the xenon propellant, and the "propellant distribution and control" system controls the flow of xenon to the ion thruster. The "gimbal" represents a two-axis pointing mechanism that points the ion thruster in the desired direction. Finally, the ion thruster itself combines the solar array power as con-

ditioned by the power conditioning unit with the propellant to produce thrust. The system in Fig. 3 shows only a single ion thruster; however, the *Dawn* ion propulsion system actually includes three ion thrusters, although only one thruster is operated at a time. The second thruster is included because the required operating time is too long to be accomplished with a single thruster. The third thruster serves as a backup.

Ion thruster operation. To produce thrust, ion thrusters must accomplish three basic tasks. They must ionize the propellant, accelerate the resulting positively charged ions into the exhaust, and then neutralize the ion exhaust (**Fig. 4**). The propellant is ionized by a current of electrons that is made to flow through the propellant gas in a low-pressure discharge inside the thruster. The electrons in this current collide with the xenon atoms, and some of these collisions knock electrons from the xenon atoms, resulting in positive xenon ions. A magnetic field is used to improve the efficiency with which the thruster ionizes the propellant by confining the electrons and increasing the chances that they will collide with xenon atoms. This magnetic field also causes the newly created xenon ions to drift preferentially toward the downstream end of the thruster, where they are extracted and accelerated electrostatically. Ion acceleration is accomplished through the use of two or more closely spaced, multiaperture electrodes, across which an accelerating voltage of order 1 kV or greater is applied. The velocity of the ion exhaust is determined by the magnitude of the applied net accelerating voltage and the charge-to-mass ratio of the ions.

Finally, electrons stripped from the propellant atoms in the ionization process are collected and injected into the positive-ion beam by an electron emitter positioned adjacent to the ion beam. This electron emitter is called the neutralizer, and its

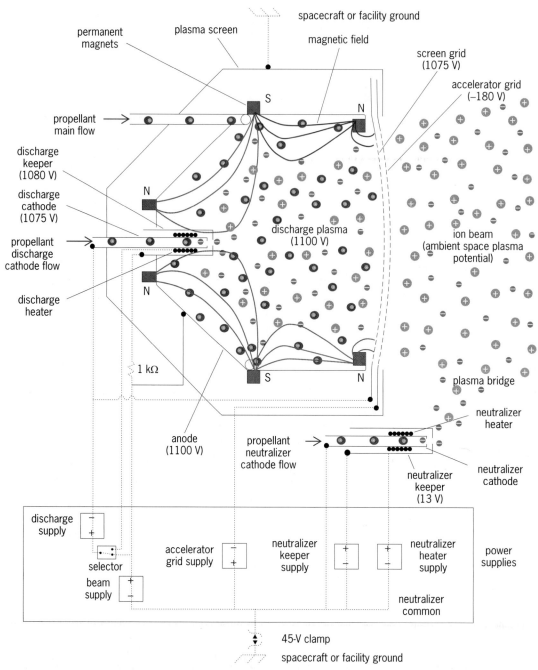

Fig. 4. *Dawn* ion thruster operation.

function is to prevent the spacecraft from accumulating a large negative charge. Without the neutralizer, the spacecraft would acquire a negative voltage equal to the net accelerating voltage (of order 1000 V or greater) in a few millionths of a second. To prevent this from happening, the neutralizer must inject electrons into the ion beam at exactly the same rate as ions are accelerated from the thruster. Fortunately, this can be accomplished fairly easily. The *Dawn* neutralizer technology was invented in the late 1960s and is based on an electron emitter called a hollow cathode. Hollow-cathode neutralizers emit a weakly ionized plasma along with lots of electrons. The plasma, in this sense, is a collection of positively and negatively charged particles, which greatly in-

creases the conductivity of the space between the neutralizer and the ion beam. The hollow cathode produces more electrons than the ion beam requires, which enables the ion beam to draw out electrons at exactly the right rate.

The electrostatic process that accelerates the ions is extremely efficient. In practice, a state-of-the-art ion accelerator system has an efficiency of converting electrical potential energy to kinetic energy of greater than 99.5%. The nearly perfect ion acceleration efficiency enables the ion engine to produce exhaust velocities more that ten times higher than the best chemical rockets while maintaining low engine component temperatures. It also results in the ion engine being the most efficient type of electric

Fig. 5. One of the *Dawn* ion thrusters mounted to the spacecraft.

thruster at specific impulses greater than approximately 2500 s.

Dawn uses the NASA 30-cm-diameter (12-in.), ring-cusp, xenon ion thruster developed in the mid-1990s and flight tested on the Deep Space 1 mission (**Fig. 5**). The maximum input power for the *Dawn* ion thruster is 2300 W. At this power level the thruster produces a thrust of 91 mN (0.020 lbf, which is approximately equal to the weight of 9 one-dollar bills). The low-thrust characteristic of ion propulsion requires long thruster operating times for typical deep-space science missions. The *Dawn* thruster design has been subjected to over 58,000 h of testing and in-flight operation, including 16,265 h on *Deep Space 1* and 30,352 h in an extended life test. The thruster can operate over an input power range from 0.5 to 2.3 kW, producing thrusts in the range from 18.8 to 91.0 mN (0.0042 to 0.0205 lbf), with specific impulses from 1740 to 3065 s, respectively. The ability to throttle thruster operation to match available input power is essential for the Dawn mission, which must operate over a factor-of-three change in solar range from 1 to 3 astronomical units.

[Acknowledgement: The research described in this article was carried out at the Jet Propulsion Laboratory, California Institute of Technology, under a contract with the National Aeronautics and Space Administration.]

For background information *see* ASTEROID; CERES; ION PROPULSION; SPACE PROBE; SPACECRAFT PROPULSION; SPECIFIC IMPULSE in the McGraw-Hill Encyclopedia of Science & Technology.   John Brophy

Bibliography. J. R. Brophy, NASA's Deep Space 1 ion engine, *Rev. Sci. Instrum.*, 73(2):1071–1078, 2002; D. M. Goebel and I. Katz, *Fundamentals of Electric Propulsion*, Wiley New York, 2008; M. D. Rayman, The successful conclusion of the Deep Space 1 mission: Important results without a flashy title, *Space Technol.*, 23(2–3):185–196, 2003; C. T. Russell et al., Dawn: A journey in space and time, *Planet. Space Sci.*, 52:465–489, 2004.

Death receptors

Apoptosis is a form of programmed cell death in which cells play an active role in their own death, ensuring that they die in a controlled and regulated manner with little disruption to the surrounding cells. Apoptosis allows unwanted cells (for example, damaged or virus-infected cells) to be efficiently removed from the body. It plays important roles in a variety of biological processes, including embryo development, the immune system, and cancer prevention. Cells undergo apoptosis in response to a wide variety of stimuli, both internal and external. Internal signals include DNA damage, cytoplasmic stress, and nutrient deprivation. By contrast, external signals can induce apoptosis by interacting with a class of cell surface receptors known as death receptors. The death receptors bind a variety of ligands known collectively as the death-inducing ligands, which can exist either as soluble, secreted proteins or as membrane-bound proteins on the surface of cells such as T lymphocytes. The binding of these ligands to the death receptors allows the apoptotic signal to be transmitted to the cell, triggering the onset of apoptosis.

Types of death receptors. Death receptors are members of the tumor necrosis factor receptor (TNFR)

Main death receptors and their ligands			
	Death-inducing ligand		
	FasL/CD95L	TNFα	TRAIL
Death receptor	Fas/CD95	TNFR1 (induces apoptosis) TNFR2 (nonapoptotic signaling)	DR4 (induces apoptosis) DR5 (induces apoptosis) DcR1 (decoy receptor) DcR2 (decoy receptor)

superfamily. (TNFs constitute a group of cytokines, which are peptides released by some cells that affect the behavior of other cells, serving as intercellular signals.) There are three main types of death receptors (see **table**) that share significant structural similarity and are grouped according to the ligand that they are activated by. The main death-inducing ligands (see table) are TNFα, Fas ligand (also known as FasL or CD95L), and TRAIL (TNF-related apoptosis-inducing ligand). In some cases, such as FasL, the ligand binds to just one type of receptor; in other cases, such as TRAIL, the ligand can bind to a number of different receptors.

The death-inducing ligand TRAIL is able to bind to at least four different receptors. The TRAIL receptors DR4 and DR5 are both able to induce apoptosis following the binding of TRAIL, whereas DcR1 and DcR2 are known as decoy receptors. These decoy receptors have extracellular domains identical to those of the DR4 and DR5 receptors, so they are able to bind effectively to TRAIL; however, they have trun-

cated intracellular domains, making them unable to send any apoptotic signals.

The typical death receptor exists as a trimerized transmembrane receptor with an extracellular domain that is capable of binding to the appropriate ligand and is often characterized by the presence of cysteine-rich motifs. The intracellular domain of the receptor consists of a cytoplasmic domain of around 80 residues that is termed the death domain. When the receptor is activated, a number of signaling molecules can bind to this death domain and subsequently activate the apoptotic signaling cascade (see **illustration**).

Death receptor signaling pathways. Binding of the appropriate ligand to a death receptor typically causes a change in the intracellular region of the receptor that results in the death domain motif becoming available. In the case of the Fas/CD95 receptor and the TRAIL receptors DR4 and DR5, an adaptor protein called FADD (Fas-associated death domain), which also possesses a death domain motif, is able to interact with the death domain of the receptor. The adaptor protein FADD also contains a protein domain known as the death effector domain (DED), which allows interaction with the equivalent DED of either pro-caspase 8 or pro-caspase 10. The caspases are a family of proteases that are responsible for breaking down key cellular components during apoptosis. They exist as inactive pro-enzymes until they are activated during apoptosis. In the case of the death receptors, pro-caspase 8 or 10 is recruited to the receptor complex by interaction between the DEDs of the pro-caspase and the adaptor protein FADD. The recruitment of the pro-caspase into the receptor complex results in the formation of the DISC, that is, the death-inducing signaling complex, the final stage of which is the autocatalytic cleavage of the pro-caspase to the active form of the caspase. Activation of the caspases is often thought to be the point of no return in the apoptotic signaling pathway, and at this stage the cell is committed to the apoptotic process. The illustration depicts the stages in the death receptor signaling pathway.

Signaling through the TNFR1 receptor is slightly different, since binding of the ligand TNFα can activate either pro-apoptotic or pro-survival signaling pathways. Which pathway is activated depends upon the precise composition of the proteins recruited to the receptor. For example, apoptotic signaling typically involves the recruitment of FADD and caspase

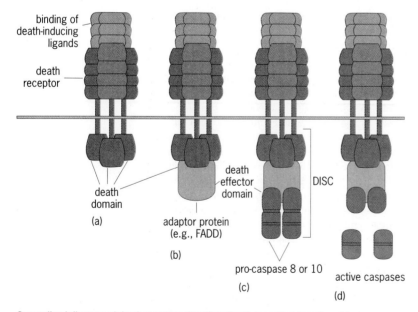

Generalized diagram of death receptor signaling. Death receptors usually exist as preformed trimers and are activated by binding of ligand trimers (*a*). Receptor activation results in a change in the structure of the receptor, revealing the presence of protein motifs called death domains. Adaptor proteins (such as FADD) that also contain these death domain motifs can then bind to the receptor (*b*). These adaptor proteins contain a protein motif called the death effector domain, which allows the recruitment of pro-caspases such as pro-caspase 8 (*c*). Recruitment of pro-caspase 8 then leads to autocatalytic cleavage of the pro-enzyme into an active caspase (*d*), which activates the caspase cascade and leads to apoptosis.

8. In contrast, survival signaling typically involves proteins such as RIP (receptor-interacting protein) and TRAF2 (TNF receptor–associated factor 2) and often leads to activation of the NF-κB signaling pathway (NF-κB, or nuclear factor κB, is an important transcription factor).

Regulation of death receptor signaling. Death receptor signaling can be modulated through a number of mechanisms. Amplification of the apoptotic signaling can occur following clustering of the death receptors. One of the earliest signaling events following binding of the ligand to the death receptor is the generation of the second messenger ceramide (a fatty acid amide of the amino alcohol sphingosine), which can promote the fusion of lipid rafts (specialized membrane microdomains enriched in certain lipids and proteins) in the plasma membrane, causing clustering of the death receptors. This clustering is able to greatly enhance the apoptotic signal. In fact, it does so to such a degree that death receptor signaling in the absence of receptor clustering is rarely able to activate the full apoptotic process.

Death receptor signaling is also able to interact with the intrinsic apoptotic pathway. The intrinsic pathway activates the apoptotic process following stimuli such as DNA damage and cell stress. Typically, these stimuli activate the bcl-2 (B cell leukemia 2) family of proteins, which are able to interact with one of the main regulators of apoptosis—the mitochondria. Mitochondria contain a number of pro-apoptotic factors, such as cytochrome c, and the bcl-2 proteins regulate the release of these factors by controlling the formation of a pore in the outer membrane of the mitochondria. When these pro-apoptotic factors are released from mitochondria, they can activate the caspase cascade and trigger the onset of apoptosis. Death receptor signaling interacts with this pathway through the cleavage of a bcl-2 protein called Bid (bcl-2 interacting domain) by the caspase 8 that is activated at the receptor complex. Cleaved Bid then translocates to the mitochondria, stimulating the formation of the pore and the release of the pro-apoptotic factors. In this manner, the intrinsic, mitochondria-related apoptosis pathway is used to amplify the apoptotic signal from the death receptors.

Apoptotic signaling from the death receptors can also be modulated by the presence of anti-apoptotic proteins (such as cFLIP), which are able to block death receptor signaling by binding to the receptor complex in place of the pro-caspases, thus preventing caspase activation.

Role in physiology and disease. Death receptor signaling is particularly important in the functioning of the immune system. Cytotoxic T lymphocytes express the death-inducing ligands on their surface and can induce apoptosis in cells that have become infected with viruses or cells that have become damaged and are thus at increased risk of developing into cancer cells. An important characteristic of many cancer cells is that they are particularly resistant to apoptosis, a feature that allows uncontrolled cell growth. Mutations in apoptotic signaling pathways are common in cancer, and much research is focused on developing ways to overcome apoptosis resistance and therefore inducing apoptosis in cancer cells.

One particularly interesting area of research involves the use of the death-inducing ligand TRAIL. Treatment of cancer cells with TRAIL has been found to induce apoptosis in many cancer cell lines, and it appears that cancer cells are more susceptible to TRAIL-induced apoptosis than normal, healthy cells. Administration of TRAIL to mice can suppress the growth of tumors and appears to produce little toxicity. Since TRAIL binds to both the active death receptors DR4/DR5 and the inactive decoy receptors DcR1/DcR2, monoclonal antibodies against the DR4/DR5 receptors that selectively activate the death receptors have been developed. Clinical trials involving either TRAIL itself or monoclonal antibodies against the DR4 or DR5 receptors have shown that these agents appear to be safe and well tolerated by patients, but the response rates are low, suggesting that they may not be potent enough. Future research is focusing on combination strategies in which TRAIL receptor targeting agents are combined with cytotoxic therapies in order to sensitize the tumor cells and increase the rate of apoptosis.

For background information *see* APOPTOSIS; CANCER (MEDICINE); CELL (BIOLOGY); CELL MEMBRANES; CYTOKINE; ENZYME; LIGAND; LIPID; MITOCHONDRIA; SIGNAL TRANSDUCTION; TUMOR in the McGraw-Hill Encyclopedia of Science & Technology.

Philip R. Dash

Bibliography. I. Lavrik, A. Golks, and P. H. Krammer, Death receptor signalling, *J. Cell Sci.*, 188:265–267, 2005; C. N. A. M. Oldenhuis et al., Targeting TRAIL death receptors, *Curr. Opin. Pharmacol.*, 8:433–439, 2008; S. Wang, The promise of cancer therapeutics targeting the TNF-related apoptosis-inducing ligand and TRAIL receptor pathway, *Oncogene*, 27:6207–6215, 2008; N. S. Wilson, V. Dixit, and A. Ashkenazi, Death receptor signal transducers: Nodes of coordination in immune signalling networks, *Nat. Immunol.*, 10:348–355, 2009.

Deployable systems

Deployable systems have the ability to change their geometries to meet different application requirements. They range from very common items like scissors to highly specialized ones like a space antenna boom. An umbrella is an example of a common deployable system that can be opened and folded.

Applications. Because of their flexible shapes, deployable systems have been widely adopted for household, exhibition, toy, civilian, humanitarian, space, and military applications. In space applications, deployable booms can be used, for example, to mount communication devices (**Fig. 1**). A deployable boom can be collapsed into a compact form for convenient transportation and then deployed for its end use (**Fig. 2**).

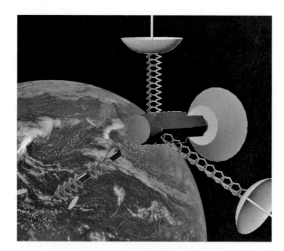

Fig. 1. Space station using deployable boom.

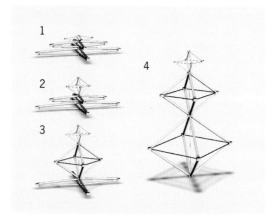

Fig. 2. Deployment of a boom tower from the folded configuration in stage 1 to the fully deployed configuration in stage 4.

Fig. 3. Deployable pneumatic shelter.

In recent years, considerable effort has been made to develop deployable structures for sheltering purposes. Such systems are also known as mobile foldable shelters, transformable structures, retractable roofs, and adaptive systems. The basic mechanisms can be telescopic, accordion, pantograph, or pneumatic (air-inflated) systems (**Fig. 3**). The use of deployable pneumatic hospitals was made popular (in thousands of m²) by Médecins Sans Frontières in their various humanitarian activities, such as in Kash-

mir (earthquake) in 2005, Sudan in 2006, Yemen in 2008, and Gaza in 2009. Deployable frames and pneumatic shelters are also commonly used to cover personnel and aircraft in various U.S. Army military operations and in research and development activities.

A deployable system may involve a complex combination of cable struts and membranes (**Fig. 4**).

Technical challenges. The special transformation features of deployable systems require designers to resolve two main challenges in terms of geometry and physical connections.

One challenge is to figure out the geometries of the folded and deployed forms and to make sure that the deployment creates a minimum of physical conflicts. This morphological study needs to consider the actual mechanism behind the geometry. The geometry of the system can be simple, like a container shape (**Fig. 5**), or complex, like an origami shelter (**Fig. 6**).

Fig. 4. Deployable tension-membrane structure.

vertical deployment

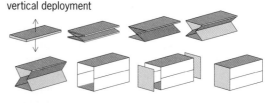

horizontal deployment

Fig. 5. Exploration of container deployment.

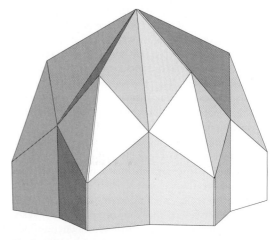

Fig. 6. Deployable origami shelter.

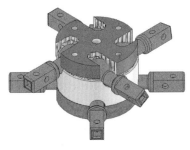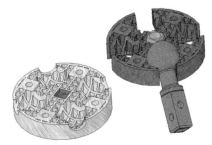

Fig. 7. Ball-joint system in deployable structure.

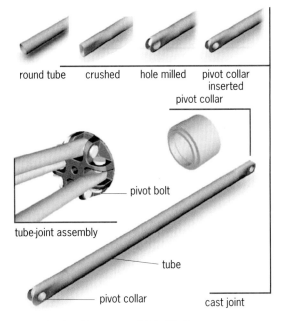

round tube   crushed   hole milled   pivot collar inserted

pivot collar

pivot bolt

tube-joint assembly

tube

pivot collar   cast joint

Fig. 8. Deployable structure joint with pin.

The second challenge is that the connections of deployable systems have to allow the transformation between the folded and deployed forms. The design of such connections requires consideration of the actual deployment requirements of the system, accounting for the self-weight effect if the structure is large in scale. It is common to find ball-type (**Fig. 7**) or pin-type (**Fig. 8**) joints used in deployable-system connections.

Tolerances must be well designed at the various points of connection. If the joint is too loose, the system cannot be deployed in accordance with the designed intention. If the joint is too tight, deformation due to self-weight can cause jamming during deployment.

Trends. A new dimension in designing deployable systems is to allow for a higher level of interaction with the environment in which the deployable system transforms and reacts to the changes in environmental conditions.

The scale of interaction can be small, like the human heat-activated shelter in **Fig. 9**, or very large, like a movable solar facade for buildings (**Fig. 10**). The interaction can be severe and require high strength for the system, as is the case with a transformable fish farm cage (**Fig. 11**) in the sea, where the wave load is severe.

Fig. 10. Movable solar facade.

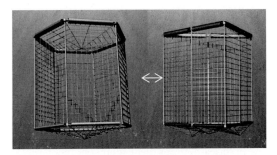

Fig. 11. Transformable fish farm cage.

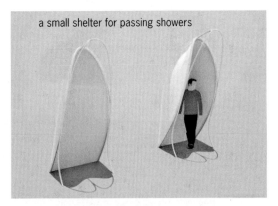

a small shelter for passing showers

Fig. 9. Deployable shelter activated by body heat. (*Courtesy of Focomoco LLP, Singapore*)

[Financial support by the RSAF/MINDEF/ Singapore is acknowledged to develop various deployable shelters for military operations, and some concepts are illustrated in the text. Contributions of photos courtesy of Ms. Ma Chenyin, Mr. Li Ya, Dr. Tran Chi Trung, Dr. Sivam Krish, Focomoco LLP, Singapore, and VRich Technology LLP, Singapore are greatly appreciated.]

For background information *see* ENGINEERING DESIGN; MECHANISM; PANTOGRAPH; SPACECRAFT STRUCTURE; STRUCTURAL DESIGN; STRUCTURE (ENGINEERING); TOLERANCE in the McGraw-Hill Encyclopedia of Science & Technology.   Jat-Yuen Richard Liew; Khac Kien Vu

Bibliography. P. Fortescue, J. Stark, and G. Swinerd (eds.), *Spacecraft Systems Engineering*, Wiley, 2003; C. J. Gantes, *Deployable Structures: Analysis and Design*, WIT Press, 2001; K. Ishii, *Structural Design of Retractable Roof Structures*, WIT Press, 2000; R. Kronenburg (ed.), *Transportable Environments*, E&FN Spon, 1998; K. Oungrinis, *Transformations: Paradigms for Designing Transformable Spaces*, Harvard Design School, 2006; F. Scheublin et al. (eds.), *Adaptables 2006*, Eindhoven University of Technology, The Netherlands, 2006; K. K. Vu, J. Y. R. Liew, and A. Krishnapillai, Deployable tension-strut structures: From concept to implementation, *J. Construct. Steel Res.*, 62:195–209, 2006.

Detection of respiratory viruses

Acute respiratory diseases account for an estimated 75% of all acute morbidities in developing countries, and most of these are caused by viruses. Upper respiratory tract infections (URTI) are among the most common infections in children, occurring 3–8 times per year in children under 5 years of age, and often causing acute asthma exacerbations or acute middle ear infections. The U.S. Centers for Disease Control and Prevention's National Vital Statistics Report indicates that there are 12–32 million episodes of URTI each year in children under 2 years of age. Viral respiratory tract infections can be caused by 18 different types of viruses. These include conventional agents such as influenza A and B, respiratory syncytial virus (RSV) types A and B, parainfluenza virus (types 1–4), adenovirus (50 types), the "common cold" viruses (including 2 coronaviruses and over 100 types of rhinovirus), and 6 viruses discovered since 2001—metapneumovirus, 3 coronaviruses [including severe acute respiratory syndrome (SARS)-associated virus], bocavirus, and avian influenza virus (H5N1).

Clinical disease. All of the 18 virus groups indicated above can cause a full range of respiratory tract infections, from the typical mild common cold with signs and symptoms of a runny nose and sneezing to more severe presentations such as pharyngitis, laryngitis, bronchitis, or pneumonia. The severity of disease varies depending on the level of immunity of the individual, and is generally more severe in immunocompromised patients [for example, those with human immunodeficiency virus (HIV); transplant recipients receiving antirejection drugs; diabetics; and the elderly]. As the immune competency of individuals and their ability to fight infection and cancer begins to decline in the fourth decade of life, the elderly patient is particularly susceptible to virus infections, which are often fatal in these circumstances.

Traditional diagnostic methods. In clinical practice, a specific virus is often not identified due to the lack of a laboratory test that is sensitive enough to detect it. As mentioned above, there have been six new viruses discovered since 2001. These have historically accounted for a significant proportion of infections where no virus could be detected in the absence of tests. The rhinoviruses and two coronaviruses discovered in the mid-1960s were "orphaned" by the medical community until recently since they were not considered to cause life-threatening illnesses. Typically, viruses were identified by their shape and size and classified into families; for example, the coronavirus has a "crown" around its surface made up from a single protein (**Fig. 1**). Virology laboratories have historically diagnosed only six conventional respiratory viruses using traditional methods. These methods include, first and foremost, virus isolation in cell culture using up to four different cell lines (not all viruses grow in all cell lines). Preformed cell cultures in 15-cm-long (6-in.) tubes are inoculated with specimens, placed in a roller drum where they are constantly rotated for 10 days, and viewed daily under a microscope for virus-induced cell damage, indicating the presence of a virus. In theory, culture can be sensitive enough for detection of a single living virus particle. However, at the same time, sensitivity can be lost when specific antibodies in the specimen neutralize the virus, preventing its growth (see **table**).

The second most important detection method has been direct fluorescent antibody (DFA) staining for the presence of virus-infected cells. This method involves collecting epithelial cells from a nasopharyngeal swab, fixing them to a glass microscope slide, staining with individual antibodies labeled with a fluorescent tag, and viewing the slide with a fluorescent microscope (**Fig. 2**). This method has sensitivities ranging from 65 to 90% for the six viruses commonly detected. Sensitivity can be compromised if the specimen is collected too late in the course of infection when the number of cells containing

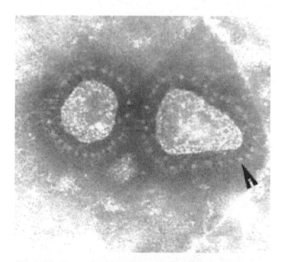

Fig. 1. Electron microscopic image of SARS coronavirus magnified 140,000 times. The arrow indicates the presence of the "corona" or crown formed by a single surface protein.

Advantages and disadvantages of various methods for detecting respiratory viruses		
Methods*	Advantages	Disadvantages
Cell culture	Low sensitivity	Not all viruses are culturable
DFA	Gold standard method	Positivity rates vary by laboratory
ELISA	Provides point of care result	Insensitive
NAAT	Highly sensitive	Expensive
M-PCR	Detects several viruses	Expensive

*Abbreviations: DFA = direct fluorescent antibody; ELISA = enzyme-linked immunosorbent assay; NAAT = nucleic acid amplification test; M-PCR = multiplex polymerase chain reaction.

viral proteins is diminished. The third most commonly used method is shell vial culture. This involves inoculating an aliquot of the specimen onto a preformed cell monolayer in a small vial containing a mixture of two susceptible cells, which is then centrifuged to enhance virus attachment and entry. The centrifugation-assisted inoculation of the cells increases the amount of viral proteins produced, allowing staining to be performed at 24–48 h and thus providing a test result to be obtained significantly earlier than the 7–10 days necessary for traditional cell culture. Rapid enzyme-linked immunosorbent assays (ELISAs), in which a monoclonal antibody conjugated to an enzyme is used to rapidly detect and quantify the presence of an antigen in a sample, have been developed as bedside tests for influenza and RSV, but these are too insensitive for routine use.

Molecular methods. The traditional methods mentioned above have been the cornerstone for diagnosis used by virology laboratories around the world for the past 25 years. The introduction of nucleic acid amplification tests (NAATs) for respiratory viruses starting in the late 1980s has heralded a new era in diagnosing respiratory virus infections. The first NAAT for respiratory viruses was developed for influenza, and used a nucleic acid amplification method called polymerase chain reaction (PCR), developed in 1983 by Kary B. Mullins, who was later awarded the Nobel Prize in Chemistry in 1993. Within a decade, NAATs were developed for all of the respiratory viruses, and most used PCR; however, other amplification schemes such as nucleic acid–sequence-based amplification (NASBA), strand displacement amplification (SDA), transcription-mediated amplification (TMA), and loop-mediated isothermal amplification (LAMP) have also been used. For all NAATs, the total nucleic acid is first extracted from the respiratory tract specimen using a variety of methods and the viral ribonucleic acid (RNA) is copied into a complementary deoxyribonucleic acid (cDNA) using an enzyme called reverse transcriptase. The cDNA is then amplified by PCR using virus-specific oligonucleotide primers, resulting in a billion copies of DNA that can be easily detected by a variety of common laboratory methods. Following the emergence of five new human respiratory viruses since 2001, there was a need for new diagnostic tests to detect these viral pathogens and NAAT filled this need. Early comparisons of molecular and traditional methods clearly indicated that the molecular methods were more sensitive than the traditional methods, often diagnosing up to 30% addi-

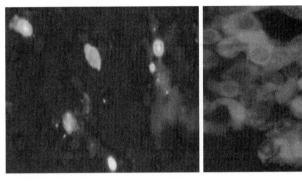

Fig. 2. Direct fluorescent antibody (DFA) staining of an influenza-positive specimen (*left*) showing the presence of virus-infected cells stained with a fluorescent monoclonal antibody, and a negative specimen (*right*) showing uninfected cells stained with a counterstain.

tional infected patients. Molecular testing methods also provided test results for clinicians often within 1 day (as compared with 2–5 days needed for traditional methods), thus improving their management of patients.

The next major advance in diagnostics was the development of multiplex PCR (M-PCR) for the detection of several different viruses in a single test. M-PCR uses multiple oligonucleotide primers, with one pair for each virus to be detected. Since M-PCR will detect several different viruses, a method is required to identify which virus is present in the specimen. This is done using a microarray (a collection of several different DNA oligonucleotides) that is either spotted onto microscope slides or cartridges (gene chips) or immobilized onto microspheres (microfluidic arrays) that are each uniquely labeled using a mixture of fluorescent dyes and identified by lasers. Each element in the array (a spot or microsphere) consists of a unique oligonucleotide (representing a unique virus) that will bind individual PCR products for each virus type or subtype. A positive specimen will generate an amplification product that is hybridized to one of the elements of the microarray and detected by a laser. One M-PCR called the xTAG™ RVP (respiratory viral panel) has recently received clearance as an in vitro diagnostic device by the U.S. Food and Drug Administration. This test was designed to identify 20 different respiratory virus types and subtypes and uses over 30 primers for target amplification and identification. Although M-PCR tests are slightly more expensive than single NAATs, they have the advantage of being able to detect many different viruses in a single test, as well as being able to detect dual infections occurring in about 10% of patients and even triple infections that are not often seen with traditional methods. NAATs using M-PCR and microarray technology offer unprecedented power for the laboratory, and are well on the way to becoming the pillars of diagnostic virology for the present century.

For background information *see* CLINICAL MICROBIOLOGY; COMMON COLD; INFLUENZA; NUCLEIC ACID; OLIGONUCLEOTIDE; PARAINFLUENZA VIRUS; RESPIRATORY SYNCYTIAL VIRUS; RESPIRATORY SYSTEM; RESPIRATORY SYSTEM DISORDERS; RHINOVIRUS; VIRUS; VIRUS CLASSIFICATION in the McGraw-Hill Encyclopedia of Science & Technology.

James B. Mahony

Bibliography. J. D. Fox, Nucleic acid amplification tests for detection of respiratory viruses, *J. Clin. Virol.*, 40(suppl. 1):S15–S23, 2007; M. Ieven, Currently used nucleic acid amplification tests for the detection of viruses and atypicals in acute respiratory infections, *J. Clin. Virol.*, 40:259–276, 2007; K. Loens et al., Detection of rhinoviruses by tissue culture and two independent amplification techniques, nucleic acid sequence-based amplification and reverse transcription-PCR, in children with acute respiratory infections during a winter season, *J. Clin. Microbiol.*, 44:166–171, 2006; J. B. Mahony, Detection of respiratory viruses by molecular methods, *Clin. Microbiol. Rev.*, 21(4):716–747, 2008.

Development and evolution

The disciplines of developmental biology (or embryology) and evolutionary biology have come together twice—once as *evolutionary embryology* in the late nineteenth century and again as *evolutionary developmental biology* (evo-devo, as it is typically known) in the late twentieth century. The current intersections of development and evolution are proving to be of paramount importance for creating a fully integrated theory of biology.

History. Prior to and into the nineteenth century, the word "evolution" had a completely different meaning from its usage today. Then, it was used to describe a particular type of embryonic or larval development—the preformation and unfolding of a more or less fully formed organism from an embryonic or larval stage (for example, a butterfly from a caterpillar, or aphids from the body of an adult female). In fact, Charles Darwin did not use the word "evolution" per se in *On the Origin of Species*, which was published in 1859, although "evolved" was the final word of his book. To reach its present-day definition, the term "evolution" had to undergo a slow transformation from *development within a single generation* to *transformation (transmutation) between generations*.

As for the field of developmental biology, in the nineteenth century, embryology (referring to the study of embryonic development) was a progressive field in biology, with researchers describing normal development and then manipulating animal embryos [*experimental* or *physiological embryol-*ogy (also known as developmental mechanics)] in search of altered outcomes that would help explain how development occurred. In addition, scientists were looking for similarities and differences in development between species, initially in a comparative context and then in an evolutionary context (as evolutionary embryology, starting in 1859). Late-nineteenth-century evolutionary embryologists such as Ernst Haeckel, Karl Gegenbaur, Francis Balfour, E. Ray Lankester, and others (including Charles Darwin) used comparative embryology to provide evidence for schemes of animal classification, to establish phylogenetic relationships among animals (common descent from a single universal ancestor), and as a major class of evidence for evolution—that is, Darwin's *descent with modification*. During this time period, embryologists observed that, from the earliest stages of development, embryos of diverse species pass through conserved embryonic stages using the same developmental processes of cell division, migration, differentiation, and morphogenesis of cells. It was also observed that the various anatomical structures arise from equivalent (homologous) germ layers. Thus, the foundation was laid for characterizing evolutionary relationships based in part on the similarities and differences in developmental processes.

Currently, experimental embryology continues to flourish through our ability to knock out (inactivate) or overexpress/misexpress (activate) genes, to transplant cells, and to manipulate differentiated cells and transform them into stem cells. Development is thus central to these twenty-first-century studies.

The late twentieth century saw the application of genetics and molecular biology to development and evolution. New molecular evolutionary theories began to emerge describing how genetics influenced life history and phenotypic plasticity [the property of a genotype to produce more than one phenotype: for example, the tadpole and adult frog, or summer and winter forms (morphs) of a single butterfly species]. At the same time, developmental biologists employed molecular biology toward understanding how genes function in the developing organism. Evolution and development emerged as two branches of biology, with most researchers in one field overlooking the progress in the other. On one side, two central pillars of evolution—natural variation and natural selection—have yet to be fully appreciated by the community of developmental biologists. On the other side, gene regulation of developmental processes is often overlooked by evolutionary theorists. In the cases where overlapping collaborations exist, the goal has been to seek and explain ultimate and proximate causes for the origin and diversity of life. As these fields begin to more fully incorporate biological models, the basis for an integrated theory of biology, that is, a twenty-first-century biology, will be created.

Relationship between development and evolution. Because natural selection operates equally on embryonic through adult stages, changes in development can mediate evolution in any life history stage. The apparent constancy of embryonic stages and the seeming lack of variation among individuals of a given embryonic stage in a given species reflect the actions of stabilizing selection (which concentrates features around a norm) and of differential survival (which removes individuals that deviate from that norm so early in development that they do not survive to be studied). Such evolutionary processes explain the conservation, stability, and seeming invariance of development [reflected in conserved embryonic stages and body plans (*Baupläne*)] in the same way that selection and differential survival and reproduction explain both the maintenance and the transformation of species.

There are a number of major development processes and evolutionary changes that need to be understood more completely in this context. These include the basis for the different body plans of multicellular organisms in the three kingdoms (plants, animals, and fungi); how transformation within and between body plans occurs; and how (or whether) major transitions, such as the origin of the limbs of terrestrial vertebrates from the fins of fish and the origin of flowers in land plants, are associated with the origin of new groups of organisms. The integrated approach to development and evolution known as evo-devo seeks a comprehensive understanding of how mechanisms operating during embryonic development mediate evolutionary change. The fundamental rationale is that no feature (character, trait) of a multicellular organism can change over evolutionary time without modification of development.

The hereditary information (that is, DNA) is transferred from generation to generation as genes, and processes such as methylation (addition of a methyl group), genomic imprinting (a phenomenon whereby one of the two alleles at a gene locus is preferentially expressed depending upon the parent of origin), chromosomal changes, and reduction in chromosome number help to modify gene action. In order for changes (mutations) to enable genetic changes that can influence evolutionary change, the genes controlling development, and therefore developmental processes, must be altered. One key intersection of molecular biology/molecular genetics with development came about through the discovery of a group of key developmental genes (homeotic/homeobox/Hox genes) that act as transcription factors and that are involved in the specification of fundamental animal features, including an anterior-posterior (A-P) axis, bilateral symmetry, and regional (often segmental) organization of body parts/segments/regions. Animals as "different" as insects, arthropods, centipedes, and crustaceans have been found to share a set of homeobox genes that were previously thought to be tied explicitly to the highly derived patterns of segmentation seen in such insects as *Drosophila*. For example, modifications of the expression patterns and of the developmental roles of three homeobox genes (designated as *engrailed*, *distal-less*, and *orthodenticle*) are associated with the origin of the different groups of echinoderms.

Other families of developmental genes (bone morphogenetic proteins, fibroblast growth factors, hedgehog genes) that act as signaling molecules provide the upstream signals that activate critical genetic pathways. Signaling by these molecules ultimately influences all aspects of development, including cell lineage decisions in the developing limbs, kidneys, teeth, heart, and skeleton. Most developmental genes are evolutionarily conserved in their cellular functions. Some have even been shown to act as master regulators of organ development conserved across the animal kingdom. One example is the *Pax-6* gene, which controls eye development in both vertebrates and insects. This was a surprising discovery, considering the morphological disparity between vertebrate and insect eyes.

Gene regulatory networks and the future of evo-devo. The future focus of evo-devo research is likely to center on a field studying the interaction of multiple genes and molecular pathways, referred to as gene regulatory networks. The developing embryo is a coordinated, highly dynamic system requiring rapid changes in cell behavior, all controlled at the molecular level. Therefore, complex molecular networks, where developmental genes and feedback networks regulate specific pathways, are present and are being discovered by scientists throughout the world. Knowledge of a gene network means that the network can be manipulated to generate altered developmental programs and thus altered morphological outcomes. This type of manipulation/experimentation is akin to evolutionary experiments and has been employed in various evo-devo studies, including research programs seeking to discover ancestral larval stages, the universal common ancestor of animals, and the dinosaur ancestor of birds. The placement of gene cascades within the context of specific lineages of cells (for example, muscle-, skeleton-, or heart-forming cells) and specific cell-to-cell or tissue-to-tissue interactions (such as those that regulate feather, limb, kidney, and tooth development) is revealing the hierarchies of interactions required to produce unique attributes of various organisms. Applying this knowledge in an evolutionary context leads to the integration of research to discover the mechanisms behind developmental (ontogenetic) and evolutionary (phylogenetic) transformations. Consequently, evo-devo constitutes one of the most exciting, vigorous, challenging, and cutting-edge sciences of the present day.

For background information *see* ANIMAL EVOLUTION; DEVELOPMENTAL BIOLOGY; DEVELOPMENTAL GENETICS; EMBRYOLOGY; GENE; GENETICS; HOMEOTIC (HOX) GENES; MACROEVOLUTION; MORPHOGENESIS; ORGANIC EVOLUTION; PLANT EVOLUTION in the McGraw-Hill Encyclopedia of Science & Technology.   Brian K. Hall

Bibliography. S. B. Carroll, J. K. Grenier, and S. D. Weatherbee, *From DNA to Diversity: Molecular Genetics and the Evolution of Animal Design*, 2d ed., Blackwell Science, Malden, MA, 2005; E. H. Davidson, *The Regulatory Genome: Gene Regulatory Networks in Development and Evolution*, Academic Press, New York, 2006; B. K. Hall, Evo-devo: Evolutionary developmental mechanisms, *Int. J. Dev. Biol.*, 47:491–495, 2003; B. K. Hall, *Evolutionary Developmental Biology*, 2d ed., Kluwer Academic, Dordrecht, the Netherlands, 1999; B. K. Hall and W. Olson (eds.), *Keywords and Concepts in Evolutionary Developmental Biology*, Harvard University Press, Cambridge, MA, 2003; B. K. Hall, R. Pearson, and G. B. Müller (eds.), *Environment, Evolution and Development: Toward a Synthesis*, MIT Press, Cambridge, MA, 2003; B. Hallgrimsson and B. K. Hall, *Strickberger's Evolution*, 4th ed., Jones and Bartlett, Sudbury, MA, 2008; B. Hallgrimsson and B. K. Hall (eds.), *Variation*, Academic Press, San Diego, 2004; M. Laubichler and J. Maienschein, *From Embryology to Evo-Devo: A History of Developmental Evolution*, MIT Press, Cambridge, MA, 2007; A. S. Wilkins, *The Evolution of Developmental Pathways*, Sinauer Associates, Sunderland, MA, 2002.

Developmental timing and oscillating gene expression

Our bodies are derived from a single cell, the fertilized egg. This cell divides and proliferates extensively, with progeny differentiating into a variety of cell types destined to form many different organs. The whole process from fertilization to birth is called embryogenesis. During embryogenesis, many events occur at predictable times. For example, in the mouse, brain and limb formation starts around 8 and 9.5 days after fertilization, respectively, and it takes about 20 days to complete embryogenesis. In each organ, cells proliferate until they reach the proper number. If proliferation continues beyond this point, the body and organs will grow disproportionately large and change shape. How embryos or the cells in embryos detect the right time is a major issue for developmental biology, and it has been suggested that some "biological clocks" serve as controllers of developmental processes. The best-known biological clock is the circadian (meaning "approximately one day") clock, which regulates day and night activities in our bodies. During embryogenesis, however, clocks with shorter time frames such as hours are required, because so many things happen in embryos in a day. The nature of such clocks remains largely unknown with one exception, the somite segmentation clock, which regulates periodic somite formation.

Somite segmentation clock. Somites (also known as metameres) are segmented blocks of tissue, which are formed transiently in a head-to-tail direction on either side of the embryonic midline (**Fig. 1***a*). Somites later differentiate into segmented tissues such as vertebrae, ribs, and skeletal muscles. The total number of somites formed varies from species to species: 44 pairs in humans, 65 pairs in mice (because of their long tail), and 300 pairs in snakes.

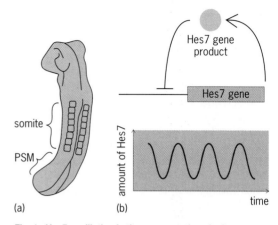

Fig. 1. Hes7 oscillation in the segmentation clock. (*a*) Somite formation by segmentation of the presomitic mesoderm (PSM). (*b*) Hes7 gene expression oscillates by negative feedback in the PSM.

A bilateral pair of somites repeatedly segments from the tail region, also known as the presomitic mesoderm (PSM) (Fig. 1*a*). The periodicity of the somite segmentation also varies from species to species: every 6–8 h in humans, every 2 h in mice, and every 60–100 min in snakes. The periodicity is so precise that the number of somites is often used to describe the embryonic stage. For example, in mouse embryos with 10 pairs of somites (the 10-somite stage), about 20 h have passed since the initiation of the first somite formation. The biological clock that regulates this periodic process (segmentation of the PSM) is called the somite segmentation clock.

The mechanism of the somite segmentation clock has been studied intensively in mouse embryos. In mouse PSM, many genes are expressed rhythmically (that is, they have oscillating gene expression). Without such rhythms, somites do not segment properly, leading to fused vertebrae and ribs. If vertebrae and ribs are fused, the neck and trunk are short, and the thoracic cavity remains small in size, restricting lung expansion and causing respiration problems. Mutations in oscillating genes also cause inherited disorders in humans such as spondylocostal dysostosis, which is characterized by loss and fusion of vertebrae and ribs. How does gene expression oscillate in the PSM? One of the most important oscillating genes is the Hes7 (hairy and enhancer of split homolog-7) gene, which encodes a transcriptional repressor. In mouse PSM, the Hes7 gene product is synthesized from the Hes7 gene and reaches peak levels after 1 h. However, this product represses its own synthesis as well as that of other target gene products (negative feedback) [Fig. 1*b*]. When new synthesis is repressed, Hes7 gene product disappears rapidly, within about 1 h, because it is extremely unstable. Disappearance of the Hes7 gene product removes the negative feedback, allowing the next round of synthesis. In this way, the amount of Hes7 gene product changes rhythmically (that is, Hes7 gene expression oscillates) with a period of about 2 h (Fig. 1*b*), and so do the amounts of other target gene products. These rhythms regulate the periodicity of somite

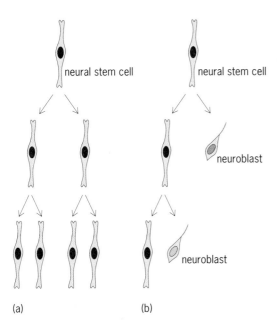

Fig. 2. Proliferation of neural stem cells. (a) Symmetric cell division. **(b)** Asymmetric cell division.

formation. If Hes7 gene is disrupted, oscillating gene expression is mostly lost, resulting in fused somites. Whether Hes7 gene expression oscillates with a period of 6–8 h in human embryos—and, if this is the case, how the periodicity difference between species is determined—remains to be examined.

Another important feature of the somite segmentation clock is the synchronicity of oscillating gene expression in thousands of the PSM cells. Oscillating gene expression occurs in phase in a group of cells that will form the same prospective somite. This synchronicity is essential for proper somite segmentation, because desynchronized oscillating gene expression results in chaotic segmentation. After PSM cells are dissociated, oscillations become irregular and out of sync, suggesting that individual PSM cells have unstable oscillators and that cell–cell communication is very important for precise and synchronous rhythms. It seems that all PSM cells send signals to each other to make the oscillating gene expression in phase.

Neural stem cells. Oscillating gene expression is a property not only of the somite segmentation clock but also of many other cell types, including neural stem cells. In neural stem cells, however, oscillating gene expression is used for maintenance of proliferation rather than the timing of certain events. In the embryonic brain, neural stem cells proliferate extensively by repeated symmetric cell division (that is, they divide into two neural stem cells) (**Fig. 2a**) and then undergo differentiation associated with asymmetric division (that is, they divide into a neural stem cell and a preneuron or neuroblast) [Fig. 2b]. By repeated asymmetric division, neural stem cells sequentially give rise to different types of neurons (Fig. 2b). Thus, maintenance of the neural stem cell population is essential to generate not only the proper number of cells but also a wide range of cell types.

Hes1, a Hes7-related transcriptional repressor, supports neural stem cell growth by repressing genes needed for neuronal differentiation (proneural genes). In the prolonged absence of Hes1 expression, proneural gene products are synthesized and induce neuronal differentiation. In such cells (differentiating neurons), proneural gene products induce synthesis of a transmembrane protein that signals neighboring cells to synthesize Hes1 and thereby to remain neural stem cells. Thus, a differentiating neuron prevents the neighboring neural stem cell from differentiating, while promoting asymmetric division into one neural stem cell and one differentiating daughter neuron. However, this mechanism of regulation poses a dilemma, because differentiating neurons are required as signal-sending cells for the maintenance of neural stem cells. This raises the question of how neural stem cells are maintained (during symmetric cell division) before neurons are formed (Fig. 2a). It was found that Hes1 is synthesized rhythmically with a period of about 2–3 h in neural stem cells (**Fig. 3**). This rhythmic expression of Hes1 is regulated by negative feedback, as is the rhythmic expression of Hes7 in the PSM. Interestingly, proneural gene products are also rhythmically synthesized in neural stem cells under the control of oscillating Hes1 expression (Fig. 3). Hes1 rhythmic expression opposes proneural gene rhythmic expression; in other words, when the Hes1 level is high, the proneural gene product level is low, and vice versa. Oscillating Hes1 and proneural gene expression leads to rhythmic synthesis of the transmembrane protein that induces Hes1 synthesis in neighboring cells (Fig. 3). It seems that all neural stem cells broadcast this signal rhythmically to prevent neuronal differentiation of the population during symmetric cell division.

Although proneural genes are known to induce neuronal differentiation, their expression must be sustained and not rhythmic to start the differentiation process. It is likely that sustained synthesis of proneural gene products is required to induce expression of all genes participating in neuronal differentiation, probably because many of these genes

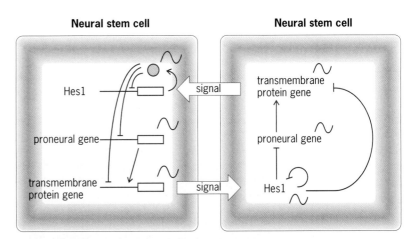

Fig. 3. Oscillating gene expression (*squiggly lines*) in neural stem cells. Neural stem cells signal each other (*large open arrows*) to induce Hes1 expression. Activation (→); repression (—|).

respond rather slowly. The current model is that the outcome depends on the mode of proneural gene expression: Oscillating expression leads to maintenance of neural stem cells (undifferentiated state), whereas sustained expression promotes neuronal differentiation (differentiated state). The mode of proneural gene expression (oscillating versus sustained) depends on the mode of Hes1 gene expression (oscillating versus absent, respectively).

If Hes1 is persistently (nonrhythmically) expressed, what happens to neural stem cells? Some cells express Hes1 persistently, and these cells proliferate very slowly or not at all. In addition, induction of nonrhythmic Hes1 expression in embryonic neural stem cells inhibits their proliferation. These cells are similar to adult neural stem cells, which are mostly dormant (nonproliferative). Could dormant cells be activated by induction of oscillating Hes1 gene expression? This question is currently under investigation. Further analyses on the gene expression dynamics will reveal the difference between embryonic and adult neural stem cells.

Conclusions. The basic mechanism of the somite segmentation clock is now known: many genes are expressed rhythmically, and this dynamic expression is regulated by negative feedback. Without such rhythms, somites are severely fused. Oscillating gene expression is also observed in many other cell types, including neural stem cells. The mode of gene expression is very important to the outcome: Oscillating versus sustained proneural gene expression leads to undifferentiated versus differentiated state, respectively, whereas oscillating versus sustained Hes1 gene expression leads to proliferative versus dormant neural stem cells, respectively. It is expected that novel strategies that control cell proliferation and differentiation will be developed with the acquisition of more detailed knowledge of the regulation of gene dynamics. Such strategies will be useful for many medical purposes such as cancer treatment and tissue regeneration.

For background information *see* BIOLOGICAL CLOCKS; CELL DIFFERENTIATION; CELL LINEAGE; DEVELOPMENTAL BIOLOGY; DEVELOPMENTAL GENETICS; EMBRYOGENESIS; EMBRYONIC DIFFERENTIATION; EMBRYONIC INDUCTION; GENE; METAMERES; MORPHOGENESIS; NEURON; PATTERN FORMATION (BIOLOGY); STEM CELLS in the McGraw-Hill Encyclopedia of Science & Technology.

Ryoichiro Kageyama; Yasutaka Niwa; Hiromi Shimojo

Bibliography. A. Aulehla and O. Pourquié, Oscillating signaling pathways during embryonic development, *Curr. Opin. Cell Biol.*, 20:632–637, 2008; M.-L. Dequéant and O. Pourquié, Segmental patterning of the vertebrate embryonic axis, *Nat. Rev. Genet.*, 9:370–382, 2008; R. Kageyama et al., Dynamic Notch signaling in neural progenitor cells and a revised view of lateral inhibition. *Nat. Neurosci.*, 11:1247–1251, 2008; Y. Niwa et al., The initiation and propagation of Hes7 oscillation are cooperatively regulated by Fgf and Notch signaling in the somite segmentation clock, *Dev. Cell*, 13:298–304, 2007.

Diesel particulate filters

One of the most important technologies in the attempt to reduce particulate matter (PM) emissions is the diesel particulate filter, or DPF. To understand the performance capabilities of diesel particulate filters, it is necessary to be familiar with the definition and composition of diesel particulates. Diesel particulate matter is composed of three main fractions: elemental carbon and ash (solid), organic fraction (liquid), and sulfate particulates (liquid).

DPF devices generally consist of a wall-flow-type filter positioned in the exhaust stream of a diesel vehicle. Wall-flow diesel particulate filters usually remove 85% or more of the soot, and can sometimes (heavily loaded condition) attain soot removal efficiencies of almost 100%. As the exhaust gases pass through the system, particulate emissions are collected and stored. The collected particulates eventually cause an excessively high exhaust-gas pressure drop in the filter, which negatively affects engine operation. Since the diesel particulate volume collected by the system eventually fills up and can even plug the filter, it is necessary to have a method to control particulate matter and regenerate the filter. In addition, regeneration of the filter systems should be "invisible" to the vehicle driver/operator and should be performed without intervention. In most cases, the diesel filter is thermally regenerated, with the collected particulates being removed from the trap by oxidation to gaseous products, primarily to carbon dioxide (CO_2).

Filter configuration and material. Conventional diesel soot filters usually consist of a wall-flow monolith. This is an extruded, usually cylindrical ceramic block that is crisscrossed by numerous small, parallel channels that run in the axial direction, while the adjacent channels are alternatively plugged at each end, thus forcing the gas to flow through the porous walls, which act as a filter medium. The flow pattern of a wall-flow filter is illustrated in **Fig. 1**.

Many different diesel particulate filter materials are available. Each one has been designed considering similar requirements, including minimum pressure drop, low cost, and durability. Moreover, because of the critical condition of use, the filter has to ensure the properties of high temperature resistance, high thermal conductivity, high mechanical resistance, and high filtering efficiency.

Wall-flow filters are easily recognized because of their characteristic checkerboard pattern, which is created by the open and plugged cells at their inlet and outlet faces (**Fig. 2b**).

The most common filter is made of cordierite (a ceramic material). Cordierite filters provide excellent filtration efficiency, are (relatively) inexpensive, and have thermal properties that make the packaging installation in the vehicles simple. The major drawback is that cordierite has a relatively low melting point, and cordierite substrates have been known to melt down during filter regeneration. The second most popular filter material is silicon carbide, or SiC. This has a higher melting point than cordierite; however,

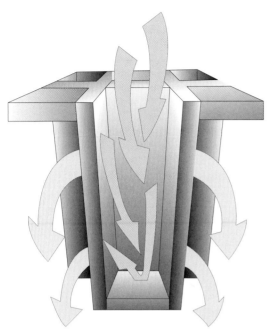

Fig. 1. Flow pattern of a wall-flow filter. (*Courtesy of Pirelli & C. Eco Technology S.p.A.*)

(a)

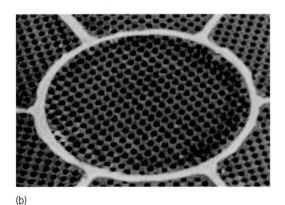

(b)

Fig. 2. Diesel particulate filters: (*a*) full-scale SiC DPF and (*b*) inlet cell pattern of a wall-flow filter. (*Courtesy of Pirelli & C. Eco Technology S.p.A.*)

it is more fragile than cordierite, making packaging an issue. Small SiC blocks are separated by a special cement in such a way that the heat expansion of the core is taken up by the cement, not by the package. SiC filters are usually more expensive than cordierite ones (Fig. 2*a*).

The DPF is packaged in a steel container that is installed in the vehicle exhaust system. Design examples of diesel particulate filters are shown in **Fig. 3**.

Filtration mechanisms. Particulate collection by any type of diesel particulate filter is based on the separation of the particles from the gas stream through deposition on a collecting surface. This separation involves the gas passing through a porous barrier that retains the particulates. This kind of filtration mechanism is typically known as shallow-bed filtration. The pore diameter of the porous barrier is usually less than the particle diameter. The particles are deposited on the media through sieving. The collected diesel particulate layer becomes the principal filter medium in surface-type filters. This layer is commonly referred to as a "filtration cake," and the process is called "cake filtration." Cake filtration is characterized by a high filtration efficiency and relatively high pressure drop. The pressure drop steadily increases with an increase in the filtration cake thickness.

Regeneration. All filter materials are designed to hold a certain quantity of soot. If the filter becomes overloaded, the particulates create an obstruction to the gas flow, which leads to an increased pressure drop and may even lead to clogging of the filter. Therefore, the filter system has to provide reliable regeneration mechanisms to ensure that it works without problems. The removal of particulates, known as filter regeneration, can be performed either continuously, during regular operation of the filter, or periodically, after a predetermined quantity of soot has accumulated. The exhaust gas temperature of many diesel engines is insufficient to regenerate the filter. For filter regeneration to work effectively, exhaust temperatures need to exceed about 500°C for noncatalyzed systems, and 250–300°C for catalyzed systems. Some diesel particulate filters use a "passive" approach and do not require an external or active control system to dispose of the accumulated soot. Passive filters usually incorporate some form of catalyst, which lowers the soot oxidation temperature to a level that can be reached by the exhaust gases during the operation of the vehicle. Another approach that can be used to facilitate reliable regeneration involves a number of "active" strategies to increase the filter temperature, including engine management to increase the exhaust

Fig. 3. Samples of canistered diesel particulate filters. (*Courtesy of Pirelli & C. Eco Technology S.p.A.*)

temperature; a fuel-borne catalyst to reduce the soot burnout temperature; a fuel burner to increase the exhaust temperature; a catalytic oxidizer to increase the exhaust temperature, with after-injection, resistive heating coils to increase the exhaust temperature, and microwave energy to increase the particulate temperature.

A computer usually monitors several sensors that measure the backpressure and/or temperature and, based on preprogrammed set points, makes decisions on when to activate the regeneration cycle.

Particulate filters also capture inorganic ash particles contained in the PM emission, with the primary source resulting from the combustion of engine lubricating oil. Engine-oil ash builds up on the surface of the inlet face of the filter and eventually clogs the pores. These incombustible particles cannot be removed from the filter through thermal regeneration. Even though the contribution of ashes to the total PM is much less than that of inorganic carbon, the accumulation of ash causes a gradual increase in the pressure drop in the filter over its lifespan. Higher backpressure is accompanied by a fuel economy penalty. If the backpressure becomes too great, an engine can stall or even be damaged. To avoid backpressure problems caused by excessive ash buildup in the filter, DPF manufacturers have prescribed maintenance schedules to clean the filters.

An alternative strategy involves the use of disposable filter cartridges, which are replaced with new units once they have filled with soot, such as during the annual vehicle inspection to ensure that they are still operating properly. Particulate traps of this kind are used in some occupational health environments. This kind of maintenance-intensive filtering system is clearly not acceptable in mobile, highway vehicle applications.

DPF performance: particle mass, number, and size distribution. For many years, particle emissions of road vehicles have been restricted in terms of mass per unit energy, or travel. Particle size distributions from internal combustion engines have received a great deal of attention since the mid-1990s because of the possible adverse health effects of fine and ultrafine particulates. Diesel emission-control strategies, based on both engine design and aftertreatment, are being examined and reevaluated with regard to their effectiveness in controlling the finest fractions of diesel particulates and particle number emissions. However, a fair performance assessment of the various control technologies will be possible only when the research community comes to an agreement on the definition of, and measurement techniques for, the smallest fractions of diesel particulates. The determination of particle sizes and numbers is much more sensitive to the measuring techniques and parameters than the quantification of particulate-mass emissions. Dilution and sampling methods are the key variables that must be taken into consideration to ensure accurate and repeatable results. On the other hand, particle sizing instruments exist that offer significantly better sensitivities than gravimetric measurements, thus presenting an attractive alternative for the PM emission measurements in future engines, provided standardized measuring methods are developed. The GRPE PMP program (Group of Experts on Pollution and Energy—Particulate Measurement Programme), set up in January 2001 and conducted under the auspices of the UNECE WP29/GRPE group (United Nations Economic Commission for Europe—Group of Experts on Pollution and Energy), has produced a recommendation for new or additional PM measurement systems to be used for European Union-type approval, testing, and development of future emission standards for both light- and heavy-duty vehicles. The new PM measurement methods include some particle size characterization parameters to address the nanoparticle emission issue.

For background information *see* AIR POLLUTION; CERAMICS; CORDIERITE; DIESEL ENGINE; DIESEL FUEL; FILTRATION; PARTICULATES in the McGraw-Hill Encyclopedia of Science & Technology.　　Debora Fino

Bibliography.　D. Fino, Diesel emission control: Catalytic filters for particulate removal, *Sci. Tech. Adv. Mater.*, 8(1-2):93–100, 2007; D. Fino et al., Catalytic filters for flue gas cleaning, in A. Cybulski and J. A. Moulijn, *Structured Catalysts and Reactors*, Chapter 16, 553–578, CRC Press, 2006; D. Fino and G. Saracco, Gas (particulate) filtration, in P. Colombo and M. Scheffler, *Cellular Ceramics: Structure, Manufacturing, Properties and Applications*, vol. 1, 416–438, Wiley-VCH (DEU), Weinheim, Germany, 2005; M. V. Twigg, Progress and future challenges in controlling automotive exhaust gas emissions, *Appl. Catal. B: Environ.*, 70:2–15, 2007.

Digital ultrasonics for materials science

The parameters that measure the resistance to compression and resistance to shear of a solid are called elastic moduli and, together with density, determine two observables: the speed of propagation of stress waves (some of which are sound waves, some of which are shear waves) and the mechanical resonances. Resonances are most easily understood by thinking of the ringing of a bell. The tones produced by the bell are determined by the shape of the bell and the material from which it is made. Generally, if the shape is known, the tones of the bell can be used to determine the elastic moduli, or stiffnesses, of the material, and vice versa. Through the elastic stiffness, the wave speeds and resonances provide a measure of two fundamental thermodynamic quantities: internal energy and free energy. Unlike most of the properties used to characterize condensed matter, the elastic moduli are tensors (there can be up to 21 different moduli) containing a wealth of detail that provide one of the most revealing probes of solids. Elastic moduli also determine the frequencies of low-frequency phonons. Phonons are the quanta of vibration in solids. They can be thermally excited, and contribute to the internal energy and heat capacity. A phonon is a resonance; the lowest-energy phonons have wavelengths that approach the size of a specimen and are measured directly by resonance ultrasound methods. Because of the connections to fundamental thermodynamic properties, ultrasound measurements in condensed matter are so important that hundreds of useful approaches have been developed.

Observables. Although the scientist interested in elastic moduli today has many measurement methods to choose from, the very highest precision and absolute accuracy come from resonant ultrasound spectroscopy (RUS) and pulse-echo (PE) methods (based on time of flight of sound pulses). At low frequencies, the resonances depend in complex ways on moduli, but are inexpensive to measure. The opposite is true for pulse-echo methods. The time of flight of an acoustic pulse during a pulse-echo measurement is determined by the wave speed (square root of the ratio of the elastic modulus for the type of wave, in the propagation direction, divided by the density). But acquiring precise timing information requires expensive hardware.

The precision of both pulse-echo methods and resonant ultrasound spectroscopy is determined by the precision with which time or frequency can be measured. These are the best measured of all physical quantities. Absolute accuracy for resonant ultrasound spectroscopy and pulse-echo methods is then limited only by the ability to measure length, perhaps the third best measured of physical quantities.

Measurements of resonances. For resonant ultrasound spectroscopy, the task is to determine the frequency of a sufficient number of mechanical resonances to enable computation of all the elastic moduli. (A simple isotropic solid has two moduli; a triclinic-structure crystal has 21.) The simplest approach, and one that is also most commonly used, is to measure the response of one transducer in contact with a rectangular parallelepiped specimen that is driven by a second transducer, also in contact, while the frequency of the driving transducer is varied. (For millimeter specimens, this frequency is typically in the range of hundreds of kilohertz to a few megahertz.)

The fundamental measurement problem is to measure and record or digitize the amplitude of the variable-frequency ac signal from the detector transducer as the frequency of the drive transducer is swept through many resonances (**Fig. 1**). When the driving frequency does not correspond to a natural vibration (resonance) of the specimen, the detected signal is very weak. As the driving frequency scans into resonance frequencies, the signal becomes much stronger.

The measurement problem is the same problem that an AM radio receiver must solve. The subtlety for both a radio and resonant ultrasound spectroscopy is that there is no simple "meter" that produces the one number required, amplitude, from a signal that is rapidly varying from positive to negative values (the ac voltage from the detecting transducer). The traditional analog resonant ultrasound spectroscopy measurement approach is to generate a reference signal at the same frequency as the signal driving the

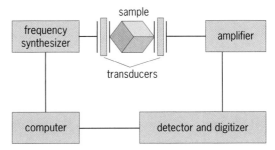

Fig. 1. Block diagram of the measurement setup for resonant ultrasound spectroscopy. A synthesizer produces a large ac signal that varies in frequency to drive the mechanical resonances of the specimen. An amplifier and phase-sensitive detector process the response of the detecting transducer for acquisition by a computer.

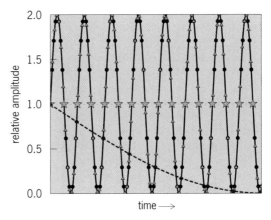

Fig. 2. Four digitization schemes of an actual signal, the solid black curve, at frequency Δf. The basic computational problem is to average a sufficient number of analog-to-digital conversions to obtain an accurate value (note zero is at bottom of plot). Open triangles are the typical digitization scheme, where the digitization frequency is unrelated to the data frequency. The average values of open triangles are different for each cycle, so multicycle averaging is required to obtain a stable value. The solid black circles are synchronously digitized. The average of the values for every cycle is identical; thus no additional averaging is required for a stable value. The stars represent the Nyquist limit. At this digitization rate, it is possible, as shown, to miss completely the presence of an ac signal. If one digitizes below the Nyquist rate, the colored diamond-shaped points would be what the digital values would reconstruct: an "alias" of lower frequency than the data.

specimen, but with fixed amplitude. A mixer, an analog device that executes the mathematical operation of instantaneous multiplication, multiplies the reference signal with the (amplified) signal from the detector. The result is the sum and difference frequencies between reference and detector signals (which, of course, are at the same frequency) combined into one output signal. The sum frequency is $2f$, where f is the drive frequency. The difference frequency is zero, or dc. This method is called homodyne detection. A low-pass analog filter (an example is a capacitor C fed by a resistor R) is used to remove the $2f$ component and reduce apparent noise. However, rapid changes in signal amplitude take time (the time "constant" is just RC) to "settle" to an accurate value, greatly limiting the rate at which the frequency can be swept. In the end, we must, as in all modern laboratories, convert this dc signal to a digital value to

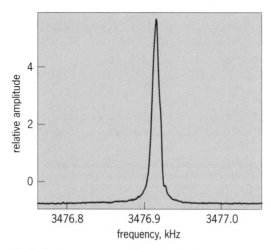

Fig. 3. Resonance measured for a diamond monocrystal. The quality factor or *Q* (frequency divided by width) at 23 K nears 500,000. The extremely narrow resonance means that the energy losses caused by the resonant ultrasound spectroscopy measurement method are extremely low.

be recorded by a computer. So why not do this conversion right off?

It is here that modern digital approaches part company with analog. In the synchronous digitization scheme for resonant ultrasound spectroscopy, dc is never generated. Instead of using a single frequency as in the homodyne scheme, three different frequencies are used in what is called heterodyne. Two of them are f, the drive, and $f + \Delta f$, the reference frequency. The frequencies f and $f + \Delta f$ are fed to the mixer. A third frequency, $n\ \Delta f$, where n is an integer, "clocks" an analog-to-digital converter to make measurements at an integer multiple of Δf. The mixer, as before, produces

two frequencies, but unlike the homodyne scheme, they are $2f + \Delta f$ and Δf. A simple fast-response filter removes the $2f + \Delta f$ component without sacrificing response speed. What has been accomplished is to transfer all the information in the detecting transducer signal coming at the variable frequency f into the same information but now at a fixed frequency Δf. This permits all the electronics to be focused on measuring only one frequency Δf, called the intermediate frequency (IF).

So far, no intrinsically digital processes have been implemented. However, when the digitizer operates on what is left, namely the Δf component, by digitizing it at a frequency $n\ \Delta f$, an intrinsically digital measurement process is implemented that has no analog counterpart. Referring to **Fig. 2**, there are several regimes in which a digitizer can be operated to acquire the amplitude of the signal at the fixed frequency Δf. A key digitization rate is the Nyquist limit (twice the frequency to be digitized), below which the frequency recorded is no longer that of the signal being digitized: An "aliased," or incorrect, lower frequency is reported by the digitized data stream. Thus it is essential to digitize at well above the Nyquist limit (oversampling). By digitizing "synchronously" at an integer multiple of the Nyquist rate, it is possible to obtain an accurate amplitude of the signal in one cycle, with no further settling time required. How this works can be seen by examining Fig. 2, where several possible digitization methods are displayed for extraction of the amplitude of the signal (the black solid curve) at the frequency Δf produced by the resonant ultrasound spectroscopy heterodyne system. For the (asynchronous) triangles, the average of the values for one cycle, a measure of signal amplitude, must necessarily fluctuate as the number of triangles and their position on the waveform vary with time. However, for the solid circles, synchronously clocked, the average is the same for any cycle (the average is stable for any given phase), and hence there is no need for additional averaging to get a stable value. No analog filter can do this. This is the powerful advantage of a synchronously clocked resonant ultrasound spectroscopy system. Frequency can be stepped very quickly with no compromise in quality or signal-to-noise ratio. **Figure 3** shows an example of resonances obtained by a synchronously digitized resonant ultrasound spectroscopy system from a diamond single crystal. The extremely "sharp" resonances were accurately acquired using synchronously digitized resonant ultrasound spectroscopy. The measure of sharpness is called the *Q*, and is the frequency divided by the width of the resonance or, equivalently, the number of oscillations to half loudness that the bell makes after it is rung.

Measurements of time of flight. A time-of-flight or pulse-echo ultrasound system launches a short (typically submicrosecond) burst of sound at some high frequency (tens of megahertz) and records the time it takes for that signal to bounce off the face of the specimen opposite the transducer and return to the transmitter. There are numerous technical

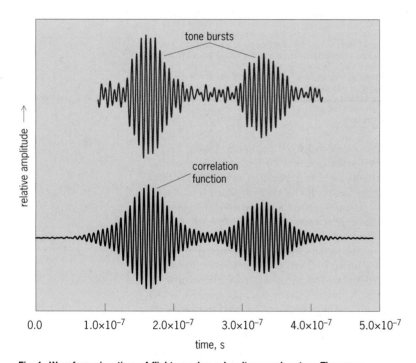

Fig. 4. Waveforms in a time-of-flight or pulse-echo ultrasound system. The upper waveform shows the ac tone bursts received by the single transducer used to both launch and detect bursts of sound so that the transit time of sound can be measured. The lower waveform is the digital correlation function of two echoes.

challenges that are common to both analog and digital approaches to the timing measurement, but the most precise analog approach, called pulse-echo overlap, works as follows. The task is to determine the time delay between successive tone bursts, shown in the upper waveform of **Fig. 4**. One way to capture conceptually what must be accomplished is to consider varying the repetition frequency of an oscilloscope sweep so that the successive sweeps occur nearly exactly at the same time intervals as do echoes from the specimen face. Then one adjusts the repetition frequency until the two echoes overlap. This can be made precise by mixing the transducer receive signal with a reference signal and adjusting the repetition rate for a minimum output from the mixer. There are problems with this successful technique, not the least of which is the unambiguous determination of when overlap occurs. Because the two signals to be overlapped are neither the same amplitude nor have the same shaped envelope, it is easy to be off by one cycle or more. However, if the signals are digitized directly, then a very special manipulation not possible with analog systems can be made. This manipulation is to digitally grab, say, the second and third echoes in software. Then one can software time delay one of the signals with respect to the other, multiply all the digitized points of the original signal with those of the time-delayed one, and plot the sum, shown in the lower waveform of Fig. 4. The result, called the correlation function, yields an unambiguous determination of best overlap if the transducer drive frequency is varied. Although we skip the details here, the process is easily programmed into the data acquisition code used to get data from the oscilloscope into the computer. Although it is in principle possible to do this in an analog system, the continuously variable delay lines and mixers required do not, in general, exist. Furthermore, using this digital approach eliminates mixers, gates, switches and more from the electronics system used for pulse-echo ultrasound measurements, thus reducing complexity and cost, while improving accuracy and precision. In the end, all that is

needed is the pulse generator and a digital oscilloscope, making this approach accessible to any motivated researcher.

Figure 5 shows a typical result, and the end product, of many sound speed studies—the bulk modulus versus temperature, in this case, for a monocrystal of rhenium diboride (ReB_2), a superhard material measured using resonant ultrasound spectroscopy.

For background information *see* ELASTICITY; FREE ENERGY; HETERODYNE PRINCIPLE; INTERNAL ENERGY; MIXER; OSCILLOSCOPE; PHONON; Q (ELECTRICITY); RADIO RECEIVER; RESONANCE (ACOUSTICS AND MECHANICS) in the McGraw-Hill Encyclopedia of Science & Technology.　　Albert Migliori; Yoko Suzuki; Jonathan B. Betts; Victor Fanelli

Bibliography. A. Migliori et al., Resonant ultrasound spectroscopic techniques for measurement of the elastic moduli of solids, *Physica B*, 183:1–24, 1993; C. Pantea et al., Digital ultrasonic pulse-echo overlap system and algorithm for unambiguous determination of pulse transit time, *Rev. Sci. Instrum.*, 76:114902, 2005.

Discrete analytic functions

Discrete analytic functions, also known as circle-packing maps, are mappings between circle packings whose properties faithfully reflect the properties that are characteristic of classical analytic functions. A circle packing is nothing more than a collection of circles with prescribed tangencies that lie on a surface. The tangency information may be encoded in a graph in which each circle is represented by a vertex, and each point of tangency at which two circles meet is represented by an edge connecting the vertices that correspond to those two circles. A discrete analytic function then may be represented by a mapping between two circle-packing graphs of this type. Though discrete analytic functions may be used to build discrete approximations of classical analytic functions of both planar domains and surfaces, their real importance derives from the discovery that their properties mark them as true analogs in the discrete setting of the familiar analytic mappings of classical complex analysis. As such, discrete analytic functions offer both the rigidity and versatility of their classical counterparts and, in particular, offer a range of manipulations of planar domains that makes them valuable for flat two-dimensional representations of surfaces in three dimensions. The theory of discrete analytic functions has been developed to the point that it stands alone as a part of a general movement in the recent development of mathematics—that of the discretization of geometry. Before presenting a more detailed description of these functions, it will be instructive to place the subject in its proper context within the recent development of mathematics.

Context. The classical theory of complex analysis—the calculus of complex-valued functions of a complex argument—is one of the pillars of pure and applied mathematics and has a distinguished

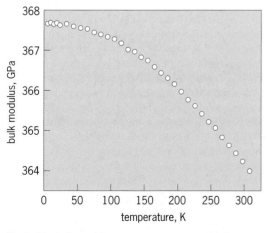

Fig. 5. The bulk modulus versus temperature of ReB₂, a superhard material, measured using resonant ultrasound spectroscopy. This is often the end goal of acoustics measurements in condensed matter.

history that stretches back to Leonhard Euler in the eighteenth century. The primary attribute of classical analytic functions that distinguishes them among all continuous complex-valued functions of a complex variable is that they are conformal almost everywhere. This means that these functions preserve angles and shape locally, so that, for example, infinitesimal circles are mapped to infinitesimal circles. The theory is rich and covers much of classical and special function theory, including polynomial and rational mappings of the Riemann sphere as well as many of the special functions of mathematical physics and number theory. Though it still is an active area of research, the classical theory reached mature development by the end of the nineteenth century. The development of discrete analytic functions, on the other hand, has occurred only over the past quarter-century and has developed to the extent that it offers a fully discrete analog of the classical theory of analytic mappings of planar domains; however, it has yet to achieve maturity in its development of discrete analytic functions on more general domains. Discrete polynomial functions are well understood, but discrete rational functions defined on the Riemann sphere are only partially understood, and the theory of discrete analytic functions defined on even more general Riemann surfaces is still in its infancy.

Discrete differential geometry. This new theory of discrete analytic functions is representative of a recent development in mathematics that is occurring with increasing sophistication. This development has its roots in applied computational science, particularly in the various sorts of two- and three-dimensional imaging problems that have become accessible only recently with the advent of powerful desktop computers. These imaging problems led to challenging mathematical problems in collecting, representing, manipulating, and deforming representations of embedded surfaces and solids in three-dimensional space. This ultimately meant that tessellations of two- and three-dimensional objects needed to be represented and manipulated as data in computations. The concern was to preserve as much as possible, in the combinatorial data that encodes the geometry of these objects, the original metric data—distances, intrinsic geodesics, angles, and intrinsic and extrinsic curvature. Often this meant calculating geodesics and curvature on combinatorial objects rather than on the original smooth ones, and the usual tools of Riemannian geometry are rendered ineffective in this combinatorial setting.

Discretization of geometry was and is driven by the habitual need to see and manipulate images. Its impetus came from a varied assortment of sources, and it became obvious to several groups of mathematical and computer scientists that the problems they faced were not going to be solved by mere approximation. What was needed was a discrete version of Riemannian geometry that was faithful in its domain to the spirit of the classical field and that represented the continuous objects whose properties it mirrored in some faithful way, but whose computational tools could stand on their own and not as an approximation to the continuous. This led to the identification of natural analogs in the discrete setting of the usual tools of differential geometry—polyhedral surfaces and spaces, discrete normal vector fields, discrete curvature operators, discrete Laplace-Beltrami operators, discrete geodesics, and discrete Gauss maps. The paradigm, though, is different from that of classical numerical approximation, where the justification for the discrete operations and calculations is approximation of the continuous. Rather, the standard in this emerging field of discrete differential geometry is that the discrete theory is a self-contained whole, with natural, exact tools leading to exact calculations, not approximations. Though the classical theory emerges in the limit of small mesh size, often this is not the overriding interest. When used as an approximation tool, this sort of discrete theory offers much more than mere approximation, because the full force of the theory is available for manipulating the approximated images.

Emergence of discrete analytic functions. The theory of discrete analytic functions has its origins in the pure mathematics of Riemann surfaces and the theory of circle packings on those surfaces. Circle-packing theory, developed originally in the 1980s, began to emerge as a separate discipline in the mid-1990s and only recently has found its natural home in the setting of discrete differential geometry. The part of discrete differential geometry that concerns the discretization, in the sense already articulated, of classical complex analysis and conformal geometry is now subsumed under the heading "discrete conformal geometry." Circle packing is one approach to discretizing parts of these classical disciplines and has both advantages and disadvantages when compared to others. Its greatest success is the topic of this article—providing a faithful discrete analog of classical complex analytic function theory in the complex plane, with discrete versions of classical analytic mappings that in their particularities share essential features and qualities of their classical counterparts. Many of the classical theorems of complex analysis have circle-packing analogs that actually imply the truth of the classical theorems in the limit as circle radii approach zero. This is not a two-way street, however, as the truth of the classical rarely implies that of the discrete. It is as if the discrete is more fundamental and more primitive, with the classical theory a derivative of the discrete.

Circle packings and mappings. To understand discrete analytic functions, one must first understand the basics of circle packing, and this requires a bit of technical discussion. We begin by considering an arbitrary triangulation K of a genus-g compact oriented surface S, that is, a surface with g handles. We stress that S carries no metric structure; it is merely a topological surface produced by adding a finite number (namely, g) of handles to a sphere. It may be proved that there is then a unique conformal structure on S and, in that conformal structure, a circle packing $C = \{C_v : v \in V(K)\}$, a collection of circles in S indexed by the vertex set $V(K)$ of K, unique up to

Möbius transformations, with C_v tangent to C_w whenever the vertices v and w of K are adjacent. To explain this, a conformal structure is not quite as rich a structure as that given by a metric, because it allows for angle measurements only, and not distance measurements. Nonetheless, a conformal structure induces on S the structure of a Riemann surface and a compatible (infinite) family of metrics. Möbius transformations are circle-preserving transformations of the plane, which together form a well-understood collection of classical analytic maps.

To unwrap this concept a bit further, when the genus $g = 0$, then S is a topological two-dimensional sphere, the surface of a ball, and up to equivalence there is only one conformal structure on S, the one identifying S as the Riemann sphere. The circles C_v are then usual circles in the Riemann sphere, or usual circles in the plane when projected stereographically. When the genus is $g = 1$, then S is a topological torus, the surface of a doughnut, and there is a continuous two-dimensional family of conformal structures available for S, no two of which are equivalent. The triangulation K chooses exactly one such structure that carries with it an essentially unique flat metric, and the circles C_v are circles with respect to this metric. The generic case is when $g \geq 2$. Then there is a continuous $(6g-6)$-dimensional family of pairwise nonequivalent conformal structures available for S and the combinatorics of K choose exactly one such structure. This structure carries a unique metric of constant curvature -1, a hyperbolic metric, and the circles C_v are hyperbolic circles. This hyperbolic metric gives the surface a noneuclidean geometric structure that locally models the noneuclidean geometry discovered in the nineteenth century by N. I. Lobachevski and János Bolyai. The packing C is unique up to hyperbolic isometries, or distance-preserving transformations. We emphasize the fact that none of the other uncountably many conformal structures with their hyperbolic metrics supports a circle packing whose tangencies are encoded in the combinatorics of K. The content of this paragraph generalizes to arbitrary surfaces, open or closed, with or without boundary, and with both finite and infinite circle packings. *See* GEOMETRIZATION THEOREM; HYPERBOLIC 3-MANIFOLDS.

A discrete analytic function f is a mapping between two circle packings that assigns to each circle of the first a unique circle of the second and preserves tangency. In more detail, given two circle packings $C = \{C_v : v \in V(K)\}$ and $D = \{D_w : w \in V(L)\}$, where V(L) is the vertex set of a triangulation L of a second surface T, a discrete analytic function f between C and D is a correspondence between these collections of circles for which $f(C_v)$ and $f(C_{v'})$ are tangent circles of D in the surface T triangulated by L whenever v and v' are adjacent vertices in K. Discrete analytic functions in general are difficult to construct, because there are topological obstructions that must be navigated. Fortunately, there are no such obstructions when the surface is simply connected, as in the special case of a topological disk. It is this special case in which the standard applications of

circle packing are used to construct discrete analytic functions that approximate classical ones of interest. For example, a smooth surface embedded in three-dimensional space is usually a topological disk, or can be transformed into one by cutting, and admits classical conformal mappings to the plane. These planar, conformal representations of the surface may be approximated by discrete conformal mappings generated by circle packing, and the theory of discrete analytic functions may be applied to manipulate these flat maps, approximately preserving the conformal structure of the flat representation.

A fitting metaphor comes from physics. The discrete theory—circle packings and their discrete analytic functions—is the quantum theory from which the classical theory of analytic functions emerges. Classical analytic functions are continuous deformations of the classical complex plane and can be very complicated, but when viewed at the atomic scale, that is, from the tangent plane, they are local complex dilations that preserve infinitesimal circles.

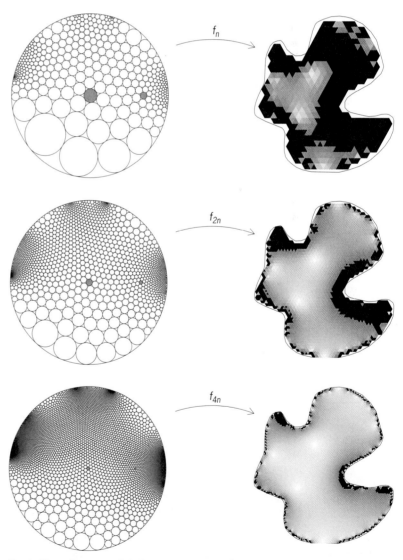

Fig. 1. Three discrete analytic functions approximating a conformal Riemann mapping of the unit disk to a planar domain. Light to heavy color encodes angular (conformal) distortion from 0 to 10%, while black encodes distortion of greater than 10%. The distortion levels improve rapidly with refinement. (*Courtesy of Ken Stephenson*)

Discrete analytic functions preserve actual circles and model the behavior of their classical counterparts. The salient large-scale features of the continuous classical functions arise from this atomic-scale circle-preserving property of the discrete functions.

Applications to approximation. The power and versatility of discrete analytic functions and circle-packing techniques will be illustrated with four examples.

Approximating the Riemann mapping. The classical Riemann mapping theorem asserts the existence of a one-to-one analytic mapping from any proper, simply connected domain in the plane onto the unit disk. Generally, these mappings are notoriously dif-

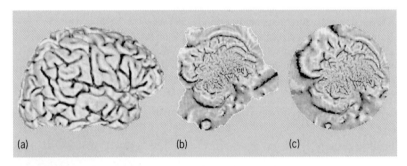

(a)   (b)   (c)

Fig. 2. Representations of the surface of the human brain by planar conformal projections. (*a*) Gray-matter surface of the right hemisphere of the brain. (*b*) Corresponding approximate conformal map in the Euclidean plane. (*c*) Corresponding map in the hyperbolic plane. The shading encodes mean curvature, with bright/white as positive curvature and dark/black as negative curvature. (*Courtesy of Monica Hurdal*)

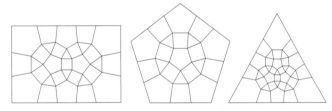

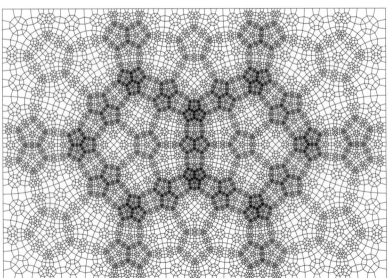

Fig. 3. The dodecahedral subdivision rule applied three times in succession to a rectangle. At each application, the rectangles are subdivided according to the pattern shown in the upper left, the pentagons according to that of the upper middle, and the triangles according to that of the upper right. Circle packing generates the embedding and a discrete conformal structure in which roundness emerges. (*Courtesy of Bill Floyd*)

ficult to compute, and various schemes have been constructed to approximate them. Circle packing provides a fast and effective algorithm for approximating these Riemann mappings by discrete analytic functions using a scheme first proposed by William Thurston and implemented in the 1980s (**Fig. 1**). This was one of the important results that initiated great interest in developing the theory of discrete analytic functions. The techniques have been generalized and developed to the point that conformal mappings between simply connected planar domains can be approximated with excellent accuracy away from the boundary of the domain. The approximation scheme is iterative and so offers a sequence of approximations that become more accurate as the sequence progresses, converging exactly in the limit, even at the boundary. Further generalizations to non-bijective mappings allow for the discrete modeling of classes of classical analytic functions, including discrete Blaschke products, discrete polynomials, and discrete trigonometric functions.

Brain mapping. One of the recent uses of discrete analytic functions is in constructing flat mappings of curved surfaces in three-dimensional space. For example, the surface of the human brain is a highly convoluted surface with complicated geometry but simple topology. Sophisticated medical imaging tools can generate precise representations of this surface as a triangulation in three-dimensional space. It is valuable for some applications to have a flat, planar representation of this surface and, much as the nonplanar globe may be represented by the familiar Mercator projection, which is conformal, the surface of the brain may be represented by planar, conformal projections (**Fig. 2**). Circle-packing tools have been developed that construct such projections as images of discrete analytic maps. Though the conformality is only approximate, the flat maps can be manipulated easily using the properties of discrete analytic functions that mirror the conformal properties of their classical counterparts. Monica Hurdal and Kenneth Stephenson have been principal developers of these applications.

Constructing conformal structures. A useful application of circle packing in pure mathematics has been in placing approximate conformal structures on cellular subdivisions of planar regions that are generated by purely combinatorial data. An example may be generated by applying the so-called dodecahedral subdivision rule to a rectangle three times in succession (**Fig. 3**). At each application, the rectangles of the subdivision are subdivided according to the pattern shown in the upper left of Fig. 3, the pentagons according to that of the upper middle, and the triangles according to that of the upper right. Circle packing generates the embedding, in which the almost roundness that the eye sees at multiple scales emerges. Technically, this is an approximate view of the shadows cast on the sphere at infinity by the first-, second-, and third-generation hyperbolic half-spaces determined by the dodecahedral tiling of hyperbolic 3-space, and is a part of the work of James Cannon, William Floyd, and Walter Parry.

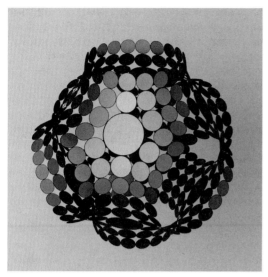

(a)

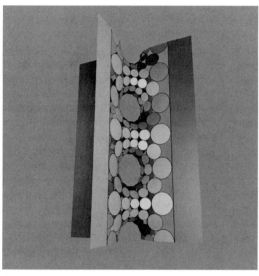

(b)

Fig. 4. Discrete (*a*) Schwarz and (*b*) Scherk minimal surfaces generated by circle patterns based on square combinatorics. (*Courtesy of Boris Springborn*)

Minimal surfaces. Minimal surfaces are surfaces embedded in three-dimensional space with zero mean curvature, and they model the physical surfaces produced by soap films suspended across wire frames. Recently, Alexander Bobenko and Boris Springborn have introduced new algorithms based on circle patterns whose tangencies are encoded using the square combinatorics of the integer lattice in the plane, rather than the triangular patterns of tangencies that are normally used. These algorithms have been successful, not only in conformal flattening, but also in an elegant discrete representation of minimal embedded surfaces (**Fig. 4**). These new techniques extend the theory of discrete analytic functions and are based on the physics of integrable systems.

For background information *see* COMPLEX NUMBERS AND COMPLEX VARIABLES; COMPUTER-AIDED DESIGN AND MANUFACTURING; COMPUTER-AIDED ENGINEERING; COMPUTER GRAPHICS; CONFORMAL MAPPING; DIFFERENTIAL GEOMETRY; MAP PROJECTIONS; MINIMAL SURFACES; QUANTUM MECHANICS; RIEMANN SURFACE in the McGraw-Hill Encyclopedia of Science & Technology.   Philip L. Bowers

Bibliography. P. L. Bowers, Review of *Introduction to Circle Packing: The Theory of Discrete Analytic Functions, Bull. Amer. Math. Soc.,* 46(3):511–525, 2009; P. L. Bowers and K. Stephenson, *Uniformizing Dessins and Belyi Maps via Circle Packing,* vol. 170, no. 805 in *Memoirs of the American Mathematical Society,* 2004; M. K. Hurdal and K. Stephenson, Cortical cartography using the discrete conformal approach of circle packings, *NeuroImage,* 23(suppl. 1):S119–S128, 2004; K. Stephenson, *Introduction to Circle Packing: The Theory of Discrete Analytic Functions,* Cambridge University Press, Cambridge, 2005.

Durability of wood–plastic composite lumber

Wood–plastic composite (WPC) lumber has been marketed as a low-maintenance, high-durability product. Retail sales in the United States were slightly less than $1 billion in 2008. Applications include decking, railing, windows, doors, fencing, siding, moldings, landscape timbers, car interior parts, and furniture. The majority of these products are used outdoors and thus are exposed to moisture, decay, mold, and weathering. WPCs are composites made primarily from wood- or cellulose-based materials and plastic(s). The wood utilized is usually in particulate form, such as wood flour or very short wood fibers. Because of the thermal stability of wood, only thermoplastics that melt at temperatures equal to or below 200°C (392°F), which is the degradation point of wood, are utilized in the production of WPCs. The most common plastics used for WPCs are polyethylene (PE), polypropylene (PP), and polyvinyl chloride (PVC). The WPC blends can vary in percentage, with wood content up to 70%, but 50% wood/50% plastic is common. It was first thought that mixing plastic and wood together would result in plastic encapsulation of wood, thereby preventing both moisture sorption and fungal decay. However, after over a decade of use in outdoor exposure, issues have surfaced that are caused by wood decay, susceptibility to mold, and polymer degradation. After the first-generation products showed these problems, the industry began to address the performance issues by improving the formulations with additives, including fungicides, mildewcides, and photostabilizers. Further improvements are continually being investigated today. Researchers around the globe are currently working on understanding the fundamental mechanisms of degradation and on ways to improve WPCs.

Moisture. Moisture is required for biological decay of wood. Air-dried wood should have no more than 20% moisture by weight to resist decay. If free water is added to attain 25–30% moisture content or higher, decay will occur in nondurable species. Wood can

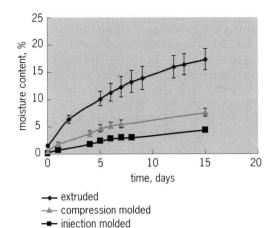

Fig. 1. Effect of processing on the moisture sorption of WPCs containing 50% pinewood flour during water soaking.

also be too wet for decay because there is not enough air for the fungi to develop.

The overall uptake of moisture in WPCs is relatively slow compared to solid wood, and the amount of moisture absorbed by WPCs can vary widely as a result of the influences of the wood flour content, wood particle size, processing methods, and additives. However, moisture uptake in the outer 5 mm (0.2 in.) of WPC commercial products has been shown to be sufficient for fungal attack to occur. Water evaporation from the WPC is also very slow because the plastic inhibits moisture movement, thus providing an opportunity for biodegradation to occur. Laboratory research and field evaluations indicate that WPCs absorb moisture to the point where they can be degraded by fungi.

There are different ways to produce WPCs. The three most common ways are extrusion, compression molding, and injection molding. WPC processing methods influence the moisture sorption, thus making them more or less susceptible to decay. Extruded composites absorb the most moisture, compression-molded ones absorb less, and injection-molded ones absorb the least (**Fig. 1**). A thin, polymer-rich surface layer forms in injection-molded composites. In addition, higher pressures are used in this process, resulting in the collapse of the wood fiber bundles, which makes a higher-density composite compared with extrusion and compression molding. The amount of moisture absorbed is also influenced by wood particle size and additives. Another variable is whether the WPC is extruded as a solid or a hollow profile. A hollow profile will create a larger surface area that allows for moisture sorption.

The hydroxyls (chemical groups in which oxygen and hydrogen are bonded and act as a single entity) of wood are primarily responsible for water sorption, which leads to wood swelling. This then can lead to warping and expansion. When WPCs are exposed to moisture, the wood fibers swell, causing stress that can result in damage to the matrix (for example, microcracks), fracture of the wood particles due to restrained swelling, and ultimately interfacial

breakdown. For solid wood, moisture alone reduces the strength and stiffness. However, fungal decay reduces these measures by the greatest amount. The mechanical properties of WPCs are more severely affected by moisture absorption than by fungal colonization. The microcracks in the matrix and the damage to the wood particles cause a loss in elasticity and strength.

Decay. Fungi have four basic growth requirements: food (for example, wood, which contains hemicellulose, cellulose, and lignin), moisture (above the fiber saturation point; about 30% moisture content), proper temperature [10–35°C (50–95°F); optimum: 24–32°C (75-90°F)], and oxygen (from the air).

Currently, the indicator of decay in laboratory testing of WPCs is by weight loss using solid wood standards. The soil-block test [American Society for Testing and Materials (ASTM) D 1413 or ASTM D 2017] is utilized in North America, and the agar-block test [European Standards (EN) 113] is used in Europe. In both types of tests, WPC specimens are exposed to a single fungus (either white-rot or brown-rot) and then mass loss of each specimen is calculated in percent, based on initial dry mass. Because WPCs vary in their percentage of wood content, the weight losses can be very low (**Fig. 2**). The test fungi commonly used are the brown-rot fungi *Gloeophyllum trabeum* and *Poria placenta*, and the white-rot fungus *Coriolus versicolor*.

Mechanical property losses can be used to evaluate decay specimens. Both three-point bending tests and dynamic mechanical analysis (DMA) have been utilized. Weight loss is a more sensitive indicator of fungal decay than flexural strength testing, except when using WPCs with high wood contents (for example, 70%).

Laboratory decay testing is helpful in predicting the long-term durability of WPCs, but field studies are needed to verify overall performance. Field exposure can encompass deterioration caused by moisture, fungi, ultraviolet (UV) light, wind, temperature, freeze/thaw cycling, wet/dry cycling, termites, and mold for both aboveground specimens (**Fig. 3**) and in-ground specimens (**Fig. 4**).

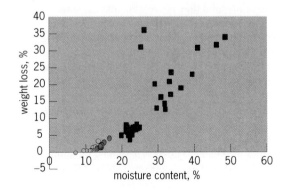

■ extruded ● injection molded ● compression molded

Fig. 2. Effect of moisture content on fungal attack of WPCs containing 50% pinewood flour after 12 weeks of exposure in the soil-block test.

Fig. 3. Aboveground WPC deck boards after 6 years of field exposure in Madison, Wisconsin.

Fig. 4. In-ground field test of WPC specimens [1.9 cm × 1.9 cm × 45.7 cm (0.75 in. × 0.75 in. × 18 in.)] in Saucier, Mississippi.

Mold and stain fungi. Fungi that grow superficially on wood are called molds, and those that cause discoloration of wood in storage and service are called stain fungi. Both mold and stain fungi are members of the Ascomycetes and Deuteromycetes (Fungi Imperfecti), yet a few of the molds are Zygomycetes (Mucorales). The aesthetic quality of wood-based products is significantly lowered as a result of discoloration by both mold and stain fungi. Stain fungi produce extracellular enzymes to utilize sugars, proteins, and extractives, which are nonstructural components of sapwood. In general, the stain fungi do not decay the structural components (that is, the wood cell wall), but some can cause soft rot. Mold and stain fungi are mainly an aesthetic problem for WPCs, but their effects are difficult to stop without periodic washing with dilute bleach solution.

Polymer degradation. WPCs exposed to weathering by solar radiation (ultraviolet light), oxidation, and rain water may experience color change as well as loss of mechanical properties. The color change affects the aesthetic quality, whereas the mechanical loss affects the performance. Weathering destroys the surface of the WPC by surface oxidation, matrix crystallinity changes, and interfacial degradation, which can lead to openings and pathways for moisture and decay to occur.

The UV resistance of WPCs is evaluated in the laboratory for minimum test durations of 2000 h of exposure in a weatherometer according to ASTM specifications. Flexural strength is determined before and after exposure. Various standards are used to reference both the plastic and the wood components. Mechanical and physical properties are determined, along with WPC degradation standards (that is, moisture, biodeterioration, UV resistance, and so on).

Methods of protection. As stated previously, fungi have four basic requirements (food, moisture, temperature, and oxygen). Temperature and oxygen are very difficult to control or limit, but moisture and nutrient exclusion can be regulated. Nutrient exclusion can be achieved by encapsulating the wood fibers in plastic, reducing the amount and size of wood in the WPC. It can also be achieved via the processing method. If the moisture content of the wood filler can be kept below 20%, then WPC decay may be prevented. This can be done not only by completely encapsulating the wood particles in plastic, but also by hydrophobation (treatments that increase water resistance) of the WPC surface or by chemical modification of the wood portion. The wood hydroxyl groups can be chemically modified to make them hydrophobic. This is done by either bulking the cell walls or cross-linking the cellulose hydroxyls. Chemical modification of the wood changes the wood fibers from hydrophilic to a permanent hydrophobic state. Anhydrides, isocyanates, and epoxides are the most common modification agents. Acetylation (bonding of an acetyl group onto an organic molecule) has been most actively investigated with solid wood because of its low toxicity and low cost, and it is currently being commercialized.

The most common way to protect WPCs from decay is via addition of additives such as chemical biocides or antimicrobials. The preservative zinc borate is often utilized because it has relatively low water solubility (making it resistant to leaching), it can withstand common extrusion temperatures, it has low toxicity (for both humans and the environment), and it has low cost. It is effective against wood decay fungi and insects, but it is not effective against mold and stain fungi at low concentrations. Other broad-spectrum treatments can be utilized against mold, mildew, and bacteria.

Photostabilizers are often added to WPCs to improve weatherability and decrease polymer degradation. For the polyolefins (plastics composed solely of carbon and hydrogen), ultraviolet absorbers (UVAs), free-radical scavengers, and pigments are generally used. Benzophenones and benzotriazoles are UVAs. Hindered amine light stabilizers (HALS) are an example of free-radical scavengers. Pigments are also utilized and protect by physically blocking the light. As an added bonus, they allow consumers to choose from a variety of colors for their product.

Moisture control is the key to protecting WPCs from biological decay and weathering, and to preserving ultimate performance. Research continues in this area with opportunities utilizing new and improved processing methods or new paints and surface coatings. Nanotechnology may also help with fundamental material characterization of the WPCs, and the use of nanoparticles may improve moisture, decay, and photodegradation resistance as well as thermal expansion and creep.

For background information *see* COMPOSITE MATERIAL; LUMBER; PLASTICS PROCESSING; POLYMER; STRUCTURAL MATERIALS; WOOD COMPOSITES; WOOD ENGINEERING DESIGN; WOOD PRODUCTS; WOOD PROPERTIES in the McGraw-Hill Encyclopedia of Science & Technology.   Rebecca E. Ibach

Bibliography. A. Klyosov (ed.), *Wood-Plastic Composites*, John Wiley & Sons, Hoboken, NJ, 2007; K. Oksman and M. Sain (eds.), *Wood-Polymer Composites*, CRC Press, Boca Raton, FL, 2008; R. M. Rowell (ed.), *Handbook of Wood Chemistry and Wood Composites*, CRC Press, Boca Raton, FL, 2005; T. P. Schultz et al. (eds.), *Development of Commercial Wood Preservatives: Efficacy, Environmental, and Health Issues*, American Chemical Society, Washington, D.C., 2008; N. Stark (ed.), *Ninth International Conference on Wood and Biofiber Plastic Composites*, Forest Products Society, Madison, WI, 2007.

Earliest humans in the Americas

The Americas were the last continents (except for Antarctica) colonized by *Homo sapiens* (that is, anatomically modern humans) and they represented the "end of the road" or final stage of the global expansion process that started in sub-Saharan Africa around 100,000 years ago. Although some researchers in the past postulated that the origin of humankind was in South America, currently all of the available data support the model that humans migrated to the American continents as *Homo sapiens* at the end of the Pleistocene (an epoch spanning about 1.8 million to 10,000 years ago and commonly characterized as when the earth entered its most recent phase of widespread glaciation). This means that no ancestors of *Homo sapiens* ever occupied or evolved in the Americas. Although this has been a highly contested debate, it seems that Neandertals lived in the Old World until roughly 30,000 years ago (when they became extinct), coexisting with the ancestors of modern humans who had expanded throughout the Old World from Africa between 100,000 and 60,000 years ago. The descendants of these modern humans then entered the Americas sometime at the very end of the Pleistocene.

Routes into the Americas. Given the strong evidence (for example, genetic and morphometric data) linking the indigenous people of the Americas with Asiatic ancestors, the main entrance route seems to be across the Bering Strait and neighboring areas (named Beringia), when the whole region emerged as a land bridge as a consequence of a drop in sea level during the Late Pleistocene [ca. 27,000–11,000 carbon-14 years before the present (^{14}C yr BP)]. [Note that all radiocarbon dates (expressed as

^{14}C yr BP) are not calibrated here, meaning that they do not match exactly with the calendar years; for example, at around 12,000 ^{14}C yr BP, the calendar years are 2000 years older (around 14,000 yr BP).] Once in Alaska, there were two possible routes: crossing a narrow ice-free corridor that was open between the Laurentide and Cordilleran ice sheets (the vast North American glaciers, which covered much of the Northwest Territories and the rest of Canada) at ca. 11,500 ^{14}C yr BP or following a coastal route along the Pacific coast of North America (see **illustration**). However, the earliest archeological sites in Alaska, such as Swan Point and Broken Mammoth (both dated at around 11,700 ^{14}C yr BP), are slightly younger than the earliest sites found in the rest of the continent. These data thus raise a problem that is still unresolved and would favor the coastal route.

An alternative model has been proposed, stating that some of the early Americans descended from Late Paleolithic people from the north coast of Spain (the "Solutrean," ca. 22,000–16,500 ^{14}C yr BP), who may have followed the edge of the ice sheet that covered the North Atlantic during the last Ice Age (see illustration). However, this provocative model is still speculative.

Evidence. Until recent decades, the main evidence for the early peopling of the Americas had come from remains at archeological sites, principally stone tools (mostly the distinctive "fluted points") associated with the bones of extinct animals (bison, mammoth, camelids, ground sloth, giant ground sloth, American horses, and so on) and eventually with charcoal from hearths. However, in the past 30 years, morphometrical studies of human skeletons, mainly the skull, have provided new tools for approaching the peopling of the Americas. More recently, genetic studies on the DNA sequences of indigenous populations as well as from ancient bones have enhanced our understanding of the process of human expansion from the Old World to the Americas. Current models of the early peopling of the Americas are combining these three lines of evidence, although giving varied degrees of importance to each of them.

The baseline for the peopling of the Americas is given by what is called the "Clovis culture," a well-established population of hunter-gatherers that inhabited the Great Plains of North America. Clovis populations exploited extinct megafauna (for example, mammoth and bison) and used a distinctive type of projectile spear point, the "Clovis point," which was discovered for the first time at the Blackwater Draw site in eastern New Mexico. Recent studies have been oriented toward carefully dating Clovis sites as well as refining the chronology of other contemporaneous and pre-Clovis sites in the Americas. Also, new sites that potentially date to the pre-Clovis period have been carefully excavated, providing new high-quality data for investigation.

In regards to Clovis, the latest analysis of ^{14}C ages indicates that the chronological span of this culture was between 11,050 and 10,800 ^{14}C yr BP. There are several sites that are a few thousand years older than

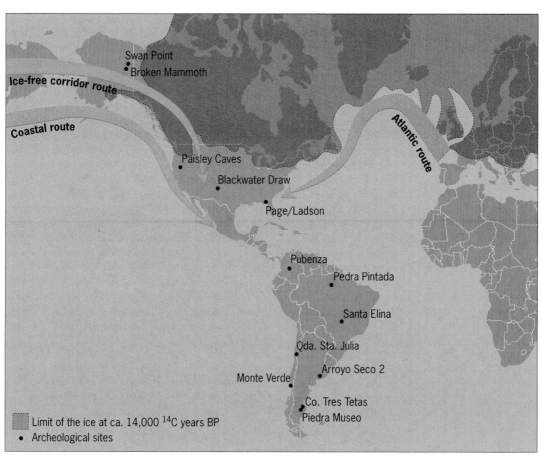

Map showing the main possible routes of entry into the Americas and the main archeological sites mentioned in the text.

Clovis that are south of the Laurentide ice sheet, and these provide strong evidence for an earlier human presence in the Americas. Some of these sites include the following: Paisley Caves (Oregon), where human coprolites (ancient excrement) have been dated to ca. 12,300 ^{14}C yr BP; Monte Verde (in southern Chile), which dates between ca. 12,800 and 12,300 ^{14}C yr BP and has been interpreted as having a forest-adapted economy based primarily on the collection of wild plant foods and secondarily on the scavenging and/or hunting of large and small animals; the Page/Ladson site (Gulf of Mexico, Florida), which dates to ca. 12,400 ^{14}C yr BP and shows a good association of extinct megafauna with tools; and Arroyo Seco 2 (Pampas of Argentina), where the oldest archeological component dates to ca. 12,200 ^{14}C yr BP. Therefore, new archeological data are pushing back the early peopling of the Americas at least 1500 years as well as confirming that there were people in both North and South America before the Clovis period. Moreover, in South America, there are sites that are contemporary with Clovis (for example, Pedra Pintada, Cerro Tres Tetas, Piedra Museo, Quebrada Santa Julia, and so on), indicating that the major environments of America were already occupied by humans with various adaptive strategies around 11,000 ^{14}C yr BP. A few sites, such as Meadowcroft Rockshelter (Pennsylvania), and other recently excavated sites, such as the Abrigo de Santa Elina (Mato Grosso State, Brazil) and Pubenza (Mag-

dalena River, Colombia), could indicate an even earlier human occupation, possibly as early as ca. 30,000 ^{14}C yr BP. However, these sites have not yet been fully investigated and their analyses have not been fully published, making it difficult to evaluate the existing evidence.

Models based on craniofacial evidence complement the picture, although there are also basic disagreements. A classic model (originally proposed in the early twentieth century by Aleš Hrdlička and more recently by Joseph Powell) stated that all Native America populations (except Aleutians and Eskimo) descend from only one original Eastern Asian (mongoloid) stock that entered into the Americas at the end of the Pleistocene. This model explains that the morphological differences are likely the result of genetic drift and natural selection rather than different migrations. In contrast, other physical anthropologists (notably, Walter Neves and Hector Pucciarelli) proposed two early entries: one premongoloid migration between 15,000 and 12,500 ^{14}C yr BP (the "Paleoamericans"), who likely became extinct, and a second mongoloid migration, between 9000 and 8000 ^{14}C yr BP, which gave rise to the modern Native Americans. However, it is essential to elucidate the probable sources of variation of craniofacial morphology, including random and nonrandom factors, and to understand the relationships between differences in morphological traits and different ancestry.

Mitochondrial DNA (mtDNA) analysis and other genetic markers (such as blood groups) show strong similarities between Native American populations and those from Central and East Asia, suggesting that the latter could be the ancestors. The mtDNA is a circular DNA duplex, generally 5–10 copies, contained within a mitochondrion. It is maternally inherited since only the egg cell contributes significant numbers of mitochondria to the zygote, thereby providing examples of DNA lineages that can be followed over long periods of time. The DNA from contemporary Native American Indians can be separated into four haplogroups, or lineages, named A, B, C, and D. A fifth haplogroup, X, is less well represented (only in some North American populations) and is known from both Central Asia and Europe. The X haplogroup is a large, diverse haplogroup with many lineages, but the lineage found in Native American Indians is different from those in Eurasia. Genetic research suggests that the four founding haplogroups may have differentiated in Asia sometime between 45,000 and 15,000 ^{14}C yr BP, giving origin to all American populations. Based on the study of ancient mtDNA, from skeletal remains recovered from archeological sites, an age of 13,500 ^{14}C yr BP has been proposed for the early populations that entered into the Americas.

Conclusions. In sum, if an Asiatic origin and a Beringia route of entry for the earliest Americans is accepted as the most probable, it is necessary to take into account that the maximum extension of the ice sheets during the Late Pleistocene was between 18,000 and 14,000 ^{14}C yr BP. During this time, the ice-free corridor did not exist (the Laurentide and the Cordilleran ice sheets were joined) and the Pacific coast of northwestern North America was probably still covered by ice. Given the fact that human presence before 18,000 ^{14}C yr BP has not been confirmed in the Americas, the most probable date of entry should be around 14,000 ^{14}C yr BP following the recently deglaciated Pacific coast. This coincides with the estimation based on contemporary and ancient mtDNA. These early populations would belong to a single Asiatic population, although later population migration waves from Asia (or even from Europe) cannot be ruled out. Once in America, these groups experienced a rapid process of expansion based on flexible and successfully adaptive strategies. By 12,500 ^{14}C yr BP, they arrived in the Southern Cone (the southernmost areas) of South America; and by 11,000 ^{14}C yr BP, the major environments of North America were already occupied by humans. At the same time, demographic growth and associated cultural processes produced the Clovis culture, one of the first groups who managed to successfully adapt to the North American landscape.

For background information see ANTHROPOLOGY; ARCHEOLOGY; DEOXYRIBONUCLEIC ACID (DNA); EARLY MODERN HUMANS; FOSSIL HUMANS; MOLECULAR ANTHROPOLOGY; PALEOINDIAN; PHYSICAL ANTHROPOLOGY; PLEISTOCENE; POPULATION DISPERSAL; POPULATION DISPERSION; RADIOCARBON DATING in the McGraw-Hill Encyclopedia of Science & Technology.   Gustavo G. Politis

Bibliography. M. Faught, Archaeological roots of human diversity in the New World: A compilation of accurate and precise radiocarbon ages from earliest sites, *Am. Antiq.*, 73(4):670–698, 2008; T. Goebel, M. R. Waters, and D. H. O'Rourke, The late Pleistocene dispersal of modern humans in the Americas, *Science*, 319:1497–1502, 2008; W. A. Neves and H. M. Pucciarelli, Extra-continental biological relationships of early South American human remains: A multivariate analysis, *Ciên. Cult.*, 41:566–575, 1989; J. F. Powell, *The First Americans: Race, Evolution, and the Origin of Native Americans*, Cambridge University Press, Cambridge, U.K., 2005; T. Schurr, The peopling of the new world: Perspectives from molecular anthropology, *Annu. Rev. Anthropol.*, 33:551–583, 2004; J. Steele and G. Politis, AMS ^{14}C dating of early human occupation of South America, *J. Archaeol. Sci.*, 36:419–429, 2009; M. R. Waters and T. W. Stafford, Redefining the age of Clovis: Implications for the peopling of the Americas, *Science*, 315:1122–1126, 2007.

EarthScope: observatories and findings

Along the western margin of North America lies a complex system of faults that accommodate the relative motions of the Pacific, North American, and Juan de Fuca plates. Sudden release of accumulated elastic strains via slip along faults produces earthquakes. Networks of EarthScope geodetic instrumentation (**Fig. 1**) are capturing these strains before, during, and after earthquakes. EarthScope also contains dense arrays of seismometers (Fig. 1) that record body and surface waves emitted by earthquakes around the world. The timing and shape of these seismic waveforms yield vital information about the structures that lie below the Earth's surface, much like medical computed tomography (CT) scans can show detailed structural variations within the human body. Ongoing detailed subsurface imaging, waveform analysis, fault-zone monitoring, and strain measurements through the EarthScope facility are revealing new insights about the Earth. Examples include the recent discovery of silent slip events (or slow-slip earthquakes) accompanied by seismic tremor activity along subduction-zone faults beneath parts of the United States, Canada, and other regions of the world, stunning images of three-dimensional geometry of subducted plates and foundering mantle drips, crustal stress changes preceding earthquakes, and samples of the San Andreas Fault brought up from nearly 3 km depth. These core samples and ongoing monitoring of the San Andreas Fault are providing the first-ever measurements of chemical and structural features of a major plate-bounding fault zone at seismogenic depths, providing scientists with an opportunity to obtain a better understanding of the physics of the nucleation, propagation, and arrest of earthquake-associated slip. EarthScope is funded by the National Science

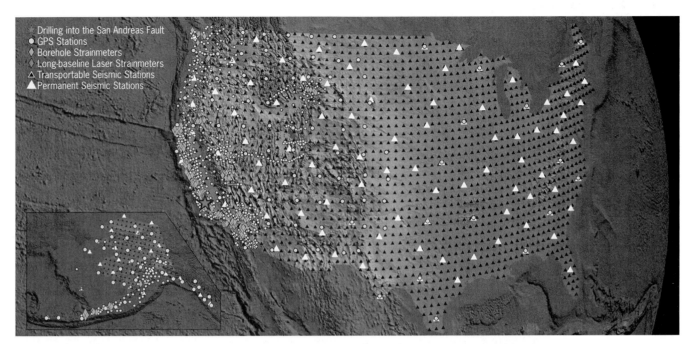

Drilling into the San Andreas Fault
● GPS Stations
◆ Borehole Strainmeters
◇ Long-baseline Laser Strainmeters
△ Transportable Seismic Stations
▲ Permanent Seismic Stations

Fig. 1. Location of EarthScope USArray and Plate Boundary Observatory (PBO) instruments, along with San Andreas Observatory at Depth (SAFOD) borehole location. (*Courtesy of EarthScope*)

Foundation in partnership with NASA and the U.S. Geological Survey (U.S.G.S.). An EarthScope national office is responsible for the coordination of scientific, educational, and outreach activities. More information about EarthScope science, data, education, and outreach activities can be obtained through its office.

EarthScope facility. Through the monumental efforts and cooperation of scientists, engineers, professional staff, private, state and federal landowners, and volunteers, the construction phase of the EarthScope facility has been an enormous success. EarthScope is now providing open data for scientific discovery, innovation, and education. EarthScope has three primary components. The first of these is USArray, consisting of the permanent, transportable, flexible, and magnetotelluric arrays. The Transportable Array is a network of 400 seismographs placed in temporary sites across the United States, with a station spacing of about 70 km. After a residence time of 18 months, each instrument is picked up and moved to a new location on the eastern edge of the array. Nearly 2000 locations will have been occupied by the Transportable Array during the measurement period of EarthScope. USArray will have covered the entire lower 48 states by 2013 and will then be moved to Alaska during years 2013–2018. The Flexible Array consists of more than 2000 portable seismometers geared for detailed studies of specific targeted areas. In collaboration with the U.S.G.S.'s Advanced National Seismic System, the USArray has upgraded and expanded the Permanent Array, with a uniform spacing of 300 km across the United States. The Permanent Array provides important baseline measurements for both the portable and flexible arrays. To date, more than 600 Transportable Array sites have been installed, and more than 2000 Flexible Array sites have been oc-

cupied. The magnetotelluric array is composed of seven permanent and 20 portable sensors that record naturally occurring electric and magnetic fields, used to image the Earth's interior. More than 170 magnetotelluric sites have been occupied, and seven backbone magnetotelluric sites have been installed. All USArray data are freely available to scientific and educational communities and to the public. More than 12 terabytes of EarthScope data have been collected and archived, and many new discoveries about the Earth's structure are emerging.

Plate Boundary Observatory. The Plate Boundary Observatory (PBO) is designed to measure the 4D strain field (that is, three spatial dimensions and time) across the active plate boundary zone system, where earthquakes are common, as well as within the North American plate interior, where signals from glacial isostatic adjustment are now being resolved. Core instrumentation consists of a network of Global Positioning System (GPS) receivers and strainmeters, along with aerial and satellite imagery. PBO observations provide temporal constraints on plate boundary deformation, ranging from seconds to millennia, enabling the understanding of the physics that govern deformation, faulting, and fluid transport within the Earth's lithosphere. PBO installation has been completed on time and on budget. A total of 1100 permanent and continuously operating Global Positioning System (CGPS) stations have been installed or added to existing (legacy) networks. These CGPS stations are deeply anchored to ensure stability and high-quality data acquisition. CGPS stations are ideal for measuring motions associated with volcanic and plate tectonic processes, strain accumulation on and near faults, earthquake-associated displacements, and postseismic deformation. These data therefore provide constraints on fault-zone and

Fig. 2. Location of EarthScope SAFOD borehole in a perspective plot with the viewer at a depth of about 3 km below the surface. The topography image represents the bottom side of the Earth's surface and the white dots are microearthquakes that have been located within the San Andreas fault zone. (*Courtesy of EarthScope*)

magma-chamber behavior on time scales of days to decades. In addition to the CGPS stations, the PBO of EarthScope also has a pool of 100 portable GPS instruments for temporary or semipermanent deployments. These instruments support exploration of specific tectonic targets, and they are available for rapid response measurements following earthquakes or volcanic events.

Strainmeter instrumentation for PBO component of EarthScope consists of 74 installed borehole tensor strainmeters (BSM) and six long-baseline laser strainmeters (LSM). The BSMs measure minute amounts of distortion of an instrument cemented within the bedrock. They are sensitive to these minute distortions of the bedrock at periods of minutes to months. The BSM instruments thus enable strain resolution over time intervals in between what seismometers and CGPS instruments can provide. The BSMs also contain short-period borehole seismometers, which provide the opportunity to measure seismic waves with high signal-to-noise characteristics. These borehole seismometers, for example, have greatly expanded the detection capability for volcanic tremor activity beneath Yellowstone caldera. The LSM instruments use a laser to measure the relative position of end monuments that are hundreds of meters apart. The LSMs are highly stable and measure precise strains from months to decades, and they are useful for detecting strain events associated with movements along fault zones.

The PBO also operates GeoEarthScope, which includes airborne light detection and ranging (lidar)

imagery for detailed mapping of the Earth's surface, satellite interferometric synthetic aperture radar (InSAR) imagery for precise mapping of surface changes during deformation events, and geochronology data to provide age constraints on prehistoric earthquakes and long-term fault offsets. These data sets and methods enable the quantification of deformation on time scales that extend back several millennia, or in the case of geochronologic data, back through Earth history. As in the case for USArray data, all PBO data are freely available to scientific and educational communities and to the public.

San Andreas Fault Observatory at Depth. The San Andreas Fault Observatory at Depth (SAFOD) is geared to measure the physical and chemical processes associated with deformation and earthquake generation within the San Andreas Fault zone. The SAFOD consists of several components that include downhole sampling, measurements, and long-term monitoring. The SAFOD consists of a 2.2-km-deep pilot hole and a main hole that was drilled through the San Andreas Fault. The main hole is vertical down to a depth of 1.5 km and then takes a 60° turn from vertical and passes through the San Andreas fault zone, ending at a depth of 2.7 km. This main hole was designed to pass through two sections of repeating clusters of microearthquakes, first identified and located using downhole instrumentation within the pilot hole (**Fig. 2**). Several sidetracks were drilled laterally off of the main hole to obtain core samples from two actively deforming traces of the San Andreas Fault. These core samples constitute the

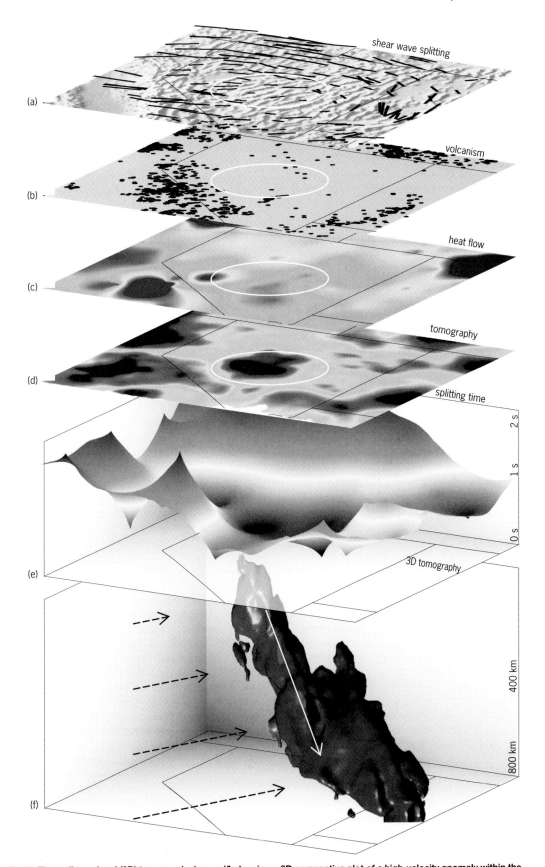

Fig. 3. Three-dimensional (3D) tomography image (f) showing a 3D perspective plot of a high-velocity anomaly within the mantle below central Nevada, determined using USArray seismometers and recorded seismic waves from earthquakes. Also shown are (a) fast directions of shear wave anisotropy within the mantle, (b) locations of past volcanic activity, (c) surface heat flow values, and (d, e) splitting-time magnitudes of anisotropy. The anomaly is thought to represent a downwelling zone of mantle flow beneath the Great Basin of central Nevada. (*Reprinted by permission from Macmillan Publishers Ltd: Nature Geoscience, J. D. West et al., Vertical mantle flow associated with a lithospheric drip beneath the Great Basin, Nat. Geosci., 2(6):438–443, 2009, copyright 2009*)

first samples ever obtained from an active fault zone at seismogenic depths. Borehole seismometers, accelerometers, and tiltmeters have been installed at depth. These instruments are yielding precise measurements of variations in deformation, fluid pressure, microseismicity, and radiated seismic energy within one of the world's major fault zones.

Scientific discovery facilitated by EarthScope data. Fault core samples collected from the San Andreas Fault have revealed the presence of the mineral serpentinite within the fault zone. Scientists have speculated for decades about whether serpentinite might be responsible for the creeping section of the San Andreas Fault. Significantly, the mineral talc was found within cuttings of serpentinite collected from the active trace of the fault. Talc has a low frictional strength at elevated temperatures, and this important discovery may explain the low resolved shear stress on the San Andreas Fault, which has been argued to be a weak fault within a strong crust for more than 20 years. Other important findings on the physical properties within the fault zone support the weak-fault hypothesis, including a lack of elevated pore fluid pressures and lack of evidence for frictionally generated heat within the fault zone. The high signal-to-noise characteristics of recorded waveforms within the SAFOD boreholes have allowed scientists, for the first time, to measure stress changes within a fault zone. Measuring the time-dependent changes in travel time, F. Niu and coworkers have determined that stress changes occurred approximately 10 and 2 h before two microearthquakes located within the San Andreas Fault zone. SAFOD samples are now being analyzed by more than 20 laboratories around the world. Sample analysis, coupled with downhole measurements within the SAFOD observatory, promise to reveal important new data about the physics that govern earthquake initiation and rupture propagation.

The USArray has already enabled important discoveries about subsurface structures, including new constraints on the geometry of the subducted lithosphere within the Pacific Northwest, new constraints on mantle flow patterns, and images of foundering lithosphere beneath the Sierra Nevada and Great Basin of Nevada (**Fig. 3**). Moreover, a revolutionary method, termed ambient noise tomography (ANT), is now being applied using USArray seismometers. This method takes advantage of the long time series of ubiquitous seismic noise recorded on USArray seismometers and, therefore, does not require earthquake-radiated energy. The dense seismic ray coverage using ANT methods is providing new high-resolution images of the velocity structure of the crust and upper mantle within regions covered by the USArray. Ambient noise tomography promises to reveal unprecedented results on the structure and velocity anisotropy of the crust and uppermost mantle within North America as the USArray moves eastward across the North American continent.

Magnetotelluric measurements provide estimates of mantle conductivity, yielding information on the water content within the mantle. Magnetotelluric measurements, coupled together with seismic estimates of wave-speed and attenuation variations, hold promise for providing new constraints on the variations of the physical and chemical state of the crust and upper mantle.

The phenomenon of episodic tremor and slip (ETS) is a relatively new discovery, first recognized during the planning stages of EarthScope. ETS is a process involving very slow slip on faults. It was detected in a deeply buried fault, detected using GPS, accompanied by seismic tremors. The slow slip and seismic tremors originate from deep subduction thrust systems beneath Cascadia, within the Pacific Northwest, where the Juan de Fuca plate is subducting beneath the North American plate. These slow slip events occur with relatively striking regularity, with typical periods of between 11 and 18 months, depending on location. ETS events throughout the Pacific Northwest region continue to be measured with increasingly higher resolution using installed PBO CGPS stations and USArray instruments. The new data have greatly contributed to an improved understanding of the earthquake rupture physics and the seismic hazards associated with subduction-zone slip events.

New PBO data have provided insight about strain within creeping and locked portions of fault zones, details of earthquake-related displacement fields, and important constraints for studies of large-scale continental dynamics. PBO measurements are thus providing critical information for describing the strain budget within the earthquake cycle as well as for understanding the longer-term forces responsible for continental-scale deformation.

PBO instruments are having a scientific impact beyond the understanding of the kinematics and dynamics of the solid earth. For example, GPS signal reflections are now being used to measure near-surface soil moisture and its changes with time. The GPS instruments within PBO can thus provide independent calibration points as a complement to planned satellite-based missions to monitor soil moisture content, which is important for understanding the effects of weather and climate change on agriculture and for mitigating the effects of drought.

All components of EarthScope have supported and developed a variety of educational materials for educators and the general public, including teacher workshops, the development of museum displays, the involvement of K–12 grades in site visits, and student internships. The open-data availability of EarthScope has facilitated cooperative and productive collaborations of scientists around the world, and has facilitated the training and development of future generations of scientists who will be accustomed to dealing with large data streams and multidisciplinary approaches to solving problems.

For background information *see* COMPUTERIZED TOMOGRAPHY; EARTH INTERIOR; EARTHQUAKE; FAULT AND FAULT STRUCTURES; GEODESY; GEODYNAMICS; LITHOSPHERE; PLATE TECTONICS; SEISMOGRAPHIC INSTRUMENTATION; SEISMOLOGY; SUBDUCTION ZONES; VOLCANO; VOLCANOLOGY in the

McGraw-Hill Encyclopedia of Science & Technology.
William Holt

Bibliography. J. R. Arrowsmith and O. Zielke, Tectonic geomorphology of the San Andreas Fault zone from high resolution topography: An example from the Cholame segment, *J. Geomorphol.*, doi:10.1016/j.geomorph.2009.01.002, 2009; G. D. Bensen, M. H. Ritzwoller, and Y. Yang, A 3-D shear velocity model of the crust and uppermost mantle beneath the United States from ambient seismic noise, *Geophys. J. Int.*, 177(3):1177–1196, 2009; O. Boyd, C. H. Jones, and A. F. Sheehan, Foundering lithosphere imaged beneath the southern Sierra Nevada, California, USA, *Science*, 305:660–662, 2004; M. R. Brudzinski, Seismology: Do faults shimmy before they shake?, *Nat. Geosci.*, 1:295–296, 2008; M. R. Brudzinski and R. M. Allen, Segmentation in episodic tremor and slip all along Cascadia, *Geology*, 35:907–910, 2007; W. L. Chang et al., Accelerated uplift and magmatic intrusion of the Yellowstone caldera, 2004 to 2006, *Science*, 318:952–956, 2007; L. M. Flesch et al., The dynamics of western North America: Stress magnitudes and the relative role of gravitational potential energy, plate interaction at the boundary and basal tractions, *Geophys. J. Int.*, 169:866–896, 2007; H. Gilbert et al., Imaging Sierra Nevada lithospheric sinking, *EOS*, 88(21):225–229, 2007; S. Hickman and M. Zoback, Stress orientations and magnitudes in the SAFOD pilot hole, *Geophys. Res. Lett.*, 31:L15S12, 2004; E. D. Humphreys and D. D. Coblentz, North American dynamics and Western U.S. tectonics, *Rev. Geophys.*, 45:RG3001, 2007; E. C. Klein et al., Evidence of long-term weakness on seismogenic faults in western North America from dynamic modeling, *J. Geophys. Res.*, 114:B03402, 2009; K. M. Larson et al., Use of GPS receivers as a soil moisture network for water cycle studies, *Geophys. Res. Lett.*, 35:L24405, 2008; M. M. Miller et al., Periodic slow earthquakes from the Cascadia subduction zone, *Science*, 295:2423, 2002; D. E. Moore and M. J. Rymer, Talc-bearing serpentinite and the creeping section of the San Andreas fault, *Nature*, 448:795–797, 2007; M. P. Moschetti, M. H. Ritzwoller, and N. M. Shapiro, Surface wave tomography of the western United States from ambient seismic noise: Rayleigh wave group velocity maps, *Geochem. Geophys. Geosyst.*, 8:Q08010, 2007; F. Niu et al., Preseismic velocity changes observed from active source monitoring at the Parkfield SAFOD drill site, *Nature*, 454:204–208, 2008; P. K. Patro and G. D. Egbert, Regional conductivity structure of Cascadia: Preliminary results from 3D inversion of USArray transportable array magnetotelluric data, *Geophys. Res. Lett.*, 35:L20311, 2008; C. M. Puskas and R. B. Smith, Intraplate deformation and microplate tectonics of the Yellowstone hot spot and surrounding western US interior, *J. Geophys. Res.*, 114:B04410, 2009; G. Rogers and H. Dragert, Episodic tremor and slip on the Cascadia subduction zone: The chatter of silent slip, *Science*, 300:1942–1943, 2003; J. B. Roth et al., Three-dimensional seismic velocity structure of the northwestern United States, *Geophys. Res. Lett.*, 35:L15304, 2008; I. Ryder and R. Burgmann, Spatial variations in slip deficit on the central San Andreas Fault from InSAR, *Geophys. J. Int.*, 175:837–852, 2008; G. F. Sella et al., Observation of glacial isostatic adjustment in "stable" North America with GPS, *Geophys. Res. Lett.*, 34:L02306, doi:10.1029/2006GL027081, 2007; K. Sigloch, N. McQuarrie, and G. Nolet, G, Two-stage subduction history under North America inferred from multiple-frequency tomography, *Nat. Geosci.*, 1:458–462, 2008; W. Szeliga et al., GPS constraints on 34 slow slip events within the Cascadia subduction zone, 1997–2005, *J. Geophys. Res.*, 113:B04404, 2008; J. D. West et al., Vertical mantle flow associated with a lithospheric drip beneath the Great Basin, *Nat. Geosci.*, 2(6):438–443, 2009; Y. J. Yang et al., Structure of the crust and uppermost mantle beneath the western United States revealed by ambient noise and earthquake tomography, *J. Geophys. Res.*, 113:B12310, 2008; G. Zandt and E. Humphreys, Toroidal mantle flow through the western US slab window, *Geology*, 36:295–298, 2008; G. Zandt et al., Active foundering of a continental arc root beneath the southern Sierra Nevada, California, *Nature*, 431:41–46, 2004; M. D. Zoback et al., New evidence on the state of stress of the San-Andreas Fault system, *Science*, 238:1105–1111, 1987; M. D. Zoback and S. H. Hickman, Preliminary results from SAFOD Phase 3: Implications for the state of stress and shear localization in and near the San Andreas Fault at depth in central California, *EOS, Trans. Amer. Geophys.*, 88(52), 2007.

Electronic medical record

The replacement of paper-based medical documentation by electronic documentation dates back to the early use of computers in the 1960s. Since then, electronic medical records (EMRs) have become more complex, and both the content and the functionality have increased substantially and will increase further in the near future. The implementation of an EMR is a long-term project and is related to a transformation process for the health-care system. Today's health-care systems are fragmented, not very efficient and ineffective, and with safety shortcomings. The development of an efficient information and communication technology (ICT) infrastructure for documentation, archiving, retrieval, communication, and processing of health-related data provides a great opportunity for improving health care and social systems worldwide. Therefore, a shift from provider-oriented documentation to patient-centered communication and integrated health care is essential. This could be achieved by the implementation of EMRs.

There are many definitions of an EMR. In principle, an EMR is a repository, communication, and processing system for all relevant data, information, and knowledge about the health status and health care of a unique identified person. The EMR is a virtual record that is stored in many different information systems and connected by a unique personal identifier. EMRs can be seen as the key element of eHealth, the electronic network for health and social care. Sometimes the EMR is considered to be an electronic document in one health-care institution, and the virtual patient-centered network of EMRs of many different health-care providers is called the electronic health record (EHR) or electronic health-care record (EHCR). Today, there are many promising implementations of different parts of the EMR, mainly on the level of single organizations or of specific functions on a regional level. However, few hospitals affirm that they have a comprehensive EMR.

Content of the EMR. Whereas most health-care providers are mandated to keep records about all diagnostic and therapeutic measures, a central question for the EMR is which part of this information should be available in the network. Selection of so-called relevant information will become one of the key functionalities of EMRs, but different professionals and specialists may require different views. Therefore, an appropriate filtering process and an ideal presentation of this information will be essential. An EMR should contain clinical data and administrative or demographic data, using standardized terminology. The clinical data should include symptoms, diagnoses, findings/results, procedures and interventions, medications, technical aids and implants, allergies and other risk factors, vital signs, laboratory test data, complications, nursing data, discharge letters, patient history, follow-up care plan, preventive measures, functional status, and immunizations. The set of administrative data can be structured in a similar manner and should contain at least the following data: patient identification, provider identification, and identification of the care episode. This can be supplemented by demographic and physiological data, for example, age, gender, blood pressure, heart rate, and other information relevant for the current care process. Additional data on patient wills, organ donation, signed documents about the permission to end care under certain circumstances, the consent for communication of the data to other health-care providers, and the use of the data in anonymous form for clinical studies are important from a legal aspect and should be included. In the near future, a standardized minimum core data set should be defined in a continuity of care record (CCR) to enable communication and cooperation among different care teams and providers. Other areas of structured documentation include vaccination (eVaccination), emergency data sets, diaries for patient with diabetes, data sets for reimbursement, and so on. In addition, with regard to patient mobility, an international agreement for a core data set will be necessary.

Functionality. The functionalities of an EMR range from data captured by writing free text to structured documentation, online recording/analysis of biosignals, storage of pictures, online ordering, scheduling, and so on, up to decision support, pattern recognition, and integration of medical knowledge. Complex EMRs can be broken down into many small but interrelated modules related to both content and functionalities. A fine example of this is the electronic prescription of medication (ePrescription), which could be implemented worldwide. To increase patient safety, an electronic ordering of a prescription for a medication by an identified provider for a unique identified person is automatically checked with regard to interactions with other medications or allergies of this patient. Thus, all medications as well as additional information on the prescription can be stored in a structured way using standard terminology.

Usability. There are many concerns from the user's point of view about the usability of EMRs. Usability is a key element for the acceptance of EMRs, especially from the health-care provider's viewpoint. The usability of the documentation process itself has to be improved, for example, by implementing user-friendly human-machine interfaces, mobile devices, or speech recognition systems. The information retrieval process has to be user-specific, which means that a selection of information that meets the specific demands of a user has to be easily accessible. Data processing and presentation have to be clearly arranged to allow a well-structured overview of the relevant information.

Interoperability and standards. Interoperability is defined as the ability of different information systems to exchange data accurately and consistently and to process the information exchanged. To guarantee national and international interoperability and to offer a sustainable pathway for the further development of EMRs, both technical and semantic standards are necessary.

Technical interoperability should assure reliable communication by containing the structure and syntax of a document. There are several different organizations, including HL7 (Health Level 7), ISO 13606, and openEHR, that provide standards. Another example is DICOM (Digital Imaging and Communication in Medicine), which is primarily used for picture archiving, processing, and communication. Integrating the Healthcare Enterprise (IHE) provides recommendations on how to store electronic documents and facilitate their sharing.

Semantic interoperability allows for the preservation of meaning when information is transferred between information systems using metadata for description of the data. Here, the main problem of EMRs is to generate shareable and computable information from different information systems. For standardization, a reference model is used to represent the generic properties of the health record information, and this can then be used to create complex data structures. The reference model contains the generic building blocks for the EMR. It specifies the way to create more complex data structures and the context for each data item to meet the legal requirements. The data can be expressed in a formal manner that enables the information to be shared between systems, thereby allowing generation of standardized electronic documents.

Privacy and security. Health-related data are very sensitive, and there exist many concerns about privacy and security. A comprehensive set of technical, organizational, and legal measures is absolutely essential to avoid the misuse of health-related data and to increase acceptance of EMRs. Security measures for the (virtual) EMR go far beyond the boundaries of local organizations because the initiating party does not know who will have access to the data in the future. Therefore, additional laws regarding privacy, especially pertaining to eHealth, are necessary. However, the basic principle for privacy is the "need-to-know principle," which means that access is permitted only when it is relevant for the care process. To achieve the highest level of privacy and security, the following requirements have to be satisfied:

1. Authentication of the person or system requiring information.

2. Authorization of the access (based on the role concept for different users).

3. Access logging.

4. Encrypted transfer.

5. Auditing about adherence.

Outlook. Because of the growing complexity and expensiveness of health and social care, an elaborate EMR provides an opportunity to improve efficiency, effectiveness, safety, and quality. Although there are only a few studies available about the cost-benefit relation of EMRs, most of them show a positive return on investment after a few years, especially if indirect costs are also considered.

To guarantee an effective and efficient patient-centered health-care system, an EMR system should be implemented in all countries worldwide in the near future, independent of specific health-care systems, thereby supporting a transformation of the health-care process. Although a lot of work on standardization has been done, further research is necessary. The successful implementation of an EMR depends on improved functionality and usability, additional measures for privacy and security of information systems, and the availability of standards and terminologies to enable technical and semantic interoperability. In addition, the implementation of the EMR is a great opportunity and a great challenge for the information technology industry.

For background information *see* CLIENT-SERVER SYSTEM; COMPUTER SECURITY; COMPUTER STORAGE TECHNOLOGY; DATABASE MANAGEMENT SYSTEM; HUMAN-COMPUTER INTERACTION; INFORMATION MANAGEMENT; INFORMATION TECHNOLOGY; MEDICAL INFORMATION SYSTEMS; MEDICINE; PUBLIC HEALTH in the McGraw-Hill Encyclopedia of Science & Technology.   Karl-Peter Pfeiffer

Bibliography. B. Blobel and P. Pharow, Analysis and evaluation of EHR approaches, *Meth. Inform. Med.*, 48:162–169, 2009; K. Häyrinen et al., Definition, structure, content, use, and impact of electronic health records: A review of the research literature, *Int. J. Med. Informat.*, 77:291–304, 2008; D. Moner et al., Archetype-based semantic integration and standardization of clinical data, *Conf. Proc. IEEE Eng. Med. Biol. Soc.*, 1:5141–5144, 2006; P. G. Shekelle, S. C. Morton, and E. B. Keeler, Costs and benefits of health information technology, *Evid. Rep. Tech. Assess.*, 132:1–71, 2006.

Enhanced ionic conductivity in oxide heterostructures

Fuel cells are electrochemical devices used to generate energy out of hydrogen. In a fuel cell, two conducting electrodes are separated by an electrolyte that is permeable to ions (either hydrogen or oxygen, depending on the fuel-cell category) but not to electrons. An electrode catalytic process yields the ionic species, which are transported through the electrolyte, while electrons blocked by the electrolyte pass through the external circuit. Polymeric membrane (PEMFC) or phosphoric acid fuel cells (PAFC) operating at low temperatures are the preferred option for transportation because of their quite large efficiencies (50%), compared with gasoline combustion engines (25%). Other uses are also being considered, such as battery replacements for personal electronics and stationary or portable emergency power. Solid-oxide fuel cells (SOFCs), operating at high temperatures, are a better option for stationary power generation because of their scalability. Here O^{2-} ions are the mobile species that travel at elevated temperatures (800–1000°C) through a solid electrolyte material to react with H^+ ions in the anode to produce water (**Fig. 1**). The high operating temperatures of solid oxide fuel cells are a major impediment to their widespread use in power generation. Thus, reducing this operating temperature is currently a major materials research goal, involving the search for novel electrolytes as well as active catalysts for electrode kinetics (oxygen reduction and hydrogen oxidation).

Among oxide-ion conductors, those of anion-deficient fluorite structures such as yttria-stabilized zirconia (YSZ), $xY_2O_3:(1 - x) ZrO_2$, are extensively used as electrolytes in SOFCs. Doping with Y_2O_3 is

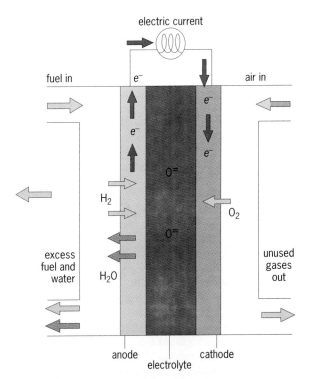

Fig. 1. Working scheme of a solid-oxide fuel cell. (*U.S. Department of Energy*)

known to stabilize the cubic fluorite structure of ZrO_2 and to supply the oxygen vacancies responsible for the ionic conduction. These materials are characterized by a large number of mobile oxygen vacancies, which are randomly distributed in the structure, and thus give rise to a completely disordered anion (oxygen) sublattice. Traditionally, the main strategy to reduce the operating temperature has been to search for novel electrolyte materials with larger oxide-ion conductivity values. Only recently has the use of artificial nanostructures appeared as a promising new direction for dramatically improved properties.

Doping. The common approach to increase the conductivity of bulk electrolytes is to chemically substitute elements that preferentially increase the concentration of the mobile species. There is often an optimum level of doping, beyond which the conductivity begins to decrease. At the optimal level, the only way to increase conductivity further is to raise the operating temperature to enhance the mobility of the defects. The figure of merit of $0.15 \, \Omega\text{cm}^2$ for the area-specific resistance (ASR, resistivity times thickness) for practical fuel-cell application requires increasing the bulk conductivity above $0.01 \, \text{Scm}^{-1}$ (the lower limit for 15-micrometer-thick electrolytes). While this limit is reached at $950°C$ for yttria-stabilized zirconia (YSZ) electrolytes, it has been achieved in the $500°C$ range using $Ce_{1-x}M_xO_{2-\delta}$ (M: Sm, Gd, Ca, Mn) or doped $LaGaO_3$ compounds in a category of fuel cells referred to as intermediate temperature (IT) fuel cells.

Nanotechnology. Nanotechnology is expected to have a major impact on the next generation of fuel cells because ionic transport is known to be strongly modified in nanomaterials. In fact, the term nanoionics has been coined to embrace the new concepts in ion transport and electrochemical storage, which result from confinement in nanostructures. Space-charge effects are an interesting family of size effects that show up when sample dimensions are comparable to the space charge (or Debye) length, the length scale of charge inhomogeneities. In ionic compounds, the accumulation of defects at surfaces or boundaries breaks charge neutrality and creates an electric field, which will be screened over the space-charge length by depletion or accumulation of mobile charges. The occurrence of space-charge effects in YSZ has been a subject of controversy in recent years. Contradictory results can be found in the literature, with some reports pointing to increased conductivity in nanostructured samples while others do not, evidencing the need for further work. In particular, the use of thin-film technology to design "artificial materials" to optimize fuel-cell materials is an interesting direction for exploring these issues.

Heterostructures. Thin-film growth techniques allow the combined growth of thin layers of different materials in heterostructures with a high crystalline perfection and very controlled properties. Several studies have reported modified ionic conductivity in thin films involving YSZ, although as in the case of bulk samples there is conflicting experimental

evidence with contradictory interpretations. Conductivity enhancements also have been reported in other related systems, such as heterostructures alternating gadolinia-doped ceria (ionic conductor) and zirconia (insulator) or superlattices alternating calcia-stabilized zirconia (CSZ) and insulating aluminum oxide. The importance of the disorder associated to structure misfit at the interfaces in promoting ion diffusion has been shown.

A strong enhancement of several orders of magnitude recently has been reported in heterostructures alternating nanometer-thick YSZ layers (with 8 mol% nominal yttria content and thicknesses ranging between 1 and 30 nm) with insulating strontium titanate ($SrTiO_3$, STO) 10-nm-thick layers grown by radio-frequency (RF) sputtering in a pure oxygen atmosphere. X-ray diffraction experiments display sharp superlattice peaks, confirming the layered structure. Scanning transmission electron microscopy (STEM) observations have shown that the layers are continuous and flat over lateral distances of a few micrometers (**Fig. 2a**). Despite

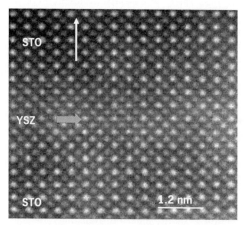

(a)

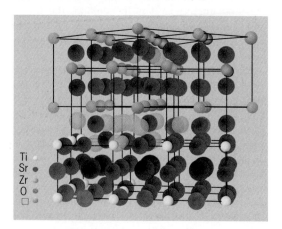

(b)

Fig. 2. Heterostructure with alternating YSZ and STO layers. (*a*) Z-contrast STEM image of the STO/YSZ interface of the [$STO_{10 \, nm}$/YSZ_{1nm}] superlattice (with nine repeats), obtained with VG Microscopes HB603U STEM. An arrow marks the position of the YSZ layer. (*b*) Solid-spheres model of the YSZ/STO interface, showing a 3D view of the interface and illustrating the compatibility of the perovskite and fluorite (rotated) structures.

the large difference in the bulk lattice constants ($a_{STO} = 0.3905$ and $a_{YSZ} = 0.514$ nm), YSZ grows epitaxially, rotated by $45°$ around the c axis and strained (7% in the $a - b$ plane) to match the STO lattice. A sketch showing the epitaxial matching of the two crystal structures is presented in Fig. 2b. Conductance of the YSZ/STO heterostructures, measured by using conventional impedance spectroscopy techniques in a lateral geometry, displays a noticeable increase as compared to the thick films or bulk samples, and the reported conductivity reaches the value of 0.01 Scm^{-1} at temperatures below $100°$C. Interestingly, conductance is essentially independent of the YSZ thickness and scales with the number of interfaces, indicating that it is an interface phenomenon which is attributed to the motion of oxygen ions. The decrease in the activation energy (0.6 eV) compared to bulk (1.1 eV) YSZ conductivity suggests that the oxygen ions at the YSZ/STO interfaces can move much more easily. By using atomic resolution electron microscopy and energy-loss spectroscopy, it has been possible to observe the atomic structure of the interface between the STO and YSZ layers, which suggested a highly disordered oxygen plane at the interface. The STO remained TiO$_2$-terminated with Ti in a 4+ charge state, and would favor the epitaxial growth of YSZ on STO despite their different crystalline structures (fluorite and perovskite, respectively).

The reported colossal ionic conductivity enhancement in strained epitaxial YSZ/STO heterostructures represents an interesting new pathway toward novel artificial electrolytes. Heterostructures between dissimilar crystal lattices may hold the potential to achieve ionic conductivity sufficiently high to enable fuel cells to operate much closer to room temperature.

For background information see ARTIFICIALLY LAYERED STRUCTURES; CATALYSIS; CRYSTAL STRUCTURE; ELECTROCHEMISTRY; EPITAXIAL STRUCTURES; FILM (CHEMISTRY); FUEL CELL; IONIC CRYSTALS; MESOSCOPIC PHYSICS; NANOTECHNOLOGY; SCANNING TUNNELING MICROSCOPE; SEMICONDUCTOR HETEROSTRUCTURES; SPACE CHARGE; SPUTTERING; SURFACE AND INTERFACIAL CHEMISTRY; X-RAY DIFFRACTION in the McGraw-Hill Encyclopedia of Science & Technology.

Javier Garcia-Barriocanal; Alberto Rivera-Calzada; Maria Varela; Zouhair Sefrioui; Enrique Iborra; Carlos Leon; Stephen J. Pennycook; Jacobo Santamaria

Bibliography. J. Garcia-Barriocanal et al., Colossal ionic conductivity at interfaces of epitaxial ZrO$_2$:Y$_2$O$_3$/SrTiO$_3$ heterostructures, Science, 321:676-680, 2008; X. Guo, Comment on colossal ionic conductivity at interfaces of epitaxial ZrO$_2$:Y$_2$O$_3$/SrTiO$_3$ heterostructures, Science, 324:465, 2009; X. Guo et al., Ionic conduction in zirconia films of nanometer thickness, Acta Mater., 53:5161-5166, 2005; C. Korte et al., Ionic conductivity and activation energy for oxygen ion transport in superlattices—the semicoherent multilayer system YSZ (ZrO$_2$ + 9.5 mol% Y$_2$O$_3$)/Y$_2$O$_3$, Phys. Chem. Chem. Phys., 10:4623-4635, 2008; I. Kosacki et al., Nanoscale effects on the ionic conductivity in highly textured YSZ thin films, Solid State Ionics, 176:1319-1326, 2005; S. J. Litzelman et al., Opportunities and Challenges in materials development for thin film solid oxide fuel cells, Fuel Cells, 8(5):294-302, 2008; B. C. H. Steele and A. Heinzel, Materials for fuel-cell technologies, Nature, 414:345-352, 2001.

Epigenetics and plant evolution

For almost a century, our understanding of evolution has been based on the modern evolutionary synthesis, also known as "neo-Darwinism." This paradigm assumes that natural selection is acting solely on the amount and structuring of heritable genetic variation, for which the ultimate origin is random mutation. Accordingly, genetic uniformity will severely constrain the adaptive flexibility of a given population or species, sooner or later resulting in evolutionary failure. During the 1990s, a challenge to this foundation was posed: several studies revealed that heritable phenotypic variation did not need to be based on variation in primary (coding) DNA sequences; instead, novel permutations of spatial and temporal patterns of gene expression could be achieved via a suite of noncoding changes, even in the complete absence of genetic variability.

Epigenetics. The term epigenetics literally means "in addition to, or outside of, conventional genetics," and it is now used to describe the study of stable (heritable) changes in gene expression that are not due to changes in DNA sequence. Epigenetic processes can involve regulatory RNAs or various biochemical modifications to DNA, as well as to the proteins closely associated with DNA that are able to control the activity of particular genes and transposons (a type of DNA that can move around within the genome, often causing mutations and changing the amount of DNA in the genome). By providing differential access to underlying genetic information, epigenetic signposts define how genetic programs can be differentially executed in both time (during development) and space (in different organs), usually by switching on and off or reducing the activity of specific genes. While the stability of epigenetic modifications through cell divisions has been demonstrated and studied extensively, there is now mounting evidence that in plants a great portion of the epigenetic signals cross generations to become part of the inherited information.

However, epigenetic states can become disrupted by random changes (mutations) during aging or under the influence of the environment (see **illustration**). How, then, does the genome integrate such developmental and environmental signals into its heritable information? Our rapidly growing understanding of epigenetic processes identifies several main classes of mechanisms, which are often interrelated and regulate gene activity in a complex, interactive fashion.

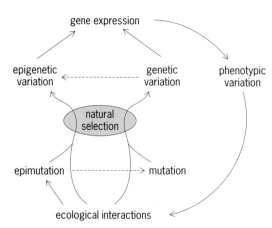

The complex interrelations between genetic and epigenetic evolutionary pathways. Dashed arrows indicate possible, but not necessary relationships. Both epigenetic and genetic variation are increased by (epi)mutations, but tend to be constantly reduced by natural selection as a direct result of ecological interactions. In addition, epimutations can also be triggered directly under the influence of the environment, which in this case will rather act toward increasing variation.

Methylation of DNA. Although all DNA nucleotides can be methylated, the most prevalent form of DNA methylation is cytosine methylation. Cytosine methylation is currently the best-described epigenetic mechanism and usually involves the addition of a methyl group to a CpG site (that is, a cytosine followed by a guanine in the DNA sequence) or to a CpNpG motif. CpG sites are often clustered in regulatory regions of genes, and their methylation results in reduced activity or even transcriptional silencing of the associated genes. Methylation is also efficient for suppression of transposon proliferation. The control of mobile elements such as transposons is a requirement, since their accumulation limits survival potential: more genomic material requires more nutrients (particularly nitrates and phosphates) for replication and slows down growth rates.

In plants, several genes with methylated epialleles (epigenetic variants of a gene) have been found to affect traits such as the morphology of flowers, vegetative and seed pigmentation, pathogen resistance,

and development (see **table**). For example, methylation and silencing of the *cycloidea* gene resulted in a peloric (radially symmetrical) variant of *Linaria* flowers, which arose approximately 250 years ago and has been inherited and maintained ever since.

In plants, transgenerational inheritance of DNA methylation appears to rely on several methyltransferase enzymes that replicate methylation patterns during both mitosis and meiosis. Several genes that encode methyltransferases and are involved in the initial establishment and maintenance of methylation patterns have been identified. These include, for example, MET (methyltransferase), CMT (chromomethylase), and DRM (domains rearranged methylase). DNA methylation is believed to be strongly tied to other epigenetic phenomena such as histone modifications and chromatin condensation (structural modification of chromatin from an expanded to a condensed form). In addition, cytosine methylation may have an impact as a mutagenic factor on the genetic level of variation because methylated cytosines have a high rate of spontaneous deamination to thymidines.

Chromatin remodeling. Remodeling of chromatin structure through chemical modification of histones (proteins closely associated with DNA) is another primary mechanism for controlling gene expression. Histones provide the core structures for chromatin packaging in eukaryotes, with each histone being wrapped in 146 base pairs (bp) of DNA to form the nucleosome and constructed from two copies each of four subunits. These subunits may be subjected to an array of covalent modifications (for example, acetylation, methylation, phosphorylation, and ubiquitination) that are interconnected at the regulatory level and make up a set of mechanisms for which concerted efforts control the condensation level of chromatin.

The most studied of the histone modifications is lysine acetylation, which is generally correlated with euchromatin (a more relaxed chromatin structure) and greater levels of gene transcription. Conversely, deacetylation results in suppression of expression as the chromatin is condensed from euchromatin to

Examples of meiotically transmitted epialleles in plants			
Organism	Locus/epialleles	Phenotype	Mechanism
Arabidopsis thaliana	*bal*	Dwarfism, but elevated disease resistance	Overexpression due to hypomethylation at some pathogen resistance genes
	fwa	Delayed flowering	Transposon-associated gain of function of a homeodomain gene
	sup	Abnormal floral organ number	Gene silencing due to DNA hypermethylation of coding sequence
Capsella bursa-pastoris	*CCA1* and *TOC1*	Flowering time variation	Differential expression of circadian core genes in the photoperiodic pathway
Linaria vulgaris	*Lcyc*	Peloric (radially symmetrical) flowers	Gene silencing due to DNA hypermethylation of coding sequence
Solanum lycopersicum	*Cnr*	Colorless fruits	Methylation in promoter region
Zea mays	*P-pr*	Reduced pigmentation	Gene silencing due to cytosine methylation of coding sequence

heterochromatin (a tightly packed form of DNA with limited transcriptional activity). Typically, these reactions are catalyzed by enzymes with histone acetyltransferase or histone deacetylase activity.

Small RNA molecules. Based on their structure and origin, as well as on the pathways in which they participate, the short (20–25 bp) noncoding regulatory RNAs are divided into two main classes in plants: small interfering RNAs (siRNAs) and microRNAs (miRNAs). Most small RNAs are processed from double-stranded RNAs (dsRNAs) by dedicated sets of enzymes and, once formed, regulate various biological processes, often by interfering with messenger RNA (mRNA) translation. They usually recognize their RNA targets by sequence-specific base pairing. When base pairing is perfect or nearly perfect, translation is repressed and the target mRNA is cleaved into fragments. In certain conditions, small RNAs may also activate gene expression; however, such mechanisms are currently not well understood. Regulatory processes mediated by small RNA molecules are much more rapid than stopping mRNA production, and they seem more flexible, as they are occasionally reversible.

Epigenetics and evolution. Epigenetic regulation is essential for development and differentiation of the various cell types in an organism, but it may have also extensive morphological, physiological, and ecological consequences at the population level. Expression levels of a gene can alone determine phenotypes that contribute to the natural variation on which selection works. A few recent studies have demonstrated the existence of some level of heritable variation in epigenetic marks within natural populations. Population bottlenecks, environmental stress, and ecological change may stimulate epigenetic modifications and novel phenotypes, partly via activation of transposons. Epigenetic processes may be highly relevant and opportune in the context of global environmental change.

By allowing for reversible phenotype variability, epigenetic variation could play a key role in natural variation, short-term adaptation, and evolution of plants in dynamic/disturbed environments. Several epialleles are expected to have ecologically significant fitness effects in the wild (for example, genes affecting pathogen resistance or vegetative pigmentation). Other epimutations may influence reproductive characteristics, such as breeding systems and pollination syndromes (for example, floral morphology and flowering time), and they may drive evolutionary divergence and speciation. Because epigenetic changes can regulate the level of gene expression, epialleles can produce continuous variation in phenotypes rather than discrete phenotypic classes. Modeling studies even suggest that epigenetic variation can facilitate jumps between fitness slopes by reducing genetic barriers represented by valleys in an adaptive landscape. Through modulation of phenotypes that trigger fitness differences and selection, as well as by affecting mutation rates and transposon mobility, epigenetic information can even have an impact on genetic variation.

This all suggests a need to reexamine most of our basic evolutionary assumptions and consider what epigenetic marks contribute to natural variation that is potentially visible to selection. However, at present, we are just beginning to understand the interrelations between epigenetic variation and phenotypic variation, selection, and ecological interactions in wild settings.

Polyploidy and hybridization. Epigenetic remodeling is especially prevalent after genomic stress, such as that caused by hybridization and polyploidization (genome doubling), when it accompanies structural changes. Polyploidy and hybridization are known to be prominent evolutionary forces and have occurred multiple times during the evolution of flowering plants, most probably starting with their origin. Most crops are more recent polyploids, and many are hybrids as well.

Epigenetic alterations after hybridization and polyploidy may include suppression/activation of transposable elements, gene silencing, and subfunctionalization (tissue-specific expression of gene copies), and they result in an effective and flexible way to control expression of redundant genes and stabilize the genome. Because of its potentially reversible nature, utilization of epigenetic regulation of duplicated gene expression could be advantageous relative to classic genetic mutations for adaptation in polyploids during evolution and development. Epigenetic effects are now also hypothesized to be an important source of phenotypic variation in invasive species, which are able to adapt to novel, sometimes stressful environments with limited genetic variation. Many of these species also have had hybridization in their background. The implications for stable epigenetic variation in plant genomes generating phenomena such as hybrid vigor and phenotypic plasticity remain to be fully appreciated.

For background information *see* ALLELE; CHROMOSOME; DEOXYRIBONUCLEIC ACID (DNA); GENE; GENETICS; GENOMICS; MUTATION; NUCLEOSOME; PLANT EVOLUTION; POLYPLOIDY; RIBONUCLEIC ACID (RNA); TRANSCRIPTION; TRANSPOSONS in the McGraw-Hill Encyclopedia of Science & Technology.   Ovidiu Paun

Bibliography. O. Bossdorf, C. L. Richards, and M. Pigliucci, Epigenetics for ecologists, *Ecol. Lett.*, 11:106–115, 2008; H. Grosshans and W. Filipowicz, The expanding world of small RNAs, *Nature*, 451:414–416, 2008; R. R. Rapp and J. F. Wendel, Epigenetics and plant evolution, *New Phytologist*, 168:81–91, 2005; E. J. Richards, Inherited epigenetic variation—revisiting soft inheritance, *Nat. Rev. Genet.*, 7:395–401, 2006; E. J. Richards, Population epigenetics, *Curr. Opin. Genet. Dev.*, 18:221–226, 2008.

Evolutionary patterns of the Ediacara biota

The animal kingdom experienced a massive diversification in the early Cambrian Period, between approximately 541 and 520 million years ago (Ma). During this 21-million-year interval, many animal phyla—each characterized by a unique body plan—diverged. This diversification event, dubbed the Cambrian explosion, has been a subject of intensive debate among bioscientists and geoscientists. In 1859, Charles Darwin speculated that the Cambrian explosion was an artifact due to the preservational failure of early animals that lacked hard skeletons. According to Darwin, animals evolved long before the Cambrian explosion, but they left no fossil record. In the past 150 years, paleontologists have thoroughly investigated many Precambrian successions in search of early animals. Among the many findings, the discovery of soft-bodied Ediacara fossils is arguably the most significant.

So far, paleontologists have described more than 120 Ediacara genera and nearly 200 species from nearly 40 localities, including the eponymous locality at Ediacara Hills in South Australia. These fossils are preserved in rocks dated between approximately 575 and 541 Ma, in the second half of the Ediacaran Period (635–541 Ma), which immediately precedes the Cambrian Period (541–488 Ma). Although evidence of life on Earth dates back to 3500 Ma and microscopic animals may have evolved in the early Ediacaran Period or possibly earlier, Ediacara fossils represent the oldest known macroscopic and morphologically complex life forms. As such, they may represent a prelude to the Cambrian explosion and may provide important insights into the early evolution of macroscopic animal body plans.

Phylogenetic interpretation. Despite the high taxonomic diversity and wide geographic distribution, the phylogenetic interpretation of many Ediacara fossils has been controversial. Traditionally, many Ediacara fossils have been interpreted as evolutionary precursors to Cambrian animals. For example, some frondose Ediacara fossils (*Charniodiscus*, **Fig. 1***a*; *Charnia*, Fig. 1*b*; *Rangea*, Fig. 1*c*) were interpreted as octocorals; segmented fossils (*Dickinsonia*, Fig. 1*d*; *Spriggina*, Fig. 1*e*) were interpreted as polychaetes; fossils with a headlike shield (*Parvancorina*, Fig. 1*f*) were interpreted as arthropods; and pentaradial fossils (*Arkarua*, Fig. 1*g*) were interpreted as echinoderms. However, a closer look at these Ediacara fossils reveals that they differ in detail from their supposed modern relatives. A radical alternative interpretation was proposed by the German paleontologist, Dolf Seilacher, who argued on taphonomic and functional morphological grounds that most Ediacara organisms were constructed from serially or fractally arranged tubular elements and that they represent an extinct kingdom (Vendobionta) that was phylogenetically distinct from the animal kingdom. Recent discoveries from Russia and Australia, though, suggest that some Ediacara organisms (for example, *Kimberella*, Fig. 1*h*) were bilaterally symmetrical, had anterior-posterior differentiation, and were capable of locomotion. These features place *Kimberella* firmly within the animal kingdom, perhaps representing early members of bilaterally symmetrical animals (bilaterians). This is significant because bilaterians represent the majority of animal diversity and are charac-

terized by important evolutionary innovations such as a central nervous system and directional locomotion. The existence of bilaterians in the late Ediacaran Period is also supported by shallow burrowing traces that were most likely produced by motile bilaterians with a muscular or hydroskeleton (fluid-filled body cavity) system.

Even though *Kimberella* may be a bilaterian animal, the phylogenetic affinity of many other Ediacara fossils remains uncertain. Some of them have been interpreted as microbes, algae, fungi, protists, or even lichens. It is plausible that the Ediacara biota represents a cross section of the biosphere in the late Ediacaran Period and may include a group of phylogenetically diverse organisms. In other words, the Ediacara biota does not need to represent a single phylogenetic clade (a taxonomic group containing a common ancestor and its descendants—for example, the Kingdom Animalia), and one cannot shoehorn all Ediacara fossils into either the Vendobionta or the Animalia. Hence, the Vendobionta as a phylogenetic concept may not be applicable to all Ediacara fossils.

Taxonomic diversity pattern. Although the Ediacara biota may not represent a clade, it is still possible to analyze its macroevolutionary patterns because such patterns often place constraints on underlying evolutionary processes and mechanisms. The most straightforward way to characterize macroevolutionary patterns is to quantify taxonomic diversity. On the basis of taxonomic composition and age, three Ediacara assemblages have been recognized: the Avalon (575–565 Ma), White Sea (565–550 Ma), and Nama (550–541 Ma). The Avalon assemblage is best represented by Ediacara fossils from the Conception Group in Newfoundland, Canada; this assemblage is restricted to deep-water environments and represents an early evolutionary stage with relatively low taxonomic richness (that is, a low total number of species). The White Sea assemblage, best seen in the Ediacara Member of South Australia and the Redkino Series in the White Sea area of Russia, shows a substantially greater taxonomic diversity and a significantly greater geographic and environmental distribution relative to the Avalon biota. The Nama assemblage is represented by fossils from the Nama Group of Namibia and shows a drop in taxonomic diversity. Most Ediacara taxa vanished before the Cambrian explosion; only a few Ediacara-like fossils have been found in the Cambrian Period. A simple compilation of taxonomic diversity shows that the White Sea assemblage has at least twice as many taxa as the other two Ediacara assemblages (**Fig. 2***a*).

The veracity of the taxonomic pattern, of course, depends on whether all Ediacara fossils have been exhaustively sampled and correctly identified. A better-sampled locality may appear to have more taxa than a poorly sampled locality. Fortunately, there are several statistical methods to correct such sampling biases, and the taxonomic pattern seems to hold even after the sampling biases are corrected (Fig. 2*a*). The taxonomic identification of Ediacara fossils, though, is not straightforward, and misidentifications are not

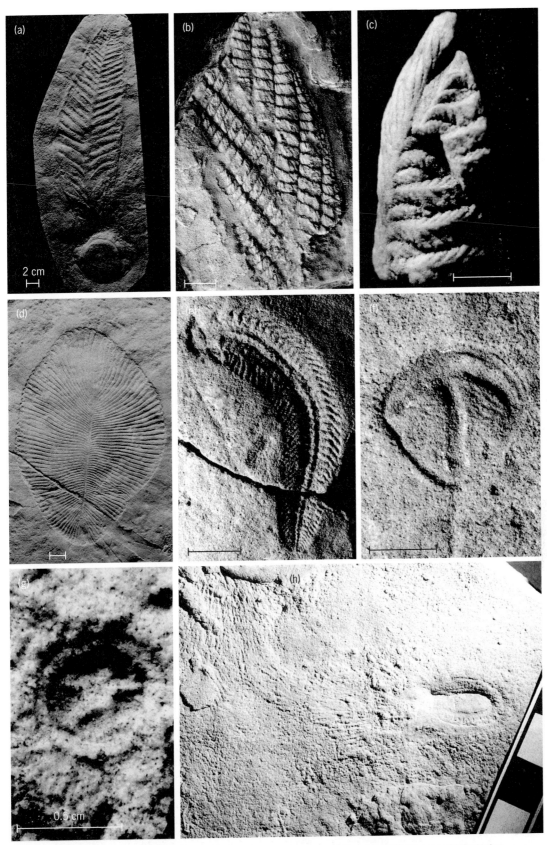

Fig. 1. Representatives of Ediacara fossils: (*a*) *Charniodiscus*, (*b*) *Charnia*, (*c*) *Rangea*, (*d*) *Dickinsonia*, (*e*) *Spriggina*, (*f*) *Parvancorina*, (*g*) *Arkarua*, and (*h*) *Kimberella*. Scale bars are 1 cm unless otherwise noted. (*Panels b and e–g courtesy of Jim Gehling and Marc Laflamme; panel h courtesy of Mikhail Fedonkin*)

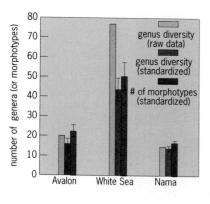

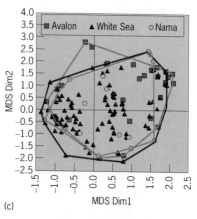

Fig. 2. Taxonomic diversity and morphospace range of the Ediacara biota. (*a*) The number of genera (raw numbers and those corrected for sampling intensity) and the number of morphotypes (corrected for sampling intensity) in the Avalon, White Sea, and Nama assemblages. (*b*) The construction of a database with rows of species occurrences and columns of characters. Entries are sorted by taxonomic names (column 1), which are used in compilations of taxonomic diversity. Species occurrences (column 2, with symbols representing the three Ediacara assemblages) are used in morphospace analyses. (*c*) The morphospace range of the three Ediacara assemblages from multidimensional scaling (MDS) analysis. Note that the three Ediacara assemblages have largely overlapping morphospace ranges.

always easy to correct. Sometimes, morphologically similar specimens that belong to the same species may be given different names if they are found in different localities and are studied by different scientists; this type of misidentification can be corrected by plotting the number of morphotypes (as opposed to named taxa), and it is encouraging that the taxonomic pattern still holds when this type of taxonomic misidentification is corrected (Fig. 2*a*). However, other types of taxonomic misidentification are more difficult to correct. For example, specimens of the same species, when preserved in different ways, may be misidentified as different taxa. Similarly, different ontogenetic (developmental) stages of the same species, when preserved in the fossil record, may be misidentified as different taxa. These taxonomic issues can be addressed only when more preservational and ontogenetic data are available. At present, these issues cannot be addressed adequately, limiting the significance of taxonomic compilations of Ediacara fossils.

Morphological pattern. Another way to analyze evolutionary patterns is to quantify the morphological history of Ediacara fossils, regardless of their taxonomic identifications. This method has certain advantages because it overcomes some of the aforementioned taxonomic problems and also because it can more objectively evaluate the "morphological distance" among Ediacara organisms. The method starts with the identification of a set of morphological characters that are used in the description of Ediacara fossils. The characters can be presence-absence characters or multistate characters. Ediacara fossils (typically at the species occurrence level) are then coded using these characters. For example, the presence or absence of a certain feature can be coded as 1 or 0, respectively, and a multistate character (for example, the number of digits) can be coded as 0, 1, 2, 3, and so on. This coding results in a data matrix (number of species occurrences by number of characters; Fig. 2*b*), which can then be analyzed using various multivariate statistical techniques (for example, multidimensional scaling analysis) to quantify the empirical morphospace (conceptual array of all possible morphologies) represented by the coded Ediacara fossils. Morphospace analysis of Ediacara fossils seems to show a pattern that is different from the taxonomic diversity pattern. Specifically, the morphological range represented by the Avalon assemblage is comparable to those occupied by the White Sea and Nama assemblages (Fig. 2*c*), which is in sharp contrast to the taxonomic diversity pattern showing that the White Sea assemblage is twice as diverse as the Avalon and Nama assemblages (Fig. 2*a*).

The veracity of the morphological pattern depends on whether Ediacara *fossils* truthfully preserve the morphologies of Ediacara *organisms* and whether the coded characters capture all morphological aspects of Ediacara organisms. Some morphological features may be distorted during preservation and others may not be preservable at all, although efforts can be made to code as many preserved morphological features as possible. Thus, the

Fig. 3. Schematic diagrams showing (*a*) the missing record of Precambrian animals (dashed lines) according to Charles Darwin and (*b*) the current phylogenetic interpretation of the Ediacara biota in the context of the Cambrian explosion of animals.

morphological pattern is significant only to the extent of preserved and coded morphological features.

If the morphological pattern of Ediacara evolution can be confirmed in future studies, it implies that the morphospace range was maximized during the Avalon explosion (575-565 Ma, approximately 30 million years before the Cambrian explosion), when taxonomic diversity was low. Furthermore, it did not expand during the White Sea taxonomic diversification; nor did it contract during the Nama taxonomic decline. Consequently, the Avalon and Nama morphospaces seem to have been more sparsely populated (species are more different from one another) than the White Sea morphospace. Intriguingly, this pattern of a rapid morphospace expansion followed by taxonomic diversification within a constrained morphospace seems to also characterize other diversification events including the Cambrian explosion.

Future outlook. The taxonomic and morphological patterns of Ediacara evolution also raise several important questions: What drove the rapid morphological expansion in the Avalon assemblage, what constrained the subsequent White Sea assemblage from further morphological expansion, and what allowed the Nama assemblage to retain the full range of morphospace despite diversity decline? Above all, what caused the extinction of most Ediacara taxa at the Ediacaran-Cambrian transition? The observation that rapid morphospace expansion followed by taxonomic diversification within a constrained morphospace may be a common evolutionary pattern indicates that the underlying evolutionary mechanisms are unlikely to be intrinsic to a specific clade. Instead, ecological feedbacks may have played an important role in driving the rapid morphospace expansion and, together with phylogenetic constraints, they may also have limited further expansion or contraction of an established morphospace until a major extinction took place. Such an extinction event may have occurred at the Ediacaran-Cambrian transition, setting the stage for the Cambrian explosion. At present, the magnitude, pattern, and drivers of this extinction are not well understood, although the necessary phylogenetic continuity of animals requires that at least some animal lineages must have crossed the Ediacaran-Cambrian boundary.

In closing, it is interesting to note that paleontologists, in their search for Precambrian animals envisioned by Charles Darwin (**Fig. 3a**), have discovered an unexpected diversity in the Ediacara biota. The Ediacara biota represents a distinct evolutionary stage characterized by many unique taxa that appear to have gone extinct at the Ediacaran-Cambrian transition. The evolution of the Ediacara biota seems to follow a general pattern of rapid initial morphological expansion and subsequent taxonomic saturation within an established morphospace. Although members in the Ediacara biota may ultimately share a distant common ancestor, they represent different branches in the tree of life, which was still in its young sapling stage. The twigs that would soon sprout into the Cambrian explosion must be buried within the Ediacara biota, among many other phylogenetically diverse branches (Fig. 3b).

For background information *see* ANIMAL EVOLUTION; BIODIVERSITY; CAMBRIAN; EDIACARAN BIOTA; FOSSIL; MACROEVOLUTION; PALEONTOLOGY; PHYLOGENY; PRECAMBRIAN; TAPHONOMY; TAXONOMY in the McGraw-Hill Encyclopedia of Science & Technology.   Shuhai Xiao

Bibliography. M. D. Brasier, *Darwin's Lost World: The Hidden History of Animal Life*, Oxford University Press, Oxford, U.K., 2009; M. A. Fedonkin et al., *The Rise of Animals: Evolution and Diversification of the Kingdom Animalia*, The Johns Hopkins University Press, Baltimore, 2007; M. F. Glaessner, *The Dawn of Animal Life: A Biohistorical Study*, Cambridge University Press, Cambridge, U.K., 1984; A. H. Knoll, *Life on a Young Planet: The First Three Billion Years of Evolution on Earth*, Princeton University Press, Princeton, NJ, 2003; G. M. Narbonne, The Ediacara biota: Neoproterozoic origin of animals and their ecosystems, *Annu. Rev. Earth Planet. Sci.*, 33:421-442, 2005; B. Shen et al., The Avalon explosion: Evolution of Ediacara morphospace, *Science*, 319:81-84, 2008; S. Xiao and M. Laflamme, On the eve of animal radiation: Phylogeny, ecology, and evolution of the Ediacara biota, *Trends Ecol. Evol.*, 24:31-40, 2009.

Exercise and cognitive functioning

Extensive research in humans and animals has established that exercise has beneficial effects on general health as well as on the health and functioning of the brain. For example, the benefits of exercise on cardiovascular health, obesity, diabetes, and cancer progression have been well described. Moreover, recent research has elucidated aspects of cognitive health and function that are improved by exercise, as well as mechanisms that might be mediating these cognitive benefits. In humans, the greatest benefits of exercise on cognitive function have been demonstrated in aged populations, in part because much human research focuses on interventions to counteract the normal cognitive decline that is associated with aging. The breadth of human exercise studies has expanded in recent years, revealing that beneficial effects of exercise on the brain extend across the life span, from the aged to young adults, adolescents, and children, and even to prenatal development.

Various aspects of brain health and function are improved by exercise. The systematic study of the benefits of exercise on brain function began in the early 1970s and revealed that older adults (65+ years of age) who participated regularly in physical activity had faster response speeds on simple tests of reaction time. These early findings suggested that more efficient processing was occurring in the brains of adults who exercised regularly, relative to their sedentary counterparts. Later studies (including randomized clinical trials) revealed that participation in exercise also benefits higher cognitive function, particularly aspects of higher cognitive function that decline with aging, such as learning and memory, and "executive control processes" (involved in multitasking, decision making, planning, attention, and dealing with distraction). In fact, while exercise training imparts relatively broad improvements across a variety of perceptual and cognitive processes, the benefits of exercise training appear to be most robust for executive control function. Interestingly, the benefits of exercise on cognitive preservation in normal aging may be more pronounced in women than in men, which is an emerging concept that is under active investigation.

In addition to improving cognitive function, recent evidence indicates that exercise has some therapeutic and preventative efficacy in depression. Randomized and crossover clinical trials demonstrate the efficacy of aerobic or resistance training exercise (2-4 months) as a treatment for depression in both young adults and older individuals. The benefits are dose-dependent, meaning that greater improvements are seen with higher levels of exercise, and appear to be similar to those achieved with antidepressant medication. However, it is likely that further research will reveal that the efficacy of exercise for treatment of depression may depend on the type and severity of depression, as well as the underlying cause of the depression.

Extensive research demonstrates that exercise also has neuroprotective effects on the brain. These effects have been best defined with respect to reducing brain injury and to delaying onset of and decline in several neurodegenerative disorders. For example, for individuals affected by stroke, poststroke therapeutic exercise accelerates functional rehabilitation. In addition, a number of studies have demonstrated that regular exercise participation throughout life delays the onset of and reduces the risk for Alzheimer's disease, Huntington's disease, and Parkinson's disease, and can even slow functional decline after neurodegeneration has begun.

Exercise has benefits across the life span. Although robust benefits of exercise on brain health and function have been documented in the aged, the effects of physical activity on human cognitive health during development and young adulthood have received relatively sparse attention. However, similar to benefits seen in the aged population, a number of studies have revealed a positive relationship between physical fitness and cognitive performance in school-age children (4-18 years of age) and young adults (18-25 years) for several cognitive categories. For example, in school-age children, aerobic fitness was positively related with achievement test performance, perceptual skills, intelligence quotient, performance on verbal and mathematics tests, and academic readiness. Interestingly, a beneficial relationship was not found between physical fitness and memory in children, in contrast to the benefits found in older adults. In young adults, there is evidence for improved cognitive flexibility (the ability to restructure knowledge in multiple ways depending on the changing situational demands) with fitness. However, unlike older adults, executive function was

not related to cardiovascular fitness in young adults. These results suggest that the relationship between greater cardiovascular fitness and some aspects of cognitive function emerges only later in life, after early adulthood. There are no human data yet on beneficial effects of exercise on prenatal development, but there is intriguing evidence from animal studies. Specifically, in rats, offspring of mothers that exercised during pregnancy showed higher brain weights, better learning and memory function, and more neurons in brain regions critical for learning and memory compared to offspring born to sedentary mothers. It will be important to determine if these animal findings extend to humans and to determine the extent of exercise participation that will have optimal effects on prenatal development.

Mechanisms for exercise benefits to brain health. Studies have been undertaken in both humans and animals to investigate the mechanisms by which exercise can provide benefit to brain health and cognitive function.

Human studies. Recent research in humans has focused on elucidating the mechanisms underlying the benefits of regular exercise participation on cognitive function and brain health. Human brain imaging studies have revealed that exercise slows the loss of brain tissue that normally occurs in aging. Specifically, cardiorespiratory fitness levels are associated with larger volumes of particular brain regions, including the hippocampus, prefrontal cortex, and parietal cortex. In addition, higher aerobic fitness levels are positively associated with whole brain volume in early-stage Alzheimer's disease, suggesting that exercise slows the loss of brain tissue in this progressive neurodegenerative disease. Intervention studies demonstrate that aerobic exercise not only slows loss of brain tissue, but it can actually reverse the decay by increasing brain volume of the prefrontal and lateral temporal regions. This finding was based on 6 months of exercise participation by healthy adults (60-79 years of age) who were previously sedentary. Importantly, the brain regions where exercise has structural effects are involved in the aspects of cognition that decline in aging patients and in patients with Alzheimer's disease, suggesting that slowing the deterioration of these brain regions slows loss of cognitive function. Imaging studies have also revealed that exercise prevents the declines in blood perfusion of the brain that occur with aging. In this case, the increased cerebral blood perfusion with cardiovascular fitness likely contributes to the preservation of brain tissue seen in exercising older adults. Imaging studies still have to be undertaken to assess whether or not similar effects of cardiovascular fitness and exercise on brain structure and cerebral perfusion occur in young adults or children.

Animal studies. In parallel with human studies, extensive animal studies have demonstrated beneficial effects of exercise on cognitive function and brain health, and have elucidated a number of mechanisms at biochemical, molecular, and electrophysiological levels that mediate the effects of exercise on the brain. Most exercise studies in animals assess the effects of several weeks of voluntary wheel running or forced treadmill running in rats or mice, using young, middle-aged, or old animals. Nearly all of the effects of exercise that occur in young animals are also seen in old animals.

One of the most reproducible effects of exercise in rodents is enhanced neurogenesis in the hippocampus, which is a brain region that is critical for learning and memory, as well as being most responsive to exercise. The hippocampus in the adult brain contains specialized stem cells that divide continuously through life, which then differentiate into neurons and other types of brain cells (neurogenesis). Exercise increases the generation and survival of new neurons in the hippocampus, and the integration of these new neurons into the hippocampal circuitry is thought to contribute to the beneficial effects of exercise on learning and memory, as well as in depression. In addition to enhancing neurogenesis, exercise stimulates synaptic complexity by increasing dendritic branching and numbers of synaptic spines (the sites of communication between neurons). Levels of many synaptic components that modulate synaptic efficacy are also increased in the brain following exercise, and electrophysiological experiments have demonstrated that exercise facilitates synaptic responses to such an extent that encoding of new information is facilitated and synaptic function becomes more efficient. These effects likely mediate the facilitation of learning that is observed in animals that have been allowed to exercise.

Increased neurogenesis and enhanced synaptic complexity and efficacy are paralleled by widespread growth of blood vessels (angiogenesis) in the hippocampus, cortex, and cerebellum, thereby providing increased nutrient and energy supply to the brain. Furthermore, exercise stimulates more efficient energy production in a number of brain regions, increasing levels of enzymes involved in glucose use and metabolism. Finally, levels of several classes of growth factors are increased in the brain in response to exercise. These growth factors modulate nearly all of the functional end points enhanced by exercise, suggesting that regulation of growth factors by exercise may constitute a common mechanism driving many of the benefits of exercise for the brain.

Healthy body, healthy mind. An emerging fundamental concept is that brain health and function are influenced by the general health of the body. Specifically, brain function is compromised by the presence of peripheral disease states such as cardiovascular disease, diabetes-related conditions (hyperglycemia, insulin insensitivity), and obesity, among other conditions. Remarkably, exercise reduces all of these peripheral risk factors for cognitive decline, in addition to the direct effects that exercise has on brain biology. Unfortunately, the prevalence of these peripheral disease states is on the rise in many countries, accompanied and exacerbated by a general decline in physical activity levels and exercise participation. This is affecting not only the adult population but also children, who are growing increasingly sedentary and unfit. The combination of increased sedentary lifestyle and poorer general health has significant deleterious consequences for brain health and function. In particular, combined sedentary lifestyle and poorer general health already at a young age is likely to compromise optimal cognitive function in children, setting the stage for earlier onset of cognitive impairments as well as faster cognitive decline later in life. Clearly, encouraging people of all ages to embrace exercise in their lifestyles is an important goal for maintaining cognitive health and function throughout the life span.

For background information *see* AFFECTIVE DISORDERS; AGING; ALZHEIMER'S DISEASE; BRAIN; COGNITION; INFORMATION PROCESSING (PSYCHOLOGY); LEARNING MECHANISMS; MEDICAL IMAGING; NEUROBIOLOGY; OBESITY; SPORTS MEDICINE in the McGraw-Hill Encyclopedia of Science & Technology.

Nicole C. Berchtold

Bibliography. S. Colcombe and A. F. Kramer, Fitness effects on the cognitive function of older adults: A meta-analytic study, *Psychol. Sci.*, 14(2):125–130, 2003; C. W. Cotman, N. C. Berchtold, and L. A. Christie, Exercise builds brain health: Key roles of growth factor cascades and inflammation, *Trends Neurosci.*, 30(9):464–472, 2007; Q. Ding et al., Exercise affects energy metabolism and neural plasticity-related proteins in the hippocampus as revealed by proteomic analysis, *Eur. J. Neurosci.*, 24(5):1265–1276, 2006; P. Heyn, B. C. Abreu, and K. J. Ottenbacher, The effects of exercise training on elderly persons with cognitive impairment and dementia: A meta-analysis, *Arch. Phys. Med. Rehabil.*, 85(10):1694–1704, 2004; C. H. Hillman, K. I. Erickson, and A. F. Kramer, Be smart, exercise your heart: Exercise effects on brain and cognition, *Nat. Rev. Neurosci.*, 9(1):58–65, 2008; A. F. Kramer, K. I. Erickson, and S. J. Colcombe, Exercise, cognition, and the aging brain, *J. Appl. Physiol.*, 101(4):1237–1242, 2006.

Experimental search for gluonic hadrons

While gluons play a very fundamental role in quantum chromodynamics, their role in ordinary matter is not easily seen. However, theoretical predictions indicate that there should exist a family of particles to which the gluons explicitly contribute to the observable behavior (gluonic hadrons). The GlueX experiment at the Thomas Jefferson National Accelerator Facility is being built to find these new particles and map out their properties. Over the coming decade, experimental measurements are expected that will directly check these predictions and test our understanding of the role of gluons in ordinary matter.

Quark confinement and QCD. Most of the visible mass of the universe is composed of protons and neutrons (nucleons), which are the particles that make up the cores of atoms. These nucleons are in turn composed of more fundamental particles known as quarks and gluons. Unfortunately, these small constituents appear to be forever confined inside the protons and neutrons. Understanding the confinement of quarks is one of the chief fundamental questions in physics. It is believed that the theory of quantum chromodynamics (QCD) can explain this confinement. However, an exact understanding of how QCD works has been extremely elusive. QCD is known to work under the extreme conditions found in high-energy particle collisions, but the current knowledge of what it is doing under conditions found in the everyday world is quite limited. Advances in high-speed computing and experimental facilities that are currently being built at the Thomas Jefferson National Accelerator Facility (JLab) in Newport News, Virginia, provide hope of making significant advances in answering this question within the next decade.

While protons and neutrons are the most obvious examples of particles made from quarks, they are only the lightest members of a rather large family of three-quark objects known as baryons. In addition to the baryons, there is a second family of particles, known as mesons, which consist of a bound state of a quark and an antimatter partner to a quark, an "antiquark." Taken together, the mesons and baryons are known as hadrons. While the study of both baryons and mesons has led to the current understanding of QCD, the simpler quark-antiquark ($q\bar{q}$) systems have historically provided a cleaner environment in which to compare theory and experiment. Thus, much of the remainder of this article will focus on what can be learned from the mesons.

Meson spectroscopy. The observation, nearly four decades ago, that mesons come in families with nine members, with each family being characterized by unique values of quantum numbers, led to the development of the quark model. In the quark model, mesons are $q\bar{q}$ bound states, while the quantum numbers represent properties of their quantum wave functions. The two relevant quantum numbers are parity, which is the correlation between the total spin of a meson and the behavior of its quantum wave function when reflected in a mirror, and charge conjugation, which relates to the interchange of a particle with its antiparticle. About one quarter of all the quantum numbers that could be observed are not possible for a simple $q\bar{q}$ system, and early observations yielded only mesons whose quantum numbers were consistent with a $q\bar{q}$ bound state. None of the non-$q\bar{q}$, or exotic, quantum numbers was observed.

The understanding of how quarks form mesons has evolved within QCD, and a richer spectrum of mesons that takes into account not only the quark degrees of freedom, but also the gluonic degrees of freedom, is now expected. Gluonic mesons with no quarks (glueballs) are expected (bound states of pure glue), but their unambiguous identification is complicated by the fact that they can mix with nearby $q\bar{q}$ states. Excitations of the gluonic field binding the quarks together can also give rise to new particles. These so-called hybrid mesons can be viewed as bound states of a quark, an antiquark, and gluons, where the gluons form a "flux tube" between the

Fig. 1. Results of a computer simulation showing the intensity of the gluonic field in a $q\bar{q}$ system. Between the quark and the antiquark, the intensity is confined to a narrow tubelike region, a flux tube.

quark and the antiquark (**Fig. 1**). The excitations of this flux tube lead to hybrid mesons. It is particularly interesting that some of these hybrid mesons can have exotic quantum numbers. The spectroscopy of these exotic hybrid mesons is simplified because they are easily separated from the conventional $q\bar{q}$ states.

GlueX experiment. In coming years, significant computational resources will be dedicated to understanding QCD, including confinement. These "lattice QCD" calculations solve QCD numerically on a small space-time grid (the lattice). To carry out these calculations requires some of the largest computer resources available, but to date that is the only known way to solve QCD and make predictions. The prediction (from lattice QCD) of the hybrid spectrum, including decays, will be a key part of this program. However, experimental data will be needed to verify these calculations. The spectroscopy of exotic mesons provides a clean and attractive starting point for the study of gluonic excitations. The Gluonic Excitations (GlueX) experiment at JLab is designed to collect high-quality and high-statistics data on the photoproduction of light mesons, with particular emphasis on identifying exotic hybrid mesons. The exotic mesons will be produced by a high-energy photon beam that collides with a proton (liquid hydrogen) target in the middle of the experiment. *See* LATTICE QUANTUM CHROMODYNAMICS.

Photon production. The accelerator at JLab produces beams of electrons. These are in turn used to

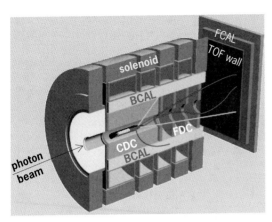

Fig. 2. GlueX detector. Components are described in the text.

produce photons by means of a process known as bremsstrahlung. The electron beam is passed through a very thin sheet of material in which some of the electrons scatter from the atoms in the material, emitting a high-energy photon in the process. For GlueX, this process is augmented by using a very thin diamond crystal. If crystal planes and the incoming electron beam are correctly aligned, an enhancement of photons at certain energies, much like a diffraction pattern in optics, occurs. Interestingly, not only is there a coherent enhancement at certain energies, but the enhanced photons have a high degree of linear polarization. It is also possible to detect the scattered electron and "tag" the emitted photon. All of these effects are needed by the GlueX experiment.

Detector. The GlueX detector is based on a solenoidal magnetic field that provides optimal detector coverage for this photoproduction experiment (**Fig. 2**). A tagged, linearly polarized photon beam with an energy of about 9 proton masses is incident on a liquid-hydrogen target. The target is surrounded by a small detector that is used to help identify when an event should be recorded. Next is the central drift chamber (CDC), used to reconstruct the paths of electrically charged particles. This is followed by a cylindrical barrel calorimeter (BCAL), which is used to reconstruct photons. Downstream of the CDC are four packages of the forward drift chambers (FDC), followed by a time-of-flight (TOF) wall. These latter, in conjunction with the CDC, allow for accurate measurements of the momentum and energy of charged particles. The most downstream element is the forward calorimeter (FCAL), which reconstructs the energy and direction of photons. Space has been reserved between the downstream end of the magnet and the TOF wall for a future particle identification system. This design provides for nearly complete detection of both charged particles and photons. This nearly complete acceptance is crucial for the analyses that will identify the exotic mesons.

Use of photons to produce exotic hybrids. At first glance, the choice of a photon beam to produce exotic hybrids seems odd. However, this choice has been driven by many factors that suggest exactly the opposite. Most models of exotic hybrids expect them to be excitations built on $q\bar{q}$ systems in which the intrinsic spins (like little tops) of the q and the $\bar{q}$ are aligned. Unfortunately, most particle beams that are available have the $q\bar{q}$ with their spins antialigned. The photon is unique in that it can momentarily fluctuate into a spin-aligned $q\bar{q}$ system. Based on this picture, models predict that it should be easier to produce exotic hybrids by using photons than with other methods (**Fig. 3**). In addition, very recent lattice QCD calculations show that in some cases the above assumption is indeed true. Finally, very little data exist from experiments using photon beams; thus, the reactions that will be explored in the GlueX experiment have not been studied before. All of these circumstances combine to make the search that will be carried out by GlueX very interesting.

Analysis of results. To complicate matters, the way in which the very-short-lived exotic hybrids fall apart (decay) into detectable particles makes the search for them difficult. Rather than appearing as narrow bumps in a distribution, they manifest themselves as a complicated distribution of particles in several dimensions. Thus, a sophisticated search is required to find the hybrids. The tools for doing these searches

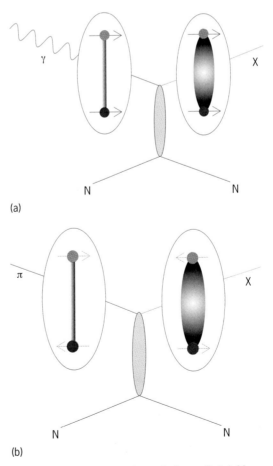

(a)

(b)

Fig. 3. Diagrams showing why producing exotic hybrids using photons should be easier than with other methods. (a) Spin-aligned $q\bar{q}$ pair in a photon interacting to produce spin-aligned $q\bar{q}$ pair in the experiment. (b) Spin-antialigned $q\bar{q}$ pair interacting to produce an antialigned pair.

are known as "amplitude analysis," but the scale of what will need to be done in GlueX dwarfs the efforts in previous experiments, both in the size of the volume of data and in the complexity of the analysis.

Expectations. Theoretically, there are expected to be three nine-member families of exotic hybrid mesons that are light enough to be accessible at JLab. The GlueX experiment hopes to map out the properties of as many of these mesons as possible. Simply being able to identify members from more than one family will provide an experimental measure of the properties of the excited gluonic field, and such results can be compared to the lattice QCD calculations. In addition to the existence of the particles, their modes of decay should provide insight regarding their internal quark structure.

Construction. Construction of the GlueX experiment started in 2009 as part of a larger upgrade to the JLab facility. Construction is anticipated to be completed in 2014, with full-scale data collection starting in 2015. Interesting results are anticipated after about a year of data collection. At the same time, computer calculations of lattice QCD are improving, and there is hope of obtaining both predictions from QCD and experimental results to compare with them by 2016.

For background information, *see* BARYON; BREMSSTRAHLUNG; ELEMENTARY PARTICLE; GLUONS; MESON; PARITY (QUANTUM MECHANICS); PARTICLE ACCELERATOR; PARTICLE DETECTOR; PHOTON; QUANTUM CHROMODYNAMICS; QUARKS; SPIN (QUANTUM MECHANICS); SYMMETRY LAWS (PHYSICS) in the McGraw-Hill Encyclopedia of Science & Technology.

Curtis A. Meyer

Bibliography. F. E. Close and J. J. Dudek, Hybrid meson production by electromagnetic and weak interactions in a flux-tube model, *Phys. Rev. D*, 69:034010, 2004; F. E. Close and P. R. Page, Glueballs, *Sci. Amer.*, 279(5):80–85, November 1998; V. Crede and C. A. Meyer, The experimental status of glueballs, *Prog. Part. Nucl. Phys.*, 63:74–116, 2009; A. R. Dzierba, C. A. Meyer, and E. S. Swanson, The search for QCD exotics, *Amer. Sci.*, 88(5):406–415, 2000; E. Klempt and A. Zaitsev, Glueballs, hybrids, multiquarks: Experimental facts versus QCD inspired concepts, *Phys. Rep.*, 454:1–202, 2007; C. Rebbi, The lattice theory of quark confinement, *Sci. Amer.*, 248(2):54–65, February 1983.

Fermi Gamma-ray Space Telescope

Cosmic high-energy gamma rays are produced by a diverse array of sources in the universe that signal the presence of natural particle accelerators. Among these are distant cores of active galaxies and the mysterious gamma-ray bursts, pulsars and cosmic-ray interactions in our Galaxy, and the Sun and Moon in our solar system. The recently launched *Fermi Gamma-ray Space Telescope*, with unprecedented sensitivity and angular resolution, is discovering hundreds of new sources of cosmic gamma rays with

the hope of probing the inner workings of the high-energy accelerators and possibly uncovering the nature of dark matter.

Fermi mission and the Large-Area Telescope. Launched on June 11, 2008, from NASA's Kennedy Space Flight Center (formerly Cape Canaveral Air Force Station), with an expected lifetime of 5–10 years, the *Fermi* telescope is exploring the universe at high energy, not only providing a deeper knowledge of what was already observed by its successful predecessor EGRET (on the *Compton Gamma-Ray Observatory*), but also opening a new and important window on a wide variety of phenomena. The Large-Area Telescope (LAT), the primary instrument on the Fermi Gamma-ray Space Telescope mission, is an imaging, wide-field-of-view, high-energy gamma-ray telescope, covering the energy range from below 20 MeV to more than 300 GeV (corresponding to wavelengths from above 6×10^{-14} m down to less than 4×10^{-18} m). The LAT was built by an international collaboration with contributions from space agencies, high-energy particle-physics institutes, and universities in France, Italy, Japan, Sweden, and the United States.

Gamma rays, photons with extremely high energy, cannot be reflected or refracted. The LAT is therefore a pair-conversion telescope, consisting of a 4×4 array of 16 modules, each of which consists of a tracker and a calorimeter (**Fig. 1**). Each tracker is a stack of trays, each containing both high-Z (high-atomic-number) material (tungsten foil) and a layer of sensitive detectors (silicon strip sensors). A segmented anticoincidence detector covers the tracker array, and shields the detector from the charged-particle flux

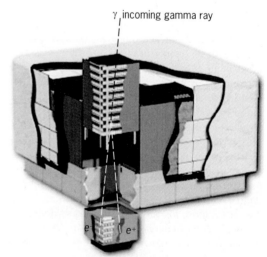

Fig. 1. Schematic view showing how the *Fermi* Large-Area Telescope (LAT) works. Each of the 16 "tower" modules consists of a tracker and a calorimeter. Each tracker is a stack of trays, containing both high-Z material (tungsten foil) to convert a gamma ray into an electron-positron pair (e^-, e^+) and sensitive detectors (silicon strip sensors) to track the pairs and reconstruct the incident gamma direction. Below each tracker, a cesium iodide (CsI) calorimeter measures the energy of the electron-positron pair. A segmented anticoincidence detector surrounds the towers and identifies cosmic charged particles among gamma rays.

that outnumbers the flux of gamma-ray photons by a factor of 10,000. An incoming charged particle releases a signal in the surrounding anticoincidence detector, and is identified by the internal trigger logic. In a gamma-ray event, the gamma ray enters the detector and probably strikes a high-Z atom, producing an electron-positron (e^+, e^-) pair. Each of these particles travels inside the tracker, giving rise to a certain signal level in each detector layer and thereby producing a track. The electron-positron pair will then end its travel in the calorimeter where the energy of the events is measured. The onboard software, combining the information from all the subsystems, does a detailed analysis of the event; candidate gamma rays are identified. A more detailed analysis is done on the ground, extracting the final information on selected events. Thus, for each event, the LAT is able to measure the direction, energy, and arrival time of the gamma ray. Thanks to the large field of view, the LAT is able to view the entire sky every two orbits (3 h), providing a complete picture of the gamma-ray sky.

Galactic sources. These include pulsars and their nebulae, x-ray binaries, and globular clusters.

Pulsars and their nebulae. The pulsed emission from rotation-powered pulsars, compact neutron stars spinning with periods between a few milliseconds and several seconds, is dominant at gamma-ray energies, and such pulsars are the most numerous gamma-ray sources in our Galaxy in the gigaelectron-volt (GeV) band. EGRET detected gamma-ray pulsations from six pulsars, using their known periods from radio and x-ray observations. Detecting sources in the Galaxy is often difficult, but pulsars radiate right into the region of *Fermi*'s greatest sensitivity between 1 and 10 GeV (corresponding to wavelengths from 1.2×10^{-15} m down to 1.2×10^{-16} m), and their pulsations act as beacons shining through the fog of the high gamma-ray background in the galactic plane. In just its first 6 months of operation, *Fermi* discovered pulsed gamma rays from an additional 40 pulsars and several new classes of gamma-ray pulsars (**Fig. 2**). Pulsars were originally discovered in the radio band (there are presently about 1800 radio pulsars). Owing to its increased sensitivity and angular resolution, *Fermi* has for the first time succeeded in discovering new pulsars through their gamma-ray pulsations alone. By June 2009, 16 new pulsars had been discovered through such a "blind frequency search." *Fermi* has also discovered a second new class of gamma-ray sources, millisecond pulsars, spinning at rates nearly high enough to cause the neutron star to break apart. Millisecond pulsars are thought to have acquired their short periods through being spun up by accretion of material from a binary companion star. For these pulsars, the pulsed gamma-ray luminosity is a large fraction of their total power output as a spinning magnetic dipole, and in its first several months of observations *Fermi* saw only the very nearby ones.

How is it that pulsars are producing such high-energy gamma rays that are radiated by particles accelerated to energies of 10^{13} eV, higher that any Earth-bound accelerator can achieve? Since neutron stars are rotating, extremely strong magnets, they should act like dynamos to generate very strong electric fields that can accelerate particles to the energy required to radiate gamma rays. One important clue that *Fermi* pulsar observations have provided is that the particle acceleration is apparently taking place far from the surface of the neutron star, where the speed of rotation approaches the speed of light. There, effects of special relativity twist the radiation into strange patterns that we see in the gamma-ray pulsations. We now think that the gamma rays are radiated into a broad-patterned beam filling nearly the whole neutron-star sky, while the radio beams are narrower. Since we are more likely to see the gamma-ray than the radio beams, many if not most pulsars may be radio-quiet for us and visible only at gamma-ray energies.

X-ray binaries. *Fermi* has discovered another new class of galactic GeV sources, binary stars emitting gamma rays that are modulated at their orbital periods. Named x-ray binaries since they were first discovered 4 decades ago at x-ray wavelengths, these sources can produce x-rays and gamma rays through either accretion of matter onto a neutron star or black hole or by interaction of the wind from a pulsar with the wind from its companion star. *Fermi* has detected periodic gamma rays from two x-ray binaries. One of them, LS I +61.303, emits strong variable emission from radio to x-ray energies and undergoes radio outbursts.

Globular clusters. *Fermi* has detected gamma-ray emission for the first time from a cluster of stars in our Galaxy (Milky Way Galaxy). The globular cluster 47 Tucanae contains many x-ray binaries and 23 known radio millisecond pulsars. *Fermi* has detected it as a bright gamma-ray source, and the emission is consistent with the combined emission from the cluster millisecond pulsars, even though the pulsed gamma rays have not yet been detected.

Cosmic-ray electrons. In addition to detecting gamma rays, *Fermi* has the ability to identify and measure the properties of the cosmic-ray leptons (electrons and positrons) that enter the detector. The first several months of operation have resulted in the best measurements of the local electron and positron cosmic-ray spectrum to date. An apparent excess at high energy may signal a local source of these particles in addition to the galactic cosmic rays that could be pulsars or a more exotic source such as dark matter.

Active galaxies. Present-day astronomy covers all wavelengths. *Fermi* data are thus combined with the observations from a series of facilities, in space and on the ground, to study the details of the emission from such sources as active galactic nuclei, galaxies that have a supermassive black hole in their core and are powered by an accretion disk.

Blazars. As largely anticipated, most of the high-galactic-latitude sources detected by *Fermi* are blazars, a type of active galactic nucleus well known to display extreme observational properties such as large and rapid variability, apparent superluminal motion, flat or inverted radio spectra, and large and

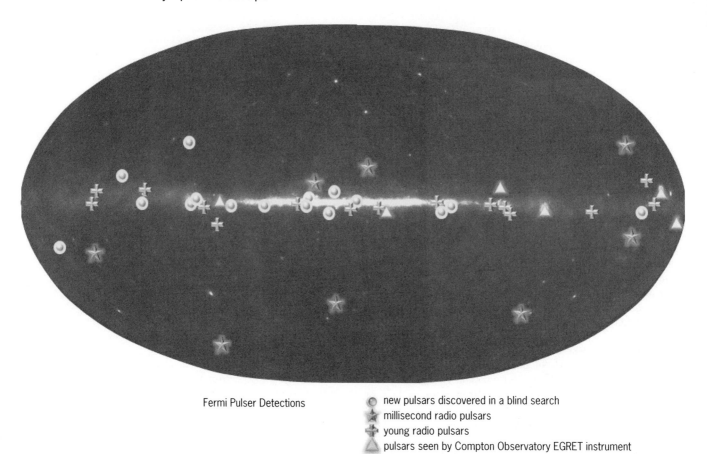

Fermi Pulser Detections

⊙ new pulsars discovered in a blind search
★ millisecond radio pulsars
✛ young radio pulsars
▲ pulsars seen by Compton Observatory EGRET instrument

Fig. 2. Gamma-ray pulsars detected by the *Fermi* telescope in its first 6 months of operation, superposed on an all-sky map of all gamma rays.

variable polarization. Blazars are thought to be objects emitting radiation from a relativistic jet that is viewed close to our line of sight, thus causing strong relativistic amplification.

Radio galaxies. These are thought to be "misaligned blazars," in which the jet structure is visible in radio-frequency emission as lobes. The *Fermi* LAT remarkably detected gamma-ray emission from the core (limited to a region of a few light-years in extent) of two radio galaxies, Centaurus A and NGC 1275. In analogy with blazars, the low-energy component is probably due to synchrotron radiation of relativistic electrons accelerated within the outflow, while Compton scattering by the same electrons is most likely responsible for the nonthermal x-ray and high-energy gamma-ray components.

Gamma-ray bursts and the Gamma-ray Burst Monitor. Gamma-ray bursts (GRBs) are highly energetic explosions signaling the death of massive stars in distant galaxies. The Gamma-ray Burst Monitor (GBM), the second instrument onboard the *Fermi* telescope, assists the LAT in the study of gamma-ray bursts, providing a full-sky monitor for this rapid transient signal in the 8 keV to 40 MeV energy range (corresponding to wavelengths of 1.5×10^{-10} m down to 3×10^{-14} m). The LAT and GBM can both independently detect gamma-ray bursts. The GBM connects the region of the spectrum in which the prompt emission from gamma-

ray bursts is well studied with the higher-energy and more poorly studied LAT range. Thanks to the LAT, the number of gamma-ray bursts detected at high energies doubled in the first few months of the mission, providing for the first time enough events at high energy to perform a detailed temporal-spectral analysis.

Within 10 months after its launch, the LAT instrument had firmly detected 8 gamma-ray bursts at energies above 100 MeV (corresponding to wavelengths less than 1.2×10^{-14} m). The extremely bright gamma-ray burst GRB 080916C, detected by the LAT, had more than 100 counts above 100 MeV. **Figure 3** shows the composite light curve of this powerful cosmic explosion. This burst represents a perfect template for studying gamma-ray burst phenomena, and should help us to understand how particles are accelerated by these powerful explosions. Photons with energies greater than 100 MeV were detected up to hundreds of seconds after the kiloelectronvolt-to-megaelectronvolt (keV-to-MeV) flux declined below detectability. Measurement of the redshift of a gamma-ray burst (which is used to derive its distance) also makes possible the determination of the isotropic luminosity, which is the energy released in gamma rays if the source were to emit with equal intensity in all directions. For this burst it corresponds to 4.6 times the solar rest energy. This high value supports the idea that the

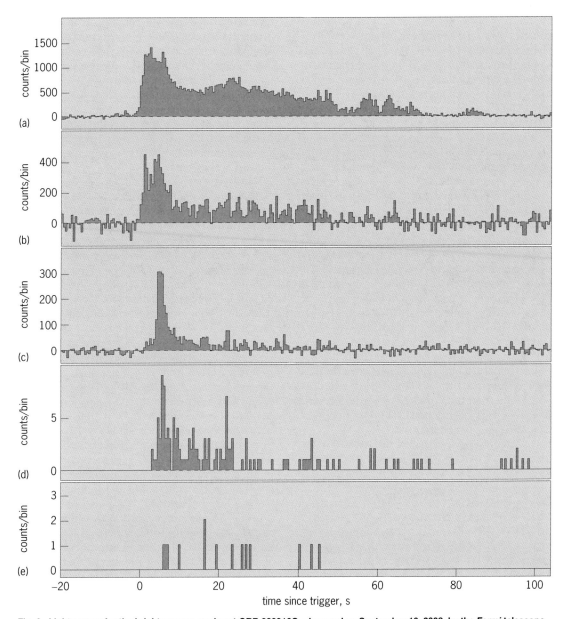

Fig. 3. Light curves for the bright gamma-ray burst GRB 080916C, observed on September 16, 2008, by the *Fermi* telescope. Each panel shows the detected signal as a function of time in a particular energy range, with the "zero" of the time scale determined by the GBM trigger. (*a*) Events in the lowest energy range of the GBM detector (8–260 keV). (*b*) Emission detected by the GBM in a higher energy range (260 keV to 5 MeV). (*c*) All events detected by the LAT. (*d*) LAT events with energy above 100 MeV. (*e*) LAT events with energy above 1 GeV. (Intensity is given in counts/bin, where the bin size is 0.5 s, so, for example, 10 counts/bin is equivalent to 20 counts/s.)

outflow powering this emission was actually collimated into a narrow jet. The high flux combined with the rapid observed variability implies the existence of a source region with high energy density, and energy from high-energy events is likely to be absorbed by pair production. Our observation of such events can be understood if the emitting regions are moving at very near the speed of light, as expected for an expanding shell of material (also called a fireball) from a massive stellar explosion. From the values of observed quantities, the motion of the relativistic expanding shell is in fact near the speed of light.

The high energy of the photons and the large distance of gamma-ray bursts can also test a prediction of some quantum gravity models that photons of different energy travel with different speeds, with high-energy photons traveling more slowly and therefore arriving later than low-energy photons. For gamma-ray bursts whose distance is known, high-energy events can be used to set a conservative upper limit on the delay, implying a robust lower limit on the mass that sets the scale of quantum gravity (which is of the same order of magnitude as the Planck mass). Using the *Fermi* observation of GRB 080916C, a lower limit only one order of magnitude smaller than the Planck mass was set, raising the previous limit obtained in this fashion by an order of magnitude.

Dark matter and exotic physics. About 80% of the mass in the universe is in a nonbaryonic form known as dark matter. The leading candidate for the dark matter is weakly interacting massive particles (WIMPs), postulated by supersymmetric particle physics theories. While WIMPs themselves would be difficult to detect, they could annihilate with each other to produce other particles, including neutral pions that decay into gamma rays. *Fermi* is searching for the gamma-ray signals of WIMP annihilation, but this poses a challenging enterprise. Most annihilation channels would produce gamma-ray continuum radiation, which would be very difficult to distinguish from other gamma-ray backgrounds. One annihilation channel would produce a gamma-ray line at the energy corresponding to the WIMP mass, which could be at several hundred GeV, in the *Fermi* sensitivity band.

Fermi is searching for the signatures of dark-matter annihilation in several of the most promising locations, including the center and halo of the Milky Way Galaxy and nearby dwarf satellite galaxies, and in the extragalactic background. The many small dwarf galaxies that surround the Milky Way Galaxy are thought to offer the best prospect since they have very low background and source confusion. Finding the continuum signal of dark-matter annihilation or the "smoking gun" annihilation-line emission will require more time than was available in *Fermi*'s first year of operation; indeed, it may take years for *Fermi* to either detect a signal from dark matter or provide a meaningful upper limit.

For background information *see* ASTROPHYSICS, HIGH-ENERGY; BINARY STAR; BLACK HOLE; COSMIC RAYS; DARK MATTER; ELECTRON-POSITRON PAIR PRODUCTION; GALAXY, EXTERNAL; GAMMA-RAY ASTRONOMY; GAMMA-RAY BURSTS; GAMMA-RAY DETECTORS; MILKY WAY GALAXY; NEUTRON STAR; PULSAR; QUANTUM GRAVIATION; RADIO ASTRONOMY; RELATIVITY; STAR CLUSTERS; SYNCHROTRON RADIATION; WEAKLY INTERACTING MASSIVE PARTICLE (WIMP); X-RAY ASTRONOMY in the McGraw-Hill Encyclopedia of Science & Technology.   Alice K. Harding; Nicola Omodei

Bibliography.   W. B. Atwood et al., The Large Area Telescope on the *Fermi Gamma-ray Space Telescope* mission, *Astrophys. J.*, 697:1071–1102, 2009.

Films of metal-containing surfactants

For many years, functional molecules with multiple controllable and responsive properties have been at the center of major research efforts. Such molecules may combine, for example, the properties of polymers with response to heat, as in thermoplastic materials, or the properties of dyes with response to electricity, as in electrochromic materials. Responsive materials that display enhanced characteristics are gaining in versatility and leading to numerous new applications. Considerable progress has been made toward supramolecular assembly at interfaces. We will discuss the formation of films of metallosurfactants. Self-assembled monolayers, metallopolymers, and dendrimers will not be discussed.

Classic surfactants. Surfactants are surface-acting agents with widely different applications due to their detergent, dispersant, emulsifying, and foaming properties. From hygiene products and cosmetics, drugs and drug-delivery vehicles, to concrete curing, automotive lubrication, and environmental remediation, surfactants are ubiquitous. What makes surfactants unique is that they contain a polar and water-soluble portion (the hydrophilic headgroup) along with an apolar and water-insoluble portion (the hydrophobic tail) [**Fig. 1***a*]. These molecules are called amphiphilic and serve a multibillion-dollar industry worldwide.

Monolayer films. Surfactants are sometimes called "schizophrenic molecules" because of their amphiphilic nature. This leads to competing forces acting to yield an energetically favorable conformation within a given environment (usually water). As a consequence, surfactants tend to self-assemble into aggregate structures, migrating to interfaces and associating with other neighboring molecules. Two limiting extremes can be drawn for these thermodynamically driven events. If the surfactants are soluble, spontaneous formation of molecular aggregates, such as micelles, takes place when a critical concentration is reached. If the surfactant is insoluble in water, a two-dimensional film will be formed at the air-water interface. Assuming that the thickness of the film is similar to that of a single molecule, it is called a Langmuir monolayer. The usual technique for evaluating these monolayers and their associated multilayers is a compression isotherm, which plots surface pressure versus the average molecular area at constant temperature and variable film area compressed between movable barriers in a trough. Figure 1*b* shows a schematic compression isotherm, from which information relative to the organization of the monolayer, the limiting area per molecule, and the area, pressure, and mechanism of monolayer collapse can be determined. Other interface-dedicated techniques, such as Brewster angle microscopy and fluorescence microscopy, allow for real-time evaluation of the monolayer compression at the air-water interface. The Langmuir monolayer can be transferred onto solid substrates, forming Langmuir-Blodgett films (Fig. 1*c–d*), which are analyzed by infrared spectroscopy, atomic force microscopy, and a host of other surface-dedicated techniques.

Metal-containing surfactants. The combination of amphiphilic behavior with the well-understood, controllable, and tunable properties of transition-metal complexes leads to new materials that exhibit interfacial organization, along with variable geometric, charge, redox, optical, and magnetic properties. Current design involves incorporation of the metal ion into the polar headgroup of the amphiphile by means of chemical bonding. The understanding of the cooperativity between transition-metal ions and amphiphilic organic scaffolds in the resulting metal-containing surfactants, or metallosurfactants, has become desirable because of its relevance to potential high-end technological applications such as catalysis, templating of mesoporous materials, analyte sensing, and molecular electronics.

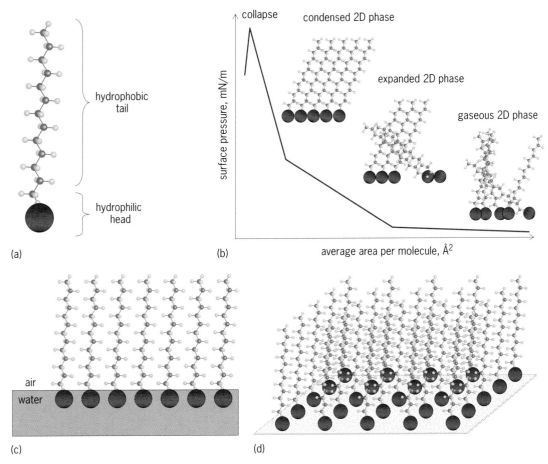

(a)

(b)

(c)

(d)

Fig. 1. Schematic representation of (a) a surfactant, (b) a compression isotherm plotting surface pressure versus area per molecule, indicating increased order of the monolayer, (c) an ordered Langmuir monolayer at the air–water interface, and (d) a Langmuir-Blodgett film transferred onto a solid substrate.

The origin of this approach unquestionably resides on the observation that inclusion of metal ions, such as sodium, calcium, and cadmium, in the aqueous subphase enhanced the order of lipid monolayers. More recently, amphiphilic ligands with specific chelating properties have been synthesized, and transition-metal complexation prior to film formation has been favored.

On one hand, the metal ion can exhibit preferential geometries that favor octahedral or tetrahedral headgroups, or foster distinctive protonation status. On the other hand, the denticity (number of donor atoms that form a chemical bond with the metal) and the rigidity of the ligand are also relevant. Therefore, the final topology and properties of a given metallosurfactant are dictated by the nature of both the metal ion and the amphiphilic ligand, influencing the properties of the resulting films.

Work on monolayer films began with metallosurfactants containing ruthenium and other inert metals. Illustrative cases can be drawn from the research by the groups of V. W.-W. Yam and M.-A. Haga. Initially interested in catalysis and optical sensor work, Yam and coworkers developed a series of ruthenium(II) and rhenium(I) surfactants of general formula $[Ru(bpy)_2L]^{2+}$ (**Fig. 2a**) and *fac*-$[Re(CO)_3(bpy)(L)]^+$, where L denotes several amphiphilic ligands containing C_{14}- and C_{18}-alkyl or

-alkoxo groups attached to pyridine, bipyridine, and diazafluorene. These systems were characterized by multiple spectroscopic techniques and had their isothermal compression and Langmuir-Blodgett film formation properties investigated. The Yam group introduced the use of second-harmonic-generation

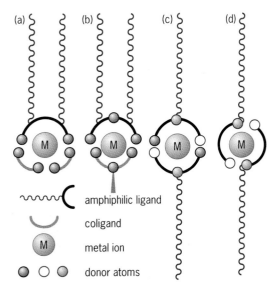

(a)   (b)   (c)   (d)

〜〜〜〜 amphiphilic ligand

◡   coligand

Ⓜ   metal ion

● ○ ●   donor atoms

Fig. 2. Schematic representation of several (a–d) metal-containing surfactants.

(SHG) spectroscopy as a tool to analyze such films and, more recently, has demonstrated the feasibility of dioxygen sensing elements containing single-layered structures of luminescent indicators of ruthenium(II) bipyridyl complexes on glass surfaces. The Haga group has also focused its approach on ruthenium surfactants based on terpyridine ligands. Taking advantage of the metal inertness, a stepwise approach allowed the generation of the terpiridine complex $[(C_{18}\text{-terpy})Ru(\text{terpy-}PO_3H)]^+$ (Fig. 2b), with enhanced hydrophobic ($-C_{18}$) and hydrophilic ($-PO_3H$) components that lead to highly ordered films. Currently, the group is exploring the self-assembly of multinuclear complexes on indium-tin oxide (ITO) electrodes. In the last few years, the C. N. Verani and coworkers has focused on the development of redox-active metallosurfactants. The redox response in these systems includes the metal center and is extended to functionalities, such as phenolates, incorporated into the design of the amphiphilic ligands. The work is dedicated to a closer look at the film formation and redox response of such metallosurfactants. Significant contributions include the establishment of archetypical modeling as a valid experimental and computational tool for understanding the coordination sphere around the metal in asymmetric metallosurfactants with $[N_{pyridine}N_{amine}O_{phenolate}]$ donors (Fig. 2c) and the demonstration that rigidity of the amphiphilic ligand in such cobalt(III) surfactants leads to distinctive collapse mechanisms. Studies focusing on copper-containing species with bidentate $[N_{amine}O_{phenolate}]$ donors (Fig. 2d) have shown a subtle equilibrium between amphiphilic and redox properties, where enhancement of one property may lead to decreased activity of the other. In a study on $[Cu(L^{PyCn})X]$ amphiphiles with L^{PyCn} pyridilmethylamines with C_{18}, C_{16}, C_{14}, and C_{10} chains, and X being monodentate chloro or bromo coligands, the unequivocal formation of patterned films was observed. The resulting patterns at the air-water interface depend on the chain length, as well as the nature of the coligand. Complimentary to the role of amphiphile rigidity, it was also demonstrated that different metals bound to the same amphiphilic ligand lead to distinctive collapse mechanisms.

More complex Langmuir-Blodgett film structures through the inclusion of metal clusters, the development of hierarchical linear materials, and the development of tridimensional networks have been reported. The development of such extended systems affords multiple metal centers in close proximity, yielding several magnetic-coupling schemes and oxidation states. These materials become relevant for efforts in molecular magnetism. Verani and collaborators published the synthesis of a family of metallosurfactants incorporating tridimensional $[Cu_4]$ clusters described as $[L_2Cu_4(\mu_4\text{-}O)(\mu_2\text{-}Y)_4]$, where L is a ligand containing an $[N_2O]$ donor set and two terminal C_{18} chains and Y is a bidentate acetate or benzoate coligand (**Fig. 3a**). These species had their structures solved by x-ray crystallography and displayed antiferromagnetic coupling, along with monolayer for-

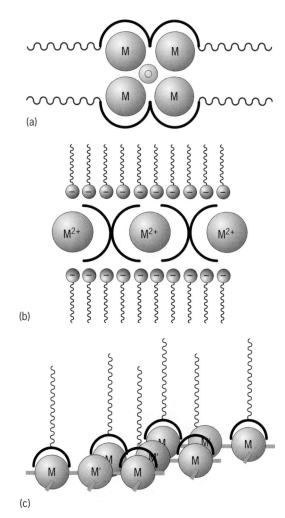

(a)

(b)

(c)

Fig. 3. Schematic representation of complex metal-containing surfactant systems. (*a*) Cluster containing surfactant. (*b*) Polyelectrolyte-amphiphile system. (*c*) Two-dimensional network.

mation. Modest collapse pressures were observed, but films resulting from transfer onto silica displayed good order when analyzed by atomic force microscopy. Another approach developed by D. G. Kurth and coworkers takes advantage of the catenation properties of ditopic bis-terpyridine ligands to build a linear $[M\text{-}L\text{-}M\text{-}L\text{-}M]^{+n}$ scaffold based on iron(II) metal ions. This positively charged, linear coordination polyelectrolyte can be neutralized with a negatively charged surfactant, such as dihexadecyl phosphate (Fig. 3b). The resulting polyelectrolyte-amphiphile complex is diamagnetic and can be transferred as a Langmuir-Blodgett film onto a solid substrate. Heating causes the alkyl chains to relax and results in distortion of the geometry around the iron centers, triggering a reversible transition to a paramagnetic state. The diamagnetic-to-paramagnetic transition is reversible and provides extensive control of structure and function from molecular to macroscopic length scales.

While polyelectrolyte-amphiphile systems take advantage of ionic interactions, the approach developed by D. R. Talham and collaborators is based on sequential coordination bonds between nickel(II) ions dissolved in the aqueous subphase and the Langmuir monolayer of a pyridine-based surfactant containing pentacyanoferrate(III) as the head group (Fig. 3c). This interaction leads to the formation of a two-dimensional cyanide-bridged iron-nickel network at the air–water interface that is transferred onto solid supports as either monolayer or multilayer films. Characterization of the films revealed a face-centered square grid structure that undergoes a transition to a ferromagnetic state at temperatures below 8 K. Subsequent work on this approach allowed for the construction of tridimensional networks.

Outlook. The formation of responsive, ordered films that result from combinations of amphiphilic designer ligands and metal ions is promising. Still in its infancy, the field has witnessed major developments, strongly indicating that the understanding of how metallosurfactants can drive complex interfacial phenomena, such as controlled morphological changes, domain formation, and collapse mechanisms, aiming at device development would constitute a major accomplishment.

For background information see COORDINATION CHEMISTRY; FILM (CHEMISTRY); LIGAND; MAGNETOCHEMISTRY; MICELLE; MONOMOLECULAR FILM; SUPRAMOLECULAR CHEMISTRY; SURFACE AND INTERFACIAL CHEMISTRY; SURFACTANT in the McGraw-Hill Encyclopedia of Science & Technology.

Cláudio N. Verani

Bibliography. Y. Bodenthin et al., Inducing spin crossover in metallo-supramolecular polyelectrolytes through an amphiphilic phase transition, *J. Am. Chem. Soc.*, 127(9):3110–3114, 2005; B. W.-K. Chu and V. W.-W. Yam, Synthesis, characterization, Langmuir–Blodgett film-forming properties, and second-harmonic-generation studies of ruthenium(II) complexes with long hydrocarbon chains, *Inorg. Chem.*, 40(14):3324–3329, 2001; J. T. Culp et al., Supramolecular assembly at interfaces: Formation of an extended two-dimensional coordinate covalent square grid network at the air-water interface, *J. Am. Chem. Soc.*, 124(34):10083–10090, 2002; J. A. Driscoll et al., Interfacial behavior and film patterning of redox-active cationic copper(II)-containing surfactants, *Chem. Eur. J.*, 14(31):9665-9674, 2008; F. D. Lesh et al., On the effect of coordination and protonation preferences in the amphiphilic behavior of metallosurfactants with asymmetric headgroups, *Eur. J. Inorg. Chem.*, pp. 345–356, 2009; K. Wang et al., Effect of subphase pH and metal ion on the molecular aggregates of amphiphilic Ru complexes containing 2,2′:6′,2″-terpyridine-4′-phosphonic acid at the air-water interface, *Langmuir*, 18(9):3528-3536, 2002.

Fischer-Tropsch synthesis

The Fischer-Tropsch synthesis is composed of three subsystems: gasification, Fischer-Tropsch synthesis (FTS), and upgrading to commercial products

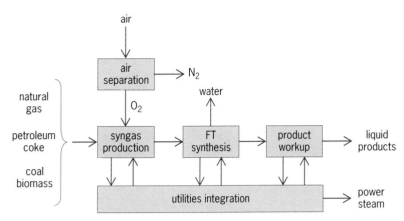

Fig. 1. Schematic of the overall Fischer-Tropsch process.

(**Fig. 1**). Based on the initial investment, the gasification system represents about 50% of the investment cost, with FTS representing 35% and upgrading 15%. The percentage of the investment for the gasification step increases to 65–70% for coal, with the remaining two systems divided in the same ratio as for natural gas. About 1.5% of the coal that is produced worldwide is gasified, and about half of this is used by Sasol (South African Synthetic Oils) FTS plants. It is expected that the cost for gasification of biomass will be similar to, or even higher than, that for coal, depending upon the biomass used.

Gasification. Today, for Fischer-Tropsch synthesis, two types of syngas [carbon monoxide (CO) and hydrogen (H_2)] are produced from coal: one produces a H_2/CO ratio near 2.0 or greater, and the other produces a H_2/CO ratio in the range of 0.6–1.2. The autothermal reforming (ATR) of natural gas produces a H_2/CO ratio of about 2.0. For example, the Sasol plants in Sasolburg and Qatar now use ATR to generate their syngas for use with iron and cobalt catalysts.

Today, only four coal gasification processes have been used for large-scale FT processes. The one that has been most widely used is the Sasol-Lurgi gasification process. This process produces about 10 wt.% coal tar products and a syngas with a H_2/CO ratio of 2.0 or higher. The Sasol iron catalyst has been widely used at the commercial level for more than 50 years, using a syngas with a ratio of H_2/CO of about 2.0. Thus, the syngas generated using the Sasol-Lurgi gasifier is appropriate for a process that uses either an iron or a cobalt catalyst.

The other gasification processes [Shell, GE (formerly known as Texaco), and Conoco-Phillips E-Gas] produce a syngas with a lower H_2/CO, in the range of 0.6 to 1.2. With these gasifiers, the syngas could be fed directly to the FT reactor following gas cleanup, since the iron catalyst exhibits sufficient water-gas-shift ($CO + H_2O \rightarrow CO_2 + H_2$) [WGS] activity to produce the additional hydrogen needed for the synthesis. However, for the cobalt catalyst, it would be necessary to install a WGS unit in front of the FT reactor to increase the H_2/CO ratio to approximately 2.15.

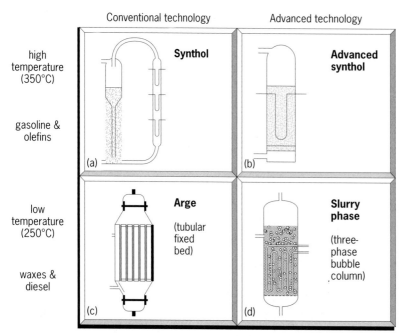

Conventional technology   Advanced technology

high temperature (350°C)

gasoline & olefins

low temperature (250°C)

waxes & diesel

Synthol

(a)

Advanced synthol

(b)

Arge

(tubular fixed bed)

(c)

Slurry phase

(three-phase bubble column)

(d)

Fig. 2. Four reactor types currently used at the commercial scale: (*a*) Synthol circulating fluidized bed, (*b*) fixed fluidized bed, (*c*) fixed bed, and (*d*) slurry bubble column.

Selectivity of Sasol processes (wt.% based on carbon)

Product	Tubular fixed-bed reactor	Synthol (fluidized bed)
CH$_4$	2	7
C$_2$ to C$_4$ olefins	2	24
C$_2$ to C$_4$ paraffins	2	6
Gasoline	17	36
Middle distillate	9	12
Heavy oils and waxes	67	9
Water-soluble oxygenates	1	6

SOURCE: B. Jager and R Espinoza, Advances in low temperature Fischer-Tropsch synthesis, *Catalysis Today*, 23(1):17–28, 1995.

Synthesis. For the Fischer-Tropsch reaction, there are two general synthesis process temperatures: low temperature (200–260°C) and high temperature (300–350°C). Today, both temperature ranges are used in commercial processes, and each has properties that are unique to it.

About 90% of Sasol's production uses the high-temperature process, as does nearly the entire production at PetroSA (formerly Mossgas). Until recently, Sasol operated circulating fluidized-bed reactors for its high-temperature FT operations, and PetroSA continues to operate this reactor type. Recently, Sasol switched its production from circulating fluidized-bed reactors to fixed fluidized-bed reactors (**Fig. 2**). Among the advantages of fixed fluidized-bed reactors are lower construction cost, higher reactor productivity, continuous use of all the catalyst inventory, lower catalyst consumption, and less abrasion, resulting in longer on-stream time. The products from the high-temperature reactors must all be in the gas phase at the reaction temperature. If this is not the case, then liquid (and solid) products will condense and cause agglomeration of the catalyst particles, which in turn causes loss of product. Thus, about 50% (molar, not weight %) of the products are in the C$_1$–C$_4$ gaseous range, and more of the liquid products are in the gasoline range and not in the more desirable diesel range.

The low-temperature fixed-bed reactor, as well as the more recent slurry bubble-column reactor, produces much heavier products (see **table**). In this process, heavy oils and waxes make up more than 65% (carbon based) of the total products, and these can be subsequently converted to transportation fuels with a ratio of about 80% diesel and 20% gasoline. Thus, the low-temperature FT produces a much higher amount of the desirable transportation fuel, diesel. The reasons that diesel is the desired transportation fuel are that it contains essentially no heteroatoms (especially sulfur) and has a very high (80 or greater) cetane number. Gasoline, while free of heteroatoms, has a low (about 65–75) octane number and is not as desirable a fuel.

There are two catalytic elements that are now considered for the FT synthesis at the commercial scale: iron and cobalt. The performance of each of these catalytic elements may be enhanced by adding other elements, called promoters. The early German work used a supported cobalt catalyst, but during World War II Germany started work to replace cobalt with iron. Following World War II, when commercial work started in South Africa and the United States, iron was the catalyst of choice.

Synthesis products have been found to follow a polymerization law that is expected for a C$_1$ monomer (**Fig. 3**). The products follow a distribution law known as the Anderson-Schulz-Flory distribution in the following equation,

$$\Phi_n = \Phi_1 \alpha^{n-1}$$

where Φ_n is the mole fraction of the product containing n atoms and α is the chain growth probability. When $\alpha = 0$, the only product is methane. When α is large, a high-molecular-weight wax is formed.

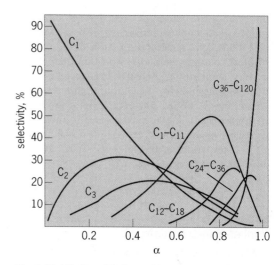

Fig. 3. Distribution of various carbon number fractions for various values of the value of α.

The perception is that the cobalt catalyst produces heavier products than the iron catalyst. However, Sasol reports an α value of 0.95 for its iron catalyst, and an unpromoted cobalt catalyst operating at normal FT conditions for the catalyst [for example, 220°C, 25 bar (2.5 MPa), and $H_2/CO = 2$] will have an α value of about 0.87, which is much lower than that for the iron catalyst. Products of the iron catalyst can be described by a so-called two-alpha plot, in which the low-α products have an α value of about 0.7 and the higher products have an α value of 0.95. Because the two-alpha value for the iron catalyst produces considerable amounts of light products, it may be possible for the cobalt catalyst to make more transportation fuel than the iron catalyst does for each quantity of carbon fed.

Upgrading to commercial products. The upgrading to fuels is accomplished by modifications of processes used in the conversion of crude oil to transportation fuel. The upgrading of the light products is illustrated by the process used at PetroSA, where the primary light FT products are produced using a high-temperature process. A silicalite (zeolite) catalyst is used to convert the gaseous olefins to heavier oligomers that fall in the gasoline and/or diesel range. For the heavy oils and waxes, some version of mild hydrocracking is used. The Qatar plant uses a hydrocracking process that was developed by Chevron, which includes a mild hydroisomerization and hydrocracking function of the catalyst under the conditions used.

Figure 4 shows that the upgrading will depend upon the FT temperature that is used. For the high-temperature FT operation, the amount of hydrocracking that is needed is minimal, if any, but for the low-temperature FT, hydrocracking will be a major operation. In contrast, the high-temperature FT oligomerization of the light gases will be a major operation, whereas it will make a minor contribution to the transportation fuels for the low-temperature FT operation.

While both upgrading processes are currently used in commercial operations, both need some improvements. For the oligomerization process, the gasoline fraction has too little branching to provide the high octane number needed for today's gasoline and too much branching to provide a high cetane number for diesel. For hydrocracking, A. de Klerk of the University of Pretoria indicated that one needs to consider the density-cetane number-yield, and that the fuel requirements for two of the three specifications for European diesel can easily be met. Thus, while the current refining processes lead to fuels that are environmentally acceptable, some areas remain where improvements are needed if FT fuels are to advance beyond being just a blending stock to meet environmental regulations.

For background information *see* CATALYSIS; CETANE NUMBER; COAL CHEMICALS; COAL LIQUEFACTION; DIESEL FUEL; FISCHER-TROPSCH PROCESS; FLUIDIZATION; GASOLINE; HYDROCRACKING; ISOMERIZATION; OCTANE NUMBER; PETROLEUM PROCESSING AND REFINING; PETROLEUM PRODUCTS; REFORMING

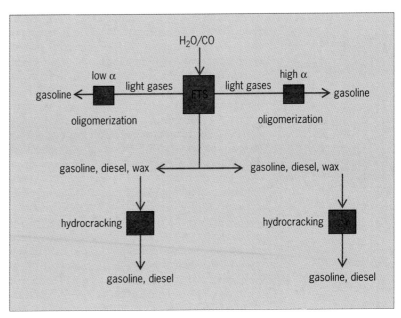

Fig. 4. Schematic for upgrading of FT products to commercial products.

PROCESSES; SYNTHETIC FUEL in the McGraw-Hill Encyclopedia of Science & Technology.   Burtron H. Davis

Bibliography. R. B. Anderson, *The Fischer-Tropsch Synthesis*, Academic, Orlando, FL, 1984; B. H. Davis and M. L. Occelli, *Fischer-Tropsch Synthesis, Catalysts and Catalysis, Studies in Surface Science and Catalysis*, vol. 163, Elsevier, Amsterdam, 2007; M. E. Dry, The Fischer–Tropsch process: 1950–2000, *Catal. Today*, 71(3-4):227–241, 2002; C. Higman and M. van der Burgt, *Gasification*, Elsevier, Amsterdam, 2003; B. Jager and R. Espinoza, Advances in low temperature Fischer-Tropsch synthesis, *Catal. Today*, 23(1):17–28, 1995; A. de Klerk, *Fischer-Tropsch Refining*, Ph.D. thesis, University of Pretoria, 2008; S. Lee, J. G. Speight, and S. K. Loyalka, *Handbook of Alternative Fuel Technologies*, CRC Press, Boca Raton, FL, 2007; A. P. Steynberg and M. E. Dry, *Fischer-Tropsch Technology, Studies in Surface Science and Catalysis*, vol. 152, Elsevier, Amsterdam, 2004; H. H. Storch, N. Golumbic and R. B. Anderson, *The Fischer-Tropsch and Related Synthesis*, John Wiley & Sons, New York, 1951.

Forensic dentistry

Forensic dentistry, or forensic odontology, can be defined as the use of dental knowledge in the judicial process. Forensic dentistry is best known for its usefulness in linking an identifying name to unknown human remains. This may be accomplished in single instances, such as when a body is discovered along a lake or river shore having presumably drowned. More than one victim may be involved, such as in a house fire or automobile accident. In some cases, referred to as multiple-fatality incidents (or mass fatalities, disaster victim identification), large numbers of deceased individuals may require identification. Recent examples include Hurricane Katrina in 2005, the Sumatra–Andaman earthquake and tsunami in 2004, the World Trade Center attack in 2001, or any of a number of common carrier incidents, including the 2009 crash of Continental Connection Flight 3704 in Buffalo, New York. Forensic dentists also assist in law enforcement cases involving suspected biting activity (animal or human) and in abuse cases involving head and neck injury. In addition, many forensic dentists act as expert witnesses in litigation involving professional liability claims against dentists and in personal injury cases involving the jaws or other oral structures.

Training. In the United States, forensic dentistry is not a specialty of dentistry recognized by the American Dental Association, such as oral surgery, orthodontics, or endodontics. Although virtually every dental graduate receives training in areas of dentistry basic to the understanding and practice of forensic dentistry, additional study and training is advised for those who label themselves as forensic dentists. Such training focuses not only on the dental aspects, but also on the various forensic sciences in general, death and crime scene investigation, and the judicial process. This training may be acquired over several years of study through continuing education courses, participation and mentoring in the field, and active participation in regional, state, and national forensic associations, such as the American Academy of Forensic Sciences. Training may be accelerated by enrolling in one of the several week-long courses focusing on forensic dentistry, such as those of the Armed Forces Institute of Pathology, the University of Texas Health Science Center at San Antonio, or the Miami-Dade Medical Examiner's Office. In addition, longer intensive programs leading to certificates or advanced degrees in the field are offered by some universities (such as the University of Texas Health Science Center at San Antonio and McGill University). The American Board of Forensic Odontology offers certification in the field of forensic odontology.

Identification. The identification of unknown human remains constitutes the majority of the cases handled by forensic dentists. As in most identification sciences, to accomplish the task, an unknown object must be compared to an object of known origin to determine the degree of similarity and to decide if the degree of similarity constitutes identification. In most cases, to establish dental identification (that is, to successfully allow the comparison to take place), first, there must be some hint as to the putative identity of the body or remains. For example, in a vehicle fire, the license number or vehicle identification number (VIN) may be researched to discover the identity of the owner, who may be presumed to be the driver or may be able to assist law enforcement in naming the possible driver. Similarly, the names of the inhabitants of a burned home are often easily obtained. In a drowning case or in the case of other human remains discovered, law enforcement can research the names of persons listed as missing, enabling the search area to be expanded to include more potential individuals. Second, in order to use this information, the names and locations of dental practitioners who may have treated the individual prior to death must be discovered. Once known, those records can be requested—the Health Insurance Portability and Accountability Act (HIPAA) exception [47 CRF 164.512(g)(1)] exists, as well as the subpoena power of the courts, coroners, and medical examiners to compel production, if required. Comparison of the dental characteristics present in the remains, including which teeth are filled, filled with which dental materials, which teeth have been extracted, and unusual positions of teeth, are compared with the dental characteristics recorded in the patient's dental record by the treating dentists. Most forensic dentists prefer to compare radiographs from the treating dentists' records to radiographs from the morgue before issuing an opinion concluding identification (**Fig. 1**). A nonidentification or exclusion also can be rendered, often with lesser amounts of dental information because certain findings are exclusive. For example, teeth that are known to have been extracted cannot reappear in the jaw, and teeth that are known to have been restored or filled cannot revert back to an unfilled or original state.

Situations in which large numbers of victims are killed may overtax the resources of the local, regional, or even state medical examiner/coroner systems. Recovery and identification of the victims is one of the top priorities in this type of disaster. Dental identification has long played a leading role because of the speed and accuracy with which identification can be established. Teeth, which consist primarily of inorganic materials, are resistant to decomposition. In addition, teeth are among the most durable structures in the human body and are able to withstand extreme thermal insult as well as physical force. As the investigating authorities determine who the missing (and therefore presumably deceased) individuals are, an organized effort to retrieve their antemortem dental information is combined with a systematic postmortem examination of the recovered dental structures. Forensic dentists analyze, often with the aid of computer comparison software, the findings to arrive at dental identifications. After Hurricane Katrina, the federal Disaster Mortuary Operational Response Team (DMORT) team was deployed to assist the state. Because of the destruction of many of the local dental offices in the storm-affected area, great difficulty was experienced in gathering antemortem dental records, resulting in fewer than 200 identifications based on forensic dentistry. Dental information, such as the presence of gold "caps" on front teeth, "crooked" front teeth, and spaces between front teeth, was often used to support tentative identifications resulting from other methods.

Age estimation. Providing an estimated dental or chronological age for an unidentified body may assist law enforcement agencies to narrow their search parameters in missing persons' or other databases. Forensic dentists can provide age estimates based on features of developing teeth and on changes in teeth that have already formed. This information may assist

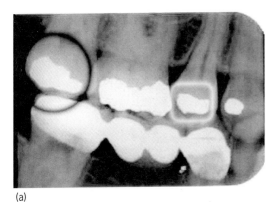

(a)

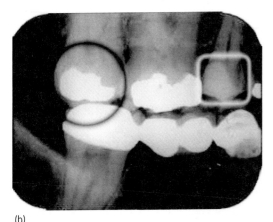

(b)

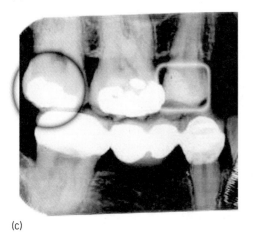

(c)

Fig. 1. A dental identification from a homicide case. (*a*) Radiograph of the decedent exposed 18 years prior to her murder. (*b*) Radiograph of the decedent exposed 1 month prior to death. (*c*) Radiograph of the skeletal remains that were thought to be her. Note that the dental restoration within the dark circle is unchanged over the 18-year span. The restoration within the light square in *a* was later replaced by a larger restoration visible within the light square in *b* and *c*. The dental bridge on the lower teeth remained unchanged throughout the 18-year span as well. In *c*, one upper tooth (to the right of the light square) was not recovered with the body; the socket for that tooth is evident.

medical examiners and others to discover the identities of those persons who may be difficult to identify because of skeletonization, decomposition, severe burning, or trauma. Forensic odontologists are also consulted by immigration and other agencies to assist in estimating the ages of persons who may be in the country illegally. In these cases the person's age is im-

portant since adults and juveniles are managed very differently in the legal system.

Recent developments in age estimation, such as determining the ratios of different forms (isomers) of amino acids found in teeth, allow for more accurate age estimations. When combined with a method of determining the levels of radioactive carbon-14 found in the enamel of teeth from past aboveground nuclear testing, the forensic dentist can give investigators good estimates of both the age at death and the date of birth of an individual. By calculation, the date of that person's death may also be discovered. Age estimation from teeth is another way that forensic odontologists assist other specialists to discover the answers to difficult scientific questions.

Bite-mark analysis. The other major field of forensic dentistry is bite-mark analysis. This has become a controversial subject recently because of several highly publicized cases in which individuals, whose conviction for crimes was based partially or entirely on bite-mark evidence, have now been exonerated through the use of DNA comparisons showing that other individuals were present at the crime scene. Bite-mark analysis is an identification discipline as are fingerprints, tool marks, and ballistics, to name a few. It is based on the knowledge and experience of the forensic odontologist, who, after studying the evidence, reaches a conclusion.

Contact with teeth or the act of biting may result in a discernable and distinct pattern injury in human skin. Teeth may also produce patterns in foodstuffs, such as chewing gum, fruit, cheese, or even nonfoodstuffs such as the steering wheel of a vehicle after an accident. Bite-mark analysis consists of answering two sets of fundamental questions. First, is the injury a result of a bite? If so, was the bite produced by human or animal teeth? And, can the patterned injury be documented sufficiently for further study, examination, and testing? If these questions can be answered positively, then a second set of questions can be addressed. Can the bite injury be linked to a particular individual as the contributor of the injury? Or perhaps just as important, can the bite injury exclude a particular individual or individuals as having been the contributor(s)? Also important for investigators to note, regardless of the answer to the second group of questions, is that if the answer to the first question is positive (that is, the injury is a human bite mark), then the area surrounding the bite mark should be subject to collection of DNA evidence, which can be used to link a suspect to the victim at the time of the infliction of the bite mark.

It is beyond the scope of this article to discuss in detail the methods of comparison and study, but in general the forensic dentist compares the teeth of the suspected biter(s) to the injury using one of several methods. Dental models (plaster or dental stone) acquired by legal means are required. The forensic dentist compares the size, shape, and arrangement of the teeth involved in the injury to the discrete injury patterns (**Fig. 2**). Overall visual comparison of the arrangement, including the absence of teeth, may be

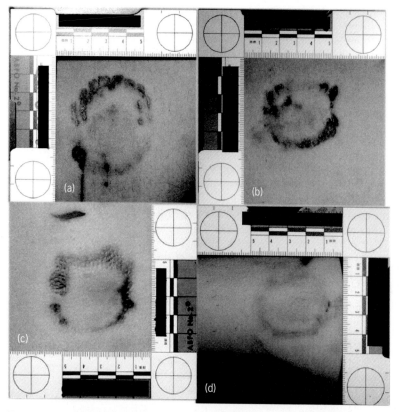

Fig. 2. Four human bite marks on human skin. The American Board of Forensic Odontology (ABFO) #2 scale is present in each photograph to assist in documenting the size of the injury as well as the positioning of the camera. Bite (*a*) exhibits a higher degree of forensic value than the others, with (*d*) having the least amount of information recorded.

For background information *see* CRIMINALISTICS; DATING METHODS; FORENSIC BIOLOGY; FORENSIC MEDICINE; RADIOCARBON DATING; TOOTH in the McGraw-Hill Encyclopedia of Science & Technology.

Robert E. Barsley; David R. Senn

Bibliography. C. M. Bowers, *Forensic Dental Evidence: An Investigator's Handbook*, Elsevier, San Diego, 2004; R. B. J. Dorion (ed.), *Bitemark Evidence*, Marcel Dekker, New York, 2005; *Doyle v. State*, 159 Texas, C.R. 310, 263 S.W.2d 779, Jan. 20, 1954; E. E. Herschaft et al. (eds.), *Manual of Forensic Odontology*, 4th ed., American Society of Forensic Odontology, 2007; P. G. Stimson and C. A. Mertz (eds.), *Forensic Dentistry*, CRC Press, New York, 1997.

Forensic science education

Forensic science has never been more popular or more popularized. From documentary TV shows such as *Forensic Files*, to their more fictionalized offspring such as the *CSI* series, forensic science has gained a solid foothold in the public's perception of its practitioners and methods. Belying that perception, however, is the reality of what it takes currently to become a forensic scientist, namely, a strong science education. The educational foundation for a forensic science degree has changed over the years in response to the increasing recognition that forensic science is a separate discipline in its own right and not merely applied chemistry, biology, or other science. The development of forensic science and the growth in the number of educational programs offering forensic science degrees have contributed to the need for accreditation of these programs. With this accreditation comes a level of assurance for students and employers of what the degree entails and what can be expected of graduates from accredited programs. A recent report from the National Academy of Sciences recommends an increased emphasis on graduate education in forensic science and, with that, additional research to improve the reliability of the science used in investigations, laboratories, and courtrooms.

The first forensic science educational program was created at Michigan State University in 1946. Currently, there are hundreds of programs with the words "forensic science" or "forensic chemistry" in their title (see **illustration**). Until recently, the growth of programs was slow and somewhat erratic because of the nature of forensic science as a profession traditionally within law enforcement and the open nature of academics. Legal investigations, criminal or civil, often have a component of sensitive or classified information; this necessitated training rank-and-file chemists and biologists "on the job" (that is, in the confines of the forensic laboratory) about the legal aspects of their work. For a time, forensic laboratory directors preferred to hire traditional scientists and train them rather than hiring graduates from forensic science programs because the quality of the education varied so greatly in the latter.

sufficient to exclude an individual as the contributor. Reproduction of the biting edges of the teeth through computer scanning and software programs can be used to produce a transparent overlay that may be compared to photographs. Measurement of the widths and breadths of the injuring teeth as well as measurement of the spaces and angles between adjacent teeth may be compared to the photograph. Articulated dental models or the suspect may be used to bite onto substrates in an attempt to reproduce the injury pattern.

There are many complicating factors in the comparison, not the least of which is, if the bite injury is on human skin, the elasticity, underlying tissues, and the dynamic position of that skin at the time of the bite injury will influence the resulting pattern of injury. There are chapters in textbooks and there is at least one textbook devoted to the analysis of bite-mark injuries. The American Board of Forensic Odontology, as part of its mission in certifying forensic odontologists, has published guidelines to enhance the reporting and dissemination of the opinions reached in bite-mark cases by its member dentists.

Bite-mark comparison has been accepted in U.S. jurisprudence since a Texas case in 1954 and continues to be accepted in U.S. courts today. However, the 2009 National Academy of Sciences report on forensic science raised questions that must be answered for all identification sciences.

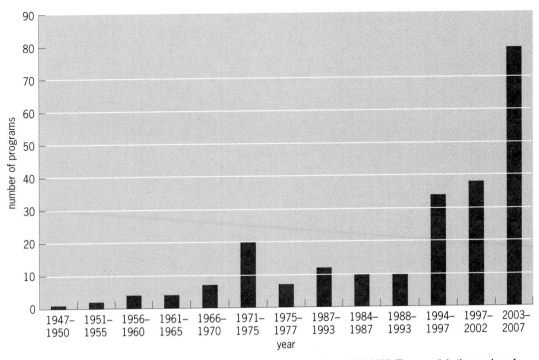

Cumulative number of forensic science programs, undergraduate and graduate, 1947–2007. The growth in the number of forensic science programs has been particularly rapid in the last few years, probably in response to the increase in popular television shows highlighting forensic science. (*From J. Peterson et al., 1977 and G. P. Jackson, 2009*)

The profession of forensic science requires a solid grounding in the natural sciences, an understanding of statistics and interpretation, an appreciation for management skills, and the perspective of the role of forensic science in the criminal justice system. The curricula in educational programs, however, have not always met these needs. Historically, a forensic science educational program was started by a scientist who had worked at a government forensic laboratory. Often, the program consisted of that one scientist and one or more adjunct faculty, few of whom had any experience in an operational forensic laboratory. An educational program thus became associated with its originating scientist and, unless a successor was found, the program faltered when that scientist left.

Accreditation and standardization of programs. With the increased recognition of forensic science as a separate discipline, additional calls from the community demanded improvement in the quality of forensic science education. Although accreditation and standardization of forensic science programs was not a new message, a confluence of effort, resources, and people was needed. A technical working group on education and training in forensic science (TWGED) was convened by the National Institute of Justice and a report was issued. The report covered four areas: what prospective students could expect from a career in forensic science, undergraduate curricula, graduate curricula, and guidelines for continuing professional development. This report provided the first coherent consensus about what a forensic science degree should require. The American Academy of Forensic Sciences (AAFS) recognized the importance of this document and in 2002 created an ad hoc committee to convert the TWGED guidelines into enforceable standards. The following year, the committee became the Forensic Science Educational Program Accreditation Commission (FEPAC) and a permanent part of the AAFS. The mission of the FEPAC is to maintain and to enhance the quality of forensic science education through formal evaluation and recognition of college-level academic programs. The primary function of the commission is to develop and maintain standards and to administer an accreditation program that recognizes and distinguishes high-quality undergraduate and graduate forensic science programs. Five educational programs volunteered to be part of a pilot accreditation project in 2004. In 2008, the FEPAC was recognized by the Association of Specialized and Professional Accreditors (ASPA). As of January 2009, 26 programs had been accredited by the FEPAC (see **table**).

Degree programs. The FEPAC accredits forensic science educational programs that lead to a bachelor's or graduate degree in forensic science or in a natural science with a forensic science concentration. The program must be housed in a regionally accredited institution of higher learning. All programs must adhere to a set of basic standards regarding planning and evaluation, institutional support, student support services, administrative practices, and student complaints. The director and the faculty must be able to fully support the program's mission and goals. Programs must have interaction with operational forensic laboratories and are responsible for keeping records of student achievement.

Undergraduate programs. Undergraduate forensic science programs have to ensure that each student obtains a fundamental education in the natural

List of FEPAC accredited forensic science programs as of January 2009	
Bachelors of science programs	Masters programs
Albany State University, GA	University of Alabama
Cedar Crest College	University at Albany (SUNY at Albany)
Eastern Kentucky University	Arcadia University
Florida International University	Duquesne University
Metropolitan State College of Denver	Florida International University
University of Mississippi	University of Illinois at Chicago
University of New Haven	John Jay College of Criminal Justice
University of North Texas	Marshall University
Ohio University	Michigan State University
The Pennsylvania State University	Oklahoma State University
Virginia Commonwealth University	Sam Houston State University
West Chester University	Virginia Commonwealth University
West Virginia University	

sciences, builds on this foundation with increasingly advanced science courses, and develops an appreciation for the interrelations of science, forensic science, and its application. Specific courses, some with laboratory components, must be taken in biology, physics, chemistry, and mathematics. An additional complement of specialized science courses, along with additional forensic science coursework, is required. Advanced courses are also required to deepen a student's understanding and knowledge of science and forensic science.

Graduate programs. Graduate programs vary in their structures, and the FEPAC graduate standards acknowledge this fact. Unlike the undergraduate standards—which are fairly proscriptive and structured in terms of specific courses and how many credits—the graduate standards are more flexible and allow programs to specialize (in forensic molecular biology, for example). A graduate seminar is required, as is a written thesis or its equivalent that is subjected to peer review.

Doctoral programs. No forensic science doctoral programs, either professional (such as an M.D. or D.D.S.) or research-based (such as a Ph.D.), currently exist in the United States or Canada. Such programs do exist in Australia, the European Union, and the United Kingdom. Doctoral degrees with an emphasis in forensic science research are offered in the United States and Canada; these degrees are centered around forensic science and tend to be housed within a non–forensic science department. The complexity and richness of forensic science as a research discipline, however, means that forensic science research Ph.D. or professional doctorates will eventually become more common. A recent report by the National Academy of Sciences recommends additional funding for forensic science graduate programs and research to improve the reliability and effectiveness of forensic science.

Other areas of interest. Beyond the scientific and curricular needs of the forensic sciences, two other areas bear mentioning.

Women in forensic science. Women constitute a majority of this field, from education (nearly 80%) to professional organizations (about 70%) to the laboratory (60%). Why this is the case is not well understood, but several factors seem to have a bear-

ing on women's career choices. An early and supported interest in science, seeing forensic science as a viable and exciting career, a sense of improving society by helping to catch criminals, and a personal experience or connection to a tragedy involving a victim of a crime all seem to influence women into forensic science as a career. More study is clearly needed to understand this phenomenon, particularly given that women make up 40% or less of other science, technology, engineering, and math disciplines.

Business of forensic science. The second area of increased interest in forensic science is that of the "business of forensic science," that is, how laboratories operate, improve their effectiveness, and enhance the quality of science they provide. Forensic laboratories occupy an interesting business niche. They have budgets but are usually not profit-oriented, so traditional measures of success (sales or revenue) do not apply. Forensic laboratories do "manufacture" products, turning evidence into knowledge (reports and testimony). The obstacles are many, as the evidence can range from unusable to excellent quality, yet the laboratory is expected to operate at better than a six-sigma level of success. The effective allocation of limited resources in a government-based scientific nonprofit that works on an "on-demand" basis is a critical challenge to the accurate, efficient, and accountable operation of a forensic laboratory. Forensic science plays a central role in the criminal justice process, and the "business of forensic science" will command greater attention in the future. Institutions have already paired Ph.D. degrees in biotechnology with M.B.A. degrees to better provide young entrepreneurial scientists with the business background they need to create start-up companies in the technology sector.

Outlook. Education and research in the forensic sciences in the next decade will require strong science skills, mathematics, statistical ability, and a creative mindset willing to ask fundamental questions.

For background information *see* CRIMINALISTICS; FORENSIC BIOLOGY; FORENSIC CHEMISTRY; FORENSIC ENGINEERING; FORENSIC EVIDENCE; FORENSIC PHYSICS in the McGraw-Hill Encyclopedia of Science & Technology.   Max Houck

Bibliography. M. Houck, Is forensic science a gateway for women?, *Forensic Sci. Policy Manage.*, 1(1):65–69, 2009; M. Houck et al., FORESIGHT: A Business Approach to Improving Forensic Science Services, *Forensic Sci. Policy Manage.*, 1(2):85–95, 2009; G. P. Jackson, The Status of Forensic Science Degree Programs in the United States, *Forensic Sci. Policy Manag.*, 1(1):2–9, 2009; C. Midkiff, Forensic science courses for a criminal justice program, in *Forensic Science*, 2d ed., G. Davies (ed.), pp. 67–76, American Chemical Society, 1986; National Academies of Science, *Strengthening Forensic Science in the United States: A Path Forward*, 2009; National Institute of Justice Special Report, *Education and Training in Forensic Science: A Guide for Forensic Science Laboratories, Educational Institutions, and Students*, June 2004; J. Peterson, D. Crim, and P. DeForest, The status of forensic science programs in the United States, *J. Forensic Sci.*, 22(1):17–33, 1977; J. A. Siegel, The appropriate educational background for entry level forensic scientists: A survey of practitioners, *J. Forensic Sci.*, 33(4):1065–1068, 1988; R. Turner, Forensic science education: A perspective, in *Forensic Science*, 2d ed., G. Davies (ed.), pp. 3–11, American Chemical Society, 1986.

Freezing tolerance and cold acclimation in plants

Freezing is a major environmental stress that inflicts injury to plant tissues and limits the productivity and geographic distribution of wild and crop species. Most tropical and subtropical species have little to no freezing tolerance. However, plants from temperate regions not only have some "constitutive" freezing tolerance (which is present all the time, typically at low levels), but they also have the genetic ability to increase this tolerance significantly when exposed to environmental cues that signal the arrival of winter, such as a period of low temperatures and/or short days. In temperate climates, such conditions are typically encountered by overwintering perennials in autumn, resulting in a seasonal increase in freezing tolerance. The ability of plants to increase freezing tolerance in response to changing environment is called cold acclimation. The term deacclimation refers to reduction/loss of freezing tolerance originally attained through cold acclimation and, in nature, happens typically in early spring with the rise of temperatures. Depending on the depth of deacclimation, it may be either irreversible or reversed by

subsequent exposure to low temperatures that may cause reacclimation, that is, restoration of at least a portion of the lost tolerance.

The timing and extent of cold acclimation and deacclimation during the annual growth and developmental cycle are of critical importance for winter survival, particularly in view of climate change, that is, unpredictable extreme weather and climate events. For example, plants may acclimate less completely to cold if exposed to milder autumn climate, and thus may be damaged by even mild sudden frosts. Alternatively, they may deacclimate prematurely as a result of unseasonable midwinter warm spells and get damaged by the cold that follows. Thus, efficient cold acclimation ability and high deacclimation resistance and reacclimation capacities are important for plants to survive freezing conditions during cold winters.

Physiology of freezing at the plant cell level. During a frost episode, ice typically forms in intercellular spaces and cell walls of plant tissues. This occurs by ice nucleation of intercellular fluid, which has a higher freezing point than the cell sap and which normally contains ice-nucleating agents, such as dust and bacterial proteins. Ice nucleation refers to the coming together of water molecules to form a stable nucleus of critical size that can trigger further crystallization of ice. Among other factors such as tissue mass and surface moisture, ice nucleation is influenced by cooling rates. During a natural freeze, temperatures generally decline gradually in the subfreezing range at rates less than $2°C/h$, promoting ice nucleation of extracellular fluid. Upon freezing, the vapor pressure of the ice state of water at a given temperature is less than that of the liquid solution, and this difference increases as temperature decreases. Thus, once extracellular ice forms, water molecules move from the protoplasm through the plasma membrane via osmosis, leading to cellular dehydration (**Fig. 1**). The net amount of dehydration depends on both the initial solute concentration of the cell sap and the subfreezing temperature that determines the vapor-pressure differential across the plasma membrane. For example, assuming an initial cell sap concentration of approximately 0.5 osmolar (Osm), freezing at $-10°C$ will remove about 90% of the osmotically active water from the cell; in contrast, the same freezing removes only about 80% of the freezable water if the initial cell sap concentration is 1 Osm (Fig. 1).

Equilibrium (extracellular) and nonequilibrium (intracellular) freezing. Slow cooling during a freezing episode (such as <2–$3°C/h$) allows the exosmosis

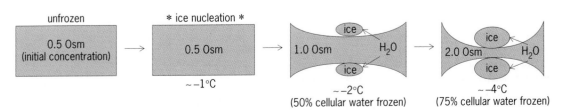

Fig. 1. Diagrammatic representation of the mechanism of cellular dehydration due to extracellular freezing.

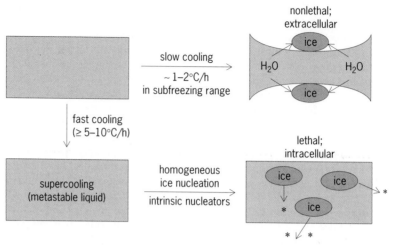

Fig. 2. Consequences of "slow" versus "fast" freezing in plant tissues.

(passage of a liquid outward through a cell membrane) of cellular water to extracellular ice at a speed sufficient to increase cell sap concentration as rapidly as the temperature drops. This allows the water potential of the cell sap to be in equilibrium with that of the extracellular ice—hence, extracellular freezing is "equilibrium freezing." In contrast, during fast cooling (such as $\geq 5-10^\circ$C/h in subfreezing range), which seldom happens in nature, the water cannot diffuse out rapidly enough for the cell sap to remain in chemical equilibrium with the extracellular ice. Consequently, the cell sap becomes supercooled—dilute aqueous solutions remaining as metastable liquids below their freezing points. During this chemical disequilibrium, the spontaneous formation of "intracellular" ice ("homogeneous ice nucleation") may be induced if the temperature is sufficiently lowered, which, unlike extracellular freezing, is instantaneously lethal as the ice crystals disrupt the integrity of the living protoplasm (**Fig. 2**).

Although extracellular freezing can be tolerated by the plant cell, physical consequences of ice formation in tissues are severe. Aside from the mechanical stress due to ice crystals, the freeze-induced dehydration causes shrinkage/collapse of vacuolated cells, increases cell osmolarity and protoplasmic concentration, and influences interactions between the plasma membrane and cell walls at attachment points.

Mechanism of freeze–thaw injury. Freezing-induced cellular dehydration is the most common cause of injury to plant tissues and is manifested in structural and functional perturbations to cell membranes. The physiological consequence of these perturbations—loss of compartmentation—is detectable as leakage of cell contents/solutes. Research with rye protoplasts (plant cells with their cell walls removed) shows that freeze-induced structural destabilization involves various lesions in membranes. The first type is the dehydration-induced loss of osmotic responsiveness and is a predominant form of injury in non-cold-acclimated protoplasts. During freezing, protoplasts contract osmotically as they dehydrate. Upon thawing, they are unable to expand, indicating im-

pairment of osmotic properties and semipermeability of the plasma membrane. Such lesion is characterized by the transition of plasma membrane from lamellar (normal) to HexII phase (abnormal) structure, which is brought about by different cellular membranes (plasma membrane, chloroplast bilayers, and so on) coming into close apposition and forming interbilayer stacking. Such stacking with hexagonal packing symmetry (hence, the HexII phase) disrupts membrane structure, causing a loss of osmotic responsiveness and solutes (**Fig. 3**).

The second form of membrane injury in nonacclimated protoplasts is known as expansion-induced lysis. Here, osmotic responsiveness is not affected during freeze-induced contraction but is lost during thaw-induced expansion. Because of the irreversible deletion of membrane in the form of endocytotic vesicles during the contraction, the protoplast bursts before regaining the original volume when melted water is drawn back during thaw. Cold-acclimated protoplasts, however, do not form such vesicles. Instead, they retain membrane surface area, allowing reexpansion of cells during thaw. It is worth noting that the aforementioned membrane lesions are observed in isolated protoplasts and may not reflect exactly the mechanisms that occur in intact plant cells.

Freeze–thaw also results in loss of membrane calcium, leading to compromised membrane integrity. Ca^{2+} forms ionic bridges with the negative charges of

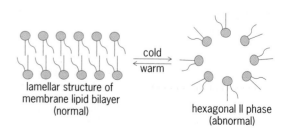

Fig. 3. Diagrammatic representation of the transition of membrane lamellar structure to HexII phase at cold temperature, and vice versa at warm temperature.

the membrane lipid bilayer, thereby providing structural packing. Disruption of these bonds destabilizes membranes and, consequently, may affect functional efficiency of membrane transport proteins (ion pumps, such as H^+-ATPase) that maintain proper ionic balance across cell membranes. Inhibition of ion pumps can lead to leakage of cell contents and loss of turgor (flaccidness) in plant tissues, which is a characteristic symptom of freeze–thaw injury. Depending on the rate of temperature change, intensity, and duration, freeze–thaw injury can be either repairable or irreversible. Research indicates that repair of the membrane transport system (such as H^+-ATPase) may be required for recovery from freeze–thaw injury. Recovery is manifested in the reuptake of leaked solutes and the regain of turgor.

Cold acclimation. Acquired tolerance to freezing, that is, cold acclimation, is the expression of the genetic potential under inductive conditions that shifts the threshold of a cell's tolerance to freeze-induced dehydration (and other stresses due to extracellular freezing) to much colder temperatures than that for nonacclimated tissues (**Fig. 4**).

Physiology of cold acclimation. Because temperature directly affects kinetic and thermodynamic processes, and thus all aspects of plant biochemistry and physiology, cold acclimation is a complex trait involving myriad changes in cell biology and metabolism as tissues transition from freeze-sensitive to freeze-tolerant or from less freeze-tolerant to more freeze-tolerant states. Because dehydration is a large component of freezing stress, there is an overlapping relationship between cold acclimation and dehydration (drought) stress tolerance in terms of cellular and biochemical adjustments.

Cold acclimation accomplishes two major functions: (1) the more universal adjustment of cellular metabolism to the biophysical constraints imposed by low temperatures and (2) the actual induction of freezing tolerance. Research validates that the cellular changes associated with cold acclimation are regulated by altered expression (upregulation or downregulation) of numerous genes—hence, it is a multigenic process. Several of these major changes are shown schematically in **Fig. 5**. The most notable changes include reduction or cessation of growth, reduction of tissue water content, changes in membrane lipid composition, increased levels of antioxidants, temporary increase in abscisic acid (ABA, a plant hormone), and accumulation of solutes (for example, soluble sugars) and "stress" proteins.

Mechanism of cold acclimation. Global transcript (gene) profiling analyses using the model plant species *Arabidopsis* indicate that the expression of hundreds of genes is altered during cold acclimation. Genes induced during cold acclimation not only produce proteins that protect cells from the injurious effects of low temperatures and freeze desiccation [often referred to as *COR* (cold-regulated) genes], but also regulate (turn "on" or "off") the expression of *COR* genes. A set of cold acclimation-responsive genes is further involved in the signal transduction of cold

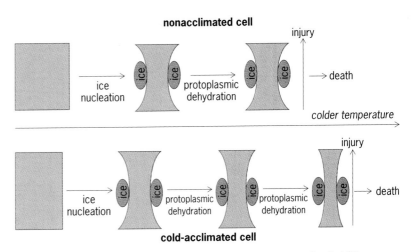

Fig. 4. Diagrammatic representation of the shifting of the "tolerance threshold" for freeze-induced cellular dehydration to relatively colder temperatures after cold acclimation.

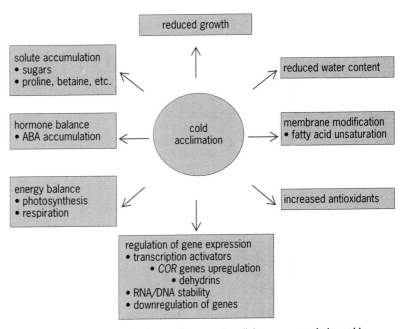

Fig. 5. Some of the commonly observed changes in cellular processes during cold acclimation. ABA = abscisic acid; *COR* genes = cold-regulated genes.

response, that is, the communication of cold signal through the cellular machinery, eventually eliciting a protective response (**Fig. 6**).

Some examples of proteins that belong to the first group are dehydrins (a family of hydrophilic proteins) and enzymes for sugar synthesis and fatty acid metabolism. Protection of cell membranes and cellular enzymes against freeze damage is achieved by mitigating the desiccation stress and/or biochemical and physical restructuring of membrane lipids. Dehydrins are believed to stabilize membranes either by direct interaction with membranes or by interacting with surrounding water (like a sponge) and keeping the membranes and enzymes sufficiently hydrated in spite of freeze desiccation. Desaturases catalyze biochemistry that increases membrane lipid unsaturation at low temperatures. This allows normal

cold

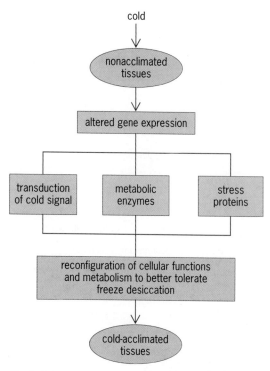

Fig. 6. Schematic illustration of molecular and biochemical components of the cold acclimation response.

functioning of membrane proteins because a membrane with a relatively high proportion of unsaturated fatty acids can maintain its characteristic fluidity even at low temperatures.

Accumulation of sugars during cold acclimation is a common occurrence. Such accumulation, through a colligative effect (that is, dependent on the number of molecules but not their nature), would reduce the amount of water lost during extracellular freezing. Consequently, upon freezing, cold-acclimated tissues experience relatively less desiccation and mechanical stress, with the latter due to less extracellular ice. Studies with *Arabidopsis* indicate that mutant seedlings impaired in their ability to cold-acclimate do not accumulate sugars, whereas those that are constitutively more freeze-tolerant than normal seedlings accumulate sugars even at warm temperatures, suggesting an important role for sugars in cold acclimation. Besides sugars, other low-molecular-weight organic solutes, such as proline, glycine betaine, and the raffinose family of oligosaccharides (RFOs), are also known to accumulate during cold acclimation. The specific role of these so-called compatible solutes in conferring freezing tolerance is not well understood. However, their specific biophysical properties and interaction with cell membranes are often emphasized in prevention and/or amelioration of membrane damage due to freeze desiccation.

The plant hormone ABA has also been shown to mediate the development of freezing tolerance and perhaps plays a central role in the signal transduction of cold acclimation. Evidence supporting such a hypothesis includes accumulation of ABA in tissues during the early stages of cold acclimation, increased freezing tolerance of tissues at normal (noninductive of cold acclimation) temperatures by exogenous ABA treatment, impaired cold acclimation ability of certain ABA-deficient mutants and its restoration upon exogenous ABA application, and induction of *COR* genes by ABA treatment.

Environmental stresses, such as freezing and drought, are long known to cause excessive accumulation of reactive oxygen species (ROS), as a result of metabolic dysfunction. ROS can cause peroxidation of membrane lipids, leading to membrane damage. Plants have evolved strategies to cope with ROS, including accumulation of antioxidant metabolites (such as vitamin C and glutathione) and activation of antioxidant enzymes (such as superoxide dismutase, catalase, and peroxidase) that function cooperatively to scavenge ROS generated during low-temperature stress.

For background information *see* CELL MEMBRANES; CELL WALLS (PLANT); COLD HARDINESS (PLANT); GENE; NUCLEATION; OSMOSIS; PHYSIOLOGICAL ECOLOGY (PLANT); PLANT ANATOMY; PLANT CELL; PLANT PHYSIOLOGY; PLANT-WATER RELATIONS; PROTEIN; TEMPERATURE in the McGraw-Hill Encyclopedia of Science & Technology.   Rajeev Arora

Bibliography.   R. Arora and J. P. Palta, *In vivo* perturbation of membrane-associated calcium by freeze thaw stress in onion bulb cells: Simulation of this perturbation in extracellular KCl and alleviation by calcium, *Plant Physiol.*, 87:622–628, 1988; C. L. Guy, Cold acclimation and freezing stress tolerance: Role of protein metabolism, *Annu. Rev. Plant Phys. Plant Mol. Biol.*, 41:187–223, 1990; C. L. Guy, Freezing tolerance of plants: Current understanding and emerging concepts, *Can. J. Bot.*, 81:1216–1223, 2003; S. Penfield, Temperature perception and signal transduction in plants, *New Phytol.*, 179:615–628, 2008; M. Uemura et al., Responses of plasma membrane to low temperatures, *Physiol. Plant.*, 126:81–89, 2006; Z. Xin and J. Browse, Cold comfort farm: The acclimation of plants to freezing temperatures, *Plant Cell Environ.*, 23:893–902, 2000.

Fungal secondary metabolites

Fungi are eukaryotic organisms known to inhabit almost all ecological niches of the Earth and have the ability to utilize various solid substrates as a consequence of diversity of their biological and biochemical evolution. Some of the solid substrates utilized by fungi are dead and decaying material, including herbivore dung (saprophytic and coprophilous fungi), live plants (endophytic, parasitic, and mycorrhizal fungi), lichens (lichenicolous and endolichenic fungi), and insects (entomopathogenic fungi). A characteristic feature of many of these fungi, especially those that exhibit filamentous growth and have a relatively complex morphology, is their ability to produce secondary metabolites. Soilborne, parasitic, and saprophytic fungal sources are relatively well investigated with regard to their secondary metabolites, and currently there is intense interest in secondary metabolites of symbiotic fungi that live in association with land plants, insects, lichens, and marine organisms. In contrast to primary metabolites such as proteins, DNA, RNA, polysaccharides, and so on, which occur universally, secondary metabolites are small-molecule organic compounds found restricted to a particular species, genus, or family. Thus, the presence or absence of certain secondary metabolites has been used successfully in the classification (chemotaxonomy) of large ascomycete genera (including *Alternaria, Aspergillus, Fusarium, Hypoxylon, Penicillium, Stachybotrys,* and *Xylaria*) and in a few genera of basidiomycetes. Many secondary metabolites are not involved directly in the normal growth, development, or reproduction of the fungus in which they occur, but they may play an important role in ecological interactions with other organisms. For this reason, many fungal secondary metabolites exhibit useful biological activities and are of interest to the pharmaceutical, food, and agrochemical industries. Production of secondary metabolites often occurs after fungal growth has ceased as a result of nutrient limitations but with an excess carbon source available, making it possible to manipulate their formation. It is intriguing that some endophytic fungi are capable of producing secondary metabolites previously known from plants. Noteworthy examples include production of two clinically important anticancer drugs, paclitaxel (Taxol®) and camptothecin, by *Taxomyces andreanae* and *Nothapodytes foetida*, respectively, and a synthetic precursor of an anticancer drug, podophyllotoxin, by *Phialocephala fortinii*.

Functional diversity. In some fungi, secondary metabolism (the process that results in the production of secondary metabolites) has been found to commence during the stationary or resting phase of their development and is often associated with sporulation and colony formation. Some well-documented functions of fungal secondary metabolites include enhancement of spore survival by acting as virulence factors and protecting against ultraviolet (UV) light, and augmentation of their fitness and competitive ability against other fast-growing organisms (see **Table 1**). Fungal metabolites associated with sporulation may activate sporulation (for example, linoleic acid analogs produced by *Aspergillus nidulans*), provide pigmentation required for sporulation structures (for example, melanins produced by *Alternaria alternata*), or have toxic properties to ward off competing organisms (for example, mycotoxins produced by some *Aspergillus* species). The relationship between production of secondary metabolites and regulation of asexual sporulation by a G-protein–mediated growth pathway in *Aspergillus* species was established over a decade ago. Also, it has been speculated that secondary metabolites in fungi function as metal chelators (combining with metal ions and removing them from their sphere of action), which is important in mineral nutrition, and that pathways leading to their formation act as safety-valve shunts that prevent the accumulation of toxic intermediates of primary metabolism under conditions of unbalanced growth. However, much remains to be learned about the functions performed by most fungal secondary metabolites in their producing organism.

Structural diversity: biosynthesis and impact of genetics and genomics. Fungal secondary metabolites encompass over 30,000 known compounds with an extremely diverse array of chemical structures. It is intriguing that all these secondary metabolites originate from a few common biosynthetic pathways utilizing precursors (small biosynthetic units or building blocks) formed during primary metabolism. The intermediates resulting from condensation of these small biosynthetic units are further elaborated ("tailored" or "decorated") by numerous enzyme-catalyzed reactions, leading to products with a diversity of structures. Thus, fungal secondary metabolites are conveniently classified based on their biosynthetic origin as polyketides, nonribosomal peptides, terpenes, and alkaloids. The majority of fungal secondary metabolites are polyketides, nonribosomal peptides, or a combination of both classes with hybrid structures. Biosynthesis of fungal secondary metabolites often involves elaborate biochemical pathways and is regulated by a group of genes known as biosynthetic genes. The insights that have been gained from recent advances in genetics, genomics, molecular biology, and bioinformatics have contributed to the understanding and manipulation of these genes for improved production, or inhibition of production, of fungal secondary metabolites.

During the biosynthesis of fungal polyketides, acetyl coenzyme A (acetyl-CoA) and malonyl-CoA are assembled by dedicated multifunctional enzymes,

---

TABLE 1. Some functionally diverse fungal secondary metabolites

Secondary metabolite*	Function	Fungal producer
Butyrolactone I	Sporulation induction	*Aspergillus terreus*
Linoleic acid analogs	Sporulation induction; affects spore development	*Aspergillus nidulans*
Melanin	Protection of spores from UV light; spore survival	*Alternaria alternata; Cochliobolus heterostrophus*
Patulin	Antibiotic—wards off competitors	*Penicillium urticae*
Zearalenone	Sporulation induction	*Fusarium graminearum*

*For chemical structures of some of these secondary metabolites, see the illustration.

referred to as polyketide synthases (PKSs). Fungal PKSs are monomodular multidomain enzymes, which act many times ("iteratively") on a growing polyketide chain, and each iteration extends the growing ketide by a two-carbon unit. The vast structural diversity of fungal polyketides results from the number of these iteration reactions, the extent to which the carbonyl groups are reduced to alcohol or hydrocarbon moieties, the cyclization of the polyketide chain, and the nature of the extender unit used. Biosynthesis of nonribosomal peptides involves the assembly of small peptides by nonribosomal peptide synthases (NRPSs), which are also dedicated multifunctional multidomain enzymes. The structural diversity of nonribosomal peptides in fungi results from the length of the peptides produced and whether they have undergone any cyclization reactions. Fungal metabolites with both polyketide and nonribosomal peptide structural moieties are assembled by hybrid NRPS–PKS enzyme systems. Many fungi capable of producing secondary metabolites are rich in genetic material dedicated to secondary metabolism and contain genes for PKSs, NRPSs, and cytochrome P_{450} oxygenases (which catalyze the addition of a hydroxyl group to an organic compound using molecular oxygen). The genome of the fungus *Aspergillus nidulans*, for example, contains 27 genes for PKSs, 14 for NRPSs, and 102 for P_{450} monooxygenases. Genome sequencing of several fungi has revealed that the number of genetic loci presumably dedicated to secondary metabolism far exceeds the number of secondary metabolites encountered to date in those fungal species. This observation suggests that the biosynthetic potential of these fungal species is currently underexplored and that a multitude of secondary metabolites remains to be discovered.

Unlike the genes involved in primary metabolism that are found scattered throughout the genome, genes involved in the biosynthesis of fungal secondary metabolites are usually arranged in clusters, which is a finding that has important implications for gene discovery, regulation, and evolution. Two main hypotheses have been advanced to rationalize gene clustering. The first states that gene clustering occurs for the purpose of coordinated gene expression through sharing of regulatory elements, and the second hypothesis is based on evolutionary evidence that supports the horizontal genetic transfer of metabolic pathways from prokaryotes to eukaryotic fungi and then from fungi to fungi. The clustered nature of the biosynthetic genes has greatly facilitated the discovery of genes involved in a biosynthetic pathway once the first gene is identified. Identification of these genes is important because they could be manipulated to direct the synthesis of new and useful secondary metabolites. In addition, the functional role played by a secondary metabolite in the biology of the fungus could be deciphered from a greater understanding of the regulation of secondary metabolism in response to environmental stimuli. Recently, a considerable number of fungal biosynthetic gene clusters have been identified. However, a corresponding increase in the number of new secondary metabolites from fungi has not been reported. It is possible that manipulating regulatory factors or modifying growth conditions combined with application of newly developed, highly sensitive mass spectroscopic techniques may lead to the identification of hitherto unknown fungal secondary metabolites. Most of the gene clusters for secondary metabolite production contain a transcription factor that acts specifically on genes within the cluster. However, recent studies suggest that secondary metabolite production in fungi is also regulated by global transcription factors encoded by genes not linked to the biosynthetic gene clusters and that such genes regulate many physiological processes and generally respond to environmental cues such as nutrition, temperature, and pH. The tendency for biosynthetic genes to cluster makes it possible to gather most, if not all, of the genes responsible for the production of a given secondary metabolite within one large fragment of DNA and introduce it into a fast-growing recipient strain, which is a process that has implications for efficient and large-scale production of fungal secondary metabolites. Eventually, application of genomics will also help in understanding many aspects of fungal secondary metabolism, including (1) why fungi produce secondary metabolites, (2) what is their ecological relevance, (3) how to diversify and improve production of biologically active compounds that will affect the pharmaceutical industry, and (4) how to avoid production of mycotoxins that may affect food safety and thus human and animal health.

TABLE 2. Some biologically active fungal secondary metabolites

Secondary metabolite*	Biological activity	Fungal source
Penicillin G	Antibacterial	*Penicillium chrysogenum* *Aspergillus nidulans*
Echinocandin B	Antifungal	*Aspergillus nidulans* var. *echinulatus*
Integric acid	Antiviral	*Xylaria* sp.
Cyclosporin A	Immunosuppressant	*Beauveria nivea*
Fumagillin	Antitumor	*Aspergillus fumigatus*
Lovastatin	Cholesterol-lowering	*Aspergillus terreus*
Aflatoxin	Mycotoxin	*Aspergillus flavus, A. parasiticus*
Fusarin C	Mutagen	*Fusarium moniliforme*

*For chemical structures of some of these secondary metabolites, see the illustration.

Polyketides

Aflatoxin B1   Butyrolactone I   Lovastatin   Zearalenone

Peptides   **Polyketide-peptide hybrid**

Penicillin G   Fusarin C

Cyclosporin A

Terpenes   **Polyketide-terpene**   **Alkaloid**

Integric acid   Fumagillin   Ergotamine

Structural diversity of fungal secondary metabolites.

Biological activity. Fungal secondary metabolites are well known for their biological activity and represent some of today's important and useful pharmaceuticals and agrochemicals (see **Table 2** and **illustration**). Among the pharmaceuticals, most noteworthy are penicillins, cephalosporins, and fusidic acid with antibacterial activity; echinocandin B, pneumocandins, griseofulvin, and strobilurins with antifungal activity; integric acid and integresone with antiviral activity; cyclosporin A and mycophenolic acid with immunosuppressive activity; fumagillin and rhizoxin with antitumor activity; lovastatin and pravastatin with cholesterol-lowering activity; and ergot alkaloids (for example, ergotamine) with antimigraine activity. Gibberellins and zearalenones are fungal secondary metabolites used in agriculture as plant growth hormones and in animal husbandry as growth promoters, respectively. Some fungal metabolites such as mycotoxic aflatoxins and mutagenic fusarin C possess potent toxic and carcinogenic activities and are therefore important in human, animal, and plant health.

For background information *see* AFLATOXIN; BIOSYNTHESIS; FUNGAL BIOTECHNOLOGY; FUNGAL ECOLOGY; FUNGAL GENETICS; FUNGAL GENOMICS; FUNGI; GENE; MEDICAL MYCOLOGY; MYCOLOGY in the McGraw-Hill Encyclopedia of Science & Technology.   A. A. Leslie Gunatilaka

Bibliography. M. A. Fischbach, C. T. Walsh, and J. Clardy, The evolution of gene collectives: How natural selection drives chemical innovation, *Proc. Natl. Acad. Sci. USA*, 105:4601–4608, 2008; E. M. Fox and B. J. Howlett, Secondary metabolism, regulation, and role in fungal biology, *Curr. Opin. Microbiol.*, 11:481–487, 2008; A. A. L. Gunatilaka, Natural products from plant-associated microorganisms: Distribution, structural diversity, bioactivity, and implications of their occurrence, *J. Nat. Prod.*, 69:509–526, 2006; N. P. Keller, G. Turner, and J. W. Bennett, Fungal secondary metabolism—from biochemistry to genomics, *Nat. Rev. Microbiol.*, 3:937–947, 2005.

Fungi and fungal toxins as weapons

Preparations of human, animal, or plant fungal pathogens as spores or fragments are a major security concern as biological weapons. Moreover, the preparation of purified toxins from toxigenic fungi is accomplished by a straightforward process, and conform to substances known as chemical weapons. A list of fungal toxins is provided in the **table**. The factors that are fundamental to producing a serviceable bioweapon are (1) efficient manufacture, (2) ease of conversion to a weapon ("weaponization"), (3) longevity of the organism or toxin in storage, (4) efficient dispersal, and (5) stability when exposed to the environment. Other factors are concealment and ability to obtain the toxin or fungus. The threat posed by these weapons may not be obvious initially. However, the economic consequences of simply reacting to an actual attack can be huge. Indeed, somber assessments of the dangers of bioweapons to the United States, for example, have been made. The effect on society could be anything from insignificant to catastrophic. It is thus much more appropriate to focus on prevention, followed by readiness and response. Current terrorist tactics have shifted worldwide attention to the protection of food supplies. A broad range of actions and programs is being developed and implemented to prevent, deter, and respond to potential attacks. These include enhanced laboratory capability, advanced tracking, increased examination, more surveillance, greater training, recovery plans, and new medical treatments.

Fungal biological weapons. Human pathogenic fungi should be given greater consideration as biowarfare and bioterrorism agents. The obvious growth (and infection) of fungi on animals is called mycosis. Such fungi are primary pathogens. *Histoplasma capsulatum*, for example, is a parasitic fungus that affects the lungs and causes histoplasmosis in humans. Nevertheless, human pathogenic fungi, with the exception of *Coccidioides* species, are not found among lists of microbes with potential for biological warfare and bioterrorism. Some fungi are comparable to other microbes—various bacteria, for example—with respect to their biological and pathogenic attributes. The current apparent indifference to fungi as biological weapons is a perception engendered by noncommunicability, lack of history of use or development, and a low (although increasing) incidence of symptomatic disease following natural infection. Awareness of weapon potential is an important consideration for greater preparedness against the threat posed by biowarfare and bioterrorism.

Furthermore, some fungi are notorious plant pathogens, making them suitable as "anticrop" bioweapons in the agricultural sector with the potential for causing famine and having economic implications. For example, in the 1950s and 1960s, the United States stockpiled more than 30,000 kg of wheat stem rust spores to spray on crops, which was more than enough to infect all of the wheat

Approximate classification of some toxins from fungi			
Metabolite	Weapon	Mycotoxin	Pharmaceutical
Aflatoxins	+	+	
Ochratoxin A		+	
Cytochalasins		+	
Beauvericin		+	
Enniantins		+	
Destruxins			
Oosporein			
Moniliformin		+	
Efrapeptins			
Beauveriolides			
Amanita phalloides toxins	+	+	
Patulin		+	+
Mycophenolic acid		+	+
Penicillin		+	+
T-2 toxin	+	+	
Ergot alkaloids		+	+

grown around the world (this supply was destroyed in 1973). Countries where crops are developed almost as a monoculture (that is, cultivation of a single crop) are at particular risk from natural pathogens. Fungal biocontrol agents also provide an interesting example that is of relevance to weapons. These are fungi used as alternatives to chemical pesticides to control pests and crop pathogens. An example is the use of *Fusarium oxysporum* to destroy coca plants in a number of Latin American countries, where the idea is to introduce disease in order to stop the manufacture of cocaine. The technology involved in this case is similar to what would be required for a bioweapon in general.

Fungal toxins. Many fungi produce toxins (see table). It is important to appreciate that they are considered to be toxins only if they are active physiologically at low concentrations. Within this group, there is a subset referred to as mycotoxins, which are derived from fungi that occur naturally in food, and it is these that are conventionally considered as potential weapons (**Fig. 1**). Some of these can also be considered as pharmaceuticals, with concentration being the crucial factor in governing when a pharmaceutical becomes toxic. T-2 toxin (a trichothecene mycotoxin produced by *Fusarium* that is highly irritating to skin and mucous membranes) and aflatoxins (toxins produced by some strains of the fungus *Aspergillus flavus*, which are the most potent natural carcinogens yet discovered; see structure below) allegedly have been researched by various

Aflatoxin B$_1$

governments with regard to their ability to be deployed as weapons. Regardless of whether these toxins were ever tested or used as weapons, their psychological effect on a potential enemy cannot be ignored. In fact, T-2 toxin appears to be a valid weapon, as exposure to a few milligrams is potentially lethal.

There has been concern since antiquity about toxins from macrofungi (for example, "toadstools") due to accidental consumption of the fruiting body. These fungi have been used as weapons in the form of poisons (even if these poisons were usually intended only for individuals). However, α-aminitin remains a major concern, because it is extremely toxic, water-soluble, and heat-stabile. This toxin is produced by *Amanita phalloides* (**Fig. 2**). Further scientific investigations of mycotoxins were undertaken following the discovery of aflatoxin from *Aspergillus flavus* in the 1960s (**Fig. 3**). Mycotoxins have been ranked as the most important chronic risk factor in the diet (for example, above pesticide residues), although chronic effects are of little interest to weapon manufacturers. They are considered to be more toxic acutely than pesticides.

The overlapping relationships among mycotoxins, pharmaceuticals, and fungal biochemical weapons are represented diagrammatically in Fig. 1. The same compound can be represented in different fields. Presently, there are only a few metabolites, out of potentially thousands, that are considered to be mycotoxins. A minute percentage of toxic fungal metabolites have been considered seriously as weapons. Aflatoxins and T-2 toxin are also obviously mycotoxins. Ergot alkaloids (isolated from the dark purple or black sclerotium of the fungus *Claviceps purpurea*) are other potential bioweapons. Also, there may be fungal metabolites more toxic than mycotoxins that are not detected normally in the environment.

Water as a vector. Water, either drinking or nondrinking, may be an effective medium for toxin and biological weapon dispersal. The threat from contaminated drinking water is obvious. In the case of nondrinking water, the toxin could be spread, for example, from a shower and then inhaled. Workplaces such as farms or car washes where high volumes of water are employed could be susceptible to toxins or fungi. Moreover, drinking water for animals may be at a considerably higher level of risk than that for humans, thereby increasing the threat to food sources.

Genetically modified fungi. It is possible to speculate about genetically altered fungal strains that would be more virulent and/or produce higher yields of toxins than the wild type. Countries with large resources could perhaps develop such strains, although their use would be limited to a worst-case scenario. It is probably only a matter of time before such fungi exist (if they do not already), considering the burgeoning developments in genetics.

Treatment and decontamination. Supportive therapies for mycotoxicosis consist of improved diet and hydration of patients. Taking superactivated charcoal (a powdered, granular, or pelleted form of amorphous carbon characterized by an extremely large

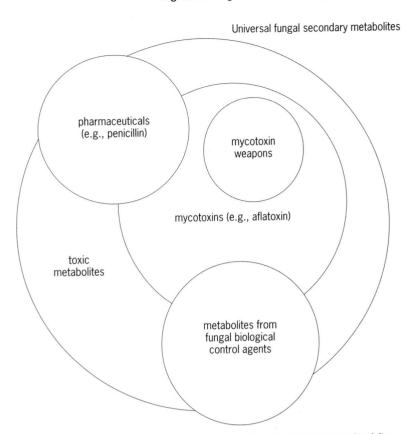

Fig. 1. Venn diagram of the relationship between fungal metabolites in terms of toxicity. The sizes of the circles are in relative proportion to the actual numbers of compounds.

Fig. 2. Fruiting body of *Amanita phalloides*, which produces α-aminitin, a dangerous toxin.

surface area per unit volume because of an enormous number of fine pores) orally may be effective if toxins are swallowed (for example, T-2 toxin). The low-calorie sweetener aspartame is very protective against ochratoxin A intoxication, and Oltipraz (an antischistosomal and anticarcinogenic agent) effectively protects against aflatoxin B_1 acute toxicity and carcinogenicity. Some strains of *Lactobacillus*

Fig. 3. Conidiophores or spore-bearing structures of *Aspergillus flavus*, the producer of aflatoxins.

effectively bind dietary mycotoxins and may be an effective treatment.

Biotoxins from fungi are difficult to remove from food and water. Washing with bleach effectively oxidizes most aflatoxins and a number of other mycotoxins. Potassium permanganate under alkaline conditions appears to be effective for a wider range of mycotoxins and for more situations than bleach.

Some priorities. Mycological and chemical sampling and detection methodologies need to be improved to address these emerging concerns. Increased investment and research into the taxonomy of fungi is necessary in order to provide unambiguous identification of fungi that produce toxins. Inactivation of mycotoxins and decontamination of food plants are other important issues that need proper research. In addition, foods need to be ranked in terms of their vulnerability to attack, and risks to people when foods are intentionally contaminated must be assessed. Research and governmental facilities that collect fungi (for example, culture collections) need to be highly organized with regard to security to prevent the organisms from being employed nefariously. Finally, there is a requirement to have a better understanding of what are normal levels of fungi and toxins in the environment.

For background information *see* AFLATOXIN; BIOTECHNOLOGY; FUNGAL GENETICS; FUNGAL GENOMICS; FUNGI; GENETIC ENGINEERING; MEDICAL MYCOLOGY; MYCOTOXIN; POISON; TOXIN in the McGraw-Hill Encyclopedia of Science & Technology.

Robert Russell Monteith Paterson;
Nelson Lima

Bibliography. J. Burnett, *Fungal Populations and Species*, Oxford University Press, 2003; CAST, *Mycotoxins: Risks in Plant, Animal, and Human Systems*, Council for Agricultural Science and Technology, 2003; R. J. Cole, B. B. Jarvis, and M. A. Schweikert, *Handbook of Secondary Fungal Metabolites*, Academic Press, 2003; C. P. Holstege et al., Unusual but potential agents of terrorists, *Emerg. Med. Clin. N. Am.*, 25:549–566, 2007; E. Latxague et al., Methodology for assessing the risk posed by the deliberate and harmful use of plant pathogens in Europe, *EPPO Bull.*, 37:427– 435, 2007; R. R. M. Paterson, Fungi and fungal toxins as weapons, *Mycol. Res.*, 110:1003–1010, 2006; R. R. M. Paterson and N. Lima, The weaponisation of mycotoxins, in M. Rai and A. Varma (eds.), *Mycotoxins in Food, Feed, and Bioweapons*, Springer-Verlag, 2009.

Geometrization theorem

Thurston's geometrization conjecture (now, a theorem of Gregory Perelman) aims to answer the question: How could one describe the possible shapes of our universe? Here we will be assuming (although, some scientists think otherwise) that our space is three-dimensional (3-D) [it has three directions: width, depth and height], closed (which means that you cannot travel infinitely far away from a given point in space, and that it does not have a boundary), connected (you can get from any point to any point), and orientable (if you travel along a path and return to the point of departure, your left and right hands will not get switched).

Constructing spaces from building blocks. In order to understand the answer to this question, we first look at the two-dimensional (2-D) case for guidance. If our universe were a 2-D surface S, we could describe its shape by combining copies of three basic building blocks:

1. The disk D^2 (given by the inequality $x^2 + y^2 \leq 1$).

2. The annulus A^2, also known as the cylinder (disk with one hole).

3. The pair of pants P^2 (disk with two holes).

Each surface S is obtained by gluing these blocks along their boundary circles. For instance, the 2-D sphere S^2 (given by the equation $x^2 + y^2 + z^2 = 1$) can be obtained by gluing together two disks (lower and upper hemispheres in S^2). The 2-D torus T^2 (the surface of a doughnut) is obtained by identifying (gluing together) boundary circles of the cylinder (**Fig. 1**). [In our description of shapes we are not concerned with the measurements of distance. For instance, instead of the disk of radius 1, we could have taken, say, an ellipse. Mathematicians define this field of study as topology.]

Every surface S can also be obtained as a connected sum of tori. This is done by taking a collection $T_1, \ldots, T_g$ of 2-D tori and attaching them to the 2-sphere S^2 by first removing a disk from each torus and g disks from S^2 and then identifying the boundary circles. Both descriptions of surfaces will be useful

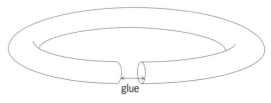

glue

Fig. 1. Making a 2-D torus.

when dealing with 3-D universes, which mathematicians call 3-D manifolds.

How can these descriptions of 2-D surfaces be generalized to dimension 3? We can start by promoting the 2-D building blocks to 3-D ones by adding an extra spatial dimension (a circle) to each. For instance, the solid torus, $D^2 \times S^1$, is obtained by rotating the 2-D disk around an axis in 3-D space that is far enough from the disk. The result is the 3-D "doughnut". Similarly, one obtains $A^2 \times S^1$ (a doughnut with a thinner concentric doughnut removed) and $P^2 \times S^1$ (a doughnut from which we have removed two thinner doughnuts). The boundary of each 3-D block we obtained consists of one, two, or three tori. Now, take a collection of these 3-D blocks and glue them together along their boundary tori. The resulting spaces are called graph manifolds. For instance, we can take the 3-D sphere S^3 given by Eq. (1) in four-dimensional

$$x^2 + y^2 + z^2 + w^2 = 1 \qquad (1)$$

(4-D) space. One can describe S^3 as a graph manifold by gluing together two solid tori: Take a small spherical cap on S^3 around the north pole, $(0, 0, 0, 1)$, then rotate this cap around the xy plane. The result of the rotation is a solid torus. Removing this solid torus from S^3 leaves us with another solid torus. As an additional example, we can consider the 3-D torus $T^3 = S^1 \times S^1 \times S^1$. It can be obtained by gluing together the two boundary tori of $A^2 \times S^1$.

By analogy with the description of 2-D surfaces above, one could have expected that all 3-D manifolds are graph manifolds. It turns out that this is far from being the case. The missing manifolds are best described by means of hyperbolic geometry.

Hyperbolic geometry. On a small scale (say, the scale of a single planet), the universe appears flat: The distances between points in space can be computed by the Pythagorean formula, or, as mathematicians would say, by means of the Euclidean (flat Riemannian) metric, which, in coordinates, is given by Eq. (2). However, on the scale of a planetary system (and beyond), the universe is curved by gravity. The distance measurements are described by means of a Riemannian metric. To define such a metric, from now on we will use the coordinates x_1, x_2, x_3; the notation x will be reserved for a point in a manifold. Then a Riemannian metric is given by a metric tensor, g_{ij}, through Eq. (3). This metric tensor is required to

$$ds^2 = (dx)^2 + (dy)^2 + (dz)^2 \qquad (2)$$

$$ds^2 = \sum_{i,j} g_{ij} dx_i dx_j \qquad (3)$$

be symmetric, that is, $g_{ij} = g_{ji}$, and positive-definite, that is, ds^2 is positive.

Hyperbolic 3-space, $\mathbf{H}^3$, can be described as the upper half-space ($x_3 > 0$) in 3-D space, with the Riemannian metric given by Eq. (4). In this descrip-

$$ds^2 = \frac{dx_1^2 + dx_2^2 + dx_3^2}{x_3^2} \qquad (4)$$

tion, hyperbolic space does not seem to have any preferred origin, but appears to have a preferred direction. To dispel this misperception, we can change the coordinates and describe the hyperbolic metric on the unit ball, $x_1^2 + x_2^2 + x_3^2 < 1$, given by Eq. (5).

$$ds^2 = 4 \frac{dx_1^2 + dx_2^2 + dx_3^2}{\left(1 - \left(x_1^2 + x_2^2 + x_3^2\right)\right)^2} \qquad (5)$$

It is then clear that such a metric is homogeneous (it looks the same at each point) and isotropic (is the same in every direction). Mathematicians describe this property by saying that the hyperbolic metric has constant curvature. There are three classes of constant-curvature metrics: positive curvature (the metric on the sphere), zero curvature (the metric of flat space), and negative curvature (the hyperbolic metric). In order to visualize what negative curvature means, we note that the volume enclosed by a sphere of (hyperbolic) radius r in hyperbolic space grows exponentially fast for large r. (In flat 3-D space, the volume grows only cubically.)

How can our universe possibly have negative curvature? At first glance, this would violate our assumption that it is closed (has finite size and is borderless). Let us reexamine the flat metric. We can start with a cube Q in the flat 3-D space, and take opposite faces of Q and identify them by parallel translations. The result has the shape of the 3-D torus T^3, which then inherits a flat Riemannian metric from Q since the translations are isometries (they preserve the metric).

How does this help to construct closed hyperbolic manifolds? Imagine that you have a finite polyhedral convex solid P in $\mathbf{H}^3$ and a collection of isometries of $\mathbf{H}^3$ identifying the faces of P. The result of identification (subject to certain conditions) is a closed hyperbolic 3-D manifold M^3. Its Riemannian metric is the one inherited from P. One of the earliest examples of this construction is the Seifert-Weber dodecahedral space, obtained from a right-angled hyperbolic dodecahedron P (**Fig. 2**).

More generally, one can take a convex finite-volume solid P in $\mathbf{H}^3$ with finitely many faces (P could have infinite diameter; **Fig. 3**) to obtain a hyperbolic 3-D manifold M^3 of finite volume. The resulting manifolds M^3 are hyperbolic, and every finite-volume hyperbolic manifold appears in this way. Even if M^3 has infinite diameter, there still exists a manifold (with boundary) N^3 of finite diameter (and a metric different from the one of M^3), so that M^3 is obtained by removing from N^3 its boundary. The boundary of N^3 is called the ideal boundary of M^3, since it appears infinitely far from the points of the manifold M^3. This boundary is a finite collection of 2-D tori T^2. The manifold N^3 (unlike M^3) is compact, that is, one cannot travel indefinitely far from a given point; therefore,

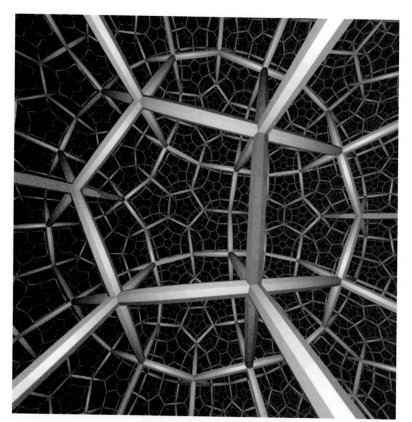

Fig. 2. Life in Seifert-Weber dodecahedral space. (*Courtesy of Geometry Center, University of Minnesota*)

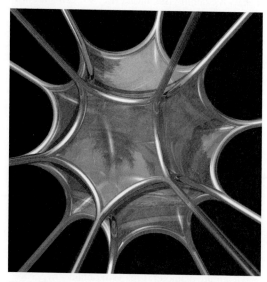

Fig. 3. Hyperbolic dodecahedron of infinite diameter. The infinite spikes are rounded off "chimneys," which are here triangular rather than rectangular. (*Courtesy of Paul Nylander, http://bugman123.com/*)

this procedure is said to compactify M^3. To visualize the compactification, imagine that the solid P consists of a finite part (a house) and an infinite part, a rectangular chimney rising from the roof (this chimney has infinite height). Identification of the faces of P is such that the opposite vertical sides of the chimney are identified by horizontal translations. After this identification, at each finite height level of the chimney we see a 2-D torus. As the height increases, these tori move away from an observer staying in the house. At the infinite height, the tori reach a limit which is the ideal boundary torus of M^3. The compactification process amounts to compressing an infinite chimney to a finite one, so that the ideal boundary torus lowers to a finite height (bringing it "from heaven to earth"). It is common to say that a 3-D manifold (with unspecified geometry) is hyperbolic if it admits a hyperbolic metric of finite volume. (Such a metric is known to be unique.) To simplify the terminology, one also refers to the compactifications N^3 as hyperbolic manifolds. The boundary tori of a hyperbolic manifold satisfy an important technical property: They are incompressible. A boundary surface S of a manifold N^3 is called incompressible if any circle on S that bounds a 2-D disk in N^3 already bounds a disk in S.

One can show that hyperbolic manifolds are never graph manifolds, so we indeed have a new class of possible universes. More generally, a manifold is called geometric if it admits a Riemannian metric that is locally homogeneous (that is, it locally looks the same at all points). *See* HYPERBOLIC 3-MANIFOLDS.

Thurston's geometrization conjecture. We are now ready to state Thurston's geometrization conjecture: Every (closed, oriented) 3-D manifold M^3 can be obtained as follows: Start with a finite collection of graph manifolds (with incompressible boundary, if there is any) and hyperbolic manifolds. Then glue these manifolds along their respective boundary tori so that the result has no boundary. It is possible that the resulting manifold M' is disconnected (you cannot reach one point from another). To remedy this, take a connected sum of the components of M': Remove small balls from distinct components and glue together the exposed boundary 2-D spheres. The result is M^3.

In other words, one can build M^3 by using the three types of basic 3-D building blocks appearing in graph manifolds, together with hyperbolic manifolds. Equivalently, one can state the Thurston geometrization conjecture by saying that M^3 can be built from a collection of geometric manifolds by gluing them along incompressible boundary tori and then taking a connected sum. (The conjecture derives its name from this formulation.)

William Thurston proved his conjecture in the 1970s under the technical assumption that M^3 is a Haken manifold. This work covered a large class of 3-D manifolds, but far from all of them.

Poincaré conjecture. An important special case of the Thurston geometrization conjecture is the Poincaré conjecture, to which Thurston's technique did not apply: If M^3 is a closed, simply connected 3-D manifold, then M^3 is the 3-D sphere. (The manifold M^3 is said to be simply connected if every circle in M^3 can be contracted to a point.)

This conjecture was first formulated by Henri Poincaré in the early twentieth century and, despite numerous efforts, remained out of reach until Perelman's work. Its analogs were proven for higher-dimensional manifolds (dimensions equal to or greater than 5 by the 1960s, and dimension 4, in the topological form, by the 1980s) through the efforts of many mathematicians (most notably, Stephen Smale, John Stallings, and Michael Freedman).

Spherical space forms conjecture. Another special case of the Thurston geometrization conjecture is the spherical space forms conjecture: Suppose that M^3 is a closed 3-D manifold with finite fundamental group. (In the context of 3-D manifolds, this property can be formulated as follows: If c is any circle in the manifold M^3, then, for some $n > 0$, the circle c^n, obtained by tracing the original circle n times, can be contracted to a point in M^3.) Then M^3 is a spherical space form, that is, it admits a metric of constant positive curvature. In particular, M^3 is a graph manifold.

Ricci flow. The Ricci flow was introduced by Richard Hamilton as a possible approach to the Thurston geometrization conjecture. Hamilton used it to prove this conjecture for manifolds of positive curvature in 1982. He also wrote a number of other papers establishing a program for proving the Thurston geometrization conjecture using the Ricci flow.

Roughly speaking, the idea is to define a flow on the space of Riemannian metrics that, starting with an arbitrary metric tensor g on a 3-D manifold, "homogenizes" g and in the limit separates the manifold into graph manifolds and homogeneous pieces. To define the Ricci flow, consider a metric tensor $g_{ij}(x)$ on a 3-D manifold M. This metric can be encoded in a 3-by-3 array of numbers, indexed by i and j, that depend on the point x in M. The curvature of g is described by means of the Riemannian curvature tensor R_{ipjq} (given by a 4-D array of numbers). This tensor has an important simplification, the Ricci tensor, given by Eq. (6), which is a symmetric ($\mathrm{Ric}_{ij} = \mathrm{Ric}_{ji}$)

$$\mathrm{Ric}(g) = \mathrm{Ric}_{ij} = \sum_{p,q} R_{ipjq} \, g^{pq} \qquad (6)$$

2-D array of numbers. Then, the Ricci flow is given by the differential equation (7). The Ricci flow can

$$\frac{\partial g(t)}{\partial t} = -2\mathrm{Ric}[g(t)] \qquad (7)$$

be regarded as an analog of the heat flow, describing the evolution of temperature in flat space. As the heat flow homogenizes the temperature distribution in space, the Ricci flow homogenizes Riemannian metrics.

By rescaling both space and time, one gets the normalized Ricci flow of metrics of constant volume, given by Eq. (8). Here $r = r(t)$ is some scalar func-

$$\hat{g}'(t) = -2\mathrm{Ric}[\hat{g}(t)] + {}^{2}\!/_{3}\, r\hat{g}(t) \qquad (8)$$

tion. Suppose that $\hat{g}(t_0)$ is a stationary point of the normalized Ricci flow, that is, $\hat{g}'(t_0) = 0$. Then $\hat{g}(t_0)$ is an Einstein metric: Its Ricci tensor is a scalar multiple of $\hat{g}(t_0)$. In dimension 3, Einstein metrics all have constant curvature, hence M^3 is geometric. This suggests

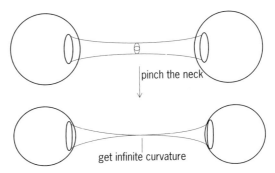

Fig. 4. Neck pinching, resulting in infinite curvature.

the strategy: Start with an arbitrary metric $g(0)$ on M^3 and let it evolve via the normalized Ricci flow, hoping that it will converge to a stationary metric (that will have to be homogeneous). Hamilton showed that this strategy actually works if $g(0)$ has positive curvature. However, it has been known since the early 1980s that in many cases 2-D spheres inside of M^3 can cause the normalized Ricci flow to blow up (the curvature becoming infinite) in a finite amount of time.

An example of this phenomenon is known as the "dumbbell": Take two copies of the round 3-D sphere and connect them by a thin neck. The neck will get pinched (under both the Ricci flow and the normalized Ricci flow) in finite time as the 2-D sphere in its cross section gets pinched into a point (and has curvature blowing up to infinity in the process) as $t \to T_1 < \infty$ and our "universe" splits in two (**Fig. 4**).

In this example the Ricci flow "finds" the 2-D spheres along which M^3 has to decompose as a connected sum (before becoming geometric). The example also suggests that one should look for these spheres in the part, $M(t)^+$, of M^3 where the curvature of $g(t)$ is "high." Hamilton was hoping that one could prove that the regions of high curvature on $M^3(t)$ [and the corresponding singularities at the first blowup time] are "nearly standard," similar to the dumbbell example.

Perelman's work. Understanding singularities of g at the first blowup time T_1 was the first among major obstacles in the way of Hamilton's program by the mid-1990s. Working in nearly total isolation and secrecy from the mid-1990s to 2002, Perelman introduced several new important tools and ideas and modified Hamilton's program; his work culminated in a sequence of three preprints in 2002–2003 proving the entire Thurston geometrization conjecture. Instead of looking at the Ricci flow as just an evolution of a metric on a 3-D manifold M^3, he analyzed the geometry of the 4-D manifold N^4 obtained from M^3 by adding the extra dimension (the time t). This 4-D geometry is specifically designed to reflect properties of the Ricci flow. Using this geometry (in combination with the Alexandrov geometry of singular metric spaces), Perelman established that singularities of the Ricci flow are, indeed, standard after appropriate rescaling and taking a limit. This allowed

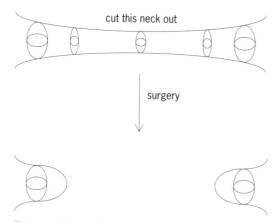

cut this neck out

surgery

Fig. 5. Ricci flow with surgery.

him to prove, without identifying the geometry of singularities precisely, that the topological type of high-curvature regions $M(t)^+$ in $M(t)$ is standard, the main examples being a cup (a 3-D ball), a spherical neck (the product of the 2-D sphere and the interval), or a spherical space form (a 3-D manifold that admits a metric of positive curvature). On the other hand, Perelman managed to understand the geometry of $M(t)^+$ well enough to be able to cut it off from the rest of M^3 along some spheres and attach nearly round spherical caps along these spheres (**Fig. 5**).

As a result, Perelman defined (realizing an earlier idea of Hamilton) the Ricci flow with surgeries, which evolves both the geometry and the topology of M^3. He proved that the surgery times T_i are well separated and the Ricci flow with surgeries thus exists for all times $t \geq 0$. Perelman also proved that (under certain assumptions) the entire manifold becomes extinct in a finite time, the curvature becoming infinitely high everywhere. (This was also independently established by Tobias Colding and William Minnicozzi.) This means that the original manifold was the connected sum of spherical space forms. In particular, this happens if M^3 is simply connected or, more generally, has a finite fundamental group; this yields the Poincaré conjecture and the spherical space forms conjecture. Lastly, extending prior work of Hamilton, Perelman proved that at $t = \infty$ the manifold $M(t)$ splits along incompressible tori into the pieces M_{thick}, where the metric becomes hyperbolic, and the pieces M_{thin}, where the metric collapses to metrics of dimension 1 or 2.

This work was described by Perelman in three preprints that he posted on the Internet in 2002–2003: "Entropy formula for the Ricci flow and its geometric applications," "Ricci flow with surgery on 3-manifolds," and "Finite extinction time for the solutions to the Ricci flow on certain three-manifolds." In order to establish the full Thurston geometrization conjecture, one also needs to show that M_{thin} is a graph manifold. A proof of this proposition (originally promised by Perelman) was soon provided by Takashi Shioya and Takao Yamaguchi, and later on by other mathematicians.

So far, Perelman's papers have not been published in a peer-reviewed journal (and he does not seem to be interested in this), but only posted on the Mathematical Archive, which is a common place for mathematicians to post preliminary versions of their work. Many arguments in Perelman's proofs are rather sketchy and some important details are missing. Soon after the appearance of Perelman's preprints, several teams of mathematicians started to work toward filling in the missing details. Three detailed accounts of Perelman's proof of the Thurston geometrization conjecture have emerged:

1. In the summer of 2003, Bruce Kleiner and John Lott started to post missing details of Perelman's proof on the Internet; their work culminated in an article covering the entire proof.

2. John Morgan and Gang Tian published a book with a self-contained treatment of the Poincaré conjecture (and spherical space forms conjecture) part of the Thurston geometrization conjecture.

3. An account of the entire proof by Huai-Dong Cao and Xi-Ping Zhu, which caused a great deal of controversy.

Mathematical significance. The proof of the Thurston geometrization conjecture solved a number of important open problems (for instance, the Poincaré conjecture). The Thurston geometrization conjecture also makes it possible to reduce many open problems about 3-D manifolds to questions about hyperbolic manifolds, which then can be attacked using geometric, analytic, or algebraic methods. For instance, using the Thurston geometrization conjecture, one can define an algorithm that will determine if two 3-D manifolds have the same topological shape. Such an algorithm is known to be impossible for manifolds of dimension equal to or greater than 4.

Relevance outside of mathematics. The Thurston geometrization conjecture deals with possible shapes of our universe. Determining the actual shape of our space is an important problem in cosmology, which is best understood in terms of the Thurston geometrization conjecture. For instance, astronomers, following a suggestion of Jean-Pierre Luminet, have recently attempted to determine, by observing the cosmic background radiation, whether our universe has the shape of a certain space form (Poincaré dodecahedral space).

For background information *see* COSMOLOGY; MANIFOLD (MATHEMATICS); RELATIVITY; RIEMANNIAN GEOMETRY; TOPOLOGY in the McGraw-Hill Encyclopedia of Science & Technology.   Michael Kapovich

Bibliography. H.-D. Cao and X.-P. Zhu, A complete proof of the Poincaré and geometrization conjectures—application of the Hamilton-Perelman theory of the Ricci flow, *Asian J. Math.*, 10:165–492, 2006; B. Kleiner and J. Lott, Notes on Perelman's papers, *Geom. Topol.*, 12:2587–2855, 2008; J. Morgan and G. Tian, *Ricci Flow and the Poincaré Conjecture*, vol. 3 of *Clay Mathematics Monographs*, American Mathematical Society, Providence, RI, 2007; W. Thurston, *Three-Dimensional Geometry and Topology, I*, vol. 35 of *Princeton Mathematical Series*, Princeton University Press, 1997.

Green computing

Green computing refers to the responsible use of computer and related resources with the object of curbing the negative effects that the increasing use of computers and their associated technologies have on the Earth's limited natural resources. Green computing is part of the broader concept of green technology (Green Tech), or clean technology (Clean Tech), which aims to use environmental and materials science to ensure that current and future technologies are socially equitable, economically viable, and do not damage or exhaust natural resources (sustainable). Green computing has been evolving over the last decade and, although there is no universal agreement as to what it may finally encompass, it generally considers all aspects of computer technology that contribute to reducing global warming and electronic waste (e-waste) through the use of alternative energy sources, power management, recycling, and biodegradable materials. Green computing also includes the implementation of national or international policies, agreements, and regulations to accomplish the above goals.

Why do we need to go green? According to the Population Institute, climate change is perhaps the most crucial environmental challenge of this century. Eleven of the world's 12 highest annual global temperatures on record have occurred since 1995. There seems to be little doubt that the world's increasing population, along with the continuous demand for more fossil-based fuels, have contributed to an increase in the concentration of greenhouse gases—carbon dioxide, methane, nitrous oxide, and fluorocarbons—which are believed to be responsible for global warming. The consequences of global warming are dire at best. The effects of global warming can result in accelerated melting of the polar ice caps, which would result in rising sea level, as well as changes in precipitation patterns, including long-term droughts. These two consequences of global warming are enough to disrupt both life and the world's economy. As a result of the 1997 Kyoto Protocol for the United Nations Framework Convention on Climate Change, most of the countries of the world began to take steps toward reducing emissions of greenhouse gases. Much of these emissions are the result of power plants that use fossil fuels to generate electricity. According to the U.S. Environmental Protection Agency (EPA), power plants are responsible for 40% of the carbon dioxide emissions in the United States. On a global scale, the United States is by far the country with the most emissions of greenhouse gases. In fact, the carbon dioxide equivalent of the United States, measured in thousands of metric tons, is about 6746, compared to more populated countries, such as China (3650) and India (1228), or the European Union (4030). As of this writing the United States has not ratified the Kyoto Protocol.

Along with the need to reduce global warming, it is also necessary to decrease the contamination of the planet with toxic waste and, in particular, e-waste. Under the Resource Conservation and Recovery Act (RCRA) issued by the EPA, televisions, monitors, computers, computer peripherals, audio and stereo equipment, VCRs, DVD players, video cameras, telephones, facsimile and copying machines, cellular phones, wireless devices, and video game consoles are primary examples of electronics and of what constitutes e-waste. According to estimates by the United Nations, each year around 20–50 million tons of computer gear and cellular phones are dumped into landfills worldwide. As a result, dangerous chemicals, such as mercury and poly(vinyl chloride) [PVC], are dumped into the environment. PVC generally contains toxic additives that leach out. Toxic waste can harm humans, animals, and plants if they encounter it buried in the ground, in stream runoff, in groundwater that supplies drinking water, or in floodwaters, as happened after Hurricane Katrina. Some toxins, such as mercury, persist in the environment and accumulate for hundreds of years. Humans or animals often absorb these toxins through food such as fish. In the long run, continuous consumption of these toxins will have a negative effect on human or animal health. In addition, hazardous materials from the world's leading economies often end up as detritus in the world's desperate places. Environmental groups say that there is a good chance electronic waste will end up in a dump somewhere in a third-world country, where thousands of laborers scavenge it for the precious metals inside, unwittingly exposing themselves and their surroundings to innumerable toxic hazards. A 1989 international treaty, known as the Basel Convention, restricts such transfers; however, the United States has not ratified this treaty. The EPA has a set of voluntary guidelines, known as R2 (responsible recycling), which include the general principles and specific practices for recyclers disassembling or reclaiming used electronics equipment, including those electronics that are exported for refurbishment and recycling. Similar, but more restrictive guidelines have been adopted in Europe and Asia.

How can green computing help? One of the main objectives of green technologies and, in particular, of green computing is to reduce the "carbon footprint" required or generated by computer technology. As indicated earlier, power plants are highly responsible for the production of greenhouse gases. To alleviate this, it is necessary to reduce the demand for electricity by computers throughout the world. According to the EPA, U.S. data centers alone consume as much power in a year as is generated by five power plants. Thus, the need for data center efficiency is a must. Several approaches to alternative sources of energy have been suggested to accomplish the goal of reducing their carbon footprint. The **table** shows some of the most promising technologies, along with some of their advantages and drawbacks.

To further reduce the consumption of energy, electronic devices, such as computers, printers, and some other peripherals, are making use of Advanced Configuration and Power Interface (ACPI). If it detects the system has been inactive for some time,

Comparison of green power technologies		
Technology	Advantages	Drawbacks
Solid oxide fuel cells (SOFCs)	Electrochemical power plants that produce no air pollutants. They make more efficient use of energy by reducing the amount of hydrocarbon-based fuel needed to generate the same amount of energy as current combustion engines.	Still rely on hydrocarbons and do not eliminate carbon dioxide (CO_2) emissions. They require a hydrogen infrastructure that still has safety concerns and requires high costs to produce, store, and transfer.
Solar energy	Eliminates air pollution and CO_2 emissions. Generates electricity from photons emitted by the Sun. This technology will play an important role in the future.	Still costly to produce. Current energy collectors only absorb energy from narrow range of the Sun's light-wave emissions.
Wind energy	Eliminates air pollution and CO_2 emissions. Wind turns blades connected to a drive shaft that turns an electric generator to produce electricity.	Negative effect on bird population and visual impact on landscape.
Nuclear	Nuclear fission of enriched radioactive isotopic material to produce electricity. This is the single greatest source of energy with no impact on global warming and capable of sustaining present and future energy needs.	Produces radioactive waste that last for thousands of years. Accidents can happen, which may have devastating consequences for human health and the environment.
Enhanced geothermal systems	Harnesses the heat generated naturally by the Earth to generate electricity. Wells are drilled into high-temperature basement rock (over 250°C) that is naturally fractured. Cold water pumped into the wells absorbs heat as it travels through the fractured rock. The heat absorbed by the water is converted into electricity via steam turbines. Water is reused and pumped again for a new cycle. This technology has zero carbon emissions.	Determining appropriate sites for drilling is difficult as a result of lack of experience with this new technology. A site is suitable for about 20 to 30 years. Maintenance of a large volume of rock (several cubic kilometers) has not been quantified nor verified properly.

usually determined by the user or system defaults, it turns off the power or switches the system to a low-power state. According to the EPA, power management is important because the energy consumption by inactive computers in the United States alone is equivalent to the energy produced by five power plants in a year. Following some basic recommendations to keep consumption of energy to a minimum, such as turning off your computer and all its peripherals when they are not in use, may result, on average, in reducing nearly half a ton of carbon dioxide and saving more than $60 a year in energy costs. As an example of the effectiveness of the use of ACPI, it was widely reported during the last quarter of 2008 that the software company Symantec will save as much as $800,000 a year by setting up power-management software that turns off employee computers overnight and on weekends. Leading computer manufacturers, such as Apple and Dell, have also announced that their new products will be more energy-efficient. For example, the new MacBook laptops are 30% more energy-efficient than previous models, use less packaging, and meet two of the "gold standards" of green computing: the Energy Star 4.0 certification and the Electronic Product Environmental Assessment Tool (EPEAT) gold-level rating. The EPEAT evaluates electronic products in relation to 51 environmental criteria. Twenty-three of these criteria are required and 28 are optional. The EPEAT's environmental criteria are contained in the IEEE (Institute of Electrical and Electronics Engineers) Standard 1680. Under the EPEAT program, manufacturers declare their products' conformance to a comprehensive set of environmental criteria. Products are rated bronze, silver, and gold based on how many criteria are met. EPEAT gold products must meet all 23 required criteria and at least 75% of the optional criteria. Energy Star is a joint program of the EPA and the U.S. Department of Energy aimed at protecting the environment through energy-efficient products and practices. The Energy Star program, which only rates energy and environmental issues of monitors, desktops, and laptops, intends to soon have a category for servers as well. The EPA is currently developing an Energy Star rating for entire data centers.

Green computing also emphasizes the use of renewable and biodegradable materials because they consume less energy in their preparation and are easier to dispose of at the end of their useful life. It is estimated that approximately 209,700,000 tons of solid waste are generated annually in the United States, with about 1% of it classified as computer and/or electronic equipment. Of the total amount of waste generated, only an estimated 134,000 tons are recycled (based on 1998 statistics, according to PC Recycler, Inc.). Therefore, there is a need to use products that can change the cycle of electronic products from "cradle to grave" to "cradle to cradle." The latter can only be achieved if electronic devices can be manufactured entirely from products that can be fully reclaimed or reused. Some manufacturers, such as Fujitsu, are currently using biodegradable plastic (bioplastic) for the main chassis of their personal computers. The bioplastic used is a type of poly(lactic acid), based on lactic acids produced from vegetable starch harvested from corn, potatoes, or wheat. The lactic acid is produced by microorganisms that act on the starch and then is chemically treated to form a plastic called polylactide (PLA). The Irish company MicroPro Computers is producing an entirely biodegradable and modular computer called iameco (pronounced "I am eco"). Iameco uses recycled wood panels for its computer case, keyboard, mouse, and monitor panels. One of the most remarkable aspects of the iameco computer is that implanted within the wood panels are seeds from native-tree species. When these computers are buried in a landfill, the wood breaks down and trees may grow. To make the computer more environmentally friendly, the iameco computer is modular so you can replace parts to keep it up to date.

Recently, some of the largest wireless communications equipment manufacturers, including NEC, Ericsson, Intel, and Ozmo, have invested time and effort to realize green communications (Green Comm) to satisfy the demanding requirements of wireless service subscribers. Several initiatives have been suggested in the cell phone industry. The idea is to make cell phones repairable and easily remodeled. This will result in phones in which the "casing" reflects fashion shifts, while preserving the guts of the phone as much as possible. A major objective in the wireless industry is reduction of the energy consumption of the transmitter/receiver (base station) towers. In fact, base stations generally consume 80% of the power used by wireless networks, with one 3G network using some 5–10 million kilowatt-hours of power each year.

Additional changes that apply not only to telecommunication equipment but also to the industry in general include the elimination of lead from printed-circuit boards. For example, in traditional manufacturing process, lead alloys are used to attach silicon chips to the inside of packages and to interconnect the electronic components of circuit boards. Newer technologies do not require lead alloys, instead using alloys consisting of tin, silver, and copper, which are not as harmful to the environment.

For background information *see* BIODEGRADATION; CLIMATE MODIFICATION; COMPUTER PERIPHERAL DEVICES; DISTRIBUTED SYSTEMS (COMPUTERS); ELECTRIC POWER GENERATION; ELECTRONIC PACKAGING; ENERGY SOURCES; GLOBAL CLIMATE CHANGE; HAZARDOUS WASTE; INDUSTRIAL ECOLOGY; MOBILE COMMUNICATIONS; POLYESTER RESINS; POLYVINYL RESINS; RECYCLING TECHNOLOGY; TOXICOLOGY in the McGraw-Hill Encyclopedia of Science & Technology.   Ramon A. Mata-Toledo; Young B. Choi

Bibliography. B. Choi Young, *From Green Computing to Green Communications: A New Application of GreenTech*, 2009.

Guard cells

Guard cells are morphologically distinct cells that are found in the epidermis (the outermost cell layer) of terrestrial plant shoots. Each pair of specialized guard cells circumscribes and defines a microscopic hole called a stoma (the plural form is stomata) [**Fig. 1**]. It is through these microscopic pores that plants lose water vapor to the atmosphere, and it is also through these microscopic pores that plants take up carbon dioxide gas (CO_2), which is then fixed into carbohydrate compounds by the internal, photosynthetic cells of the leaf (the mesophyll cells). This exchange of gases with the atmosphere is limited to the stomata because the rest of the epidermis is covered with a waxy layer, the cuticle (Fig. 1), which is essentially impervious to these gases.

If plants lose too much water, they will desiccate and die. Conversely, if plants do not capture enough CO_2 in the form of carbohydrates, they will not have sufficient energy resources to sustain life. Regulation of both water vapor loss and CO_2 uptake rates are controlled by the guard cells, which adjust the size of the stomatal pores by changing their volume in response to both external and internal signals. Therefore, guard cells are essential in regulating the opposing priorities of the terrestrial plant, to maximize CO_2 uptake while minimizing water vapor loss. This central importance has resulted in the evolution of complex sensory systems within the guard cells that perceive and respond to a great diversity of signals, including environmental stimuli, such as light, CO_2 concentrations, humidity, and pathogens, and internal stimuli, particularly plant hormones.

Stomata are typically found in highest density on the underside of leaves, but they can also be found

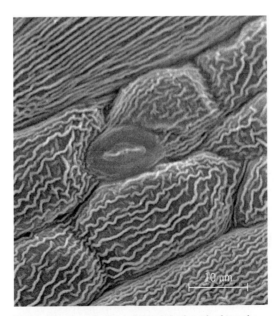

Fig. 1. Scanning electron micrograph of a pair of guard cells (with closed stoma) surrounded by larger epidermal cells. The wavy white lines prominent on the epidermal cells are wax deposits in the cuticular covering.

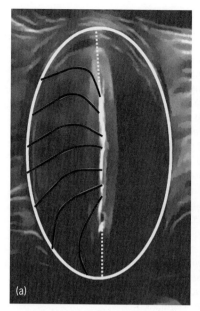
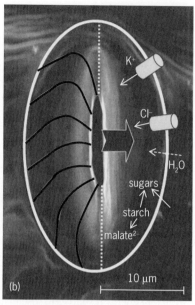

Fig. 2. Schematic diagram of processes leading to opening of a stoma (a → b). An increase in intracellular solutes drives water influx, which causes the guard cells to swell and bow out longitudinally (black arrow, b). Radially oriented cellulose microfibrils that aid in this directional expansion are depicted on the left guard cell.

on the top surface of leaves and on flowers and young stems to varying degrees depending on the plant species. Their distribution is controlled by a number of genes that regulate their placement and number, and can be influenced by environmental factors present during development such as CO_2 concentration. Regardless of the guard cell location and plant species, the fundamental biophysical mechanism that results in a change in guard cell volume remains constant: Water uptake by the guard cells results in increased pressure within the guard cells. This increased turgor pressure against the guard cell wall, which can exceed the pressures found in car tires, causes swelling and separation of the two guard cells, thereby opening the stoma (**Fig. 2**). Physically, separation of guard cells is the result of an outbowing (outwardly curving) expansion. This anisotropic (unequal along the axes) expansion of the guard cells can be traced to their specialized cell walls. The arrangement of reinforcing cellulose microfibrils in the guard cell wall (Fig. 2) favors expansion along the longitudinal axis of the cell. In addition, the dorsal cell wall (facing the neighboring epidermal cells) is thinner than the ventral cell wall (facing the pore). Accompanied by the greater density of cellulose microfibrils in the pore region, this cell wall arrangement then promotes the outbowing of the guard cells as cell volume increases, thus resulting in a widening of the stomatal pore.

Mechanism of guard cell volume change. Water uptake is driven by an increase in intracellular solute concentration resulting from import of potassium (K^+) and chloride (Cl^-) ions across the cell membrane of the guard cell and/or by increases in sugars and other organic molecules such as $malate^{2-}$ as a result of metabolic activity within the guard cell. Adenosine triphosphate (ATP)-powered pro-

ton pumps (H^+-ATPases) actively transport protons across the cell membrane out of the cell, causing a more negative membrane potential. This increased negativity of the guard cell provides the energetic driving force for uptake of positively charged K^+ ions. K^+ uptake occurs through ion channels in the cell membrane that are defined by specific proteins and structured so as to be selective for the K^+ ion. This influx of K^+ ions is electrically neutralized by Cl^- ion uptake and $malate^{2-}$ synthesis from starch breakdown. The increase in solute concentration drives water into the cell in order to equalize osmotic concentration. Increases in sugars from metabolic activity, photosynthetic production, import, or breakdown of osmotically inactive starch to osmotically active sugars also cause water uptake and swelling in order to maintain osmotic equilibrium. Conversely, when H^+ pump activity is inhibited and/or anion-selective channels in the cell membrane that mediate Cl^- and $malate^{2-}$ efflux open, the membrane potential of the guard cell becomes less negative, driving K^+ efflux. Sugar and/or $malate^{2-}$ metabolism may also occur, removing these osmotically active solutes. The net decrease in solute concentration leads to efflux of water and reduction of turgor, deflating the guard cells and closing the stomatal opening between them.

Environmental response and signaling. Guard cell volume changes rapidly and reversibly in response to a number of environmental stimuli, including light, CO_2 levels, humidity, and pathogen attack. Stomatal opening is triggered by light, particularly blue and red light. This allows the CO_2 uptake needed for photosynthesis, because red and blue are also the main colors (wavelengths) of light that are absorbed by chlorophyll. Blue light is absorbed by guard-cell chlorophyll, but it is also specifically perceived by photoreceptor molecules in the guard cells that are called phototropins. This specific blue light perception results in a signal transduction cascade that ultimately activates H^+-ATPases, leading to stomatal opening. Red light is perceived by guard-cell chloroplasts (plastids occurring in the green parts of plants, containing chlorophyll pigments, and functioning in photosynthesis and protein synthesis), but whether red light also activates guard-cell H^+-ATPases is still under debate. Alternatively, light-induced stomatal opening may be driven by an increased sugar content within the guard cell resulting from photosynthesis, starch breakdown, and/or sucrose import. A reduction in leaf intercellular CO_2 concentration also stimulates stomatal opening, so light-driven photosynthetic CO_2 fixation by the mesophyll cells of the leaf also triggers guard-cell volume increases and stomatal opening. Moreover, because intercellular CO_2 levels are linked to atmospheric CO_2 levels, global elevation of atmospheric CO_2 concentrations as a result of the burning of fossil fuels may have particular impact on stomatal apertures and gas-exchange regulation.

Drought induces stomatal closure via guard-cell response to increases in a hormone, abscisic acid

(ABA). As a key regulator of plant water status, ABA is produced in the roots when drought conditions are sensed. It is then transported through the vascular system of the plant and accumulates in the shoot, where it inhibits stomatal opening and promotes stomatal closure, thereby reducing water loss from the plant. ABA controls stomatal aperture by inhibiting the H^+-ATPases and activating plasma membrane anion channels; the resultant changes in membrane potential then drive K^+ efflux. ABA can also induce a rise in cytosolic Ca^{2+} concentration, an important intracellular signal that both inhibits inward K^+ channels and activates anion channels, further ensuring stomatal closure.

Stomata also close rapidly in response to a reduction in atmospheric humidity: Drier air increases the driving force for water loss from the plant and thus is another environmental condition that would promote plant desiccation in the absence of any countermeasure by the guard cells. Although the mechanisms by which guard cells sense water vapor and CO_2 concentrations are not known explicitly, the ensuing response pathways within guard cells utilize some of the same signaling intermediates and ion-transport molecules as have been identified to function in ABA responses.

As natural openings in the plant surface, stomata provide pathogens with an entry into the leaf interior. However, plant guard cells have evolved the ability to sense and respond to such invaders: Stomatal closure is triggered by at least some species of pathogenic bacteria, as well as by purified bacterial molecules such as PAMPs (pathogen/microbial-associated molecular patterns). As a counterdefense, some bacterial virulence factors, such as the phytotoxin coronatine, suppress this stomatal closure response. Severe outbreaks of bacterial infection in crop plants often follow periods of heavy rain and high humidity. These atmospheric conditions undoubtedly facilitate pathogen dispersal, but they also favor open stomata, and it has been hypothesized that bacterial infections may capitalize on this effect to achieve entry into the leaf.

From guard cell to globe. The multisensory nature of guard-cell signal perception and response has made guard cells a favored model for the study of overlapping signaling systems in plant cells. The unique properties of guard cells also make them a favorite model for electrophysiological investigations into the nature of ion-transport systems of plants. The essential role that guard cells play in balancing water loss and CO_2 uptake also dictates the key importance of guard-cell responses to such issues as whole-plant water status, drought tolerance and crop productivity, vegetative responses to and impacts on atmospheric CO_2 concentrations, and even the global water cycle of the past, present, and future. For example, studying guard-cell densities of fossilized plants has contributed to knowledge about prehistoric atmospheric conditions. Consequently, continued research into the environmental cues and intrinsic interactions within the guard cell that con-

trol stomatal movements is not only fundamental science but has far-reaching implications on such crucial concerns as world hunger and climate change.

For background information *see* ABSCISIC ACID; CARBON DIOXIDE; DROUGHT; EPIDERMIS (PLANT); LEAF; PHOTOSYNTHESIS; PLANT ANATOMY; PLANT CELL; PLANT HORMONES; PLANT ORGANS; PLANT PHYSIOLOGY; PLANT TISSUE SYSTEMS; PLANT-WATER RELATIONS in the McGraw-Hill Encyclopedia of Science & Technology.

Laura Ullrich Gilliland; Sarah M. Assmann
Bibliography. N. A. Campbell and J. B. Reece, *Biology*, 7th ed., Pearson/Benjamin Cummings, 2005; L. M. Fan, Z. Zhao, and S. M. Assmann, Guard cells: A dynamic signaling model, *Curr. Opin. Plant Biol.*, 7:537–546, 2004; J. M. Kwak, P. Mäser, and J. I. Schroeder, The clickable guard cell, version II, in *The Arabidopsis Book*, American Society of Plant Biologists, Rockville, MD, 2008; R. M. Roelfsema and R. Hedrich, In the light of stomatal opening: New insights into "the Watergate," *New Phytol.*, 167:665–691, 2005; L. Taiz and E. Zeiger, *Plant Physiology*, 4th ed., Sinauer Associates, Sunderland, MA, 2006.

Hadrosaurid (duck-billed) dinosaurs

Hadrosaurids, popularly known as duck-billed dinosaurs, comprise a highly diverse natural group (a clade) of plant-eating dinosaurs that evolved in the Late Cretaceous period (approximately 100–65 million years ago) [**Fig. 1**]. Hadrosaurids have the distinction of being the best-known dinosaur group in terms of the completeness of their fossil record. Hadrosaurids are known from hundreds of nearly complete fossil skeletons, including juvenile and embryonic skeletons, and thousands of isolated bones and teeth. Skin impressions and other soft tissue traces (including the original soft tissues themselves), footprints and trackways, and stomach contents have been described. Intact eggs and nests, as well as eggshells, are also relatively well known. Because of this incredible record, hadrosaurids have

Fig. 1. Skeleton of *Gryposaurus incurvimanus*, a representative hadrosaurid dinosaur.

Hadrosaurinae

Lambeosaurinae

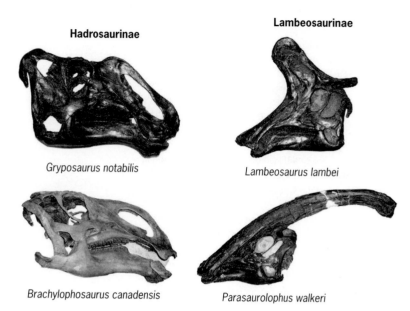

Gryposaurus notabilis

Lambeosaurus lambei

Brachylophosaurus canadensis

Parasaurolophus walkeri

Fig. 2. Skulls of representative hadrosaurine (left) and lambeosaurine (right) hadrosaurids.

contributed significantly to the overall understanding of dinosaur biology.

Description. Hadrosaurids are ornithischian dinosaurs and form part of a more inclusive group, called the Ornithopoda, sharing a similar general body plan. The most important differences between hadrosaurids and other closely related dinosaurs are concentrated in the skull. Hadrosaurids are characterized by a specialized feeding system that consists of a broad, toothless beak and hundreds of highly specialized teeth that are tightly packed to form complex dental batteries. These dental batteries form large grinding surfaces that were used to mechanically break down fibrous plant material. Chewing was accomplished via a jaw movement mechanism termed pleurokinesis (in which a portion of the upper jaw slid outward over the lower teeth to grind the food). Direct evidence of diets can be found by analyzing fossilized gut contents and dung (called coprolites). Gut contents associated with *Brachylophosaurus* and *Corythosaurus*, together with coprolites attributed to *Maiasaura*, indicate that their diet included leaves and other fibrous plant material, and even rotting wood.

The oldest definitive hadrosaurid fossils are approximately 95–85 million years old, and the last hadrosaurid fossils coincide closely with the end of the Cretaceous period, which marked the end of the Mesozoic Era and the "Age of Dinosaurs." Although the group probably existed for little more than 30 million years, they quickly diversified and expanded their geographic range to become the most abundant large dinosaurian herbivores in Late Cretaceous terrestrial ecosystems of the Northern Hemisphere. There are over 50 species of hadrosaurids currently known, with 6 new species named from 2007–2008. The average size of a duck-billed dinosaur is 10 m (33 ft) in length and over 3 m (10 ft) high at the hips, with an estimated body mass be-

tween 2 and 4 metric tons (4400–8800 lb). Although the majority of species reached this size, at least two hadrosaurids (*Shantungosaurus* and *Zhuchengosaurus*) attained the size of large sauropods, with mass estimates of 10–20 metric tons (22,000–44,000 lb) or more.

Hadrosaurids are often considered facultatively bipedal, meaning that they could walk on either two or four legs, depending on their preference. Most hadrosaurids were probably predominantly quadrupedal, opting for a four-footed gait when walking and foraging. Many fossilized trackways of hadrosaurids indicate that they used their hands when walking; however, when moving at speed, they may have opted for a bipedal gait.

Evolutionary relationships. Hadrosauridae is a clade nested within a series of larger, more inclusive clades, including Hadrosauriformes and Ornithopoda, which include related herbivores with a similar body plan. Hadrosauridae is composed of two major subclades recognized at the subfamily level: Hadrosaurinae and Lambeosaurinae (**Fig. 2**). Hadrosaurines are characterized by broad beaks and long low skulls adorned by a crest made up of solid outgrowths of the nasal bone. This clade includes the genera *Maiasaura*, *Brachylophosaurus*, and *Gryposaurus*. Lambeosaurines are characterized by a hollow crest that encloses an expanded nasal cavity. This group includes *Parasaurolophus*, *Hypacrosaurus*, and *Olorotitan*. Below the subfamily level, the relationships of genera within each clade are poorly understood, with multiple different and contradictory hypotheses of relationship being proposed. There is particularly poor consensus with regard to the interrelationships of hadrosaurines, and some have suggested that this group may even be paraphyletic (that is, a taxonomic group that does not contain all of the descendants of its most recent common ancestor). There is more consensus on the interrelationships among lambeosaurines. A series of poorly known Asian taxa form successive sister taxa to a clade comprising "parasaurolophs" and "corythosaurs." The parasaurolopha group contains the tube-crested *Parasaurolophus* and *Charonosaurus*, whereas the corythosaur group includes *Hypacrosaurus*, *Corythosaurus*, *Lambeosaurus*, and *Olorotitan*.

Distribution and migration. Hadrosaurids were widely distributed during the Late Cretaceous. Their remains have been found on every continent except Australia and Africa, but they are most common in western North America and Eurasia. Notable occurrences of hadrosaurid fossils occur in Alaska, the Canadian Arctic, Siberia, and Antarctica, which were similarly positioned in the Late Cretaceous as they are today. Although there were no polar ice caps at this time, the high paleolatitude of these localities implies extreme seasonal variation in sunlight, temperature, and vegetation availability. It was once thought that hadrosaurids must have made long seasonal migrations into and out of the polar regions to take advantage of the high plant productivity in the summer and avoid the dark winter. A recent calculation suggests that migrations were no more than 2600 km (1600 mi), and stable carbon and oxygen isotopic ratio analyses suggest that hadrosaurids had relatively small geographic ranges inconsistent with large-scale migration. The recent recovery of hadrosaurid eggshells from northern Siberia indicates that hadrosaurids nested in the northern polar region and strongly suggests that the hadrosaurids stayed in the region throughout the year. This implies a robust metabolism tolerant to the relatively harsh polar conditions.

Exceptional preservation. Patches of skin impressions have been found associated directly with many hadrosaurid skeletons. Some of the most spectacular dinosaur finds in history include several virtually complete duck-bill "mummies," that is, complete skeletons that are almost totally encased in skin impressions. These specimens of *Edmontosaurus*, *Brachylophosaurus*, and *Corythosaurus* provide unparalleled views into what a living dinosaur would have looked like. The skin was composed of mostly small, nonoverlapping, pebblelike scales. Larger ridged scales typically augmented the background of smaller scales. A segmented, keratinous epidermal frill similar to that of modern iguanas ran down the middle of the back. Even natural casts of the keratinous covering of the beak at the front of the mouth have been recovered in several specimens. In 2009, true soft tissue remains were described for the hadrosaurid *Brachylophosaurus*. Demineralization of a 78-million-year-old thighbone resulted in the recovery of soft microscopic structures resembling blood vessels, bone and blood cells, and other proteinaceous material. The preserved collagen material was molecularly characterized, and molecular phylogenetic analyses confirmed the close relationship of nonavian dinosaurs to modern birds.

Life history and social behavior. Hadrosaurid eggs and nests are also known from numerous sites in Europe, Asia, and North America. Their spheroolithid-type eggs are spherical and have a characteristic microstructure. Clutches consisted of approximately 15–20 eggs that were apparently deposited in a shallow nestlike structure. Despite their close relationship, the eggs of the lambeosaurine *Hypacrosaurus* are four times bigger in terms of volume than those of the hadrosaurine *Maiasaura*. Multiple preserved nests in one Montana site suggest that *Maiasaura* nested in colonies and returned to the same nesting site year after year.

Histological investigations have revealed poor development of the joint ends of the limb bones and calcified cartilage structures that would have impaired locomotion in newly hatched hadrosaurids. This led to the hypothesis that hadrosaur hatchlings required parental care until they at least doubled their hatching size. Like other dinosaurs, hadrosaurs grew more rapidly than living reptiles. Bone microstructure also reveals that it probably took 10 to 12 years for *Hypacrosaurus* to reach its full size, and that it reached sexual maturity early, at only 2 or 3 years of age.

Duck-billed dinosaurs differ from their close relatives in the development of cranial crests, which are associated with elaboration of the nasal vestibule. The bony crest is formed by the premaxilla and nasal bones, and differs between species in size and shape. In lambeosaurines, the crest is small in juveniles and becomes more prominent with increasing size and maturity. As such, the crests are hypothesized to represent visual display structures important in social and sexual signaling. The prominent crests of lambeosaurines are hollow and contain enlarged and convoluted nasal passages. These unique and bizarre crests have presented a classic problem in functional anatomy. A large number of functional hypotheses have been put forward since the early 1920s, but recent studies have reconstructed the soft tissue divisions of the nasal cavity and used digitally reconstructed brain casts to narrow the possibilities and explore the function of the crest. This work confirms the most probable hypothesis: specifically, that the elaborate nasal cavity within the lambeosaurine crest functioned as a resonation chamber to produce species-specific calls for communication. Bone beds dominated by the remains of a single species (*Maiasaura*, *Edmontosaurus*, or *Brachylophosaurus*) and containing the remains of hundreds of individuals provide additional evidence that some hadrosaurids were social animals that lived together in large herds.

For background information *see* ANIMAL EVOLUTION; CRETACEOUS; DINOSAURIA; EXTINCTION (BIOLOGY); FOSSIL; ORNITHISCHIA; PALEONTOLOGY; REPTILIA; TAPHONOMY in the McGraw-Hill Encyclopedia of Science & Technology.   David C. Evans

Bibliography. D. C. Evans, Nasal cavity homologies and cranial crest function in lambeosaurine dinosaurs, *Paleobiology*, 32:109–125, 2006; H. C. Fricke, R. R. Rogers, and T. A. Gates, Hadrosaurid migration inferences based on stable isotope comparisons among Late Cretaceous dinosaur localities, *Paleobiology*, 35:270–288, 2009; P. Godefroit et al., The last polar dinosaurs: High diversity of latest Cretaceous Arctic dinosaurs in Russia, *Naturwissenschaften*, 96:495–501, 2009; J. R. Horner, Dinosaur reproduction and parenting, *Annu. Rev. Earth Planet. Sci.*, 28:19–45, 2000; J. R. Horner, D. B. Weishampel, and C. A. Forster, Hadrosauridae, in D. B. Weishampel, P. Dodson, and H. Osmólska (eds.), *The Dinosauria*, 2d ed., pp. 438–463, University of California Press, Berkeley, 2004; M. H. Schweitzer et al., Biomolecular characterization and protein sequences of the Campanian hadrosaur *B. canadensis*, *Science*, 324:626–631, 2009.

Hardening of wood

Wood is a very versatile biomaterial. It is used in construction because of its availability, sustainability, strength, and low cost, and in furniture because of its warmth and beauty. For many applications, wood is used without any modification. In some cases, however, there is a need to improve its performance properties to compete with other higher performing and more expensive materials.

When wood is used where hardness is a requirement (such as in flooring), there are several ways to consider which wood to use. A wood that is naturally hard is the first choice, although the hardness of wood varies greatly. For the most part, hardness is a function of density, that is, the more dense the wood, the harder it is. The density of dry wood varies widely based on the volume of void space (lumens and vessels) in the wood. For example, the density of dry balsa wood ranges from 100 to 200 kg/m^3, with a typical density of about 140–170 kg/m^3 (about one-third the density of other hardwoods), whereas a wood such as lignum vitae has a density of 1280–1370 kg/m^3 (this wood does not float).

The Janka scale of hardness measures the force required to embed a 1.128-cm (0.444-in.) steel ball to half its diameter in wood. It is the industry standard for determining the ability of various species to resist denting and wear. The Janka hardness of lignum vitae ranks highest of the trade woods, 4500 pounds-force (lbf) [or 20 kilonewtons (kN)], meaning that it takes 4500 pounds of force (or 20 kilonewtons) to embed the steel ball halfway into the wood's surface. **Table 1** shows the Janka hardness of several wood species. It is not surprising that oak, maple, and hickory are species of choice for

lower cost, readily available flooring materials. Some of the Brazilian hardwoods are also used for flooring, but they are much more expensive.

If additional hardness is desired and wood is the selected resource for the application, then methods to make softer woods harder should be employed. There are two major methods that have been used to increase the hardness of wood: compression and chemical impregnation.

Compression. Wood can be compressed under very high pressure to give a product with the hardness nearly doubled and the scratch resistance improved by over 200%. Scratch resistance is measured by dragging a stylus with a given weight on it across the wood and measuring the depth of the scratch. There are many ways of achieving hardness and scratch resistance by compression, including the use of microwaves, heat, steam, and pressure. The objective of a process of wood deformation is to achieve both densification of the wood and permanent fixation of the densification in the product. Mechanical properties such as bending and tensile strength are improved by 30–70%. One method of compressing wood involves softening the wood at high temperature and high water vapor temperature, and then compressing the wood to reduce it by one-third to half of its original thickness. The compression is fixed by maintaining wood in a compressed state for a predetermined period of time. The procedure is applicable for most kinds of wood, provided the wood source is free of larger knots and has low heartwood content.

An early application of this technology was the production of a product known as Staypak®. By compressing the wood under conditions that cause the cell wall lignin (the cementing material between fibers) to flow, the internal stresses resulting from the compression relax and the wood takes on a new structure in its new compressed state. A temperature range of 150–170°C is used and the wood is compressed while heated. The density is increased 25–40%, there is an increase in tensile and flexural strength proportional to the increase in density, there is an increase in impact strength and toughness, and the hardness is 10–18 times that of the noncompressed wood.

Chemical impregnation. Wood can be impregnated with various chemicals to improve hardness. A solution of phenol-formaldehyde can be impregnated into wood and cured, resulting in a 15–25% density increase, an increase in compressive strength in proportion to the density increase, and an increase in hardness that is more than proportional to the density increase.

The phenol-formaldehyde polymers are the oldest commercial synthetic polymers, first introduced around the beginning of the twentieth century. Their inventor, Leo Baekeland, worked out conditions to produce a tough, light, rigid, chemically resistant solid from two inexpensive ingredients. The actual chemistry is complicated and still not completely understood. The polymers are thermosetting (cannot be melted or dissolved), and the main reaction (**1**)

TABLE 1. Janka hardness of several wood species		
Wood species	Janka hardness, lbf	Janka hardness, kN
Eastern white pine	380	1.7
Basswood	410	1.8
Chestnut	540	2.4
Douglas fir	660	2.9
Southern yellow pine	690	3.1
Sycamore	770	3.4
Cedar	900	4.0
Black cherry	950	4.2
Teak	1000	4.4
Black walnut	1010	4.5
Yellow birch	1260	5.6
Red oak	1290	5.7
American beech	1300	5.8
Ash	1320	5.9
White oak	1360	6.0
Hard maple	1450	6.4
Birch	1470	6.5
Brazilian oak	1650	7.3
Locust	1700	7.6
Rosewood	1780	7.9
Hickory	1820	8.1
Purple heart	1860	8.3
African rosewood	1980	8.8
Mesquite	2345	10.4
Brazilian cherry	2350	10.5
Brazilian rosewood	3000	13.3
Ebony	3220	14.3
Brazilian teak	3540	15.7
Brazilian walnut	3680	16.4
Lignum vitae	4500	20.0

(1)

is the production of methylene bridges between aromatic rings. Many side reactions also occur, and some of these give the phenol-formaldehyde polymer its dark color.

The phenol-formaldehyde-impregnated wood can be compressed, making one of the hardest woods known. In this case, the density is increased 40% and the hardness is increased 10–20 times that of the starting wood. Master automotive dies have been made from this type of wood, as well as laminated airplane propellers used during the Second World War. Wood sawdust mixed with a phenol-formaldehyde polymer is known as Bakelite™, and was one of the very early plastics used in the United States.

Wood can also be treated with acrylic monomers, with the monomers being polymerized in the wood voids to increase hardness. For example, some industrial flooring is treated with acrylics to reduce damage done by high traffic. The major monomer used in these acrylic-impregnated flooring products is methyl methacrylate (MMA) [$H_2C=C(CH_3)—C(=O)—OCH_3$]. Methyl methacrylate [Chemical Abstracts Service (CAS) number 80-62-6; International Union of Pure and Applied Chemistry (IUPAC) name: methyl 2-methylprop-2-enoate] is a colorless liquid and is the methyl ester of methacrylic acid. MMA polymers and copolymers are used for waterborne coatings such as latex house paint, are used in plates that keep light spread evenly across liquid crystal display (LCD) computer and TV screens, and also are used to prepare corrosion casts of anatomical organs, such as coronary arteries of the heart. The principal application of MMA is the manufacture of transparent polymethyl methacrylate (PMMA) acrylic plastics (Plexiglas™). MMA is also used for the production of the copolymer methyl methacrylate-butadiene-styrene (MBS), used as a modifier for polyvinyl chloride (PVC).

In the wood industry, MMA polymers are not only used in hardened engineered wood flooring, but they have also been used to produce furniture (desk writing surfaces, tabletops), decorative products (knife handles, clock faces, plaques), musical instruments (bagpipe chanters, finger boards for stringed instruments, instrument bodies, mouthpieces for flutes and trumpets), and sports equipment (gold club heads, baseball bats, hockey sticks, laminated skis, gun stocks).

The treating procedure is rather simple, and **Fig. 1** describes the steps in the process. The wood is placed in a stainless steel reactor and weighted down with a stainless steel weight (so the wood does not float in the monomer). A vacuum is applied to remove the air from the wood and the monomer is introduced under vacuum with nitrogen gas into the reaction chamber. A nitrogen pressure can be applied, depending on the thickness of the wood to be treated. Usually, only a thin veneer is treated that will be laminated on top of a sub-base of plywood, particle board, or high-density fiber board.

The monomer solution can contain a few percent of a cross-linking agent such as trimethylolpropane trimethacrylate. The free radical reaction is catalyzed using a Cobalt-60 source, peroxides (including benzoyl peroxide), or Vazo® catalysts. Vazo® is a registered trade name of nitriles (RC≡N cyanides derived by removal of water from an acid amide) that are used to cure acrylics. These catalysts generate

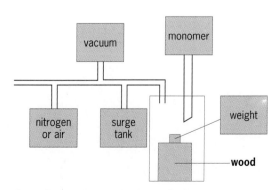

Fig. 1. Equipment setup for the acrylic treatment process.

the free radicals that initiate the polymerization reaction (2).

$$\text{methyl methacrylate} \quad \xrightarrow[\text{vinyl polymerization}]{\text{free radical}} \quad \text{poly(methyl methacrylate)} \qquad (2)$$

methyl methacrylate   poly(methyl methacrylate)

Polymerization of MMA is exothermic and much heat is generated during the polymerization, which needs to be controlled. After the curing is complete, the reactor is drained, flushed with nitrogen, and cooled, and the treated wood is removed. There may be residual polymer on the surface that must be removed. Not all woods treat the same, so the time of monomer impregnation, vacuum, and pressure, as well as the level of loading and the cure time, will vary depending on the species.

As with all polymerization reactions, the volume of the monomer shrinks upon polymerization. In the case of MMA to PMMA, the shrinkage is about 20%. The left-hand image in **Fig. 2** shows the open structure of oak before MMA-PMMA treatment. The right-hand image shows the polymer in the lumens after treatment, which is easily seen in the void structure of the wood.

The hardness of the PMMA-treated wood is increased 50–100%, depending on the wood treated. Indentation resistance is increased 50–70% and the static bending properties of the PMMA-treated wood

TABLE 2. Static bending properties of acrylic lumen-filled wood

Property*	Untreated, MPa	Treated, MPa
MOE	9.3	11.6
FSAPL	44.0	79.8
MOR	73.4	130.6
MCS	44.8	68.0

*Abbreviations: MOE = modulus of elasticity; FSAPL = fiber stress at proportional limit; MOR = modulus of rupture; MCS = maximum crushing strength.

are greatly improved (**Table 2**). Dyes can be added to change the color of the impregnated wood, with darker browns being the most popular (**Fig. 3**). Oak and maple are often dyed to resemble walnut. The PMMA is resistant to aliphatic hydrocarbons, cycloaliphatic compounds, fats, and oils, and also to weak acids and bases at temperatures up to 60°C. The resistance to weathering of PMMA is very good. PMMA has good insulating properties, high dielectric strength, and high tracking resistance. PMMA is naturally transparent and colorless. (The transmission for visible light is 92%, and the refractive index is 1.492 for PMMA.) There are types that transmit ultraviolet rays, and types that absorb them almost completely, as a result of which sensitive dyes on painted surfaces are protected from fading.

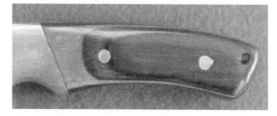

Fig. 4. Veneers can be treated with MMA and a dye in order to obtain various colors that are laminated into custom knife handles.

Hobbyists also employ these various methodologies to treat wood laminates to be used for knife handles. Veneers are treated with MMA and a dye to give veneers of different colors that are laminated into custom knife handles (**Fig. 4**).

For background information *see* POLYMER; POLYMERIZATION; STRENGTH OF MATERIALS; VENEER; WOOD COMPOSITES; WOOD DEGRADATION; WOOD ENGINEERING DESIGN; WOOD PROCESSING; WOOD PRODUCTS; WOOD PROPERTIES in the McGraw-Hill Encyclopedia of Science & Technology.

Roger M. Rowell

Bibliography. R. E. Ibach and W. Dale Ellis, Lumen modifications, in R. M. Rowell (ed.), *Handbook of Wood Chemistry and Wood Composites*, Chap. 15, pp. 421–446, Taylor and Francis, Boca Raton, Florida, 2005; M. Inoue and M. Norimoto, Permanent fixation of compressive deformation in wood by heat treatment, *Wood Res. Tech. Notes*, 27:31–40, 1991; M. Inoue et al., Steam or heat fixation of compressed wood, *Wood Fiber Sci.*, 25(3):404–410,

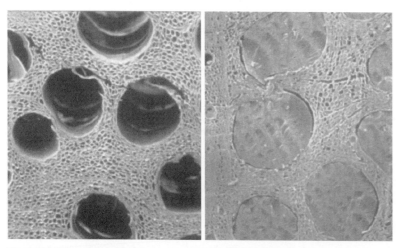

Fig. 2. (*Left*) The open structure of oak before MMA-PMMA treatment. (*Right*) The polymer in the lumens after treatment.

Fig. 3. Various dyes can be added to change the color of the impregnated wood.

1993; J. A. Meyer, Industrial use of wood-polymer materials: State of the art, *Forest Prod. J.*, 32(1):24–29, 1982; J. A. Meyer, Treatment of wood-polymer systems using catalyst-heat techniques, *Forest Prod. J.*, 15(9):362–364, 1965; J. A. Meyer, Wood polymer materials: State of the art, *Wood Sci.*, 14(2):49–54, 1981; J. A. Meyer, Wood-polymer materials, in R. M. Rowell (ed.), *Chemistry of Solid Wood*, Advances in Chemistry Series No. 207, Chap. 6, pp. 257–289, American Chemical Society, Washington, D.C., 1984; R. M. Rowell and P. Konkol, Treatments that enhance physical properties of wood, General Technical Report 55, USDA Forest Service, Forest Products Laboratory, Madison, WI, 1987; M. H. Schneider and A. E. Witt, History of wood polymer composite commercialization, *Forest Prod. J.*, 54(4):19–24, 2004; USDA, Specialty treatments, in *Wood Handbook: Wood as an Engineering Material*, General Tech. Rep. 113, Chap. 19, pp. 405–418, USDA Forest Service, Forest Products Laboratory, Madison, WI, 1999.

Histone modifications, chromatin structure, and gene expression

Most cells of the human body typically contain DNA having about 6,000,000,000 base pairs, which would measure nearly 2 m in length if stretched out end to end. Yet this genomic DNA is packaged into cellular nuclei that are roughly 10 micrometers, or 0.00001 m, in width. This remarkable compaction is aided by the wrapping of the genome into a complex between DNA and proteins known as "chromatin."

Compaction of DNA. The first level of compaction into chromatin is the winding of 147 base pairs of DNA around a group of 8 positively charged proteins called the histones (there are typically two copies each of the histone proteins named H2A, H2B, H3, and H4). This unit of spooled DNA and its associated histones is called the nucleosome, whereas the free DNA that links one nucleosome to another is called "linker" DNA. Nucleosomes and intervening linker DNA appear in microscopic images as "beads on a string," with the nucleosomes being the beads and the linker forming the string. The beads on a string are further folded many times to form the fully compacted state of the genome, but this further folding is still fairly mysterious.

The compaction of DNA into nucleosomes poses a conundrum, as DNA is tightly wrapped around the histones, making access to this DNA difficult for other cellular proteins. Yet every process that occurs on DNA—turning genes on and off, copying the genome, repairing DNA damage, and so on—must necessarily occur in the context of this extensive wrapping. How do large protein machines such as RNA polymcrase access DNA that is wrapped up in nucleosomes? A huge variety of factors are involved in regulating access of the cellular machinery to nucleosomal DNA, such as cell machines that use energy to move the histones off a piece of DNA.

Modification of histones. One of the more extensive systems involved in regulating chromatin structure involves the role of covalent modification of the histone proteins in regulating chromatin-based processes. Covalent modification of proteins means the linkage, by covalent bond, of a chemical group to an amino acid of the protein. A wide range of covalent modifications occurs on various proteins in cells, and a high fraction of these occur on the histone proteins (which are but a handful of the tens of thousands of cellular proteins). Modifications described on histones to date include phosphorylation of serines, acetylation of lysines, methylation of lysines and arginines, ubiquitination (addition of ubiquitin) of lysines, and many others. Not only are histones decorated with many different types of chemical groups, but a given chemical group can be attached to many different sites. For example, the end or "tail" of the histone H4 can be acetylated on lysine 5 (abbreviated as K5), lysine 8 (K8), lysine 12 (K12), or lysine 16 (K16) (or any combination of these).

One implication of all of this modification is that the packaging proteins associated with the genome, the histones, have the potential to carry a huge amount of information beyond that carried in the genome itself [see part (*a*) of the **illustration**]. How, then, is it possible to make sense of this? Two important types of questions are typically asked. First, how, when, and where are histones modified in a given way? Second, what is the outcome of the modification in question?

How are histone modifications placed on the genome? A wide range of enzymes that deposit modifications on the histones has been identified. Lysine acetylation is carried out by histone acetyltransferase enzymes, and can be removed by enzymes called deacetylases. Histone methylation is carried out by methyltransferases, and recently several families of demethylases have been identified. As a result of these competing on-and-off enzymes, histone modifications are much more dynamic than DNA sequences—an acetyl group in chromatin typically persists for minutes, whereas methyl groups may last

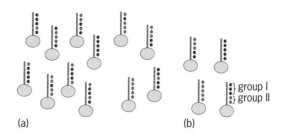

(a)   (b)   group I
                                                  group II

Combinatorial usage of histone modifications. Over 100 covalent histone modifications have been described, raising the question of why so many exist. In principle, it could be imagined that each of the modifications could occur independently, as depicted in part (*a*) [here, a modification is schematically shown as a small dark circle, and nucleosomes are shown as large ovals]. This means that many billions of different combinations could occur, and each might mean something different to the cell. However, it appears that, in the cell, many modifications occur together in groups, as shown in part (*b*). This suggests that less information is carried by histones than is possible, but raises the question of why so many modifications are used by the cell if they often occur in groups.

for hours. Conversely, a given DNA base is stable to mutation for millions and millions of cellular generations.

An interesting difference between various modifying enzymes is the way they operate on histones. Some enzymes prefer to modify histones that are fully assembled into nucleosomes, and appear to operate in situ on the chromosome. In other words, these enzymes are recruited to a stretch of chromatin in the cell, where they modify a small patch of nucleosomes, perhaps as a signal to other proteins. On the other hand, some modifying enzymes prefer to modify "free" histones, which have not been wrapped with DNA. In this case, the modifications in question enter into the genome only when those free histones are assembled into chromatin, and thus they can be imagined to go everywhere that new histones are incorporated (although this is not necessarily the case).

Hence, where do histone modifications occur along the genome? After all, each nucleosome on the genome can in principle be modified independently of the nucleosomes around it. This means that the approximately 30 million nucleosomes in a single cell have the potential to carry quite a bit of information of their own, beyond the information in the genome that they package.

Mapping studies. By using modern mapping technologies, scientists can identify each and every nucleosome in a cell that carries a certain modification, such as methylated lysine 4 (K4) in its histone H3 molecules. Several important facts are revealed by these mapping studies. Most notably, a large fraction of histone modifications appear to be deposited during the process of transcription (the process of making an RNA copy of a gene). For example, some modifications (many acetylation states and some methylation states, such as the aforementioned K4 methylation) are found at the front end of genes, whereas others are found over the body of genes. In at least some of these cases, this is probably a consequence of the relevant modifying enzyme being carried by RNA polymerase, the enzyme that makes RNA copies of genes. In other words, the act of making an RNA copy of a gene leaves an imprint on the gene's packing material—"this bit of the genome has been copied at least once." Other marks occur in broad domains, the most notable being the marking of an entire chromosome in the process of dosage compensation. Men carry one X chromosome, whereas women carry two. In order to equalize expression of genes on the X chromosome, one X chromosome is inactivated in women, and this inactive X is characterized by huge domains of silencing histone marks.

Another interesting discovery in mapping studies is that many histone modifications occur together [see part (b) of the illustration]. In other words, groups of modifications exist where a nucleosome with modification 1 will generally be found to also carry modifications 2, 5, 12, and 56 (for example). Why is this? *How* it occurs is sometimes understood—some of the enzymes that modify histones prefer to work on a histone that already carries

some other modification, so the two modifications in question will tend to co-occur. The question of what purpose is served by having large groups of modifications that occur together (instead of just one representative per group), on the other hand, is a vexing one and the subject of intensive current research.

Together, mapping studies indicate a great deal about what the average chromatin state of the genome looks like in a given cell type. This has many uses. For example, determining the boundaries of genes in complicated genomes like the human genome from DNA sequence alone is still difficult. However, by mapping histone marks, it is possible to "ask" the cell where the gene's boundaries are. This leads to a picture in which the histones that package the genome act as a sort of annotation system for the cell to take notes about the status of the genomic stretch underneath. However, while the landscape of histone modifications is very informative about some things, it often indicates very little about why the cell puts those marks there. In other words, what do histone modifications do?

Effects on chromatin structure. How do covalent modifications change chromatin structure? Generally speaking, a given modification can have effects through either of two mechanisms. First, modification of amino acids can change their basic properties in ways that affect chromatin packaging—acetylation of lysine, for example, results in neutralization of its positive charge. The decrease in charge is believed to "loosen" the wrapping of DNA around the histones, since a significant contributor to the tightness of DNA around the histones is the attraction of the negatively charged DNA to the positively charged histones. Highly acetylated nucleosomes are then easier to unwrap, allowing cellular proteins such as RNA polymerase to have easier access to the underlying DNA sequences.

Second, modified amino acids are shaped differently from unmodified amino acids, and can be recognized by specific binding partners that can bring protein machines to particular locations along the chromosome. This mode of regulation is particularly interesting, since different binding partners often exist for each different modified amino acid. In other words, there are proteins that bind to histone H3 methylated on lysine 4 (K4), whereas other proteins bind to the same histone when it is methylated on lysine 27 (K27). The recruited proteins often play roles in aspects of gene regulation such as initiation of transcription or repair of DNA damage. For example, in embryonic stem cells, "master regulator" genes involved in cell differentiation are wrapped in nucleosomes that are methylated on K27; this, in turn, recruits a complex called Polycomb, which then represses these genes. This repression is a key aspect of what allows embryonic stem cells to maintain the ability to become any cell type in the body.

Functional effects. As for the functional effects of histone modifications, these vary from mark to mark, depending on how they change the structure of the chromatin fiber or what type of protein complex they recruit. Some histone marks are very broadly involved in controlling some process—phosphorylation of the H2A variant H2A.X, for instance, is a signal for initiation of a DNA damage repair pathway. Other marks are more subtle—trimethylation of H3 lysine 36 is involved in preventing the cell from making RNA copies starting at incorrect locations inside the body of genes. The range of functions spans virtually every DNA-based process, but recent years have pointed toward an unexpected role for histone modifications in fine-tuning, rather than overt control, of major processes such as transcription.

Future outlook. The world of histone modification is complex in detail, but simplifying themes have emerged. Histone marks are dynamic, often being deposited and removed by competing enzymes many times in a single cell's lifetime. They provide the cell with a way to mark where major events have occurred, such as transcription or genomic replication, and thereby allow the cell in a sense to keep notes of events that have occurred throughout the genome. They are involved in the control of most processes that occur on DNA, in ways that range from initiating entire processes to fine-tuning of rates, levels, or distances. Modern genomewide methods are allowing scientists to map these marks, thereby enabling these investigators to see the notes that the cell leaves for itself. This complex choreography of decoration occurring throughout the genome has begun to seem understandable, and the future holds promise for a better understanding of the role of histone modification in normal cells, and the contribution of errors in these processes to diseases such as cancer.

For background information *see* BIOCHEMISTRY; CHROMOSOME; DEOXYRIBONUCLEIC ACID (DNA); ENZYME; GENE; HUMAN GENOME; NUCLEOPROTEIN; NUCLEOSOME; PROTEIN; RIBONUCLEIC ACID (RNA); TRANSCRIPTION in the McGraw-Hill Encyclopedia of Science & Technology.   Oliver J. Rando

Bibliography. T. Kouzarides, Chromatin modifications and their function, *Cell*, 128(4):693–705, 2007; C. L. Liu et al., Single-nucleosome mapping of histone modifications in *S. cerevisiae*, *PLoS Biol.*, 3(10):e328, 2005; T. S. Mikkelsen et al., Genomewide maps of chromatin state in pluripotent and lineage-committed cells, *Nature*, 448(7153):553–560, 2007; O. J. Rando, Global patterns of histone modifications, *Curr. Opin. Genet. Dev.*, 17(2):94–99, 2007; K. E. Van Holde, *Chromatin*, Springer-Verlag, New York, 1989.

Human fossils from Omo Kibish

In 1967, paleontologists made a series of major fossil discoveries of early humans in the ancient sedimentary layers of the Kibish Formation along the banks of the Omo River in southwestern Ethiopia (**Fig. 1**). Omo I, the most complete specimen discovered, consists of numerous fragments of the skull, teeth, and much of the skeleton, including several limb bones, whereas Omo II preserves only a nearly com-

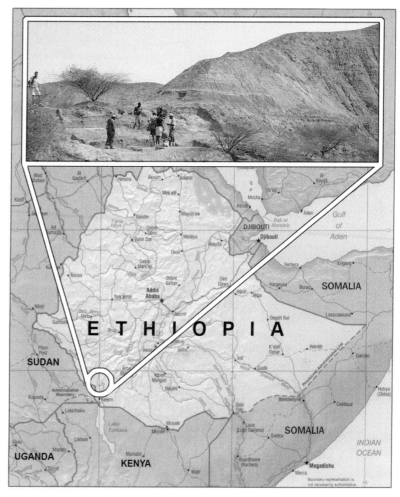

Fig. 1. Map of the Horn of Africa. The circle indicates the general location of the Kibish Formation along the Omo River in southwestern Ethiopia. The inset shows recent excavations at KHS (Kamoya's Hominid Site), where new bones from the Omo I early *Homo sapiens* fossil were discovered. (*Photo courtesy of J. Fleagle*)

plete neurocranium, the portion of the skull that surrounds the brain. While the modern appearance of Omo I was quickly accepted by most paleoanthropologists (that is, specialists in the study of human evolution), the Omo II skull was described as more primitive, with many similarities to more ancient members of the human lineage. In the years following these discoveries, the importance of the Kibish fossils for understanding the origins of our own species, *Homo sapiens*, was surrounded by considerable speculation and confusion about the age and exact provenance of these fossils. Motivated by this, researchers recently returned to the Kibish Formation to clarify the many ambiguities surrounding the fossils found decades earlier, as well as to search for new fossil and archeological material. Their efforts yielded new fossils and a wealth of stone tools, and also provided an age of approximately 195 ka (thousand years) for the fossils, securing their place as the earliest evidence of our own species yet recovered. Thus, in the decades since their initial discovery, the human fossils from the Omo Kibish continue to occupy a critical role in our understanding of modern human origins and provide insight into the last 200,000 years of human evolution.

Human fossils and localities. The Kibish Formation is a geological unit located along the Omo River in southwestern Ethiopia. The formation is subdivided into four distinct blocks of successive sedimentary layers called "Members," with Member I at the base of the sequence and Member IV at the top. The importance of the Kibish Formation to studies of human origins first became clear in 1967 when a team from the National Museums of Kenya, under the direction of Richard Leakey and working as part of the International Paleontological Research Expedition to the Omo River, made a series of major fossil discoveries at several neighboring localities in the Kibish Formation. The marked anatomical differences between the two most complete skulls led to considerable debate among specialists about the antiquity and stratigraphic provenance of these fossils.

Omo I is the most complete fossil known from the Kibish Formation. It was discovered at the site of KHS (Kamoya's Hominid Site) in the upper layers of Member I during the course of the first expedition to the region, although renewed fieldwork at KHS has yielded additional fossil remains (Fig. 1). Several new bones, including two manual phalanges (finger bones), a large portion of the pelvis (hip bone), a

part of the right femur (thighbone), and a talus (anklebone), have now been added to the fragments of the skull, jaw, teeth, and major limb bones discovered in 1967 (**Fig. 2**). All of the bones from KHS almost certainly derive from a single associated skeleton, since no bone is duplicated, all appear to come from an adult individual, and the newly discovered portion of the thighbone fits perfectly with a broken thighbone fragment found in 1967. Today, the Omo I individual is known from much of its skeleton, albeit in a very fragmentary condition with few complete bones. This individual is modern in overall appearance, although a few of the limb bones do exhibit some more primitive features. This is consistent with the interpretation of Omo I as an early representative of modern humans (*Homo sapiens*) and supports archeological and genetic evidence in establishing an early African origin for our species.

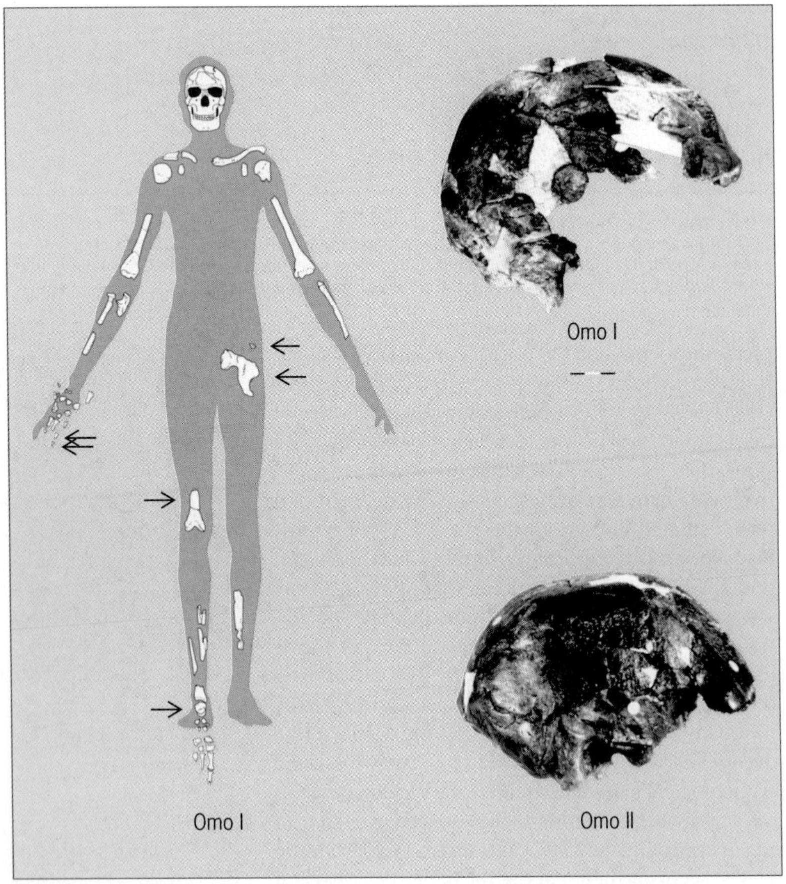

A second fossil locality (PHS, Paul's Hominid Site) was discovered a few kilometers away from KHS on the opposite side of the Omo River in the upper reaches of Member I of the Kibish Formation. The Omo II fossil preserves only the neurocranium. Although this fossil was initially assigned to *Homo sapiens*, subsequent work has consistently described it as more primitive in appearance, retaining many similarities to *Homo erectus*, a more ancient human ancestor (Fig. 2). Another specimen (Omo III) preserves only a small portion of the upper face and the front part of the neurocranium. This fossil was also recovered in 1967, although it received very little attention as a result of its fragmentary preservation and uncertain age.

In 2003, more human fossils were discovered at a new locality in Member I of the Kibish Formation: AHS (Awoke's Hominid Site). A nearly complete tibia (large shinbone) and a small portion of the fibula (small shinbone) from this site may belong to the same adult individual. The AHS fossils are considered broadly contemporaneous with the Omo I and Omo II individuals. Recently, a few other fragmentary human fossils have been found higher up in younger parts of the Kibish Formation (Members III and IV), including two skull fragments from Chad's Hominid Site (CHS). The thickness of the parietal fragment (a bone that forms part of the side of the braincase) from CHS confirms that it belonged to an adult individual, although the fragment is too small to preserve much anatomical detail. The thickness of the second fossil from CHS, an occipital fragment from the back of the skull, indicates that it belonged to a young child, probably between 2 and 4 years of age.

With its modern appearance, Omo I has long occupied a central place in discussions of modern human origins, providing support for an African origin of our species. In contrast, the more primitive-looking Omo II skull does not fit so neatly into models of modern human origins. Paleoanthropologists have still not reached a consensus on the biological and evolutionary significance of the anatomical differences between the Omo I and Omo II skulls. One interpretation suggests that this diversity reflects the range of physical variation found within the early human

Fig. 2. Major fossil *Homo sapiens* discoveries in the Kibish Formation. (*Left*) The bones of Omo I, the earliest *Homo sapiens* fossil known to science, are shown in anatomical position surrounded by the outline of a modern equatorial African. Arrows indicate the newly recovered fragments attributed to this individual. (*Right*) The fossil skulls of Omo I (top) and Omo II (bottom) discovered in Member I of the Kibish Formation exhibit clear differences in morphology, notably in the rear of the skull (to the left of the image) and along the forehead. Scale bar = 3 cm (1.2 in.). (*Left image courtesy of O. Pearson; right images courtesy of M. Day*)

Fig. 3. Chert (silica) foliate (leaf-shaped) biface stone tools from Member III of the Kibish Formation. These stone tools were discovered at a new locality in the vicinity of the Omo II site during recent expeditions to the Omo River Valley. (*Photo courtesy of J. Shea*)

population. Alternatively, these distinct skulls might indicate that two very different peoples coexisted along the Omo River.

Stone tools. Recent fieldwork yielded a large number of stone tools from both surface collections and excavations at KHS and other sites in the Kibish Formation. These artifacts help to put the Kibish human fossils into a broader archeological and behavioral context. As a group, the stone tools recovered from the Kibish Formation exhibit a low number of retouched tools (tools that have been reworked along their edge) and a high occurrence of bifaces or tools worked on both sides (**Fig. 3**). These characteristics align the Kibish assemblages with archeological industries found across equatorial Africa during the Middle Stone Age, the period of African prehistory that began around 300,000 years ago and lasted until about 50,000 years ago.

Animal fossils. The recent expeditions also yielded large numbers of fossil mammals, birds, and fish from the Kibish Formation. All of the mammalian and bird fossils represent extant animals, many of which can still be found in the Omo River Valley today. However, a few of the mammals [for example, the giant forest hog (*Hylochoerus meinertzhageni*), the largest wild member of the pig family; and the duiker (*Cephalophus* sp. indet.), a small antelope species] are found only in less arid parts of Africa today, suggesting that the Omo River Valley may have been wetter during the time of Omo I and II compared to the present day. Likewise, the fossil fish recovered from the Kibish Formation are species such as Nile perch and catfish that can still be found in the local rivers, although many of the fossil fish are larger than their modern counterparts.

Geologic age. Initially, the absolute age of the Kibish fossils was difficult to establish because of the limitations of dating methods available to researchers in the late 1960s. Two carbon-14 (^{14}C, radiocarbon) dates in excess of 39 ka and a uranium-thorium date of ca. 130 ka were obtained from Nile oyster shells collected just above the fossil-bearing layer of Member I. However, because of problems associated with the application of uranium dating to shells, these dates offered little chronological resolution beyond confirming that the Member I fossils were older than the limits of radiocarbon dating (that is, older than approximately 30,000 to 40,000 years ago). The modern faunal community suggested by the animal fossils also did not help to resolve the antiquity of the Kibish human fossils.

The recent fieldwork in the Omo River Valley served to clarify the geological provenance of the Kibish fossils and shed light on the absolute age of the fossils through the application of modern dating methods. The evidence gathered in the course of the new expedition supports the original interpretation that both Omo I and II came from the upper part of Member I in the Kibish Formation. Therefore, these specimens are approximately contemporaneous in age. By applying argon-argon dating to rock-forming minerals known as feldspars collected from tuffs, or layers of volcanic air-fall material, sandwiched in Member I and Member III of the Kibish Formation, workers were able to bracket the age of the Kibish fossils between approximately 196 ± 2 ka and 104 ± 1 ka. To further narrow the age estimate for the fossils recovered from Member I (including Omo I and II and the new AHS finds), researchers took a closer look at the nature of the sediments preserved in the Kibish Formation. They found evidence for the rapid deposition and formation of Member I. In addition, they were able to link the formation of local sapropels (dark-colored sediments rich in organic matter) to dated sapropel phases documented in the Mediterranean Sea. Together, these diverse lines of evidence allowed researchers to narrow the age estimate to roughly 195 ± 5 ka. This secures their place as the earliest-known *Homo sapiens*, and contributes to the wealth of paleontological, archeological, and genetic evidence establishing the origin of modern humans in Africa over 150,000 years ago.

For background information *see* ANTHROPOLOGY; ARCHEOLOGY; DATING METHODS; EARLY MODERN HUMANS; FOSSIL; FOSSIL HUMANS; MOLECULAR ANTHROPOLOGY; PHYSICAL ANTHROPOLOGY; PREHISTORIC TECHNOLOGY; SEQUENCE STRATIGRAPHY; TAPHONOMY in the McGraw-Hill Encyclopedia of Science & Technology.   Danielle Royer

Bibliography. J. G. Fleagle (ed.), Paleoanthropology of the Kibish Formation, Southern Ethiopia, *J. Hum. Evol.*, 55:359–530, 2008; D. C. Johanson and K. Wong, *Lucy's Legacy: The Quest for Human Origins*, Harmony Books, New York, 2009; I. McDougall, F. H. Brown, and J. G. Fleagle, Stratigraphic placement and age of modern humans from Kibish, Ethiopia, *Nature*, 433:733–736, 2005; P. Mellars, K. Boyle, O. Bar-Yosef, and C. Stringer (eds.), *Rethinking the Human Revolution: New Behavioural and Biological Perspectives on the Origin and Dispersal of Modern Humans*, McDonald Institute for Archaeological Research Monographs, Cambridge, U.K., 2007; P. R. Willoughby, *The Evolution of Modern Humans in Africa: A Comprehensive Guide*, AltaMira Press, Lanham, MD, 2006.

Humboldt squid beak biomimetics

Living organisms make robust materials such as silk, shell, bone, and wood out of abundant yet intrinsically weak building components. These materials are adapted to perform multiple functions in vivo, exhibit unique combinations of properties, and are processed using efficient biosynthetic pathways under mild ambient conditions. The perception of biological materials as sources of "bio-inspiration" for novel, multifunctional materials has grown significantly in popularity. Limited resources and the need for more recyclable materials have added momentum to this perception. Manufacturing materials with minimal environmental impact is also becoming increasingly urgent. "Biomimetics" is roughly defined as abstracting useful design from living organisms. The first step in biomimetics research necessarily explores the fundamental relationships between material structures and their properties, which are investigated at multiple length scales, from macroscopic to the molecular. The ultimate goal is to transfer abstracted design concepts to the next generation of materials and materials fabrication (**Fig. 1**).

Squid beaks and biomimetics. One of the hardest and stiffest organic materials known is the beak of the Humboldt squid (jumbo squid, *Dosidicus gigas*). Thus, squid beaks represent an excellent case study in biomimetics. The beaks have a sharp hard tip, yet are completely organic and devoid of mineral (unlike hard human tissues, such as teeth and bones). This raises the first intriguing question about squid beaks: How does nature contrive to design an organic material to be as hard and stiff as high-performance synthetic polymers? A second question pertains to avoiding contact damage that occurs whenever hard and soft materials share a common interface. This subtlety can be appreciated by a cursory inspection of squid anatomy. The razor-sharp beak is connected to a compliant golf ball–sized muscular bundle called the buccal mass, which controls the biting action of the beak. How does nature manage to cut with a sharp instrument that is mounted in tissue with the consistency of a gelatinous substance?

The organic matter of the beak consists of a blend of proteins and chitin (a fibrous polysaccharide). The chitin fibers are embedded in a matrix of proteins that undergo cross-linking (that is, covalent interactions between themselves or with chitin) to various degrees. The beak tip (rostrum) is stiffer and harder

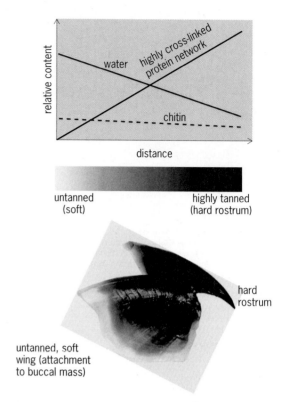

Fig. 2. **Biomolecular and mechanical gradients in squid beak.**

than the strongest synthetic polymers, with a cross-linking density approaching 20% by weight. The rostrum also exhibits a layered microstructure, which toughens the beak against crack propagation and endows it with significant fracture energy (typically in the range of the toughest polymers). A partial answer to how nature makes an organic "knife" thus includes a multilayer structure having preferentially oriented fibers embedded in a highly cross-linked protein matrix.

Actually, the beak properties and composition vary in the squid beak. The relative proportions of chitin, proteins, and cross-link density are graded from the base to the tip and are closely correlated with a gradient of mechanical properties. The proximal base region in contact with the buccal mass is soft and composed largely of highly hydrated chitin (up to 70 wt% water). Toward the distal tip of the beak, the relative chitin and water contents decrease, whereas that of protein increases (**Fig. 2**). This is accompanied by a large, 100-fold increase in hardness and stiffness. Mechanical gradients play an important role in the structural integrity of many junctions between tissues and are particularly impressive in squid beaks (see **table**). This exquisite tuning of structural properties allows proper chewing and biting activity by the hard tip without inflicting self-damage to the surrounding soft tissue. The gradient distribution of properties makes the structure particularly tolerant to abrasion at the tip, whereas the overall design makes it an attractive model system for engineering biomaterials in restorative applications.

Biochemical characterization of the beak is still incomplete and is limited to three areas: macroscopic

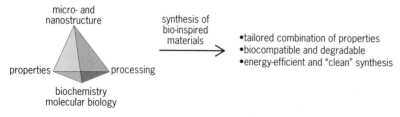

Fig. 1. **The biomimetic approach. Interrelations in biological materials between the function of the tissues, their properties, their structure at multiple length scales (from the macroscopic scale to the nanoscale), and the biochemistry of the building blocks are first elucidated at the fundamental level. Those principles are then applied toward design and synthesis of novel materials.**

Selected examples of elastic mismatch in tissues and joints of various biological materials			
Structure, junction	E_1 (stiff domain), GPa	E_2 (compliant domain), GPa	E_1/E_2
Bone/tendon	25	1	25
Enamel/dentin	70	20	3.5
Mollusk shell	70 (mineral)	1.5 (matrix)	47
Mussel threads	0.5	0.05	10
Squid beaks	6	0.05	120

mapping of protein, chitin, and water distribution; protein amino acid composition; and identification of covalent cross-links between beak proteins. Beak proteins are rich in glycine and alanine, which are common to several structural proteins (for example, silk fibroin and elastin). However, the high histidine content in the beak proteins is more distinctive, and this is increasingly emerging as a common theme of chemically sclerotized hard tissues. The presence of dopa (3,4-dihydroxyphenylalanine), which is better known as a precursor for the biosynthesis of the neurotransmitter dopamine or for its role in marine adhesives, in the beak proteins is further evidence of chemical sclerotization, because oxidized dopa forms strong covalent bonds with histidine residues during maturation to the robust structure (**Fig. 3**). Physicochemical interactions between the proteins and chitin in the beak are believed to control mechanical properties of the structure in the following way: Chitin is a hydrophilic polysaccharide and, in its native state, absorbs water molecules to become a gel-like material. When dehydrated, its stiffness increases dramatically such that its mechanical properties can potentially be varied over two orders of magnitude between the fully hydrated and fully dried states. Squid beak biosynthesis thus involves a closely regulated manipulation of hydration at the molecular level to control stiffness and strength gradients.

Current research and development. Mimicking the design of a biological load-bearing structure such as squid beak is a highly interdisciplinary undertaking. Even defining the paradigm, a necessary first step, involves (1) biochemical characterization of the macromolecules and their interrelationships in the beak, (2) biomechanical studies at different length scales, (3) investigating molecular architecture to the nanometer scale, and (4) insights into the dynamic process of beak assembly (Fig. 1).

Biochemical characterization. At one level, squid beak resembles the multiple-part epoxy adhesive available from a hardware store. Before mixing and hardening, the epoxy components (resin and catalyst) are neatly and separately stored, and they can be quite easily chemically characterized. After mixing, however, the cured epoxy cannot be redissolved into its component parts by any solvent. The durability of squid beak to heat, solvents, and even enzymes is well known, but (unlike epoxy) the chemical identity of its soluble precursors is not known with any precision. Given this, it is necessary to extract gram quantities of beak with mild acid and disruptive reagents

(for example, 8 M urea) to obtain micrograms of soluble protein for further characterization. The working assumption with such low-yield extractions is that the harvested soluble protein is the residue of a major beak protein that has escaped cross-linking.

Current technology allows at least partial characterization of microgram amounts of protein as long as each protein in a crude mixture is adequately separated from others. Gel electrophoresis is the most popular method for protein separation, after which each protein can be proteolytically digested in situ and the released peptides subjected to mass spectrometric analysis for amino acid sequencing. Even though the protein sequence is a "fingerprint" of its identity, the process involved with sequencing of proteins from one end to the other is prohibitively expensive and time-consuming. Instead, a few peptides are sequenced and these are exploited to provide the templating primers for a molecular biology approach in which the gene copy for the protein in question is found and amplified. The protein sequence is then deduced from the "gene" sequence. This may seem indirect, but it is usually cheap, fast, and reliable. The molecular approach has another advantage that relates to processing—that is, beak cells whose protein expression is differentially regulated are ultimately responsible for producing the observed macromolecular gradients. Knowing the gene sequence of each protein allows one to determine the concentration and location of each protein being made.

There is at least one aspect of beak biochemistry that does not lend itself to the molecular approach—that is, cross-linking. Cross-links are not revealed by gene sequences and must be characterized directly. Cross-link characterization usually involves controlled hydrolysis of the beak, leading to breakage

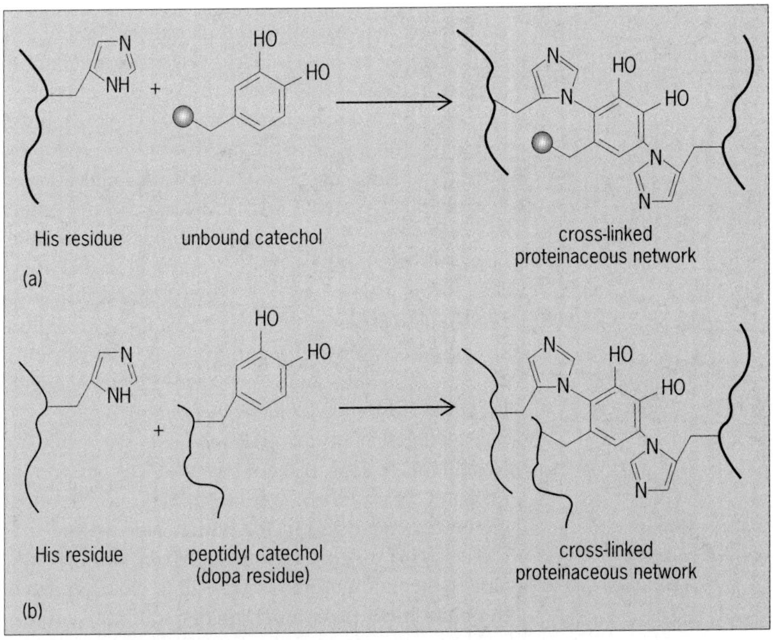

Fig. 3. Molecular cross-links formed in tanned (hard) portions of the beak. Current research aims at elucidating whether dominant cross-link mechanisms involve (a) unbound catechols or (b) peptide-bound catechols (dopa).

only of peptide bonds in proteins and *O*-glycosidic bonds in chitin. The cross-linked amino acids can then be identified by mass spectrometric and nuclear magnetic resonance analyses of the various chromatographically separated hydrolysis products. So far, the cross-links most often encountered in squid beak are between histidine and dopa or related catechols (*ortho*-dihydroxybenzenes) [Fig. 3]. To date, no cross-links have been detected between chitin and protein.

Biomechanics. The most important biomechanical issue in beak pertains to the role played by dehydration in determining beak properties and, more generally, chitin-based biocomposites. Two physicochemical models are currently being explored. The first model states that hardening is related to cross-linking density in beak proteins (as occurs in the hardening of epoxy resins). The second model proposes that dehydration of fibrous chitin networks is caused by percolation of hydrophobic proteins into this network and by the displacement of water molecules bound to chitin fibers by hydrophobic interactions. Abrasion, friction, and wear at the nanometer scale are other characteristics that are being studied to corroborate the predicted superiority of the hard tip region over current synthetic thermoset polymers used in abrasion-resistance applications.

Structural studies. Structural investigations are needed to explore the architectural arrangement of building blocks at the nanoscale. So far, most insights about beak microarchitecture have been obtained using x-ray scattering and high-resolution and scanning electron microscopy. They have shown that the nanometer-size chitin fibers are preferentially oriented at the tip of the beak, whereas no preferred orientation is evident in the proximal region. However, our knowledge of chitin/protein interfaces at the nanoscale is still at a very early stage, and this is also true with respect to the gradual organization of the biomolecules along the proximal-to-distal axis. The limitation here comes from our ability to visualize interfaces between biomacromolecules at a very small scale using traditional electron microscopy techniques. Recent progresses in surface-probe microscopy instrumentation for biological applications, such as atomic force microscopy, is anticipated to help elucidate the intimate spatial interactions between the protein and chitin. These techniques also hold much promise for measuring the interaction strength between chitin and purified beak proteins at the single-molecule level and are receiving increasing attention in many biomimetic studies.

Inspiration for novel materials. Are we ready to mimic the lessons of squid beaks? The answer is probably yes and no. Insect cuticles are the closest biochemical system for which bio-inspired materials based on chitosan (deacetylated chitin) have already been attempted. Chitosan is closely related to chitin, but it has a much higher solubility and excellent properties as a templating scaffold for fabrication of various composites. The graded structure of cephalopod beaks, though, represents an even greater potential, if the underlying chemical subtleties governing gradients can be discovered. Because many manufactured goods involve interfaces between mechanically mismatched materials that undergo contact deformation and failure in time, many promising innovations are possible. Control of water gradients by appropriate adjustments in the levels of cross-linking or hydrophobic polymers would allow creation of multifunctional materials. This could include genetic engineering techniques in order to biosynthesize the protein precursors of the hard tissue. In bio-implants, for example, this could notably lead to the production of graded structures exhibiting cartilage-like properties at one end of the material and hard, wear-resistant properties at the other, and could find application in engineered durable joints between soft and hard materials. With regard to the all-organic composition of cephalopod beaks, applying the bio-inspired principles of beak maturation and hardening could lead to a new generation of degradable high-performance composites. Hardened squid beak has much in common with the lignin/cellulose based chemistry of hardwoods and is likely to be decomposed to humins by soil microbes.

For background information *see* CEPHALOPODA; COLEOIDEA; ENGINEERING DESIGN; MATERIALS SCIENCE AND ENGINEERING; POLYMER; POLYSACCHARIDE; PROTEIN; SQUID; STRENGTH OF MATERIALS; TEUTHOIDEA; TISSUE in the McGraw-Hill Encyclopedia of Science & Technology.

Ali Miserez; J. Herbert Waite

Bibliography. P. Fratzl and J. Aizenberg (eds.), *Adv. Mater.*, 21:379–492, 2009; M. A. Meyers et al., Biological materials: Structure and mechanical properties, *Prog. Mater. Sci.*, 53:1–206, 2008; A. Miserez et al., The transition from stiff to compliant materials in squid beak, *Science*, 319:1816–1819, 2008; T. Muller, Biomimetics: Design by nature, *Natl. Geogr.*, April, pp. 68–90, 2008; A. C. Neville, *Biology of Fibrous Composites: Development Beyond the Cell Membrane*, Cambridge University Press, Cambridge, U.K., 1993.

Hydrogen-powered cars

Surprisingly, hydrogen was first used for automotive propulsion almost 2 centuries ago. Its use has not been widespread in part because it is an energy carrier, not a fuel, requiring energy input to extract it from water or other chemical compounds. Because of environmental issues relating to greenhouse gas emissions and criteria pollutants, and concerns about energy security, hydrogen has become an attractive alternative to fossil fuels for several reasons. First, hydrogen is available in vast amounts terrestrially, as it is bound in water and other chemical compounds. Second, combustion of hydrogen (H_2) with oxygen (O_2) yields only water vapor and heat. As a result, no carbon dioxide (CO_2), the major greenhouse gas of concern, is produced. Third, kilogram for kilogram, hydrogen in the molecular form (H_2) contains about three times the energy of gasoline, although

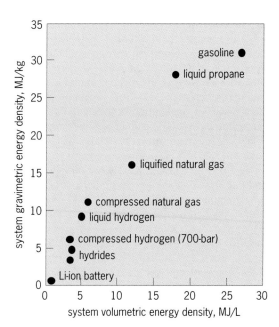

Fig. 1. Volumetric and gravimetric energy densities of the entire fuel delivery system for a variety of fuels and fuel-storage methods, including the volume and weight of the fuel, storage tank, and fuel delivery hardware.

gasoline is more attractive on a system basis (**Fig. 1**). And fourth, splitting water, the most abundant source of hydrogen, can be done via noncarbon energy sources in a variety of ways.

History. In 1783, Antoine Lavoisier gave the name hydrogen ("water creator") to the light, flammable gas produced by reacting metals with strong acids. Remarkably, the first hydrogen-fueled ground vehicle would follow barely 30 years later. In 1813, François Isaac de Rivaz tested a prototype that burned hydrogen in a spark-ignition internal combustion engine (ICE), thus gaining the distinction of building the world's first ICE-powered automobile. Supplied by compressed hydrogen gas in an attached balloon and using a Volta cell to trigger each spark by hand, his test drive covered about 100 m (330 ft).

Experiments continued with hydrogen-fueled ICE vehicles, such as Etienne Lenoir's 1860 "Hippomobile." By the end of the nineteenth century, mainstream automotive development had turned to petroleum, and hydrogen cars were limited to occasional demonstration vehicles throughout most of the twentieth century. Practical vehicles appeared by necessity during the 1941 siege of Leningrad, when petroleum shortages prompted the conversion of 200 GAZ-AA Soviet Army trucks to run on hydrogen.

General Motors built the first hydrogen fuel-cell vehicle, the GM Electrovan, in 1966 (**Fig. 2**). A converted GMC Handivan, its entire cargo space was occupied by 32 5-kW alkaline fuel cells (AFCs) and, like spacecraft, cryogenic tanks for storing both liquid hydrogen and liquid oxygen [AFCs require high-purity CO_2-free oxygen].

Since then a variety of hydrogen-powered test vehicles have been built. Early versions mostly relied on burning hydrogen in an ICE, or occasionally AFCs. Liquid hydrogen storage was preferred because of its volumetric density, but compressed H_2 gas was also common. Other vehicles, such as a transit bus (1976) and converted Cadillac Seville (1977) built by Roger Billings, GM's H2-4 Chevrolet (1978), and a Mercedes Daimler Benz van (1984), used metal-hydride beds.

By the 1990s, renewed interest in hydrogen-powered vehicles was stimulated by growing concerns over petroleum dependency and greenhouse gas emissions. Technical advances in polymer electrolyte membrane (PEM) fuel cells (FC) have today made them the most practical and potentially cost-effective power plants for mobile applications. Almost every automotive manufacturer has demonstration vehicles.

In this decade, emphasis has shifted from experimental platforms to development of practical, consumer-ready vehicles, with performance and features comparable to conventional gasoline ICE vehicles. BMW continues to favor ICE power plants with liquid hydrogen storage, for example, in their Hydrogen 7 automobile (2006). Most manufacturers are focused on PEM fuel cells. Storage systems are trending away from liquid storage toward high-pressure compressed-hydrogen tanks at 200–700 bar (20–70 MPa), as illustrated by GM's HydroGen 1 (2000) and early HydroGen 3 (2001), which used liquid H_2 storage, while a later HydroGen 3 (2002),

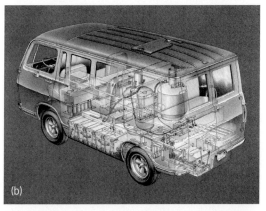

Fig. 2. General Motor's 1966 Electrovan, the first hydrogen fuel-cell vehicle. (*a*) Vehicle shown on the right, beside the recent HydroGen 1. (*b*) Interior diagram, showing liquid H_2 and liquid O_2 storage tanks. (*Photos courtesy of General Motors Company*)

Fig. 3. GM Equinox hydrogen fuel-cell vehicle. More than 100 of these vehicles are being driven by the public to assess real-world performance. (*Photo courtesy of General Motors Company*)

Sequel (2005), Chevy Equinox FC (2006; **Fig. 3**), and HydroGen 4 (2008) all use compressed gas.

Power plants. Engines for extracting the chemical energy from hydrogen for vehicles fall into two classifications: internal combustion engines, in which the thermal energy from burning hydrogen is converted to mechanical energy, and fuel cells that use electrochemical energy conversion to power electric drive motors (**Table 1**). ICEs are attractive because the underlying technology has propelled vehicles for more than a century. ICEs, however, are ultimately limited to the efficiencies of heat engines, while electrochemical fuel cells are not limited by thermal efficiencies.

The theoretical thermal efficiency η_{th} [Eq. (1)] and

$$\eta_{th} = 1 - \frac{1}{(V_1/V_2)^{\gamma-1}} \qquad (1)$$

temperature rise from T_1 to T_2 during the adiabatic compression stroke of an Otto cycle ICE [Eq. (2)]

$$T_2 = T_1 \left(\frac{V_1}{V_2}\right)^{\gamma-1} \qquad (2)$$

are governed by the compression ratio (V_1/V_2), where V_1 and V_2 are the uncompressed and compressed volumes, respectively, and γ is the specific heat ratio c_p/c_v. A higher compression ratio improves the thermal efficiency, subject to the constraint that T_2 must be less than the autoignition temperature.

Hydrogen's higher autoignition temperature and larger γ (1.4 for H_2, 1.1 for gasoline) make hydrogen ICEs inherently more efficient than gasoline. Hydrogen ICEs have several drawbacks, including (1) the high autoignition temperature makes a self-igniting diesel cycle difficult, (2) the energy required to ignite hydrogen is about an order of magnitude smaller than that for gasoline, making hydrogen ICEs much more prone to pre-ignition (knock), and (3) high-temperature combustion in air still produces NOx pollutants, although these can be minimized by adjusting the air/hydrogen ratio.

Gasoline ICEs can be modified to run with hydrogen, but these engines are not optimized for hydrogen. The power output per stroke is only about 85% of that obtained with gasoline because of the much higher volumetric fuel/air ratio of hydrogen compared to gasoline. For stoichiometric combustion, gasoline occupies less than 2% of the cylinder volume (1 gasoline molecule for every 12.5 O_2 molecules, which in turn comprise 20% of air, Table 1), while H_2 fills 30% of the cylinder ($H_2/O_2 = 2/1$). Equivalently, with hydrogen the effective cylinder displacement is 30% smaller. The power output can be boosted to about 120% of that with gasoline by injecting high-pressure hydrogen after the cylinder is full of air and the intake valve closes. Hydrogen-specific engines typically have larger cylinders and higher compression ratios to optimize the power per stroke and engine efficiency. Typical gasoline engines have an overall efficiency of 25–30%. Recently, BMW demonstrated 42% engine efficiency for an optimized hydrogen ICE.

Early automotive fuel cells used alkaline fuel-cell (AFC) technology, in which two electrodes sandwiched a porous separator saturated with an alkaline solution, such as potassium hydroxide (KOH). Electrons generated at the anode supplied current through the external circuit, balanced by an OH^- counterflow from the cathode through the separator (Table 1). AFC applications require pure oxygen or purified air to prevent poisoning by CO_2 through carbonate formation. Although they have the potential to reach 70% efficiency, the inability to use air directly limits large-scale application of AFCs for transportation.

More recent vehicle applications have converged on the proton exchange membrane (or polymer electrolyte membrane) fuel cell. Here the separator is a thin (about 50-μm) nonconducting polymer membrane (for example, Nafion™) permeable to protons but not other gases. Hydrogen is catalytically

TABLE 1. Comparison of gasoline, ethanol, and hydrogen power plants

Power plant*		Principal reaction	CO_2 produced, g/MJ	H_2O produced, g/MJ
Gasoline ICE		$C_8H_{18} + 12.5O_2 \rightarrow 8CO_2 + 9H_2O$ (octane)	65.3	30.1
Ethanol ICE		$C_2H_5OH + 3O_2 \rightarrow 2CO_2 + 3H_2O$	63.7	39.1
H_2 ICE		$2H_2 + O_2 \rightarrow 2H_2O$		
Alkaline FC	Anode	$H_2 + 2OH^- \rightarrow 2H_2O + 2e^-$		
	Cathode	$O_2 + 2H_2O + 4e^- \rightarrow 4OH^-$	0	63.4
PEM FC	Anode	$H_2 \rightarrow 2H^+ + 2e^-$		
	Cathode	$4H^+ + O_2 + 4e^- \rightarrow 2H_2O$		

*ICE = internal combustion engine; FC = fuel cell; PEM = polymer electrolyte membrane.

TABLE 2. Comparison of selected transportation technologies*

Fuel	Power plant†	TTW mpg equivalent, relative to gasoline ICE	Fuel source	WTW energy use	Fraction fossil energy, %	Fraction petroleum energy, %	Total greenhouse gases	NOx
Gasoline	ICE	1	Petroleum	1	~100	89	1	1
	Hybrid ICE/EV	1.24	Petroleum	0.81	~100	89	0.81	0.88
Diesel	ICE	1.21	Petroleum	0.80	~100	91	0.87	0.85
	Hybrid ICE/EV	1.45	Petroleum	0.67	99	91	0.68	0.77
E85 ethanol/ gasoline‡	ICE	1.00	Corn	1.29	54	22	0.82	1.94
			Cellulosic	1.67	18	17	0.28	2.10
	Hybrid ICE/EV	1.24	Corn	1.05	53	22	0.66	1.64
			Cellulosic	1.35	18	17	0.23	1.76
H₂	ICE	1.20	Electrolysis	2.38	86	2	2.36	5.84
			Methane	1.15	98	1	0.93	1.46
	Hybrid ICE/EV	1.48	Electrolysis	1.94	86	2	2.01	4.80
			Methane	0.93	98	1	0.76	1.26
	Hybrid PEM/EV	2.63	Electrolysis	1.09	86	2	1.12	2.44
			Methane	0.52	98	1	0.42	0.47
	PEM FC	2.38	Electrolysis	1.20	86	2	1.24	2.69
			Methane	0.58	98	1	0.46	0.52
			Renewable	0.54	0	0	0	0

*Comparison of tank-to-wheels (TTW) efficiency and well-to-wheels (WTW, including contributions from fuel production) total energy use, greenhouse gas release, and NOx as a representative pollutant. All values are normalized to those for a conventional gasoline internal combustion engine vehicle. For each WTW energy the fractions derived from fossil sources (oil, coal, natural gas, etc.), and more specifically from petroleum, are also given. (*Adapted from http://www.transportation.anl.gov/pdfs/TA/339.pdf*)

†ICE = internal combustion engine, EV = electric vehicle, PEM = polymer electrolyte membrane, FC = fuel cell.

‡Mixture of 85% ethanol and 15% gasoline by volume.

split at the anode, and the external electron flow is balanced by protons permeating through the membrane, where they react with oxygen at the cathode (Table 1). Honda's FXC fuel-cell concept vehicle reportedly has an efficiency of 60%.

Table 2 compares the relative tank-to-wheels (TTW) vehicle efficiency and well-to-wheels (WTW) energy use, greenhouse gas production, and pollution for selected transportation technologies. These estimates are based on a specific vehicle (a pickup truck) and may differ for other vehicle platforms, but the relative values provide insight into the performance of different vehicle technologies. WTW includes the energy use and emissions required to produce and deliver the fuel to the vehicle as well as those of the vehicle itself. For ethanol-based (E85) fuel, WTW effects are further broken down according to whether the ethanol is produced from corn or from nonfoodstuff cellulose. For hydrogen vehicles, the fossil fuel energy use and CO_2 emissions are entirely attributable to hydrogen production and transport, and depend on whether the H_2 was extracted by electrolysis of water using primary energy from the electric grid, by reacting (reforming) methane with high-temperature steam, or by any of a number of noncarbon renewable or sustainable energy sources, all of which are nearly petroleum-independent. As can be seen in Table 2, vehicles based on PEM fuel cells using hydrogen from completely renewable energy sources provide the most attractive WTW pathway.

Hydrogen storage. Nominally 4–8 kg (9–18 lb) of hydrogen, depending on vehicle size, are required for a driving range approaching 480 km (300 mi), which is clearly a challenge for hydrogen as a low-density gas. Drivable prototype hydrogen cars have been built that store hydrogen as compressed gas, as liquid hydrogen, or in a metal hydride, or use hydrogen generated on-board either by cracking a liquid hydrocarbon fuel or by hydrolysis of sodium borohydride ($NaBH_4$).

Liquid hydrogen has been a popular storage option because, given its density of 0.07 g/cm³, 5 kg (11 lb) of liquid hydrogen corresponds to about 70 liters (18.5 gal), which is only modestly larger than a typical gasoline tank. Liquid H_2 boils at 20 K ($-424°F$), necessitating a superinsulated cylindrical cryogenic tank. The technology suffers several drawbacks: (1) there is a 30% energy penalty for liquefaction relative to the energy content of the hydrogen (mostly from its latent heat), (2) storage and cryogenic transfer at the fueling station, and (3) the tank gradually empties because of boil-off from small heat leaks, an issue referred to as dormancy.

Compressed gas is the option of choice for the near term. Early compressed storage systems used multiple conventional tanks at pressures at or below 15 MPa (150 bar), but with limited vehicle range (80–290 km or 50–180 mi). To achieve useful driving range with tanks of realistic volume, modern compressed gas storage uses 30–70-MPa (300–700-bar) Type IV pressure vessels, consisting of an inner one-piece permeation-resistant high-molecular-weight polymer liner, overwrapped with multiple layers of high-tensile-strength carbon fiber/laminate composite, with a protective outer impact-resistant shell. GM's Sequel vehicle (2005), using three tanks totaling 8 kg (18 lb) of 70-MPa (700-bar) hydrogen, has demonstrated a driving range of 480 km (300 mi).

Solid hydrides can reversibly release and absorb hydrogen if the hydrogen is weakly bound (enthalpy

---

TABLE 3. Selected hydrogen vehicles

Vehicle	Type	Power plant*	Storage	Implementation
BMW Hydrogen 7 car	Dual fuel	ICE	Liquid H_2/gasoline	100 lease/demo vehicles
Chrysler ecoVoyager minivan	Plug-in hybrid	45-kW PEM, Li-ion battery	Compressed, 350 bars	Concept
Ford Focus FCV car	Hybrid	PEM, NiMH battery	Compressed, 350 bars	30-vehicle trial
General Motors Equinox SUV	Hybrid	115-kW PEM, 35-kW NiMH battery	Compressed, 700 bars (3 tanks)	100+ consumer trial
General Motors Sequel crossover	Hybrid	73-kW PEM, 65-kW Li-ion battery	Compressed, 700 bars (3 tanks)	Concept
Honda FCX Clarity car	Hybrid	100-kW PEM, Li-ion battery	Compressed, 350 bars (2 tanks)	Lease intent for 200 vehicles
MAN city bus		ICE	Compressed, 350 bars (10 tanks, roof mounted)	14+ in operation
Mazda 5 Hydrogen RE minivan	Dual-fuel mild hybrid	Rotary ICE, Li-ion battery	Compressed, 350 bars/gasoline	Government lease vehicles
Quantum Hydrogen Hybrid Prius car	Hybrid	ICE, battery	Compressed, 350 bars; or 10 bars, metal hydride	50+ converted Prius vehicles
Toyota FCHV SUV	Hybrid	90-kW PEM, 21-kW NiMH battery	Compressed, 700 bars	~25 demonstration vehicles
VW Passat Lingyu car	Hybrid	55-kW PEM, Li-ion battery	Compressed	20 demonstration vehicles

*ICE = internal combustion engine, PEM = polymer electrolyte membrane fuel cell.

---

of hydrogen release $18 < \Delta H < 35$ kJ/ mole H_2). Hydrogen is liberated by a modest increase in temperature or decrease in hydrogen pressure. "Interstitial" hydrides, such as $LaNi_5$-based hydrides, provide the best engineering performance, but their weight penalty is prohibitive, as the hydrogen content is only 1–3% by weight. Advanced high-capacity hydrides offer reduced weight and volume, but practical and cost-effective materials for solid hydride storage remain in the research stage.

On-board hydrogen generation has been implemented by extracting hydrogen from liquid hydrocarbon fuels. Chemical reformers catalytically convert methanol (CH_3OH) to H_2 and CO_2. Numerous prototype cars using methanol, or gasoline as in GM's Chevrolet S-10 pickup (2001), showed that the cost and complexity of on-board liquid fuel reforming, and the fact that they still produce the greenhouse gas CO_2, make these impractical for mass-market vehicles. Millennium Cell Inc. pioneered on-board hydrogen generation by catalytically reacting sodium borohydride ($NaBH_4$) with water to form sodium borate ($NaBO_2$) and H_2. Daimler-Chrysler and Peugeot have built several $NaBH_4$-based demonstration vehicles. The need to exchange an alkaline sodium borate slurry with fresh $NaBH_4$ during refueling, and the cost and energy of reprocessing the spent borate, make this system commercially impractical.

Other storage mechanisms are being explored. Cryoadsorption uses weak van der Waals binding of H_2 molecules on very high specific–surface-area materials (2500–3500 m^2/g) such as activated carbon or metal-organic frameworks (MOFs) to store up to 7 wt% hydrogen near liquid nitrogen temperature (77 K, $-321°F$). Converting between a hydrogen-rich and hydrogen-deficient liquid organic ring compound releases hydrogen, although it cannot be reversed onboard. Cryocompression envisions a superinsulated high-pressure tank that uses low temperature to increase the hydrogen gas density. While

active subjects of research and development, no vehicles have yet been built that use these principles.

Recent vehicles. Several hundred distinct hydrogen-fueled experimental vehicles have been made, including passenger cars, vans, SUVs, pickup trucks, buses, motorcycles, scooters, utility vehicles, mining vehicles, boats, submarines, and aircraft. Numerous cities worldwide now run small fleets of prototype hydrogen-fueled public buses. **Table 3** lists a selection of recent vehicles, some of which are being made available to the public in small numbers as learning platforms aimed toward future production. Hydrogen vehicles can and must be designed to meet all applicable safety standards. As with gasoline or any fuel, hydrogen must be handled with respect for its large chemical energy content. It is otherwise nontoxic and rapidly disperses, if released.

Most current hydrogen vehicle designs are hybrids and include a high-performance battery to supplement the fuel cell in high-demand situations (load leveling) and to recapture electrical energy by regenerative braking. Hydrogen-powered fuel-cell vehicles have been developed that have commercial performance and driving range, and can be refueled in about 5 min.

GM's Autonomy concept vehicle shows the potential of redesigning the automobile around hydrogen fuel cells from the ground up. The entire power train is compressed into a flat "skateboard" chassis with a detachable passenger compartment, using drive-by-wire instead of mechanical linkages for vehicle control. GM's Sequel uses a skateboard chassis that is only 28 cm (11 in.) thick, illustrating concepts that may well represent the future of automotive transportation.

Down the road. If a hydrogen fuel-cell transportation era with its tantalizing advantages is to be realized, three major hurdles must be overcome.

Light, robust, and economical fuel-cell propulsion systems must be developed. Significant progress continues to be made, and there is confidence within the automotive industry that this objective can be achieved.

A safe, light, compact, cost-effective, and durable system for storing hydrogen onboard a vehicle needs to be developed. A great deal of international effort is focused on this challenge.

A hydrogen production and distribution infrastructure needs to be established. The leadership of national, state, and local governments will be required to enable construction of refueling facilities. The distribution infrastructure will be dependent to large degree on the means for production and storage, and could be either centralized (pipelines or delivery trucks), decentralized (electrolyzers, photocatalyzers, or reformers at homes and at fueling stations), or a combination of both. Currently, H_2 is produced chiefly by reforming natural gas, but in the future we can look to dissociating water, our greatest source of hydrogen, by means of noncarbon energy sources. Among the possibilities are electrolysis using electricity from solar, wind, hydroelectric, and nuclear resources; direct generation of H_2 by photocatalysis using sunlight and semiconductors; nuclear and solar thermochemical cycles; and biological and bio-inspired methods. The full potential of hydrogen-powered vehicles can be realized if H_2 is produced by completely renewable energy sources.

For background information *see* ALTERNATIVE FUELS FOR VEHICLES; AUTOMOBILE; AUTOMOTIVE ENGINE; FUEL CELL; HYDRIDE; HYDROGEN; INTERNAL COMBUSTION ENGINE; METAL HYDRIDES in the McGraw-Hill Encyclopedia of Science & Technology.

Frederick E. Pinkerton; Jan F. Herbst

Bibliography. L. D. Burns, J. B. McCormick, and C. E. Borroni-Bird, Vehicle of change, *Sci. Amer.*, 287(4):64–73, October 2002; N. Brinkman et al., Well-to-wheels analysis of advanced fuel/vehicle systems: A North American study of energy use, greenhouse gas emissions, and criteria pollutant emissions, 2005; F. E. Pinkerton and B. W. Wicke, Bottling the hydrogen genie, *Indus. Phys.*, 10(1):20–23, February-March, 2004.

Hyperbolic 3-manifolds

Hyperbolic geometry is the non-Euclidean geometry discovered by János Bolyai, N. I. Lobachevsky, and K. F. Gauss about 200 years ago. Work of George Daniel Mostow, William Thurston, and Gregory Perelman shows that hyperbolic 3-manifolds are the most important class of three-dimensional (3-D) manifolds. Therefore, it is of great importance to come to a deep understanding of hyperbolic 3-manifolds. The first part of this article begins with basic ideas, but ultimately focuses on the most natural tool for analyzing a hyperbolic 3-manifold: its volume. The second part of the article investigates the geometric rigidity of hyperbolic 3-manifolds.

Notion of a manifold. A two-dimensional (2-D) manifold (or 2-manifold) is a topological space that locally

looks like the *xy* plane. Here, "looks like" is a statement about shape, not measurements: We are in the realm of rubber-sheet topology, where two things look the same if one can be stretched and pulled to get the other.

The simplest example of a 2-manifold is the *xy* plane ($\mathbf{R}^2$) itself. A more interesting example is the 2-D torus, obtained by taking a (rubber) cylinder and gluing the opposite circle boundaries together. A better way to think of the torus is to start with a square in $\mathbf{R}^2$ and identify opposite sides. The first such identification, when carried out, produces a cylinder; performing the second identification produces a torus. But we really do not have to perform the identifications to realize that we have a torus. In fact, the square with opposite sides identified completely describes a torus.

Note that $\mathbf{R}^2$ is infinite in extent, but that the torus is finite in extent. A major goal of low-dimensional topology is to understand compact, orientable manifolds. ("Compact" is the topological version of "finite in extent," and an "orientable" manifold has no mirror-reversing paths; traveling along a mirror-reversing path in a 3-D space will have the effect of changing you from right handed to left handed.) In two dimensions, the list of all compact, orientable 2-manifolds has been known for over 100 years. The list consists of the sphere, the torus, the 2-holed torus, the 3-holed torus, and so on (**Fig. 1**).

Any 2-manifold can be decomposed into triangles, a process called triangulation. The 2-manifold can then be cut along the triangles (keeping track of the cuts) to produce a description of the manifold as a collection of triangles with the edges identified in pairs. Thus, all 2-manifolds can be gotten by appropriate gluing of triangles.

A 3-D manifold or 3-manifold is a space that locally looks like $\mathbf{R}^3$. A good example is the 3-torus, which is obtained by taking a unit cube and identifying opposite faces as follows: A given point on one face is identified to the point on the opposite face that is the end point of the line segment running between these two faces and perpendicular to the faces at both ends. The study of 3-manifolds is considerably more difficult than the study of 2-manifolds, and powerful techniques are needed. Thurston revolutionized the study of 3-manifolds by showing that topological questions about the shape of a 3-manifold could be fruitfully analyzed by introducing a natural geometry to the manifold, and then using this geometry to discover topological insights.

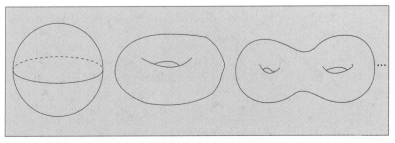

Fig. 1. Compact, orientable 2-manifolds: sphere, torus, 2-hole torus, ...

The most important such geometry is hyperbolic geometry.

Geometry. Euclidean geometry will be briefly discussed, in preparation for describing hyperbolic geometry. Euclidean geometry in two dimensions can be constructed from the xy plane with the infinitesimal metric $ds^2 = dx^2 + dy^2$ (an infinitesimal version of the Pythagorean theorem). Straight lines are locally length-minimizing paths with respect to the metric. A triangle is formed from three straight line segments, and all triangles have angle sum π radians ($180°$). An infinitesimal square parallel to the axes has edges of "length" dx and dy, hence infinitesimal area $dxdy$, and the area of an object A in the Euclidean plane can be computed by integrating $dxdy$ over A.

We describe hyperbolic geometry in a similar fashion. Consider the upper half of the xy plane [that is, all points (x,y) in $\mathbf{R}^2$ with $y > 0$] together with the infinitesimal metric $ds_{\mathrm{H}}^2 = (dx^2 + dy^2)/(y^2)$. The shortest paths ("geodesics") with respect to ds_{H} are fundamentally important; they turn out to be Euclidean circles perpendicular to the x axis and vertical Euclidean straight lines. A hyperbolic triangle is formed from three geodesic segments, and all such triangles have angle sum less than π radians ($180°$). An infinitesimal square parallel to the axes has edges of "length" dx/y and dy/y, hence infinitesimal area $dxdy/(y^2)$, and the area of an object A in our upper-half-plane model of hyperbolic geometry can be computed by integrating $dxdy/(y^2)$ over A. The area so computed of a hyperbolic triangle turns out to be the difference between π and the sum of its angles (measured in radians).

Euclidean geometry and hyperbolic geometry naturally generalize to 3-D versions.

Geometric 2-manifolds. A Euclidean 2-manifold is a 2-manifold together with a means of measurement that locally produces exactly the same calculations as in the Euclidean plane. A torus can be given a Euclidean structure: Think of the torus as a square with opposite edges identified. All points in this torus locally produce exactly the same calculations as the Euclidean plane: The interior points of the square have little disk neighborhoods that are simply pieces of the Euclidean plane; the edge points are identified in pairs, and each point of the pair has a half-disk neighborhood with the two half-disks glued together to form a disk in the Euclidean plane; the four vertex points are all identified, and each has a quarter-disk neighborhood with the four quarter-disks glued together to form a disk in the Euclidean plane.

It can be shown that identifying opposite edges of an octagon produces a 2-holed torus. Does it admit a Euclidean structure? The obvious attempt would start with a regular octagon and analyze the various points. The interior points and the edge points work fine, but the eight vertices of the octagon are all identified, and after gluing together little neighborhoods of the eight vertices the angle sum at the (one) vertex is too large. One can prove that the 2-holed torus does not admit a Euclidean geometric structure.

However, a 2-holed torus does admit a hyperbolic structure. (The definition of a hyperbolic 2-manifold is the same as the definition of a Euclidean 2-manifold but with "Euclidean" replaced by "hyperbolic.") This can be seen from the viewpoint of triangles: Put a point at the center of a regular octagon and then connect the point by geodesics to the eight vertices of the octagon, thereby creating eight congruent triangles. Here the needed 2π angle condition at the one vertex on the boundary of the octagon (after identifying) becomes the condition that all the triangles have angles $\pi/8$, $\pi/8$, and $\pi/4$ radians ($22.5°$, $22.5°$, and $45°$). Indeed, triangles of that type exist in hyperbolic space. We can compute the area of such a triangle to be $[\pi - (\pi/8 + \pi/8 + \pi/4)] = \pi/2$. So, the area of this octagon, hence the area of this hyperbolic 2-holed torus, is $8(\pi/2) = 4\pi$. A similar argument can be used to show that, for all $n > 1$, the n-holed torus admits a hyperbolic structure of area $2\pi(2n - 2)$.

Low-volume Hyperbolic 3-Manifolds

A hyperbolic 3-manifold is a 3-manifold together with a means of measurement that locally produces exactly the same calculations as in hyperbolic 3-space. Hyperbolic 3-manifolds are of paramount importance in the theory of 3-manifolds. How can we come to a good understanding of hyperbolic 3-manifolds? The most natural approach is to use the hyperbolic structure to measure the size, or volume, of the space. Does the volume provide useful information about hyperbolic 3-manifolds? Building on work of Palle Jorgensen and Michael Gromov, Thurston proved (in 1978) a structure theorem for the volumes of hyperbolic 3-manifolds. There is a compact, orientable hyperbolic 3-manifold of least volume v_1, and a next lowest volume v_2, and a next lowest volume v_3, and so on, limiting on the least limiting volume v_ω. Then there is a next lowest volume, and a next lowest volume after that, and so on, limiting on the second lowest limiting volume. This procedure can be continued without end. The non-limiting volumes are volumes of compact orientable hyperbolic 3-manifolds. The limiting volumes have a geometric meaning, as well. They are volumes of certain types of noncompact hyperbolic 3-manifolds called cusped hyperbolic 3-manifolds. A cusp is essentially an infinitely high rectangular chimney with the opposite faces identified. A cross section is a square with opposite edges identified, that is, a torus. A more physical description of the cross section is that of the surface of a bagel.

We cannot claim to understand hyperbolic 3-manifolds if we do not have a good understanding of the low-volume ones. In particular, what is the minimum volume v_1, and which hyperbolic 3-manifolds have this volume? This is a hard problem which, as will be discussed, has been solved recently. The solution proceeds on two fronts. On the first front, one can investigate lots of likely hyperbolic 3-manifolds and their volumes, and put forth the winner as the leading candidate for the low-volume manifold. So, the volume of v_1 is certainly less than or equal to that of this leading candidate. On the other front, one can argue by theoretical methods that v_1 must be greater than or equal to some determined value (typically, by finding useful geometric objects in the manifolds and analyzing their volumes). That is, one can come up with lower bounds for the value of v_1. If the lower bound is equal to that of the low-volume candidate, then v_1 has been found.

SnapPea and the first front. Among surfaces, the topologically simple manifolds (the sphere and the torus) do not admit hyperbolic structures, whereas the complicated manifolds (all the rest) do admit hyperbolic structures. Work of Thurston and Perelman shows that a similar situation occurs for hyperbolic 3-manifolds: They form the largest and most complicated class of 3-manifolds.

For many years, there was slow progress in finding hyperbolic 3-manifolds, and this is not surprising given that hyperbolic 3-manifolds are topologically complicated. The situation changed dramatically in the late 1970s when Thurston developed a general process for constructing infinite families of hyperbolic 3-manifolds. He started with the 3-D sphere and removed a knotted curve. If the knot is sufficiently complicated, then Thurston showed how the 3-D space that remains, called the knot complement, can be given a cusped hyperbolic 3-manifold structure and its volume is one of the limiting volumes described above (similarly for links, which are collections of disjoint knotted curves). Further, Thurston proved that there are an infinite number of distinct ways of gluing the removed knot back that produce a new (compact) manifold with a hyperbolic structure. The process of gluing the removed knot back (generally, in a new way) is called Dehn filling. Using the chimney description of a cusp, we can think of Dehn filling as chopping off the top of the chimney and filling in the exposed bagel surface (that is, a torus) with the rest of the bagel. The tricky point is that this can be done in many fundamentally different ways.

In the 1980s, Jeffrey Weeks developed a computer program (SnapPea) that carried out Thurston's knot-based procedure. Draw a knot for SnapPea, and it will determine whether the knot complement admits a hyperbolic structure, and will explicitly describe the structure. SnapPea's method is a natural generalization of the triangulation-based method for determining if a 2-manifold admits a hyperbolic structure. SnapPea starts by decomposing the knot complement into tetrahedra. Thus, the knot complement can be described as a collection of tetrahedra together with gluing data. SnapPea then attempts to find specific hyperbolic tetrahedra that will produce an angle of 2π radians (360°) at each edge after the gluings are carried out. If SnapPea can accomplish this, then all points in the manifold obtained by gluing together the tetrahedra will locally look like hyperbolic 3-space, and the knot complement will have been given a hyperbolic structure. SnapPea is amazingly efficient at carrying out this procedure.

SnapPea has many other remarkable capabilities. It can take a cusped hyperbolic 3-manifold and ana-

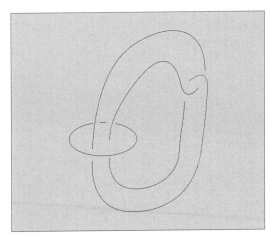

Fig. 2. Whitehead link. The lowest-volume closed hyperbolic 3-manifold is a certain Dehn filling of the complement of this structure in the 3-dimensional sphere.

lyze all the possible Dehn fillings for hyperbolicity, thereby carrying out Thurston's Dehn-filling result described above. Given an explicit description of a hyperbolic 3-manifold in terms of hyperbolic tetrahedra, it is a simple matter for SnapPea to compute the volume of this manifold. The basic idea is that, in analogy with the 2-D situation where there is a simple formula for the area of a hyperbolic triangle in terms of its angles, there is a formula for the volume of a hyperbolic tetrahedron in terms of the angles along the edges. (The angle at an edge is made by the two faces meeting at that edge.) This tetrahedral formula involves an intriguing integral that can be numerically approximated to desired accuracy.

Weeks and others used SnapPea to analyze huge numbers of hyperbolic 3-manifolds, including the calculation of their volumes. After these extensive analyses, it was reasonable to put forth the winner as a candidate for the lowest-volume closed hyperbolic 3-manifold: It is obtained by a certain Dehn filling of the complement of the Whitehead link (**Fig. 2**) in the 3-D sphere, and its volume is 0.9427....

Second front. Lower bounds for v_1 were first obtained by finding objects in hyperbolic 3-manifolds and using their volumes to provide volume bounds. For example, a short geodesic loop (a loop is a curve that begins and ends at the same point; and a geodesic loop is a loop that is locally length-minimizing) in a hyperbolic 3-manifold is at the core of a tube (**Fig. 3**). It turns out that the shorter the core geodesic, the larger the tube and, in fact, the larger the volume of the tube. Using this sort of idea, a lower bound for volume of 0.001 was produced by around 1980. About 20 years later, Nathaniel Thurston (William Thurston's son) teamed with David Gabai and Robert Meyerhoff to prove that the shortest geodesic in a hyperbolic 3-manifold has a much larger tube than previously realized. In fact, it is large enough to get the lower bound on volume up to 0.166.

Gabai, Meyerhoff, and Nathaniel Thurston used a rigorous computer analysis of a six-dimensional (6-D) real parameter space (three complex dimensions) to

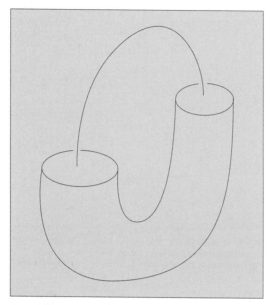

Fig. 3. Cutaway view of geodesic loop at the core of a solid tube.

prove their theorem. The parameters included the length of a shortest geodesic, the size of a largest tube around it, and a parameter describing how the tube would bump into itself if it were blown up beyond its largest size. Given specific values for the parameters, the question posed was whether a hyperbolic 3-manifold existed that had a shortest geodesic of that length, with a tube of that size, sitting in that way inside it. The computer program broke the parameter space up into roughly 10^9 boxes of varying sizes and analyzed each box separately.

The Gabai-Meyerhoff-Thurston (GMT) result stimulated considerable research and the lower bound was quickly pushed to 0.33. In particular, Ian Agol developed tools for comparing the volume of a closed manifold and the volume of a particular cusped manifold from which it is obtained via Dehn filling. (Agol's approach critically uses the large tubes of GMT.) It is considerably easier to get lower bounds for volumes of cusped manifolds than closed manifolds, because the neighborhood of the cusp is a natural object of considerable volume sitting in the manifold. In fact, by the time of Agol's work, the cusped minimum volume v_ω had already been found by Chun Cao and Meyerhoff.

Next, Agol and Nathan Dunfield used Perelman's work on Ricci flow to improve Agol's closed-versus-cusped volume control, and the lower bound thereby produced was 0.66. Further, their result showed that any cusped manifold with volume greater than 2.848 would only produce (via Dehn filling) closed manifolds of volume greater than the volume of the Weeks manifold. *See* GEOMETRIZATION THEOREM.

The final breakthrough was achieved by Gabai, Meyerhoff, and Peter Milley, who found all cusped hyperbolic 3-manifolds of volume less than 2.848. Loosely, the method of Gabai-Meyerhoff-Milley is to start with a cusp and then expand the cusp neighborhood as far as possible. At some point, it will bump

into itself, but one can continue to expand the cusp neighborhood past the various bumpings. If there are few early bumpings, then there is a lot of volume. If there is extensive early bumping, then this is important topological information. We are led to a dichotomy: either there is little early bumping and the volume can be shown to be at least 2.848, or there is substantial bumping and the possible topological bumping information leads to possible cusped manifolds realizing the bumping data. The analysis of these possibilities requires significant (and rigorous) computer assistance. In particular, all such cusped manifolds with volume less than 2.848 are found (there are ten of them).

Milley then analyzed all compact hyperbolic manifolds obtained from Dehn filling on these ten low-volume cusped manifolds (again with rigorous computer assistance) and found that the Weeks manifold had the lowest volume, thereby proving that the Weeks manifold is the unique compact orientable hyperbolic 3-manifold of minimum volume.

Robert Meyerhoff

Geometric Rigidity of Hyperbolic Manifolds

To investigate the geometric rigidity of hyperbolic 3-manifolds, we will begin in two dimensions, as before.

Situation in two dimensions. As previously discussed, abstractly gluing together opposite edges of an octagon produces a 2-holed torus, as shown in **Fig. 4**. A pair of opposite edges in Fig. 4*a* should be glued with their arrows pointing in the same direction. Then, in Fig. 4*b*, each pair of similar edges becomes a single edge with an arrow.

Figure 5 shows that it is possible to build an octagon in the upper half-plane such that all the edges are the same length, and all the angles are $\pi/4$ radians (45°). At first glance this pair of pictures may be surprising. It should be remembered that the edges are hyperbolic geodesics, that is, hyperbolic straight lines, which means that they are either vertical lines or portions of semicircular arcs hitting the *x* axis perpendicularly. Also, no vertex of the octagon lies on the *x* axis, though 5 of them lie close to it.

The hyperbolic length of a parametrized geodesic, or any path, $(x(t), y(t))$ defined for $a \leq t \leq b$, is given by the integral below, where $\dot{x}(t)$ [respectively $\dot{y}(t)$

$$\int_\alpha^b \frac{\sqrt{\dot{x}(t)^2 + \dot{y}(t)^2}}{y(t)} dt$$

is the derivative of $x(t)$ [respectively $y(t)$]. Length is difficult to see in Fig. 5*a* and *b*, but with work one can verify that the edge lengths agree, and are all about 3.06. Abstractly gluing the opposite edges forms a 2-holed torus, as before. These four gluings identify all eight vertices to a single point in the 2-holed torus. To verify that this is a hyperbolic 2-holed torus, one must check that the total angle around the vertex is 2π radians (360°). As $8 \cdot \pi/4 = 2\pi$, this is correct.

How unique is this hyperbolic structure on the 2-holed torus? More concretely, are there other

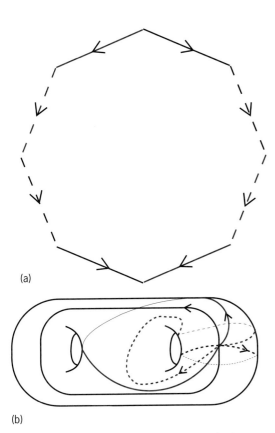

(a)

(b)

Fig. 4. Gluing opposite sides of an octagon to obtain a 2-holed torus. (*a*) Octagon. A pair of opposite edges should be glued with their arrows pointing in the same direction. (*b*) 2-holed torus. Each pair of similar edges has become a single edge with an arrow.

distinct octagons in the hyperbolic upper half-plane such that (1) opposite edges have equal length, and (2) the sum of the vertex angles is 2π radians (360°)? By distinct we mean that we are not interested in other congruent octagons, namely those with identical interior angles and edge lengths. For example, one could simply move the octagon of Fig. 5 horizontally to obtain many congruent hyperbolic octagons. These do not interest us. Are there truly different octagons in the hyperbolic upper half-plane satisfying conditions (1) and (2)?

Briefly, the answer is yes, there exists an infinitude of such octagons. One is shown in **Fig. 6**, though from the picture alone the desired properties are not obvious. It has four edges of length roughly 3.11, and four of length roughly 4.79. The vertex angles are also no longer the same, changing from 45° to approximately 128.6°, 16.1°, and 19.2°. (Each angle occurs at multiple vertices. It is uncommon to write approximate angles in radians.) The lesson to learn from this example is that 2-D hyperbolic manifolds are not rigid. Our 2-holed torus can support many different hyperbolic structures. This flexibility has been known to mathematicians (in different guises) for at least 150 years.

Situation in three dimensions. The situation in three dimensions is very different and more complex. The upper half-plane is replaced with the upper half-space. We can replace our octagon with a dodecahedron (**Fig. 7**). [The obvious generalization to an

octahedron does not quite work.] There are many ways abstractly to glue up the faces of a dodecahedron. Here we choose to identify opposite faces by a $3\pi/5$-radian (108°) twist, that is, 3/10 of a full rotation. Gluing together these six pairs produces a 3-manifold called the Seifert-Weber dodecahedral space. After gluing, the 30 edges of the dodecahedron become just 6 edges in the 3-manifold. In other words, 5 edges of the dodecahedron glue up to a single edge in the 3-manifold. How can we put a hyperbolic structure on this object?

To answer this question, we must think a bit about hyperbolic polyhedra in the upper half-space. A geodesic in the upper half-space is again either a vertical line or a semicircular arc hitting the horizontal xy plane orthogonally. By analogy, a hyperbolic

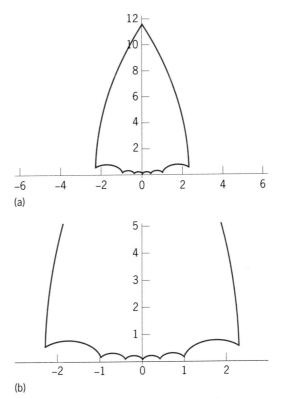

(a)

(b)

Fig. 5. Hyperbolic octagon with angles $\pi/4$ radians (45°). (*a*) The whole octagon in the upper half-plane. (*b*) Magnified view of the part to the octagon near the *x* axis.

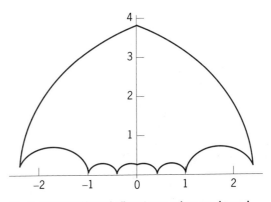

Fig. 6. Deformed hyperbolic octagon, whose angles and edge lengths are changed from those of the octagon in Fig. 5.

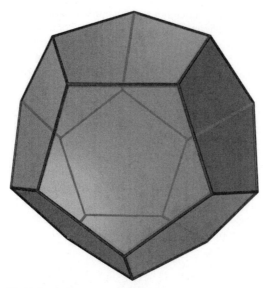

Fig. 7. Euclidean dodecahedron. Gluing together opposite faces with a $3\pi/5$-radian ($108°$) twist produces a 3-manifold called the Seifert-Weber dodecahedral space.

plane in the upper half-space is either a vertical Euclidean plane or a hemisphere hitting the horizontal xy plane orthogonally. A hyperbolic polyhedron is formed by polygonal pieces of hyperbolic planes.

Rigidity of compact hyperbolic 3-manifolds. With this we can return to our dodecahedron. Remarkably, there is a hyperbolic dodecahedron with 12 congruent faces and dihedral angles all equal to $2\pi/5$ radians ($72°$). (The dihedral angle is the interior angle between two faces.) The edges of this hyperbolic dodecahedron are shown in **Fig. 8**. When visualizing the faces, it should be remembered that each is a piece of a hemisphere. When forming the Seifert-Weber dodecahedral space, five edges of the hyperbolic dodecahedron are identified. Each of these five edges has a dihedral angle of $2\pi/5$ radians ($72°$), so the five dihedral angles glue up to form a 2π-radian ($360°$) "angle" about an edge of the Seifert-Weber dodecahe-

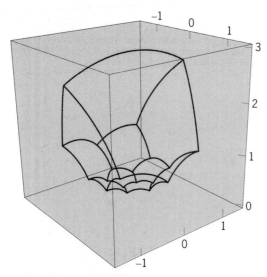

Fig. 8. Edges of a hyperbolic dodecahedron, with 12 congruent faces and dihedral angles all equal to $2\pi/5$ radians ($72°$).

dral space. We conclude that this 3-manifold admits a hyperbolic structure given by our hyperbolic dodecahedron with the above gluing instructions; it is a hyperbolic 3-manifold. (The punctilious reader may worry that we have not considered the vertices of the Seifert-Weber dodecahedral space. Perhaps surprisingly, this is not necessary.)

Is this hyperbolic structure on the Seifert-Weber dodecahedral space unique? Making this question precise is difficult. Sacrificing a little rigor for clarity, let us approximate it by the following explicit question. Do there exist other distinct hyperbolic dodecahedra such that all the dihedral angles remain $2\pi/5$ radians ($72°$)? By distinct we mean that we are not interested in congruent dodecahedra, those with identical dihedral angles and edge lengths. We are interested in dodecahedra whose edge lengths are not all identical. For example, one could simply move or rotate the dodecahedron of Fig. 8 horizontally to obtain many congruent hyperbolic dodecahedra. As before, these do not interest us. Are there truly different dodecahedra in the hyperbolic upper half-space with all dihedral angles equal to $2\pi/5$ radians ($72°$)?

The answer is no. Try as we might, it is not possible to wiggle the faces of our hyperbolic dodecahedron without changing the dihedral angles. This negative answer to our second question hints that the hyperbolic structure on the Seifert-Weber dodecahedral space should be unique. This hint is correct; it is unique. When the hyperbolic structure is unique we say that the space is rigid. This rigidity, and the negative answer to our second question, are consequences of the same underlying theorem, called the Mostow rigidity theorem. In the 1960s, Mostow proved that the hyperbolic structure on a compact 3-manifold is always unique. (Intuitively, a noncompact manifold goes on forever in some direction, for example, an infinite cylinder without ends.) This remarkable theorem implies that the floppy topological "shape" of a 3-manifold uniquely determines its hyperbolic structure, a hard geometric object. In fact, Mostow proved something far more general, and his achievement is surely one of the twentieth century's best theorems.

What does it mean that the Seifert-Weber dodecahedral space is rigid? The "shape" of this space can be described completely: Take a dodecahedron and glue opposite faces with a $3\pi/5$-radian ($108°$) twist. This description does not specify the twelve points in the upper half-space of Fig. 8. It does not even specify the appropriate lengths and dihedral angles of the dodecahedron. From Mostow's rigidity theorem we know that these hard geometric properties are determined by the shape. Rigidity also implies that the hyperbolic volume of the Seifert-Weber dodecahedral space is determined by its shape, as was discussed in the first part of this article.

Deformability of noncompact hyperbolic 3-manifolds. What happens when we drop the requirement that our hyperbolic 3-manifold is compact? In other words, what if we let our space have some directions that go on forever? For example, consider the hyperbolic "octahedron" of **Fig. 9**. The checkerboard is the *xy* plane, where each square has length-one sides. The colored lines are edges of the hyperbolic octahedron. Only two of the octahedron's hemispherical faces are shown. Two others are not shown. The four other faces are pieces of vertical planes, each formed by two vertical edges and a semicircular edge.

An octahedron has six vertices. However, only five vertices are visible in Fig. 9, and they lie in the *xy* plane, which is not actually part of hyperbolic space. One may contrast this with the five lower vertices of the compact hyperbolic octagon, which are slightly above the *x* axis (Fig. 4*b*). The sixth vertex of the octahedron has disappeared completely. It is "the point at infinity," thought to live infinitely far up from the *xy* plane. This means that the octahedron is not compact; it goes on forever vertically.

All the original dihedral angles are $\pi/2$ radians (90°), which is visible in the picture. For example, the four vertical faces form a square chimney going up forever. The two displayed hemispherical faces also intersect in a right angle. Is this noncompact hyperbolic octahedron still rigid? More precisely, is it possible to wiggle the faces of this polyhedron while preserving all the dihedral angles of the original? Now the answer is yes. One example is shown in **Fig. 10**. Notice how all the edges present in the original polyhedron still have $\pi/2$-radian (90°) dihedral angles. There are, however, two new edges in Fig. 10. Each new edge appears as a vertex lifts vertically away from the horizontal plane. In the picture, the vertex nearest us and its opposite both move up, forcing two pairs of faces to intersect along two new edges. For reasons we cannot adequately explain here, this sort of deformation producing new edges is allowed.

With more geometric theory, one can take the behavior of our deformed octahedron and apply it to deform hyperbolic structures on noncompact 3-manifolds. Loosely speaking, this means abstractly gluing up the faces of our octahedra and keeping track of total angles around the edges in the glued-up manifold. William Thurston pushed this phenomenon very far, proving a general theorem roughly stating that noncompact hyperbolic 3-manifolds always admit deformations of their hyperbolic structures. The example of the octahedron is not excep-tional, but rather representative. These deformations were also discussed in the section on SnapPea and the first front, in the first part of this article.

The study of geometric rigidity does not stop here. It is also interesting to study deformations of infinite-volume hyperbolic manifolds. This subject has grown tremendously over the past 30 years, and has recently experienced tremendous break-throughs. Currently, relatively little is known about deformations of hyperbolic structures in higher dimensions, and similar questions in other settings are actively studied. The deformation theory of Einstein manifolds, or various types of projective structures, and the remarkable work of Gregory Margulis all fall into this vibrant field of mathematics.

Fig. 9. Hyperbolic octahedron. The colored lines are its edges, and two of its four hemispherical faces are shown. The four other faces are pieces of vertical planes, each formed by two vertical edges and a semicircular edge. The checkerboard is the *xy* plane.

Fig. 10. Deformed hyperbolic octahedron, whose faces are modified from those of the octahedron in Fig. 9, while preserving all the dihedral angles.

[Prof. Storm was partially supported by NSF grant DMS-0741604 and the Roberta and Stanley Bogen Visiting Professorship at Hebrew University.]

For background information *see* MANIFOLD (MATHEMATICS); NONEUCLIDEAN GEOMETRY; RELATIVITY; RIEMANNIAN GEOMETRY; TOPOLOGY in the McGraw-Hill Encyclopedia of Science & Technology.

Peter A. Storm

Bibliography.   C. Adams and R. Franzosa, *Introduction to Topology*, Pearson, Upper Saddle River, NJ, 2008; L. Ji, A summary of the work of Gregory Margulis, *Pure Appl. Math. Q.*, 4(1):1–69, 2008; D. Mackenzie and B. Cipra, *What's Happening in the Mathematical Sciences*, vol. 6, American

Mathematical Society, 2006; G. D. Mostow, *Strong Rigidity of Locally Symmetric Spaces*, vol. 78 in *Annals of Mathematics Studies*, Princeton University Press, Princeton, NJ, 1973; W. Thurston and S. Levy (eds.), *Three-Dimensional Topology and Geometry*, Princeton University Press, Princeton, NJ, 1997; J. Weeks, *The Shape of Space*, Marcel Dekker, New York, 2002.

Hypersonic test facilities

Systematic research in the field of hypersonic flight in the Earth's atmosphere began in the 1950s. This was due to the development of intercontinental ballistic missiles and the beginning of the space age. Understanding hypersonic flight required new theoretical concepts based on reliable experimental results. However, in the 1950s, scientists did not have data readily available to assist them in the aerodynamic design of high-velocity flying vehicles. As a result, various test facilities were built to reproduce specific features of hypersonic flight under ground-based conditions.

Ballistic ranges. The first hypersonic facilities were ordinary cannons, which shot simple models with velocities of about 1000 m/s (3300 ft/s). The model flight within the measurement section was studied by optical methods. Acceleration schemes were improved, and the flight velocity of moderate-size models was increased to 6000 m/s (20,000 ft/s). Such facilities called ballistic ranges are still in use. For instance, the Arnold Engineering Development Center (AEDC) Range/Track G in the United States uses a two-step light-gas gun that allows an 8-kg (18-lb) model to be accelerated to a velocity of 4000 m/s (13,000 ft/s). The possibility of increasing the velocity up to 10,000 m/s (33,000 ft/s) has been considered. The main drawbacks of ballistic ranges are the moderate admissible mass of accelerated models and their single application. One version of ballistic ranges is a facility where the model is accelerated on a sled with rocket boosters (also called rocket sled test tracks). The largest facility of this type, the Holioman High-speed Test Track (HHSTT) in the United States, allows real objects with masses of 100 and 400 kg (220 and 882 lb) to be accelerated over a length of 6157 m (20,200 ft) to velocities of 3000 and 1500 m/s (9800 and 4900 ft/s), respectively.

Wind tunnels. Specialists in the field of experimental aerodynamics are more familiar with facilities where the motionless model is exposed to a flow of air (or another test gas) with a given velocity. Such facilities are called wind tunnels. They are divided into the following classes: subsonic (the Mach number, M, is less than 1), transonic (M is approximately equal to 1), supersonic (M is greater than 1.5, but less than 5), and hypersonic (M is greater than 5) wind tunnels. Here the Mach number, M, is the ratio between the flow velocity, V, and the velocity of sound, a, at the flow point considered (that is, $M = V/a$). In any case, the test gas with an elevated pressure first enters a settling chamber, where the flow is decelerated and the flow parameters are smoothed to uniform, and then the gas flow is accelerated to a necessary velocity in a special contoured nozzle owing to the difference in pressure.

It was found rather rapidly that the difference in pressure, that is, the ratio of the gas pressure P_s in the settling chamber to the pressure at the nozzle exit P_e should be very large for the gas flow to be accelerated to hypersonic velocities. For instance, a ratio of $P_s/P_e = 530$ is needed to generate a flow with $M = 5$, and a ratio of $P_s/P_e = 4.78 \times 10^6$ should be applied to generate a flow with $M = 20$. Usual multistage compressors can provide limited pressures of the gas (up to approximately 40 MPa or 5800 lb/in.²); therefore, an inevitable feature of hypersonic wind tunnels is exhaustion of the gas into a vacuum tank with pressures of 1–10 Pa (1.45×10^{-4} to 1.45×10^{-3} lb/in.²) to achieve the needed pressure ratio.

In addition, for high velocities to be obtained, the initial gas should possess an extremely large internal energy (high temperature). It follows from thermodynamic considerations that the main part of energy of the gas accelerated in the nozzle is converted to kinetic energy of the flow; correspondingly, the remaining internal energy of the moving gas decreases. The temperature, T_e, and the velocity of sound of the moving gas, which should correspond to the real flight values of temperature and velocity of sound, decrease in proportion to the decreasing internal energy. Thus, to reproduce conditions of flight with a velocity of 1470 m/s (4823 ft/s), $M = 5$, at an altitude of 20 km (12.4 mi), where the air temperature is 217 K ($-69°$F), the initial temperature of the gas in the settling chamber of the wind tunnel should be $T_s = 1300$ K (1880°F). As the modeled flight velocity is increased to 2950 m/s (9678 ft/s), $M = 10$, the required initial temperature of the gas reaches $T_s = 3800$ K (6380°F), and the value of T_s necessary for a velocity of 4600 m/s (15,092 ft/s), $M = 15$, exceeds 10,000 K (17,540°F).

As it is difficult to provide the above-mentioned high values of the gas temperature, a less rigorous condition is used in most cases to model a real flight velocity. Only the Mach number is reproduced in experiments, which is one of the commonly used criteria of similarity for aerodynamic problems. In this case, the flow temperature, the velocity of sound, and the absolute velocity can be substantially lower than the real flight values. The physical limit for temperature reduction is the beginning of test-gas condensation, which implies the emergence of solid particles in the flow and distorts the flow pattern around the model. If the test gas is air, whose condensation starts at a temperature T_e less than 65 K ($-343°$F), the required values of temperature are $T_s = 390$ K (242°F) for $M = 5$, $T_s = 1350$ K (1970°F) for $M = 10$, and $T_s = 3000$ K (4940°F) for $M = 20$. Using a gas with a low condensation temperature (for example, helium, whose condensation temperature is 4 K or $-452°$F), one can obtain a hypersonic flow with $M = 19$ with the initial temperature of the gas equal to room temperature, $T_s = 293$ K (68°F). However, the thermodynamic properties of helium are different than air.

In addition to solving the principal problem of generating a hypersonic flow, there are some important operational requirements for hypersonic wind tunnels. For instance, the size of the generated flow should be consistent with the size of the models being tested. Also, the duration of the test regime (running time) should be sufficient for a steady flow to be obtained. And, finally, the resultant flow should have the minimum possible degree of mechanical and chemical contamination. All existing hypersonic wind tunnels, and those being designed, reflect the history of long-term research and design activities aimed at satisfying the above-listed requirements and restrictions.

Blowdown wind tunnels. A blowdown wind tunnel releases stored high-pressure gas for a short period of time. **Figure 1** shows the structure of a typical hypersonic blowdown wind tunnel, the S4 ONERA, in Modane, France. ONERA (Office National d'Etudes et Recherches Aérospatiales) is the French national aerospace research center. The test gas (air) is supplied from 29 m³ (1024 ft³) upstream storage tanks, where it is stored at a pressure up to 270 bar (3916 lb/in.²). Then, the air flow passes through a heater, which is heated before the experiment up to 1850 K (2870°F) by propane combustion products. After a fast-response valve is opened, the gas enters the settling chamber, is accelerated to a specified velocity in the nozzle, and flows around the model mounted in the test section. After the test section, the air flow passes over a pipeline into a vacuum tank with a volume of 8000 m³ (2.8 × 10⁵ ft³). The initial pressure in the vacuum tank is 100 Pa (1.45 × 10⁻² lb/in.²). The maximum pressure of the gas is $P_s = 120$ bar (1740 lb/in.²), and the maximum temperature is $T_s = 1800$ K (2780°F). The running time is several tens of seconds and is limited by the increase in pressure in the vacuum tank. **Table 1** shows the main characteristics of the facility.

Simple considerations show that running time reduction leads to a proportional decrease in energy necessary for conducting the experiment. Such a method is sometimes used in blowdown wind tunnels. Thus, the largest hypersonic wind tunnel 9 in AEDC has a running time of 0.3–6 s. The main advantage of reducing the test duration to 1 s or less, however, is the fact that this method offers a number of new structural solutions that allow record-beating parameters of the test gas to be obtained.

Shock wind tunnels. The first representatives of the new activities were shock wind tunnels. The oper-

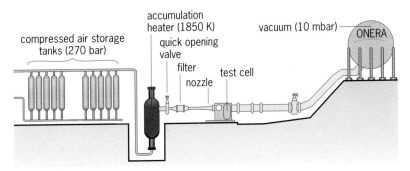

Fig. 1. Hypersonic blowdown wind tunnel, the S4 ONERA in Modane, France.

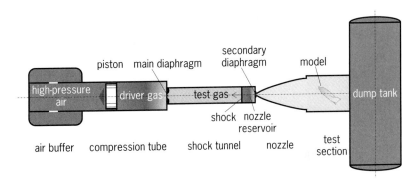

Fig. 2. Shock tunnel with a heavy piston. The shock has been reflected and is moving in the direction opposite to the gas flow.

ation process of shock wind tunnels is based on instantaneous breakdown of a diaphragm between the high-pressure and low-pressure gases in a cylindrical tube. After that, a shock wave passes over the low-pressure gas with a moderate-size region of the test gas with high temperature and pressure formed behind the shock wave. The test gas obtained under such conditions flows out into the nozzle and forms a flow with required parameters. A typical duration of existence of the working flow in shock tunnels varies from 1 to 10 ms.

Despite a long history of shock tunnels, improvements are still implemented, and shock tunnels are intensely used for experimental research. Notable among currently active shock tunnels are the Large Energy National Shock Tunnel (LENS) at the Calspan-University of Buffalo Research Center (CUBRC), United States; the NASA HYpersonic PULSE (HY-PULSE) facility, United States; and the Aachen Shock Tunnel, TH2, Germany.

The main difference in the structure of various shock tunnels is the method of obtaining the driving gas. The driving gas parameters are increased by using an electric heater, detonation of an explosive gas mixture, or adiabatic compression by a moving

TABLE 1. Operational characteristics of the S4 wind tunnel					
Mach number, M	Nozzle diameter, mm	Pressure, bar	Temperature, K	Reynolds number ($Re_l \times 10^{-7}$), m⁻¹	Running time, s
6.4	685	10–42	500–1800	Less than 2.8	40–90
10.0	994	15–120	1000–6000	Less than 0.78	25–85
12.0	994	30–120	1300–1550	Less than 0.32	25–85

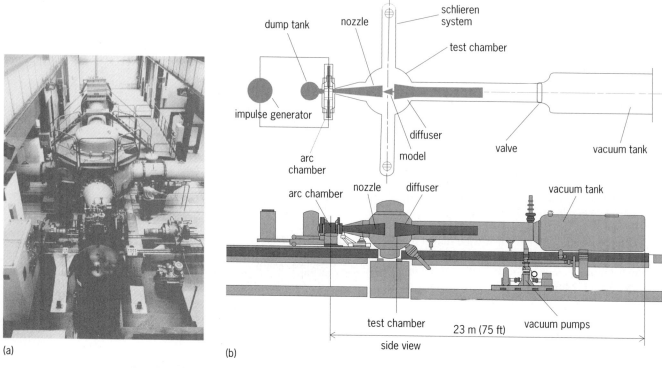

(a)   (b)

Fig. 3. **ONERA F4, a high-enthalpy wind tunnel. (***a***) Researchers in France use the F4 to explore hypersonic winds. (***b***) Diagram depicts the internal structure of the F4.**

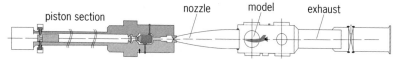

Fig. 4. **The U-11, a hypersonic wind tunnel with adiabatic compression, at the TsNIIMASh facility in Russia.**

piston. Adiabatic compression by a piston ensures the strongest shock waves and the highest parameters of the generated flow. Two large shock tunnels with a heavy piston (**Fig. 2**) are the High Enthalpy Wind Tunnel Gottingen (HEG), Germany, and the High Enthalpy Shock Tunnel (HIEST), National Aeronautical Laboratory, Kakuda, Japan. These facilities generate a flow with a Mach number, M, of 10, stagnation temperature up to 10,000 K (17,540°F), and stagnation pressure up to 1500 bar (21,750 lb/in.^{2}). (Stagnation pressure and temperature are the conditions when a high-velocity gas is brought to rest, that is, stagnated.) The running time in these facilities is less than 2 ms.

Hot-shot wind tunnels. In hot-shot wind tunnels, the operation process is based on energy supply in the form of an electric discharge to a motionless test gas in the settling chamber, which triggers the working process in the form of quasisteady exhaustion of the gas from the settling chamber. The pressure in the settling chamber after the discharge reaches 200 MPa (29,000 lb/in.^{2}) and the temperature reaches 4000 K (6740°F). In contrast to shock tunnels, the operation process here is not limited to the motion of shock waves, and almost the entire amount of the accumulated gas is used here for flow generation. Therefore, the running time in hot-shot wind tunnels increases to hundreds of milliseconds. A typical hot-shot facility is the ONERA F4 wind tunnel in France (**Fig. 3***a*). There are approximately 20 wind tunnels of this type in the world. The nozzle-exit diameter (Fig. 3*b*) ranges from 0.3 to 0.5 m (1 to 1.6 ft). All of these wind tunnels are designed to generate high-velocity flows (M equals 8 to 20 and higher).

Adiabatic compression. In wind tunnels with adiabatic compression, the test gas is heated by means of adiabatic compression by a rapidly moving piston. The piston is driven by a driving gas with a comparatively low pressure equal to 20–40 MPa

TABLE 2. Characteristics of TsNIIMASh facilities		
Parameter	U-11	U-7
Stagnation pressure	Up to 250 MPa	Up to 250 MPa
Stagnation temperature	Up to 4000 K	Up to 3000 K
Range of Mach numbers	4–20	10–15
Range of Reynolds numbers per meter	0.1×10^6 to 300×10^6	0.1×10^6 to 30×10^6
Nozzle-exit diameter	0.4 m; 0.8 m	0.4 m; 0.8 m
Running time	0.1–3.0 s	0.1–1.0 s
Test gas	Air, nitrogen, helium, and carbon dioxide	

(2900–5800 lb/in.²). During piston acceleration, almost all of the energy of the driving gas becomes converted into kinetic energy of piston motion and then to internal energy of a small mass of the test gas. It is the large ratio of the masses of the driving gas and test gas, as well as the high efficiency of energy conversion, that ensure unique characteristics of wind tunnels with adiabatic compression. The main representatives of this class of wind tunnels are the piston-driven gas-dynamic tunnels, U-11 (**Fig. 4**) and U-7, at the TsNIIMASh facility in Russia (**Table 2**).

For background information *see* AERODYNAMICS; GAS DYNAMICS; HYPERSONIC FLIGHT; MACH NUMBER; REYNOLDS NUMBER; SHOCK WAVE; WIND TUNNEL in the McGraw-Hill Encyclopedia of Science & Technology.   Valerie Zvegintsev

Bibliography. J. D. Anderson, Jr., *Hypersonic and High Temperature Gas Dynamics*, 2d ed., American Institute of Aeronautics and Astronautics, Reston, VA, 2006; F. Lu and D. Marren (eds.), *Advanced Hypersonic Test Facilities*, vol. 198 in *Progress in Astronautics and Aeronautics*, American Institute of Aeronautics and Astronautics, 2002; A. Pope and K. L. Goin, *High-Speed Wind Tunnel Testing*, 1965, Krieger, reprint 1978.

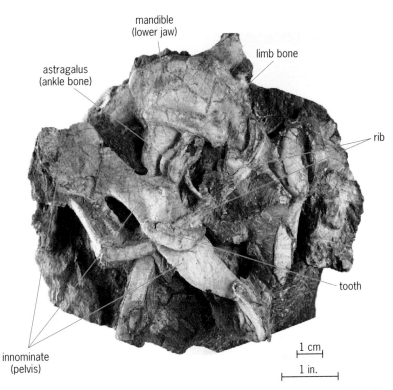

Fig. 1. A block of sediment showing several disarticulated bones of *Indohyus*. At least 25 individuals of various ages were preserved in this death assemblage.

Indohyus: the origin of whales

For the last 2 decades, the origin of cetaceans (whales, dolphins, and porpoises) has become one of the best case studies in mammalian evolution. A rich collection of fossil cetaceans, mostly from Asia, Africa, and North America, traces the transition of cetaceans from a terrestrial to a completely aquatic environment. Cetaceans initially took to the seas about 50 million years ago in an ancient shallow seaway, the Tethys Sea, which was located between the Indian subcontinent and Asia. Within about 10 million years in the aquatic environment, cetaceans evolved specialized ears for hearing underwater, reduced the size of their hindlimbs, and evolved tail flukes for propulsion. The cetacean body plan has so radically evolved from terrestrial artiodactyls (even-toed hoofed mammals) that modern cetaceans are no longer able to bear their weight on land and are obligatorily aquatic.

Cetaceans are the only lineage of artiodactyls that became fully aquatic. However, recent fossil findings near the Kashmir region along the India-Pakistan border show that another artiodactyl family, Raoellidae, is closely related to cetaceans. Raoellid artiodactyls initially had aquatic habits as well, but never lost their ability to bear weight on land. They probably spent most of their lives wading in shallow aquatic habitats. These recent findings indicate that several fossil artiodactyls occupied shallow-water habitats, but only a single lineage evolved a body plan that allowed for the penetration of deep marine environments.

Closest relatives of cetaceans. One family of artiodactyls, Raoellidae, was for decades known solely on the basis of isolated teeth and fragmentary skull elements. Within the past few years, paleontologists uncovered new skeletal elements (skulls, jaws, and numerous postcranial elements) of the raoellid *Indohyus* in the Kashmir region along the India-Pakistan border, enabling the full skeleton of this creature to be accessible for study for the first time. These skeletal elements were preserved in fossil mudstones in an archaic Eocene streambed. (The Eocene epoch occurred about 55–33 million years before the present.) Many individuals died there, and their bodies decomposed in such a manner that the bones were stacked on top of one another (**Fig. 1**). Researchers estimated that at least 25 individuals of varying ages were buried together in this death assemblage.

From the analysis of this fossil assemblage, it was determined that *Indohyus* was about the size of a raccoon (**Fig. 2**). It had a long tail, long limbs, and an elongated snout. The middle ear of *Indohyus* had a thickened medial tympanic wall (**Fig. 3**), called an involucrum, which is characteristic of cetaceans. Along with the presence of an involucrum, both *Indohyus* and cetaceans also display crushing basins in the center of their molars (Fig. 3). Using analyses

Fig. 2. Reconstruction of *Indohyus*. (*Illustration by Carl Buell; copyright by J. G. M. Thewissen*)

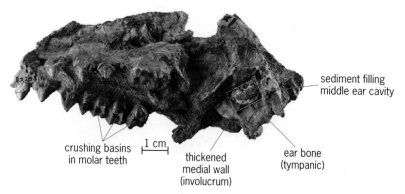

Fig. 3. The skull of *Indohyus* (RR 208) has a middle ear cavity that was filled with air in life. This cavity has been filled in with sediment during the fossilization process. The ear is broken in this specimen, showing the thickness of the medial wall. Both cetaceans and *Indohyus* have a thickened medial wall, called an involucrum.

of characteristics of bone shape and molecular data, researchers found *Indohyus* to be the closest fossil relative to cetaceans. In addition, *Hippopotamus* was found to be the closest relative to cetaceans still living today (**Fig. 4**).

The fossil record indicates that *Indohyus* lived about 48 million years ago in Asia, whereas the earliest fossil relatives of *Hippopotamus* are from 15-million-year-old sediments in Africa. This leaves about 35 million years of evolution between these two close cetacean relatives, on separate continents. As such, future fossil finds may better resolve the *Indohyus*-hippopotamid relationship and biogeographical history.

Skeletal indicators of habitat. Mammals that wade and swim in shallow bodies of water, such as *Hippopotamus*, otters, and manatees, have limb bones

with thick cortices (outer, superficial layers) and reduced marrow (medullary) cavities. These thickened bones act as ballast to weight the skeleton and counteract the buoyancy effect of water. Their marrow cavities are typically less than 55% of the bone diameter, whereas terrestrial mammals have thinner cortices and a larger marrow cavity that is approximately 60–70% of the diameter. The limb bones of *Indohyus* display a small marrow cavity that was only 42% of the total bone diameter. This dramatic increase in cortical thickness at the expense of the medullary cavity is termed osteosclerosis. Although the dimensions of the long bones of *Indohyus* indicate that it occupied an aquatic niche and was an amphibious wader, the earliest cetaceans displayed a comparatively extreme form of osteosclerosis in that the medullary cavity was only 10% of the total bone diameter. It could be that the extremely thickened bones of cetaceans afforded them a decreased buoyancy compared to *Indohyus*, a characteristic that may have ultimately lead to their successful transition to a completely aquatic lifestyle.

Further evidence was gathered to pinpoint the habitat of *Indohyus* by analyzing stable oxygen isotopes in the fossilized tooth enamel. Enamel oxygen-18 values are most affected by ingestion of environmental water, as well as through physiological processes such as sweating and respiration. Aquatic mammals have enamel values that are typically much less (2–3 parts per thousand) than those of terrestrial mammals. The oxygen-18 value for *Indohyus* was much less (4 ppt) than that of terrestrial mammals, indicating these isotopes had a very strong aquatic signature. Therefore, both analyses of the chemical

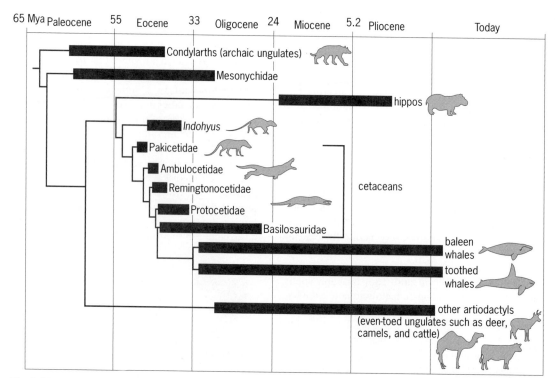

Fig. 4. Diagram showing that *Indohyus* is the closest fossil relative to cetaceans, with *Hippopotamus* being the closest living relative.

composition of teeth and bone thickness indicate that *Indohyus* occupied an aquatic habitat. Interestingly, analyses of the oxygen-18 stable isotopes of archaic cetaceans also indicate that they were aquatic.

Why did artiodactyls take to the water? It could be that the ancestors to cetaceans, hippopotamids and raoellids, initially took to the water for shelter from predators. The water chevrotain (*Hyemoschus aquaticus*) lives in modern-day western and central Africa, and will flee to small bodies of water in order to escape predation. It is plausible that, from this initial wading step, cetacean lineages spent successively more time in water, and eventually began feeding in water similar to modern-day *Hippopotamus*.

For background information *see* ANIMAL EVOLUTION; ARTIODACTYLA; BONE; BUOYANCY; CETACEA; FOSSIL; HIPPOPOTAMUS; MACROEVOLUTION; MAMMALIA; ORGANIC EVOLUTION; PHYLOGENY; SYSTEMATICS in the McGraw-Hill Encyclopedia of Science & Technology.   Lisa Noelle Cooper; J. G. M. Thewissen

Bibliography. J. H. Geisler and J. M. Theodor, *Hippopotamus* and whale phylogeny, *Nature*, 458:E1–E4, 2009; N. M. Gray et al., Sink or swim?: Bone density as a mechanism for buoyancy control in early cetaceans, *Anat. Rec.*, 290:638–653, 2007; M. A. Taylor, Functional significance of bone ballast in the evolution of buoyancy control strategies by aquatic tetrapods, *Hist. Biol.*, 14:15–31, 2000; J. G. M. Thewissen et al., Whales originated from aquatic artiodactyls in the Eocene epoch of India, *Nature*, 450:1190–1195, 2007; E. M. Williams, Synopsis of the earliest cetaceans: Pakicetidae, Ambulocetidae, Remingtonocetidae, and Protocetidae, pp. 1–28, in J. G. M. Thewissen (ed.), *The Emergence of Whales: Evolutionary Patterns in the Origin of Cetacea*, Plenum Press, New York, 1998.

Intervehicle communications

Vehicle-to-vehicle (V2V) and vehicle-to-roadside (V2R) communication allow both passenger safety and driving comfort to be improved significantly. For example, V2V communication could allow a vehicle detecting an icy road to inform following vehicles and thereby prevent accidents. Similarly, V2R communication could be used near construction sites to warn vehicle drivers about a reduced number of lanes or to give advice for an alternative route. For improved traffic efficiency, vehicles could exchange the latest traffic flow information.

The advantages of vehicle-to-wherever (V2X) communication have been recognized in many regions of the world. Research initiatives worldwide have produced valuable results, and additional projects are in progress to solve various issues. Many regional efforts have been undertaken to consolidate the results from research projects into industrial standards. Unfortunately, the regional variance of boundary conditions for the development of V2X communication makes it difficult to establish an easy worldwide harmonization. Still, the mutual awareness of the worldwide initiatives is growing. An increasing number of companies are trying to bridge the gaps between regional developments.

Technical and regional challenges. Acronyms such as V2V, V2R, and V2X describe communication patterns in the area of intervehicle and vehicle-to-roadside communication. These acronyms cover a wide variety of applications that range from hazard notification and traffic flow optimization, to passenger entertainment.

The V2X applications and technology have specific requirements and characteristics resulting in several technical challenges. For example, one technical challenge is due to many applications relying on position awareness of the nodes, to indicate an incident's location, and to judge how to react to a received message. Another technical challenge is derived from the mobility of vehicles that form the network, since applications cannot rely on sessions or stable routing.

In addition to technical challenges, regional influences for these V2X communication systems have to be considered. Radio frequencies need to be available, installation of supporting roadside infrastructure has to be negotiated with road administrations, and agreements on interoperable standards have to be achieved. Finally, there are significant challenges to create a robust business case for the market introduction of V2X technology. Usually, new technologies are introduced first into the premium vehicle segment. Since safety related V2V communication requires a rapid market penetration of equipped vehicles, introducing V2X technology for limited use in the premium market is not a viable economic option.

In order to address the aforementioned challenges, many regional efforts have been undertaken to consolidate results from research projects in industrial standards. Industry consortia have been formed, namely the Car2Car Communication Consortium (C2C-CC) in Europe, the Vehicle Safety Communication Consortium (VSCC) in the United States, and the Advanced Cruise-Assist Highway System Research Association in Japan (AHSRA). The outcome of these efforts is the Institute of Electrical and Electronics Engineers (IEEE) 1609.x and IEEE 802.11p standards in the United States. In parallel, the International Standards Organization (ISO) is already trying to tie together a standard for a continuous Communications Air Interface for Long and Medium Range (CALM). In 2007, the European Telecommunications Standards Institute (ETSI) established a technical committee for standardization in the area of intelligent transportation systems (TC ITS). Furthermore, field operational tests (FOTs) and proof-of-concept tests are on the way to support the consolidation of V2X technology. These include the Safe Intelligent Mobility–Test Area Germany (SIM-TD), the Japanese Smartway Project, the European project PRE-DRIVE Car to Infrastructure (C2X), and the United States Crash Avoidance Metrics Partnership (CAMP) Cooperative Intersection Collision Avoidance Systems-Violation (CICAS-V) FOT.

V2X communication applications. In general, V2X communication applications can be categorized into three classes: active safety applications, also called cooperative driver assistance or safety-related applications; decentralized floating car data applications, which are used to control traffic flow optimization applications; and user communication and information services, which are used in convenience applications.

Active safety applications. The purpose of active safety applications is to support vehicle drivers in potentially dangerous situations. This is based on the exchange of position information, sensor data, and other information between vehicles. The active safety application class covers any kind of emergency notification for the driver, for example, hazard warnings, accident warnings, or traffic jam warnings. Furthermore, the class covers driver assistance applications such as automatic lane merging or (intersection/forward) collision warning.

Decentralized floating car data applications. The aim of decentralized floating car data applications is to collect and distribute data on traffic flow. The vehicles act as sensors (sources of information), information processing units, and sinks of information. The collected data can be used to determine the current traffic situation, which is, along with other data relevant for vehicles on the same road, passed to following vehicles. These vehicles can adapt their itinerary accordingly, if appropriate, from vehicles further ahead. Or, slightly differently, vehicles could query traffic flow, weather, and road conditions on their route from these distant vehicles.

User communication and information services. User communication and information services include passenger communication via Internet protocols as well as transmission of other user data. Example applications are online gaming or chat between passengers on different vehicles, vehicle-to-Internet communication for info-fueling (for example, music or map download), and any other vehicle to infrastructure communication (for example, tolling or advertisements).

V2X communication technology. The idea behind V2X communication is that vehicles and roadside infrastructure are able to communicate among each other directly. For that purpose, they use wireless communication systems that are based on technology similar to a wireless local-area network (WLAN) such as IEEE 802.11a/b/g.

Several aspects make V2X communication networks different from other communication networks. Developed networks and protocols need to be reliable (for example, error-free transmission and reliable reception of messages); deliver safety messages quickly; prioritize information (since different messages are sent simultaneously over the shared communication medium); be available all of the time; and be secure (warnings must be trustworthy).

Major conclusions from these requirements are that V2X communication networks need specifically tailored protocols and dedicated frequencies. They cannot be run in a license-free band such as WLAN.

Worldwide research and standardization. The regional variance of boundary conditions for the development of V2X communication poses a difficult challenge to worldwide harmonization. Since the United States, Japan, and Europe are heavily involved in V2X research and standardization, it is crucial that they have mutual awareness of the worldwide initiatives to implement V2X technology. As such, governments and private industries are working together to help bridge the gaps between regional developments.

V2X in the United States. The driving forces in the United States are the Department of Transportation (DOT) and vehicle manufacturers. The main efforts are concentrated in the DOT-funded projects such as the Vehicle Infrastructure Integration (VII) Program and CICAS. Industry consortia that support those efforts are the Vehicle Infrastructure Integration Consortium (VIIC) and CAMP.

In 1999, the Federal Communications Commission (FCC) allocated 75 MHz in the 5.9-GHz band as Dedicated Short-Range Communication (DSRC) spectrum, exclusively for automotive use with the primary purpose of safety use. **Figure 1** contains an overview on the resulting frequency and channel allocation, comparing it to the (planned) allocations in Japan and the European Union.

The communication protocols for V2X communication are standardized within IEEE. The physical and medium access control (MAC) layers are defined in IEEE P802.11p. Higher communication layers are defined by the IEEE 1609.x series, and together they form what is referred to as Wireless Access for the Vehicular Environment (WAVE).

V2X in Japan. In Japan, research, standardization, and development of V2X communication systems are mainly driven by public authorities such as the Ministry of Land, Infrastructure, and Transportation (MLIT), the Ministry of Internal Affairs and Communication (MIC), Ministry of Economy, Trade, and Industry (METI), and the National Police Agency (NPA).

Japan has developed multiple standards for V2X communication. One of them is the Association of Radio Industries and Businesses (ARIB) STD-T75 dedicated short-range communications (DSRC) standard (5.8 GHz) for V2R communication. Another basis for V2R communication is the Vehicle Information and Control System (VICS), which includes infrared beacon units located on roadside. Furthermore, Japan has developed a communication standard for the 700-MHz band that will support V2R and V2V communication.

Public-funded V2X communication research in Japan is conducted in the projects Advanced Safety Vehicle (ASV), Advanced Highway Systems (AHS), and Driving Safety Support Systems (DSSS). The difference among these projects is that ASV focuses mainly on equipping vehicles with intelligent safety systems, while AHS and DSSS focus on V2R communication. The AHS program investigates the use of roadside sensors such as cameras and infrared so that AHS is able to provide information on the traffic situation to vehicles via 5.8 GHz DSRC. The DSSS program focuses on V2R communication using infrared beacons together with 5.8 DSRC.

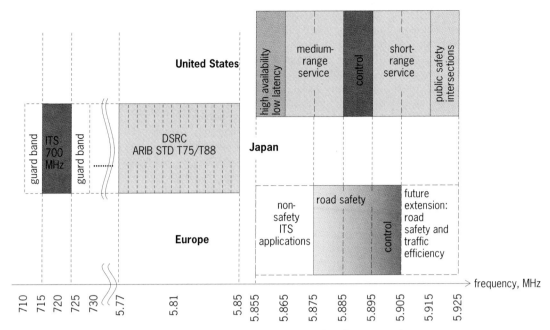

Fig. 1. Frequency and channel allocation in the United States, Japan, and the European Union.

V2X in Europe. The European V2X communication activities are mainly driven by two entities, public-funded research projects and the C2C-CC. The upcoming generation of research projects is now targeting FOTs to validate proposed solutions suggested by previous research. One example is the aforementioned SIM-TD project in Germany.

The C2C-CC is the organization in Europe that drives the standardization process. It is an industry consortium where vehicle manufacturers, suppliers, and research institutes consolidate the input of the different research projects. Furthermore, the C2C-CC prepares and supports activities like the frequency allocation process. Those activities and the standardization are carried out in the ETSI.

The current status of results of the C2C-CC activities can generally be summarized as follows. Regarding frequency allocation, a dedicated bandwidth of 30 MHz will be available for traffic safety applications in the 5.9 GHz band (Fig. 1). An additional bandwidth of 20 MHz might be available as future extension, for road safety and traffic efficiency. Non-safety ITS applications might use 20 MHz in the ISM (industrial, scientific, and medical) band below 30 MHz.

With respect to communication protocols, the most important difference of the protocols proposed for the European Union compared to other V2X communication protocols as they are proposed in Japan or the United States is the strong orientation towards direct V2V communication with less focus on vehicle-to-infrastructure (V2I) communication. This is reflected in the C2C-CC transport and network protocols that provide features such as multihop message dissemination and geographic routing.

Typical equipment for V2X communication. Owing to the early stage of technology development, typical equipment for V2X communication consists mainly

Fig. 2. DENSO Wireless Safety Unit (WSU), used in automobiles. (*DENSO AUTOMOTIVE Deutschland GmbH*)

of prototypes that are designated to support research and standardization. An example for such a prototype is the DENSO Wireless Safety Unit (WSU; **Fig. 2**). The WSU is DENSO's second-generation feasibility test platform for communication protocol evaluation and application prototyping. It is specifically designed for automotive environments (that is, various temperatures, shock, and vibration) and its primary focus is on safety-related applications.

In conjunction with the WSU, DENSO provides a communication software stack that implements 802.11p, P1609.3, and P1609.4. This stack can be accessed from applications via a well-documented application programming interface (API). Furthermore, the provided software contains a set of dedicated test tools.

In addition, the WSU can be used in combination with the "AKTIV" Communication Unit and its followup development, the OpenWAVE Engine (a C2C-CC conformant communication solution for active safety). The development was initiated by BMW Forschung und Technik GmbH, and implemented by the partners Cirquent and Philosys, in the context of the German AKTIV (Adaptive and Cooperative Technologies for Intelligent Traffic) project.

Outlook. The development of V2X communication reflects the strengths and weaknesses of where it will be located. For example, the progress of development in Europe is slowed down by the large number of member countries and fragmented activities, projects, and players. However, with the activities started in the ETSI, the motivation for cooperation has accelerated in order to push standardization. In Japan and the United States, there are only a few focused projects, and, hence, research results are quickly brought into standards. Probably the most supportive factor in Japan is the already existing toll systems that are used as a base for development.

Although the developed protocols reflect the boundary conditions in the different regions around the world, there are also commonalities. One of these commonalities is that there is a strong possibility that some of the communication standards worldwide will be based on technologies similar to IEEE P802.11p. This allows for a common communication hardware technology based on that standard. Then, only regional adjustments, such as communication driver parameters and MAC protocols will have to be made to support regional standards.

Another common development in V2X worldwide is the trend to start nationwide or even larger FOTs. Based on recent hardware developments and intensive research, smaller-scope field tests with few vehicles have been performed. Hence, the next step is to gain practical experiences via large-scope implementations.

With this in mind, and knowing that the envisioned applications worldwide are also similar, there is the motivation for companies to work together and tackle the remaining issues (such as security) on a global basis.

For background information *see* AUTOMOBILE; CONTROL SYSTEMS; HIGHWAY ENGINEERING; RADIO SPECTRUM ALLOCATION; TRAFFIC CONTROL SYSTEMS; TRANSPORTATION ENGINEERING; WIRELESS FIDELITY (WI-FI) in the McGraw-Hill Encyclopedia of Science & Technology.

Tim Leinmüller; Robert K. Schmidt; Bert Böddeker; Roger W. Berg; Tadao Suzuki

Bibliography. W. Franz, H. Hartenstein, and M. Mauve, *Inter-Vehicle-Communications Based on Ad Hoc Networking Principles: The FleetNet Project*, Universitaetsverlag Karlsruhe, 2005; B. Gallagher, H. Akatsuka, and H. Suzuki, Wireless communications for vehicle safety: Radio link performance and wireless connectivity methods, *IEEE Veh. Tech. Mag.*, 1(issue 4):4–24, December 2006; R. Lasowski, T. Leinmüller, and M. Strassberger, *OpenWAVE Engine/WSU—A Platform for C2C-CC*, 15th World Congress on Intelligent Transport Systems, New York, November 16–20, 2008; T. Leinmüller, E. Schoch, and C. Maihöfer, *Security Issues and Solution Concepts in Vehicular Ad Hoc Networks*, 4th Annual Conference on Wireless On Demand Network Systems and Services (WONS 2007), Obergurgl, Austria, 2007.

Iron-based superconductors

The phenomenon of superconductivity has a rich and interesting history, starting in 1911 when Kamerlingh Onnes discovered that upon cooling elemental mercury to very low temperatures, the electrical resistance suddenly and completely vanished below a critical temperature T_c of 4 K ($-452°$F). This resistanceless state enables persistent currents to be established in circuits to generate enormous magnetic fields, and to store and transport energy without dissipation. Superconductors have other unique properties, such as the ability to expel and screen magnetic fields and quantum oscillations controlled by the magnetic field that provide extraordinary measurement sensitivity. Over the intervening years, the number of superconducting materials has grown, with higher critical temperatures and improved metallurgical properties, and these have found their way into a number of technological applications, such as magnetic resonance imaging (MRI) systems for the health-care industry. The field was shocked in 2008 by the surprise discovery of a completely new class of superconductors based on iron. These iron-based superconductors have initiated a flurry of activity as researchers try to understand the origin of the superconductivity in these new materials and develop them for potential use in devices. In this latter context, the new materials have quite high (relatively speaking) superconducting transition temperatures (T_c) and rather favorable current-carrying capabilities that should make them useful in practical applications.

How does a metal become superconducting? In metals, electrons are free to move and provide electrical conduction, but collisions with other electrons, lattice vibrations, and impurities and defects in the material cause resistance, and thus energy dissipation, in the system. Superconductors circumvent this problem by binding two electrons together into pairs, which must all move together in a coordinated fashion. If the temperature is low enough, there is insufficient thermal energy to break apart and disrupt these pairs, so collisions are not possible and the pairs can move through the material without any interference—resistanceless conductivity. The extraordinary thing about this superconducting pairing is that electrons have the same charge and therefore strongly repel each other. So how can this bound state exist? It took half a century to unravel this mystery, but the pairing occurs for two electrons with equal and opposite speeds, rather than as two electrons "glued together." This unusual state of two electrons in a bound state is called a Cooper pair and is a fundamental property of all known superconductors, including the iron-based ones.

A quantitative and complete theory of superconductivity was developed in 1957, which explained that the pairing interaction originated from the lattice vibrations of the solid. This theory provided a thorough understanding of all the superconductors known at the time and all the "conventional" superconductors discovered since then. One cornerstone of this understanding is that any magnetic atoms in the lattice tend to break the Cooper pairs; therefore, magnetism is very detrimental to the superconductivity. However, in 1986, a new class of "high-temperature" superconductors was discovered that completely contradicted this rule. These were oxides in which the crystal structure contained sheets of copper and oxygen, called cuprates. Oxides typically are not even conductors, let alone superconductors, but more surprising was that copper ions carry a magnetic spin (like a compass needle), which is the kiss of death for conventional superconductors. Moreover, it turns out that in the cuprates, the magnetism not only is tolerated, but appears to play a key role in the Cooper pairing. Up until 2008, all known high-temperature superconductors exhibited two essential ingredients: copper-oxygen planes of atoms and magnetic moments on the copper. Hence, it was thought that these two properties were essential to achieve high-T_c superconductivity.

Discovery of iron-based high-T_c superconductors. The atomic structure and bonding of a material control its properties. For the iron systems, there are four different structure types that have been identified so far, typified by LaFeAsO, $SrFe_2As_2$, LiFeAs, and Fe(Te-Se). The structure for the first two types, which have the highest T_cs, are shown in **Fig. 1**. The common structural feature is a layer of Fe and As atoms (like the Cu-O layer for the cuprates), which is separated by a noniron layer such as La-O (for LaFeAsO) or Sr (for $SrFe_2As_2$). These materials undergo a small structural distortion below room temperature, along with the development of magnetic order, and are metals but not superconductors. Like the cuprates, superconductivity is achieved by chemically substituting, or "doping," the system to change the electronics. In $SrFe_2As_2$, for example, we can start with the Sr (or Ba or Ca) and dope that site with K. For the LaFeAsO, we can substitute fluorine for oxygen. An example of how the properties change and the superconductivity develops with doping is shown in **Fig. 2** for the CeFeAsO system as fluorine is substituted for oxygen. As the F content increases, the temperature at which magnetic order develops is lowered, and the magnetic order is completely suppressed before superconductivity develops. However, fluctuating iron magnetic moments are still present in the superconductivity regime. Similar phase diagrams are found for the other types of iron-based materials, and this general type of phase diagram is also found for the cuprate systems.

The iron-based superconductor T_cs are too high to be explained by the conventional theory, and have a number of additional features in common with the cuprates. They contain iron-arsenic (or selenium) layers of atoms, the iron atoms have magnetic moments, the superconductivity is established by Cooper pairs, and magnetism is known to play a key role in the superconducting state. But they have a number of important differences as well. The undoped "parent" materials, in both cases, exhibit magnetic order. However, the iron-based systems are metals, while the cuprates are insulators, which means that there

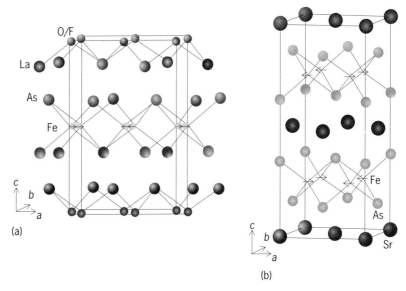

Fig. 1. The basic building blocks of the atomic structure for the two types of iron-based superconductors with high critical temperatures T_c: (a) LaFeAsO and (b) $SrFe_2As_2$. The feature common to the iron superconductors is a metallic layer of iron bonded to arsenic atoms; the iron atoms form a square array, with the arsenic atoms above and below the iron plane. The iron-arsenic layer is sandwiched between layers of La-O or Sr. This basic structural unit is then repeated in all three directions ad infinitum to form the macroscopic crystal structure of the material. Below room temperature, the structure distorts slightly so that the iron lattice is no longer exactly square, and the iron moments order in an antiferromagnetic arrangement, which means that half point in one direction, and half in the opposite direction. The iron magnetism is a key ingredient for all aspects of these materials.

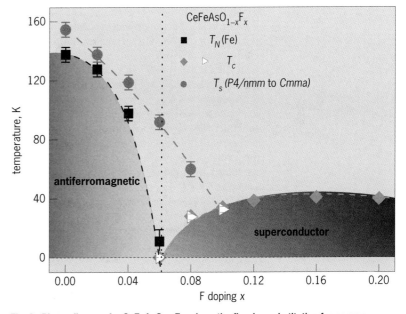

Fig. 2. Phase diagram for $CeFeAsO_{1-x}F_x$, where the fluorine substitution for oxygen changes the electronics. With increasing fluorine content x, the magnetic order disappears, and then the superconductivity regime appears. The basic behavior with doping is the same for all the iron-based high-T_c materials, and also has some similarities to the phase diagrams for the copper-oxide superconductors.

are fundamental differences in the electronics of these materials. For the cuprates, the Cu and O atoms are in the same thin layer, and this renders the superconducting properties highly anisotropic, being very good within the layer and poor along the direction between the layers. This makes it difficult (and thus expensive) to fabricate wires. The structures in Fig. 1 show that the iron-arsenic layers are

Examples of superconducting materials for the three different types of superconductors: the conventional materials where the Cooper pairing originates from lattice vibrations, the layered copper-oxide (cuprate) materials, and the new iron-based superconductors. Representative critical fields quoted are in tesla—for comparison the Earth's magnetic field is about 0.00005 tesla (about $^1/_2$ gauss); (1 tesla = 10,000 gauss). Then for Hg, H_{c2} = 450 gauss.

Material	Type	T_c, K (°F)	Pairing	Critical field (H_{C2}), T	Date
Hg	Conventional	4 (−452)	Lattice	0.045	1911
Nb_3Sn	Conventional	18 (−427)	Lattice	30	1954
Nb(Ti) wire	Conventional	10 (−442)	Lattice	15	1962
$La_2CuO_{4.1}$	Cuprate	42 (−384)	Magnetic	100	1986
$YBa_2Cu_3O_7$	Cuprate	90 (−298)	Magnetic	250	1987
MgB_2	Conventional	39 (−389)	Lattice	20	2001
LaFeAs(O,F)	Iron-based	26 (−413)	Magnetic	56	2008
SmFeAs(O,F)	Iron-based	56 (−359)	Magnetic	250–300	2008
$(Ba,K)Fe_2As_2$	Iron-based	37 (−393)	Magnetic	75	2008

thick, with the As atoms positioned well off from the plane of iron atoms, and this makes the superconducting properties much closer to isotropic. This is especially the case when high magnetic fields or large currents are required, which is a very important advantage for applications. The different natures of the anisotropy originate from a fundamental difference in the pairing—for the cuprates, the Cooper pairs prefer to be in the Cu-O planes and are highly anisotropic, while for the iron-based systems, they are almost isotropic.

The superconducting transition temperature is one important parameter of a superconductor, but it is not the only property of interest. The size of the magnetic field that a superconductor can support (or the related maximum current) before the superconducting state collapses is another vital property—the higher the critical field (called H_{C2}), the better. For applications, the cost of the raw materials is important, as are the metallurgical properties, which dictate the ease of fabrication and consequent cost-effectiveness. For example, cuprate superconductors, with their high transition temperatures and large critical fields, have been around for 23 years (see **table**), but because of the difficulty and cost of making wires, the superconducting magnets used in MRI systems are still made from conventional Nb(Ti) wire, which must be cooled with (expensive) liquid helium to a very low temperature (4 K, −452°F).

A few examples of prototype materials for the three classes of superconductors are listed in the table. The iron-based superconductors exhibit values of T_c that exceed those of all superconductors except some of the cuprates, while their critical magnetic fields are unsurpassed, making them particularly attractive for applications requiring large magnetic fields and large currents. In the short time since they were discovered, there are already four different general types of iron systems, with many chemical substitutions possible. Perhaps more types of materials will be discovered and the T_cs could be higher, but the chemical versatility that is already available is one of the dramatic strengths of the new iron-based materials. This flexibility is opening up new research avenues to probe the origin of the superconductivity and has fostered hopes that this may ultimately lead to a complete theoretical understanding of both classes of high-T_c superconductors. The chemical flexibility will also allow scientists and engineers to tailor the properties for specific commercial technologies, which may include high-magnetic-field applications like medical MRI imaging, high magnetic fields for scientific research, energy storage technologies, and more efficient transfer of electricity over regional power grids. The discovery of this new class of superconductors based on iron has tremendously revitalized the field of superconductivity, and should provide many more surprises and promises for the future.

For background information *see* HIGH MAGNETIC FIELDS; IRON; LOW-TEMPERATURE PHYSICS; MAGNETISM; SUPERCONDUCTING DEVICES; SUPERCONDUCTIVITY in the McGraw-Hill Encyclopedia of Science & Technology.			Jeffrey W. Lynn

Bibliography. Y. Kamihara et al., Iron-based layered superconductor La[$O_{1-x}F_x$]FeAs (x = 0.05–0.12) with T_c = 26 K, *J. Am. Chem. Soc.*, 130:3296–3297, 2008; C. Kittel, *Introduction to Solid State Physics*, 8th ed., Wiley, Hoboken, N.J., 2005; J. W. Lynn et al., *High Temperature Superconductivity*, Springer-Verlag, New York, 1990; J. W. Lynn and P. Dai, Neutron studies of the iron-based family of high T_C magnetic superconductors, *Physica*, C469:469–476, 2009.

Large Kuiper Belt objects

Orbiting the Sun at distances beyond the orbit of the planet Neptune are millions of small icy bodies occupying a region of space called the Kuiper Belt. To date, over 1200 Kuiper Belt objects (KBOs) have been discovered. Although the vast majority of these known objects are small (less than 500 km or 310 mi in diameter), there are now several objects known with diameters greater than 1000 km (620 mi), including the well-known dwarf planets Pluto and Eris. Many of these large objects have atmospheres, moons, and surfaces that show evidence of collisions, making them subjects of intense scientific interest. Details about the largest Kuiper Belt objects are given in the **table**.

Pluto. Pluto, discovered by Clyde Tombaugh in 1930, is the second largest Kuiper Belt object, with

The largest Kuiper Belt objects

Name	Discoverer	Year	Size[†], km	Semimajor axis, AU	Orbital period, yr	Known moons
Eris	BTR*	2005	2400	67.7	557	1
Pluto	Clyde Tombaugh	1930	2300	39.5	248	3
Haumea	BTR	2004	2000 × 1500 × 1000[‡]	43.1	283	2
Sedna	BTR	2003	1500	526	~12,000	
Makemake	BTR	2005	1500	45.8	310	
Quaoar	BTR	2002	1200	43.6	288	1
Orcus	BTR	2004	1000	39.2	245	1
Varuna	Spacewatch Survey	2000	1000	43.1	283	

*BTR = Michael Brown, Chadwick Trujillo, and David Rabinowitz (Palomar Survey).
[†]Pluto's size is well known from a relatively precise determination of the diameters of Pluto and Charon enabled by mutual events in the 1980s, during which Charon passed directly in front of and behind Pluto as seen from Earth. The other diameters are all uncertain by hundreds to several hundreds of kilometers due to the indirect methods (thermal measurements) that were used to determine them. Although the diameters of Quaoar and Eris were measured directly with the *Hubble Space Telescope*, thermal measurements from the *Spitzer Space Telescope* disagree with the *Hubble* measurements somewhat, so the error bars for them are likely to be just as large as for the other Kuiper Belt objects. 1 km = 0.62 mi.
[‡]Haumea is not spherical (see Fig. 2).

a diameter of 2300 km (1430 mi), and was the first Kuiper Belt object to be discovered. It orbits the Sun in an eccentric (elliptical) orbit that brings it 30 astronomical units (AU) from the Sun at its perihelion (closest point to the Sun) and 49 AU from the Sun at its aphelion (farthest point from the Sun). Pluto has a thin atmosphere composed of molecular nitrogen (N_2), methane (CH_4), and carbon monoxide (CO). These molecules exist as ices on its surface and sublime to form Pluto's thin atmosphere. This atmosphere may freeze out onto the surface once Pluto moves farther away from the Sun (it is currently near its perihelion).

Pluto has three known moons (Charon, Nix, and Hydra). Charon is slightly greater than half Pluto's size and orbits Pluto every 6 days. Nix and Hydra are much smaller (less than 5% of Pluto's diameter) and orbit outside the orbit of Charon. The *New Horizons* spacecraft (launched in 2006) will flyby the Plutonian system in 2015 and is expected to provide detailed images and spectral information on Pluto and its moons.

Eris. Eris, discovered in 2005, is the largest Kuiper Belt object known, with a diameter slightly larger than Pluto's (5%). It also has a small moon (Dysnomia) orbiting it. By tracking the position of Dysnomia relative to Eris and finding its orbital period, astronomers were able to calculate Eris's mass using Kepler's laws. The discovery that Eris was larger and more massive (25%) than Pluto led ultimately to the demotion of Pluto from planet to dwarf planet status by the International Astronomical Union in 2006. In order to keep Pluto as a planet, Eris (and perhaps hundreds of other large Kuiper Belt objects) would also have had to be called planets.

Eris and Pluto can be thought of as twins—they have similar masses and compositions (they are composed of a mixture of ice and rock)—but Eris is currently more than three times farther away from the Sun than Pluto (**Fig. 1a**). Its surface is therefore much colder than Pluto's and its atmosphere has completely frozen out onto its surface. In AD 2257, Eris will come to its closest point to the Sun (38 AU) and may become warm enough for its frozen volatile ices to sublime and form a thin atmosphere similar to Pluto's.

Haumea. Haumea is the third largest Kuiper Belt object, after Pluto and Eris. Soon after its discovery in 2004, astronomers noticed that the brightness of Haumea increased and decreased every 2 h. This variation, it was concluded, is due to Haumea's strange shape and fast rotation. Haumea is shaped like an American football (**Fig. 2**) and rotates end over end every 4 h. When the long end is visible, Haumea appears brighter, but when only the short end is visible, it appears fainter. The two peaks in its brightness are slightly unequal; one of the ends must be brighter than the other, presumably because of their albedos. The odd shape is actually caused by its fast rotation—the rapid spinning causes it to stretch out into the football-like shape.

Haumea has two small moons orbiting it, Hi'iaka and Namaka. Their orbits enabled astronomers to determine that Haumea's mass is about one-third that of Pluto's. Haumea has a very high density (2.7 g cm^{-3}) compared with all other Kuiper Belt objects (it is very massive for its size). This means that, unlike most large Kuiper Belt objects, which are made up of about half water ice (low density, 1 g cm^{-3}) and half rock (high density, 3 g cm^{-3}), Haumea is made almost purely of rock. Interestingly, though, spectral observations of the light coming off of Haumea's surface have revealed that it is covered not with rock, but with nearly pure water ice. Haumea is therefore made up of rock with only a thin coating of pure water ice on its surface.

In the outer solar system, most large objects are thought to have a rocky core surrounded by a thick water-ice mantle. The observation that Haumea has only a very small amount of a water-ice mantle suggested that it likely experienced a large collision with an object of similar size early in its history. This collision must have been so violent that it stripped off most of Haumea's icy mantle, left it spinning rapidly, and formed its two moons. Recently, Michael Brown and colleagues discovered a cluster of small objects of nearly pure water ice in orbits very similar to that of Haumea. They concluded that these icy objects are pieces of the original icy mantle of Haumea that were ejected in the massive collision. In addition, Haumea's two moons also show the same water-ice spectral signatures. Haumea, its moons, and the small

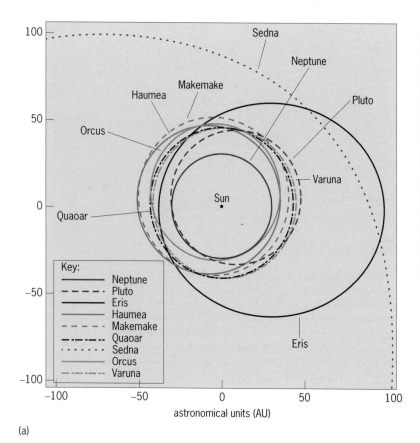

(a)

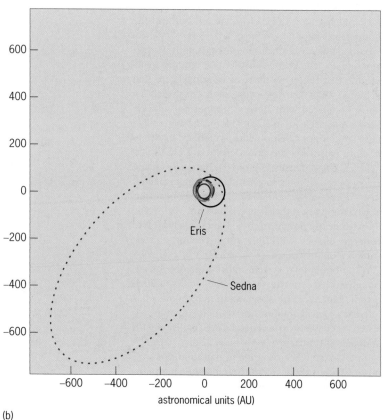

(b)

Fig. 1. The outer solar system (*a*) at a scale of about 100 astronomical units (AU), showing the orbits of the largest Kuiper Belt objects and the planet Neptune, and part of the orbit of Sedna; and (*b*) at a larger scale, showing the entire orbit of Sedna.

Fig. 2. The shape of the Kuiper Belt object Haumea. (*Michael Busch*)

icy Kuiper Belt objects in similar orbits are the first known collisional family in the Kuiper Belt.

Makemake. With a diameter of approximately 1500 km (930 mi), Makemake (pronounced Makay-Makay) is the fourth largest Kuiper Belt object, after Pluto, Eris, and Haumea. Unlike most of the other large Kuiper Belt objects, no moon has been discovered orbiting Makemake, so its mass is unknown. Spectral observations of the surface of Makemake reveal a surface with spectral features of solid methane, similar to the spectra of Pluto and Eris.

Sedna. Sedna has a diameter of approximately 1500 km and an orbit that never enters the classical Kuiper Belt (30–50 AU; Fig. 1). It is discussed here because of its similar size to the large Kuiper Belt objects and because it is likely the first of a new class of objects in the outer solar system that orbit beyond the Kuiper Belt (Fig. 1*b*). Sedna was discovered near its perihelion, and it is currently at a distance from the Sun of 93 AU. Whereas while most Kuiper Belt objects have perihelia between 30 and 40 AU and are therefore under the gravitational influence of Neptune, Sedna's perihelion is far beyond Neptune's region of gravitational influence. Because of its distance and faintness, not much is known about its surface. Spectra acquired with the largest telescopes in the world, such as Keck and Gemini, have revealed the presence of water ice and tentative detections of methane ice and other dark organic compounds. Sedna's surface is so cold that all its ices are completely frozen to its surface and do not sublime to form an atmosphere.

Quaoar, Orcus, and Varuna. These objects have diameters of approximately 1000 km and orbits in the Kuiper Belt. Quaoar and Orcus are each known to have one moon. The surfaces of these objects are covered with varying amounts of water ice and dark material likely composed of organic compounds that have been irradiated over billions of years. All of these objects except Quaoar have insufficient mass to prevent the evaporation of volatile molecules such as CH_4, N_2, or CO that could form thin atmospheres.

Search for large objects in the outer solar system. Objects in our solar system appear to move relative to the fixed background stars: The closer an object is to the Earth, the faster it appears to move across the sky. To find Pluto in 1930, Clyde Tombaugh painstakingly imaged vast areas of the sky multiple times, manually blinked images of the same part of the sky taken at different times, and searched for points of light that appeared to move. It was not until 1992 that David Jewitt and Jane Luu found the next Kuiper Belt object, 1992 QB1, but this object was at least ten times smaller than Pluto. In 2000–2005, Brown, Chadwick Trujillo, and David Rabinowitz searched the northern sky using the 1.2-m (48-in.) telescope at Palomar Observatory to look for bright moving objects. This survey found many of the large objects discussed above, including Eris, Haumea, Makemake, Quaoar, Orcus, and Sedna. Other surveys, such as Spacewatch and the Deep Ecliptic Survey, have also discovered large numbers of Kuiper Belt objects.

Over the next decade, the number of objects discovered in the Kuiper Belt is likely to increase by over an order of magnitude as a result of searches by two dedicated survey telescopes, PanSTARRS and the Large Synoptic Survey Telescope (LSST). These telescopes will search the sky to discover even fainter and more distant objects and will likely find several more large Kuiper Belt objects. Even before then, a Southern Hemisphere search for Kuiper Belt objects now scheduled may turn up some large examples.

For background information *see* ASTRONOMICAL SPECTROSCOPY; COMET; HUBBLE SPACE TELESCOPE; KEPLER'S LAWS; KUIPER BELT; NEPTUNE; PLANET; PLUTO; SPITZER SPACE TELESCOPE in the McGraw-Hill Encyclopedia of Science & Technology.

Emily L. Schaller

Bibliography. A. M. Barucci et al. (eds.), *The Solar System Beyond Neptune*, University of Arizona Press, Tucson, 2008; A. Stern and J. Mitton, *Pluto and Charon—Ice Worlds on the Ragged Edge of the Solar System*, 2d ed., Wiley, Hoboken, NJ, 2005; A. Stern and D. Tholen (eds.), *Pluto and Charon*, University of Arizona Press, Tucson, 1998.

Lattice quantum chromodynamics

Quantum chromodynamics (QCD) is the theory of quarks and gluons. Quarks are the fundamental constituents of nuclear particles such as the proton and the neutron. Isolated quarks have never been observed, and it is thought that they never will be. Quarks interact via forces mediated by particles called gluons. These forces have the unusual property that, although they appear feeble in high-energy collisions, they become very strong at the energies typical in nuclear particles. This property, called asymptotic freedom, explains why the constituents of protons appeared almost free in early high-energy collision experiments and yet have never been observed. They are permanently confined inside protons, neutrons, and the other hadrons (particles that contain quarks and gluons).

This unusual property means that methods that had been used to solve other quantum field theories were insufficient to solve QCD. QCD is very similar mathematically to quantum electrodynamics (QED), the quantum field theory of electromagnetism. Quantum electrodynamics can be solved by calculating a series of Feynman diagrams, which results in a convergent series of terms that approximates QED to very high accuracy. Unlike QED, in which there is only one kind of charge, quark charges come in three different kinds, which have been whimsically named "colors." Gluons change a quark of one color into a quark of another color. Gluons can also interact with each other, unlike the photons that mediate the electromagnetic force. These self-interactions cause the color force to become large at hadronic energies, and the Feynman diagram series, which succeeds at very high energies, fails. This led to a search for new methods for approaching QCD, and resulted in the invention of lattice QCD and the most productive approach to solving it: large-scale computer calculations.

Principles. Lattice QCD approximates the real world by defining fields representing quarks and gluons on a four-dimensional space-time lattice (**Fig. 1**). The quarks live on the sites of the lattice, and the gluons live on the links connecting the sites. In the real world, quarks follow continuous paths through space-time, but in this approximation they hop from site to site on the lattice in discrete steps. The predictions of the theory are obtained in the limit that the lattice spacing approaches zero, the continuum limit. Lattice QCD as a quantum field theory is defined with the path integral formalism. Quantum amplitudes are defined as sums over all possible configurations of the fields weighted by a certain function of the fields, the classical action. Monte Carlo methods are used to create a set of configurations of the gluon fields that approximates the complete path integral to higher and higher accuracy as more configurations

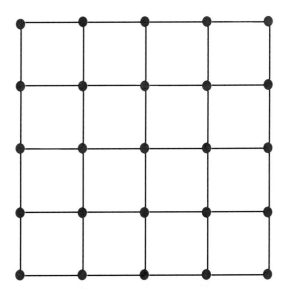

Fig. 1. The fields of lattice quantum chromodynamics are defined on a four-dimensional space-time lattice. The quarks live on the sites of the lattice; the gluons live on the links that connect the sites. The physical theory is obtained in the limit that the lattice spacing approaches zero.

are added. Large computer calculations are used to create these sets of configurations. Because the number of sites on the lattice becomes large as the lattice spacing becomes small, taking the continuum limit is computationally very expensive. The largest calculations might take a period of months on one of today's largest computers, performing hundreds of trillions of arithmetic operations per second.

Applications. Applications of lattice QCD include inferring the parameters of the standard model, understanding the structure and interactions of nuclear particles, and studying the behavior of nuclear matter in extreme environments.

Parameters of the standard model. Lattice QCD is used to understand physics at distance scales both larger and smaller than the hadronic distance scale. Physics at the shortest distances that are now understood is described by the standard model of particle physics. Quarks are among the elementary constituents of matter in the standard model. Their masses and the strength with which one decays into another are fundamental parameters of the standard model. These are unpredicted, free parameters in the standard model. Most particle physicists believe that they will be understood someday in terms of more fundamental theories that will be discovered as we investigate physics at shorter distances and higher energies than we can reach today. Accurately determining them is one of the most important goals of current research. Because quarks and gluons are confined within hadrons, these masses and couplings cannot be directly observed. Experiments can determine directly only the masses and other properties of hadrons. Lattice QCD calculations, which determine the properties of hadrons in terms of the properties of quarks and gluons, enable us to infer the masses and couplings of the quarks and gluons from the observed properties of hadrons. Some of these quantities, such as the masses of the lightest quarks, can be determined only with the aid of lattice QCD. **Figure 2** illustrates one of these processes: the decay of a *B* meson (consisting of a massive bottom flavor of quark and an up or down antiquark) into a pion, an electron, and a neutrino. The electron and the neutrino are elementary particles too, and do not

feel the strong force. The *B* meson and the pion are made of quarks. In this decay, the bottom quark in the *B* meson decays into an up quark to form a pion. Lattice calculations derive the strength of the bottom-to-up quark coupling from the rate of the *B* meson decay.

Structure and interactions of nuclear particles. Understanding the structure and properties of protons, neutrons, and the other hadrons is another goal of lattice QCD. Lattice calculations have been used to determine previously poorly known quantities such as the size, shape, and distribution of electromagnetic charge and current within hadrons. They have determined how the spin of the proton arises from the spins and orbital angular momenta of the quarks and gluons. The known masses of the observed hadrons are being calculated with ever-increasing accuracy to test the theory.

Nuclear physics, the study of atomic nuclei with many nuclear particles, developed before QCD was understood. The goal of deriving nuclear physics from first principles has not yet been achieved. Mathematical models of nuclear structure incorporate scattering parameters that are in principle calculable from QCD. With lattice QCD, this quest has begun. The most important scattering parameters describing the scattering of two mesons such as pions or kaons have been calculated to good accuracy. Pions and kaons are especially simple hadrons, built from a quark and an antiquark. The goal now is to achieve comparable accuracy for the protons and neutrons that are contained in nuclear matter, which are built of three quarks and are more complicated to calculate.

QCD in extreme environments. When nuclear matter is brought to extremely high temperatures and pressures, it is believed to behave quite differently from the way it behaves in ordinary nuclei. Quarks are normally confined in a particular nucleon and do not hop from one nucleon to the next. When the nucleons are heated to high enough temperatures or pushed tightly enough together, however, confinement disappears and the quarks from neighboring nuclei mix together freely. This is thought to happen in sufficiently dense neutron stars and in quark-gluon plasma formed in the high-energy collisions of heavy nuclei, and also to have happened in the early universe. One of the parameters used in models describing the evolution of the quark-gluon plasma in heavy nuclei collisions has been calculated already: the temperature at which quarks become deconfined. Other parameters, such as the plasma viscosity, are expected in the future.

Future goals. Lattice QCD calculations are still in their adolescence, with much more remaining to be accomplished. In the future, we expect to see calculations from first principles of the parameters of nuclear structure models, of new properties of the quark-gluon plasma, and of the parameters of the standard model to higher and higher precision. Most particle physicists believe that the standard model is an approximation to a yet-to-be-discovered, more fundamental theory, and experiments are searching for clues to it. New interactions in these theories are

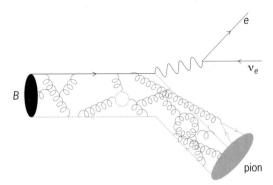

Fig. 2. The strength of the coupling of the bottom quark contained in a *B* meson into a lighter up quark is inferred with lattice QCD from the decay rate of a *B* meson into a pion, an electron, and a neutrino. The heavily colored line represents the heavy *b* quark, the fully colored lines represent light quarks, and the curly gray lines represent gluons holding the quarks together.

likely to require new types of lattice theories for their solutions.

For background information *see* ACTION; FEYNMAN DIAGRAM; GLUONS; MONTE CARLO METHOD; NUCLEAR STRUCTURE; QUANTUM CHROMODYNAMICS; QUANTUM ELECTRODYNAMICS; QUANTUM FIELD THEORY; QUARK-GLUON PLASMA; QUARKS; STANDARD MODEL; SUPERCOMPUTER in the McGraw-Hill Encyclopedia of Science & Technology.   Paul B. Mackenzie

Bibliography. M. Creutz, *Quarks, Gluons, and Lattices*, Cambridge University Press, Cambridge, U.K., 1983; T. DeGrand and C. DeTar, *Lattice Methods for Quantum Chromodynamics*, World Scientific, Singapore, 2006; D. H. Weingarten, Quarks by computer, *Sci. Amer.*, 274(2):116–120, February 1996.

Leaf senescence and autumn leaf coloration

When viewed by astronauts, Earth is the blue planet. When imaged by Earth observation satellites, the planet is green. This reflects the fact that water and chlorophyll are the signatures of life in the solar system. Satellite images show the green color of vegetation ebbing and flowing with the seasons (**Fig. 1**). A whole season of the year in temperate regions is named for the fall of leaves that follows the seasonal replacement of chlorophyll with the yellows, oranges, reds, purples, and browns of autumn foliage. Such a global-scale biological event must surely have a purpose and a mechanism. We are now learning the *how* of color change in plants; the *why* is more elusive, but new insights from genetics and evolutionary biology are providing possible answers.

Light is essential for green plants, but can be harmful too. The small ephemeral weed *Arabidopsis thaliana* is botany's laboratory rat: a universally studied model organism that does just about everything a flowering plant should do, in an experimentally tractable way. Like those of trees and crop species, its leaves develop into the green solar panels that power growth through photosynthesis, then, when their productive job is done, they initiate senescence, turn yellow, and eventually die. There are many *Arabidopsis* mutants with alterations in genes controlling development, physiological function, and

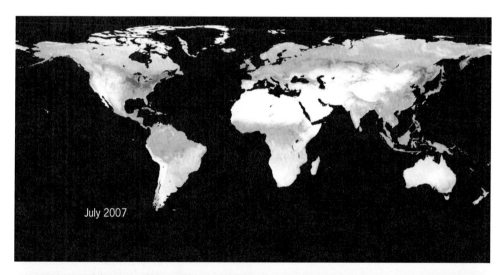

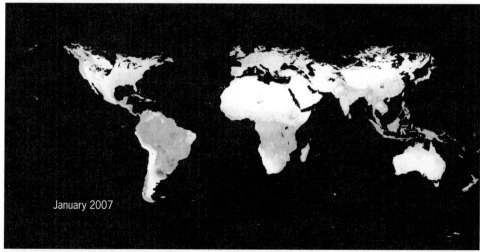

Fapar Index
1.0
0.8
0.6
0.4
0.2
0.0

Fig. 1. Photosynthetic pigments of foliage on a global scale, measured by the Envisat satellite in Fapar units (Fapar = fraction of absorbed photosynthetically active radiation). The images are monthly averages comparing coverage in January and July 2007, showing the extent of seasonal variation in chlorophyll, particularly in northern temperate regions. (*Courtesy of European Space Agency Envisat satellite, MERIS data*)

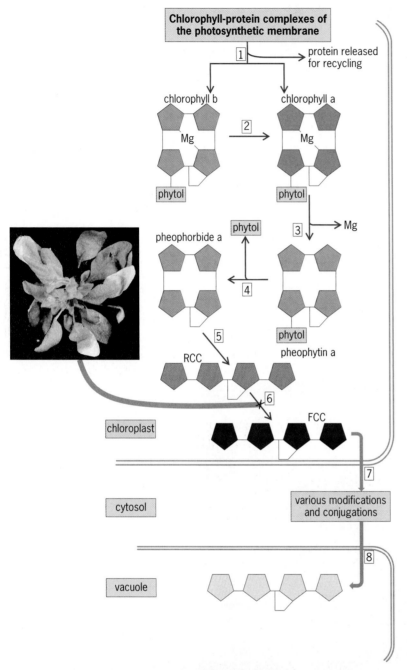

Fig. 2. The cellular mechanism of chlorophyll breakdown in plant senescence. The enzymes and transporters responsible for the sequence of biochemical transformations are as follows: [1] stay-green; [2] chlorophyll b reductase; [3] dechelation reaction; [4] pheophytinase; [5] pheophorbide a oxygenase; [6] RCC reductase; [7] ATP-dependent catabolite transporter; and [8] ABC transporter. The *Arabidopsis* mutant, *acd2*, is blocked at reaction [6] and suffers from light-dependent leaf lesions as a consequence of the anomalous buildup of the normally transient phototoxic intermediate RCC. (FCC = fluorescent chlorophyll catabolite.)

destined to die: chlorophyll and its derivatives, if not carefully processed by organized biochemical pathways in green cells, can be toxic when illuminated. The mutant's name, *acd2*, stands for *accelerated cell death*-2 and clearly expresses the consequences of disrupting pigment metabolism in senescing green tissues.

The behavior of *acd2* can be understood in evolutionary terms by considering how certain single-cell algae dispose of chlorophyll. If cultures of *Chlorella protothecoides* growing photosynthetically in the light are transferred to darkness and deprived of a nitrogen source, they turn yellow by a process that resembles the yellowing of senescing leaves. At the same time, the culture medium accumulates a red pigment that chemical analysis shows to be a chlorophyll derivative. A similar red chlorophyll catabolite (RCC) is an intermediate in the metabolism of chlorophyll during leaf senescence (Fig. 2). RCC is potentially harmful because it generates reactive oxygen species and free radicals that destroy the fabric of living cells. Single-cell aquatic plants deal with the threat from RCC by pumping it out of the cell. Multicellular land plants do not have this option and have evolved a detoxification mechanism that begins with the conversion of RCC to a colorless product by RCC reductase. In the mutant *acd2*, RCC reductase is missing. As a consequence, when chlorophyll is degraded, RCC builds up and mediates cell death in the light. The ultimate destination of the product of the RCC reductase reaction is the central vacuole, the large membrane-limited aqueous compartment that occupies most of the volume of the green cell. Sequestering products out of harm's way in the vacuole is a further detoxification measure employed by land plants. It seems that the progression from green to yellow that is characteristic of foliar senescence is a highly visible symptom of toxin disposal.

Recycling is a way of life for green plants. If eliminating chlorophyll is a hazardous business requiring stringent detoxification measures, why doesn't the plant simply abandon the pigment by dropping it unchanged within falling leaves? The answer is that the removal of chlorophyll is associated with, and necessary for, the salvage of protein nitrogen from the senescing leaf to support the development of young organs or storage tissues. In many crops and natural plant communities, nitrogen is an important limiting factor for growth and productivity. Senescence evolved as a way of liquidating the investment of nitrogen in green tissues when they become old and unproductive and reinvesting it in younger, better-adapted structures.

Most of the salvageable nitrogen in green cells is located in the chloroplasts (cell plastids occurring in the green parts of plants, containing chlorophyll pigments, and functioning in photosynthesis and protein synthesis). A large fraction of this protein is bound to chlorophyll. Chlorophyll (and protein) degradation begins with the controlled unpacking of pigment-protein complexes (Fig. 2). Observations on mutants indicate that nitrogen from pigment-binding

environmental response. Among these mutants is *acd2*, a genetic variant that suffers from lesions on older leaves in the light (**Fig. 2**). When the *acd2* gene was finally cloned and identified in 2001, it turned out to be an inactive form of a gene encoding red chlorophyll catabolite (RCC) reductase, an enzyme of chlorophyll breakdown. This provides a clue as to why plants take the trouble to expend physiological energy on changing the colors of leaves that are

proteins cannot be released and relocated to destination tissues unless chlorophyll is removed. For example, a mutant gene called *stay-green* (*sgr*) renders plants that carry it unable to deliver chlorophyll from pigment-protein complexes to the chlorophyll breakdown machinery. Such plants cannot mobilize chlorophyll-associated protein, which remains more or less unchanged in the leaf even as the nitrogen from other proteins is being recycled during senescence.

Yellow pigments are revealed during senescence. The picture that we are developing of color changes in senescing leaves is of the controlled detachment of chlorophylls from their nitrogen-rich binding proteins in chloroplast membranes, followed by a series of transformations via transient green and red intermediates, culminating in colorless end products that accumulate in the cell vacuole (Fig. 2). In ripening crops and many other species, green is replaced by golden yellow. Carotenoids, the chemical group responsible for the yellow and orange pigments of senescing foliage, are already present in green leaves and are unmasked as chlorophyll is removed. Some carotenoids are built into chlorophyll-protein complexes. As the complexes and chloroplast membranes are taken apart, carotenoids coalesce in intensely colored lipid-rich globules within the senescing chloroplast (**Fig. 3**). Leaves tend not to make more carotenoids during senescence; however, in other senescencelike processes such as the ripening of tomatoes, bell peppers, and other colorful fruit, substantial amounts of new carotenoids may be synthesized. Carotenoids are potent antioxidants, retained during leaf senescence as part of the cellular equipment defending against photodamage.

Some senescing leaves turn red. In many regions of the world, such as the forests of New England in the United States, senescing leaves turn spectacularly red. Anthocyanins, the chemical constituents commonly responsible for the fiery colors of autumnal foliage, are water-soluble pigments that (in contrast to the fat-soluble carotenoids) are actively synthesized by senescing photosynthetic tissues and accumulated in the vacuoles of leaf cells (Fig. 3). Genes for carotenoid synthesis are extremely ancient, exhibiting a high degree of conservation from flowering plants all the way to unicellular organisms resembling those present at the origin of life. On the other hand, recognizable genes for metabolism of anthocyanin-like compounds do not occur until the algae evolved, became integrated into development only once land plants began to appear, and are characteristic of the relatively recent (in evolutionary terms) flowering plant groups.

Autumn colors evolved to defend against a hostile environment. It has been suggested that anthocyanins are sunblockers, defending vulnerable tissues from the harmful effects of excess light. There is also evidence that anthocyanins can act as antioxidants. An alternative or additional explanation for the origin of foliar anthocyanin is coevolution. According to this hypothesis, autumn coloration is a signal of

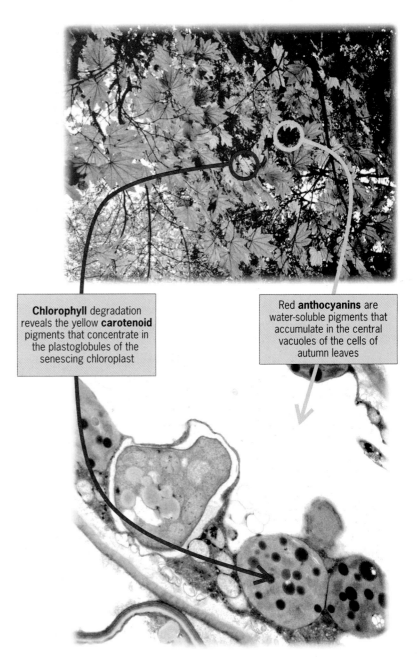

Chlorophyll degradation reveals the yellow **carotenoid** pigments that concentrate in the plastoglobules of the senescing chloroplast

Red **anthocyanins** are water-soluble pigments that accumulate in the central vacuoles of the cells of autumn leaves

Fig. 3. Detail of the ultrastructure of a senescing cell showing the different locations of the yellow (carotenoid) and red (anthocyanin) pigments of autumn leaves.

quality, directed at insects that migrate to the trees in autumn. The red color of foliage may be symptomatic of the plant's unsuitability as a host because of high levels of chemical defenses, of poor nutritional status, of imminent leaf fall, or of any other characteristic that would induce a lower fitness in the insects. Mathematical models based on signaling theory generally support the coevolution explanation and emphasize the significance of insect visual systems and behavior and of the nature of the link between leaf color, defense status, and plant vigor.

Outlook. In conclusion, researchers studying the meaning of color in senescing leaves are able to enjoy the pleasures of engaging with biological events that are both scientifically challenging and esthetically satisfying.

For background information *see* ABSCISSION; CAROTENOID; CHLOROPHYLL; COLOR; DECIDUOUS PLANTS; LEAF; PHOTOSYNTHESIS; PIGMENTATION; PLANT ANATOMY; PLANT PHYSIOLOGY; PLANT PIGMENT in the McGraw-Hill Encyclopedia of Science & Technology.   Howard Thomas

Bibliography. M. Archetti et al., Unravelling the evolution of autumn colours: An interdisciplinary approach, *Trends Ecol. Evol.*, 24:166–173, 2009; H. Ougham et al., The control of chlorophyll catabolism and the status of yellowing as a biomarker of leaf senescence, *Plant Biol.*, 10(suppl. 1):4–14, 2008; H. J. Ougham, P. Morris, and H. Thomas, The colors of autumn leaves as symptoms of cellular recycling and defenses against environmental stresses, *Curr. Top. Dev. Biol.*, 66:135–160, 2005.

Local structural probes

The deeply penetrating ability of x-rays and their short wavelength, comparable to atomic size, makes them an ideal probe for structural characterization of materials. In addition, many elemental absorption edges are located in the x-ray part of the electromagnetic spectrum. As such, x-rays are especially useful for a variety of spectroscopic measurements, providing chemical or elemental specificity, and sensitivity to orbital and magnetic ordering.

Focusing x-ray beams. After Wilhelm Conrad Röntgen discovered x-rays in 1896, he immediately searched for ways to focus them. Since the wavelength of x-rays is about 5000 times shorter than that of visible light, x-ray beams could be focused down to subnanometer dimensions. This is well below the limit imposed by diffraction effects on visible light, on the order of 200 nm. However, until recently, most x-ray–based characterization techniques were carried out in a nonlocal, global averaging mode, where the diffraction or spectroscopic information is obtained via spatial averaging over the macroscopic dimensions (measured in millimeters) over the entire sample. Availability of nanoscale-focused x-ray beams enables a wide range of x-ray diffraction or spectroscopy-based techniques that can be performed in an imaging mode, providing local information about relevant inhomogeneities (that is, chemical, elemental, structural, electronic, and magnetic) with nanoscale spatial resolution.

It took more than 100 years after Röntgen's discovery of x-rays to achieve sub-100-nm x-ray beam focusing, primarily due to developments of high-brightness synchrotron radiation sources and advances in the synthesis of materials and devices with tailored structural properties at the nanoscale level. In general, x-ray beams can be focused using three types of optics: reflective, refractive, and diffractive (**Fig. 1**).

Reflective optics. Reflective optics (Fig. 1*a*), that is, mirrors, make use of the fact that the refractive index for x-rays in matter is slightly less than unity and x-rays incident at the surface (that is, at grazing angles that are lower than the so-called critical angle of "total internal reflection") will be reflected with nearly 100% reflectivity. Since the critical angles at which these parabolic mirrors operate are on the order of a few milliradians (less than 0.1°), using reflective mirrors to focus requires relatively long focal lengths. One of the key advantages of the reflective approach is that focusing properties depend primarily on the shape of the mirror, and not on the wavelength of the incoming beam, so that polychromatic beams (a mixture of a wide spectrum of x-ray wavelengths or energies) can be focused in the same spot. Focusing these polychromatic (so-called pink) beams is important for many research applications, for example, microdiffraction studies of polycrystalline materials, or Laue diffraction studies of crystallized proteins and other biological molecules. High efficiency (almost 100% reflectivity at below critical angles) is another key advantage of using the reflective optics approach. The main technical challenge in producing high-quality reflective optics is the ability to produce "ideally shaped" parabolic or elliptical mirrors with errors as low as a few tens of nanometers, over long distances (on the order of a meter).

Refractive optics. Refractive optics (Fig. 1*b*) uses the same principle as standard glass lenses commonly found in reading glasses, optical microscopes, and

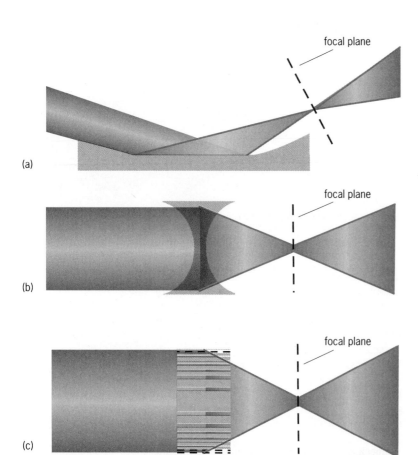

Fig. 1. Typical x-ray focusing approaches. (*a*) Reflective optics. (*b*) Refractive optics. (*c*) Diffractive optics.

cameras. However, since the refractive index is less than unity, focusing is achieved with biconcave-shaped rather than biconvex-shaped lenses (commonly called compound refractive lenses). And since the refractive index for x-rays typically deviates from unity by a very small factor, on the order of one part in 10,000, a stacked assembly of many thick lenses is often required. Unlike reflective optics, which benefit from coatings made of materials with a high atomic number, Z, such as platinum or gold ("high-Z materials"), refractive lenses are made of materials with a low Z to reduce adsorption effects, such as beryllium or silicon ("low-Z materials"). Similarly to reflective optics, the nonideality in the shape of the lenses, on the order of 10-100 nm, is the crucial parameter that defines the dimensions of the focal spot.

Diffractive optics. The use of diffractive optics (Fig. 1c) to focus x-ray beams is usually achieved with Fresnel zone plates, featuring alternating transparent and opaque concentric rings, or Fresnel zones. These rings (zones) are geometrically designed to produce maximum constructive interference for a specific focal length point and a given wavelength. The widths of the zones decrease in inverse proportion to their radii. The Fresnel zone plate operates on a similar principle as a diffraction grating. Since the focal length is strongly wavelength-dependent for diffractive optics, and nonvanishing bandwidth introduces spatial broadening of the x-ray beamspot, diffractive optics is ideal for experiments requiring highly monochromatic radiation. The degree of focusing achieved with diffractive optics is typically defined by the dimensions of the outer (and narrowest) zones. Advances in the synthesis of nanostructures make it possible to make zone plates producing beams focused to 20–30 nm, and focusing of hard x-ray beams down to 1 nm is considered technologically achievable (using the multilayer Laue lens approach) in the near future. However, the efficiency of the zone plates, defined as the ratio of the transmitted and focused x-ray intensity to the x-ray intensity incident on the zone plate, is typically low (of the order of 10–40%). Opaque zones need to be greater than 100 nm along the beam-propagating direction to adsorb most of the high-energy x-rays, but also narrow (approximately 1–10 nm) in transverse direction for the outer zones. This means that zone plates with a high efficiency for hard x-rays (above 10 keV) require delicate high-aspect-ratio nanostructures made of highly adsorbing, high-Z materials (for example, gold or tungsten) that are difficult to synthesize. Despite these difficulties, diffractive optics allow for short focal lengths that are required to achieve nanometer-scale focusing that approaches the ultimate diffraction limit for x-rays.

Scanning and full-field imaging x-ray microscopy.

Regardless of the optics of choice, there are two modes in which the focusing elements are commonly used. These modes are scanning microscopy and full-field imaging (**Fig. 2**). In scanning microscopy, the focusing optic is placed upstream of the sample, with the sample placed in the focal spot and rastered to

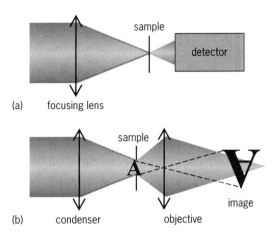

(a)   focusing lens

(b)   condenser   objective

Fig. 2. X-ray microscopy modes. (*a*) Scanning. (*b*) Full-field.

provide a detailed microscopic image of the sample with spatial resolution defined by the dimensions of the focused x-ray spot. An alternative approach is the so-called full-field imaging, where the focusing optics is placed downstream of the sample, and serves as an objective, with another focusing element (condenser) placed upstream of the sample. This is the same approach used in optical microscopes, providing a large field of view of the sample with spatial resolution defined by the quality of the objective optics. Scanning microscopy is considerably slower, as it requires detailed scanning of the specimen, but also provides a much higher overall efficiency and ease of interpretation. It is often a preferable option for many x-ray microscopy applications, including microdiffraction and fluorescence microscopy.

Lensless coherent x-ray diffractive imaging.

An alternative approach to using lenses for focusing in scanning microscopy or as an objective in full-field imaging is the so-called lensless or coherent x-ray diffractive imaging (CXDI) approach (**Fig. 3**). This technique relies on mathematical phase-retrieval

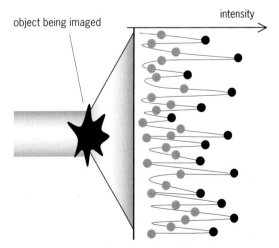

Fig. 3. Coherent x-ray diffractive (lensless) imaging. Although phases of the diffraction pattern are lost during measurements, phases can be retrieved via oversampling, that is, measurements of additional diffraction intensities at higher spatial frequencies (colored points).

algorithms to obtain a real-space image of the object with nanoscale resolution from the diffractive coherent pattern. The computer algorithm effectively acts as a lens, so that the ultimate spatial resolution is limited by the total coherent flux (directly related to the brightness of the x-ray source), bypassing the need for difficult-to-manufacture optical elements.

The main idea behind lensless imaging is the use of oversampling, that is, measurements of coherent diffraction pattern intensities at a spatial frequency at least twice the inverse of the imaged object dimensions, to retrieve phase information lost in experimental measurements and obtain full reconstruction of the illuminated object. This coherent x-ray diffractive imaging has been used to produce three-dimensional (3D) images of samples, typically microscopic metal or biological particles, with spatial resolution as low as 5 nm. This substantially exceeds state-of-the-art spatial resolution that can be obtained with various focusing optical elements. Coherent x-ray diffractive imaging in transmission geometry is used to obtain 3D electron density distribution within the sample, whereas similar measurements performed in high-angle Bragg diffraction geometry provide information about 3D distribution of strain and crystallographic defects with nanoscale spatial resolution. A combination of CXDI and scanning microscopy approaches, a so-called ptychographical iterative engine, has been demonstrated to be successful in imaging extended, rather than small isolated, objects. Ptychography refers to folding, and is the oversampling in real space (by imaging overlapping regions) and oversampling in reciprocal space (by measuring diffraction patterns at least twice the Nyquist spatial frequency, that is, the inverse of the total size of the object).

Applications of x-ray microscopy. In addition to structural properties of materials, x-ray microscopy can be used to provide valuable information about chemical composition by using fluorescence or near-edge spectroscopy techniques. X-ray microscopy can also couple to magnetic and orbital ordering using resonant scattering. Sensitivity to elemental, chemical, orbital, and magnetic properties, in addition to structural (x-ray diffraction microscopy), makes x-ray microscopy an important tool for a variety of disciplines, such as condensed matter and materials physics, chemistry, biology, medicine, geosciences, astrophysics, and environmental sciences. Because the measurements can often be performed in a variety of environments, and since the deep penetrating power of hard-x-ray samples do not require sectioning, many systems can be investigated in situ, or in time-resolved mode to study nanoscale dynamics.

For background information *see* DIFFRACTION; DIFFRACTION GRATING; FLUORESCENCE MICROSCOPE; MICRORADIOGRAPHY; REFRACTION OF WAVES; X-RAY DIFFRACTION; X-RAY MICROSCOPE; X-RAY TELESCOPE; X-RAYS in the McGraw-Hill Encyclopedia of Science & Technology.   Oleg G. Shpyrko

Bibliography. C. Jacobsen and J. Kirz, X-ray microscopy with synchrotron radiation, *Nat. Struc. Biol.*, 5:650–653, 1998; H. C. Kang et al., Nanometer linear focusing of hard x rays by a multilayer Laue lens, *Phys. Rev. Lett.*, 96:127401, 2006; P. Kirkpatrick and V. Baez, Formation of optical images by X-rays, *J. Opt. Soc. Am.*, 38:766, 1948; B. C. Larson et al., Three-dimensional X-ray structural microscopy with submicrometre resolution, *Nature*, 415:887–890, 2002; J. Miao et al., Extending the methodology of X-ray crystallography to allow imaging of micrometre-sized non-crystalline specimens, *Nature*, 400:342–344, 1999; H. F. Poulsen, *Three-Dimensional X-ray Diffraction Microscopy, Mapping Polycrystals and Their Dynamics*, Springer Tracts in Modern Physics, vol. 205, Springer, Berlin, 2004; I. K. Robinson and J. Miao, Three dimensional coherent x-ray diffraction microscopy, *MRS Bull.*, 29:177–181, 2004; A. Snigirev et al., A compound refractive lens for focusing high-energy X-rays, *Nature*, 384:49–51, 1996.

Location-based decision support

Over the years, we have become a mobile information society. Increased mobility has affected many areas of our daily lives, such as travel, communication, consumerism, social behavior, and the environment. Mobile people face challenging problems in space and time. These problems need to be solved on the spot, such as navigating an unfamiliar city or deciding on the fastest public transportation mode to a destination. Location-based decision support facilitates people's mobile decision making. It is based on location-based services (LBS)—information services that are sensitive to the location of a mobile user. These services allow a user to query a location from a mobile terminal, such as a mobile phone or personal digital assistant (PDA), and to exploit spatial information about the user's surrounding environment, such as his or her proximity to other entities in space.

Location-based decision services. Personalization is an essential aspect of making mobile decisions that are valuable to the user. It concerns the personal management of space through user preferences and characteristics. Customizing and adapting LBS to users is important because people differ in their spatial and cognitive abilities, and their information needs depend highly on the personal and situational context. Disabled people require different wayfinding instructions, for example. And route elements for people using wheelchairs must not include stairs.

Recent research has focused on the development and design of mobile location-based decision services (LBDS), which provide personalized spatial decision support to users. These services are built on the integration of multicriteria decision analysis (MCDA) and can provide analytic evaluations of the attractiveness of alternative destinations and choices being offered. MCDA is based on the idea that humans use multiple decision criteria to determine the best solution to a problem. The following description of a mobile hotel finder service illustrates how a LBDS functions.

The hotel finder service features multicriteria location-based decision support for the task of finding suitable hotels in an unfamiliar environment, depending on the user's location and preferences. It integrates the ordered weighted averaging (OWA) decision rule, which allows users to choose a decision strategy as part of their decision-making preferences. This leads to different answers by the location-based decision service depending on the decision maker's level of risk taking. Decision strategies range from optimistic (that is, risk taking) to pessimistic (that is, cautious), and allow for full trade-off to no trade-off between the different decision criteria. For example, with the optimistic strategy, the decision maker focuses on the higher outcomes, thus incurring the risk of accepting an alternative with excellent values on some criteria but potentially poor values on other criteria. Users navigate through the steps of an MCDA process that includes determining decision alternatives (hotel destinations), selecting decision criteria (such as room rate and Wi-Fi access), standardizing the criterion values for all alternatives, determining importance weights for the criteria, and using a multicriteria decision rule to aggregate the weighted standardized criterion values to an evaluation score and rank for each alternative. The user interface of the mobile device provides both the functionality for displaying the geographic data and a dialogue component to elicit the user's input of MCDA parameters. **Figure 1** shows the service for the city of Toronto, Canada. The map window can be provided by map servers such as Google Maps or Microsoft Virtual Earth.

Benefits and critical issues. The widespread adoption of location-based services is expected to lead to great benefits by providing large segments of the population with real-time, location-based decision support for purposes ranging from trivial (navigation and friend-finder services) to critical (emergency response). A major advantage of LBDS is that such decision support can be tailored to the user of the service, thereby providing an individual with the optimal information, as needed. Further benefits include real-time (24/7) data access and potential time savings compared to traditional decision-making approaches. The gradual reduction in the cost of mobile devices and application access over time will broaden the accessibility and distribution of these services.

On the other hand, one must be aware of the security and privacy issues that modern technology brings with it, and location-based decision services are no exception. In extreme cases, the capacity for real-time integration of location information and personal data can lead to so-called geoslavery—monitoring and exercising control over the physical location of an individual. As a response to this potential danger, various initiatives have called for the implementation of location privacy protection methods and laws to regulate and restrict the use of existing human tracking systems. Finding the right

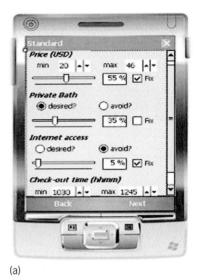

(a)   (b)

Fig. 1. User interface for the Toronto hotel finder, showing (*a*) standardization and weighting of criteria and (*b*) presentation of the results, with the additional option of calculating the route to the optimal hotel.

balance between customer service and privacy invasion will be a major goal for the future.

Applications. The application areas for location-based decision support are many. Currently, the most popular services are mobile guides and navigation services. Mobile guides provide users with a wealth of information about their surrounding environment, and many expect that they will gradually replace traditional analogue tour guides. Wayfinding and navigation services provide route instructions to both car drivers and pedestrians. One of the challenges in providing optimal wayfinding instructions for pedestrians is the representation of the navigable space, because, unlike car drivers, pedestrians are not bound to street networks, which leads to more complex calculations. Recent research has focused on the integration of analog and digital media, such as the combination of static paper maps and digital displays. This integration can be accomplished through the use of mobile phones that are equipped with cameras. In this way, digital information, such as the location of nearby automated teller machines or restaurants, can be displayed on top of an analog map. Future mobile guides will be able to access knowledge from online repositories and use this content to create educational audio tours starting and ending at stationary city maps (**Fig. 2**).

Emergency services are another key application of location-based decision support. Automatic positioning methods and communication technology help to save critical time during rescue operations, such as in car accidents when injured people are unable to report their location. Mobile emergency services can help rescue teams improve their response operations by receiving location-sensitive information and instructions from an emergency operations center. Such a center locates and coordinates its emergency crews in the field through Global Positioning System (GPS) technology and provides up-to-date information and decision-making parameters to them.

Fig. 2. WikEar, a novel mobile guide that generates location-based audio tours. (*Copyright © Johannes Schöning, Institute for Geoinformatics, University of Münster, Germany*).

Location-based decision support can be used in a wide range of applications for businesses and administrations. Mobile commerce allows people to make transactions on the move and receive location-based advertisements, such as electronic discount coupons for restaurants in the surrounding area. Commercial enterprises use LBS to calculate optimal delivery routes for shipping goods based on their customers' locations and up-to-date traffic information. Administrations are supported by location-based technology in the areas of asset management and local commerce.

Recently, a novel form of location-based services has emerged in the area of social networking. These services can determine the locations of friends and family members, and a user is notified by the service when one of his or her friends comes within a certain geographic proximity. Decision support includes the selection and suggestion of meeting points and activity locations, such as a restaurant to spontaneously meet for dinner.

For background information *see* AUTOMATED DECISION MAKING; DECISION SUPPORT SYSTEM; DECISION THEORY; MOBILE COMMUNICATIONS in the McGraw-Hill Encyclopedia of Science & Technology.

Martin Raubal

Bibliography. A. Brimicombe and B. Li, *Location-Based Services and Geo-Information Engineering*, John Wiley & Sons, Chichester, U.K., 2009; J. Dobson and P. Fisher, Geoslavery, *IEEE Tech. Soc. Mag.*, 22(1):47–52, 2003; A. Küpper, *Location-Based Services: Fundamentals and Operation*, John Wiley & Sons, Chichester, U.K., 2005; J. Malczewski, *GIS and Multicriteria Decision Analysis*, John Wiley & Sons, New York, 1999; J. Raper et al., Applications of location-based services: A selected review. *J. Location Based Serv.*, 1(2):89–111, 2007; C. Rinner and M. Raubal, Personalized multi-criteria decision strategies in location-based decision support, *J. Geogr. Inform. Sci.*, 10(2):149–156, 2004.

Lunar and planetary mining technology

Sustained human life and activities rely on a wide variety of minerals and rocks as fuel and raw material for production of goods. Mining and processing methods have evolved through millennia of human experience with terrestrial deposits. If a Moon base is established to learn more about the lunar environment and the Solar System, to prepare for Mars missions, to intercept the asteroids, and to launch deep-space missions, materials will need to be mined from the local environment, for example, to provide water and probably air.

Lunar and planetary mines will use technology derived from current state of practice and adapted for new environments. These technologies will allow the desired material to be removed from the unwanted material with which it is mixed, using the least energy expenditure possible.

At present, humanity is confined to a single planet and faces many dangers and challenges. The United States plans to return humans to the Moon by 2020, after a nearly 50-year hiatus. India, China, Japan, Russia, the European Union, and other countries have programs underway focused on exploring the Moon. Mars exploration is ongoing and human landings will follow. The asteroid Apophis will approach Earth in 2029 and again in 2036. A significant increase in human interplanetary capabilities may be needed to control its behavior.

Geology of the Moon and Mars. Mars formed in basically the same way and at the same time as the Earth did, when the original solar nebula coalesced to form the Sun, the rest of the planets, the asteroids, and the comets. Mars is smaller than the Earth. Mercury, Venus, Earth, and Mars are all rocky planets (unlike the gas giants Jupiter and Saturn and the ice giants Uranus and Neptune), but only Earth has permanent liquid water. Mars has water ice at its poles (sampled by the *Phoenix* lander) and in permafrost elsewhere; even Mercury has water ice in perpetually shadowed regions at its poles.

The surface of Mars consists largely of basalt flows and the sand worn from them by billions of years of winds and meteoroid impacts. It is different from the Moon in that it has an atmosphere, albeit a very thin one, that protects it from smaller meteoroids. The presence of liquid water as groundwater and occasionally as surface water in its past has also created a different, wider suite of minerals than is found on the Moon. Some of these minerals, such as the clays, could be mined.

The Moon came into being when a proto-planet collided with the Earth only a few hundred million years after the Earth formed. Many of the pieces from that collision coalesced into a molten ball that cooled slowly and crystallized, almost like the Earth had done. Unlike the Earth, though, the Moon's surface consists of large smooth areas of basalt, called the maria, and rough areas of anorthosite, called the highlands. Both are covered by a layer of pulverized rock (regolith) that formed when meteoroids hit the Moon's surface, breaking the underlying rock into

Moon mining challenges	
Issue	Implications
Nearly absolute vacuum	Space suits required
Excessive radiation	Many problems with electronic devices/communication systems
Extreme temperature (-180 to 120°C) gradient	Regular material (rubber, lubricants, etc.) cannot be used in the design of the machine; excessive fatigue and thermal design issues
Low gravity, one-sixth of Earth	Need for larger mass to provide for reaction forces
Potential for impact by meteorites	Dangerous work conditions; may need additional shields for impact prevention
No cooling or flushing medium	Cleaning of the hole for drilling is a problem; also, overheating of the cutting tools
Dust suspension over extended time	A major problem, affecting equipment design, wear of the equipment, visibility, seals and air locks, etc.
Limited source of power	Requires very energy-efficient systems; may limit the size of equipment to be used
Difficulty in repair/maintenance	Equipment should be semiautonomous or fully automated, very rugged, simple, and durable
Cost of over $200k/kg to send material to the Moon	Equipment should be lightweight for transportation

extremely fine (dust-sized) pieces over a very long period of time. Quite a few pebbles, rocks, and boulders also litter the surface, from recent impacts. Over time, all are reduced to regolith. The Moon may also have water ice between the regolith grains in completely shadowed polar craters, but that has not been confirmed.

Mining targets. Mining of raw materials has been proposed for outer-space bases to support space missions. Because of the extremely high cost of transporting raw materials from Earth, all the fuels, air for breathing, or construction materials should be mined from the Moon or Mars, or an asteroid or comet, to be cheaper and less energy-intensive. This would also allow more room in launched payloads for people and for scientific instruments.

Some of these materials can be extracted from the regolith and some from the rock as well, though that will be harder to get to. Water ice is particularly valuable. Hydrogen incorporated in other minerals is almost as good, because oxygen occurs in most of the minerals on the Moon and Mars. Hydrogen implanted in regolith grains by the solar wind, for example, is concentrated 100 times less than water ice that would produce the same amount of hydrogen for fuel production. The key is to mine enough of the raw material to produce the needed volume of each product.

Particular constituents of the regolith other than ice that are of interest include volcanic glass beads and helium-3 (^{3}He), an isotope of helium that could produce clean power for civilization from nuclear fusion with relatively few radioactive byproducts. Helium-3 is considered a safe, environmentally friendly fuel candidate for fusion reactors. Futuristic power plants have been demonstrated in proof-of-concept but are likely decades away from commercial deployment. These generators are being considered as a replacement for fossil fuels. Helium-3 exists on the Moon in greater quantities than on the Earth, and is a potential candidate for commercial mining and eventual export to Earth. Scientists estimate that there are about 10,000 tons of helium-3 on the Moon, which is enough to power

the world for thousands of years. The equivalent of a single space shuttle load, or roughly 40 tons, could supply the entire United States' energy needs for a year.

Challenges and mining system requirements. Most of the extraterrestrial bodies, such as Mars, the Moon, and asteroids, have much lower gravities than Earth. In turn, the maximum excavation forces that can be applied to a digging tool by a lander or a rover will be much lower. On the Moon, for example, an excavator will weigh six times less and apply lower digging forces. On Mars, the available force will be three times lower than on Earth. To deal with this limitation, an approach that relies less on the mass of the landing vehicle is required. In the space arena, when a trade-off must be made between lower mass and lower power, lower mass always wins. In general, mining activities on the moon pose special challenges (see **table**).

Moon dust is one of the major challenges to any activities on the Moon, including mining. Dust can impair visibility and interfere with many functions of the machinery, including its electronics and mechanical parts. Transportation of the material from the mine site to the base is a bit more straightforward and can be done by using modular systems such as rovers. A system for planetary mining should be mobile, lightweight, low-power/efficient, low-maintenance, remotely operated or fully automated, flexible with variation in ground conditions, simple and modular in design for replacing parts, and capable of performing in extreme temperature conditions.

Drilling. A number of approaches to drilling equipment and techniques for use on Mars and the Moon are being developed and tested in the United States, Canada, and Europe. Most of these light exploration drills operate in either rotary or percussive modes, and fragment rock by grinding instead of chipping. More energy (per unit material produced) is needed to grind rock than to chip, but grinding requires less downward force on the bit to initiate, and the small exploration rovers used to date can generate only low drilling forces. Heavier rovers and drills will be able to generate greater drilling forces;

Fig. 1. Photos of (*a*) proposed rock abrasion tool (RAT) [*Honeybee Robotics, Inc.*] and (*b*) coring bit tested for the Jet Propulsion Laboratory's FIDO (field integrated design and operations) rover (*NASA Jet Propulsion Laboratory*).

alternatively, drills can be anchored directly to the ground. **Figure 1** shows some drills used in planetary exploration.

Excavating. Several excavator concepts have been suggested for the Moon and Mars. Each of these designs can work well in soft regolith, consolidated regolith, or intact rock of a particular strength range, but will lose significant capability (often becoming inoperable) when applied to different material or to

Fig. 2. Rotary cutting test with prototype low-energy planetary excavator (LPE) cutterhead: (*a*) contracted configuration for hard materials; (*b*) expanded configuration for soft materials. (*NASA*)

mixed ground (combinations of soft and hard soils and rocks).

An initial focus on the less cohesive portions of the lunar regolith will reduce the scope of this problem, although one innovative cutterhead prototype is able to change the spacing of its bits in response to material strength and hardness variations. **Figure 2** shows the cutterhead of the low-energy planetary excavator (LPE) in expanded and contracted configurations, for soft and hard materials, respectively.

Several teams have built and tested prototypes of bucketwheel and bucket ladder excavators. These designs can achieve high production rates with continuous material removal (such as conveyors), though efficiency can drop rapidly in mixed ground or stronger material. Drum-type excavators have similar limitations. Drag scrapers, as shown in **Fig. 3**, are less complex and more flexible, but need to be automated to reproduce the operator skill required for effective operation. Some of the concepts for surface material excavation are shown in **Fig. 4**.

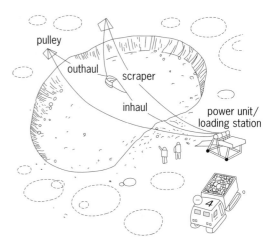

Fig. 3. Schematic view of a lunar mine using a drag scraper, an old-school but effective technology. (*From R. E. Gertsch et al., 1983*)

Excavating and handling very fine powders efficiently is difficult, yet almost 50% by weight of the lunar regolith is smaller than 50 μm across. Handling material this size is difficult to test on Earth, because the atmosphere affects how grains interact with each other. Regolith grains tend to cling together more on the Moon, which has no atmosphere. A drilling and excavation method that uses small jets of gas to mobilize the grains is being developed for the fine fraction of the regolith.

Beneficiation and processing. Beneficiation and processing transform the raw material from the mine into products that can be used. **Figure 5** shows a rover during 2008 field testing that excavates and processes regolith onboard to create oxygen. This size machine is useful for exploration and site characterization. Larger modules will be needed for full-scale production to supply a lunar outpost.

Outlook. Several spacecrafts have recently fired impactors into the surface of the Moon to study the

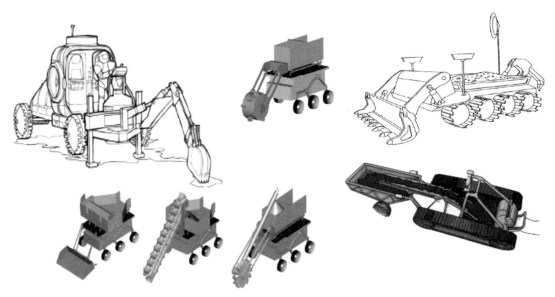

Fig. 4. Various concepts for excavating loose regolith.

Fig. 5. A high-tech regolith drilling and processing rover. (*NASA*)

regolith that was thrown out. Among other things, these missions are mapping the lunar surface and determining whether the tantalizing hydrogen signals from the Moon's polar regions are water ice.

Meanwhile, Mars is under scrutiny with a series of orbiters and rovers from the European Union and the United States. The exceptionally long-lived Spirit and Opportunity rovers, preceded by the small Sojourner rover, were equipped with small grinding-type drills. They will be succeeded by the Mars Science Laboratory, with rock-coring capability. In brief, the future efforts in the study and development of planetary mining systems will focus on material characterization by remote sensing, development of exploratory drilling systems (core drills), and refinement of the excavators for mining and construction activities.

For background information *see* ASTEROID; BASALT; COMET; MARS; MINING; MOON; REGOLITH; SOLAR SYSTEM; VOLCANIC GLASS in the McGraw-Hill Encyclopedia of Science & Technology.

Jamal Rostami; Leslie Gertsch

Bibliography. H. Benaroya and L. Bernold, Engineering of lunar bases, *Acta Astronaut.*, 62(4-5):277-299, 2008; J. Billingham, W. Gilbreath, and B. O'Leary (eds.), *Space Resources and Space Settlements*, NASA SP-428, 1979; L. Gertsch, J. Rostami, and R. Gustafson, Review of lunar regolith properties for design of low power lunar excavators, *Proceedings of 6th International Conference on Case Histories in Geotechnical Engineering*, Arlington, VA, Aug. 11-16, 2008; R. E. Gertsch, A method for mining lunar soil, in *Space Manufacturing 1983*, *Proceedings of the Sixth Conference*, Princeton, NJ, May 9-12 (A84-29852 12-12), San Diego, CA, Univelt, Inc., pp. 337-346, 1983; G. L. Kulcinski and H. H. Schmitt, Fusion power from lunar resources, *Fusion Technol.*, 21(4):2221-2229, 1992; M. F. McKay, D. S. McKay, and M. B. Duke (eds.), *Space Resources*, NASA SP-509, 1992; R. P. Mueller and R. H. King, Criteria for lunar outpost excavation, *Space Resources Roundtable: SRR IX*, Golden, CO, Oct. 26, 2007; C. R. Neal, The Moon 35 years after Apollo: What's left to learn?, *Chem. Erde: Geochem.*, 69(1):3-43, 2009; J. Rostami et al., Design and preliminary testing of low energy planetary excavator, *SME 2009 Annual Meeting*, Feb. 27-29, Denver, CO, Preprint #09-134, 2009.

Marine transportation and the environment

The transportation system affects the natural environment in many ways. Land, vegetation, and natural habitats are impacted by the physical construction of the facilities used to move people and goods, such as highways, airport runways, and seaport facilities. The vehicles or vessels that use the transportation system also impact the surrounding environment, for example, through pollutant emissions from engines, discharges into water bodies, and noise and vibration disturbances from vehicle operations. Marine transportation, as one part of the transportation system, impacts the natural environment in a variety of ways, often resulting in governmental policies and vessel design or technology strategies aimed at reducing such impacts. With more than 90% of global trade carried by sea, and with the amount of goods loaded aboard ships reaching 7.4 billion metric tons (8.2 billion short tons) in 2006, marine transportation's impact on the environment can be widespread.

Vessel-related impacts. Marine vessels can have a number of effects on the physical environment, ecosystems, and human health.

Air-pollutant emissions. The most important air pollutants associated with marine transportation are sulfur oxides (SOx), nitrogen oxides (NOx), and particulate matter (PM) originating primarily from diesel fuel. Marine vessels accounted for approximately 8.8% of all NOx emissions from the U.S. freight transportation sector in 2002, and 8.5% of all particulate matter at the 10-micrometer level (PM10). Although these percentages do not seem particularly high, the health-related problem is that many of these emissions are usually concentrated in certain locations (for example, at ports or on-board vessels).

Air pollutant emissions can affect human health in many ways. For example, the Health Effects Institute concluded that long-term exposure to diesel exhaust in a variety of occupational circumstances is associated with a 1.2- to 1.5-fold increase in the relative risk of lung cancer compared with workers classified as unexposed. Recent literature relates the health risk from diesel particulate matter to the time and proximity of exposure. Lung cancer incidence rates have been found to be higher for ship engine personnel than for deck officers, suggesting that reduced exposure in open or ventilated locations on ships, generally forward of the engine exhaust and outside of machinery spaces, reduces the risk to similar worker cohorts on the ship.

In order to reduce marine-related emissions, the U.S. government regulates marine NOx and PM emissions through engine emission standards, and SOx through fuel quality standards. Based on these standards, it is expected that PM and NOx emissions will be reduced by 80–90% over current levels by 2020. A 500 parts per million (ppm) sulfur standard was set in 2007, and a 15 ppm sulfur standard is to begin in 2012. For international shipping, an agreement under the auspices of the International Maritime Organization, signed by over 100 countries since 1998, sets emissions standards for ships involved in international trade.

International shipping is an important source of greenhouse gas emissions (primarily carbon dioxide, CO_2). Global shipping CO_2 emissions are almost 10^9 metric tons per year, and would rank sixth in the world if compared across national inventories—representing nearly twice Germany's CO_2 emissions and about 19% of annual CO_2 emissions from the United States.

Water quality, discharges, and spills. Discharging wastes or other contaminants into the natural water system can be a significant environmental problem. Water pollution in ports may originate from ship ballast water and waste, the use of ship antifouling paints (which contain chemicals to prevent marine plants and invertebrates from attaching themselves to the hull), ship oil spillage, and waterway dredging. Large ships use ballast tanks to control their displacement and trim under different loading conditions. The tanks take in seawater when additional weight is needed, and carry it over an entire voyage before discharging it in another part of the world. As noted in a report by the Organization for Economic Cooperation and Development, "the routine discharge of ballast water from marine vessels, if ballast is not segregated from cargo, introduces oil pollution at sea and in coastal waters, and can lead to introduction of nuisance species transported from the boat's origin to its destination. Shipping is a source of oil and chemical spills at port, in coastal waters, and more rarely at sea. The routine maintenance dredging of ports and inland waterways stirs up toxic sediment and frequently leads to the disposal of dredged material in the open ocean. . . ."

International and national water pollution laws and regulations establish standards and processes that must be followed in marine transportation. For example, the 1973 International Convention for the Prevention of Pollution from Ships (referred to as *MARPOL 73/78*) established a set of discharge standards and equipment requirements designed to prevent operational oil pollution.

Invasive species and habitat disruption. Invasive species are defined as "an alien species whose introduction does or is likely to cause economic or environmental harm or harm to human health." Ballast water discharge is considered a major pathway for the transfer of aquatic invasive species. Invasive species can affect aquatic ecosystems directly, or indirectly by affecting the land in ways that harm aquatic ecosystems. Invasive species represent the second leading cause of species extinction and loss of biodiversity in aquatic environments worldwide.

Large commercial and military ships may contain over a million gallons (more than 3.7 million liters) of water with up to 300 species, and it is estimated that 100 million metric tons of ballast water with exotic plankton are released daily in U. S. waters. Species attached to a hull or living in or on other species are transported among harbors. In the United States, of the most recent 171 invasive species introduced due to shipping, more were linked to hull fouling than ballast water. The risk of introducing invasive species due to hull fouling is a function of several factors: vessel speed, harbor residence time, voyage duration, surface area, last cleaning, and areas on the vessel not subject to shear, including intakes and sea chests (recessed hull areas in which intakes are located).

Besides the introduction of invasive species, additional habitat disruption can be caused through toxic discharges, vessel wakes, noise, and changes in sedimentation. In sensitive environmental areas, environmental analyses are often undertaken to determine different ways of mitigating such impacts, such as vessel speed restrictions, and regulations on discharges. With respect to invasive species, new technologies are emerging for antifouling paints, such as less toxic compounds and Teflon® hull coatings that do not allow organisms to attach themselves to the hull. Other strategies include using alternate ballast water exchange areas away from sensitive environmental areas, and treating ballast on-board with cyclonic separation, deoxygenation, filtration, ultraviolet radiation, and chemicals; or on shore by connecting to existing treatment systems.

Noise, vibration, and vessel wake. Vessel-related noise and vibration can affect a vessel's crew, sensitive receptors along the shoreline, and natural habitats. The formation and collapse of vapor-filled bubbles (cavitation) cause noise, vibration, and often rapid erosion of propeller material, especially in fast, high-powered vessels. Underwater sound propagation can be affected by many factors, including water temperature, pressure, salinity, and air bubble occurrence. The attenuation losses due to absorption in water are less than the attenuation of sound in air. In most cases, sound in water will travel farther than sound in air. Underwater sound pressure levels have been shown to cause disruption of activities in marine mammals, including but not limited to migration, breeding, care of young, predator avoidance or defense, and feeding.

The reference level for sound pressure underwater is 1 micropascal (μPa). In the United States, the Marine Mammal Protection Act (MMPA) identifies two levels of harassment of marine mammals. The MMPA Level A harassment involves injury and is triggered when received underwater sound pressure levels are at or above 180 dB relative to the 1 μPa reference level. Level B harassment refers to disturbance without injury and is triggered when the received underwater sound pressure levels are at or above 160 dB.

Vessel wakes can cause erosion of shorelines and in some cases disruption of sedimentation patterns. Primary mitigation strategies include managing vessel operations, such as speed management, as well as improved design of propeller blades and hulls.

Energy consumption. Marine transportation is the most energy efficient mode of moving commodities as measured in kilojoules per ton-kilometer, with energy efficiency increasing with size of vessel. Nevertheless, approximately 4% of the world's oil demand (146 million tons of oil equivalent) is consumed by marine transportation, with 27,000 slow-speed cargo vessels accounting for 60% of the fuel consumed. Fuel costs range from 35 to 45% of the total cost of operating a container ship, and approximately 25% of a double hull tanker's total costs. The consumption of oil for ship propulsion therefore is a concern to both ship owners and to society as a whole, and is an important sector for strategies to reduce oil consumption.

The types of strategies being considered to reduce petroleum-based energy consumption include the use of alternative fuels such as natural gas and hydrogen; alternative engine technologies, for example, electric, fuel cell, and dual fuel engines; operating strategies such as optimal speed based on fuel consumption, improved hull cleaning, and shore electric hookups when in port; and improvements in vessel design such as more energy efficient hull designs.

Port operations. Although vessel design, operations, and maintenance are important in mitigating maritime environmental impacts, the environmental consequences of port operations also are significant and may be mitigated, for example, through improved loading and unloading operations, use of shore power management and electrification, and ballast treatment.

Port operations have become an important environmental issue in many cities around the world. The primary concerns have related to the community disruption that occurs with large volumes of trucks and trains that often serve the port, as well as the environmental impacts associated with port operations (especially port-related pollutant emissions). These concerns have focused on several key points: land-use compatibility and zoning with surrounding communities that can constrain port expansion; increasing volumes of traffic serving the port (and often traversing local communities); socioeconomic impacts on the community; and potential impacts on cultural resources, such as historic structures, archeological sites and landscapes, and traditional cultural properties. In many port cities, these types of impacts are studied as part of environmental impact analyses that are undertaken when major expansion projects are being contemplated.

The types of strategies that have been considered to mitigate such impacts include improved terminal operations to reduce queues and delays, the use of barges and short sea shipping to replace trucks, the use of hydraulic hybrid port equipment, electrification of cranes, port automation, natural resource monitoring, the use of environmental management systems, and the use of pricing strategies to encourage cargo movement on more energy and environmentally efficient transportation modes. The Port of Los Angeles's water quality strategy is a good example of the types of programs that might be considered as part of a comprehensive consideration of water resources for ports. The strategy includes the following elements.

Water resources action plan. The Ports of Los Angeles and Long Beach, California, are cooperating on an action plan to protect water and sediment quality in the harbors. Both ports will integrate their existing programs and adopt new approaches, especially those that exceed regulatory requirements.

Clean water program. The port has invested in water circulation and quality models as well as in studies to determine how to improve water quality at nearby beaches. Storm drains and sewer lines have been repaired, and a proactive beach management strategy has been implemented.

Consolidated slip remediation. The port is working with the local water quality board to clean up a toxic hotspot in the harbor. Remediation may include capping sediment or the removal of sediment to a confined site.

Oil spill prevention. The port participates in a state reduction program as well as helps manage a shared rapid response network and program.

Sediment quality improvement programs. For many years, the port has remediated contaminated areas by sequestering the contaminants in confined disposal facilities or removing them to a special upland disposal area.

Watershed and storm water management. The port has conducted a water quality modeling study focusing on storm water contamination from the major storm water channel feeding into the harbor. It is the first port on the U.S. West Coast to implement a storm water treatment system at a container terminal.

Water quality monitoring. The port has monitored water quality at 31 stations in the harbor since 1967. Samples are tested monthly for dissolved oxygen, biological oxygen demand, and temperature.

Similar types of comprehensive environmental mitigation strategies are being implemented in many ports around the world that focus on the environmental impacts of marine transportation.

For background information *see* AIR POLLUTION; ENVIRONMENTAL MANAGEMENT; INVASION ECOLOGY; GLOBAL CLIMATE CHANGE; HARBORS AND PORTS; MARINE ECOLOGY; MERCHANT SHIP; PROPELLER (MARINE CRAFT); TRANSPORTATION ENGINEERING; WATER POLLUTION in the McGraw-Hill Encyclopedia of Science & Technology.   Michael D. Meyer

Bibliography. International Maritime Organization (IMO), *Annex VI of MARPOL 73/78 and NOx Technical Code*, IMO-664E, International Maritime Organization, London, 1998; K. Nauss, *Diesel Exhaust: A Critical Analysis of Emissions, Exposure, and Health Effects*, DieselNet Technical Report, Summary of a Health Effects Institute (HEI) Special Report by the HEI Diesel Working Group, 1997; H. Saarni, J. Pentti, and E. Pukkala, Cancer at sea: A case-control study among male Finnish seafarers, *Occup. Environ. Med.*, 59:613–619, 2002; K. O. Skjølsvik et al., *Study of Greenhouse Gas Emissions from Ships (MEPC 45/8 Report to International Maritime Organization on the outcome of the IMO Study on Greenhouse Gas Emissions from Ships)*, MARINTEK Sintef Group, Carnegie Mellon University, Center for Economic Analysis, and Det Norske Veritas, Trondheim, Norway, 2000.

Metal ions in neurodegenerative diseases of protein misfolding

Metal ions are essential nutrients involved in diverse biological and biochemical processes. These processes include respiration, photosynthesis, muscle contraction, cell division, and biomineralization. Metal ions are key components in electrolytes, cellular compartments, and bones. A very large percentage of proteins and nucleic acids (DNA and RNA) require metal ions to function properly.

In general, there are two classes of physiologically relevant metal ions: those that transfer electrons, such as iron, copper, molybdenum, nickel, and manganese, and those that do not, such as calcium, zinc, sodium, potassium, and magnesium. To ensure optimal levels, the body regulates all metal-ion trafficking, but levels of metal ions that participate in electron transfer (reduction-oxidation reactions) are especially tightly controlled because of the capacity of these metal ions to generate cell-damaging species. Metal ions that transfer electrons are called redox-active metal ions, and these will be discussed in this review.

Amyloidoses: Diseases of protein misfolding. Amyloid diseases that are associated with progressive brain dysfunction, such as Parkinson's, Alzheimer's, and prion disease, involve proteins that adopt a three-dimensional structure other than that required for their normal activity. A general feature of these amyloid diseases is that the misfolded protein self-associates into small clusters, which eventually deposit as solid plaques (**Fig. 1**). These protein plaques,

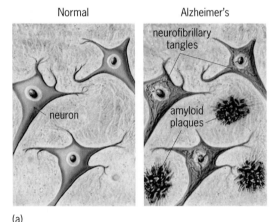

(a)

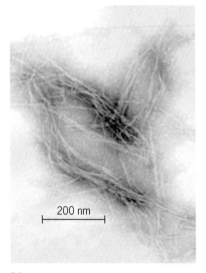

(b)

Fig. 1. Illustration of (a) amyloid plaques found in AD patients (*from http://www.ahaf.org*) and image of (b) amyloid fibrils with copper formed in vitro (*from J. W. Karr et al., Amyloid-beta binds Cu²⁺ in a mononuclear metal ion binding site, J. Am. Chem. Soc., 126(41):13534–13538, 2004*).

Amyloid diseases and associated proteins	
Disease (abbreviation)	Protein (abbreviation)
Alzheimer's (AD)	Amyloid-β (Aβ)
Parkinson's (PD)	α-synuclein (α-syn)
Prion	Prion protein (PrP)

clearly visible by pathologists, have been the hallmarks of amyloid diseases. Each amyloid disease creates plaques that contain a disease-specific protein. In Parkinson's and prion disease, it is unclear whether the plaques or precursors of the plaques, called oligomers, are the main culprit when it comes to initiating brain cell damage. In Alzheimer's disease, oligomers have been shown to be the most dangerous.

Relevant to our review are Alzheimer's disease, Parkinson's disease, and prion diseases such as Creutzfeld-Jakob disease and Kuru (humans), mad-cow disease, scrapie (sheep), and chronic wasting disease (elk, deer). Each disease is associated with a specific misfolded protein (see **table**), the physiological function of which is essentially unknown and, as a result, widely debated.

Metal ions and neurodegenerative amyloid diseases. It is known that proteins that bind or transport redox-active metal ions are involved directly in neurodegenerative diseases, including Menkes, Wilson's, and amyotrophic lateral sclerosis (ALS). Defects in these proteins lead to a metal-ion excess (or deficiency) or an inability of the brain to regulate oxidative stress. Oxidative stress is a condition in which the body cannot effectively manage (reduce) all of the oxidized species being produced. Cell damage that results from this excess of oxidized species has been implicated in heart disease, cancer, and several amyloidoses.

Although oxidative stress is not caused simply by metal-ion misregulation, metal ions contribute to it by generating reactive oxygen species (ROS), which, in turn, use up a body's antioxidants more rapidly than they can be replaced. The metal ions that participate in ROS generation include copper (Cu) and iron (Fe). An electron is transferred when the metal ion cycles between a lower (Cu^{1+} or Fe^{2+}) and a higher (Cu^{2+} or Fe^{3+}) oxidation state. The ROS (peroxide, superoxide, or hydroxyl radicals) damage DNA, proteins, and even cell membranes, destabilizing cells and leading to cell death.

The role of metal ions in Alzheimer's disease, Parkinson's disease, and prion diseases still is not entirely clear. Early work showed that metals are associated with the plaques found in Alzheimer's disease and in Lewy bodies in Parkinson's disease–afflicted brains. What is not clear, however, is whether this co-localization is a cause or an effect of the disease. There is evidence to suggest that metal ions are intimately involved in disease initiation and progression. In the case of prion disease, copper induces endocytosis of the prion protein, a process whereby the cell engulfs the prion protein so that it can be

recycled. When copper levels are changed, this recycling process does not occur, potentially leading to increased concentrations of the protein that can then be converted to the toxic form of the protein (called PrP^{Sc}). Copper levels fluctuate as we age, and decreases in copper have been hypothesized to play a role in Alzheimer's disease. This idea has been supported by experiments showing increased neuron survival in a mouse model of Alzheimer's disease when copper availability is increased. In patients with an increased risk of Parkinson's disease, it has been shown that neuron damage occurs subsequent to increases in brain iron levels, suggesting that iron misregulation is the initiating step in the disease. It is important to note that copper and iron homeostasis are intimately connected, so that disruptions in levels of one metal ion affect the other.

Molecular-level interactions. Here we will focus on copper.

Alzheimer's disease. Early work by A. I. Bush and coworkers showed that Cu^{2+} binds to amyloid-β (Aβ) protein. The identities of the protein amino acids that bind to the Cu^{2+} ion have been the center of contention for well over a decade. The general consensus is that Cu^{2+} binding occurs in the N-terminal section of the peptide. The species that directly bind Cu^{2+} include the N-terminal amine group, two histidine residues, and another species that binds via an oxygen atom (**Fig. 2***a*). Recent studies have shown that the Cu^{2+} coordination environment does not change as Aβ forms oligomers. This is consistent with work showing that the N-terminal section of Aβ is not critical for Aβ self-association. These experiments also demonstrated that Cu^{2+} could be added to and removed from Aβ fibrils (Fig. 1*b*), showing that Cu^{2+} in neuritic plaques may be exchangeable with copper in the extracellular space.

Recent work has demonstrated that Cu^{1+} also binds to Aβ. In this case, Cu^{1+} binds two imidazoles, most likely from two adjacent histidine residues, in a linear fashion (Fig. 2*b*). There are three histidines in the N-terminal region of the Aβ peptide, two of which are sequential. Considering that both the Cu^{1+} and Cu^{2+} coordination environments involve two

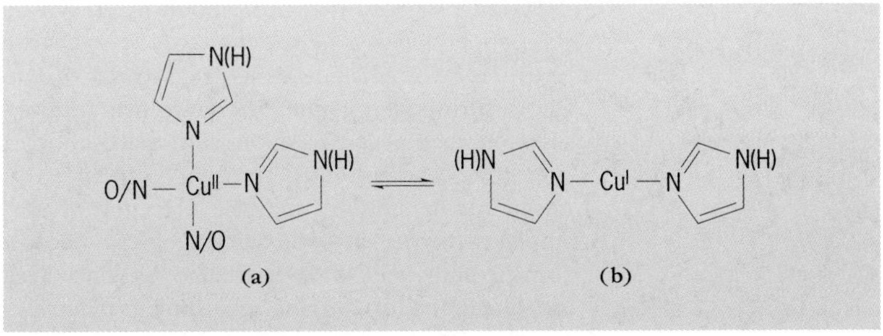

Fig. 2. The proposed Aβ binding site for (*a***) Cu^{2+} and (***b***) Cu^{1+}. The Cu^{2+} binding site consists of two imidazole nitrogens and coordination from two additional nitrogen- and/or oxygen-containing amino acids. The Cu^{1+} binding site contains a linear coordination of two imidazole nitrogens. (***J. Shearer and V. A. Szalai, The amyloid-beta peptide of Alzheimer's disease binds Cu(I) in a linear bis-his coordination environment: Insight into a possible neuroprotective mechanism for the amyloid-beta peptide, J. Am. Chem. Soc., 130(52):17826–17835, 2008)***

histidines, it is likely that the same two histidines are involved in both cases.

Understanding the Cu^{1+} binding site is important because metal-mediated redox chemistry has been suggested to be the source of oxidative damage to the neocortex of Alzheimer's disease patients. Early studies showed that incubation of $A\beta$ with Cu^{2+} and oxygen (O_2) produced ROS. These ROS would then be responsible for the damage to surrounding tissue. Contrary to this neurotoxic role, neuroprotective roles for $A\beta$ also have been proposed for a number of years. For example, it has been shown that when the $A\beta$-Cu^{2+} complex is chemically reduced to the Cu^{1+} form with ascorbate (an antioxidant found in large amounts in the brain), reoxidation of $A\beta$-Cu^{2+} by O_2 is kinetically slow compared to that of unbound copper ions. This has led to the idea that copper binding to $A\beta$ slows the rate at which Cu^{2+}/Cu^{1+} can cycle back and forth to create ROS, which in turn decreases the amount of ROS, leading to a neuroprotective role for $A\beta$.

The $A\beta$ peptide is a cleavage product of the amyloid precursor protein (APP), a transmembrane protein that is thought to be involved in copper transport. APP coordinates Cu^{2+} in an approximately square planar geometry via two histidine residues, one tyrosine residue, and one water molecule. An axial water molecule also has been proposed to coordinate the bound Cu^{2+} center in APP. Upon reduction to Cu^{1+}, the binding site does not change dramatically, which is surprising given the fact that a tetrahedral coordination environment is more likely for Cu^{1+}. APP equilibrates between monomeric and dimeric forms. The APP dimer form is hypothesized to increase production of $A\beta$ oligomers. Cu^{2+} binding to APP induces the formation of another type of APP dimer, but this APP dimer decreases $A\beta$ oligomer amounts.

Parkinson's disease. Parkinson's disease is another neurodegenerative disorder that is associated with a loss of copper homeostasis. Parkinson's disease is characterized by the presence of an increased amount of neurodegenerative protein deposits, which predominantly comprise the α-synuclein (α-syn) protein, which also contains copper. As with Alzheimer's disease, the exact role of copper in Parkinson's disease is poorly defined, although it has been proposed that Cu^{2+} modulates α-syn deposition, induces a new neurotoxic α-syn structure, or is involved in metal-mediated oxidative damage to surrounding tissue. There are many reports showing that oligomerization and aggregation of α-syn is accelerated in the presence of Cu^{2+}. One recent report indicates that the presence of Cu^{2+} amplifies the toxicity of α-syn oligomers. Because copper and iron levels fluctuate as one ages, the likelihood of forming neurotoxic metal-ion–protein complexes increases, which may explain why the elderly are more affected by Parkinson's and Alzheimer's diseases.

Prion disease. Copper has also been implicated in prion diseases. The prion protein (PrP) contains a highly conserved octarepeat region containing four

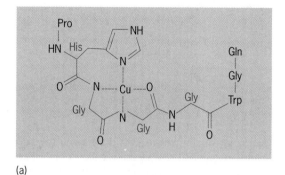

(a)

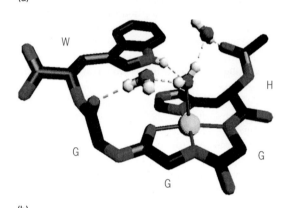

(b)

Fig. 3. Proposed Cu^{2+} binding site in PrP. (*a*) Model showing equatorial binding of Cu^{2+} within the conserved sequence of the octarepeat region (P, Pro, proline; H, His, histidine; G, Gly, glycine; G, Gly, glycine; G, Gly, glycine; W, Trp, tryptophan; G, Gly, glycine; Q, Gln, glutamine). (*b*) Three-dimensional model showing how amino acids in the minimal binding sequence bind Cu^{2+} (sphere in lower right). (*G. L. Millhauser, Copper and the prion protein: Methods, structures, function, and disease, Annu. Rev. Phys. Chem., 58:299–320, 2007*)

copies of the PHGGGWGQ amino-acid sequence, each of which binds a single Cu^{2+} ion. Spectroscopic techniques have shown that the HGGGW sequence is the minimal segment needed to coordinate a single Cu^{2+} ion in the native coordination environment. Using complementary crystallographic experiments, G. L. Millhauser and coworkers have proposed that each Cu^{2+} binding site within the octarepeat region contains an imidazole nitrogen from histidine, two backbone amides from the subsequent glycine residues, and a backbone carbonyl from the second glycine residue (**Fig. 3a**). Tryptophan is essential to recreate the native Cu^{2+} binding site because it stabilizes an axial water molecule coordinated to the Cu^{2+} ion (Fig. 3b). This is a unique model for a protein's metal-binding site because most proteins coordinate the metal ion through side-chain residues rather than backbone interactions. A fifth and even a sixth Cu^{2+} binding site has been observed in PrP outside of the octarepeat region. These additional Cu^{2+} binding sites are significant because they might be involved in cooperative binding of Cu^{2+} ions. Even though the exact biological role of PrP is still not known, these findings lead to the idea that PrP is a "cuprostat" responsible for regulating the amount of copper in the central nervous system. Additionally, recent work has shown that under biologically relevant conditions, Cu^{2+} inhibits the formation of PrP fibrils, suggesting that perhaps copper regulates the structural conversion of normal PrP to its neurotoxic scrapie form.

Outlook. The redox-active metal ions copper and iron are essential for the body's normal biological functions. When misregulated, these metal ions can have disastrous effects. The biochemical events that lead to misregulation of these redox-active metal ions have not been fully delineated, but such misregulation is thought to precipitate neurotoxic events in several of the most prevalent neurodegenerative diseases plaguing society. In all of these diseases, further work to understand where and how metal-ion binding affects the structure and function of proteins, both in a test tube and in animal model systems, will bring us closer to unraveling the molecular-level involvement of metal ions in Alzheimer's disease, Parkinson's disease, and prion diseases and designing therapies to combat them.

[Disclaimer: The Alzeimer's Association funded some of the research described here.]

For background information *see* ALZHEIMER'S DISEASE; AMYLOIDOSIS; BIOINORGANIC CHEMISTRY; COPPER; PARKINSON'S DISEASE; PRION DISEASE; PROTEIN; PROTEIN FOLDING in the McGraw-Hill Encyclopedia of Science & Technology.

Jesse W. Karr; Veronika A. Szalai

Bibliography. D. Berg, Disturbance of iron metabolism as a contributing factor to SN hyperechogenicity in Parkinson's disease: Implications for idiopathic and monogenetic forms, *Neurochem. Res.*, 32(10):1646–1654, 2007; O. V. Bocharova et al., Copper(II) inhibits in vitro conversion of prion protein into amyloid fibrils, *Biochemistry*, 44:6776–6787, 2005; X. Huang et al., Cu(II) potentiation of Alzheimer Aβ neurotoxicity: Correlation with cell-free hydrogen peroxide production and metal reduction, *J. Biol. Chem.*, 274(52):37111–37116, 1999; J. W. Karr, L. J. Kaupp, and V. A. Szalai, Amyloid-β binds Cu^{2+} in a mononuclear metal ion binding site, *J. Am. Chem. Soc.*, 126(41):13534–13538, 2004; J. W. Karr and V. A. Szalai, Cu(II) binding to monomeric, oligomeric, and fibrillar forms of the Alzheimer's disease amyloid-β peptide, *Biochemistry*, 47(17):5006–5016, 2008; G. K. Kong et al., Structural studies of the Alzheimer's amyloid precursor protein copper-binding domain reveal how it binds copper ions, *J. Mol. Biol.*, 367(1):148–161, 2007; G. L. Millhauser, Copper and the prion protein: Methods, structures, function, and disease, *Annu. Rev. Phys. Chem.*, 58:299–320, 2007; A. E. Oakley et al., Individual dopaminergic neurons show raised iron levels in Parkinson disease, *Neurology*, 68(21):1820-1825, 2007; J. Shearer and V. A. Szalai, The amyloid-β peptide of Alzheimer's disease binds Cu^I in a linear bis-his coordination environment: Insight into a possible neuroprotective mechanism for the amyloid-beta peptide, *J. Am. Chem. Soc.*, 130(52):17826–17835, 2008; J. A. Wright, J. X. Wang, and D. Brown, Unique copper-induced oligomers mediate alpha-synuclein toxicity, *FASEB J.*, 23(8):2384–2393, 2009.

Methane on Mars

The atmospheres of planets in the solar system vary greatly in composition. The gas giants (Jupiter, Saturn, Uranus, and Neptune) are dominated by hydrogen (H_2) captured during their formation, so the principal trace elements in their atmospheres appear as fully hydrogenated gases (such as CH_4, NH_3, and H_2O). Atmospheres of the terrestrial planets (Venus, Earth, and Mars) are deficient in hydrogen and thus are strongly oxidizing. Consequently, hydrocarbon gases should be absent from their atmospheres if atmospheric processes alone govern their production and destruction.

The presence of abundant methane in Earth's atmosphere ($\sim$1.6 parts per million) requires sources other than atmospheric chemistry. Living systems produce more than 90% of Earth's atmospheric methane; the balance is of geochemical origin. On Mars, methane has been sought for nearly 40 years because of its potential biological significance, but it was detected only recently. Its distribution on the planet is found to be patchy and to vary with time, suggesting that methane is released recently from the subsurface in localized areas, and is then rapidly destroyed.

The atmosphere of Mars is composed primarily of carbon dioxide (CO_2, 95.3%) along with minor amounts of nitrogen (N_2, 2.7%), carbon monoxide (CO, 0.07%), oxygen (O_2, 0.13%), water vapor (H_2O, 0–300 ppm), and radiogenic argon (1.6%); other species and fully hydrogenated gases such as methane (CH_4) are rare. Methane production by atmospheric chemistry is negligible, and its lifetime against removal by photochemistry is estimated to be several hundred years. Thus, the presence of significant methane would require recent release from subsurface reservoirs. The ultimate origin of this methane is uncertain, but either biological or geochemical production is possible.

Before 2000, searchers obtained sensitive upper limits for methane by averaging over much of Mars's dayside hemisphere, using data acquired with Mars-orbiting spacecraft (*Mariner 9*) and Earth-based and space-based observatories (Kitt Peak National Observatory, *Infrared Space Observatory*). These negative findings suggested that methane should be searched at higher spatial resolution since the local abundance could be significantly larger at active sites. Since 2001, searches for methane have emphasized spatial mapping from terrestrial observatories and from Mars orbit (*Mars Express*).

Ground-based spectroscopic observations. The first definitive detections and spatial maps were achieved using high-dispersion infrared echelle spectrometers at three ground-based observatories (NASA's IRTF and the W. M. Keck Observatory, both on Mauna Kea in Hawaii, and Gemini South, in Chile). For a typical observation, the spectrometer's long entrance slit is held to the central meridian of Mars (**Fig. 1a**) while spectra are taken sequentially in time. For each snapshot in time, spectra are acquired simultaneously at contiguous positions along the entire slit length,

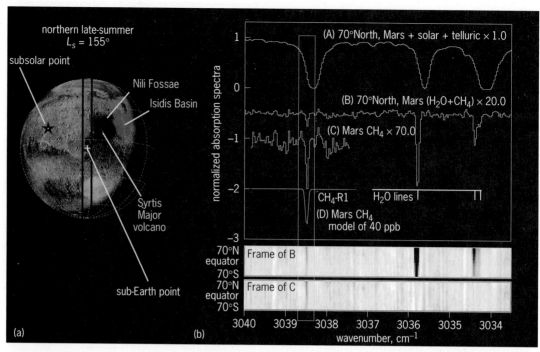

Fig. 1. Observation of Mars in Northern Summer, 2003, when the solar longitude (L_s) was 155°. (*a*) Aspect of the planet. The subsolar and sub-Earth points and several conspicuous geologic features are marked, and the spectrometer entrance slit (NASA-IRTF) is shown to scale (vertical lines). With this geometry, spectra are acquired simultaneously for 35 spatial footprints along the slit. (*b*) Spectrum acquired at 70°N latitude is shown (top, A). Spectral signatures of Mars methane (CH_4) and water (H_2O) appear after subtraction of terrestrial and solar signatures. The depth of the spectral lines is shown in a grayscale representation extending over the latitude range 70°N to 70°S on the planet (bottom, two panels). Spectral lines of water (three lines) and methane (one line) are strongest in the northern hemisphere (frame of B). The latitudinal distribution of methane alone appears (frame of C) after subtracting a synthetic model of H_2O. Two other spectral lines of methane (R0 and P2) and several other lines of water were detected and mapped at nearby spectral settings.

sampling latitudinally resolved spatial footprints on the planet (35 footprints, for the geometry shown in Fig. 1).

The most sensitive searches are made at wavelengths near 3.3 micrometers, where methane has a strong vibrational spectral band and Mars is seen mainly in reflected sunlight. The collected spectra contain (solar) Fraunhofer lines, spectral lines of Mars's atmospheric gases (mainly H_2O, CH_4, and CO_2), and lines of gases in Earth's atmosphere. Removing the solar and terrestrial components from the composite spectrum reveals spectral lines of gases in Mars's atmosphere (Fig. 1*b*). One line of CH_4 and three distinct lines of H_2O are seen in this example. [Two other lines of methane (R0 and P2) and several other lines of water were detected at nearby spectral settings.] The spatial distributions of methane and water are then mapped in latitude by combining these signatures in a grayscale pattern (Fig. 1*b*, bottom). Combining a temporal sequence of these individual snapshots forms latitude-longitude maps.

A series of such measurements taken in Mars's Northern Summer in 2003 found that methane varied significantly in abundance over the surface of Mars, and in time (**Fig. 2**). The most compelling results from these searches are (1) the unambiguous detection of multiple spectral lines of methane, (2) evidence for spatial variations that imply active release in discrete regions, and (3) evidence for seasonal variations that imply a CH_4 lifetime of months rather than the 300-plus years implied by photochemistry. The short lifetime may be the result of reaction with strong oxidants such as peroxides located in the soil or on airborne dust grains.

Methane appears notably enriched over several localized areas (Fig. 2): A (east of Arabia Terra, where water vapor is also greatly enriched), B_1 (Nili Fossae), and B_2 (the southeast quadrant of Syrtis Major). Unusual enrichments of hydrated minerals (phyllosilicates) were identified in Nili Fossae from *Mars Express* and from the *Mars Reconnaissance Orbiter* (Fig. 2). The observed morphology and mineralogy of this region suggests that these bedrock outcrops might be connecting with reservoirs of buried material rich in volatile species. The characteristic arcuate ridges in the southeast quadrant of Syrtis Major were interpreted as consistent with catastrophic collapse of that quadrant, perhaps from interaction with a volatile-rich substrate. This could provide conduits connecting the substrate with the atmosphere. When averaged over all latitudes and the Mars year, spectral data from *Mars Express* also imply an enhancement in methane in this longitude range.

Geochemical production of methane. Methane is produced in Earth's crust by several processes, including (1) oxidation of peridotite (a magnesium-iron silicate mineral) to serpentine, when hot rock combines with water; and (2) the production of pyrite (an iron sulfide). Both processes release abundant H_2 that can combine with carbon (from CO_2 or carbonate), releasing methane. Process 1 is observed in mud volcanoes and in low-temperature hydrothermal vents. Process 2 is observed in the laboratory under anoxic conditions. A third process, the reaction of subducted CO_2 and H_2O in Earth's mantle magma, produces methane that emerges as a minor effluent in volcanic emissions (in which H_2O, CO_2, and SO_2 are the principal gases released). There is no independent evidence that such processes are active on Mars today; for example, no excess SO_2 emission and no geothermal anomalies (hot spots) are known. However, the release of methane could be the signature of such processes, whether currently active or representing stored methane produced in an earlier epoch.

Habitable biomes. On Earth, cold-loving and cold-tolerant microorganisms have been discovered and in some cases resuscitated from cores into permafrost, subpermafrost, and ice deposits that range in age from 40,000 to millions of years. Some of the microorganisms are methanogens, and evidence suggests that they are actively producing CH_4 at temperatures below the freezing point of water. Experiments performed on other isolates recovered from these deposits demonstrate that they are metabolically active at temperatures down to -15 to $-20°C$ (5 to $-4°F$). When coupled with field and laboratory studies of permafrost and hydrates and with Mars's crustal properties, these findings suggest the Martian crust can be subdivided vertically into four potentially habitable biomes.

Surface photic zone. At depths of less than 1 m (3.3 ft), this zone is potentially habitable by microorganisms if they can withstand the withering ultraviolet and cosmic-ray flux. Such organisms could release methane gas.

Subphotic dissociation zone. At depths from 1 m to about 100 m (330 ft), this zone is potentially habitable by rock-eating microorganisms that utilize atmospheric gases (such as H_2 and CO as electron donors, and O_2 and CO_2 as electron acceptors) and produce CH_4 and water ice. Gas hydrates of CH_4, H_2S, CO_2, O_2, and N_2 are not stable in this zone but dissociate into ice and gas bubbles that diffuse through the photic zone and thence into the atmosphere, so the subphotic dissociation zone could also contribute a flux of biomarker gases (for a 100-m-thick zone the gas transport time is on the order of one Martian year).

Permafrost zone. In this zone, extending from greater than 1 m to less than 2000 m (6600 ft), habitable oases may be restricted to lenses of hypersaline solutions isolated from the Martian atmosphere and in lithostatic pressure equilibrium. Although methanogens are known on Earth that could survive in such an environment, on Mars the CH_4 they produce would be trapped within the permafrost zone as hydrate.

Subpermafrost zone. This zone, extending from greater than 2000 m to greater than 6000 m (19,700 ft), is acknowledged as having the great-

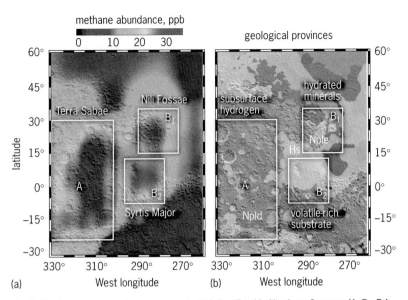

Fig. 2. Regions where methane appears notably localized in Northern Summer (A, B₁, B₂), and their relationship to mineralogical and geomorphological domains. (*a*) Observations of methane near the Syrtis Major volcanic district. Region A (Terra Sabae) is rich in subsurface hydrogen, B₁ (Nili Fossae) displays minerals that formed in water, and B₂ (southeast quadrant of Syrtis Major shield volcano) overlaps a region that appears to have collapsed. (*b*) Geologic map superimposed on topographic shaded relief. Distinct terrain types are coded by shading. The most ancient terrain [dissected (Npld) and etched (Nple) plateau material] is Noachian in age (~3.6–4.5 billion years old, an era when Mars was wet), and is overlain by volcanic deposits of Hesperian age (light gray) from Syrtis Major (Hs, ~3.1–3.6 billion years old). (*After M. J. Mumma et al., Strong release of methane on Mars in Northern Summer 2003, Science, 323:1041–1045, 2009*)

est potential for habitability because liquid water is readily available to potential life forms. The lower permafrost boundary may vary in depth according to local conditions. Even though this zone is completely sealed from Mars's atmosphere (except perhaps at scarp faces under special conditions), radiolytic, mineral hydration/oxidation, and hydrothermal processes could liberate hydrogen and thus sustain a biosphere of anaerobic microbes. In some Earth-analog sites, radiolysis of ice and water provides a source of H_2, H_2O_2, and O_2 at a rate that sustains a subsurface biosphere.

Biomarkers. Trace gases in the Martian atmosphere that could potentially represent biomarkers could originate from the surface photic and the subphotic dissociation zones, particularly at low latitudes, from the permafrost zone wherever tectonic or impact related erosion and depressurization has occurred recently or brine seepage is now occurring, or from the subpermafrost zone wherever it has been breached by recent fracturing due to impacts or by geothermal heating. Seasonal release at scarps may occur at favorable locations (**Fig. 3**).

Pointers to habitable subsurface regions. Three major indicators help guide the search for regions of possible extant life on Mars: The remanent magnetic field (**Fig. 4a**) identifies regions where signatures of Mars's magnetic dynamo are still preserved. The magnetic signature is prominent in the ancient highlands, but is missing in regions subjected to flood volcanism (Tharsis, Elysium, northern plains) and in areas modified by major cratering events (Isidis, Argyre, Hellas). If established in early Mars, methane

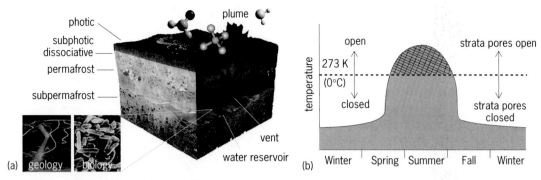

Fig. 3. Subsurface structure envisioned for Mars. (*a*) The subpermafrost region may vent gases as pores open and close seasonally at scarps. (*b*) Schematic showing possible mechanism for controlling seasonal release of gases. (*After M. J. Mumma and NASA*)

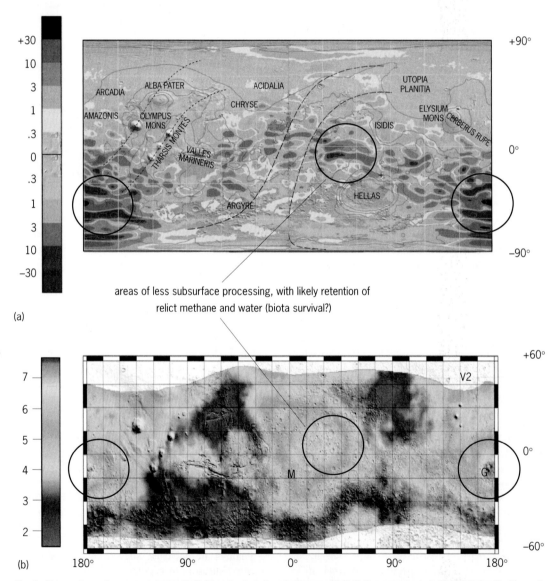

Fig. 4. Comparison of remanent magnetic field and subsurface hydrogen. (*J. E. P. Connerney et al., Tectonic implications of Mars crustal magnetism, Proc. Natl. Acad. Sci., 102(2):14970–14975, 2005*) (*a*) Remanent magnetic field measured by *Mars Global Surveyor* (remanent magnetic signature from the first few kilometers of the crust), which generally coincides with Noachian terrain, and may mark the least modified subsurface domains on Mars. Shades encode the change in radial magnetic field (in nanotesla per latitude degree) observed within the latitude/longitude bin (1° × 1°) represented by each pixel. (*W. P. Boynton et al., Concentration of H, Si, Cl, K, Fe, and Th in the low-and mid-latitude regions of Mars, J. Geophys. Res., 112, E12S99, doi:10.1029/2007JE002887, 2007*) (*b*) The abundance of subsurface hydrogen measured by *Mars Odyssey* is coded as water-equivalent concentration by mass (%). Locations of the *Viking* (V1, V2), *Pathfinder* (PF), *Spirit* (G), and *Opportunity* (M) landed spacecraft are shown. Black circles identify regions showing high hydrogen, high magnetization, and extreme vertical relief. These may be especially favorable locations for active vents where gases of biogenic and/or geochemical origin may be released into the atmosphere. (*After M. J. Mumma and NASA*)

reservoirs may survive in favored places (such as those showing high remenant magnetic signatures), along with any coexisting local biocommunities. The presence of subsurface hydrogen concentrations on Mars inferred by *Mars Odyssey* is often interpreted as buried water ice, but complex hydrocarbons cannot be excluded (Fig. 4*b*).

The regions of high subsurface hydrogen are roughly congruent with those of high magnetization (black circles, Fig. 4), suggesting that both signatures survived intact since Hesperian times (3.5–1.8 billion years ago), making them prime regions to search for vents where methane, water, and other gases may be released seasonally. The congruent regions are characterized by areas of strong vertical relief where scarps penetrate to depths of several kilometers, or deeper. Deep canyons and craters are of special interest because they might access the subpermafrost habitable zone (Fig. 3).

The regions of methane release (Fig. 2) date from the earliest epochs of Mars history, Noachian (4.5–3.5 billion years ago) and Hesperian (~3.1–3.6 billion years ago), when the planet was wet and volcanism was active. However, the release of methane is itself not sufficient to identify its origin. Quantitative release rates for methane and higher hydrocarbons (C_nH_{2n+2}) and their (C, H) isotopologues, along with sulfuretted species (H_2S, CH_3SH, SO_2) and other gases (N_2O, NH_3), are needed to assess biogenic versus geologic origins.

For background information *see* ASTROBIOLOGY; ASTRONOMICAL SPECTROSCOPY; MARS; METHANE; METHANOGENESIS (BACTERIA); PERMAFROST; PLANET; ROCK MAGNETISM in the McGraw-Hill Encyclopedia of Science & Technology.   Michael J. Mumma

Bibliography. S. K. Atreya, P. R. Mahaffy, and Ah-S. Wong, Methane and related trace species on Mars: Origin, loss, implications for life, and habitability, *Planet. Space Sci.*, 55:358–369, 2007; B. K. Chastain and V. Chevrier, Methane clathrate hydrates as a potential source for martian atmospheric methane, *Planet. Space Sci.*, 55:1246–1256, 2007; A. Geminale et al., Methane in Martian atmosphere: Average spatial, diurnal, and seasonal behaviour, *Planet. Space Sci.*, 56:1194–1203, 2008; H. Hiesinger and J. W. Head III, The Syrtis Major volcanic province, Mars: Synthesis from Mars Global Surveyor data, *J. Geophys. Res.*, 109:E01004, 2004; M. J. Mumma et al., Strong Release of Methane on Mars in Northern Summer 2003, *Science*, 323:1041–1045, 2009; J. Mustard et al., Hydrated silicate minerals on Mars observed by the Mars Reconnaissance Orbiter CRISM instrument, *Nature*, 454:305–309, 2008.

Methicillin-resistant Staphylococcus aureus in the horse

Staphylococcus aureus is a gram-positive bacterium (**Fig. 1**) that can be found commonly in or on various animal species, including humans. It is an opportunistic pathogen, that is, a bacterium that normally coexists peacefully with its host, living on the skin or

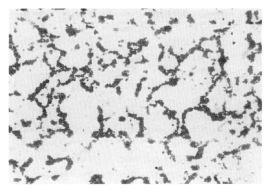

Fig. 1. Gram stain appearance of *Staphylococcus aureus*. Note the dark color of the small round bacteria and the presence of clumps of bacteria. (*Image: Dr. Richard Facklam, U.S. Centers for Disease Control and Prevention*)

in areas such as the nasal passages or intestinal tract without causing problems, but one that can cause disease in certain situations.

The presence of *S. aureus* at a body site without causing problems is termed colonization. Colonized individuals are also called carriers. This is a normal state and individuals can be colonized with *S. aureus* for long periods of time. Colonization differs from infection, which involves penetration of host barriers by the bacterium, and may cause a state of disease. The main site of *S. aureus* colonization in humans and horses is the nose, but it can also be found in the intestinal tract or on the skin.

One major problem with *S. aureus* is its tendency to become resistant to antimicrobials (antibiotics), and some *S. aureus* strains can be resistant to most available antimicrobials. One particularly concerning aspect is methicillin-resistance. Methicillin (also called meticillin) is an antimicrobial in the beta-lactam family. This family includes penicillins, cephalosporins, and carbapenems, which are very important drug classes commonly used to treat various infections. Methicillin-resistant *S. aureus* (MRSA) was identified in people not long after methicillin was released. Resistance to methicillin itself is not the concern, though, as this drug is no longer in use. However, in addition to being resistant to methicillin, MRSA is resistant to all other beta-lactam antimicrobials, which greatly limits treatment options. Furthermore, MRSA strains often acquire other genes that make them resistant to additional antimicrobial classes, with some strains being resistant to almost all available drugs.

MRSA is a tremendous problem in human medicine, and is a leading cause of infection and death in hospitalized people. Over the past 10–15 years, MRSA has emerged in the general population as well, and MRSA is now an important cause of infections in people outside of hospitals. It is perhaps this movement of MRSA into people in the general population that has resulted in the emergence of MRSA in companion animal species such as horses.

While MRSA is called a "superbug" and elicits a lot of concern, it is still *S. aureus*. Just like its methicillin-susceptible relatives, MRSA can colonize the nose and other sites, and live there uneventfully. Colonization is still of concern, though, because colonized individuals can transmit MRSA to others, and colonized individuals are at increased risk of MRSA infection in certain situations.

Emergence of MRSA in horses. Despite MRSA being recognized in humans since the 1960s, it was not until 1997 that MRSA was first reported in horses. Since then, the number of reports of MRSA in horses has steadily increased, and it is clear that many horses become infected every year. Infected horses have been reported in the United States, Canada, the United Kingdom, Ireland, Austria, Germany, France, Belgium, and Japan, and it is likely that MRSA is actually present in horses in most developed countries.

As in people, horses can be carriers of MRSA without any sign of infection. While colonized horses are more likely to develop an MRSA infection in certain situations, such as after admission to a veterinary hospital, most do not ever develop problems. As with humans, colonized horses pose a risk to others (humans and horses) because they are able to transmit MRSA. Studies of MRSA colonization in healthy horses have reported rates of 0–11%. Colonization rates are likely influenced by various factors, including the type of horse, farm infection control practices, antibiotic use, contact with the veterinary healthcare system, contact with livestock, and contact with people. Higher rates can be encountered on individual farms, with over 50% of horses carrying MRSA in some instances.

MRSA can cause a wide range of infection in horses, from mild skin or soft tissue infections to serious (and sometimes fatal) infections such as pneumonia, joint infection, and bloodstream infections. Infections occur in horses in veterinary hospitals and on farms, but outbreaks are probably more common in veterinary hospitals. Infections caused by MRSA are no different than those caused by methicillin-susceptible *S. aureus* and the majority of infections are treatable; however, prompt recognition of MRSA is critical to ensure proper treatment since initial treatment choices for infections would often involve antimicrobials to which MRSA is resistant. Risk factors for MRSA colonization or infection have not been well explored, but a few studies have reported antimicrobial treatment as a risk factor. This is not surprising and is consistent with what is known in humans, but it is a good indicator of the need for prudent use of these drugs in horses.

Comparison of human and equine MRSA. Molecular fingerprinting tools have been very useful for helping to understand the emergence and spread of MRSA in animals. Most initial reports of MRSA in horses involved one specific strain. This strain has various names depending on the typing method used and the geographic location, but it is often called CMRSA-5 or USA500 (**Fig. 2**). This is a recognized human strain that has been found in horses from numerous countries. It is suspected that this strain is somehow better adapted for survival in horses than other strains. Recently, another strain of MRSA has been found to predominate in horses in some European countries.

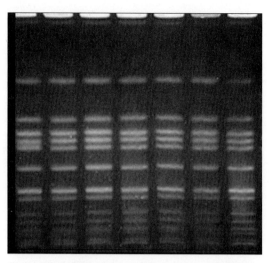

Fig. 2. Pulsed-field gel electrophoresis image of MRSA isolated from humans and horses on the same farm. Note that the banding patterns in each column, which represent a single MRSA isolate, are the same.

This strain, called ST398, has been linked to pigs and cattle, and is spreading to humans.

Transmission of MRSA between humans and horses. A major concern regarding the emergence of MRSA in horses has been recognition that the bacterium can be transmitted between humans and horses. The fact that the predominant MRSA strain in horses is a recognized human strain strongly suggests that equine MRSA was ultimately human in origin. However, while it is also clear that horses may have first been infected by humans, horses are able to spread MRSA back to humans. The first evidence that MRSA could be transmitted between species was a report where the same strain of MRSA was found in horses and a veterinary surgeon that had operated on them. The thought at that time was that MRSA was spread from the surgeon to the horses. Later investigations revealed that MRSA could be spread in both directions, both on farms and in veterinary clinics. An interesting aspect of one study of horses and their owners in Canada and the United States was the following: on every farm where MRSA was found in a horse, the same strain was found in at least one person. That study also reported MRSA colonization in approximately 10% of horse owners. Considering it is estimated that only 1–3% of people in the general population are MRSA carriers at any given time, these results indicate that there is a close link between MRSA in humans and horses and that horse ownership may be a risk factor for MRSA colonization and perhaps infection.

Further support of the potential for horses to transmit MRSA to humans has been provided by studies of equine veterinarians. The first such study reported MRSA colonization in 15% of equine veterinarians at a veterinary internal medicine conference in the United States. A follow-up study reported a similarly high (10%) colonization rate in equine veterinarians at an international equine veterinary conference, while another study reported colonization in 17% of people at a veterinary surgery specialty conference.

provided by the complementary microorganism are made available from another source. Under these conditions, the two microorganisms can function independently. Thus, there is no sense of obligation in cooperative relationships.

Commensalism is important in many aspects of ecology. An important example is nitrification, that is, the oxidation of ammonium ion to nitrite by microorganisms such as *Nitrosomonas* and the subsequent oxidation of the nitrite to nitrate by *Nitrobacter*. Another important commensalistic process occurs when anaerobic microorganisms oxidize ethanol to acetate, releasing hydrogen that can be used by hydrogen-utilizing methane producers, as shown below. This commensalistic process, involv-

$$
\begin{array}{ccc}
\text{ethanol} & & CO_2 \\
\downarrow & e^- + H^+ & \downarrow \\
& (H_2) & \\
\text{acetate} & & CH_4
\end{array}
$$

ing interspecies hydrogen transfer, can also have cooperative aspects: if the hydrogen produced during ethanol oxidation is utilized by the second organism, this decreases the end product inhibition that would occur if the hydrogen produced by the first organism were to accumulate, speeding up the entire process.

Symbiotic interactions with negative effects include parasitism, predation, amensalism, and competition. Parasitism is one of the more complex microbial interactions, and the line between parasitism and predation is often difficult to establish. Parasitism usually involves a longer-term interaction, where one member of a pair, the parasite, benefits from resources obtained from the host. Bacterial parasites also can manipulate their hosts; these include bacteria that control the sex of insects (*Wolbachia* is an excellent example). The viruses provide excellent examples of varied degrees of parasitism. As a part of their role in nature as "parasites," viruses are now recognized as playing major roles in facilitating microbial evolution by gene transfer. When the phage genome remains within the bacterium without causing lysis in what now is described as a "prophage" state, it can confer new attributes on the host bacterium. Toxin production by *Corynebacterium diphtheriae* and *Prochlorococcus* maintenance of photosynthesis-related genes are important examples.

Predation involves one organism living off another, usually involving attack or capture that leads to the death of the prey; this usually occurs over short time periods. Predation, however, can have many positive effects on the populations of both the predator and the prey. This includes driving the microbial loop (mineral recycling in waters and soils) as protozoa graze bacteria and release their nutrients, protection of ingested prey from heat and damaging chemicals, and actually increasing the pathogenicity of disease-causing microbes such as *Legionella*. Predatory bacteria also are active in nature. Several

of the most interesting examples include *Bdellovibrio*, *Ensifer*, *Vampirococcus*, and *Daptobacter*. Predation also can involve active acquisition of homologous DNA to improve genetic fitness and facilitate serotype conversion (that is, conversion between various serological types of intimately related microorganisms, distinguished on the basis of antigenic composition), as has been observed with *Streptococcus pneumoniae*, an important human pathogen.

In addition, predation can have important secondary effects. The surviving prey usually are smaller, less palatable, and less active. An interesting example of predation is kleptochloroplasty, which is the "stealing" of algal chloroplasts (cell plastids needed for photosynthesis) by protozoa. After capturing and digesting the alga, the protozoan maintains the chloroplast to provide valuable photosynthate (a chemical product of photosynthesis) for its growth.

Amensalism describes the negative effect that one microorganism can have on another. A classic example of amensalism is the production of antibiotics that result in the inhibition or death of a susceptible microorganism; these are widely used (and often misused) to attempt to control infections in humans and animals. The attine ant–fungal mutualistic relationship involves an amensalistic interaction: antibiotic-producing bacteria (actinomycetes) are maintained in the fungal garden system to control *Escovopsis*, which is a persistent parasitic fungus that can destroy the fungal garden upon which the attine ant depends.

Other important amensalistic relationships involve microbial production of bacteriocins (antibacterial substances produced by various strains of gram-negative bacteria that may inhibit the growth of other strains of the same or related species) and more specific colicins (bacteriocins produced by coliform bacteria, such as *Escherichia coli*), as well as the high-molecular-weight antimicrobial protein "killer factors" released by yeasts and protozoans. Antibacterial peptide effector molecules, called cecropins in insects and defensins in mammals, also play significant roles in innate immunity.

Competition can lead to two possible outcomes: (1) one of the interacting organisms will dominate in the use of the particular limiting resource (a nutrient, space, and so on), leading to the exclusion of the other organism; or (2) both organisms will coexist at lower levels with the need to share a single limiting resource.

It is important to recognize that the quality of these symbiotic relationships can quickly change. What had been a positive interaction can suddenly turn negative, and the relationship will then be ended or markedly curtailed. Often, these interactions can allow "cheaters" to make a living—these microbes benefit from interactions without making sufficient contributions "to carry their own weight." Cannibalism and fratricide can also occur when things no longer go well for some of the interacting microbes.

Energy sources and microbial interactions. The energy sources that are available also affect the

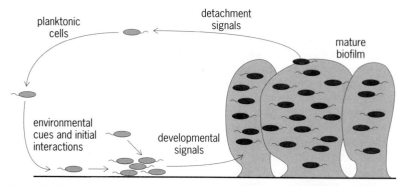

Fig. 2. Biofilms as a site for microbial interactions: gradients of nutrients, waste products, and physical conditions. Microbes in surface-attached states have undergone profound changes from those found in the planktonic state. (*Reproduced with permission from G. O'Toole, H. B. Kaplan, and R. Kolter, Biofilm formation as microbial development, Annu. Rev. Microbiol., 54:49–79, 2000*)

distribution of microbes and their interactions. In marine environments, where light is the main energy source available in surface waters, the photosynthetic microbes at the base of the food web normally either are separated and isolated, or are present as small, dispersed microflocs (tiny masses of nonsettleable microorganisms and particles) or assemblages with limited direct interactions. In biofilms (microbial communities, enveloped by the extracellular biopolymer that these microbial cells produce, that adhere to the interface of a liquid and a surface), extensive physical structures of the attached microbes can form on living and nonliving surfaces, creating distinct nutrient, waste product, light, and chemical gradients. Within biofilms, the entire range of symbiotic interactions can occur (**Fig. 2**), often involving quorum sensing, where microbes sense their own population by measuring levels of autoinducer signal molecules.

Microbial interactions can also occur inside plants and animals; every cell of a plant or animal can be penetrated by some type of microbe. In plants, such cell penetration processes can lead to important beneficial associations, such as mycorrhizal (fungal-plant) interactions; in addition, bacterial penetration of plant cells can lead to plant tissue destruction, as occurs with *Erwinia* and related plant pathogens. In animals, these interactions can involve the ingestion of invading bacteria by neutrophils, monocytes, macrophages, and dendritic cells. Dendritic cell capture of microbial pathogens is an important part of cellular immunological responses.

Studying interactions in plants, animals, and microbes. The techniques being used to study microbial interactions are changing. It is now possible to measure physiological changes in individual microbes and to measure changes in the physical environment on the scale of individual cells. Direct observational techniques can be combined with chemical, enzymatic, and molecular measurements to provide new views of what actually is happening at the level of individual microbes. This is useful because individual microbes have unique characteristics, termed microheterogeneity, that were missed by the population-level analyses employed in the past.

Techniques are now available to study microbial interactions at the same scale of resolution as used with plants and animals—the individual organism.

For background information *see* BACTERIA; BACTERIAL PHYSIOLOGY AND METABOLISM; BACTERIOLOGY; ECOLOGICAL COMPETITION; ECOSYSTEM; ENVIRONMENT; GENOMICS;; MICROBIAL ECOLOGY; MICROBIOLOGY; MUTUALISM in the McGraw-Hill Encyclopedia of Science & Technology.   Donald A. Klein

Bibliography. A. Casadevall and L.-A. Pirofski, Host-pathogen interactions: Concepts of microbial commensalism, infection, and disease, *Infect. Immun.*, 68:6511–6518, 2000; D. A. Klein, Seeking microbial communities in nature: A postgenomic perspective, *Microbe*, 2:591–595, 2007; M. M. M. Kuypers and B. B. Jorgensen, The future of single-cell environmental microbiology, *Environ. Microbiol.*, 9:6–7, 2007; L. Margulis and M. J. Chapman, Endosymbiosis: Cyclical and permanent in evolution, *Trends Microbiol.*, 6(9):342–346, 1998; M. Travisano and G. J. Velicer, Strategies of microbial cheater control, *Trends Microbiol.*, 12:72–78, 2004; C. W. Waters and B. L. Bassler, Quorum sensing: Cell-to-cell communication in bacteria, *Annu. Rev. Cell Dev. Biol.*, 21:319–346, 2005; J. B. Xavier and K. R. Foster, Cooperation and conflict in microbial biofilms, *Proc. Natl. Acad. Sci. USA*, 104:876–881, 2007.

Microbial survival mechanisms

Microorganisms have long intrigued, terrified, and baffled human beings. Although most microorganisms are not pathogenic to humans, a few are, and they receive most of the attention. More and more researchers now believe that it is how these microorganisms live that can really matter in terms of the diseases they cause.

During the course of evolution, microbial organisms have evolved ways of surviving and evading host defense mechanisms. Many of these ways are found throughout the microbial world. Currently, several are receiving a great deal of investigation by microbiologists.

Evasion of host defenses by viruses. In most cases, the pathology arising from a viral infection is due to either (1) the host's immune response, which attacks virus-infected cells or produces various hypersensitivity reactions, or (2) the direct consequences of viral multiplication within host cells. The ways that viruses have evolved to survive the defense mechanisms of their hosts are currently being elucidated through modern genomics and the functional analysis of specific gene products.

For example, the influenza virus can mutate and change antigenic sites (antigenic drift) on the virion proteins, or decrease the number of viral cell proteins being expressed (presented) on the surface of the virus. The human immunodeficiency virus (HIV) infects T cells and greatly diminishes their numbers. The HIV, measles virus, and cytomegalovirus can cause fusion of host cells. This fusion allows the virus to move from infected to uninfected cells without being exposed to the host's defenses. The herpes simplex viruses can infect neurons that express little or no major histocompatibility complex molecules. (The major histocompatibility complex comprises a family of genes that encode cell-surface glycoproteins that regulate interactions among cells of the immune system, some components of the complement system, and perhaps other related functions connected with intercell recognition.) The common cold virus produces proteins that inhibit the major histocompatibility complex from functioning. Finally, hepatitis B virus–infected cells produce large amounts of extraneous antigens that are not associated with the complete virus. These extraneous antigens then bind all of the available neutralizing antibody so that there is insufficient free antibody to bind with the complete virus.

Evasion of host defenses by bacteria. Bacteria have evolved many mechanisms for protection against host defenses. A number of common ones are listed below.

Evading the complement system. The complement system is a system of plasma proteins that play a major role in an animal's defensive immune response against bacteria. To survive, some bacteria produce capsules that prevent complement activation and subsequent destruction. Some gram-negative bacteria can lengthen the O chains in their outer lipopolysaccharide coat to prevent complement activation. Others, such as *Neisseria gonorrhoeae* (the bacterium that causes gonorrhea), generate serum resistance that interferes with the formation of the membrane attack complex during the complement cascade (the sequential activation of complement proteins resulting in lysis of a target cell).

Resisting phagocytosis. Before a phagocytic cell can engulf a bacterium, it must come into contact with the bacterium's cell surface. Bacteria such as *Streptococcus pneumoniae* (the cause of pneumonia), *Neisseria meningitidis* (a cause of bacterial meningitis), and *Haemophilus influenzae* (a cause of meningitis in children) can produce slimy capsules that prevent phagocytic cells from coming into contact with the bacteria. Other bacteria evade phagocytosis by producing specific proteins on their cell surface that prevent adherence between a phagocytic cell and the bacterium.

Other mechanisms are also used to resist phagocytosis. *Staphylococcus* species can produce leukocidins (toxic substances that kill leukocytes) that destroy phagocytes before phagocytosis can occur, and *Streptococcus pyogenes* can release a protein that breaks down one component part of the complement system that attracts phagocytes to an infected area.

Survival inside phagocytic cells. Some bacteria have evolved mechanisms for survival inside macrophages, monocytes, and neutrophils. The bacteria *Listeria monocytogenes* (the cause of listeriosis), *Shigella* species (a cause of dysentery), and *Rickettsia* species escape from the phagosome (a closed intracellular vesicle containing material captured by phagocytosis) before it fuses with a lysosome [a specialized cell organelle surrounded by a single membrane and containing a mixture of hydrolytic (digestive) enzymes] and are thus not digested. Other bacteria, such as *Mycobacterium tuberculosis*, can resist the digestive enzymes of the lysosomes because of their waxy outer layer. Finally, *Chlamydia* prevents fusion of the lysosome with the phagosome.

Evading the specific immune response. To evade the specific immune response, some bacteria (such as *Streptococcus pyogenes*) produce capsules that are not antigenic because they resemble normal host tissue components. *Neisseria gonorrhoeae* can evade the specific immune response and survive by two mechanisms: (1) It makes genetic variations in its pili (proteinaceous appendages) so that specific antibodies are useless against the new pili and adherence to host tissue can occur; and (2) it produces specific proteases that destroy secretory antibodies, thereby allowing adherence to host tissue.

Hiding within a host cell. Some bacteria survive the host's defenses by remaining inside host cells, out of the reach of the phagocytes, antibodies, and complement. For example, *Shigella* species and *Listeria* species move into host epithelial cells in the intestine. Once inside the cell, they move laterally from cell to cell and are never exposed to the host's defenses.

Bacterial biofilms resist key host defenses. Biofilms consist of communities or groups of microorganisms that attach to surfaces of animate objects such as heart valves, bones, or tissues, or to inanimate objects such as artificial heart valves, prosthetic implants, or catheters. The most common biofilm on human tissue is dental plaque, which consists of a community of microorganisms that attaches to the surfaces of teeth.

In natural environments, biofilms likely serve as an important survival factor for many different bacteria. These biofilms allow bacteria to adapt to changing environments collectively instead of as single cells.

In human hosts, many bacterial pathogens form biofilms that cause chronic infections and resist standard antibiotic treatments. These same biofilms are also able to withstand several host defense measures, such as phagocytosis. One specific example is the pathogen *Pseudomonas aeruginosa*, which is associated with the lungs of cystic fibrosis patients. By the time cystic fibrosis patients are 16–20 years of age, this pathogen will have colonized the lungs of more than 90% of them. This accounts for the chronic morbidity and increased risk of early mortality in cystic fibrosis patients. The survival mechanisms involve the formation of an exopolysaccharide called alginate, which protects the bacterium through a process called "frustrated phagocytosis." During frustrated phagocytosis, neutrophils and macrophages encounter but cannot engulf this bacterium.

Recent research on outsmarting host defenses. Numerous microbiologists have become increasingly aware that many pathogens are continually evolving and developing mechanisms to subvert the host's defenses. A number of research papers have been published recently on these newly discovered mechanisms. Recent examples include the fact that many microorganisms can develop resistance to defensins (microbicidal and cytotoxic peptides made within the body by neutrophils and macrophages). In addition, *Legionella* (the cause of Legionnaires' disease) can develop resistance to phagosome–lysosome fusion, *Listeria* can escape from phagosomes, and the uropathogenic strain of *Escherichia coli* can adhere to and invade the urinary bladder epithelium to avoid being cleared from the body. Another mechanism that is quite clever is observed with *Salmonella enterica*, a major cause of diarrheal disease. *Salmonella enterica* engages in a close-contact duel with phagocytic cells to see which can kill the other first, with *Salmonella* winning in the majority of cases.

For background information *see* ANTIBODY; ANTIGEN; BACTERIA; BACTERIAL PHYSIOLOGY AND METABOLISM; BACTERIOLOGY; BIOFILM; COMPLEMENT; HERPES; INFLUENZA; MICROBIAL ECOLOGY; MICROBIOLOGY; PHAGOCYTOSIS in the McGraw-Hill Encyclopedia of Science & Technology.

John P. Harley

Bibliography.   T. J. Foster, Immune evasion by staphylococci, *Nat. Rev. Microbiol.*, 3:948–958, 2005; J. G. Leid, Bacterial biofilms resist key host defenses, *Microbe*, 4(2):66–70, 2009; M. J. Pallen and B. W. Wren, Bacterial pathogenesis, *Nature*, 449:835–842, 2007.

Mitosis and the spindle assembly checkpoint

Mitosis is the process that produces new cells through the replication and division of existing cells. Multicellular organisms such as humans require mitotic cell division for development and growth, and to continually replenish dying cells. Prior to mitosis, the cellular genetic material (DNA), which is organized into chromosomes, is replicated to produce two sister chromatids that remain joined until late mitosis. During mitosis, each sister chromatid pair (that is, replicated chromosome) aligns at the cell equator. Then, the two sister chromatids are pulled apart, thus forming two daughter chromosomes, and are segregated to opposite sides of the cell. The cell is finally halved, resulting in the production of two identical cells from one. Mitosis must be tightly regulated to ensure that the resulting daughter cells have exactly one copy of each chromosome, because missegregated chromosomes are implicated in the initiation and progression of cancer, as well as in the formation of birth defects in humans. The safety mechanism responsible for monitoring mitotic division is known as the spindle assembly checkpoint (SAC) and has evolved to ensure that chromosome segregation occurs faithfully.

Mechanics of mitosis. The mechanics of mitosis consist of a number of important coordinated processes and actions.

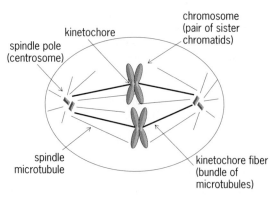

Fig. 1.   Schematic representation of a mitotic cell.

The mitotic spindle. Segregation of chromosomes occurs on the mitotic spindle, a structure comprising highly organized arrays of microtubules. Microtubules are hollow rods that continually grow and shorten from their ends. These dynamic polymers are made up of individual tubulin dimer subunits such that the addition of tubulin dimers causes microtubules to grow, and loss of tubulin dimers causes microtubules to shorten. Tubulin dimers have a certain polarity, and they can only add to a microtubule in one orientation. This confers polarity to polymerized microtubules, giving the two ends different growth characteristics: fast-growing "plus" ends and slow-growing "minus" ends. The minus ends of spindle microtubules are focused onto two points at opposite sides of the cell known as spindle poles (or centrosomes in many organisms) to form a "bipolar" mitotic spindle (**Fig. 1**).

Kinetochore–microtubule attachment. Each pair of sister chromatids uses the bipolar spindle to align at the cell equator in preparation for partitioning into two daughter cells. This is achieved by chromosomes generating attachments to spindle microtubules, thereby allowing the microtubules to push and pull chromosomes throughout the mitotic spindle. Chromosome–microtubule attachments are made at specialized regions on mitotic chromosomes called kinetochores (Fig. 1), which are very large protein structures comprising over 100 proteins. Kinetochores are built on each sister chromatid at the primary constriction site on mitotic chromosomes, which houses a region of specialized, highly compacted DNA called centromeric DNA. Proteins that specifically bind centromeric DNA dictate where the kinetochore is to be built during mitosis and serve as a platform for kinetochore construction. Distal to the centromeric platform, at the kinetochore outer domain, microtubule-binding proteins reside that serve to connect the plus ends of spindle microtubules to chromosomes.

Chromosome biorientation. Successful chromosome alignment requires that each chromatid pair biorient on the mitotic spindle, meaning that each sister chromatid must attach to microtubules emanating from opposite poles. After nuclear envelope breakdown, in prometaphase (the stage between prophase and metaphase in mitosis), the bipolar spindle forms and the plus ends of microtubules dynamically probe, via cycles of plus-end growth and shortening, for binding sites within the kinetochore outer domain. In most cases, a chromosome first becomes mono-oriented, where one kinetochore of the pair attaches to microtubules from one pole. Initially, this attachment is made by a single microtubule; however, over time, multiple microtubules bind the kinetochore to form a thick bundle called the kinetochore fiber (Fig. 1). Eventually, microtubules emanating from the opposite pole capture the opposing sister kinetochore to produce a bioriented chromosome. Microtubule plus ends that are linked to kinetochores remain dynamic as they continue to gain and lose tubulin subunits. This results in chromosome movement that is driven by the growth and shortening of attached microtubules within a kinetochore fiber. Once all chromosomes have bioriented and aligned at the spindle equator, the cell is considered to be in metaphase. An abrupt separation of the sister chromatids then marks the onset of anaphase. Sister chromatids disjoin and move toward opposite sides of the cell while remaining attached to shortening microtubules. The spindle poles move farther apart from each other, and eventually the cell is divided into two new cells during the process of cytokinesis.

Spindle assembly checkpoint. To avoid the catastrophic consequences of chromosome missegregation, eukaryotic cells have evolved a surveillance system to ensure that cells do not enter anaphase until all chromosomes have properly bioriented. This fail-safe mechanism is referred to as the spindle assembly checkpoint (SAC). The SAC, contrary to what its name implies, monitors the attachment status of kinetochores to microtubules rather than the assembly status of the mitotic spindle. In fact, the SAC is able to detect a single unattached kinetochore and will, in turn, delay anaphase onset and mitotic exit until proper attachment is achieved.

SAC targets. Prior to all chromosomes achieving biorientation, the SAC is active, and it inhibits the anaphase-promoting complex/cyclosome (APC/C). The APC/C is an E3 ubiquitin ligase responsible for eliminating proteins whose destruction is required for anaphase onset and mitotic exit. Specifically, the APC/C ubiquitinates the mitotic proteins cyclin B and securin [that is, APC/C adds ubiquitin (a small protein containing 76 amino acid residues) to these proteins], marking them for destruction by the proteasome (a large proteolytic particle found in the cytoplasm and nucleus of all eukaryotic cells that is the site for degradation of many intracellular proteins). Elimination of securin activates the protein separase, which cleaves the cohesin complex of proteins (**Fig. 2**). It is these cohesin proteins that hold duplicated sister chromatids together. Thus, upon their destruction, sister chromatids are detached from one another and are actively separated by forces from attached microtubules. Destruction of cyclin B re-

sults in the inactivation of the master mitotic kinase, CDK1 (cyclin-dependent kinase 1), driving cells to exit mitosis. Once all chromosomes are properly bioriented, the SAC is silenced, allowing for APC/C activation and anaphase onset.

Activation of the SAC. Early studies suggested that unattached kinetochores were the key activators of the SAC by correlating the timing of anaphase onset with capture of the last unattached kinetochore in living cells. Subsequent studies showed that the destruction of the one remaining unattached kinetochore in cells silenced SAC signaling and initiated anaphase onset. Interestingly, it has been suggested that the SAC may not only detect direct microtubule binding at kinetochores, but may also detect the physical tension between sister kinetochores that is generated upon chromosome biorientation. Micromanipulation experiments have shown that imposing a pulling force on a single remaining unattached kinetochore in cells inactivates the SAC and signals anaphase onset. In addition, spindle poisons (for example, paclitaxel) that relieve tension across sister kinetochores but do not induce microtubule detachment can stimulate SAC activation. Whether the SAC directly detects microtubule attachment at kinetochores or the tension produced as a result of that microtubule attachment remains a debated topic.

SAC signaling. The SAC detects chromosomes that have not bioriented and responds by generating a signal that inhibits the APC/C. Two genetic screens in budding yeast originally identified the core SAC genes. Mutations in these genes prevented host yeast cells from arresting in mitosis in response to drugs that destroyed spindle microtubules and thus prevented normal chromosome biorientation. These include "MAD" genes (mitotic arrest deficient: MAD1, MAD2, and MAD3), "BUB" genes (budding uninhibited by benomyl: BUB1 and BUB3), and MPS1 (monopolar spindle 1). These genes are conserved throughout eukaryotic evolution and encode for SAC proteins that coordinate to inhibit the APC/C. MAD2 and MAD3 (also known as BUBR1) bind an APC/C-activating protein called CDC20 (cell division cycle 20 homolog), thereby preventing APC/C activation and subsequent promotion of anaphase. A complex referred to as the MCC (mitotic checkpoint complex), containing the proteins MAD2, BUBR1, BUB3, and the APC/C activator CDC20, has been suggested to be the functional SAC effector, because this complex inactivates the ubiquitin ligase activity of the APC/C much more effectively than MAD2 or BUBR1 alone. Importantly, unattached (or improperly attached) kinetochores are thought to be required to catalyze the formation of the MCC, or functional APC/C inhibitor. This was suggested, in part, by the observation that unattached kinetochores accumulate high levels of SAC proteins, while SAC protein levels at attached kinetochores are significantly depleted. In the case of MAD2, an inactive form of the protein is present throughout the cytoplasm of mitotic cells, and only when MAD2 binds

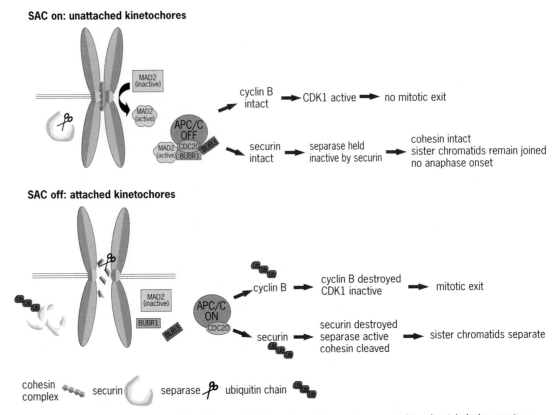

Fig. 2. Signaling pathways of the SAC. (*Top*) Unattached kinetochores (those not connected to microtubules) generate an "active" form of MAD2, which is essential in generating the MCC complex. The MCC complex binds to the APC/C-activating protein CDC20, thereby preventing CDC20 from turning on the APC/C. When the APC/C is off, both cyclin B and securin are protected from degradation. (*Bottom*) Attached kinetochores no longer generate the form of MAD2 capable of producing functional MCC complexes. This results in activation of APC/C by CDC20. Active APC/C polyubiquitinates both cyclin B and securin, targeting them for destruction by the proteasome. Securin destruction results in activation of separase, which cleaves the cohesin complex, that is, the "glue" holding sister chromatids together. Destruction of cyclin B results in the inactivation of CDK1, which drives cells to exit mitosis.

unattached kinetochores does it undergo a conformational change that renders it capable of binding CDC20 and inhibiting the APC/C. SAC proteins are not limited to the MAD and BUB families, and multiple additional proteins in higher eukaryotes participate in SAC signaling. Similar to the MAD and BUB proteins, perturbation of their function results in abrogation of the SAC and progression into anaphase with unaligned chromosomes.

Silencing the SAC. For cells to initiate anaphase after all chromosomes have bioriented, the SAC-inhibitory signal must be silenced. This occurs, in part, by the removal of SAC proteins from attached kinetochores via the minus end directed microtubule motor dynein (a large enzyme complex that hydrolyzes adenosine triphosphate to produce force for translocation along microtubules). Upon microtubule attachment to kinetochores, the SAC proteins MAD1 and MAD2 are transported along microtubules from kinetochores to spindle poles by cytoplasmic dynein. Interfering with dynein function results in accumulation of MAD1 and MAD2 at properly attached kinetochores and metaphase arrest. Microtubule binding into kinetochores may also contribute directly to checkpoint inactivation by modulating the activity of checkpoint kinases. For example, in certain organisms, microtubule binding has been shown to inactivate the checkpoint kinase BUBR1 through

interaction with the outer kinetochore protein CENP-E (centromere-associated protein E). Checkpoint silencing also likely involves activation of the protein p31[COMET], which prevents MAD2 from transitioning from its inactive form to its active, inhibitory form. The ultimate goal of these quenching pathways is shared: Silence the "wait anaphase" signal upon proper kinetochore-microtubule attachment to initiate exit from mitosis.

For background information *see* BIOCHEMISTRY; CELL (BIOLOGY); CELL CYCLE; CELL DIVISION; CELL NUCLEUS; CENTROSOME; CHROMOSOME; DEOXYRIBONUCLEIC ACID (DNA); MITOSIS; PROTEASOME in the McGraw-Hill Encyclopedia of Science & Technology.

Jennifer G. DeLuca

Bibliography. M. A. Hoyt, L. Totis, and B. T. Roberts, *S. cerevisiae* genes required for cell cycle arrest in response to loss of microtubule function, *Cell*, 66:507–517, 1991; G. J. Kops, B. A. Weaver, and D. W. Cleveland, On the road to cancer: Aneuploidy and the mitotic checkpoint, *Nat. Rev. Cancer*, 5:773–785, 2005; R. Li and A. W. Murray, Feedback control of mitosis in budding yeast, *Cell*, 66:519–531, 1991; A. Musacchio and E. D. Salmon, The spindle-assembly checkpoint in space and time, *Nat. Rev. Mol. Cell Biol.*, 8:379–393, 2007; C. E. Walczak and R. Heald, Mechanisms of mitotic spindle assembly and function, *Int. Rev. Cytol.*, 265:111–158, 2008.

Molecular modeling of polymers and biomolecules

The structure and properties of polymers and biomolecules are directly connected to their molecular architecture. Biomolecules, such as proteins, present a remarkably rich variety of complex three-dimensional arrangements based on the sequence of their building blocks (amino acids). Formation of different structures, because of subtle and often unknown changes in this sequence, can lead to improper functioning and numerous diseases. In materials design, custom synthesis of polymers with different functional groups aims at driving spontaneous self-assembly to desired patterns with nanometer-length features. Given the multitude of potential chemical groups that can be added through innovative chemistry, an a priori prediction of the thermodynamic and dynamic properties would significantly reduce expensive trial-and-error procedures. Molecular modeling provides such predictions by establishing a relationship between the chemical architecture and observed macroscopic behavior.

Starting from a specific chemical constitution and knowledge of forces between pairs of atoms, we need to predict the behavior of thousands of molecules to capture the macroscopic properties that are commonly observed. Classical statistical mechanics has long provided a rigorous mathematical framework that connects interactions between atoms to macroscopic properties (through the evolution of microscopic degrees of freedom such as atomic positions and momenta). Molecular modeling, as a form of applied statistical mechanics or "computer experiment," calculates approximate numerical solutions through sampling of realistic atomic positions and velocities. Transport properties, constitutive laws, and equations of state can all be extracted for the majority of simple molecular models with moderate computational requirements. Furthermore, molecular simulations can now efficiently address the properties of complex materials, such as polymers and assemblies of biomolecules.

Methods. A typical molecular simulation requires extended input about molecular geometry and the nature of interactions between the atoms of the molecule studied (such as hydrogen and oxygen; **Fig. 1**). During the last decades, the development of accurate parameters for atomic interactions (force fields) has flourished, aided by advances in experimental techniques (such as x-ray and neutron diffraction and nuclear magnetic resonance spectroscopy), quantum calculations on small compounds, and accurate knowledge of phase diagrams. Force fields available today are sufficient to build atomistic models for simulations of most materials far from extreme thermodynamic conditions.

Molecular models estimate the properties of a material by iteratively generating a sample of microscopic states (positions and momenta of atoms). The actual sampling procedure progresses either by a stochastic process (involving generation of random

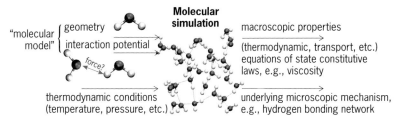

Fig. 1. Typical input and output in a molecular simulation. Liquid water as an example system.

numbers) or by a deterministic method (time integration of Newton's equations of motion). Monte Carlo simulations are typical stochastic methods applied to study thermodynamic properties. Molecular dynamics follow the deterministic evolution of the system with time, providing additional information about dynamic observables, such as transport properties. A more detailed description of methods can be found in the Bibliography. The output of the simulation consists of estimates of macroscopic properties (for example, viscosity) for the given model and exclusive detailed insight into the microscopic

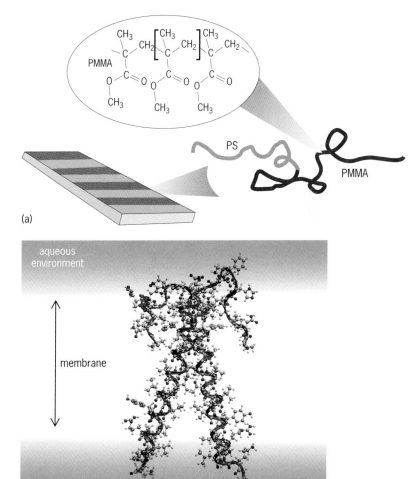

Fig. 2. Self-assembly of (*a*) synthetic copolymers [for example, composed by poly(methyl methacrylate), or PMMA, and polystyrene, or PS, units] to manufacture ordered patterns with nanometer length-scale features. (*b*) Dimer of Glycophorin A, formed by the assembly of two helical proteins spanning a lipid membrane (membrane not shown for clarity).

mechanism responsible for the observed behavior (such as liquid structure and hydrogen bonds).

Challenges. Synthetic macromolecules and biopolymers (such as proteins and DNA) consist of a sequence of repeating units (from a small number to thousands) that have distinct physicochemical properties. Special Monte Carlo and parallel molecular-dynamics algorithms can perform efficient atomistic simulations of these complex molecules. However, despite remarkable achievements in algorithmic development and the constant increase of computational power, atomistic modeling of these systems is still severely limited. Macromolecules present characteristic structures and dynamics over an extremely broad range of time and length scales. While a repeat unit has

dimensions of a few nanometers, thousands of such units are linked together to form a single molecule extending to hundreds of nanometers. Furthermore, details of repeating units are often important in reproducing long-range structure. For example in **Fig. 2a**, block copolymers synthesized with two different sequences of repeating units assemble to ordered patterns because of unfavorable interactions between the building blocks. Membrane proteins fold into a native structure that can range from an individual helical conformation of hydrophobic amino acids (secondary structure) to an overall complex produced by a well-defined assembly of different helices (quaternary structure). An example of two helices forming a dimer in a membrane is presented in Fig. 2b. In both examples (copolymers and proteins), atomic details of the chemical structure govern the overall configuration and need to be considered to successfully reproduce macroscopic behavior.

The spatial range involved in atomistic modeling of macromolecules is not the only challenge. Characteristic times associated with the motion of these molecules span an extreme range, from picoseconds (10^{-12} s) to seconds. Modern all-atom molecular dynamics can typically trace trajectories only over a few nanoseconds (10^{-9} s). Protein molecules typically require 10^{-6}–10^{-3} s to form their structures. In polymer melts, macromolecules diffuse at slow rates that decrease steeply with molecular weight (or N, number of repeating units) according to power laws (proportional to N^{-1}–$N^{-2.4}$). Because of all these limitations, research today employs molecular modeling of polymers and biopolymers as a part of a greater hierarchical scheme, where all-atom simulations serve as an intermediate rather than the final stage toward predicting macroscopic properties (a scheme is shown in **Fig. 3a**).

Hierarchical modeling. Molecular simulations can serve as a critical step toward designing accurate mesoscopic or coarse-grained models and performing quantitative theoretical calculations. Coarse-grained simulations are performed with simpler models, where groups of atoms are represented by a single interaction site. Figure 3b demonstrates how a lipid and a protein molecule can be modeled with such simpler descriptions. As a result, the computational requirements are drastically reduced compared to fully atomistic simulations. However, there are multiple grouping schemes that can be envisioned, and effective interactions between such coarse-grained sites remain to be determined. Atomistic modeling can be done to resolve such issues and guide the design of accurate mesoscopic simulations, which greatly extend the accessible spatial and temporal length scales. Phenomenological models, on the other hand, had in the past been the exclusive theoretical toolbox in polymer physics research. Calculations based on such models are extremely attractive, not only because of their minimal computational effort, but also because of their success in capturing qualitatively the observed macroscopic behavior. With the power of molecular simulation, such

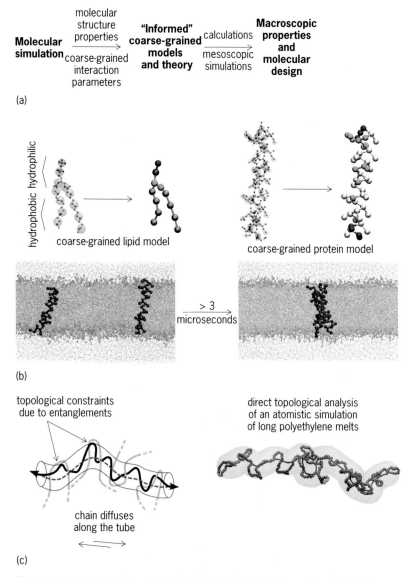

(a)

(b)

(c)

Fig. 3. Molecular simulation as part of (a) a hierarchical modeling scheme. (b) Building simpler models of biomolecules extends the accessible spatial and temporal length scales from a few nanoseconds to over microseconds. Using such an approach, the factors controlling the association of two protein molecules in a lipid membrane can be examined. (c) The reptation model consists of a simple description of molecular diffusion along a tube defined by topological constraints imposed by neighboring chains (left). Atomistic simulations can evaluate the tube parameters (right) required to perform quantitative theoretical predictions.

models are now being revisited, critically examining approximations as well as improving and extending them to perform quantitative predictions. Within such a multiscale-modeling scheme (Fig. 3a), molecular simulations today maximize their potential in polymer and biomolecular research. To demonstrate this approach, two examples focusing on the structure and transport properties of macromolecules are presented.

Modeling the structure of membrane proteins. Membrane proteins account for 25% of the open reading frames encoded in genomes and participate in critical biochemical processes. Because of experimental challenges, the structure of only a few membrane proteins has been resolved in fully atomistic detail. Atomistic modeling cannot follow the formation of this structure, which occurs over relatively long time scales (microseconds). Simpler consistent coarse-grained models of lipids and proteins (Fig. 3b) can reduce computational requirements, while maintaining critical characteristics (such as solubility in water). Lipid molecules preserve their amphiphilic character (hydrophilic and hydrophobic parts) and self-assemble into a membrane in an aqueous environment. Protein molecules insert into this bilayer of lipids and associate, forming dimers, in agreement with experimental findings. Using this approach, we are now starting to evaluate the factors that promote such dimerization events, which play a role in numerous diseases, including some forms of cancer.

Rheology of polymer liquids. In polymer melts, there are two main theories that describe molecular diffusion and dynamics, the Rouse model and the reptation (tube) model. The Rouse model has been successful in describing transport properties at moderate molecular weights. However, when considering long macromolecules above a critical number of repeating units N, Rouse dynamics fail to capture macroscopic behavior, such as the viscosity of melts. In such systems, individual polymers experience topological constraints (entanglements) induced by the uncrossability of neighboring segments (consider the motion of the chain being restricted by the neighboring molecules in Fig. 3c). The reptation model suggests that chains can translate only along the main axis (primitive path) of a tube with a finite diameter. This phenomenological model provides a mathematical formalism to predict rheological properties, such as shear modulus, based on microscopic information as the length of the primitive path passing through the center axis of the tube. Recently, algorithms employing topological analysis have extracted estimates of essential parameters such as the length of the tube and the number of entanglements per chain. Combining the theoretical calculations with the simulation-predicted values allows for quantitative predictions that are not feasible with either theory or simulations alone.

Outlook. In the field of macromolecules, both synthetic and biological, the challenges faced by the complexity of the structures require the design of a hierarchical approach capable of addressing phenomena over a broad range of time and length scales.

Molecular modeling will provide a unique and detailed view of collective molecular behavior and will serve as a valuable research tool to design new materials.

For background information *see* BIOPOLYMER; CELL MEMBRANES; COMPUTATIONAL CHEMISTRY; COPOLYMER; MACROMOLECULAR ENGINEERING; MOLECULAR MECHANICS; MOLECULAR SIMULATION; MONTE CARLO METHOD; POLYMER; PROTEIN; STATISTICAL MECHANICS; SUPRAMOLECULAR CHEMISTRY in the McGraw-Hill Encyclopedia of Science & Technology.   Manolis Doxastakis

Bibliography. M. P. Allen and D. J. Tildesley, *Computer Simulation of Liquids*, Oxford University Press, New York, 1987; H. J. C. Berendsen, *Simulating the Physical World*, Cambridge University Press, Cambridge, U.K., 2007; P. J. Cummings, Molecular simulation, in *McGraw-Hill Encyclopedia of Science & Technology*, 10th ed., vol. 11, pp. 359–362, 2007, DOI 10.1036/1097-8542.YB011080; D. L. Nelson and M. M. Cox, *Lehninger Principles of Biochemistry*, Freeman, New York, 2004; M. Rubinstein and R. H. Colby, *Polymer Physics*, Oxford University Press, Oxford, U.K., 2003; B. Smit and D. Frenkel, *Understanding Molecular Simulation*, Academic Press, San Diego, 2001.

Molecular shape and the sense of smell

The nose is a chemical detector. It houses the olfactory sensory neurons (OSNs), which are specialized cells that display on their surface genomically encoded protein receptors that have evolved to respond to airborne odorant molecules. In 1991, these protein receptors, now known as the olfactory receptors (ORs), were identified by Linda Buck and Richard Axel as belonging to a class of well-known proteins, the G-protein-coupled receptors (GPCRs). [GPCRs are dependent on guanosine triphosphate (GTP) for their function; in the nonstimulated state, these receptors are bound to a complex of three different proteins (a heterotrimer), named alpha (α), beta (β), and gamma (γ), which together constitute the G-protein.]

Like other GPCRs, the ORs adhere to the cell surface by floating in the cell's lipid bilayer outer membrane, which is a native environment that unfortunately makes structural analysis of the ORs difficult. By analogy with the few GPCRs that have been studied structurally, olfactory GPCRs are predicted to contain seven helical segments that span the OSN cellular membrane, in addition to segments exposed to the extracellular and intracellular environments (shown schematically in **Fig. 1**). The rotation of chemical bonds, in a protein or any other flexible molecule, produces differently shaped structures known as conformers. By analogy with other GPCRs, the conformational shape of an OR is predicted to alternate between two extremes: a quiescent conformer that does not trigger intracellular events, and an activated conformer that triggers intracellular events. The two forms are in

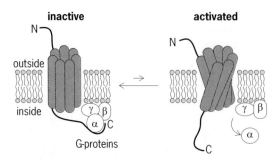

Fig. 1. Olfactory receptors are membrane-bound proteins that alternate between at least two forms: one quiescent (inactive) and one activated. Equilibrium favors the inactive form when no odorant is present. N and C refer to protein termini.

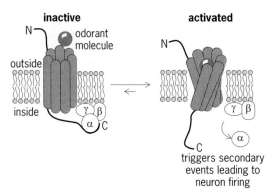

Fig. 2. Above threshold concentrations, an odorant specific for the receptor tips the equilibrium in favor of the activated form, releasing the α-subunit G-protein and leading to an action potential.

equilibrium with each other. In the absence of an activating molecule, or "agonist" in pharmacological parlance, the equilibrium favors the quiescent form, and the OSN is silent. When present, however, an agonist will upset this equilibrium by binding to and stabilizing the activated conformer (**Fig. 2**). Shifting the GPCR to its activated form sets off a series of intracellular events that culminates in the firing of action potentials by the OSN. Many GPCR agonists are small molecules. The OR agonists are odorants, that is, molecules that we can smell.

Further analogy to the few GPCRs whose structures have been elucidated has led to insights into how the odorants might be "recognized" by the ORs. Like all proteins, the ORs are polymers, that is, molecular chains of the 20 amino acids linked covalently in a precise order as specified by the genome. Within the membrane, the helical segments of each OR are predicted to converge, thereby forming an odorant binding pocket lined by some of the helical segment's amino acids. Most mammalian genomes encode a family of about 1000 ORs related by their amino acid sequences. (Humans have approximately 855 ORs, but about 465 have become disabled through mutations that are thought to disrupt the integrity of the GPCRs, converting them into nonfunctional pseudogenes.) Each OR family member is made unique by having amino acid sequence differences scattered throughout its chain. The OR family shows extra variability in the amino acid se-

quences predicted to form the binding pocket. This variability provides most of the 1000 ORs with the potential to form a differently shaped binding pocket so that each OR can be stabilized in its activated conformation by odorants having different chemical features. Furthermore, each OSN chooses only one OR from the genome to be displayed on its surface, so each OSN is dedicated to sensing the type of chemical odorant features preferred by that OR. In this way, the OSNs divide up the labor of sensing all possible molecules that may be encountered.

Odorant shape. The chemical features for OR activation among encountered odorants are sometimes described as molecular shape. However, this description is an oversimplification. While it is true that, to stabilize an OR conformer, the shape of the odorant must in some way be complementary to that of the binding site of the activated OR, shape alone is not sufficient. This is because binding between the odorant and the OR is mediated by a variety of weak noncovalent intermolecular forces. These forces are the same attractive forces that mediate molecular recognition in other biological and organic systems: hydrogen bonding, dipole-dipole attraction, van der Waals dispersion forces (attractive forces between two atoms or nonpolar molecules that arise because a fluctuating dipole moment in one molecule induces a dipole moment in the other, and the two dipole moments then interact), and the energetic benefit of excluding water from hydrophobic surfaces. Although two molecules may have similar shapes, they may have different sets of weak intermolecular forces available to them, and thus they may activate different combinations of ORs.

Many odorants are flexible molecules, that is, they are able to alternate among many different conformers by rotating around single bonds in their structure. These molecules have no single conformational shape, but rather alternate among a range of different shapes. Does the OR in these cases recognize all shapes, or does it recognize just one or a few? This question was recently tested in a closely related series of eight carbon aldehyde odorants in which a flexible linear alkyl (hydrocarbon) chain was constrained by incorporation into a series of cycloalkyl rings. Cycloalkyl (circular hydrocarbon) ring systems have fewer degrees of conformational freedom than their linear counterparts, and thus have better-defined shapes. Surprisingly, all the aldehydes, whether linear or cyclic, bound to the OR, but only a subset activated it to fire an action potential. This result indicates that an OR is not necessarily activated by all of the shapes that its odorant can adopt. Although the OR studied may bind many different shapes indiscriminately, it appears to wait to see if the odorant can adopt a key shape (conformation) that shifts the OR equilibrium in favor of activation. Shape is often likened to the static lock-and-key theory of enzyme substrate recognition first proposed by Emil Hermann Fischer. However, this recent study places the importance of molecular shape more in the context of induced fit theory as later proposed by Daniel E. Koshland: the binding molecules induce each other to adopt the best shape for binding. These experiments support the conformational shape of the odorant as a key determinant of OR activation.

Olfaction and pharmacology. Pharmacologists study the binding of receptors to their natural ligands, such as hormones and neurotransmitters, as well as to chemically synthesized molecules, or drugs. Studies in this area result in a pharmacological profile of the types of small molecules that will bind a given receptor. In the case of the GPCRs, this list includes both agonists and nonactivating, competing molecules, known as antagonists. Antagonists bind the quiescent conformation of a GPCR and, when bound, competitively block agonists from activating the receptor. Prior to 2000, all odorants were thought to be OR agonists, but in recent years odorant antagonists have been discovered, with interesting implications for the olfactory code (see below). Nonolfactory GPCRs and other pharmacologically interesting receptors often occur in families of related proteins. For example, there are at least 12 related serotonin GPCRs encoded in the human genome. All of them bind the neurotransmitter serotonin, but each is also activated or blocked by different sets of drugs. Prior to the discovery of the ORs, a receptor subtype family of 12 receptors was considered large. Placed in this context, the size of the OR gene superfamily, at 1000 genes, is staggering and by far the largest in the human genome. However, the list of activating odorants remains unknown for 95% of human ORs, and antagonism has barely been examined. Considered as a pharmacological problem, the correlation between odorant structure and odor percept, the sensory output of the sum of all ORs, can be considered to be the most complex pharmacological problem yet encountered in nature. In order to understand the relationship between structure—that is, shape—and perceived odor character, it will be necessary to study the pharmacological profile of all human functional ORs, which is a formidable challenge, since the ORs are technically difficult to study.

Olfactory code. There are too few ORs and too many odorant shapes for each OR to be dedicated to detecting a single odorant. It has therefore been hypothesized that the ORs have different but overlapping pharmacological profiles and that they pool their sensory inputs to create unique neural signals for the brain to interpret. Experiments have shown that individual ORs respond to multiple chemically related odorants, some of which are also detected by other ORs. Based on the few that have been profiled, some ORs are found to be "narrowly tuned," or quite specific—for example, being activated only by aldehyde compounds. Others appear to be "broadly tuned," being activated by a wide variety of functional groups, such as alcohols, aldehydes, and acids. Thus, little can be understood about the relationship between chemical shape and perceived odor quality by looking at any single OR profile. How many ORs work in parallel to contribute to the perception of a single odorant? A single flexible odorant, octanal, has been observed to activate approximately 55 rodent ORs. This number is probably close to the upper limit for a single odorant, since less flexible octanal analogues activate fewer ORs. The concentration of an odorant is also an important factor because ORs can have different dose-response profiles for the same odorant. Natural fragrances occur in mixtures, where each component chemical may activate a different set of ORs. Simultaneously, some mixture components may suppress activation of other ORs by antagonism. The number of possible output signals sent to the brain for interpretation seems limited only by the number of odorants and odorant mixtures possible.

Vibration theory. By analogy with nonolfactory GPCRs, odorant shape—including conformation, functional group identity, and overall size—is now accepted to be the major determinant of primary olfactory sensing. Nevertheless, it has to be admitted that, because of the experimental difficulty of studying ORs, some of what is known about activation mechanisms comes from analogous studies on "better behaved," but nonolfactory GPCRs. An interesting pre-GPCR hypothesis, maintaining that odor character arises from the vibration frequencies of the odorant, was recently revived. Like pharmacological or "shape" theory, the vibration hypothesis posits that odorants bind ORs and give rise to codes that the brain must interpret. However, the unique features of the vibration theory do not lend themselves readily to biochemical experimentation, and no direct experimental evidence has yet been adduced to support this theory.

For background information *see* ALDEHYDE; CHEMICAL SENSES; CHEMORECEPTION; MOLECULAR BIOLOGY; MOLECULE; NEUROBIOLOGY; NEURON; OLFACTION; PHARMACOLOGY; POLYMER; PROTEIN; SIGNAL TRANSDUCTION in the McGraw-Hill Encyclopedia of Science & Technology.   Kevin Ryan; Xiaozhou P. Ryan

Bibliography. L. Buck and R. Axel, A novel multigene family may encode odorant receptors: A molecular basis for odor recognition, *Cell*, 65:175–187, 1991; C. Burr, *The Emperor of Scent*, Random House, New York, 2002; S. Firestein, A code in the nose, *Science Signaling* (formerly, *Science STKE*), 2004:pe15, 2004; S. Firestein, How the olfactory system makes sense of scents, *Nature*, 413:211–218, 2001; Z. Peterlin et al., The importance of odorant conformation to the binding and activation of a representative olfactory receptor, *Chemistry and Biology*, 15:1317–1327, 2008.

Morotopithecus

Morotopithecus bishopi is a fossil ape species from northeastern Uganda dated to be more than 20.6 million years old. The fossils of *Morotopithecus* suggest that it was capable of modern apelike behaviors, including use of vertical trunk postures (orthogrady), slow climbing, and arm hanging. Other fossil apes of similar age are known from East Africa. However, unlike *Morotopithecus*, these apes would have walked on the tops of branches on all fours with a horizontally oriented torso (pronogrady). *Morotopithecus* thus represents the oldest record of modern apelike

Fig. 1. Map of ape-bearing fossil localities in East Africa circa 20 million years ago. (*Courtesy of Laura MacLatchy*)

locomotor behavior in the fossil record. Paleontologists are currently divided over whether the anatomical features associated with upright posture and suspension in the modern apes are due to inheritance or independent evolution. This debate has important implications for interpreting the evolutionary position of *Morotopithecus* and other fossil apes, as well as for reconstructing the pattern and timing of the emergence of modern ape adaptations.

Age of Moroto. Mount Moroto is an extinct volcano in northeastern Uganda (**Fig. 1**). Two fossil sites, Moroto I and Moroto II, are located about 15 km (9.3 mi) north of the summit of the mountain. Fossiliferous sediments at both sites are overlain by basalt lava deposits that have been estimated to be 20.6 million years old using the argon-40/argon-39 radiometric dating technique. This means that the fossils underlying the lava are more than 20.6 million years old. The fossil assemblage includes rodents, primates, hyraxes, elephants, rhinoceroses, ruminants,

and piglike and hippopotamuslike mammals, as well as turtles, crocodiles, and birds. The same species of mammals are found at other sites of similar age, supporting the radiometric date for the sites.

Fossils. The *Morotopithecus* ape fossils from Moroto II include a maxilla (upper jaw) and face of a single individual, parts of two mandibles (lower jaws), an isolated upper canine, lumbar (lower back) and thoracic (chest) vertebrae, and two femora (thighbones). With regard to Moroto I, there is a fragment of a scapula (shoulder blade) and a finger bone that have been assigned to *Morotopithecus* (**Fig. 2**).

The face, maxilla, and mandible specimens from Moroto II were found before any postcranial bones and initially were attributed to other, roughly contemporaneous apes, including *Proconsul*, known from sites in Uganda and Kenya, and *Afropithecus*, from Kenya. However, unlike *Proconsul*, *Morotopithecus* has larger anterior teeth, a relatively larger back molar, and more closely spaced eyes. Features separating *Morotopithecus* from *Afropithecus* include a higher face, a broader nose, and the presence of a wide fossa (hole) in the palate. More important, the primitive nature of the postcranial remains attributed to or inferred for *Proconsul* and *Afropithecus* differs from the nature of the known postcranial elements of *Morotopithecus*.

Most significant among the postcranial fossils is a middle lumbar vertebra. Its anatomical features are compatible with a stiff lower back, enabling an individual to keep its torso upright. All other large-bodied apes that have been dated to be 20–13 million years old (for example, *Proconsul*, *Afropithecus*, *Sivapithecus*, *Kenyapithecus*, *Equatorius*, *Nacholapithecus*, and *Otavipithecus*) are thought to have been pronograde quadrupeds. Features of the *Morotopithecus* vertebra that are shared with modern ape vertebrae, but not with quadruped vertebrae, include transverse processes (bony lever arms) that arise from the neural arch, increasing the mechanical advantage of muscles in the back that keep the torso from falling forward. However, the vertebra is not completely modern. For instance, the shortening of the vertebral body that characterizes living apes is not present.

The scapula fragment has a broad, oval-shaped shoulder socket, compatible with using the arm in side-to-side, not just fore-and-aft, movements. The socket is also gently curved in all directions, permitting a wide range of rotational shoulder movements. Monkeys and other quadrupeds tend to have shoulder sockets that are narrower and that are very curved in the craniocaudal (top-to-bottom) direction, but less curved side to side, favoring fore-and-aft (flexion and extension) over rotatory movements. Thus, the *Morotopithecus* shoulder is postulated to have been loaded (that is, adapted to withstand the forces incurred on the joint) over a wide range of movements, such as would occur during forelimb-suspensory and forelimb-dominated climbing behaviors, marking the earliest record of such behaviors in apes.

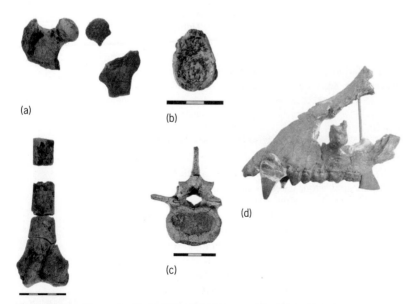

Fig. 2. *Morotopithecus* fossils, including the (a) femora, (b) scapula, (c) lumbar vertebra, and (d) maxilla. (*Courtesy of Laura MacLatchy*)

The left femur (thighbone) is quite fragmentary, but the right femur is about 75% complete and preserves both ends. Primitive features include a relatively small femoral head (the "ball" component of the hip joint). The head also lacks extensive articular surface on the posterior side, suggesting that the hip was not as mobile as those of modern apes, although still able to permit climbing. The right femur bone is extraordinarily thick in the shaft. Orangutans and other slow climbing primates also have this character, and it is possible that it represents an adaptation to strengthen the shaft in the face of the sustained muscular contractions that occur during slow and cautious climbing activities. Features of the distal femur also support climbing behaviors, including a knee joint that is frequently loaded while abducted (moved away from the midline of the body), a broad knee, and a broad, shallow groove for the kneecap.

Body mass has been estimated from the femora, scapula, and vertebra. Most estimates fall within a range of 35 to 45 kg (77 to 99 lb), which is about the size of a small female chimpanzee or orangutan.

In summary, the known postcranial remains of *Morotopithecus* suggest a stiff lower back, a nonstereotypic (varying) pattern of loading the shoulder joint, moderate hip mobility, a knee often used in abducted postures, and strongly built thighs (**Fig. 3**). Some combination of behaviors, including forelimb suspension and slow-speed brachiation (swinging below branches using only the forelimbs), cautious and deliberate climbing or clambering, vertical climbing, and quadrupedalism, was probably employed by this taxon, displaying a repertoire with greater similarities to that of the living apes than to its fossil contemporaries.

Calibration of the ape–monkey split. At more than 20.6 million years old, *Morotopithecus* is the oldest securely dated fossil ape. The oldest *Proconsul* site is Meswa Bridge in Kenya, but it is not securely dated radiometrically, although the fauna suggest that it is probably more than 20 million years old. Other *Proconsul* sites include Napak, Uganda, which is approximately 20 million years old, and Songhor and Koru, Kenya, which are 19.5 million years old. The timing of evolutionary branching events is often estimated using both genetic and fossil-based approaches. The age of *Morotopithecus* implies that apes and monkeys diverged at least 20.6 million years ago, and this age thus represents a minimum estimate for the branching event. However, because there are almost no primate fossil localities from the time period just prior to this, the fossil record cannot currently help refine this estimate. Genetic data suggest that the split may have occurred as long as 30 million years ago. In combination, fossil and genetic data indicate that the split between Old World monkeys and apes likely occurred between 30 and 20 million years ago.

Evolution of adaptively significant locomotor features. The lumbar vertebral, scapular, and femoral fossil remains of *Morotopithecus* suggest a locomotor and postural repertoire that included orthogrady, deliberate climbing, and suspension, which are behaviors found in all living apes. In contrast, *Proconsul*, *Afropithecus*, and other early apes are reconstructed as above-branch quadrupeds.

Some paleontologists think that the living apes, including gibbons, orangutans, and African apes (gorillas, bonobos, and chimpanzees), acquired their locomotor adaptations independently through parallel evolution. If this is the case, then *Morotopithecus* may represent the earliest known instance of a transition (that is, from quadrupedalism to orthogrady and suspension) that occurred several times in ape evolution. When evaluating the role of parallelism in ape evolution, it should be noted that modern apes, especially great apes (gorillas, chimpanzees, and orangutans), diverge from one another in their locomotor specializations, although they closely resemble one another anatomically. It thus seems plausible that shared ancestry played some role in causing knuckle walkers (for example, *Gorilla*) and slow suspensory climbers (for example, *Pongo*) to be very similar in form. An improved fossil record and more detailed anatomical and developmental studies of shared features that may be due to inheritance are needed.

Some anatomical data bolster the view that the living apes and *Morotopithecus* resemble one another postcranially because they inherited these features from a common ancestor. For example, studies assessing the reliability of cranial, postcranial, and soft-tissue data sets for inferring evolutionary history

Fig. 3. A reconstruction of *Morotopithecus*. (*Courtesy of Stephen Nash*)

suggest that the latter two may be superior to the former in apes and monkeys.

Ape evolution is of interest in part because many aspects of ape biology are seen as preadaptations related to human emergence. For example, a vertical trunk is necessary for bipedality, whereas large size and long life span are important for learning and social flexibility and complexity. *Morotopithecus* indicates that the traits of large body size and locomotor innovation were in place in at least one ape lineage by more than 20.6 million years ago. As more fossils are uncovered and a more detailed understanding of anatomical variation in the apes is gained, it will be possible to assess whether *Morotopithecus* represents one of the many times that these features evolved.

For background information *see* ANIMAL EVOLUTION; APES; DATING METHODS; FOSSIL APES; FOSSIL PRIMATES; MOLECULAR ANTHROPOLOGY; MONKEY; PALEONTOLOGY; PHYLOGENY; PHYSICAL ANTHROPOLOGY; PRIMATES; TAPHONOMY in the McGraw-Hill Encyclopedia of Science & Technology.

Laura MacLatchy

Bibliography. D. L. Gebo et al., A hominoid genus from the early Miocene of Uganda, *Science*, 276:401–404, 1997; T. Harrison, Late Oligocene to middle Miocene catarrhines from Afro-Arabia, in W. C. Hartwig (ed.), *The Primate Fossil Record*, pp. 311–338, Cambridge University Press, Cambridge, 2002; L. M. MacLatchy, The oldest ape, *Evol. Anthropol.*, 13(3):90–103, 2004; L. MacLatchy et al., Postcranial functional morphology of *Morotopithecus bishopi*, with implications for the evolution of modern ape locomotion, *J. Hum. Evol.*, 38:159–183, 2000; W. J. Sanders and B. E. Bodenbender, Morphometric analysis of lumbar vertebra UMP 67-28: Implications for spinal function and phylogeny of the Miocene Moroto hominoid, *J. Hum. Evol.*, 26:203–237, 1994; N. Young and L. M. MacLatchy, The phylogenetic position of *Morotopithecus*, *J. Hum. Evol.*, 46:163–184, 2004.

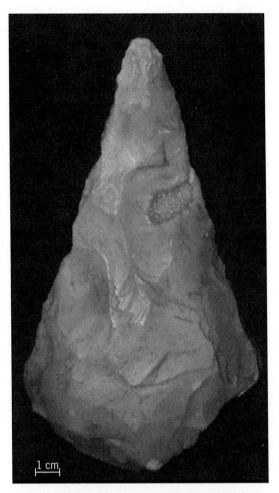

Fig. 1. An Acheulean hand axe from the Somme valley, northern France. Hand axes show variation in form, being oval, triangular, or "teardrop" in shape. Note the symmetrical form of this example and the large number of flake scars, typical of many examples from west of the Movius Line.

Movius Line

The earliest cultural traces currently date to ~2.6 Ma (millions of years ago) and are found at the site of Gona in Ethiopia, East Africa. The earliest hominin behavioral evidence is in the form of worked stone, which is referred to as Oldowan technology (named by Louis and Mary Leakey from their work at Olduvai Gorge in northern Tanzania). The Oldowan stone tool kit (a repertoire of tools of different sizes, shapes, materials, and functions) comprises simple core and flake tools, usually from locally available raw materials and expediently made. Sometime after ~1.6 Ma, a new stone tool form appeared in Africa: the Acheulean. (The term Acheulean is derived from the town of Saint-Acheul, in northern France, which is the original find site for these types of tools.) The Acheulean is represented by bifacially (worked on two sides) flaked hand axes, cleavers, and picks. The most emblematic artifact of the Acheulean is

the so-called hand axe. These are triangular, oval, or teardrop-shaped pieces that often display properties of symmetry, with a cutting edge that extends around a large proportion of their edge (**Fig. 1**). One of the major behavioral changes between the Oldowan and the Acheulean is the uniformity of design often evident in the latter stone tool technology. Technologically speaking, the Acheulean is considered to be a more advanced form of lithic (stone) technology than the Oldowan. The Acheulean eventually disappears as Levallois (prepared core) technology (named after a type site near Paris, France) became more prevalent, some 250 ka (thousands of years ago).

Demarcation by the Movius Line. Based on current paleoanthropological data, the earliest hominin dispersals out of Africa and into Asia date some time after ~2.0 Ma. It appears that these early hominins carried Oldowan technology with them. Bifaces have been found in Early Pleistocene sites outside of Africa, such as 'Ubeidya in the Jordan Rift Valley (~1.4 Ma). By at least 500 ka, Acheulean hand axes were being made across Africa, western Europe, and western Asia. In the 1940s, the eminent Paleolithic archeologist Hallam Movius noted that no hand axes were present in the eastern Old World, in either East or Southeast Asia, despite the widespread appearance of hand axes in Africa, the Levant, and India. This demarcation line (**Fig. 2**) dividing the hand axe–rich west and the hand axe–absent east became known as the "Movius Line."

Although Movius was a keen observer of archeological patterning, he perhaps went one step too far and tried to add interpretative value to the Movius Line. In fact, Movius suggested that the reason why hand axes were absent from East and Southeast Asia was that the region was a cultural backwater and culturally stagnant when compared to the western Old World. This interpretation served to ignite a controversy in East Asian Paleolithic research; in turn, it influenced a generation of indigenous archeologists to go forth and find hand axes in East Asia to disprove Movius' hypothesis. Within a few decades of Movius' observations, hand axes began to be reported from different regions of China and Korea. In particular, trihedral picks were found in Dingcun in central China, and hand axes and cleavers were reported from Chongokni in Korea. Bifacial implements have also since been reported from southern China in the Baise (Bose) Basin, and from northern China in the Luonan Basin. Bifacially worked implements have yet to be reported from Japan or Southeast Asia. The discovery of hand axes and other bifacially worked stone tools led some archeologists to argue for the abolition of the Movius Line, although this is still being debated.

Nevertheless, recent studies of the Movius Line have led to at least five important differences that still appear to be present between the sites and stone tool assemblages found east and west of the Movius Line. For example, it has been noted recently that the number of hand axe sites in East Asia is still substantially lower than the number in Africa or South Asia. Hand axes only represent a small percentage (<5%) of the lithic assemblage composition. The morphology of the East Asian hand axes is quite different from similar lithics found west of the Movius Line. In particular, the East Asian hand axes tend to be thicker, have fewer invasive flakes removed, and do not display the levels of symmetry seen in many contemporaneous western examples (**Fig. 3**). It has been shown that Levallois technology probably developed from Acheulean technology. Accordingly, if a true Acheulean technology is present in East Asia, then Levallois prepared core technology should also be present. However, Levallois technology only appears sporadically in northern China and appears rather later than in western Acheulean regions. Furthermore, the Levallois has not been reported at any of the four primary regions in East Asia where bifaces have been reported. Finally, a fifth characteristic to be noted is that none of the hand axes east of the Movius Line has been proposed to be older than ~800 ka, with most of the assemblages dating to the latter part of the Middle Pleistocene (~780–127 ka) and Late Pleistocene (~127–10 ka), although hand axes from west of the line date from ~1.6 Ma.

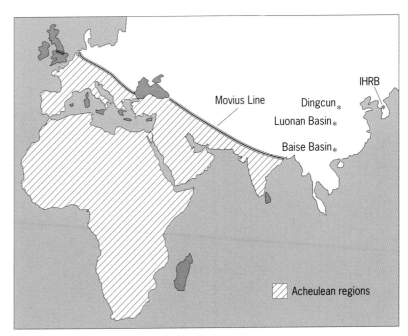

Fig. 2. The Movius Line. Although Acheulean hand axes are frequently found in large numbers in Africa, the Near East, the Indian subcontinent, and western Europe, they are found only rarely in East Asia. Small numbers of East Asian hand axes have been found at sites in the Baise Basin, in the Luonan Basin, and in Dingcun (all in China), and in the Imjin/Hantan River basins (IHRB) in Korea. This relative scarcity suggests that the Movius Line is still valid.

Explaining the Movius Line. Since Movius published his observations, a number of hypotheses have been put forth to best explain the absence or paucity of bifacially worked implements east of the Movius Line. The primary explanations are related to cultural isolation, raw material constraints, environmental barriers, and small population sizes.

1. *Hominin dispersals predated the advent of Acheulean technology.* The proposed explanation here is that the early hominin dispersals into Asia predated the development of Acheulean technology (>~1.6 Ma). Thus, the initial wave of hominin dispersal would have been by hominins wielding Oldowan stone tools. Hominin presence at sites that are possibly older than ~1.6 Ma, such as Dmanisi (Republic of Georgia), Yuanmou (China), and Mojokerto (Indonesia), lends support for this hypothesis.

2. *Raw material constraints.* In East Asia, most lithic assemblages are dominated by core and flake tools produced on poor-quality quartz and quartzite. The argument is that quartz and quartzite are not conducive to producing refined Acheulean bifacial implements. Although quartz and quartzite are probably the most common raw materials in Early and Middle Pleistocene lithic assemblages, Acheulean-like hand axes have been produced from them. A good example is the bifacial implement excavated from Zhoukoudian Locality 1 (China).

3. *"Bamboo Hypothesis."* It has been noted that at different times during the Pleistocene much of East Asia would have been covered with bamboo forests. Bamboo is a good-quality raw material for producing tools that can replicate many of the functions of Acheulean hand axes. It was thus suggested

Fig. 3. Hand axe from Chongokni, Korea. East Asian hand axes tend to show less extensive flaking and are typically less symmetrical compared with examples from west of the Movius Line (compare with Fig. 1). With a fewer number of invasive flakes removed, East Asian examples are also often thicker than those from west of the Movius Line.

that early hominins dispersing into East Asia dropped their reliance on Acheulean bifaces and began using the more readily available bamboo instead.

4. *Environmental barriers.* Some scientists suggest that mountain ranges (for example, the Himalayas) and wide river systems (for example, Ganges-Brahmaputra) served as effective barriers that prohibited wide-scale hominin dispersals from India into China and Southeast Asia.

5. *Loss of the tradition due to small population sizes.* Regular incidences of social contact between individuals play a vital role in the continued learning of a skilled tradition such as hand axe manufacture. Cultural transmission theorists thus suggest that fluctuations in population size may have influenced the disappearance of the Acheulean tradition in East Asia. Small hominin population sizes, possibly due to point four above, would have made continuing a cultural tradition difficult, if not impossible. It has often been noted by scientists working in East Asia that population densities there were likely smaller than those in contemporaneous regions of Africa or India. Cultural traditions can disappear from the archeological record in as little as one to two generations.

Future outlook. More than a half century ago, Hallam Movius made a number of noteworthy observations regarding the nature of the Paleolithic record in the eastern Old World, some that are today considered controversial and debatable. Because of the presence of Acheulean-like bifacial implements in East Asia, some researchers have argued that the Movius Line is outdated and needs to be discarded. However, after more than 50 years of research, the Early Paleolithic stone tool kits east and west of the Movius Line still display marked differences in several key properties. Current and future research projects are being designed to further assess the variability of the stone tool kits between Africa, South Asia, and East Asia and to determine the potential causes of such variability.

For background information *see* ARCHEOLOGY; ASIA; DATING METHODS; EARLY MODERN HUMANS; FOSSIL HUMANS; MOLECULAR ANTHROPOLOGY; PALEOLITHIC; PHYSICAL ANTHROPOLOGY; PREHISTORIC TECHNOLOGY in the McGraw-Hill Encyclopedia of Science & Technology.

Christopher J. Norton; Stephen J. Lycett

Bibliography. J. D. Clark, The Acheulean industrial complex in Africa and elsewhere, in R. S. Corruccini and R. L. Ciochon (eds.), *Integrative Paths to the Past*, pp. 451–469, Prentice-Hall, Englewood Cliffs, NJ, 1994; S. J. Lycett and C. J. Norton, A demographic model for Palaeolithic technological evolution: The case of East Asia and the Movius Line, *Quaternary Int.*, 2009, in press; H. L. Movius, The Lower Palaeolithic cultures of southern and eastern Asia, *Trans. Am. Phil. Soc.*, 38:329–426, 1948; C. J. Norton, K. Bae, J. W. K. Harris, and H. Lee, Middle Pleistocene handaxes from the Korean peninsula, *J. Hum. Evol.*, 51:527–536, 2006; K. D. Schick, The Movius Line reconsidered, in R. S. Corruccini and R. L. Ciochon (eds.), *Integrative Paths to the Past*, pp. 569–596, Prentice-Hall, Englewood Cliffs, NJ, 1994.

Natural circulation in nuclear systems

Natural circulation is a process by which fluid motion is driven by a density gradient and no external source of energy is required. In general, a heat source, a heat sink, and the pipes connecting them form the essential hardware of a natural circulation system. The pipes are connected to the source (heater) and sink (cooler) in such a way that they form a continuous circulation path (**Fig. 1**). When the flow path is filled with working fluid, circulation can set in automatically following the activation of the heat source under the influence of a body field such as gravity. With both the source and sink maintained at constant conditions, a steady condition is expected to be achieved that can continue indefinitely, if the integrity of the closed loop is maintained.

Natural circulation will occur in a loop system whenever the buoyant forces caused by differences in loop fluid densities are sufficient to overcome the flow resistance of the loop components (for

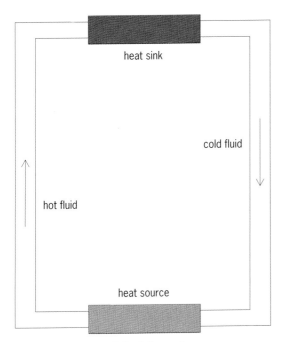

Fig. 1. A simple natural circulation system.

example, steam generators). Fluid density differences occur as a result of heating the fluid in the source region (causing the liquid to become less dense) and cooling the fluid in the sink region (causing the fluid to become denser). At the heat source, the fluid absorbs heat and heated fluid becomes lighter than the fluid surrounding it, and thus rises. At the heat sink, the fluid transfers heat and the nearby fluid becomes denser as it cools, and is drawn downward by gravity. The buoyant forces resulting from those density differences cause the fluid to circulate through the loop from the heat source to the heat sink and back again. The flow rate through the loop is limited by the sum of the resistances in the components and interconnecting piping. Because of their simplicity, natural circulation loops are widely used in energy conversion systems. To take advantage of the natural movement of warm and cool fluids, the heat source and heat sink must be at the proper elevations. Usually, the heat sink is located above the source to promote natural circulation. Such loops in which the fluid circulation is caused by the thermally induced buoyancy force are also known as natural circulation loops, thermosyphon loops, or natural convection loops.

The primary function of a natural circulation loop is to transport heat from a heat source to a heat sink. The main advantage of the natural circulation system is that the heat transport function is achieved without the aid of any fluid-moving machinery, such as pumps. The absence of moving or rotating parts to generate the motive quantity of force (motive force) for flow makes the system less prone to failure, reducing the maintenance and operating costs. The motive force for the flow is generated within the loop simply because of the presence of the heat source and heat sink. Because of this, natural circulation loops

find several engineering applications in conventional as well as nuclear industries. Notable among these applications are solar water heaters, transformer cooling, geothermal power extraction, cooling of internal combustion engines, gas turbine blades, computer cooling, and nuclear reactor cores. In addition to these industrial applications, natural circulation is also observed as weather systems, ocean currents, and household ventilation.

Passive safety systems. An important feature of several advanced reactor designs is the incorporation of passive safety systems. The International Atomic Energy Agency (IAEA) conference on *The Safety of Nuclear Power: Strategy for the Future*, convened in 1991, recommended that for new plants, "The use of passive safety features is a desirable method of achieving simplification and increasing the reliability of the performance of essential safety functions and should be used wherever appropriate." In addition to the IAEA, Europe, the United States, and other countries have established some basic goals and requirements for future nuclear power plants. In Europe, for example, the major utilities have proposed a common set of nuclear safety requirements known as the European Utility Requirements (EUR). One of the common requirements for new nuclear plants is the use of forgiving design characterized by simplicity and passive safety features, where appropriate. Consequently, nuclear plant designers select active safety systems, passive safety systems, or combinations of them for meeting their ability to perform the required safety functions with sufficient reliability as well as for their impact on plant operation and cost. A number of passive systems incorporated in advanced reactors use natural circulation as the mode of energy removal, underlining the importance of natural circulation in nuclear reactor design.

After the Three Mile Island no. 2 reactor accident in 1979, increased safety requirements and the introduction of effective and transparent safety functions led to growing consideration of passive safety systems for advanced light-water reactors (ALWRs). In the evolutionary designs, attempts have been made to reduce the complexity of the emergency core cooling system and of the long-term decay heat removal system by increased use of passive systems. Following the IAEA definitions, a passive component is a component that does not need any external input to operate, and a passive system is either a system that is composed entirely of passive components or a system that uses active ones in a very limited way to initiate subsequent passive operation. Passive safety systems are characterized by their full reliance upon natural forces, such as natural circulation and gravity, to accomplish their designated safety functions. They are also making safety functions less dependent on active systems and components, such as pumps, diesel generators, and electromotor-driven valves. ALWR design incorporates passive safety features to perform safety-related functions.

Design features proposed for the passive ALWRs include the use of passive, gravity-fed water supplies for emergency core cooling and natural circulation

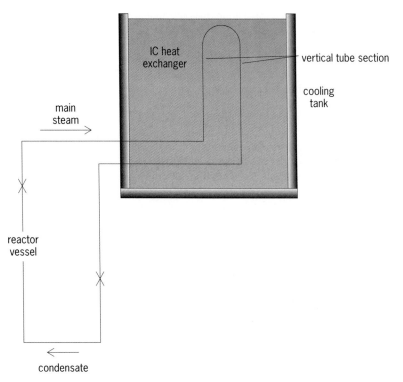

Fig. 2. Isolation condenser (IC) cooling system.

decay heat removal, and natural circulation cooling within the core for all conditions. These types of nuclear plants also use automatic depressurization systems (ADSs), the operation of which is essential during a range of accidents to allow adequate emergency core coolant injection from the lower-pressure passive safety systems. The low flow regimes associated with these designs will involve natural circulation flow paths not typical of current operational light-water reactors. Passive ALWR designs emphasize enhanced safety by means of improved safety system reliability and performance. These objectives are achieved by means of improved safety system simplification and reliance on natural forces for system operation.

Example systems. There are a number of passive safety systems considered in different plant designs. To enhance the understanding of these systems, three typical examples will be given and their operation will be explained briefly.

Isolation condenser system. The first example is the operation of the isolation condenser system (ICS). Passively cooled core isolation condensers are designed to provide cooling to a boiling water reactor (BWR) core subsequent to its isolation from the primary heat sink—the turbine/condenser set. During power operations, the reactor is normally isolated from the isolation condenser (IC) heat exchanger by closed valves (**Fig. 2**). In the event that the core must be isolated from its primary heat sink, the valves are opened and main steam is diverted to the IC heat exchanger where it is condensed in its vertical tube section. Heat is transferred to the atmosphere through the heat exchanger and isolation

condenser system/ passive containment cooling system (ICS/PCCS) pool (cooling tank). The condensate returns to the core by gravity, draining inside the tubes.

Gravity-driven cooling system. The gravity-driven cooling system is the second example case. Under low-pressure conditions, elevated tanks filled with cold borated water can be used to flood the core by the force of gravity. In some designs, the tank is capable of injecting large volumes of water into the depressurized reactor pressure vessel to keep the reactor core covered for at least 72 h following a loss-of-coolant accident. The tank has sufficient capacity to flood the entire reactor cavity. Operation of the system requires that the isolation valve be open and that the driving head of the fluid exceed the system pressure, plus a small amount, to overcome the cracking (opening) pressure of the check valves (**Fig. 3**). The performance of the gravity-driven cooling tank may be limited to a very short time, under core uncovery (loss of coolant) conditions due to steam production in the core region.

Containment passive heat removal/pressure suppression systems. The third example is of containment passive heat removal/pressure suppression systems. This type of passive safety system uses an elevated pool as a heat sink. Steam vented in the containment will condense on the containment condenser tube surfaces to provide pressure suppression and containment cooling. Two different zones of the containment, typically characterized by different pressures in case of an accident (pressure is the same during normal operation), are connected with the rising and the descending side of a pool-type steam-condenser heat exchanger (**Fig. 4**). In this case, the steam-air mixture is the working fluid, with condensate in the descending leg. The containment pressure is low (slightly above atmospheric pressure), driving forces may be low, and the working conditions may not be stable over a certain limited range. Positive driving forces may be low in this case and careful system engineering is needed.

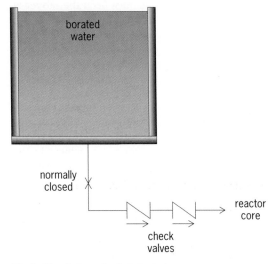

Fig. 3. Elevated gravity drain tank.

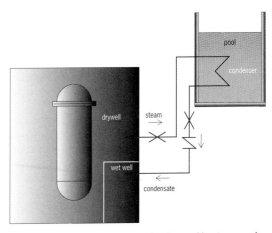

Fig. 4. Containment pressure reduction and heat removal following a loss-of-coolant accident (LOCA), using an external steam-condenser heat exchanger.

Thermal-hydraulic phenomena. As can be observed from these examples, natural circulation phenomena are very much coupled with the type of passive systems, their combinations, and also the type of the nuclear plant considered. Thermal-hydraulic phenomena and related parameter ranges that characterize the performance of passive systems do not differ, in general, from phenomena that characterize the performance of systems equipped with active components. This is specifically true for transient conditions occurring during safety-relevant scenarios. For example, friction pressure drops or heat-transfer coefficients are affected by local velocity and void fraction and not by the driving force that establishes those conditions, such as gravity head or a centrifugal pump. The same is true for more complex phenomena like two-phase critical flow or countercurrent flow limiting. A large number of thermal-hydraulic phenomena that are expected to occur in passive systems during an accident are classified in documents of the Nuclear Energy Agency of the Organization for Economic Cooperation and Development (OECD/NEA). The OECD/NEA list of phenomena for passive systems was upgraded and modified in the IAEA Coordinated Research Program on *Natural Circulation Phenomena, Modeling and Reliability of Passive Safety Systems that Utilize Natural Circulation*, considering the passive systems recently proposed by the industry. Twelve phenomena influencing natural circulation were identified and characterized for passive systems based upon the key layout of the nuclear plant designs considered: (1) behavior in large pools of liquid, (2) effect of noncondensable gases on condensation heat transfer, (3) condensation on the containment structures, (4) behavior of containment emergency systems, (5) thermofluid dynamics and pressure drops in various geometrical configurations, (6) natural circulation in closed loops, (7) steam-liquid interaction, (8) gravity-driven cooling and the behavior of the accumulator (a prepressurized reactor core flooding tank), (9) liquid temperature stratification, (10) behavior of emergency heat exchangers and isolation condensers, (11) stratification and mixing of boron, and (12) behavior of the core make-up tank (an elevated tank filled with borated water to provide coolant injection at system pressure to the reactor vessel in the event of an emergency). As can be seen from this list of identified phenomena, the field of natural circulation phenomena in nuclear plants is quite broad, and is related to other groups of phenomena that need to be considered in combination.

Outlook. The use of passive safety systems, such as accumulators, condensation and evaporative heat exchangers, and gravity-driven safety injection systems, eliminates the costs associated with the installation, maintenance, and operation of active safety systems, which require multiple pumps with independent and redundant electric power supplies. However, considering the weak driving forces of passive systems based on natural circulation, careful design and analysis methods must be used to ensure that the systems perform their intended functions, in addition to the experimental elaborations. As a result, passive safety systems are being considered for numerous reactor concepts and may potentially find applications in the future reactor concepts, as identified by the Generation IV International Forum (GIF). Another motivation for using passive safety systems is the potential for enhanced safety through increased safety system reliability.

For background information *see* BORON; CONVECTION (HEAT); COUNTERCURRENT TRANSFER OPERATIONS; DENSITY; FLUID-FLOW PRINCIPLES; FLUIDS; GRAVITY; HEAT PIPE; HEAT TRANSFER; NUCLEAR POWER; NUCLEAR REACTOR; PIPE FLOW; TWO-PHASE FLOW in the McGraw-Hill Encyclopedia of Science & Technology.   Nusret Aksan

Bibliography. N. Aksan et al., *A separate effects test matrix for thermal-hydraulic code validation: Phenomena characterization and selection of facilities and tests*, vol. I, OECD/NEA/CSNI Rep. no. (94) 82, Paris, 1993; N. Aksan and F. D'Auria, *Relevant thermal-hydraulic aspects of advanced reactor design*, OECD/NEA/CSNI Rep. no. (96)22, Paris, 1996; International Atomic Energy Agency, *IAEA Training course on natural circulation in water cooled nuclear power plants: Phenomena, models, and methodology for system reliability assessments*, IAEA-TECDOC-1474, IAEA, Vienna, 2005; International Atomic Energy Agency, *Natural circulation in water cooled nuclear power plants: Phenomena, models, and methodology for system reliability assessments*, IAEA-TECDOC, IAEA, Vienna, 2010 (in press); International Atomic Energy Agency, *Passive safety systems and natural circulation in water cooled nuclear power plants*, IAEA-TECDOC, IAEA, Vienna, 2010 (in press); International Atomic Energy Agency, *Safety related terms for advanced nuclear power plants*, IAEA-TECDOC-626, IAEA, Vienna, 1991; International Atomic Energy Agency, *Status of advanced light water cooled reactor designs*, IAEA-TECDOC-1391, IAEA, Vienna, 2004; International Atomic Energy Agency, *The safety of nuclear power: Strategy for the future*, IAEA Proceedings, Vienna, 1991.

Natural-gas-powered vehicles

Many options exist to power light-duty vehicles, which include cars, SUVs, and small trucks. Most vehicles are now powered by the combustion of gasoline, and diesel fuel is becoming increasingly common. As the transportation industry moves toward using cleaner fuels, natural gas is receiving serious consideration. Natural gas as well as biofuels and electricity from batteries or fuel cells serve the goals of improving energy efficiency, reducing or eliminating greenhouse gas emissions, reducing the cost of operating vehicles, and diversifying the available energy sources.

Powering cars with gasoline or diesel fuel, which are made from petroleum, has several major disadvantages. The price of petroleum is high, fluctuates, and is subject to international political trends. The manipulation of the international petroleum market can be a serious problem for the United States, which spent $700 billion on imported oil in 2007. As concern over global climate change increases, the combustion of gasoline and diesel fuel is recognized as both inefficient and a major contributor of greenhouse gases.

One option for reducing our dependence on imported oil is to power vehicles with natural gas, a mixture of hydrocarbons, primarily methane (CH_4). Compressed natural gas (CNG) is typically transported by pipeline. It is currently used to power buses and cars that can run directly on CNG, or by conversion of a gasoline engine.

While CNG can power vehicles, its use in the United States is rare. Only 120,000 of the 250 million cars and light-duty trucks are powered by natural gas in the United States. However, urban transit systems have increasingly turned to CNG to power their buses. As a result, nearly 25% of new transit buses on order in the United States will be CNG powered. Other nations use CNG on a wider scale, with about 7 million of the 800 million vehicles powered by natural gas.

This article emphasizes natural gas as a resource, looking both at the annual production and the reserves or resources available (the total amount extractable from the earth). Technology is often intimately linked to the availability of a particular resource, and science can be driven by needs of technology. An example is the need for efficient batteries or fuel cells for powering cars, which is stimulating research in several areas of chemistry, physics, catalysis, and engineering.

Operating principles. The energy density of several fuels is shown in **Table 1**. Gasoline and diesel fuel have exceptionally high energy density per weight and volume, thus their widespread use in transportation.

The volume density of compressed natural gas (CNG) at a pressure of 250 bars (approximately 3600 psi or pounds per square inch, or 25 MPa) is about one-third that of gasoline, so for the same range three times the volume of CNG is required compared with gasoline.

Two items should be noted. (1) The 250-bar pressure is high; scuba tanks are typically filled to only 200 bar. (2) A tank to hold CNG is typically a cylinder with rounded ends, approximately the shape of a scuba tank. It has to be strong and typically weighs much more than a tank to hold a liquid such as gasoline.

Advantages. Compressed natural gas is potentially useful where domestically produced natural gas is available. Given the limited driving range possible with storage of a reasonable quantity of CNG onboard the vehicle, and given the limited infrastructure available to supply CNG in most locations, the best applications are for fleet use, such as taxis, delivery trucks, and buses, where the vehicle returns daily to a central source for CNG refueling.

There is a premium paid for purchase of a CNG-powered car in the United States. The price of natural gas fluctuates with time, as do the prices of petroleum and gasoline. The price of CNG is generally less than that of an equivalent quantity (in energy content) of gasoline, but changes in tax and regulatory polices could affect that price relationship.

The major advantage often cited for the use of CNG as a transportation fuel in the Unites States is that it lessens the dependence on imported petroleum. However, the United States is a net importer of natural gas and the advantage of lessening the need for imported petroleum could be offset by the need to import more natural gas. This balance depends on domestic production, which is increasing in the United States as shale gas sources are being developed. World natural gas reserves are shown in **Table 2**.

Although natural gas is flammable, its flammability range is relatively narrow. Natural gas will rise into the atmosphere if a tank is punctured, which makes it safer than gasoline, which is a liquid.

Disadvantages. There are several issues concerning widespread use of CNG to power light-duty vehicles, including limited domestic production, the availability of natural gas as a resource, the very limited distribution infrastructure in the United States, and the large onboard storage tanks required for a vehicle to achieve a reasonable range.

Only 1500 locations for distribution of CNG for vehicles currently exist in the United States, and only half of them can be used by the public. By comparison, there are about 175,000 service stations selling gasoline. That lack of availability makes it nearly impossible at present to use a CNG-powered

Fuel	Volume density	Weight density
Gasoline	34.6 MJ/L	47.5 MJ/kg
Diesel	38.6 MJ/L	45.8 MJ/kg
CNG (compressed natural gas) at 250 bars	11.2 MJ/L	43.6 MJ/kg
LNG (liquefied natural gas) at $-160°C$ ($-256°F$)	26.0 MJ/L	43.6 MJ/kg

TABLE 1. Volume and weight densities of gasoline, CNG, and LNG

TABLE 2. World natural gas reserves by country as of January 1, 2009

Country	Reserves, trillion cubic feet*	Percent of world total
World	6,254	100.0
Top 20 Countries	5,674	90.7
Russia	1,680	26.9
Iran	992	15.9
Qatar	892	14.3
Saudi Arabia	258	4.1
United States	238	3.8
United Arab Emirates	214	3.4
Nigeria	184	2.9
Venezuela	171	2.7
Algeria	159	2.5
Iraq	112	1.8
Indonesia	106	1.7
Turkmenistan	94	1.5
Kazakhstan	85	1.4
Malaysia	93	1.3
Norway	82	1.3
China	80	1.3
Uzbekistan	65	1.0
Kuwait	63	1.0
Egypt	59	0.9
Canada	58	0.9
Rest of world	581	9.3

SOURCE: Worldwide look at reserves and production; *Oil & Gas J.*, 106(48):22–23, December 22, 2008.
*1 Tcf = 10^{12} ft^3 = 2.83×10^{10} m^3.

car in the United States for anything other than local trips.

However, since a flex-fuel car can run on either gasoline or CNG, it is possible to have two storage systems (CNG and gasoline) onboard, but the two storage systems add weight to a vehicle, cutting its efficiency. There is also a potential for fueling a car's tank with natural gas at home, taking advantage of the large infrastructure supporting home use of natural gas for heating. The market potential remains to be explored.

Honda Civic GX. The Honda Civic GX is the only CNG-powered car commercially available in the United States. It runs like a traditional gasoline-powered Honda Civic LX, but with less horsepower and more limited range. While they both have about the same mileage results, the GX holds the equivalent of 8 gal (30 liters) of gasoline, versus 13 gal (49 L) in the LX. Emissions from the GX (compressed natural gas) are considerably lower than from the LX (gasoline).

Natural gas has an "octane" rating of 130, considerably higher than that of the usual formulations of gasoline (about 86–92). This high octane rating allows use of a higher compression ratio (14:1) than gasoline, with a potential increase of power of 15%. (The Honda Civic GX has a compression ratio of 12.5:1 versus 10.5:1 for LX.) However, for the same displacement engine the GX has 113 hp (84 kW), while the LX has 140 hp (104 kW), according to Honda.

Supply issues: annual production. The United States is a net importer of natural gas and is likely to remain so for the foreseeable future. The annual consumption of natural gas is nearly 30 trillion cubic feet (1 Tcf = 10^{12} ft^3 = 2.83×10^{10} m^3), of which the

United States imports approximately 6 Tcf, primarily from Canada and Mexico.

Annual natural gas production globally is about 115 Tcf, according to U.S. Energy Information Administration's (EIA) *International Energy Outlook 2009*, with Russia being the largest producer (24 Tcf) followed by the United States at 20 Tcf, and the Middle East at 18 Tcf.

Supply issues: reserves. World reserves (Table 2) are estimated to be about 6300 Tcf, a number that has risen steadily as new sources have been discovered. Significant natural gas discoveries made in 2007, including some in the United States, are not included in this total. Three countries have more than half the world's total reserves: Russia, 27%, Iran 16%, and Qatar 14%. Smaller reserves are in Saudi Arabia, the United Arab Emirates, the United States, Nigeria, Venezuela, and other countries.

The United States has natural-gas reserves of at least 211 Tcf, and maybe more. Estimates of natural gas yet to be discovered (total resource) vary by large factors. The U.S. EIA cites 210–240 Tcf; the U.S. Geological Survey cites 364 Tcf, including coal-bed methane. There are claims of even higher reserves from unconventional deposits, such as from shale and possibly from methane clathrates (hydrates), the latter of unknown potential.

The annual domestic consumption in the United States is about 30 Tcf, which means that with the EIA's proven reserves estimate of 210–240 Tcf, the country has less than a 10-year supply of natural gas. This estimate will be revised as new discoveries are verified.

Energy efficiency and greenhouse gas emission. According to the U.S. Environmental Protection Agency (EPA), natural gas produces 60–90% less smog-generating pollutants than gasoline, and 30–40% less greenhouse gas (GHG) emissions. However, because natural gas is methane, and methane itself is a potent greenhouse gas, any leaks into the atmosphere reduce the clean fuel advantage of burning natural gas.

Comparison of fuel economy for CNG and gasoline versions of the 2008 Honda Civic shows that the gasoline-fueled Civic gets 26 miles per U.S. gallon (mpg) or 9 liters/100 km city and 34 mpg (7 liters/100 km) highway, while the CNG-powered version with a similar engine size gets 26 mpg (9 liters/100 km) city and 31 mpg (7.5 liters/100 km) highway. Fuel economy is thus slightly worse for the CNG version than for the gasoline version.

The EPA reports that emission of carbon dioxide is generally 25% less for a CNG-powered car than for one powered by gasoline. As always, the actual miles-per-gallon achieved depends on driving profiles and other variables. For example, the Honda Civic gasoline car shows 6.3 tons per year emission versus 5.4 tons per year for the CNG version. The EPA also claims a reduction of emissions of carbon monoxide of greater than 90% and a twofold reduction in the emissions of nitrogen oxide. With CNG, there is very little emission of particulate matter, according to the EPA.

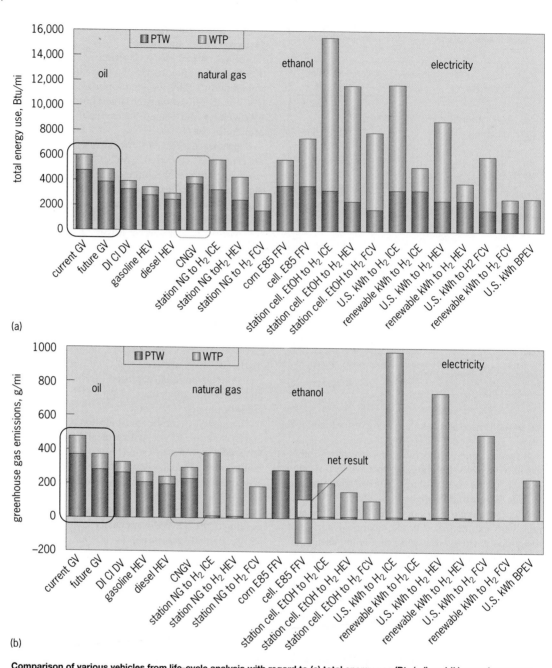

Comparison of various vehicles from life-cycle analysis with regard to (*a*) total energy use (Btu/mi) and (*b*) greenhouse gas emissions (g/mi). WTP is "well-to-pump," PTW is "pump-to-wheels," and total height of bar is "well-to-wheels" or total energy use. GV is gasoline-powered vehicle (black box) and CNGV is compressed natural gas vehicle (gray box). Other abbreviations: BPEV = battery-powered electric vehicle, cell. = cellulosic, DI CI DV = direct-injection compression-ignition diesel vehicle, E85 = mixture of 85% ethanol and 15% gasoline (by volume), EtOH = ethanol, FCV = fuel-cell vehicle, FFV = flexible fuel vehicle, HEV = hybrid electric vehicle, ICE = internal combustion engine, kWh = kilowatt-hour (electrical energy), NG = natural gas, station = service station. 1 Btu/mi = 0.656 kJ/km. 1 g/mi = 0.621 g/km. (*Argonne National Laboratory Transportation Technology R&D Center*)

The "well-to-wheels" energy efficiency for various vehicles has been studied at Argonne National Laboratory. Results for total energy use (Btu/mi) and greenhouse-gas emissions (g/mi) are shown in the **illustration**. As illus. *a* shows, a CNG vehicle (CNGV) has somewhat overall lower energy use than even a future gasoline vehicle (future GV), but not as low as a gasoline or diesel hybrid vehicle. Illustration *b* shows a similar result for greenhouse-gas emissions.

Prospects. Powering cars and light-duty trucks with CNG is a viable option for replacing gasoline as fuel, with some caveats. A CNG-powered car has fuel economy that is similar to a gasoline-powered car, and emits less carbon dioxide and pollutants. However, CNG cars cost more.

CNG may cost less than gasoline per mile traveled at the present time. This difference is not great and is subject to change. The range of a CNG-powered car is typically no more than half that of a gasoline-powered car, and there are very few CNG fueling stations in the United States, and currently no viable home-fueling options. Development of a national CNG fueling infrastructure would be hugely expensive and is unlikely in the foreseeable future.

Overall, the outlook for powering cars with electricity, once more efficient and inexpensive batteries are developed, is brighter. Compared to CNG-powered vehicles, battery-powered electric vehicles will have much higher energy efficiency, as well as carbon dioxide emissions that are potentially much lower, possibly even zero. Much of the distribution infrastructure is already in place for electricity. Although electricity must be produced from primary sources, some of which are highly polluting, as the country moves to cleaner and renewable sources for electricity, its advantage over CNG will only increase.

For background information *see* AIR POLLUTION; ALTERNATIVE FUELS FOR VEHICLES; AUTOMOBILE; COALBED METHANE; DIESEL FUEL; FUEL GAS; GASOLINE; GLOBAL CLIMATE CHANGE; METHANE; NATURAL GAS; SMOG in the McGraw-Hill Encyclopedia of Science & Technology. Alfred S. Schlachter

Neutron-rich atomic nuclei

Atomic nuclei consist of neutrons and protons bound by the strong force. Guided by the quantum-mechanical nature of the nuclear many-body system—driven, for example, by the Pauli exclusion principle and the concept of isospin—the attraction between a neutron and a proton is stronger than the attraction between two protons or between two neutrons. As a consequence, most stable, light nuclei are composed of equal numbers of protons and neutrons. Starting at about calcium (20 protons), the Coulomb repulsion between the positively charged protons reduces the binding of the nuclear system so that the most stable isotopes of an element have more neutrons than protons. Of the 3000 nuclei that have been produced in laboratories, only about 300 are stable. The other nuclei are short-lived and prone to decay until a stable nucleus is reached. The study of these so-called exotic nuclei that often exist only for fractions of a second has proven crucial for the understanding of the nuclear force and continues to provide important input for nuclear astrophysics in the quest to explain the isotopic composition of the universe.

Limits of existence. In the nuclear chart, nuclei are arranged by plotting the number of protons (Z) versus the number of neutrons (N) [**Fig. 1**]. The chart is limited on the left and on the right by the proton and neutron drip lines, respectively. For each element, the proton drip line connects the nuclei with the smallest possible neutron number—with one neutron less, the system becomes unbound. Correspondingly, the neutron drip line outlines the maximum number of neutrons that can be bound to a given number of protons; an additional neutron cannot be bound anymore. The proton drip line proceeds much closer to the stable nuclei than the neutron drip line because the Coulomb repulsion limits the number of protons that can be bound by a certain number of neutrons. As a consequence, the proton

drip line has essentially been reached up to $Z = 90$ in nuclear physics experiments, whereas the neutron drip line is firmly established only for the lightest elements—up to oxygen ($Z = 8$). A fundamental question in nuclear physics is how many neutrons a given number of protons can bind.

Experimental quest. Experiments aimed at establishing the existence of the most neutron-rich nuclei around the neutron drip line for $Z = 8$–16 are extremely challenging, because these nuclei are produced with only very small cross sections if they can be reached at all. The experimental task is threefold: The exotic, short-lived nuclei have to be produced, detected, and unambiguously identified. Fragmentation of high-energy stable nuclei impinging on stable targets, followed by in-flight separation of the reaction residues and their detection in less than 1 μs, is at present the only approach available to accomplish this task.

In the United States, this capability exists at the National Superconducting Cyclotron Laboratory (NSCL) at Michigan State University. Recently, a highly energetic beam of stable calcium-48 (^{48}Ca; $Z = 20, N = 28$) accelerated by the laboratory's coupled cyclotron facility was fragmented upon collision with a 970-μm-thick natural tungsten target. The reaction products, fully stripped of their electrons and thus positively charged ions, were separated by the A1900-S800 two-stage magnetic fragment separator and guided onto a detection system that records, event by event, the energy loss, total kinetic energy, flight time, and position of each reaction residue transmitted by the system. In NSCL's fragment separator, the production target is followed by superconducting quadrupole magnets that focus the ions produced, similar to light collection by an optical lens. Those focusing magnets are combined with dipole magnets that bend and filter ions in the same way that

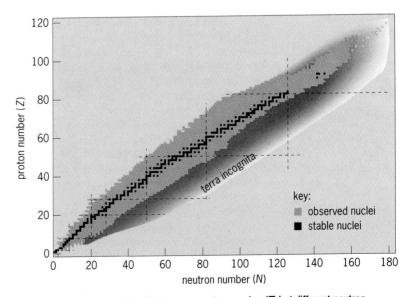

Fig. 1. Nuclear chart. Nuclei with the same proton number (**Z**) but different neutron numbers (**N**) are called isotopes of an element. (**Z** = 12 corresponds to the element magnesium, and ^{20}Mg and ^{40}Mg, for example, are two of the magnesium isotopes.) Of all nuclei, only about 300 nuclei are stable, about 3000 have been produced in laboratories, and many more are predicted to exist.

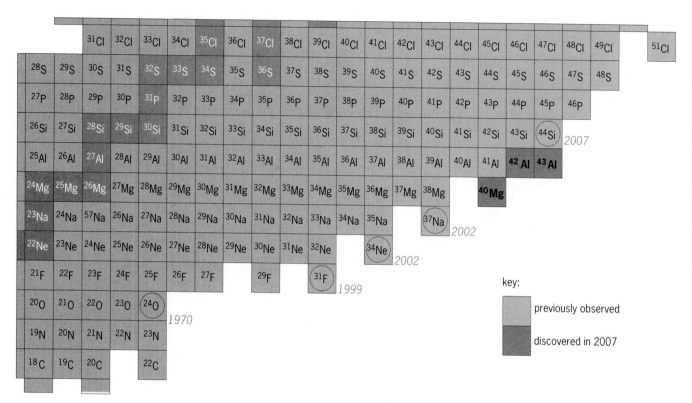

Fig. 2. Part of the nuclear chart from carbon (C; $Z = 6$) to chlorine (Cl; $Z = 17$). The neutron drip line is established only up to oxygen (O); for all heavier elements, more neutron-rich nuclei than presently observed could exist. Marked are the most neutron-rich nuclei above oxygen, along with their year of discovery. The time between discoveries is indicative of the challenge of the corresponding experiments. Nuclei corresponding to ^{39}Mg ($Z = 12$, $N = 27$), ^{36}Na ($Z = 11$, $N = 25$), and ^{33}Ne ($Z = 10$, $N = 23$), for example, have not been found to exist, for the reason discussed in the text. (*Adapted from T. Baumann et al., Discovery of ^{40}Mg and ^{42}Al suggests neutron drip-line slant towards heavier isotopes, Nature, 449:1022–1024, 2007*)

glass prisms bend light and split it into its components. The additional filtering provided by the analysis section of the S800 spectrograph works according to the same principle. A detection system composed of two thin plastic timing scintillators behind the A1900 fragment separator and in the S800 beamline and a stack consisting of a position-sensitive parallel-plate avalanche counter, seven silicon PIN diodes, and a thick plastic scintillator was used to unambiguously identify the transmitted nuclei on an event-by-event basis from their flight time, energy loss, and position. Over a total period of 7.6 days and with an average rate of 5.0×10^{11} ^{48}Ca beam particles per second, three events of magnesium-40 (^{40}Mg; $Z = 12$, $N = 28$) were unambiguously detected and identified among the zoo of less exotic reaction residues. Also, the existence of aluminum-42 (^{42}Al; $Z = 13$, $N = 29$) could be proven, and one candidate event of the even heavier nucleus aluminum-43 (^{43}Al) was identified.

Figure 2 displays the section of the nuclear chart around the nuclei discussed in this article. The nuclei corresponding to magnesium-39 (^{39}Mg; $Z = 12$, $N = 27$), sodium-36 (^{36}Na; $Z = 11$, $N = 25$), and neon-33 (^{33}Ne; $Z = 10$, $N = 23$) have not been found to exist, while their more neutron-rich neighbors have been observed. The reason for the likely nonexistence of these nuclei is the attractive pairing interaction that makes isotopes with even numbers of neutrons more stable than isotopes that have an unpaired neutron.

Previous attempts to probe the existence of ^{40}Mg at laboratories in France and Japan had been unsuccessful. Aside from the high intensity of the ^{48}Ca beam at NSCL, the success of the measurement can be attributed to the experimental approach of combining NSCL's two magnetic fragment separators—the A1900 fragment separator and the analysis section of the S800 spectrograph—to reach the high degree of suppression of unwanted nuclei that is crucial in the regime of very rare events. **Figure 3** shows the event-by-event identification of all nuclei produced and transmitted to the detector setup in the measurement described above.

Implications of the results. While most theories that attempt to predict the limits of nuclear existence calculate ^{40}Mg to be bound, the existence of ^{42}Al came as a surprise for many nuclear models—with potentially far-reaching consequences. The existence of ^{42}Al indicates that the neutron drip line might be much farther out than previously suspected, meaning that many more neutron-rich nuclei might exist than expected. This argument is related to the shell structure of the nucleus.

In a simple independent-particle picture, each neutron or proton occupies and moves in a bound quantum-mechanical orbit within an average potential that is generated by all the others. This "shell-model" concept has been instrumental for the description of systematic as well as detailed properties of atomic nuclei. In particular, it predicts that nuclei

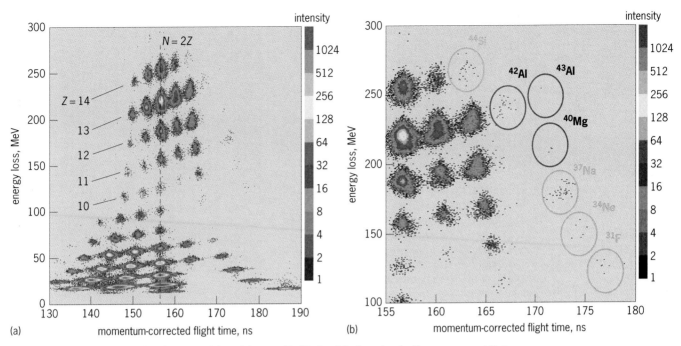

Fig. 3. Identification of the produced neutron-rich nuclei transmitted to the detector setup for the measurement that discovered ^{40}Mg, ^{42}Al, and potentially ^{43}Al. Plotted is the energy loss measured in silicon PIN diodes versus the measured flight time. All species can be separated. Gaps in the regular pattern indicate the unbound nuclei ^{39}Mg, ^{36}Na, and ^{33}Ne discussed in the text. The intensity is illustrated by a gray scale; the most neutron-rich species were produced less abundantly than the isotopes closer to stability. (*a*) All nuclei observed in the experiment. (*b*) Nuclei around the region of ^{40}Mg. (*Adapted from T. Baumann et al., Discovery of ^{40}Mg and ^{42}Al suggests neutron drip-line slant towards heavier isotopes, Nature, 449:1022–1024, 2007*)

are more stable when they have certain numbers of neutrons or protons to exactly fill given orbitals or "shells." At those "magic numbers"—2, 8, 20, 28, 50, 82, and 126—a large gap in energy to the next available level occurs, and as a result, the nucleus is more stable. For ^{42}Al ($Z = 13$, $N = 29$), neutrons are expected to start to fill the $p_{3/2}$ orbit, which can hold 4 neutrons. Calculations predict that the energy of this orbital will remain constant as neutrons are added. This indicates that—with the observation of ^{42}Al—one might expect aluminum nuclei out to aluminum-45 (^{45}Al; $Z = 13$, $N = 32$), where the $p_{3/2}$ orbit is fully occupied, to be bound. Even heavier aluminum isotopes might be bound if the energy splitting between the $p_{3/2}$ and $p_{1/2}$ orbitals becomes small, as suggested by some calculations that include forces with tensor terms.

The nuclear potential is well understood for stable nuclei, and shell-model calculations successfully describe many properties of stable nuclei, for example, energy excitation spectra and decay properties. However, for neutron-rich nuclei, modifications to the nuclear structure as known from stability have been observed: New magic numbers develop and conventional shell gaps break down. The driving forces behind those modifications are, for example, parts of the interactions between protons and neutrons that depend on the imbalance of N and Z in the nucleus (spin-isospin–dependent terms).

An understanding of the nuclear structure of exotic nuclei is not only of fundamental interest in nuclear physics, it is also crucial for the field of nuclear astrophysics, where the properties of short-

lived exotic nuclei are important input to nucleosynthesis models that describe the origin of the elements in the cosmos. In the quest to measure properties of neutron-rich nuclei—with the proof of existence being the most fundamental—experimentalists are joined by theorists in a worldwide effort to arrive at a theory for the structure of the atomic nucleus with predictive power across the nuclear chart.

For background information *see* EXCLUSION PRINCIPLE; EXOTIC NUCLEI; I-SPIN; ISOTOPE; MAGIC NUMBERS; NUCLEAR STRUCTURE; NUCLEOSYNTHESIS; PARTICLE ACCELERATOR; RADIOACTIVITY in the McGraw-Hill Encyclopedia of Science & Technology.

Alexandra Gade

Bibliography. T. Baumann et al., Discovery of ^{40}Mg and ^{42}Al suggests neutron drip-line slant towards heavier isotopes, *Nature*, 449:1022–1024, 2007; B. A. Brown, The nuclear shell model towards the drip lines, *Prog. Part. Nucl. Phys.*, 47:517–599, 2001; R. F. Casten, *Nuclear Structure from a Simple Perspective*, 2d ed., Oxford University Press, 2000; M. Thoennessen, Reaching the limits of nuclear stability, *Rep. Prog. Phys.*, 67:1187–1232, 2004.

New drug to control tuberculosis

Tuberculosis is a disease that infects approximately one-third of the world's population. Many individuals have latent tuberculosis, in which they are infected with *Mycobacterium tuberculosis*, but the bacterium is dormant in the lungs, causing no signs of infection. Latent infection is difficult to treat because the bacteria are not dividing while they exist in this dormant state. Antimicrobial drugs work to kill bacteria when they are actively growing and multiplying; latent tuberculosis cannot, therefore, be treated effectively until the bacteria begin to multiply. A new drug, PA-824, produces nitric oxide (NO) gas when it is metabolized inside the bacterial cell. This gas can kill the bacterium from within, "exploding" it and causing bacterial death even when the bacterium is not multiplying, which may in turn lead to new and effective treatments for latent tuberculosis.

Background. Tuberculosis (TB) is caused by *M. tuberculosis*, which is an acid-fast rod-shaped bacterium. This species has an unusual cell wall structure, with waxy mycolic acids (a type of complex fatty acids) in lieu of the typical Gram-positive cell wall structure. *Mycobacterium tuberculosis* divides slowly, with 16 to 20 h between each cell division. In contrast, fast-growing bacteria such as *Escherichia coli* can reproduce and divide in approximately 20 minutes. This slow growth is why treatment for TB consists of months of antibiotics use in order to kill all bacteria in a patient with TB.

The species *M. tuberculosis* has existed within humans for thousands of years. Recently, DNA consistent with five different strains of *M. tuberculosis* has been detected in remains dated approximately 9000 years old. In addition, high-performance liquid chromatography (HPLC) detected mycolic acid lipid biomarkers that are found in mycobacteria, confirming the presence of this disease in prehistoric times.

Mycobacterium tuberculosis is among the most difficult of bacteria to kill, both within and outside the body. It is very resistant to drying, disinfectants, and strong acids, and it can survive for long periods of time outside the body. The disease is spread when individuals inhale airborne organisms from a person with TB. These organisms pass into the lungs, where they are taken up by macrophages (large phagocytic cells of the immune system). They survive inside these host cells by inhibiting the production of enzymes necessary for the destruction of bacterial cells. In addition, *M. tuberculosis* possesses a mechanism that protects it from nitric oxide, which is a toxic substance produced by the macrophage cells. After 2 weeks, the host develops a delayed hypersensitivity response. This response surrounds the bacteria with more macrophages and B and T lymphocytes, forming a walled-off area called a granuloma. These granulomas are referred to as tubercles (**Fig. 1**) and are characteristic of the disease. The mycobacteria present in the tubercles remain alive but do not multiply, causing a latent infection that is difficult to treat.

It is estimated that 10 million Americans are infected with *M. tuberculosis*; most of these are likely to have latent TB and show no symptoms of the disease. Reactivation TB can occur in these patients when an individual's immune system becomes impaired by some other factor, such as advanced age or acquired immune deficiency syndrome (AIDS). In these cases, the bacteria resume growing and cause

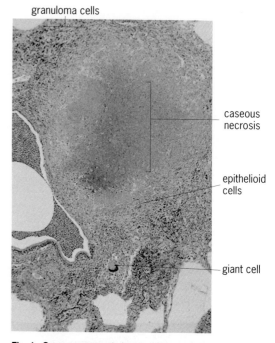

Fig. 1. Caseous necrosis has occurred in the center of this tubercle. Granuloma cells, including fibroblasts, lymphocytes, and epithelioid cells, surround the center of the caseous lesion. (*Courtesy of M. K. Cowan and K. P. Talaro, Microbiology: A Systems Approach, 2d ed., McGraw-Hill, New York, 2008*)

symptomatic active TB. Macrophages become activated to destroy the bacteria, releasing enzymes into the infected tissues. This causes death of the tissue, leading to caseous necrosis, in which dead tissue forms a cheeselike mass in the lungs. This can lead to the development of lung cavities. Individuals with active TB can transmit the organisms to other people by coughing or spitting.

Tuberculosis is prevented in many parts of the world through the use of a vaccine containing a living attenuated mycobacterium known as Bacillus Calmette-Guérin (BCG). BCG is not used in the United States because the preferred diagnosis method is the tuberculin or Mantoux skin test. Individuals who receive the BCG vaccine will test positive for TB, which eliminates the possibility of diagnosing infection early in the disease when it is easiest to treat. BCG is also unsafe for use in immunocompromised patients, as the attenuated mycobacterium can spread throughout the patient because his or her immune system cannot contain it. This has caused disease in a number of patients who have received the vaccine, including AIDS patients and children born with inherited immune deficiency diseases.

In the United States, when an individual has a positive tuberculin test (**Fig. 2**), it indicates that he or she has been infected with *M. tuberculosis*, but it does not necessarily indicate active infection. Many individuals who test positive for the bacteria have the latent form of the disease. If an individual tests positive, an x-ray of the lungs is used to detect tubercles. Patients with active TB are treated, as are patients who are identified as having latent TB or being at high risk for disease with TB. These include babies, young children, and the elderly, as well as those with human immunodeficiency virus (HIV) infection, diabetics, and drug abusers. If any members of these groups are exposed to TB, they are at high risk and should be treated to prevent disease in the future.

Probably the most significant issue with TB today is the development of antimicrobial resistance. One of the most dangerous forms of TB is extensively drug-resistant TB (XDR-TB), which is resistant to almost all drugs used to treat TB. Other strains of the mycobacteria with limited resistance to a number of drugs may cause multiple-drug-resistant TB (MDR-TB).

Because mutations can occur frequently in *M. tuberculosis*, two or more medications are used in combination to treat TB. Rifampin and isoniazid (isonicotinic acid hydrazide, INH) are often used together to treat both active and latent TB. These two drugs are referred to as first-line drugs, meaning that they are the best drugs available. Each drug kills actively growing organisms as well as metabolically inactive organisms found in tubercles. As a result, treatment must be prolonged, with at least 6 months of medication to cure the disease. In patients with compromised immune systems, this treatment period can last as long as a year. When a patient has MDR-TB or XDR-TB, rifampin and INH may not be effective. In these cases, second-line drugs, which are generally more toxic and less effective, must be used. This, in turn, leads to the need for the development of new and more effective antituberculosis medications.

PA-824 and its mechanism of action. PA-824 is a new drug that is a member of the bicyclic nitroimidazoles, which are a group of drugs currently in human trials. These drugs are active against both replicating and nonreplicating bacteria, with the latter being most likely responsible for latent TB. The mechanism of bacterial killing by the bicyclic nitroimidazoles, though, has taken time to elucidate. PA-824 and other members of this group disrupt the formation of mycolic acids in the cell envelope. However, because these are not really formed under intracellular conditions when the bacteria are not replicating, this cannot be the killing mechanism of the drug.

A team of biochemists has determined the mechanism by which PA-824 works to kill nonreplicating bacteria. The activity of this drug appears to depend on the presence of three specific bacterial molecules: a deazaflavin cofactor, F_{420}; an F_{420}-dependent glucose 6-phosphate dehydrogenase, known as Fgd1; and a deazaflavin (F_{420})–dependent nitroreductase, known as Ddn. (Note that deazaflavins are flavin coenzymes in which a nitrogen has been replaced by carbon.) It was found that mutants deficient in any of these three molecules were strongly resistant to the effects of PA-824.

Ddn reduced PA-824 to three reduction products, with the most significant being des-nitroimidazole (des-nitro). Formation of des-nitro generated reactive nitrogen species, including nitric oxide. NO is produced by macrophages after they engulf bacteria for destruction and plays a significant role in the immune response to mycobacterial infections. It has

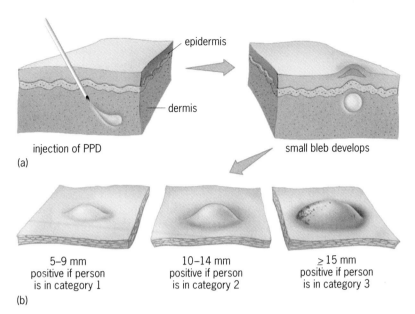

injection of PPD

epidermis

dermis

small bleb develops

(a)

(b)

5–9 mm
positive if person
is in category 1

10–14 mm
positive if person
is in category 2

≥ 15 mm
positive if person
is in category 3

Fig. 2. Skin testing for tuberculosis, using the Mantoux test. (*a*) Tuberculin PPD (purified protein derivative) is injected into the dermis. A small bleb develops, but is absorbed within a short time. (*b*) After 48 to 72 hours, the skin reaction is read and given a rating based on the size of the raised area. A reaction of less than 5 mm is negative in all patients. Category 1 includes, for example, individuals who have had contact with actively infected TB patients and HIV-positive patients; category 2 includes HIV-negative intravenous drug users, persons with medical conditions that put them at risk for progressing from latent TB to active TB, and children who have contact with members of high-risk adult populations; category 3 includes individuals who do not meet criteria in the other categories. (*Courtesy of M. K. Cowan and K. P. Talaro, Microbiology: A Systems Approach, 2d ed., McGraw-Hill, New York, 2008; and John D. Cunningham/Visuals Unlimited*)

been shown that mice lacking the ability to produce NO are extremely susceptible to infection with *M. tuberculosis*. The production of NO through the breakdown of PA-824 by bacterial enzymes means that the drug may be able to cause self-destruction of the bacteria through its own metabolism. Because humans do not produce F_{420} or any enzyme equivalent to Ddn, PA-824 has no effect on human cells, which bodes well for use of this drug as a therapeutic agent. Indeed, because many bacteria have enzymes in the same family as Ddn, it is possible that this drug and related molecules may be useful in other types of bacterial infections. In light of increasing antimicrobial resistance worldwide, PA-824 and its potential for therapeutic applications is of significant medical interest and importance.

For background information *see* ANTIBIOTIC; DRUG RESISTANCE; EPIDEMIC; INFECTIOUS DISEASE; MYCOBACTERIAL DISEASES; NITRIC OXIDE; PUBLIC HEALTH; TUBERCULOSIS; VACCINATION in the McGraw-Hill Encyclopedia of Science & Technology.

Marcia M. Pierce

Bibliography. A. Carter et al., Tuberculosis and the city, *Health Place*, 15(3):777–783, 2009; M. K. Cowan and K. P. Talaro, *Microbiology: A Systems Approach*, 2d ed., McGraw-Hill, New York, 2008; I. Hershkovitz et al., Detection and molecular characterization of 9000-year-old *Mycobacterium tuberculosis* from a Neolithic settlement in the Eastern Mediterranean, *PLoS ONE*, 3(10):e3426, 2008; E. Nester et al., *Microbiology: A Human Perspective*, 6th ed., McGraw-Hill, New York, 2008.

New malaria vaccine

A new malaria vaccine recently tested in Kenya and Tanzania has halved the risk of developing the disease in children. The vaccine, known as RTS,S, was developed by GlaxoSmithKline and is designed to prevent the destructive lysis of red blood cells after the liver is infected by *Plasmodium falciparum*. This species is responsible for the highest mortality and morbidity among victims of malaria and is considered more virulent than other members of the genus. Prevention of malaria in populations living in endemic areas could lead to the control and possibly even the elimination of malaria in some regions.

History of malaria. Malaria has caused death and illness in humans since prehistoric times. Early descriptions of malarial fevers have been found in ancient Chinese medical writings, dating back as far as 2700 B.C.E. The very name, malaria, is from the Latin *malus aria*, or bad air, indicating the belief that the disease originated from the air coming from swamps and marshes.

One of the earliest effective treatments for malaria, developed in China during the second century B.C.E., was derived from the Qinghao plant, or *Artemisia annua*. This plant is more commonly known as the annual or sweet wormwood, and its active ingredient, artemisinin, is an effective antimalarial drug that is still in use today. Quinine, another substance that is still used today to treat malaria, was discovered by the Spanish in the New World in the 1600s, where indigenous peoples used the bark of the cinchona tree as a medicine. This bark became known as Jesuit's bark in Europe because it was first discovered by Jesuit missionaries in South America. The bark was steeped in wine to dissolve the alkaloids present, and then administered to patients.

The parasite responsible for causing the disease was not discovered until 1880, when a French surgeon, Charles Laveran, studied the blood of a patient suffering from a fever for 15 days and found the parasite inside red blood cells. Studies by other scientists led to the discovery and naming of four different species of the malarial parasites: *Plasmodium malariae*, *P. vivax*, *P. falciparum*, and *P. ovale*.

The mode of transmission of malaria remained in contention until 1897, when Ronald Ross, a British medical officer, demonstrated that the parasite could be transmitted to mosquitoes from infected humans. Further evidence was gained shortly thereafter by the Italian scientist Giovanni Grassi, who sent mosquitoes that had fed on infected patients in Rome to London, where they fed on two volunteers, who subsequently developed malaria.

Epidemiology of malaria. Malaria is transmitted to a patient through the bite of the female *Anopheles* mosquito. The incubation period varies, ranging from 7 to 30 days before signs and symptoms of disease begin. Symptoms of malaria range from relatively mild fever, chills, headaches, nausea, and vomiting, to extremely dangerous forms of the disease, including cerebral malaria involving the brain, anemia due to lysis of red blood cells, and hemoglobin-

uria, or hemoglobin in the urine. Severe disease is most often associated with *P. falciparum* infection; severe malaria should be treated as a medical emergency because of its ability to cause severe illness and death.

The life cycle of *Plasmodium* species must include both the vertebrate (humans) and the invertebrate host (mosquitoes) [**Fig. 1**]. The first stage occurs when the *Anopheles* mosquito injects sporozoite forms of the parasite into the human host. These travel to the liver, where they infect liver cells and mature into the schizont form of the parasite. The schizont then ruptures, releasing merozoites, which enter red blood cells and multiply. In the red blood cell, the merozoite develops into either a trophozoite or a gametocyte. The trophozoite can then develop into a schizont, which will cause rupture of the blood cells. The cycle of recurrent fever and chills seen in malaria is due to the synchronous rupturing of the schizonts in the red blood cells.

The male and female gametocytes produced in the red blood cells are taken up by *Anopheles* mosquitoes during a blood meal and serve as the starting point for the sexual cycle. The gametocytes undergo sexual reproduction, forming zygotes in the mosquito's stomach; these then elongate, differentiate, and invade the midgut wall. Following this, they develop into oocysts. These grow larger until they rupture, releasing sporozoites, which travel to the salivary glands of the mosquito. The mosquito then transmits the sporozoites to humans during a blood meal, thus perpetuating the malaria cycle.

Modern efforts to control malaria. Malaria is controlled and prevented using one of two major strategies: (1) preventing infection by avoiding bites by parasite-carrying mosquitoes, and (2) preventing disease by taking antimalarial drugs prophylactically. These drugs do not prevent infection; rather, they prevent the parasites from multiplying in the blood, thus suppressing infection. Because nearly all attempts at developing a successful antimalarial vaccine have failed, efforts to control the disease have been limited to these two strategies. Particular focus has been directed toward preventing infection in developing countries through the use of insecticide-treated bed nets (ITNs) [**Fig. 2**]. These bed nets allow individuals to sleep protected from mosquitoes, thus lowering their risk of infection with *Plasmodium*. It is estimated that approximately 730 million bed nets impregnated with long-lasting insecticides would need to be distributed throughout endemic regions by the end of 2010 to successfully reduce malarial infection.

In addition to ITNs, other initiatives call for indoor spraying with insecticide for 172 million households and preventative treatment for 25 million pregnant women annually. In order to be effective, approximately 1.5 billion diagnostic tests for identifying malaria infection as well as millions of doses of drugs to treat *P. falciparum* and *P. vivax* infections will also be needed.

RTS,S vaccine. The RTS,S vaccine was developed to inhibit infection caused by *P. falciparum*. The

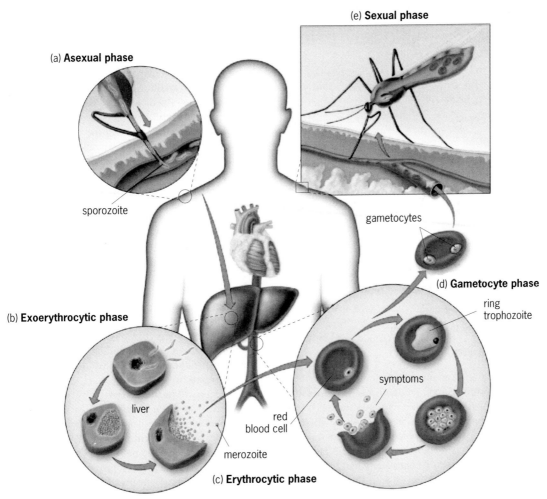

(e) Sexual phase

(a) Asexual phase

sporozoite

gametocytes

(d) Gametocyte phase

ring
trophozoite

(b) Exoerythrocytic phase

symptoms

liver

red
blood cell

merozoite

(c) Erythrocytic phase

Fig. 1. The life and transmission cycle of *Plasmodium*. (*a*) In humans, sporozoites enter a capillary during the blood meal of a mosquito. (*b*) Sporozoites invade the liver and develop into schizonts, which then rupture, releasing merozoites into the blood. (*c*) Merozoites in the blood infect red blood cells and form schizonts, which then release additional merozoites. (*d*) Some merozoites develop into gametocytes in red blood cells, which are then taken up by a feeding mosquito. (*e*) The sexual phase of fertilization and sporozoite formation occurs in the mosquito, completing the cycle. (*Courtesy of M. K. Cowan and K. P. Talaro, Microbiology: A Systems Approach, 2d ed., McGraw-Hill, New York, 2008*)

vaccine targets a protein found in the sporozoite stage of the parasite, which is the stage that infiltrates and multiplies in the liver. In the RTS,S study, rabies vaccine was used as a placebo vaccine. Children in Kenya and Tanzania were randomly assigned to two study groups (447 per group), with one group receiving the RTS,S vaccine and the second group receiving the placebo vaccine. Each vaccine was administered intramuscularly in the deltoid (the large triangular shoulder muscle), with one dose given monthly for three months. Children were monitored for the development of clinical malaria, which was defined as a temperature of 37.5°C (99.5°F) with a density of more than 200 *P. falciparum* parasites per microliter.

Of the more than 400 children receiving the new vaccine, 32 developed clinical malaria, whereas 66 children receiving the placebo vaccine developed infection. Statistical analysis determined that the efficiency of the vaccine was therefore approximately 53%. In addition, it took longer for disease to develop in the children receiving RTS,S.

The immunological basis for protection due to RTS,S is unknown. A wide variation in the pro-duction of antibodies to the circumsporozoite protein (a cell surface protein of the sporozoite) occurred among the vaccinated children. No correlation could be drawn between the production of a high level of antibodies to circumsporozoite protein and protection from infection, although this does not rule out the possibility of a role for these antibodies. The effect of cell-mediated immunity resulting from the vaccine was undetermined and was still being studied in the children receiving the RTS,S vaccine at the time that the study was published.

Others in the field have raised questions about the long-term effectiveness of the new vaccine, as well as about the variability in the risk of infection seen in the test children. In a different study, administration of the vaccine in adults was shown to give relatively short-lived protection. In addition, transmission rates for malaria in the study areas were falling at the time of the study. However, further clinical investigations are in progress and should address these and other questions that may arise about the vaccine.

Health professionals have been seeking to develop an effective malaria vaccine for a great many years.

Fig. 2. Personal bed nets are impregnated with insecticide in Benin, West Africa. (*Courtesy of M. K. Cowan and K. P. Talaro, Microbiology: A Systems Approach, 2d ed., McGraw-Hill, New York, 2008; and Roll Back Malaria Partnership*)

A combination of an effective vaccine together with the other control methods will lead to a possible elimination of malaria from endemic regions of the world. The ultimate goal is to eradicate this disease from the world's population. Although this may seem impractical at the present time, the enhancement of prevention and control can limit this disease and hopefully decrease its effect on the developing world, leading to a better future for the people of these regions.

For background information *see* AFRICA; DRUG RESISTANCE; EPIDEMIC; HAEMOSPORINA; IMMUNITY; MALARIA; MEDICAL PARASITOLOGY; MOSQUITO; PUBLIC HEALTH; QUININE; SPOROZOA; VACCINATION in the McGraw-Hill Encyclopedia of Science & Technology.   Marcia M. Pierce

Bibliography. P. Bejon et al., Efficacy of RTS,S/AS01E vaccine against malaria in children 5 to 17 months of age, *N. Engl. J. Med.*, 359(24):2521–2532, 2008; W. E. Collins and J. W. Barnwell, A hopeful beginning for malaria vaccines, *N. Engl. J. Med.*, 359(24):2599–2601, 2008; M. K. Cowan and K. P. Talaro, *Microbiology: A Systems Approach*, 2d ed., McGraw-Hill, New York, 2008; R. D. Gosling and D. Chandramohan, RTS,S/AS01E vaccine against malaria, *N. Engl. J. Med.*, 360(12):1253–1254, 2009; E. Nester et al., *Microbiology: A Human Perspective*, 6th ed., McGraw-Hill, New York, 2008.

Noninvasive diffuse optics for brain mapping

Functional neuroimaging of healthy adults has enabled the mapping of brain functions and revolutionized cognitive neuroscience. Increasingly, functional neuroimaging is being used as a diagnostic and prognostic tool in the clinical setting and to view brain development. Its expanding application in the study of disease and development necessitates new, more flexible tools. The logistics of traditional functional brain scanners [for example, magnetic resonance imaging (MRI) systems] are not well suited to subjects who are in an intensive-care unit or who might otherwise require sedation for imaging, such as infants and young children. Diffuse optical imaging (DOI)—an emerging, noninvasive technique with unique portability and hemodynamic (blood circulation) contrast capabilities—can record evoked brain function in clinical environments. However, despite its unique strengths, DOI as a standard tool for functional mapping has been limited by low spatial resolution, limited depth penetration, and a lack of reliable and repeatable mapping. To address these weaknesses, there recently has been significant interest in applying tomographic optical approaches to the problem of imaging brain function.

Technology. DOI technology is based on the same tissue sampling principles that underlie pulse oximetry, which is a common noninvasive clinical tool to monitor oxygen levels in blood. DOI uses light in red to near-infrared wavelengths to map changes in blood oxygenation and blood volume in the brain. Measurements are made using pairs of light sources and detectors, typically separated by about 1–5 cm (0.4–2 in.) and often coupled to the head surface using fiber optic bundles. The majority of previous optical neuroimaging studies have been performed using solely time traces from individual source-detector pairs or images made with simple two-dimensional back-projections. These techniques are often referred to as functional near-infrared spectroscopy (fNIRS). Because such systems require source-detector separations of around 3 cm (1.2 in.) in order to sample the cortex, their spatial resolution is thus restricted. In addition, as all measurements are a mixture of hemodynamics in the scalp, skull, and brain, data are often obscured by superficial and systemic hemodynamic artifacts. Diffuse optical tomography (DOT) seeks to overcome the limitations of fNIRS through a variety of methods. Time-resolved DOT systems rely on very high temporal resolution (approximately 1 ns) to gate the detected photons into groups that have traveled to different depths. These systems require complex electronics and photomultiplier tubes, resulting in a particular set of trade-offs among measurement density, frame rate, and field of view.

An alternative DOT strategy is to use high-density (HD) grids of sources and detectors, with each detector sensing light from many sources at different distances. These measurements overlap laterally and in depth, allowing the construction of an inversion problem, which enables higher-resolution tomographic image reconstructions of brain activity. However, the light levels seen by a detector decrease exponentially with distance from the source. Thus, in order for light from multiple source-detector pairs to be measured, the sources, detectors, and all the supporting circuitry must have very high dynamic range and low cross talk (interference). Recently, a

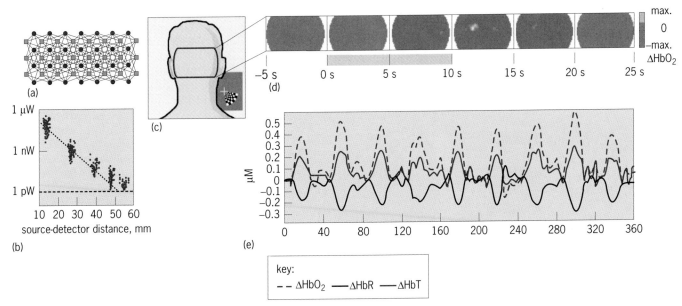

Fig. 1. Schematic of the high-density imaging grid (*a*), with 24 sources (squares) and 28 detectors (circles). Source-detector pair (SD-pair) measurements are represented by gray lines. The first-nearest neighbor (1st-nn) and second-nearest neighbor (2nd-nn) SD-pairs are at separations of 13 mm (0.52 in.) and 30 mm (1.2 in.), respectively. (*b*) Detected light level versus SD-pair separation on a human subject. All 1st-nn and 2nd-nn SD-pairs are detected simultaneously and with levels 100 times greater than the noise floor (dashed line). (*c*) Schematic showing placement of activation images on a human subject. Inset: For data in this figure, the visual stimulus presented is a reversing radial grid in the center lower right of the visual field. (*d*) Time course showing the temporal response of oxyhemoglobin (ΔHbO$_2$) due to the stimulus, which occurred during $t = 0$–10 seconds. (*e*) The high-density grid allows measurement of functional activations from individual stimuli, shown here for nine sequential stimuli (ΔHbO$_2$: dashes; ΔHbR: solid black line; ΔHbT: solid gray line). (*Adapted with permission from B. W. Zeff et al., Retinotopic mapping of adult human visual cortex with high-density diffuse optical tomography, Proc. Natl. Acad. Sci. USA, 104:12169–12174, 2007*)

HD-DOT system was developed that relies on avalanche photodiodes (which cumulatively achieve internal photocurrent multiplication) with independent 24-bit analog-to-digital converters and digital domain encoding/decoding of the source illumination. This strategy yields the high dynamic range and low cross talk necessary for using multiple source-detector distances. In addition, since the light-emitting diode (LED) sources are individually digitally encoded in software, the system is easily reconfigurable and expandable.

System configuration and results. An example of a system configured for imaging the visual cortex is shown in **Fig. 1**. The system has 24 source positions, with each having LEDs at two near-infrared wavelengths (750 and 850 nm). The sources are interleaved with 28 detectors [Fig. 1*a*, overall dimensions of 13.2 × 6.6 cm (5.2 × 2.6 in.)]. The optical fibers were coupled to the head using a flexible plastic cap molded to fit the head. Each detector samples light from all sources. For functional imaging of brain dynamics, light levels with signal-to-noise ratios (SNRs) of >100 are required. With this system, the first-nearest neighbor (1st-nn) [13 mm (0.52 in.)] and second-nearest neighbor (2nd-nn) [30 mm (1.2 in.)] source-detector pairs (SD-pairs) can be sampled simultaneously with sufficient SNRs for a total of 212 measurements (Fig. 1*b*). These data are reconstructed using finite-element modeling software and DOT inversion routines.

To demonstrate the feasibility of imaging human brain function with the new DOT system, a visual ac-

tivation study in adult humans has been conducted. Whereas previous optical imaging studies of the visual cortex have focused on discriminating contralateral activations, functional features of the human visual cortex (for example, visual field eccentricity, which is the angular distance from the point of fixation out to peripheral visual field locations) that were previously inaccessible to optical imaging have been reported. These retinotopic mappings are repeatable and consistent with previous fMRI and positron emission tomography (PET) studies. As a result of the dense spatial sampling and the removal of global and superficial signals, functional activations have high contrast-to-noise ratios (CNRs; for example, in Fig. 1, CNR = 12:1 without block-averaging and CNR = 34:1 with block-averaging), and activations due to even single stimuli can be imaged. Furthermore, discrimination of activations due to stimuli of different polar angles (**Fig. 2**) and eccentricities (**Fig. 3**) are possible.

The polar angle visual paradigm was an angularly swept radial reversing grid (10-Hz reversal, 10°/s). The 30 full sweep cycles (36 s each) were averaged together and down-sampled to 1 Hz to create 36 images with 10° phase separation. Four time points, separated by 90° phase, were chosen to yield a set of four activations, symmetric across the midline (Fig. 2). The four stimuli shown, also separated by 90° phase shifts, represent a 4.5-s shift relative to the functional activations with which they are matched. Half-maximum contours of the four activations are shown.

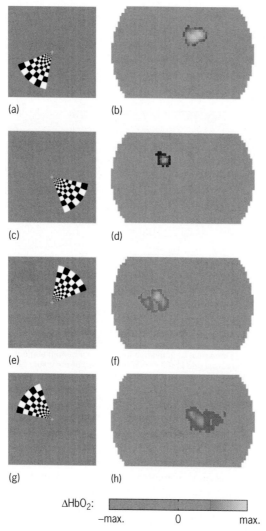

ΔHbO₂:

−max.   0   max.

Fig. 2. Examples of activations due to a counterclockwise rotating wedge stimulus. (*a*) One frame of the counterphase flickering checkerboard in the lower left visual field. (*b*) An activation due to this stimulus. In order to match the stimulus and response for this figure, a 6-s lag between stimulation and maximal response has been assumed. Note that the maximum response is in the upper right visual cortex. (*c–h*) Additional frames from the rotating stimulus. Note that the response is always maximal in the opposite visual quadrant from the stimulus.

Eccentricity within the visual field was mapped with expanding and contracting rings: minimum radius of 1°, maximum radius of 8°, width of 1.4° (three checkerboard squares), and 18 positions with 2 s per position, for a total cycle of 36 s. Four frames from a movie of the response to the expanding ring stimulus show that it is possible to locate responses to multiple visual field eccentricities (Fig. 3). Since this stimulus appears in both the right and left visual hemifields (the two halves of a sensory field), activations can be seen corresponding to the left and right visual hemispheres. As the stimulus moves outward in the visual field, both activations move upward in the field of view. Repeated studies in five subjects demonstrated that these signals were reproducible within subjects over multiple imaging sessions. Furthermore, this reproducibility was sufficient to de-

tect significant intrasubject variability, which has important consequences for more detailed cortical mapping problems (for example, identifying borders of visual cortical areas).

The inherent strengths of optical neuroimaging are well established. Cap-based imaging is suitable for a wide range of imaging situations, and the ability to measure changes in oxyhemoglobin (HbO₂; hemoglobin in combination with oxygen, present in arterial blood), deoxyhemoglobin (HbR; hemoglobin without oxygen), and total hemoglobin (HbT) can produce a more complete picture of brain function. However, the use of DOI has been hindered by insufficient spatial resolution, limited depth-sectioning capabilities, and the greatly increased complexity of larger arrays. As demonstrated by the human visual cortex results, new high-density DOT technology provides a significant step forward

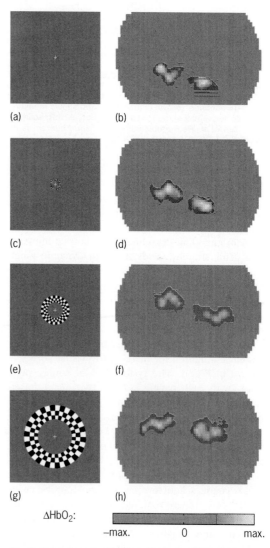

ΔHbO₂:

−max.   0   max.

Fig. 3. Examples of activations due to an expanding ring stimulus. (*a*) One frame of the expanding ring checkerboard in the central visual field. (*b*) An activation due to this stimulus. Note that the maximal response is bilateral and in the lower visual cortex. (*c–h*) Additional frames from the expanding stimulus (spaced evenly) and the resulting response (from the same session). Note that the hemodynamic response moves upward as the stimulus moves outward.

in resolving many of these obstacles to widespread use of DOI in neuroimaging, providing high contrast-to-noise functional signals, improved lateral resolution, and more accurate volumetric localization. It is anticipated that these advancements in image quality will open the path to a wide range of new studies concerned with the developing brain and the diseased brain.

For background information *see* BRAIN; COMPUTERIZED TOMOGRAPHY; ELECTRON TOMOGRAPHY; HEMOGLOBIN; MEDICAL IMAGING; NEUROBIOLOGY; OPTICAL COHERENCE TOMOGRAPHY; OXIMETRY; RADIOGRAPHY; RADIOLOGY in the McGraw-Hill Encyclopedia of Science & Technology.

Brian R. White; Joseph P. Culver

Bibliography. W. Colier et al., Simultaneous near-infrared spectroscopy monitoring of left and right occipital areas reveals contra-lateral hemodynamic changes upon hemi-field paradigm, *Vis. Res.*, 41:97–102, 2001; A. G. Yodh and D. A. Boas, Functional imaging with diffusing light, in T. Vo-Dinh (ed.), *Biomedical Photonics*, vol. 21, pp. 21–45, CRC Press, Boca Raton, 2003; B. W. Zeff et al., Retinotopic mapping of adult human visual cortex with high-density diffuse optical tomography, *Proc. Natl. Acad. Sci. USA*, 104:12169–12174, 2007.

Optical Ethernet

Optical Ethernet is the physical layer of the Local Area Network (LAN) communications protocol for sending data over fiber-optic cable. It is used for connecting Internet servers and switches inside equipment racks, within data centers, and between metropolitan data centers. Today the most widely used LAN data rate is 1 Gb/s, with increasing adoption of 10 Gb/s. These rates are insufficient to support core networking requirements, such as switching, routing, and aggregation in large data centers, Internet exchanges, and service provider peering points; and high-bandwidth applications, such as video on demand and high-performance computing environments. (A peering point is a central office where multiple service providers have assigned facilities, which then allows them to easily interconnect with one another.) To support these applications, the IEEE 802.3ba Task Force is in the process of standardizing a 100-Gb/s LAN data rate, which is the subject of this article. Formal adoption is expected in 2010, although most of the optical and electrical interface specifications are now complete and available on the Task Force Web site. For distances up to 100 m (330 ft), multimode fiber (MMF) cables are specified. For distances up to 10 km (6.2 mi), or up to 40 km (25 mi), single-mode fiber (SMF) cables are specified. The Task Force is also specifying higher LAN protocol layers and a 40-Gb/s LAN data rate as an interim step between 10- and 100-Gb/s. These, however, will not be discussed here. **Table 1** lists and spells out the acronyms used in this article.

100-Gb/s SMF technologies. The design of SMF optical transceivers is constrained by fiber-optic cable limitations, as well as limitations in the available components such as optical devices, integrated circuits, packaging, and interconnect. The two major fiber limitations in LAN applications are loss and chromatic dispersion (CD). In the 1310-nm fiber transmission window used for LAN, loss is approximately 0.4 dB/km and CD is -5 to 5 ps/nm-km. The communication link budget degradation in decibels due to loss is linear with distance, and degradation in decibels due to CD is linear with distance and quadratic with symbol rate (baud). Fiber and component limitations lead to design trade-offs between the major system parameters: baud, modulation format, channel (lane or fiber) count, and coding.

Prior SMF Ethernet standards all use the simplest and lowest implementation cost parameters: non-return-to-zero (NRZ) modulation format, single lane over a single fiber in each transmission direction (for duplex operation), and no coding. In every standard, baud matches the data rate, including at 100 Mb/s, 1 Gb/s, and 10 Gb/s.

Simply matching baud to the 100-Gb/s data rate is no longer feasible because the CD penalty prevents closing the link budget over useful distances. To retain low-cost NRZ modulation, the IEEE-adopted four lanes or wavelengths, each carrying data at 25 Gb/s (4 × 25G) as the basis for the new standard. These

---

TABLE 1. Acronyms and abbreviations in Optical Ethernet technology

Acronym	Term
CAUI	100-Gb/s Attachment Unit Interface
CD	Chromatic dispersion
CDR	Clock and data recovery
CFP	100-Gb/s Form-factor Pluggable
CMU	Clock multiplier unit
CW DFB	Continuous-wave distributed-feedback
CXP	100-Gb/s Extended-capability Pluggable
DeMux	Demultiplexer
DLL	Delay-locked loop
DML	Directly modulated laser
EML	Electroabsorption modulator laser
I/O	Input/output
LA	Limiting amplifier
LAN	Local Area Network
LD	Laser driver
MD	Modulator driver
MMF	Multimode fiber
MPO	Multifiber push-on
MSA	Multi-Source Agreement
MT	Mechanical transfer
Mux	Multiplexer
N. C.	Not connected
NRZ	Non-return-to-zero
OIF	Optical Interface Forum
PCB	Printed circuit board
PD	Phase detector
PIC	Photonic integrated circuit
PIN	*p*-intrinsic-*n*
PLL	Phase-locked loop
PMD	Physical Medium Dependent
REFCLK	Reference clock
SerDes	Serializer-deserializer
SMF	Single-mode fiber
SOA	Semiconductor optical amplifier
TEC	Thermoelectric cooler
TIA	Transimpedance amplifier
TWA	Traveling-wave amplifier
VCSEL	Vertical-cavity surface-emitting laser
VCO	Voltage-controlled oscillator
WDM	Wavelength-division multiplexing

TABLE 2. Optical LAN WDM Ethernet lanes

Lane	Center frequencies, THz	Center wavelengths, nm	Wavelength ranges, nm
L0	231.4	1295.56	1294.53–1296.59
L1	230.6	1300.05	1299.02–1301.09
L2	229.8	1304.58	1303.54–1305.63
L3	229.0	1309.14	1308.09–1310.19

lanes are wavelength-division multiplexed (WDM) over a single fiber in each transmission direction. The exact optical lane assignments are listed in **Table 2** and are referred to as the LAN WDM grid. They are located near the zero-CD wavelength of standard SMF.

Figure 1 shows the architecture for distances up to 10 km (100GBASE-LR4) and 40 km (100GBASE-ER4). The two 4 × 25G WDM optical interfaces are specified by the IEEE on the Task Force Web site. The lane rate of the electrical interface is 10 Gb/s, determined by I/O rates in mainstream CMOS technologies used for the media access controller (MAC) integrated circuit, which connects to the 100-Gb/s optical transceiver. The resulting 10 × 10G electrical interface is referred to as CAUI (100-Gb/s Attachment Unit Interface).

To translate between 10- and 25-Gb/s lane rates requires 10:4 SerDes (serializer deserializer) integrated circuits. The transmitter uses a quad modulator driver (MD) and quad electroabsorption modulator laser (EML) with four different EML wavelengths combined in a WDM multiplexer (Mux). The receiver uses a WDM demultiplexer (DeMux), quad PIN (*p*-intrinsic-*n*) photodiode, and quad transimpedance amplifier (TIA). A limiting amplifier (LA) function is required either in the TIA or SerDes integrated circuits. For 40-km distances, the architecture uses an optional semiconductor optical amplifier (SOA).

100-Gb/s SMF transceivers are packaged in the CFP (100-Gb/s Form-factor Pluggable) module, as specified by the CFP MSA (Multi-Source Agreement). This is a pluggable form factor (82 × 145 × 13.6 mm) with SMF optical connectors in the front. To minimize cost, surface-mount components and printed circuit board (PCB) transmission-line RF interconnect is used. First-generation transceivers use discrete optical components connected with fiber.

SerDes. The process alternatives for SerDes integrated circuits are CMOS and silicon-germanium (SiGe) BiCMOS. The serializer integrated circuit performs three functions: it (1) receives input data, (2) multiplexes 10 input lanes to 4 output lanes, and (3) low-pass filters and multiplies the input reference clock (REFCLK) for transmitting data. The integrated circuit is also referred to as the Mux/CMU (clock multiplier unit).

The receive function recovers the sampling phase of each lane. There is no need to recover the clock frequency since it is provided as a reference. A common topology is a delay-locked loop (DLL), which tracks the phase wander of the electrical interface.

The CMU is a narrow-band phased-locked loop (PLL) frequency synthesizer made up of a phase detector, loop filter, voltage-controlled oscillator (VCO), and digital divider. The loop filter removes high-frequency reference clock jitter so that the output jitter is determined by the CMU intrinsic phase noise. A low-phase-noise VCO uses an *LC*-tank structure.

The deserializer integrated circuit performs three functions: it (1) receives input data, (2) demultiplexes 4 input lanes to 10 output lanes, and (3) generates the clock for the input data. Both the clock frequency and phase are unknown and require CDR (clock and data recovery) for each 25-Gb/s input. This integrated circuit is also referred to as a CDR/DeMux.

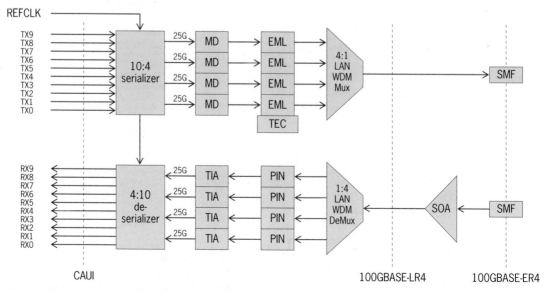

Fig. 1. 100GBASE-LR4 and 100GBASE-ER4 transceiver architecture.

A CDR is a wide-band PLL that can track fast phase variations of the input signal. Typically a fast-tuning ring oscillator is used for the VCO. A phase detector (PD) recovers phase error from a noisy and distorted signal. Either proportional (linear) or bang-bang (limiting or Alexander) PD circuits are used.

Transmitter. An EML transmitter consists of two sections: a CW DFB (continuous-wave distributed-feedback) laser diode, and an electroabsorption modulator. EMLs are typically cooled by a thermoelectric cooler (TEC) to permit optimization of the laser and modulator sections at a single temperature. MD integrated circuits provide a drive swing around 3 V with 25-GHz bandwidth. MD circuits are either multistage TWA (traveling–wave amplifier) GaAs or InP topologies, or simpler SiGe cascode and driven-cascode topologies. MD integrated circuits can be singles or quad per die and can be packaged separately or together with the EMLs.

WDM Mux and DeMux. The Mux and DeMux use wavelength-selective structures instead of simple power combiners. This requires a narrow filter to match the 800-GHz spaced wavelength of each multiplexer lane to the corresponding laser. For the LAN WDM grid, active wavelength control is required, either through process control or wavelength tuning by temperature adjustment. An additional constraint for the DeMux comes from the random polarization state of the incoming light, which requires implementation of low polarization-dependent loss and frequency shift.

Receiver. Broadband PIN receiver diodes are vertically illuminated, and require coupling with mirrors from the DeMux. For the 40-km application, an SOA is used prior to the DeMux to provide additional signal gain. The TIA integrated circuit sets the noise performance of the receiver, and can be either SiGe or CMOS. The TIA converts the photodiode single-ended input current to a differential output voltage (hence transimpedance). Because the signal is small at the TIA input, it is very susceptible to crosstalk. Techniques are required to minimize adjacent lane, supply, and substrate-coupled noise. The TIA integrated circuits must be packaged together with the PIN diodes.

Future 100-Gb/s SMF technologies. Future low-cost high-volume 100-Gb/s transceivers will be based on use of photonic integrated circuits (PICs). **Figure 2** shows architecture for use in applications below 10 km. The transmitter uses a quad laser driver (LD) and quad directly modulated laser (DML) array. The DML has a cost advantage over an EML because it consists only of a laser section, resulting in a smaller die area, lower power, and higher yield. The optical interface (100GBASE-LR4) is still specified by the IEEE on the Task Force Web site for interoperability with EML-based interfaces. The 4 × 25G electrical interface (CEI-28G) is now under definition by the Optical Interface Forum (OIF). CDRs with equalization are sufficient to retime the electrical input/output (I/O), and are simpler and less expensive than SerDes integrated circuits. The die count of a fully integrated transceiver becomes similar to today's 10-Gb/s SMF

transceivers. If high PIC yield can be achieved, comparable cost will be reached. Integration and lower electrical I/O count also lead to a smaller form factor than the CFP.

100-Gb/s MMF technologies. MMF transceivers at 100-Gb/s are for interrack and intrarack interconnections over distances below 100 m. MMF transmitters use 850-nm VCSELs (vertical-cavity surface-emitting lasers), and cost significantly less than SMF transmitters using edge-emitting EMLs or DMLs. VCSELs operate reliably up to 10 Gb/s. Faster rates require a high bias current that unacceptably degrades the VCSEL lifetime. This leads to the selection of 10 lanes for 100-Gb/s (10 × 10G).

Figure 3 shows a MMF architecture using a single 24-fiber ribbon cable, terminated by a multifiber push-on (MPO) connector based around the mechanical transfer (MT) ferrule. It has the advantage of aligning connecting fibers by referencing internal rather than external features, which minimizes thermal effects. The implementation has 12 lanes so that applications other than 100-Gb/s Ethernet, such as 12 × 12.5G InfiniBand, can be supported with common hardware. For 100-Gb/s applications, two lanes are not connected (N.C.).

CDRs. When implemented in the CFP form factor, retiming CDRs are required to support a CAUI electrical interface. This permits interchangeability with SMF transceivers. Eliminating the CDRs leads to a low-cost and small-size CXP (100-Gb/s Extended-capability Pluggable) form factor (about 15% the size of CFP), which is specified around the CXP host connector by the InfiniBand Trade Association. The resulting unretimed electrical interface is referred to as a PMD (Physical Medium Dependent) Service Interface, and is also defined by IEEE 802.3ba. The down side is that electrical requirements on the host are more stringent then with the CAUI electrical interface. In either case, the optical interface (100GBASE-SR10) is the same and is defined on the Task Force Web site.

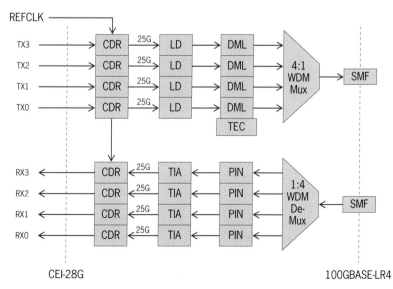

Fig. 2. Future 100GBASE-LR4 transceiver architecture.

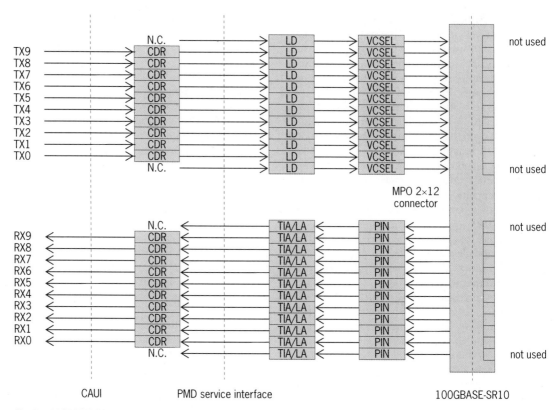

Fig. 3. **100GBASE-SR10 transceiver architecture.**

Transmitter. Arrays of 12 × 10G VCSELs are fabricated using the same fabrication process as a single 10-Gb/s VCSEL. Additionally, the top masks are optimized for alignment requirements during assembly. VCSEL array yields are inversely logarithmic with element count, so 12-element arrays are less expensive than 12 discrete elements. Single 10-Gb/s LDs are used in high volume, with array design facing power dissipation challenges. Large lane count favors eliminating discrete components such as capacitors used in single-lane designs, so LDs are DC-coupled to the VCSELs.

Receiver. Arrays of 12 × 10G PIN diodes are fabricated using the same process as single vertically illuminated 10-Gb/s PIN diodes. They also use top-level masks to aid in alignment during assembly. 10-Gb/s TIAs are used in high volume, with the major array design challenge coming from crosstalk mitigation.

Assembly. VCSEL and PIN arrays require special alignment techniques to the parallel fiber in the MPO connector. The fibers are first butt-coupled to a lens array, while the optics and integrated circuit arrays are placed on flex circuits. The two subassemblies are then precisely aligned to each other. Vertical emission of the VCSEL array can be used to view the emission aperture during alignment to the fiber/lens array.

Outlook. The highest Optical Ethernet data rate is 100 Gb/s, defined for both SMF and MMF cables by the IEEE 802.3ba Task Force. Deployment will start in the next few years for high-end data center and metro applications. Over time, as cost and size is reduced, 100-Gb/s Ethernet will migrate into high-volume applications.

For background information *see* DATA COMMUNICATIONS; FIBER-OPTIC CIRCUIT; INTEGRATED CIRCUITS; INTERNET; LASER; LOCAL-AREA NETWORKS; OPTICAL COMMUNICATIONS; PHASE-LOCKED LOOPS in the McGraw-Hill Encyclopedia of Science & Technology.   Chris Cole

Bibliography. C. Cole et al., Photonic integration for high volume low cost applications, *IEEE Commun. Mag.*, pp. 16–22, March 2009; T. Satake et al., MT multifiber connectors and new applications, *44th Electronic Components and Technology Conference*, pp. 994–999, IEEE, 1994.

Origin and evolution of echolocation in bats

Bats (order Chiroptera = "hand-wing") are among the most common living mammals. They represent nearly one-fifth of all mammalian species and live virtually everywhere, except in the harshest of polar climates. Most bats are active only during the evening and at night, making their activities mysterious to most people. Bats are among the most important seed dispersers in the tropics and, along with birds, are a major factor in controlling insect populations worldwide.

Bats are unique mammals. They are very small creatures, with most bat species weighing far less than 50 g (2 oz). The bumblebee bat (*Craseonycteris*) is the smallest living bat (weighing only 2 g) and is one of the smallest living mammals. Bats are one of three vertebrate groups (birds and pterosaurs being the others) and are the only mammals to have achieved powered flight. Their bodies reflect this aerial lifestyle most obviously by the presence of wings formed by flight membranes stretched over elongate hand and arm bones. Most living bats lack claws on all fingers except the thumb. Many other modifications of the skeleton have taken place in bats, including reduction in bone density and alterations in the shoulder and chest in order to accommodate flight muscles. Modifications of the hips, lower limbs, and feet have enabled bats to adopt their unusual upside-down roosting posture.

Biosonar. Perhaps the most remarkable attribute of bats is the ability of most species to echolocate, that is, to orient themselves and navigate precisely using a highly developed biosonar system. Most Old World fruit bats ("flying foxes") lack the ability to echolocate, whereas horseshoe bats have very sophisticated and highly sensitive echolocating abilities. The distinction between echolocating and nonecholocating bats was once used to classify living bats into two suborders, Megachiroptera (non-echolocators) and Microchiroptera (echolocating forms). It has since been discovered that bat relationships are not so simple. Nonecholocating forms are now thought to be closely related to one particular group of echolocating bats, Rhinolophoidea, which interestingly includes horseshoe bats, which possess one of the most sophisticated echolocating systems (**Fig. 1**). Together, this grouping is often referred to as Yinpterochiroptera (see **table**).

There is a great deal of variation among living bats with regard to echolocating abilities. In general, fruit bats cannot echolocate and instead rely on keen eyesight to navigate. However, one fruit bat, the cave-dwelling *Rousettus*, has developed an echolocating system involving tongue clicks to produce outgoing sound pulses. *Rousettus* is capable of navigating quite well through obstacle courses in low light. It is not certain whether simple tongue clicking was the initial form of echolocation in primitive bats or whether *Rousettus* developed this system independently from a nonecholocating ancestry.

Echolocating bats include all species other than fruit bats. Unlike the tongue clicks of *Rousettus*, all other bats produce tonal echolocating sounds in the larynx, which are normally emitted through the

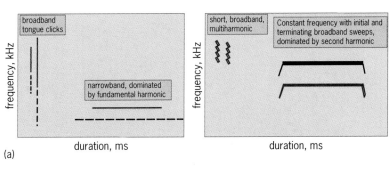

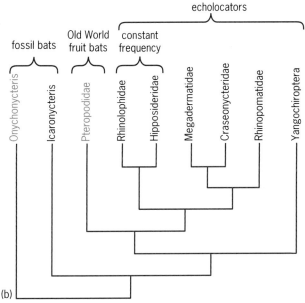

Fig. 1. Echolocation systems. (*a*) Spectrograms depicting some different echolocation sound emissions utilized by bats. (*b*) Relationships of living bats and the two oldest known fossil bats (*Onychonycteris* and *Icaronycteris*). Gray type indicates nonecholocating bats. (See the table for definitions of terms.)

mouth, but also through the nostrils in some forms. Emitted sounds range from narrowband fundamental harmonic only, to broadband multiharmonic, to constant-frequency signals often beginning and ending with broadband sweeps (Fig. 1). All signals can be further modified by duration, and bats may vary their signal depending on the complexity of the environment through which they are flying.

Definition of terms used in text and figures	
Term	Definition
Broadband	Echolocation sound emissions with broad frequency range, often spanning several harmonics
Chiroptera	Mammalian order that includes all bats
Fundamental harmonic	First harmonic of emitted echolocation sound; other harmonics are multiples of the first harmonic (for example, the second harmonic of a fundamental harmonic of 20 kHz equals 40 kHz); the fundamental harmonic varies from species to species
Harmonics	Differing sound frequencies produced by echolocating bats; the second and third harmonics are often most energetic
Narrowband	Echolocation sound emissions within narrow frequency range, often restricted to the fundamental harmonic
Pteropodidae	Family that contains all Old World fruit bats, including the rudimentary echolocating *Rousettus*
Rhinolophoidea	Superfamily that contains all echolocating bats other than those included in Yangochiroptera; families in Rhinolophoidea include Rhinolophidae, Hipposideridae, Megadermatidae, Craseonycteridae, and Rhinopomatidae
Yangochiroptera	Subordinal higher-level group that contains all living echolocating bats other than those included in Rhinolophoidea
Yinpterochiroptera	Subordinal higher-level group that contains Rhinolophoidea and Pteropodidae

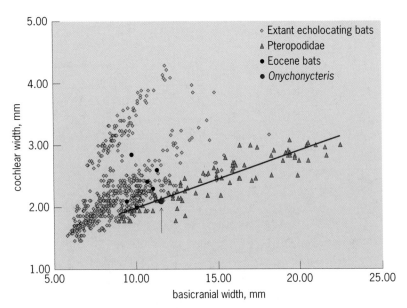

Fig. 2. Relationship between cochlear width and skull width in bats. The arrow indicates the position of *Onychonycteris* within the distribution of Pteropodidae (nonecholocating bats). (*Reproduced with permission from N. B. Simmons et al., Primitive early Eocene bat from Wyoming and the evolution of flight and echolocation, Nature, 451:818–822, 2008*)

In general, narrowband simple harmonic signals are more useful in open spaces when hunting for flying insects or commuting to and from roosts. Broadband, multiharmonic signals improve the ability to discriminate objects and to navigate in cluttered environments. Constant-frequency signals (no bandwidth) of relatively long duration (on the order of 30 ms) enable bats to classify food targets (insects) in terms of size and movement. The use of constant-frequency signals also allows for adjustments in frequency as a response to flying speed, thereby enabling these bats to compensate for Doppler shifts [that is, the change in the frequency of a wave observed at a receiver whenever the source or the receiver is moving relative to each other or to the carrier of the wave (the medium)] caused by variable flight speeds. Most bats differentiate emitted pulse from returning echo based on time separation; however, by altering constant signals based on flight speed, differentiation can also be achieved by frequency separation.

Anatomy of echolocation. Is it possible to know if an animal is capable of echolocation based solely on skeletal indicators? Fortunately, besides soft-tissue indicators of echolocating ability such as enlarged and complex external ears, there are internal skeletal structures that also can be used to determine whether an animal possesses the anatomy that will allow echolocation. The fact that such echolocating structures can be found in hard tissues is important because it means that similar features may be found in fossil bats, thereby making it possible to trace their development through time.

Three features of the bony middle and inner ear of bats can be related directly to the ability to echolocate. The first feature is the relative size of the cochlea. The cochlea of the inner ear consists of three fluid-filled canals that are housed in a bony structure called the promontorium. In bats that echolocate, the cochlea is enlarged (measured across the first half-turn) relative to the width of the skull (**Fig. 2**). Nonecholocating bats have smaller cochleae similar in relative size to those of other mammals that lack echolocating abilities. A second bony feature associated with echolocation is found on one of the small ear ossicles called the malleus. The malleus is the first of a chain of three bones (the incus and stapes being the others) that connect the eardrum (tympanic membrane) with the inner chamber of the ear. A relatively large, knobby process called the orbicular apophysis (OA) is present on the malleus of all echolocating bats, but it is absent in nonecholocating forms. The exact function of this structure is not well understood, but the tensor tympani muscle attaches to the malleus near the base of the OA. The tensor tympani muscle serves to freeze the ear ossicles in place just prior to emission of a sound pulse and aids in preventing self-deafening during sound production. The OA may improve the mechanical advantage of the tensor tympani by shifting its insertion point more posteriorly. A final bony feature found in echolocating bats is the shape of the cranial tip of the stylohyal. The stylohyal is one of the bones of the hyoid complex, which is a delicate set of bones found surrounding the larynx and tongue and extending to the base of the skull in most mammals. The cranial tip of the stylohyal is often enlarged and forms an elongate or axe-shaped plate that drapes over the auditory bulla (the bony covering of the middle ear) in echolocating bats. The stylohyal is a simple curved bar lacking any cranial expansion in nonecholocating bats.

Fossil echolocators. When did bats first begin to use echolocation, and what was the order of acquisition of the key bat characteristics of powered flight and echolocation? The fossil record of bats is anomalous in some ways because the earliest known fossil bats are also among the best known, with many being represented by virtually complete skeletons (**Fig. 3**). The two earliest known fossil bats are *Icaronycteris* and *Onychonycteris* from the Green River Formation in southwest Wyoming. The lake sediments that preserve these skeletons are early Eocene in age (53 million years old). Both bats are represented by exquisitely preserved skeletons that show minute details of the anatomy of these animals. Based on comparisons with modern bats, *Onychonycteris* is the most primitive bat now known, whereas *Icaronycteris* is somewhat less primitive.

Detailed examination of the skeletons reveals that both *Onychonycteris* and *Icaronycteris* were fully capable of powered flight. Each has elongate forelimbs and specializations of the shoulder that make it clear that each could fly. *Icaronycteris* was somewhat more primitive than modern bats because it still had tiny claws on the tips of its second and third fingers, whereas *Onychonycteris* was more primitive still in having small but clearly recognizable claws on all of its hand digits. *Onychonycteris* was also more primitive than *Icaronycteris* in the

Fig. 3. Skeleton of *Onychonycteris finneyi*, the most primitive known fossil bat, from the Green River Formation, Wyoming (52.5 million years old). (*Reproduced with permission from N. B. Simmons et al., Primitive early Eocene bat from Wyoming and the evolution of flight and echolocation, Nature, 451:818–822, 2008*)

relative proportions of front and hind limbs, having relatively shorter wings and relatively longer legs than any other known bat. Nonetheless, *Onychonycteris* could clearly fly.

Looking closely at the features of the ear in both Green River Formation bats reveals some important differences. *Icaronycteris* has a moderately enlarged cochlea, a well-developed orbicular process of the malleus, and an expanded cranial tip of the stylohyal. Each of these features is not as well developed as in most echolocating bats living today, but the evidence indicates that *Icaronycteris* was capable of at least a rudimentary form of echolocation. *Onychonycteris* does not have an enlarged cochlea, has a very small and modest orbicular process of the malleus, and has a simple, curved stylohyal lacking cranial tip expansion, which are features that are found in nonecholocating bats that live today.

At present, the best available evidence indicates that bats developed the ability to fly before they developed the ability to echolocate, because the most primitive known member of Chiroptera, *Onychonycteris*, could fly but was not capable of echolocation. However, in an evolutionary sense, bats developed echolocation relatively quickly after they took flight. Perhaps this was the result of competition with birds, which led to bats exploiting the same skies at night that birds dominated during the day.

For background information *see* ANIMAL EVOLUTION; BIOACOUSTICS, ANIMAL; CHIROPTERA; DOPPLER EFFECT; EAR (VERTEBRATE); ECHO; ECHOLOCATION; FLIGHT; NEUROBIOLOGY; PHONORECEPTION; SONAR in the McGraw-Hill Encyclopedia of Science & Technology.   Gregg F. Gunnell

Bibliography. G. F. Gunnell and N. B. Simmons, Fossil evidence and the origin of bats, *J. Mamm. Evol.*, 12:209–246, 2005; G. Jones and E. C. Teeling, The evolution of echolocation in bats, *Trends Ecol. Evol.*, 21:149–156, 2006; N. Simmons, Taking wing, *Sci. Am.*, December, pp. 96–103, 2008; N. B. Simmons et al., Primitive early Eocene bat from Wyoming and the evolution of flight and echolocation, *Nature*, 451:818–822, 2008.

Origin of the flowering plants

The most familiar flowers of the angiosperms (flowering plants) are colorful, with the colors serving to attract insect or other animal pollinators. However, many plants, including grasses and many broad-leaved trees of the Northern Temperate Zone, have inconspicuous flowers. These are pollinated by wind, so colorful structures are not needed and are not produced. The essential features that define a flower are the reproductive structures at the center of the flower, not colorful petals. Stamens are the male structures of flowers. Stamens surround the female structures, the carpels. Stamens and carpels have many technical features that distinguish them from the reproductive organs of non-flowering plants. The nonflowering plants include the gymnosperms, which make seeds but do not make flowers, and the ferns, lycopods, mosses, and liverworts, which reproduce by means of tiny spores rather than by seeds.

The female structures of angiosperms show the largest number of differences from nonflowering plants. The carpel consists of a flat structure rolled or folded inward to surround the ovules. At the tip of the carpel is the stigma, which is a specialized surface where pollen germinates and grows (forming pollen tubes) through the carpel tissues to reach the ovules deep inside, fertilizing them. The pollen tube brings two sperm. One fuses to make the zygote (and embryo), whereas the other fuses to make endosperm, which is the storage tissue for the embryo. The ovules develop into the seeds. The carpels develop into the fruit. A flower may have a single carpel, multiple separate carpels, or several carpels fused into a single pistil.

Gymnosperms also have ovules and make pollen. However, at the critical time when pollen is distributed, the ovules are, at least briefly, exposed to the air. In the most familiar examples, the conifers, scales of the young cones separate slightly so that air can reach the ovules deep inside. The ovules secrete a tiny drop of liquid that catches the airborne pollen grains. The ovules then absorb the liquid, bringing the pollen inside, and a short pollen tube grows to meet the egg.

The angiosperm stamen appears deceptively simple. It typically consists of a thin stalk bearing four pollen sacs at its top. The pollen sacs are fused in pairs and have a special tissue between the fused pollen sacs that breaks to allow the pair of sacs to open and distribute the pollen.

Gymnosperms also have pollen sacs, but in most they are borne on flat modified leaves grouped into male cones. They lack the special opening mechanism of angiosperm stamens; instead, each pollen sac cracks open, releasing pollen.

Source of the evolutionary mystery. It has long been clear that angiosperms evolved from some group of gymnosperms because they share numerous similarities, including the ovule and pollen, as well as features outside the flower. The problem is to determine which group of gymnosperms comprises the closest relatives of angiosperms and what the series of evolutionary changes was that resulted in flowers.

Three interrelated problems have made this a mysterious issue: (1) the great morphological differences between flowers and the reproductive structures of gymnosperms; (2) uncertainty in the relationships among flowering plants; and (3) uncertainty in the relationships among the various gymnosperm groups, especially when fossil as well as living gymnosperms and angiosperms are considered. These problems are interrelated because inferences of relationship, especially among fossils, are based on comparison of morphological structures, incorporating hypotheses of how structures may have evolved from something resembling one organism into a morphology resembling another. These are summarized as homologies, that is, inferences regarding what structures in one organism are comparable to which structures in another. For example, the front legs of reptiles are homologous to the wings of birds, because bird wings ultimately evolved from the front legs of their reptilian ancestors. Without clear homologies between flower structures and features of gymnosperms, it becomes problematic to identify the closest gymnosperm relatives of angiosperms. Conversely, if evidence identified the closest gymnosperm relatives, then homologies to flowers could be deduced.

The second problem (uncertainty in flowering plant relationships) has now been resolved. Previously, there was no consensus on what the earliest flowers were like: Some views of flowering plant relationships suggested that the earliest flowers were tiny, simple, and unisexual, consisting of only a single carpel or only a single stamen or group of stamens. Other inferences placed plants with bisexual flowers among the earliest-diverging angiosperms, suggesting the earliest flowers were bisexual and complex, with multiple flower parts. Without knowing the sort of flower that the theory needed to explain, there was little progress in understanding flower origins.

Relationships within flowering plants. Over the last decade, DNA sequence analyses have resolved the relationships among flowering plants. The earliest diverging group of angiosperms consists of a single species, *Amborella trichopoda* (a shrub from the Pacific island of New Caledonia). The second group to diverge contains the water lilies and also *Trithuria* (tiny submerged aquatics from Australia, New Zealand, and India). The third group contains several genera, including *Illicium*, which is the source of the Asian spice star anise. All flowers in these groups have multiple parts, and most are bisexual or bisexual-derived (because vestigial structures of the opposite sex are present in the functionally male and female flowers). The purely unisexual *Trithuria* is so outnumbered by bisexual flowers that the earliest flowers were clearly bisexual. They also had sepals and/or petals (often termed tepals, if not strongly differentiated), multiple stamens (each with four pollen sacs), and multiple separate carpels that were bucket-shaped (rather than folded over); these carpels were sealed shut by a secretion, not by actual fusion of the carpel edges.

Relationships of seed plants. Relationships of the gymnosperm groups to each other and to flowering plants remain controversial. In the mid-1980s, cladistics, which was then a new and much improved approach to analyze relationships, was applied to morphological features of living and fossil seed plants. This suggested that one group of living gymnosperms, the Gnetales, along with several fossil groups were closely related to angiosperms. One of the fossils and also some Gnetales had arrangements of parts reminiscent of flowers—flat things surrounding male structures with female structures in the middle. This suggested that this arrangement was homologous in these groups, even though the individual organs were different. However, recent DNA studies strongly suggest that the four groups of living gymnosperms (conifers, cycads, Gnetales, and *Ginkgo*) are related to each other, so none is close to flowering plants. This remains controversial because DNA studies cannot be done on fossil plants, and it is widely accepted that including more groups of organisms, as in morphological studies that include fossils, can be especially effective. Some morphological cladistic studies continue to group Gnetales with angiosperms, although other studies suggest that the grouping is not a robust result. Morphological studies place several fossil gymnosperm groups close to angiosperms, but it is not clear which is the closest. Moreover, these fossils have different reproductive structures, so it is not clear what sort of gymnosperm structures would have evolved into flowers.

Evo-devo studies. Investigations utilizing the evolution of development (evo-devo) combine knowledge of gene function with knowledge of morphological evolution to elucidate evolutionary mechanisms. Recent discoveries of genes controlling flower development make this approach powerful. This can be applied in three ways: (1) Most simply, expression of similar genes in different structures supports homology of the structures; (2) such analyses may directly suggest scenarios for evolutionary change; and (3) knowledge of gene function allows mechanistic explanations of how specific gene changes could have resulted in morphological changes.

The *LEAFY* gene, studied in several angiosperms, makes a shoot develop into a flower by turning on genes that specify the individual flower organs. Living gymnosperms have two copies of *LEAFY*. Early work suggested that one copy was more active in the male reproductive unit, whereas the other was more active in the female reproductive unit. Hence, the two *LEAFY* genes might specify male versus female reproductive units. The single flowering plant *LEAFY* is most closely related to the gymnosperm *LEAFY* thought to specify male units. This led to the "mostly male" theory, which states that the flower arose from the male reproductive unit of the ancestral gymnosperm. Ovules, originally borne separately in that ancestor, would have jumped, as a result of changed gene expression, into the male unit, generating the precursor of the carpel. The theory is consistent with much morphological data, but recent evidence questions the proposed roles of the gymnosperm *LEAFY* genes.

Theories called "out of male" and "out of female" suggest that either the ancestral male or female reproductive structures could have been modified to become bisexual, giving rise to the flower. These theories do not favor one source over the other, nor do they explain the origin of the carpel from gymnosperm structures.

A mechanistic theory seeks to identify changes in gene expression that could cause a unisexual reproductive unit to become bisexual. In flowers, the stamens are specified by the joint action of the two B-class genes along with action of the C-class gene (for identification purposes, floral organ genes are divided into A, B, and C classes). Carpels are specified by the C gene without B genes. The theory suggests that greatly increased expression of the C gene could mask effects of the B genes, rendering a male unit partly female, and thus bisexual.

Fusion of the second sperm to make storage tissue seemed strange, until it was found that the second sperm of gymnosperm pollen tubes often fuses with another nucleus, in addition to the first sperm that fuses with the egg nucleus. In gymnosperms, this second fusion appears to generate a second embryo. This second embryo could have evolved into the storage tissue of angiosperms.

Future prospects. The near future is likely to bring much progress. Paleobotanists are studying mesofossils, that is, fossils a few millimeters across, and intermediate in size between the microscopic fossils (for example, pollen) and macrofossils that have long been studied. Among these mesofossils are flowers and flower buds, often exquisitely preserved after being charred in ancient fires, swept out to sea, and buried in sediments. Studies of mesofossils have revolutionized knowledge of angiosperm diversification. Perhaps gymnosperms that are obvious close relatives to angiosperms will be discovered in such deposits. In fact, recent macrofossil studies demonstrate the great diversity of gymnosperms in the period around the likely origin of angiosperms.

New technology will greatly facilitate evo-devo studies by making genetic studies applicable on nonmodel organisms chosen for their interesting morphologies or their critical phylogenetic positions. Complete genome sequencing will soon be affordable even for small scientific projects. New methods to inactivate genes should allow gene function to be revealed in nonmodel organisms as convincingly as mutations do in standard genetics. Other technologies that allow measurement, under controlled conditions, of the binding of proteins to DNA will facilitate study of genetic control networks in nonmodels. The future of evolutionary studies will be extremely interesting.

For background information *see* FLOWER; FOSSIL; GENE; GENOMICS; MAGNOLIOPHYTA; PALEOBOTANY; PLANT ANATOMY; PLANT EVOLUTION; PLANT PHYLOGENY; PLANT REPRODUCTION; POLLEN; POLLINATION; SEED in the McGraw-Hill Encyclopedia of Science & Technology.   Michael W. Frohlich

Bibliography. R. M. Bateman, J. Hilton, and P. J. Rudall, Morphological and molecular phylogenetic context of the angiosperms: Contrasting the "top-down" and "bottom-up" approaches used to infer the likely characteristics of the first flowers, *J. Exp. Bot.*, 57:3471–3503, 2006; J. A. Doyle, Seed ferns and the origin of angiosperms, *J. Torrey Bot. Soc.*, 133:169–209, 2006; M. W. Frohlich, Recent developments regarding the evolutionary origin of flowers, pp. 63–127, in D. E. Soltis, J. H. Leebens-Mack, and P. S. Soltis (eds.), *Developmental Genetics of the Flower: Advances in Botanical Research*, vol. 44, Academic Press, San Diego, 2006; B. Glover, *Understanding Flowers and Flowering: An Integrated Approach*, Oxford University Press, Oxford, U.K., 2008; J. Jernstedt (ed.), Special issue: Darwin bicentennial, *Am. J. Bot.*, 96(1), 2009.

Oxygen sensing in metazoans

The abundance of molecular oxygen (O_2) in the atmosphere has varied enormously during Earth's history. O_2 first appeared as a chemical by-product of photosynthesis over 2 billion years ago and reached its zenith of about 30% during the Carboniferous period approximately 300 million years ago. Many geological and biological processes regulate atmospheric O_2 levels, and it is interesting to note that increased metazoan body sizes, as well as the appearance of placental mammals, correlate with a rise in O_2 levels from 10 to 20% over the past 200 million years. (Note that metazoans are the multicellular animals that make up the major portion of the animal kingdom, whose cells are organized in layers or groups as specialized tissues or organ systems.)

Essentially all animal cells use O_2 to convert sugars, lipids, and other substrates into biochemical energy in the form of adenosine triphosphate (ATP). Cellular O_2 delivery is achieved through multiple strategies. For example, insects have remarkably elaborate single-celled tracheal tubes that convey atmospheric O_2 directly to each cell. Mammals, in contrast, evolved red blood cells and highly complex cardiovascular systems to provide an adequate supply of O_2 to all tissues. Once transported into the recipient cell, O_2 is used primarily to accept electrons from cytochrome c in the terminal step of the mitochondrial electron transport chain (ETC), in which O_2 is reduced to water ($O_2 + 4e^- + 4H^+ \rightarrow 2H_2O$). This makes O_2 an essential factor in regulating cellular bioenergetics, but also poses a threat, as O_2 is highly

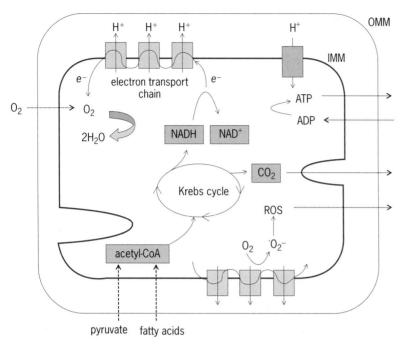

Fig. 1. Most O_2 is reduced to H_2O by the mitochondrial electron transport chain (ETC), using electrons (e^-) from NADH (reduced or hydrogenated nicotinamide adenine dinucleotide) and other substrates, generated from acetyl-CoA (acetyl coenzyme A) by the Krebs cycle (a sequence of enzymatic reactions involving oxidation of a two-carbon acetyl unit to carbon dioxide and water to provide energy for storage in the form of high-energy phosphate bonds). The ETC also pumps H^+ ions into the mitochondrial intermembrane space, and transport of H^+ ions by the F1/F0-ATPase drives production of ATP. However, O_2 can also pick up unpaired electrons from the ETC to generate superoxide ($^{\cdot}O_2^-$) and other reactive O_2 species (ROS). OMM, outer mitochondrial membrane; IMM, inner mitochondrial membrane.

electronegative and can easily recruit unpaired electrons to generate reactive oxygen species (ROS) such as superoxide (**Fig. 1**). Although relatively low ROS levels function in normal cellular signaling pathways, elevated concentrations of ROS can cause severe cellular damage by oxidizing proteins, lipids, and deoxyribonucleic acid (DNA) [thereby causing mutations]. Consequently, metazoans have evolved complex mechanisms to carefully regulate O_2 delivery at the organismal level, as well as to respond to changes in local O_2 concentrations at the cellular level.

Short-term adaptation. Anyone who has traveled to high altitude knows that immediate physiological responses to decreased O_2 levels are rapid and profound. Respiration and heart rates are quickly elevated, and, in severe cases, pulmonary edema and other dangerous pathological conditions can arise. Similarly, sleep apnea and other respiratory disorders can result in systemic O_2 deprivation (hypoxia), which may contribute to cardiovascular disease, stroke, and other maladies. How are changes in O_2 levels perceived at the organismal level, and how are these rapid physiological responses mediated?

Blood O_2 levels (pO_2) are continually monitored by the carotid body, which is a small neurosecretory organ situated near the bifurcation of the carotid artery. Oxygenated blood from the lungs returns to the heart and then flows past the carotid body on its way to the brain. Glomus cells in the carotid body sense minute changes in blood pO_2 and pCO_2 (carbon dioxide levels), and respond by releasing

neurotransmitters (including dopamine and acetylcholine) that stimulate chemoreceptor afferents to the brain, resulting in altered respiratory and cardiovascular responses. The precise molecular mechanisms by which glomus cells sense blood pO_2 are controversial, and multiple models have been proposed. However, regardless of the exact immediate sensing mechanisms, it is clear that secretion of neurotransmitters by glomus cells is preceded by elevated intracellular calcium (Ca^{++}) levels, possibly in response to inhibition of potassium (K^+) channels and opening of the voltage-dependent Ca^{++} channel. It is important to note that these neurosecretory events are extremely rapid and generally do not require new protein synthesis. Although many molecular details remain to be elucidated, it is likely that glomus cells have more than one strategy for detecting changes in pO_2. Additional organs and cell types, including the aortic body, neuroepithelial cells in the lung, and chromaffin cells of the fetal adrenal medulla, also sense changes in local O_2 concentrations and release neurotransmitters to regulate vasoconstriction or dilation accordingly.

Long-term (chronic) cellular adaptation. All metazoan cells have innate molecular mechanisms to alter their metabolic and genetic programs in response to changes in O_2 levels. Some of these responses are more gradual than those described in the previous section, and many require new gene expression and protein synthesis. For example, people who spend extended periods at high altitude develop an elevated hematocrit (red blood cell number), but this response typically occurs over weeks and reflects increased synthesis of the hormone erythropoietin (EPO). How then do cells sense O_2, and how do they adapt to changes in local O_2 levels?

Investigations into the regulation of EPO expression led to the identification of a transcriptional regulator that responds directly to changes in O_2 levels. This hypoxia-inducible factor-1 (HIF-1) is a heterodimeric DNA-binding protein complex consisting of α and β subunits. HIF-1 binds to specific sequences in the promoters and enhancers of target genes, such as EPO, and modulates their expression in response to altered O_2 concentrations (**Fig. 2**). HIF-1 is regulated in an unusual fashion: it accumulates in cells exposed to low O_2 (approximately <3–5%), but is rapidly degraded in the presence of abundant O_2 (approximately >8–10%). It is worth noting that most mammalian tissues are typically exposed to O_2 concentrations in the 1–8% range, called "physiological hypoxia," whereas hypoxic tumors or ischemic tissue can become nearly anoxic (oxygen-depleted; 0%). Interestingly, HIF-1 levels are controlled dynamically in an inverse relation to pO_2: the lower the O_2 level, the more HIF-1 protein accumulates. The discovery of HIF-1 provided a direct molecular mechanism to explain how variations in microenvironmental O_2 concentrations could alter gene expression.

Many of the genes directly regulated by HIFs encode proteins that confer obvious adaptive advantages to hypoxic cells. For example, HIF-1 increases the expression of glycolytic enzymes, glucose transporters, angiogenic factors to recruit blood vessels, and EPO. In addition, HIF-1 regulates genes involved in iron metabolism, cell growth and survival, cell movement, and many other processes, indicating that O_2 availability influences a broad array of physiological responses. Work from many laboratories has now revealed a complex signaling network in which multiple HIF complexes (containing either HIF-1α or the highly related protein HIF-2α) regulate stem cell function, tissue development and homeostasis, and disease progression.

Exactly how, though, do cells sense changes in O_2 levels? The answer to this question remains quite controversial, but again it appears that multiple mechanisms are involved. Under "normoxic" (abundant O_2) conditions, HIF-α subunits are hydroxylated on two conserved proline residues by specific HIF prolyl hydroxylase domain (PHD) enzymes that use O_2 as a substrate for the hydroxylation reaction (Fig. 2). These hydroxyproline residues constitute a "destruction" tag that triggers HIF-α proteolysis, thereby keeping HIF-α levels low in normoxic cells. However, under hypoxic conditions, the PHDs are inhibited and HIF-α accumulates to form complexes with HIF-1β, activating gene expression. Consequently, one elegant and relatively simple model of O_2 sensing posits that PHD enzymes become substrate-limited under hypoxic conditions, thereby stabilizing HIFs.

The picture is clearly more complex, though, as a wealth of data implicate other cellular signaling molecules in O_2 sensing. For example, cells experiencing moderate hypoxia (0.5–5% O_2) actually accumulate ROS generated from the mitochondrial ETC, and abundant evidence indicates that these ROS can, in turn, regulate HIF-α accumulation (possibly by inhibiting PHDs) as well as other signaling pathways. In addition, nitric oxide (NO) has been proposed to play a role in oxygen sensing by inhibiting the mitochondrial ETC, thereby making O_2 available for other cellular reactions. Clearly, the molecular strategies for cellular O_2 sensing are complex, and different mechanisms (including those that are HIF-independent) may predominate in particular cell types or under specific conditions.

When oxygen sensing goes awry. Cellular O_2 sensing mechanisms have been highly conserved in metazoan evolution; for example, the HIF/PHD pathway is found in animals ranging from invertebrates (nematodes, insects) to humans. Given the importance of oxygen sensing in cellular physiology, what are the consequences of disabling these pathways?

Mice with mutations in HIF-1α or HIF-1β genes die early in embryonic development and display striking defects in morphogenesis and cardiovascular development. This suggests that naturally occurring O_2 gradients act as important signals in the early embryo, and HIF-dependent adaptations are a critical component of normal development. Similarly, deletion of HIF or PHD genes in specific cell types in adult mice compromises blood vessel function, inflammation, and other processes, indicating that

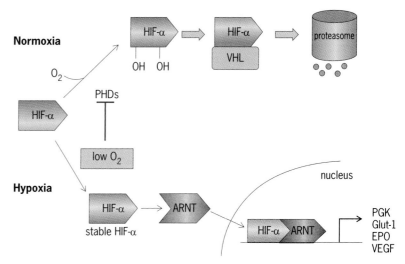

Fig. 2. Under normoxic (abundant O_2) conditions, HIF-α subunits are hydroxylated on conserved proline residues by specific HIF prolyl hydroxylase domain enzymes (PHDs). These modified residues bind an E3-ubiquitin ligase complex (VHL) that targets HIF-α for degradation in the 26S proteasome (a large proteolytic particle found in the cytoplasm and nucleus of all eukaryotic cells that is the site for degradation of most intracellular proteins), keeping total HIF levels low. Under hypoxic (limited O_2) conditions, PHDs are inhibited and HIF-α accumulates to form an active complex with ARNT (HIF-1β) and to transcribe adaptive genes. PGK, phosphoglycerate kinase; Glut-1, glucose transporter-1; EPO, erythropoietin; VEGF, vascular endothelial growth factor.

O_2 continues to influence physiology in adult tissues. Intriguingly, HIF proteins accumulate to abnormally high levels in a variety of human cancers, suggesting that unregulated hypoxic adaptations (altered metabolism, increased blood vessel recruitment, etc.) may contribute directly to tumor progression. This supposition has been borne out in mouse tumor models (although specific effects vary depending on the choice of model), and the HIF/PHD pathway has been proposed as a possible therapeutic target. Inhibition of PHD enzymes is also an attractive strategy for treating severe anemia, since the consequent elevation of HIF activity increases the expression of many genes involved in red blood cell production (including the genes that encode EPO).

Finally, recent data suggest that HIFs can influence other, apparently unrelated, molecular signaling pathways implicated in controlling normal development and cancer progression. For example, HIF-1α can bind directly to a complex containing the oncoprotein c-Myc and can inhibit its ability to promote cell growth and division. Interestingly, HIF-2α also binds the c-Myc complex, but instead potentiates c-Myc activity. Thus, despite their structural similarity and conserved regulation by PHDs, this suggests that HIF-1α and HIF-2α can oppose each other in regulating c-Myc activity in particular tumor settings. HIF-α subunits have also been reported to interact directly with components of the Notch and Wnt signaling pathways (which help to regulate cell-to-cell interactions during embryogenesis), apparently expanding the portfolio of HIF's physiological activities.

In summary, metazoans have evolved complex mechanisms to sense and respond to changes in O_2 levels at the organismal and cellular levels. These responses are essential for normal embryonic development and adult physiology, and are intimately involved in pathological contexts.

For background information *see* ADENOSINE TRIPHOSPHATE (ATP); ATMOSPHERE, EVOLUTION OF; BIOLOGICAL OXIDATION; CANCER (MEDICINE); ENERGY METABOLISM; ENZYME; FREE RADICAL; HYPOXIA; METAZOA; OXYGEN; RESPIRATION; RESPIRATORY SYSTEM in the McGraw-Hill Encyclopedia of Science & Technology.   Brian Keith

Bibliography. P. G. Falkowski et al., The rise of oxygen over the past 205 million years and the evolution of large placental mammals, *Science*, 309(5744):2202–2204, 2005; W. G. Kaelin, Jr., and P. J. Ratcliffe, Oxygen sensing by metazoans: The central role of the HIF hydroxylase pathway, *Mol. Cell*, 30(4):393–402, 2008; S. Lahiri et al., Oxygen sensing in the body, *Progr. Biophys. Mol. Biol.*, 91(3):249–286, 2006; G. L. Semenza, Targeting HIF-1 for cancer therapy, *Nat. Rev. Canc.*, 3(10):721–732, 2003; M. C. Simon and B. Keith, The role of oxygen availability in embryonic development and stem cell function, *Nat. Rev. Mol. Cell Biol.*, 9(4):285–296, 2008.

Personalized medicine

Personalized medicine may be considered an extension of traditional approaches to understanding and treating disease. Physicians have always used observable evidence to make a diagnosis or prescribe a treatment tailored to each individual patient. In the modern conception of personalized medicine, the tools provided to the physician are more precise, probing not just the visually obvious, such as a tumor on a mammogram or the appearance of cells under a microscope, but the very molecular makeup of each patient. A profile of a patient's genetic variation can guide the selection of drugs or treatment protocols that minimize harmful side effects or ensure a more successful outcome. It can also indicate susceptibility to certain diseases before they become manifest, allowing the physician and the patient to set out a plan for monitoring and prevention.

The ability to profile the structure, sequence, and expression levels of genes, proteins, and metabolites is redefining how we classify diseases and select treatments, allowing physicians to go beyond the "one size fits all" model of medicine to make the most effective clinical decisions for each patient. It is an approach that is well suited to the medical challenges faced in the twenty-first century. Although we have prevailed over many of the diseases that have plagued humanity throughout the ages, what remains are diseases of greater complexity, including diabetes, cancer, heart disease, and Alzheimer's disease. They are not caused by a single gene or a single event but by a combination of genetic and environmental factors, and they tend to be chronic, placing a heavy burden on the health-care system. Personalized medicine provides the tools needed to better manage chronic diseases and treat them more effectively.

We can now point to real-world applications of almost every aspect of personalized medicine's promise. For example, genetic profiles can better discern different subgroups of breast cancer, guiding physicians to select the best treatment protocol or, in some cases, forgo the expense and risks of chemotherapy altogether; and tests that detect variations in the way that individuals metabolize blood-thinning drugs can help predetermine the right dose for a patient, navigating the narrow therapeutic passage between reducing the risk of clots and triggering internal bleeding. Moreover, a test for mutations in the genetic coding for an enzyme can help physicians select the most effective drug for a specific cancer patient from an expanding pharmacopoeia of choices, avoiding a costly and protracted trial-and-error approach that can leave the patient suffering needlessly from adverse effects or losing precious time in battling the disease. As evidence of the benefits continues to grow, an infrastructure of laws, policy, education, and clinical practice is building around personalized medicine to support its use:

1. Medical institutions across the country have announced their commitment to putting personalized medicine into practice through dedicated centers or statewide initiatives.

2. Personalized medicine approaches are becoming best practice in hospitals, in order to ensure that patients with serious conditions (for example, cancer) are given optimum therapy from the start.

3. The regulatory system is integrating genetic testing into the labels of pharmaceutical products, ensuring that a drug is administered in a way that minimizes the risk of adverse effects and improves the chances of effective treatment.

4. Nearly every major pharmaceutical development project is incorporating information on genetic variation and its effects on the safety and effectiveness of the candidate drug.

5. Personalized medicine applications have extended beyond cancer to improve treatments in cardiovascular disease, infectious diseases, psychiatric disorders, and transplantation medicine.

6. Several of the nation's leading medical schools are launching genomics-based medical education programs to train the next generation of care providers.

7. Various health organizations have advocated policies encouraging genetic testing and preventive care, while several large private insurers have begun paying for genetic tests identifying presymptomatic high-risk populations.

8. Wide-ranging policy recommendations for personalized medicine have been made by governmental and nongovernmental groups; in addition, a genetic privacy law has been passed in the United States, and other legislation supporting personalized medicine has been introduced in the U.S. Congress.

Clinical applications. Ultimately, the success of personalized medicine will rise or fall on its ability to demonstrate its value to the health-care system, to the industries that develop its products, and to patients. The promise of personalized medicine, for which tangible evidence already exists, includes the ability to (1) shift the emphasis in medicine from reaction to prevention, (2) enable the selection of optimal therapy and reduce trial-and-error prescribing, (3) make the use of drugs safer by avoiding adverse drug reactions, (4) increase patient compliance with treatment, (5) reduce the time and cost of clinical trials, (6) revive drugs that are failing in clinical trials or were withdrawn from the market, and (7) reduce the overall cost of health care.

Shift the emphasis in medicine from reaction to prevention. Personalized medicine introduces the ability to use molecular markers that signal the risk of disease or its presence before clinical signs and symptoms appear. This information underlies a health-care strategy focused on prevention and early intervention, rather than a reaction to advanced stages of disease. Such a strategy can delay the onset of disease or minimize symptom severity. One example is a test used to look for genetic mutations in BRCA1 and BRCA2 (breast cancer 1 and 2, respectively) that indicate a hereditary propensity for breast and ovarian cancer. Women with BRCA1 or BRCA2 genetic risk factors have a 36–85% lifetime chance of developing breast cancer, compared with a 13% chance among the general female population. For ovarian cancer, women with certain BRCA1 or BRCA2 gene mutations have a 16–60% chance of disease, compared with a 1.7% chance among the general population. The BRCA1 and BRCA2 genetic test can guide preventive measures, including increased frequency of mammography, prophylactic surgery, and chemoprevention.

Over 1300 genetic tests exist that signal inherited susceptibility to conditions as wide-ranging as hearing loss and sudden cardiac arrest. Although not every test has a therapeutic option, a genetic diagnosis often permits targeted prevention or mitigation strategies.

Select optimal therapy. On average, a drug that is on the market works for only 50% of the people who take it. The consequences in terms of quality and cost of care are significant, leaving patients to contend with their disease and their medical bills as they switch from one drug to another until they find an effective therapy. Studies have linked differences in response to the differences in genes that code for the drug-metabolizing enzymes, drug transporters, or drug targets. The use of genetic and other forms of molecular screening allows the physician to select an optimal therapy the first time and avoid the frustrating and costly practice of trial-and-error prescriptions.

Make drugs safer. Overall, about 5.3% of hospital admissions are associated with adverse drug reactions (ADRs). Many ADRs are the result of variations in genes coding for the cytochrome P450 (CYP450) family of enzymes and other metabolizing enzymes. These variants may cause a drug to be metabolized more quickly or more slowly than in the general population.

As a result, some individuals may have trouble eliminating a drug from their bodies, leading in essence to an overdose as it accumulates, whereas others eliminate the drug before it has a chance to work. The consequences of not considering variation in these genes when dosing can range from futility to unpleasant or even fatal side effects.

Increase patient compliance with treatment. Patient noncompliance with treatment leads to adverse health effects and increased costs. When personalized therapies prove more effective or present fewer side effects, patients will be more likely to comply with their treatments. The greatest impact could be for the treatment of diseases such as asthma and diabetes, in which noncompliance commonly exacerbates the condition. Inherited forms of hypercholesterolemia (high cholesterol) can increase the risk of myocardial infarction before the age of 40 more than 50-fold in men and 125-fold in women. Conventional monitoring of cholesterol levels can catch the condition early, but genetic testing offers additional benefits. In addition to detecting the condition before there are observable signs of disease, knowledge of a genetic predisposition for hypercholesterolemia provides patients with a powerful incentive to make lifestyle changes and to view their condition seriously. Patients with a genetic diagnosis have shown more than 86% adherence to their treatment program after two years compared to 38% prior to testing.

Reduce time, cost, and failure rate of clinical trials. Developing a new drug is a costly and lengthy process. Theoretically, the use of pharmacogenomic data, or information about how patients' genes affect their drug responses, could reduce the time and cost of drug development, in addition to reducing the rate of drug failures by allowing researchers to focus on subsets of patient populations. Using genetic tests, researchers could preselect patients for studies, using those who are most likely to respond or least likely to suffer side effects. Enriching the clinical trial pool, as this approach is called, could reduce the size, time, and expense of clinical trials.

Reduce the cost of health care. The cost of health care in the United States and elsewhere is on an unsustainable upward climb. Incorporating personalized medicine into the fabric of the health-care system can help resolve many embedded inefficiencies, such as trial-and-error dosing, hospitalization of patients who have severe reactions to a drug, late diagnoses, and reactive treatment. Specific examples of personalized medicine are generating tangible results about their economic benefit, which can be enormous in potential health-care cost savings.

For background information *see* CHEMOTHERAPY AND OTHER ANTINEOPLASTIC DRUGS; CLINICAL PATHOLOGY; DISEASE; DRUG DELIVERY SYSTEMS; GENE; GENETICS; GENOMICS; MEDICINE; PHARMACOLOGY; POLYMORPHISM (GENETICS); PUBLIC HEALTH in the McGraw-Hill Encyclopedia of Science & Technology.   Wayne A. Rosenkrans, Jr.

Bibliography. Blue Cross Blue Shield Technology Evaluation Center, Special report: Genotyping for cytochrome P450 polymorphisms to determine drug-metabolizer status, Blue Cross Blue Shield Association, Chicago, 2004; J. A. DiMasi, R. W. Hansen, and H. G. Grabowski, The price of innovation: New estimates of drug development costs, *J. Health Econ.*, 22:151–185, 2003; C. Kongkaew, P. R. Noyce, and D. M. Ashcroft, Hospital admissions associated with adverse drug reactions: A systematic review of prospective observational studies, *Ann. Pharmacother.*, 42(7):1017–1025, 2008; L. M. Mangravite, C. F. Thorn, and R. M. Krauss, Clinical implications of pharmacogenomics of statin treatment, *Pharmacogenomics J.*, 6(6):360–374, 2006; H. D. Nelson, L. H. Huffman, and R. Fu, Genetic risk assessment and brca mutation testing for breast and ovarian cancer susceptibility: Systematic evidence review for the U.S. Preventive Services Task Force, *Ann. Intern. Med.*, 143:362–379, 2005; S. Paik et al., Gene expression and benefit of chemotherapy in women with node-negative, estrogen receptor–positive breast cancer, *J. Clin. Oncol.*, 24(23):3726–3734, 2006; K. A. Phillips et al., Potential role of pharmacogenomics in reducing adverse drug reactions: A systematic review, *JAMA*, 286:2270–2279, 2001; B. B. Spear, M. Heath-Chiozzi, and J. Huff, Clinical application of pharmacogenetics, *Trends Mol. Med.*, 7(5):201–204, 2001; M. A. Umans-Eckenhausen et al., Long-term compliance with lipid-lowering medication after genetic screening for familial hypercholesterolemia, *Arch. Intern. Med.*, 163(1):65–68, 2003.

Phoenix Mars mission

The Phoenix mission was launched by the National Aeronautics and Space Administration (NASA) to Mars from Cape Canaveral on August 4, 2007. The spacecraft passed an uneventful journey of 400 million miles to Mars, entering the planet's upper atmosphere on May 25, 2008. The "seven minutes of terror" during entry, descent, and landing (**Fig. 1**)

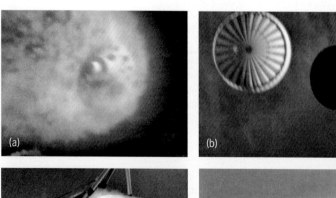

Fig. 1. The four phases of entry, descent, and landing: (*a*) heat shield; (*b*) parachute; (*c*) thrusters; (*d*) landing legs. (*Jet Propulsion Laboratory*)

Holy Cow

Fig. 2. The view under the spacecraft, taken with the robotic-arm camera. The powerful thrusters have scoured the surface, removing about 5 cm (2 in.) of top soil and revealing an underlying layer that is now known to be water ice. Some of the material has splashed onto the left strut, leaving mysterious blobs. (*University of Arizona*)

captivated viewers around the world, and first images of the alien plains were returned 2 hours later. The landing site, at 68°N latitude and 233°E longitude, was in a broad, shallow valley on the ejecta blanket of a modern crater named Heimdal (after the Norse guardian of the bridge between the real world and the spirit world).

The Phoenix mission was selected by NASA from a pool of candidate missions as the first Scout mission, a lower-cost category than core NASA missions. The spacecraft, originally *Mars Surveyor 2001*, was built by Lockheed Martin in Denver; the company was placed under contract by Jet Propulsion Laboratory (JPL) to bring the spacecraft back to flightworthiness and prepare it for its mission to the Martian arctic.

Landing. As is the case for entering the Earth's atmosphere, a heat shield (Fig. 1*a*) protected against the heat of friction and a parachute (Fig. 1*b*) slowed descent to the surface. However, Mars's atmosphere has only about 1% of the terrestrial atmospheric pressure, so a parachute slows the spacecraft to only 160 km/h (100 mi/h). Upon release from the parachute, the spacecraft's final approach to the surface was accomplished by firing 12 hydrazine thrusters (Fig. 1*c*) guided by radar and an attitude control system. Touchdown was at a speed of 8 km/h (5 mi/h), with the shock absorbed by crushable landing legs (Fig. 1*d*).

Local geology. After the landing, one of the first activities that was necessary to achieve full control of the spacecraft was to assess the stability of the lander on its three footpads. To see those footpads, hidden under the deck, the robotic arm used its camera to look underneath the lander. The first image revealed that the 12 thrusters had blasted away the topsoil, exposing a bright, flat substance that looked like an ice table. This feature was unexpected, and joyous scientists informally named it Holy Cow (**Fig. 2**).

The terrain surrounding the lander shows no sign of ice on its surface. The impression is of an undulating surface similar to patterned ground in polar regions on the Earth. As the underlying ice expands and contracts during seasonal temperature extremes, the surface is shaped into polygonal features, with mounds surrounded by troughs. The depths of the

Fig. 3. Undulating ground topped by clusters of small cobbles. This image illustrates the mounds and troughs that typify polar terrain on Mars and Earth. The structures are formed in winter by the contraction of ice, forming cracks that fill with small particles. In the summer, the expanding ice cannot resume its original shape. (*Texas A&M University, M. Lemmon*)

troughs are only 30 cm (1 ft), and the diameters are 2–3 m (7–10 ft; **Fig. 3**).

A few days later, the arm used its scoop to dig the first of a dozen trenches. Again, icelike hard material was found at a depth of 5 cm (2 in.). Other alternatives besides ice were proposed, such as rock or salt deposits. However, as a trench that was informally named the Dodo trench was being excavated, several small whitish chunks were scraped off. Four days later they disappeared; sublimation of ice to vapor is strong evidence that the material is water ice. (Temperatures are too warm to allow carbon dioxide ice to be stable in this season.)

Another trench named Snow White was positioned at the center of a nearby polygon chosen to represent the average surface when seen from orbit. Again, a hard surface was found at a depth of 5 cm (2 in.), but this time no white material was apparent; the ice appeared to fill the pore spaces in the soil. Over several days, as the ice sublimed, a thin blanket of dry soil (a lag deposit) overlaid the icy soil. Soil samples were taken for analysis from nearby surface materials ("Rosy Red") and from the dry soil lag in the Snow White trench. Three trenches with different expressions of the ice table are seen in **Fig. 4**.

Samples were delivered to a microscope station for examination. High-resolution images (4 μm/pixel) revealed two size distributions: a group of rounded, sand-sized particles, 50–150 μm in diameter; and orange-colored dust particles, a few micrometers in size (**Fig. 5**). The differences between the two groups of particles in size and color imply a different origin for each group. The particles are strongly magnetic, indicative of iron mineralogy. The larger particles display an assortment of colors and textures; some are even transparent. The global dust storms that frequently occur on Mars may have transported the finer particles to this site; the larger particles were probably rounded by saltation (the near-surface hopping of grains forced by strong winds).

Chemistry of the soil. The TEGA (Thermal and Evolved Gas Analyzer) instrument received samples from several locations. The sublimation lag from Snow White contained 2–3% ice; signatures were seen both in the calorimetry experiment as ice vaporized during the heating of the oven above 0°C (32°F) and in the positive water identification in the mass spectrometer from the gas released. All soil samples showed clear carbon dioxide peaks at temperatures above 750°C (1400°F), leading to the conclusion that calcium carbonate composes 3–5% of the soil. This discovery implies that the soil was once wet, because this compound forms in the combination of atmospheric carbon dioxide (CO_2) with liquid water creating a weak acid that leaches calcium from the soil.

Support for this conclusion came from the MECA (Microscopic, Electrochemical, and Conductivity Analyzer) instrument. Four beakers were flown that could each receive a soil sample and add water to make a solution. If the ground ice were to melt, then the properties of the wet soil would match those in a MECA cell. The solutions created from three independent samples all had a pH near 8; the addition of an acid did not change this value. The implication is that calcium carbonate is buffering the solution. Chemistry of this type is familiar on Earth; our oceans have a pH of about 8.2 and are buffered by calcium carbonate.

Other discoveries from MECA include small amounts of salts and a completely unexpected substance. The salts contain potassium, magnesium, sodium, chlorine, and calcium, but the surprise came in finding a large amount (1–2%) of perchlorate in solution. Perchlorate, an anion consisting of a chlorine atom with four oxygen atoms attached, is rare

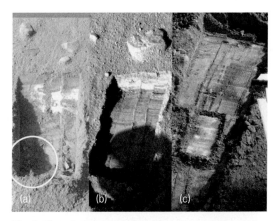

Fig. 4. Three trenches showing different expressions of the ice layer. (*a*) Trench with white ice that is determined, through spectral imaging, to be 99% pure; the chunks that have been scraped off (inside the white circle) sublimed away in less than 4 days. (*b*) Trench with chatter marks that were formed as the scoop blade attempted to dig through the frozen soil. Ice content is estimated at about 40%, an amount consistent with atmospheric water vapor diffusing into the soil and freezing in the pore spaces between soil grains. (*c*) Early morning image of Snow White trench. A thin frost has formed on the trench bottom. Scraping and rasping marks are visible where the robotic arm attempted to gather ice samples.

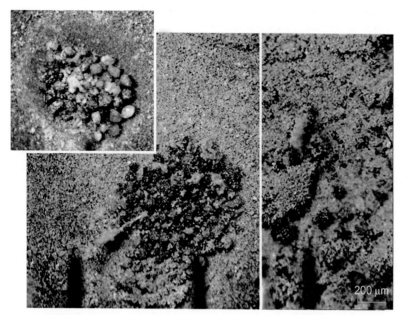

Fig. 5. **Three microscopic views of surface samples. All images are the same scale. The images show the Martian soil on a strong magnet in a vertical orientation; nonmagnetic particles tend to fall off.**

on Earth, found only in hyperarid environments because of the ease at which it goes into solution and migrates into the ground water. In drinking water, it is considered hazardous and can adversely affect the thyroid. (The solid rockets on the launch vehicle used ammonia perchlorate as the oxidizer.) Perchlorate also bonds easily with water and is often used as a desiccant in industry; its ability to attract water may control the local humidity above the soil.

Climate. Temperatures typically reached a peak of −30°C (−22°F) in the afternoon and dropped to −80°C (−112°F) before dawn. Daily temperatures repeat fairly consistently, but the pressure has a complex pattern and decreased throughout the mission as CO_2 in the atmosphere froze down to the Martian surface in the southern polar region. Despite the variability, the pressure was always above the triple point of water, between 700 and 850 Pa. If temperatures rise above the melting point, then liquid water is stable; this can happen at lower temperatures if brines are present.

Early in the mission, the atmosphere was made dusty up to about 5 km (3 mi) above the surface by turbulent mixing in the afternoon. But after 90 sols (Martian solar days), the dust cleared and water ice clouds were observed. These clouds showed precipitation late at night and were observed to release snow, which at first fell part way before subliming and then later in the season fell to the surface. (**Figure 6** shows fall streaks, defined as ice crystals that fall from the bottom of a cloud and sublime before they reach the surface, on Mars and on Earth.) In addition, frost and snow were seen on the surface by the Stereo Surface Imager (SSI) camera.

Habitability of site. The MECA instrument detected a variety of nutrients in the form of salts. These ions are important to life as seen on Earth. However, at least two important nutrients were not detected:

nitrogen and phosphorus. These two in the form of oxides are crucial to life as we know it.

The most likely type of life to be found on Mars is from the class of microbes that exist on chemical-energy sources, because intense solar ultraviolet radiation reaches the surface (Mars is not protected by an ozone layer in the upper atmosphere) and renders it uninhabitable by Earth standards. However, on Earth life adapts to subsurface environments using chemical energy sources instead of photosynthesis. A multitude of terrestrial microbes live on perchlorate, using it as an energy source as part of a redox couple. (Bioremediation is the method used to control perchlorate contamination in Earth's drinking water.)

Although energy sources and nutrients are important components in any habitable zone, the primary requirement for habitability as we know it is liquid water. On Mars, the temperatures are too cold and the humidity too low to support liquid water in the current epoch. Several lines of evidence provide hope that conditions were markedly different in the recent past. First, the Martian soil is cloddy and sticky; whatever is sticking the soil together is likely to need water to create the bonds. Second, the calcium carbonate component of the soil (3–5%) is typically formed in wet environments. Other

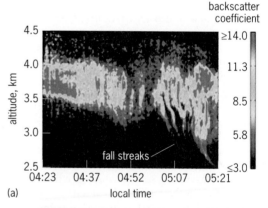

(a)

(b)

Fig. 6. **Fall streaks on Mars compared to those on Earth.** (a) **Results of LIDAR experiment that sent vertical laser pulses into a water ice cloud on sol 99. By timing the receipt of the reflected light, an altitude-versus-cloud density map was made in the early morning. The contour plot gives the backscatter coefficient, as keyed on the scale bar to the right, in units of $10^{-6}\,m^{-1}\,sr^{-1}$, as derived from the lidar backscatter signal at 532-nm wavelength. Fall streaks are observed starting at 5 a.m. local time.** (b) **Similar features in Australia. (York University, J. Whiteway)**

aqueous minerals are detected by their release of water vapor at 300 and 680°C (570 and 1260°F). But as a caveat, the upper soil levels could have been created at some other location and arrived during the global dust storms occasionally observed on Mars. Even so, the conclusions hold for the source region.

The likely scenario is that Martian obliquity (the tilt of the spin axis with respect to the orbital plane) was much greater than the current 25.2° several million years ago. When the tilt exceeds 30°, the polar cap is unstable and rapidly releases water vapor into the atmosphere. Because the summer insolation to the polar region is also increased, the temperatures are then warmer and the humidity higher. Between the subsurface ice, the nighttime snow, and the perchlorate in the soil, there are many opportunities during a warmer period to wet the soil

If the periodic variations in obliquity and orbital dynamics change conditions for liquid water from unfavorable to favorable, then we have an environment that is periodically habitable. Can microbes survive long-term cold, dry spells? This is an active area of research. Microbes in the permafrost regions on Earth are known to survive several hundred thousand years in cold conditions, lying dormant in icy cages.

Mars has a higher radiation flux as a result of its thinner atmosphere, and it may not be possible for microbes to repair their DNA codes, which are being continually damaged by high-energy cosmic rays. The only way to know for sure that this is possible is to find convincing evidence that microbes do exist on Mars. This is the goal of the next lander, the *Mars Science Laboratory*, which is scheduled to be launched in 2011 and land in 2012 in a region that was once water-rich. In the meantime, the search for clues from orbit continues using the *Odyssey, Mars Reconnaissance Orbiter*, and *Mars Express* spacecraft.

Administration. As Principal Investigator, Peter H. Smith of the University of Arizona was in charge, working with Project Manager Barry Goldstein, from JPL.

For background information *see* ASTROBIOLOGY; BUFFERS (CHEMISTRY); CARBONATE MINERALS; MARS; PERMAFROST; SPACE PROBE; SUBLIMATION; TRIPLE POINT in the McGraw-Hill Encyclopedia of Science & Technology.   Peter H. Smith

Bibliography. R. E. Arvidson et al., Mars exploration program 2007 Phoenix landing site selection and characteristics, *J. Geophys. Res.*, 113:E00A03, 2008; A. Chaikin, *A Passion for Mars*, Abrams, 2008; W. K. Hartmann, *A Traveler's Guide to Mars: The Mysterious Landscapes of the Red Planet*, Workman, New York, 2003; B. M. Jakosky, F. Westfall, and A. Brack, Mars, in W. T. Sullivan III and J. A. Baross (eds.), *Planets and Life*, chap. 18, Cambridge University Press, 2008; W. Sheehan, *The Planet Mars: A History of Observation and Discovery*, The University of Arizona Press, Tucson, 1996; P. H. Smith et al., Introduction to special section on the Phoenix mission: Landing site characterization experiments, mission overviews, and expected science, *J. Geophys. Res.*, 113:E00A18, 2008.

Phylogenetic classification of Fungi

Fungi constitute one of the major branches of the tree of life. There are about 100,000 described species in the Fungi, but it has been estimated that there are as many as 1.5 million extant fungal species. Fungi have evolved a remarkable diversity of forms and lifestyles. Some produce mushrooms, which are sexual reproductive structures, but most fungal species are very inconspicuous and generally go unnoticed. Single-celled forms called yeasts occur in multiple fungal groups, but most species grow as a network of microscopic filaments (hyphae), termed a mycelium, by which the fungus explores its environment and captures resources. Some fungi are aquatic and produce swimming cells with typical eukaryotic flagella. However, most fungi are terrestrial and lack flagellated cells, which appear to have been lost, perhaps multiple times, during fungal evolution.

All fungi are heterotrophic and have absorptive nutrition, meaning that they must feed themselves by absorbing organic molecules from the environment. Beyond this, however, fungal nutritional strategies are highly divergent. Saprotrophic fungi cause plant debris and other dead organic matter to decay, whereas biotrophic fungi obtain nutrition from living organisms. Biotrophic fungi include mutualistic forms, in which both organisms appear to benefit from the association, as well as many parasites and pathogens. Examples of mutualistic fungal symbioses include lichens, mycorrhizae (associations of fungi and plant roots), and the partnership of leafcutter ants and their cultivated fungi. Pathogenic and parasitic fungi attack animals, plants, algae, and other fungi, and they cause serious diseases in humans and agricultural crops. At the same time, fungi provide food and are used as experimental organisms in genetics and biotechnology. They are also sources of useful biochemicals, including dyes and stains, antibiotics, and enzymes that are being applied to the production of biofuels and other "green" technologies. Thus, fungi play critical roles in ecosystems and have a profound impact on human affairs.

Biological classifications. To comprehend, study, and use fungal diversity, it is essential to have a system of well-accepted names for species and more inclusive groups. Taxonomy provides the scientific names (in Latin form) that are applied to all organisms, without which discourse in biology would be severely impaired. Ever since Darwin, taxonomy has been inextricably linked to evolutionary biology. Biological classifications reflect hypotheses about phylogenetic (evolutionary) relationships, with each formally named group of organisms being equated with a clade (a group that contains an ancestor and all of its descendants; that is, a single and complete branch of the tree of life). Phylogenetic classifications have predictive value because the distributions of many attributes of organisms are correlated with their evolutionary relationships. For example, a clade of Fungi called the Polyporales produces lignin peroxidases, which are wood-decaying enzymes that have potential applications in biodegradation and biofuel production. Applied scientists working in these areas can use the classification to select species of Polyporales that might be sources of useful enzymes.

Evolutionary history and classification can be represented in phylogenetic tree diagrams that show patterns of genealogical relationships (see **illustration**). Phylogenetic trees can be used to reconstruct various aspects of evolution, such as changes in organismal forms and ecological characteristics, or the shifts in geographic ranges that occurred during the evolution of a group (biogeography). Thus, taxonomy is not merely a cataloguing exercise; it is a dynamic science concerned with understanding the history of life.

Prior classifications of Fungi. Fungi are anatomically simple, and they have a limited fossil record, which has made it difficult to reconstruct their evolutionary relationships. Recent advances in molecular phylogenetics (the study of evolutionary relationships using DNA and protein sequences) have helped to overcome these limitations. Research in fungal molecular phylogenetics began in the early 1990s with analyses of genes encoding ribosomal ribonucleic acids (rRNAs, nucleic acids that are components of ribosomes), and has expanded to include analyses of multiple genes (which provide improved resolution of relationships) and, recently, whole genomes. As the outlines of fungal relationships have emerged, older classifications based on morphology (physical characteristics) have been revised. However, the integration of molecular perspectives occurred in a piecemeal fashion, resulting in classifications in which some taxa were clades, but others were not. Furthermore, there were significant differences among the names used in several major classifications. The lack of a single broadly accepted, phylogenetically accurate fungal classification has been a source of confusion, especially for students, nonspecialists, and applied scientists.

Deep Hypha and the AFTOL classification. Two projects provided a structure for the fungal taxonomic community to work toward a unified phylogenetic classification of Fungi. Deep Hypha (2001–2006) supported conferences and research planning, whereas AFTOL (Assembling the Fungal Tree of Life; 2003–2006) supported a network of laboratories that analyzed molecular and morphological data for phylogeny reconstruction. Through the work of AFTOL and many other independent laboratories, a large body of phylogenetic data was developed. Together, a group of fungal taxonomists representing diverse classification projects worked to translate the new phylogenetic results into a classification that would reflect current understanding of evolutionary relationships and unify the disparate classifications that were then in use.

The AFTOL classification uses the Linnaean classification system. The Fungi is treated as a kingdom, and the classification includes other taxa at the rank of subkingdom, phylum, subphylum, class, subclass, and order. It is important to realize, however, that taxa placed at the same rank are not necessarily equivalent in age, number of species, or degree of difference from other taxa; the ranks are merely

nomenclatural devices employed for classification purposes.

Position of Fungi within eukaryotes and basal fungal lineages. Fungi are traditionally studied by botanists, but they are now known to be more closely related to animals (Metazoa) than to plants. Fungi and animals are contained in a clade called the Opisthokonts, along with several groups of unicellular "protists," including the nucleariids, which are ameboid organisms that appear to be the closest relatives of the Fungi. The AFTOL classification excludes several groups of funguslike eukaryotes that are now known to be outside of the Fungi, such as the Oomycetes (including the potato blight pathogen *Phytophthora infestans*) and the slime molds (Mycetozoa). Perhaps the most controversial aspect of the AFTOL classification is the inclusion of Microsporidia (as a phylum), which are highly reduced intracellular parasites that lack mitochondria. Several phylogenetic studies have suggested that Microsporidia are nested within the Fungi, although their precise placement is ambiguous.

Despite the application of multigene data sets, the deepest divergences within the Fungi remain poorly resolved. Consequently, the AFTOL classification includes a "polytomy" (an inadequately resolved division) with 10 clades diverging at the base of the Fungi. This polytomy reflects uncertainty about the branching order and is not meant to suggest that the 10 clades evolved simultaneously. One of the clades in the basal polytomy is the Microsporidia. Three others include taxa with flagellated cells that had previously been grouped as the Chytridiomycota. In the AFTOL classification, these organisms are dispersed among three phyla: Blastocladiomycota, Neocallimastigomycota, and a restricted Chytridiomycota. Two other species of "chytrids" (an informal term for fungi with flagellated cells at some stage of the life cycle), *Olpidium brassicae* and *Rozella allomycis*, are of uncertain placement and are probably outside of these groups. Some analyses suggest that *R. allomycis* could be the closest relative of the Microsporidia. Possession of flagellated cells is clearly a primitive trait of the Fungi as a whole. Overemphasis on this characteristic in prior taxonomy led to the false recognition of the chytrids as a clade.

Five other clades in the basal polytomy include taxa that had previously been grouped as the Zygomycota (a taxon not included in the AFTOL classification), based on the shared absence of flagellated cells, possession of a filamentous habit (usually without septa dividing the filaments into discrete cells), and the nearly universal lack of multicellular fruiting bodies (present in one group, the Endogonales). Like the chytrids, the "zygomycetes" appear not to form a clade, although this remains controversial. Ecologically, zygomycetes are tremendously diverse, including saprotrophic taxa, pathogens, mycorrhizal symbionts, and various forms that parasitize or trap insects and other small animals.

Dikarya. The last clade arising from the basal polytomy, the Dikarya, includes the vast majority of the described fungi. The Dikarya consists of two groups, the Ascomycota and Basidiomycota, which are

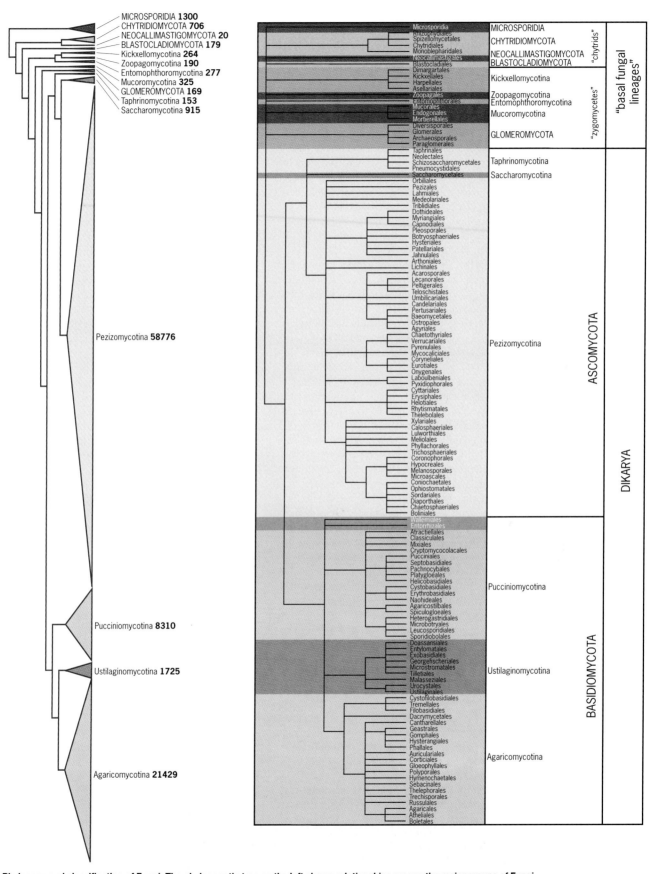

Phylogeny and classification of Fungi. The phylogenetic tree on the left shows relationships among the major groups of Fungi based on a study by T. Y. James et al. (*Reconstructing the early evolution of the fungi using a six gene phylogeny, Nature, 443:818–822, 2006*). The areas of the terminal clades (triangles) are proportional to the relative diversity of the groups, based on estimates of the number of species in each group (indicated by the numbers following group names) from the *Dictionary of the Fungi* (*P. M. Kirk et al., Dictionary of the Fungi, 10th ed., CAB International, Wallingford, U.K., 2008*). The tree on the right represents the AFTOL classification, with all of the 129 orders that are the terminal groups in the classification. Every resolved branch in the tree corresponds to a named clade, but not all intermediate-level taxa are labeled.

dominated by filamentous taxa (there are also yeasts in both groups), lack flagellated cells, and frequently produce multicellular fruiting bodies. The group is named for the dikaryon, which is a uniquely fungal structure in which the haploid nuclei (having half of the diploid or full complement of chromosomes) resulting from sexual fusion (plasmogamy) remain separate and undergo coordinated mitotic divisions, resulting in a network of cells, each with two genetically distinct nuclei. Nuclear fusion (karyogamy) occurs immediately before meiosis, which occurs in typically sac-shaped cells called asci (singular: ascus; Ascomycota) or typically pedestal-shaped cells called basidia (singular: basidium; Basidiomycota). Thus, in most Ascomycota and Basidiomycota, there is only a single truly diploid cell in the life cycle (there are exceptions, though).

Ascomycota are divided into three subphyla, the Taphrinomycotina, Saccharomycotina, and Pezizomycotina, whereas the Basidiomycota are divided into the Pucciniomycotina, Ustilaginomycotina, and Agaricomycotina. The diverse forms and lifestyles in these groups collectively comprise 110 orders, including virtually all of the fungi that are familiar to nonbiologists, such as yeasts, molds, lichens, and mushrooms. Dikarya are almost all terrestrial, and they represent (along with plants and animals) one of three clades of eukaryotes that have evolved complex multicellular forms and have radiated extensively on land. Some Dikarya are secondarily aquatic, however, and many of these have evolved appendaged spores that promote dispersal without the aid of flagella.

Future of fungal classification. The AFTOL classification represents a synthesis of current knowledge, but it is merely an intermediate step toward the development of a complete classification of Fungi. To complete the classification, it will be necessary to reconstruct the deepest divergences in the Fungi and resolve the basal polytomy. It will also be necessary to discover and describe the estimated 93% of fungal species that remain unknown. Technological advances, including recently developed methods of rapid DNA sequencing, and improvements in computer hardware and algorithms, will play a major role in this work. Modernization of nomenclatural practices, including the abandonment of rank-based taxonomy, may also accelerate progress. However, the experience of Deep Hypha and AFTOL suggests that coordination and cooperation among fungal taxonomists will be of equal, if not greater, importance in the development of a comprehensive phylogenetic classification of the Fungi.

For background information *see* BIODIVERSITY; CLASSIFICATION, BIOLOGICAL; EUMYCOTA; FUNGAL ECOLOGY; FUNGAL GENOMICS; FUNGI; MYCOLOGY; PHYLOGENY; SYSTEMATICS; TAXONOMIC CATEGORIES; TAXONOMY; YEAST; ZYGOMYCETES in the McGraw-Hill Encyclopedia of Science & Technology.

David S. Hibbett

Bibliography. D. L. Hawksworth, The magnitude of fungal diversity: The 1.5 million species estimate revisited, *Mycol. Res.*, 105:1422–1432, 2001; D. S. Hibbett et al., A higher-level phylogenetic classification of the Fungi, *Mycol. Res.*, 111:509–547, 2007; T. Y. James et al., Reconstructing the early evolution of the fungi using a six gene phylogeny, *Nature*, 443:818–822, 2006; P. J. Keeling et al., The tree of eukaryotes, *Trends Ecol. Evol.*, 20:670–676, 2005; P. M. Kirk et al., *Dictionary of the Fungi*, 10th ed., CAB International, Wallingford, U.K., 2008.

Phylogenetics: predicting rarity and endangerment

Phylogeny is the evolutionary or ancestral history of organisms. Phylogenetic trees (branching diagrams used to indicate phylogenetic relationships) depict the shared evolutionary histories of species. For well-studied groups, for example, mammals, we can estimate the Tree of Life that connects all species. In cases where there is a well-sampled fossil record, it is also possible to place species that are now extinct along these evolutionary branches. Although we are most often restricted to information on relationships among living species, such phylogenetic trees may still allow us to infer the evolutionary processes that have determined the distribution of species among the major limbs of the tree. Frequently, we find that closely related taxa of similar ages differ greatly in species richness, indicating variation in rates of diversification (speciation minus extinction) over the evolutionary history of the clade (a taxonomic group defined by a common ancestor). Of more pressing concern is the increasing rate of extinction observed within recent time periods, perhaps several orders of magnitude greater than background rates, depending on the taxa of study. Identifying the drivers of extinction is complicated by the nonindependence of species, which is a product of their shared evolutionary history. Phylogenetic methods can correct for evolutionary nonindependence and offer a powerful tool for discriminating among variables that covary with phylogeny, as is the case for many species' traits associated with elevated extinction risk.

The extinction crisis. We are entering an extinction event on a scale similar to the mass extinctions recorded in the fossil record. The International Union for Conservation of Nature (IUCN) documents 76 mammal extinctions within recent time periods; the number for birds is higher (134 species), but it is lower for amphibians (38 species). It is difficult to estimate how many extinction events have gone undocumented, especially for less charismatic groups (that is, organisms lacking widespread popular appeal), but of course these unrecorded events can only intensify current assessments of the severity of the extinction crisis. Identifying processes that are driving species toward extinction and identifying which species are most vulnerable are critical for efforts to reduce rates of species loss.

With the extinction of species, we lose not only their particular contributions to ecosystem service and function, but also their unique evolutionary histories, which can be represented by the branches of the phylogenetic tree from which they subtend

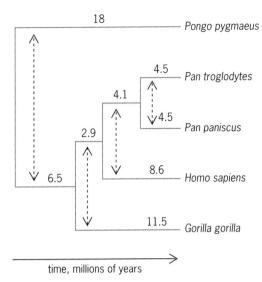

Fig. 1. Phylogenetic tree of the family Hominidae [consisting of humans (*Homo sapiens*) and the great apes—orangutan (*Pongo pygmaeus*), common chimpanzee (*Pan troglodytes*), pygmy chimpanzee (*Pan paniscus*), and gorilla (*Gorilla gorilla*)], with branch lengths drawn proportional to time (the number above each branch indicates its length in millions of years). Nodes represent the intersection of branches. Clades represent the species membership below each node. Orangutans split from the other hominids approximately 18 million years ago; therefore, orangutans capture a greater amount of unique evolutionary history (18 million years) than any other single hominid species. Double-headed dashed arrows indicate potential phylogenetically independent contrasts between sister clades.

(**Fig. 1**). Species descending from long evolutionary branches have no close relatives, and hence may be both genetically and phenotypically distinct. We might want to maximize the preservation of this phylogenetic diversity because (1) there may be little redundancy in the services provided by evolutionarily distinct species and (2) maximizing feature diversity is a sensible strategy in the face of an uncertain future.

Taxonomic selectivity. The IUCN Red List provides a categorical index of a species' "conservation health." Species for which there are sufficient data are placed into one of the following seven categories, in order of increasing extinction risk: least concern, near-threatened, vulnerable, endangered, critically endangered, extinct in the wild, and extinct. In mammals, 25% of assessed species are considered to be at risk of extinction (**Fig. 2**); the equivalent figures for birds and amphibians are 14% and 45%, respectively, suggesting taxonomic differences either in sensitivity or in extinction drivers. If extinction were a stochastic (random) process (sometimes likened to a field of bullets, in which survival is not contingent on inherent species' attributes but is due simply to chance—that is, the bullets are not targeted), we would not predict significant differences in risk between groups. However, even within mammals, strong taxonomic selectivity is clearly demonstrable. As an example, only 11% of opossums (order Didelphimorphia) are considered at risk of extinction, whereas over 60% of assessed primates are at risk (Fig. 2). Because extinction risk is thus clustered on the Tree of Life, if one species is at high

risk of going extinct, then it is likely that its close relatives are also vulnerable. As a consequence, the potential loss of phylogenetic diversity through extinction may be exacerbated—not only will we lose the unique branch lengths subtending from each species, but we might additionally lose the internal branches that connect them. The tendency for closely related species to share similar vulnerabilities has provided the stimulus to search for species attributes that predispose some species, but not others, toward extinction.

Correlates of extinction risk. At first, it might seem counterintuitive that extinction risk should covary with phylogeny. The principal drivers of extinction—for example, habitat destruction and transformation, introduction of nonnative species, and, increasingly, climate change—are of anthropogenic origin and do not evolve along the branches of the phylogeny. Critically, however, many traits (for example, body size, reproductive rate, and dispersal ability) that have been hypothesized to determine species' susceptibility show a phylogenetic signal, indicating some type of phylogenetic structure in extinction risk. Without considering phylogeny, it is not possible to assess whether a trait that correlates with

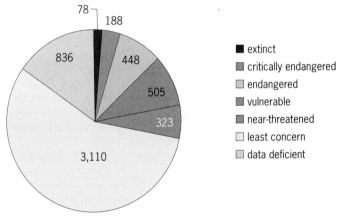

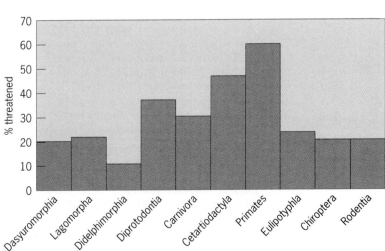

Fig. 2. (*Top*) Number of mammal species within various IUCN Red List categories. (*Bottom*) Percent of threatened (critically endangered + endangered + vulnerable) species within each major mammalian order; primates, including the great apes, contain the highest proportion of threatened species (approximately 60%), whereas Didelphimorphia (opossums) have the fewest (approximately 11%). (*Data from the 2008 IUCN Red List of Threatened Species*)

extinction risk, such as body size, influences species' vulnerability or simply covaries with a third trait as a result of shared evolutionary history, such as generation time, which may be the true determinant of risk. Because we do not have a priori information as to which traits determine species' responses, models of extinction risk that treat species as independent might therefore mislead us. Phylogeny must be considered when there is a phylogenetic signal in any variable that is not included in the analysis but that might additionally influence species' responses, even when the response (in this case, extinction risk) is not directly heritable.

One robust approach for identifying correlates of extinction risk uses phylogenetically independent contrasts to look at differences between sister clades (Fig. 1). At each node in the phylogeny, contrasts between the response (extinction risk) and predictor variables (species traits) are calculated, and statistical correlations can be evaluated. Because nodes deeper in the tree have longer independent evolutionary histories than those toward the tips, the raw differences may have unequal variance and might need to be rescaled assuming some approximation of how divergence increases through time. In mammals, several traits have been demonstrated to correlate with extinction risk—most significantly, the size of a species' geographic range. Small range size might reflect narrower environmental tolerances and increasing susceptibility to perturbations. Additionally, small-ranged species may be more vulnerable because a single localized threat might affect all individuals simultaneously. Body size is also thought to be important, because larger species tend to have longer generation times, smaller litter sizes, and lower abundances; hence, a longer time is necessary to recover following population declines. Small population sizes put species at high risk through demographic stochasticity, even if the risk is equal at an individual level. In contrast, species that reproduce rapidly can rebuild population sizes quickly and, as a result, spend less time in the danger zone of low numbers.

The size of a species' geographic range has been found to be a key predictor of extinction risk in other taxonomic groups, including birds and amphibians. However, even among mammals, the particular attributes that influence vulnerability differ across species. For instance, for Carnivora (one of the larger orders of placental mammals, including dogs, bears, and cats), bats, and marsupials, ecology (including home range size and population density) is the main predictor of extinction risk; for ungulates (hoofed mammals), though, life history traits (for example, weaning age and litter size) are more important. Furthermore, critical traits and drivers can also vary geographically: In South America, human population density is the most important determinant of a species' well-being, regardless of its biological traits; however, in Africa, it matters more how big a species is, reflecting differences in threats.

Future goals. Although detailed models for other taxa are lacking, it is likely that the patterns of ex-

tinction risk will be equally complex. It is urgent, therefore, to extend studies to include other groups; in particular, our knowledge of the distribution of rarity and vulnerability in plants is woefully poor. Preliminary work suggests that plants might demonstrate a very different phylogenetic pattern of extinction risk than mammals, with highly threatened species disproportionately represented among more species-rich taxa. Perhaps the Tree of Life for plants might prove more robust to the vagaries of the extinction process. Predictive models of extinction risk will allow us to identify at-risk species before they decline, providing opportunity for preemptive conservation strategies. Preventative conservation will likely be much cheaper than remedial actions, which is an important consideration when conservation budgets are limited. However, future threats may be different from those that have occurred historically. Climate change is considered likely to be an increasingly important future threat, yet we currently lack predictive models to project forward because we do not have substantial information on past trends.

For background information *see* BIODIVERSITY; ECOLOGICAL SUCCESSION; ECOLOGY; ECOSYSTEM; ENDANGERED SPECIES; EXTINCTION (BIOLOGY); PHYLOGENY; POPULATION ECOLOGY; SPECIATION; SYSTEMATICS in the McGraw-Hill Encyclopedia of Science & Technology.   T. Jonathan Davies

Bibliography. J. E. M. Baillie, C. Hilton-Taylor, and S. N. Stuart (eds.), *A Global Species Assessment*, IUCN, Gland, Switzerland, 2004; T. J. Davies et al., Phylogenetic trees and the future of mammalian biodiversity, *Proc. Natl. Acad. Sci. USA*, 105:11556–11563, 2008; G. M. Mace, J. L. Gittleman, and A. Purvis, Preserving the Tree of Life, *Science*, 300:1707–1709, 2003; A. Purvis, J. L. Gittleman, and T. Brooks (eds.), *Phylogeny and Conservation*, Cambridge University Press, Cambridge, U.K., 2005; D. M. Raup, *Extinction: Bad Genes or Bad Luck*, W. W. Norton & Company, New York, 1991.

Picocells and femtocells

Traditional cellular communication base stations are usually mounted on towers or buildings. They cover large (macro) areas ranging from a fraction of a mile up to a few miles (1 mi = 1.6 km), depending on their transmit power level and antenna directionality. To complement these macro coverage networks, picocells and femtocells provide users with local extensions of cellular communications services. The cells are used for many purposes, most often for closing gaps in service coverage and for providing service in specific areas, such as within buildings. This is done through the use of low-power base stations connecting the local mobile traffic to the cellular operator's network through the existing wired backhaul links such as home cable or digital subscriber line (DSL) connections. This article examines the characteristics of picocells and femtocells, the benefits of using the technology in wireless networks, and examples of current and future picocell and femtocell deployments

Characteristics. Picocells can typically cover enterprises with buildings up to 2800 m^2 (30,000 ft^2) in area, or approximately one-half the area of a soccer field, and can be installed either indoors or outdoors.

Outdoor picocells can be used to fill coverage gaps at transportation stations, shopping centers, or busy or remote cellular traffic areas. The picocell base stations usually have much lower transmit power than cellular base stations, on the order of a few watts. They are supported by wired backhaul links to relay traffic back to the main cellular network. Typically, they can be found mounted on the sides of low buildings or on lampposts.

Indoor picocells provide more limited coverage in the form of "hot spots" in airports, hotels, corporate offices, and campuses. Their coverage area can be as small as 10 m (33 ft) in an office setting. These installations typically have much less than 1 W transmit power. They are likely to be mounted on walls, and will have a wired backhaul, as well.

Femtocells are the smallest of these network extenders. They generally provide indoor coverage for residential and small-office user locations. Their transmit power is much less than 0.25 W. In cellular systems, it is usually suggested that most femtocells be operated in the 1–10-mW range to minimize the potential for interference to the macrocells using the same or nearby operational frequencies in the same area. They can be mounted on a wall, or be free-standing on a desktop. Similar to the picocells, they connect to the service provider's network via DSL or cable broadband installations.

Benefits. It is a common question why one would choose to use a picocell or femtocell instead of a Wi-Fi access point. At a high level, both serve a similar function by wirelessly connecting the user to the remote communication networks. However, the most obvious benefit of picocells or femtocells is that they allow the user's cellular phone to be used on the home or indoor network with the same capabilities as on the macrocellular network. The user can maintain the same phone number and the same user equipment while enjoying the advantages of the incremental additional coverage provided by these networks.

Femtocell and picocell base stations are essentially the low-capacity and low-power versions of cellular base stations, designed to provide limited local coverage. As a result of the deployment of these local networks, mobile operators can benefit from significant service coverage improvements. These networks can provide better coverage in difficult areas such as a remote building or urban street, or in a subscriber's home.

They also provide additional capacity to the cellular network operator at the user locations. Current designs can support more than 25 mobile phone users to connect simultaneously to the network. As a result, they reduce the capacity requirements for the macro coverage networks and, therefore, the operators can reduce the required number of macro base stations.

Picocells can also offer higher throughput to their users than the macrocellular networks. Picocell base stations or user terminals can be occupied with single or multiple "diversity" antennas. Using a multiple antenna configuration, picocells can greatly increase their capacity. It can be demonstrated that capacity levels of 14 b/s/Hz (bits/second/hertz) and 16 b/s/Hz can be achieved in 80% of the cases at a signal-to-noise ratio (SNR) of 20 decibels (dB) with 4×4 space diversity. In this example, the 4×4 configuration represents four transmit antennas at the base station and four receive antennas at the user terminal. This throughput increase will also come with improved quality of service (QoS) to the individual users owing to the enhanced local coverage, where the link quality (that is, dropped call) problems of large macro networks will not be a significant issue for these local coverage networks.

Local coverage also benefits the mobile battery life. Mobiles will consume less battery power to reach the nearby picocell or femtocell base stations, as compared to the macrocellular base stations that may be located a mile away. In addition, owing to all the benefits discussed above, these local networks present the opportunity to offer new bandwidth-hungry services, at reduced operating cost.

Issues. While the benefits of picocells are numerous, there are certain limitations on how widely they can practically be deployed. Inexpensive picocells deployed at a density of 8–15 cells per square kilometer (20–40 cells per square mile) can provide better coverage and capacity than a few macrocells covering the same area with 30 dB higher transmit power. However, such a picocell network solution can be quite costly and impractical for general-public areas owing to the relatively high cost (per subscriber) of the equipment and the availability of the installation locations and leases. As a result, it is expected that picocells will be a more applicable solution for localized dense urban areas, instead of a large-area mobile (portable or vehicular) coverage model with connected picocells.

One of the current issues for picocell and femtocell deployments is the lack of a standard architecture. This circumstance creates problems in the operational compatibility of the equipment from different vendors or operators, as well as in interference management.

Mobiles operating in different nearby picocell or femtocell networks can create interference with each other at rapidly varying levels owing to the small size and challenging operational environments of these networks. These interference issues must be mitigated for the successful operation of these networks and their harmonious coexistence with the macro base station networks. Therefore, it is important that the picocells and femtocells go through proper frequency and cellular design planning in order to avoid any interference issues and maximize capacity. Perhaps, this can easily be achieved by allocating these networks their unique operational spectrum, as was done in the case of the Clearwire venture discussed below.

If local coverage networks share operational frequencies with other systems operating in the same area, interference issues may occur. For example, individual femtocell or picocell mobiles operating in a single-frequency code-division multiple access (CDMA) network can act as a high-power interferer within the entire macro base station area, where they can block the general uplink traffic intended for the macro base station.

Similar interference conditions can also be observed in time-division duplex (TDD) networks where the uplink and downlink transmission times can be different for adjacent cells using the same frequency. To resolve this case of interference, picocell networks have been proposed that use a combination of frequency-division duplexing (FDD) and TDD schemes, while also increasing the capacity up to 90%.

Another significant issue is in the regulatory domain. Picocells and femtocells are required to be operated at their registered location. They must operate within the licensed area of the wireless network operator, and must reside at the user's address registered with the operator in order to properly receive emergency services. Unauthorized operation of these devices outside the registered location will create complex and hard-to-mitigate interference issues to other operators, and impair emergency services. In order to enforce the operation at the registered location, for example, Verizon's Wireless Network Extender uses the Global Positioning System (GPS) to verify the location of the equipment.

Deployments. An early example of picocell and femtocell designs appeared in March 1999 in documentation from Alcatel. This document provided a description of a GSM (Global System for Mobile Communications) base station that was designed to provide home coverage to existing GSM phones. In the following years, various related research projects appeared from different companies. By 2005, many small- and large-scale wireless system and equipment providers joined the development of picocell and femtocell networks, and wireless operator trials started in early 2007.

Most of the early developments focused on deploying femtocells as an extension of the GSM-based UMTS (Universal Mobile Telecommunication System) networks. However, the femtocell concept is practically the same for all cellular technologies. Indeed, the first commercial system deployment in the United States was on a cdma2000 network. This system was launched by Sprint in Denver in September 2007, and the service was called AIRAVE. It aims to provide improved in-home wireless coverage and low-cost unlimited calling. The user equipment includes a compact base station that works with any Sprint CDMA phone. It accesses to the backhaul network through the user's broadband Internet connection. Verizon also started its own femtocell service in January 2009 to complement its CDMA network

with a unit that has the size of a conventional Wi-Fi router, and can cover up to 465 m^2 (5000 ft^2) of domicile while supporting up to three simultaneous calls.

Similar to the Sprint and Verizon implementations, WiMAX (Worldwide Interoperability for Microwave Access) networks also plan to use femtocells as a part of their service. For example, Comcast is planning to use femtocells as part of its Clearwire joint venture. Clearwire aims to offer WiMAX-based cellular broadband services at 2.5 GHz. Comcast plans to extend the Clearwire WiMAX signal availability in the home, and provide backhaul through its Comcast cable modem network. In addition to use at home, Comcast subscribers can leave their home and remain connected on the main Clearwire WiMAX network. WiMAX femtocell networks have 5 MHz of dedicated spectrum in each Clearwire market. Such femtocell networks are expected to open significant application opportunities in the home, including high-definition television (HDTV) and voice over Internet Protocol (VoIP).

Outside the United States, the Japanese Personal Handy-Phone (PHS) system is a good example of a commercially viable picocell implementation with only 1% the emission levels of a conventional cell phone. The NTT-standard PHS phone outputs just 10–20 mW, while a phone connecting to a macrocellular network can output up to a fraction of a watt. To support future technologies, femtocells are also under development for the Long-Term Evolution (LTE) networks as a priority area, and it is expected that the initial deployments will start in 2012. LTE is designed to provide data rates between 15 and 100 times faster than the current 3G cellular networks, at up to downlink and uplink peak rates of 100 and 50 Mbps, respectively.

Outlook. Although the majority of cellular usage is outside the home and in vehicular environments, cellular use at home or work still represents a significant portion of network traffic. The increased volume of home and in-building cellular use will require the capacity and coverage of cellular networks to be supplemented by picocells and femtocells. These networks will service bandwidth-hungry future applications in addition to conventional voice calls at home, work, indoors, and outdoors. Although the technology is still in its infancy, it is forecast that there will be 32 million femtocell access points and more than 100 million users in the world by 2011.

For background information, *see* MOBILE COMMUNICATIONS; MULTIPLEXING AND MULTIPLE ACCESS; VOICE OVER IP; WIRELESS FIDELITY (WI-FI) in the McGraw-Hill Encyclopedia of Science & Technology.

Riza Akturan

Bibliography. H. R. Anderson, *Fixed Broadband Wireless System Design*, John Wiley & Sons, Chichester, England, 2003; S. G. Glisic, *Advanced Wireless Communications*, 2d ed., John Wiley & Sons, Chichester, England, 2007; Y. K. Kim and R. Prasad, *4G Roadmap and Emerging Communication Technologies*, Artech House, 2006.

Piezoelectric fiber composites

Producing electric power from otherwise wasted mechanical energy has created many new applications for piezoelectric materials, ranging from structural health monitoring of bridges and aircraft to powering wireless sensor networks. Wireless sensors have been used in a variety of applications such as surveillance, real-time data sharing, and condition-based monitoring of structures. Recent advances in piezoelectric materials and designs has made it possible to use piezoelectric generators as a self-sufficient, carefree power source for generating electric energy from the environment to power sensors, which can eliminate sensor batteries in remote or inaccessible places. These materials are able to produce electric energy from the vibrations available in structures such as aircraft, cars, bicycles, bridges, or in naturally occurring phenomena such as wind, water current, and waves.

Piezoelectricity. Piezoelectricity is the ability of certain crystalline materials to develop an electric charge proportional to a mechanical stress, and a geometric deformation (strain) proportional to an applied voltage. It was first discovered in 1880 by Pierre and Jacques Curie, who found that certain crystals, such as quartz and Rochelle salt, develop an electric charge when compressed in particular directions. All crystals can be classified in 32 crystal classes (point groups) according to the symmetry elements they possess. Only 20 point groups have one or more polar axes, which exhibit the piezoelectric effect. Of the 20 piezoelectric classes of crystals, 10 have a unique polar axis, called pyroelectrics. In addition to the piezoelectric effect, pyroelectrics develop electric charges as a result of a change in the magnitude of the dipoles with temperature.

Among pyroelectric crystals, those in which the spontaneous polarization can be reversed by an electric field are called ferroelectrics. The creation of dipoles can be further explained by looking into the barium titanate ($BaTiO_3$) crystal structure. At room temperature, $BaTiO_3$ has an elongated cubic structure (tetragonal) with barium at the corners, oxygen on the sides, and titanium in the center (**Fig. 1**). Large Ba cations cause the Ti ion to move off-center, producing a permanent dipole moment. The titanium ions are surrounded by six oxygen ions in an octahedral configuration, called TiO_6 octahedral, which can join from the corners to cause an effective dipole coupling within the unit cell. Application of a large electric field can cause the dipoles to align, a process called polarization or poling. In ferroelectric materials, such as barium titanate, most of the dipoles remain aligned when the field is removed, which results in a permanent polarization. These dipoles can attract surface charges on the electrodes, where the charge balance can change by movement of the dipoles as a result of any external pressure, creating electric power.

Piezoelectrics are most known for two different effects: direct and converse. The converse effect refers to the ability of piezoelectric materials to produce a strain, s, proportional to the applied electric field, E ($s = dE$, where d is the piezoelectric strain coefficient). This behavior is used in piezoelectric actuators and makes possible quartz watches, ultrasonic cleaners, and numerous other important products. On the other hand, application of a mechanical stress, σ, on piezoelectric materials creates an electric charge ($D = d\sigma$, where D is dielectric displacement, which is the charge per unit area, and d is the piezoelectric charge coefficient). This is called direct piezoelectric effect and is used for sensing applications such as in microphones, undersea sound-detecting devices, pressure transducers, and energy harvesters, which makes possible to produce electric energy from any mechanical disturbance. In both effects, the proportionality constant d is identical numerically. It should be noted that the piezoelectric coefficients are directional. For example, in the case of direct piezoelectric effect, d_{ij}, the i and j represent the directions of applied mechanical pressure and developed electric charge, respectively. The symmetry of a material may require that certain coefficients be equal or zero. For example, the nonzero piezoelectric coefficients of lead zirconate titanate (PZT) are transverse d_{31}, longitudinal d_{33}, and shear d_{15}. These coefficients are very important and help engineers to design actuators and energy-harvesting elements for specific applications based on the directions of applied stress, electric field, and polarization.

Piezoelectric generator designs. In some applications, such as powering wireless sensor networks and health monitoring of infrastructure and bridges, a high output power at low vibration amplitudes is required. Piezoelectric materials are well suited for such environments; however, their output power level is low because of the relatively low values of their piezoelectric coefficients. Therefore, many novel generator designs have been developed to amplify the stress-induced electric power. In these designs, the structure of generators is engineered in such a way that the transverse, longitudinal, or shear piezoelectric coefficients, or a combination of these coefficients, is effective. For example, in multilayer structure (**Fig. 2a**) the longitudinal coefficient (d_{33})

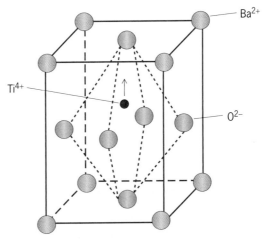

Fig. 1. Crystal structure of BaTiO₃.

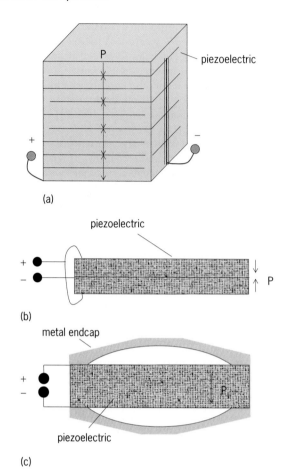

Fig. 2. Schematics of piezoelectric generator designs: (a) multilayer, (b) parallel bimorph; (c) moonie. The arrows show the polarization directions (P) in the piezoelectric layers.

and in bimorph structure (Fig. 3b) the transverse coefficient (d_{31}) are used, whereas in moonie transducers (piezoelectric ceramic between two metal endcaps; Fig. 2c) a combination of d_{31} and d_{33} coefficients is used. Besides the piezoelectric coefficients, other important material property criteria include mechanical stress and temperature tolerance, high piezoelectric coupling coefficient, k_{33}, and low dielectric loss, tan δ. Other dielectric properties such as capacitance and impedance are also important. For example, multilayer generators consist of a stack of thin piezoelectric ceramic sheets with internal electrode layers. The thin layers create high capacitance ($C = \varepsilon A / t$, where C is capacitance, ε is permittivity, A is area, and t is thickness of the layer), which in turn creates more charge $Q = CV$, where V is the voltage. The generator's impedance (Z) also should be designed to be as low as possible to maximize the power output. The impedance decreases with increasing of capacitance ($Z = 1/j\omega C$, where $j = e^{j\pi/2}$ and ω is angular frequency). The higher the number of piezoelectric layers and the magnitude of the applied stress, the greater is the output power. However, the most significant drawback in the multilayer design is its low threshold for cracking. Microcracks can initiate and propagate at the electrode–ceramic interface, especially at high-frequency vibration conditions, causing the device to fail. Because of the

low fracture toughness of bulk ceramics, composite generators have been developed that are more flexible, resistant to fracture, and protected from environmental elements.

Piezoelectric fiber composite. Generally, piezoelectric fiber composites are manufactured in two different form factors, as shown in **Fig. 3**. These are called 1-3 and 1-1 type composites, where the numbers refer to the connectivity of the piezoelectric and nonpiezoelectric materials (generally a polymer). For example, in a 1-3 composite the piezoelectric fiber is continuous in only one direction (z), whereas the polymer matrix is continuous in the x, y, and z directions (Fig. 3a). This type of composite is most suitable for hydrophones, underwater listening devices, pressure sensors, and ultrasound applications. The 1-1 composite comprises unidirectionally aligned piezoelectric fibers embedded in a polymer matrix, which increases fracture toughness and protects the fibers from mechanical and electrical breakdowns. Because of this fracture tolerance, fiber composites can be exposed to a higher vibration level to produce more power compared with those of the bulk materials. The polarization and electric output is transferred through a set of electrodes that extend perpendicular to the fiber direction. These so-called interdigital electrodes, placed on the top and bottom of the fibers, are based on piezoelectric d_{33} coefficient, which provide alternating poling directions along the fiber length (Fig. 3b). In a vibration scenario, the stress components could be from several directions, activating other piezoelectric coefficients such as shear and transverse modes. The 1-1 type composite is used as an actuator, sensor, and generator. In a bimorph structure, the 1-1 type produces significantly greater electric power compared to a single layer, as a result of the extension and contraction of piezoelectric layers and higher-frequency vibration characteristics. A bimorph generator consists of two poled piezoelectric layers that are held together with a bonding agent.

The fiber composite consists of many smaller segmented generators located between the electrode

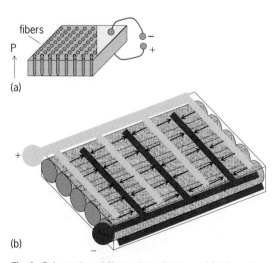

Fig. 3. Schematics of fiber composite types: (a) 1-3, and (b) 1-1.

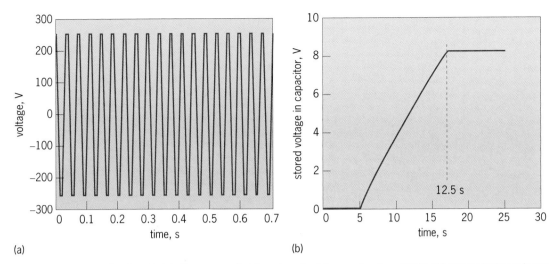

Fig. 4. Performance of a piezoelectric fiber composite bimorph under 2.4-*g* acceleration at 27 Hz: (*a*) unregulated ac voltage; (*b*) regulated dc output.

fingers connected electrically in series and parallel. Because of this electrical configuration, the generator can produce power even if a fiber segment or parts of the composite are damaged. Under mechanical vibration, a piezoelectric fiber composite bimorph can generate several hundreds of volts (**Fig.** 4*a*). This alternating voltage is too large for many applications; therefore, the level needs to be rectified and regulated to a lower voltage (Fig. 4*b*). In this example, the capacitor started charging as soon as the vibration started, and it took only 12.5 s to charge the capacitor to 8.2 V under 2.4-*g* acceleration at 27 Hz. The output energy can be either stored in a capacitor or battery, or it can be used directly to energize other low-power electronic devices. Depending on the magnitude of vibration force, amplitude, frequency, size, and number of the piezoelectric composites, low to high milliwatts of power can be generated, which is sufficient to run wireless devices, sensors, and a number of electronics without a battery or can extend battery life by charging it. Advances in electronics toward lower-power-consumption components as well as developments in materials and designs to generate more power are making possible the application of more self-powered devices.

For background information *see* CERAMICS; COMPOSITE MATERIAL; CRYSTAL STRUCTURE; DIPOLE; DIPOLE MOMENT; FERROELECTRICS; HYDROPHONE; MICROPHONE; PIEZOELECTRICITY; PYROELECTRICITY; TRANSDUCER; ULTRASONICS in the McGraw-Hill Encyclopedia of Science & Technology.

Farhad Mohammadi

Bibliography. W. G. Cady, *Piezoelectricity: An Introduction to the Theory and Applications of Electromechanical Phenomena in Crystals*, McGraw-Hill, New York, 1946; L. L. Hench and J. K. West, *Principles of Electronic Ceramics*, John Wiley & Sons, New York, 1990; B. Jaffe, W. R. Cook, and H. Jaffe, *Piezoelectric Ceramics*, Academic Press Limited, 1971; F. Mohammadi, A. Khan, and R. B. Cass, Power generation from piezoelectric fiber composites, *Mater. Res. Soc. Symp. Proc.*, vol. 736, Paper D5.5.1, 2003; S. Priya and D. J. Inman, *Energy Harvesting Technologies*, Springer-Verlag, 2009.

piRNAs (PIWI-interacting RNAs)

piRNAs are a subset of small ribonucleic acids that are expressed primarily in germline cells. Their length ranges from 23 to 31 nucleotides, and they show extreme sequence diversity. The origins of piRNAs include transposons (transposable elements), other classes of intergenic repetitive sequences, and noncoding, nonrepetitive genomic elements, called piRNA clusters. piRNAs function in RNA silencing, that is, the pathways that negatively control selected target genes. Of those, the best known are transposons. piRNAs also, rarely, target protein-coding genes. piRNAs alone are not able to silence genes. Rather, they associate with protein partners—PIWI (P element–induced wimpy testes) proteins—to perform their functions. Hence, this is the origin of their name, PIWI-interacting RNAs (piRNAs). PIWI proteins are a subgroup of the Argonaute (AGO) proteins, which are the key factors in RNA silencing. PIWI proteins are expressed primarily in germline cells, as are piRNAs. Presumably, for a given animal to maintain its reproduction system, multiple PIWI proteins are expressed that show peptide sequence similarity. However, piRNAs that bind to each PIWI member have unique characteristics. These observations have led to the proposal of two models for piRNA production: the amplification and the nonamplification loop pathways. Loss of PIWI proteins causes overexpression of transposons and failure of germline development, indicating that piRNAs, together with PIWI proteins, are necessary for germline survival and for maintaining the lineage of a species.

Origins. piRNAs were first discovered by a comprehensive small RNA profiling study in *Drosophila melanogaster* (fruit fly) testes and embryos. In this study, piRNAs were originally referred to as repeat-associated small interfering RNAs (rasiRNAs).

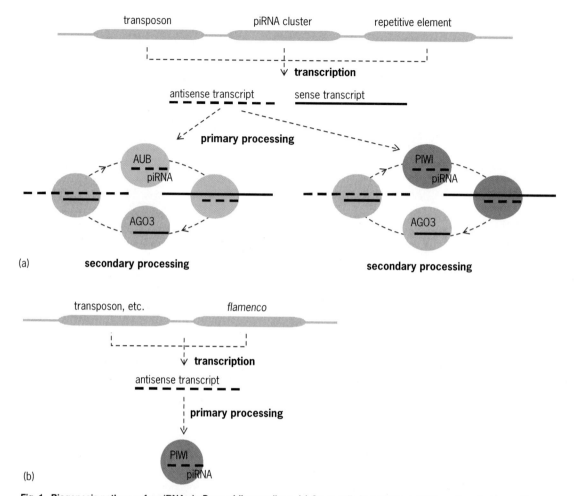

Fig. 1. Biogenesis pathways for piRNAs in *Drosophila* germlines. (*a*) Germ cells in germlines. (*b*) Somatic cells in germlines. The origins of piRNAs include transposons, other classes of intergenic repetitive regions, and piRNA clusters. piRNAs are processed by two pathways: the nonamplification loop system (indicated as primary processing) and the amplification loop system (indicated as secondary processing).

Mapping of their sequences on the *Drosophila* genome clarified their origins to be various intergenic repetitive regions, including transposons (**Fig. 1**). Subsequent sequencing of piRNAs in flies and mice identified nonrepetitive, noncoding genomic elements, which were previously unannotated and of unknown functions, as the sources of piRNA. These regions are collectively called piRNA clusters (Fig. 1). Nevertheless, in *Drosophila*, transposons and other kinds of intergenic repeats are the main sources for piRNAs in both ovaries and testes. By contrast, in mice, far fewer transposon-originating piRNAs have been found. Only a few piRNAs have been found in ovaries, whereas piRNAs are abundantly expressed in testes. Mouse piRNAs can be divided into two subgroups, prepachytene and pachytene piRNAs (pachytene is the third stage of meiotic prophase), depending on the cell type in which the piRNA is expressed and the time of expression during spermatogenesis (**Fig. 2**). Prepachytene piRNAs include prenatal piRNAs and neonatal piRNAs. They resemble fly piRNAs and are derived mainly from mobile genomic elements. Pachytene piRNAs, which are expressed mostly in adult testes, are less likely to be derived from transposons, with only 12–17% being known to correspond to transposons. The origins of piRNAs in other

organisms, including *Xenopus* (clawed frogs), fish, silkworms, nematodes, and planaria, have also been determined.

Biogenesis. A perspective that piRNA biogenesis pathways must be distinct from those of miRNAs (microRNAs) and siRNAs (small interfering RNAs) [which also act in regulation of gene expression or gene silencing] originally came from an observation in *Drosophila* ovaries that the cleaving enzymes called Dicers (DCR-1 and DCR-2) are unnecessary for piRNA production, whereas they are necessary for miRNA and siRNA production. Currently, two models for piRNA biogenesis have been suggested and are widely recognized (Fig. 1).

One such system is the amplification loop pathway, where it is thought that the endonuclease (or Slicer) activity of the PIWI proteins is involved in determining and forming the 5′ ends of piRNAs. Slicer is the activity required for the cleavage of target RNAs in RNA interference (RNAi, the process by which foreign, double-stranded RNA is recognized and degraded by specialized protein complexes within many eukaryotic cells, believed to be an evolutionarily conserved defense mechanism against RNA viruses and transposable elements), or in similar RNA silencing pathways. The idea originated from sequence analysis of piRNAs that are associated

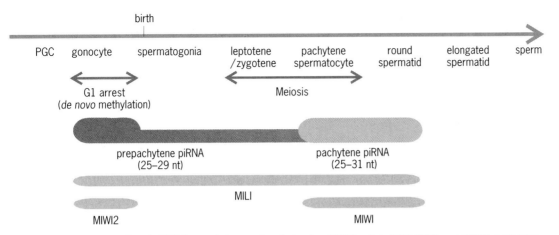

Fig. 2. Summary of the timing of piRNA (prepachytene and pachytene) and PIWI protein (MILI, MIWI, and MIWI2) expression during mouse spermatogenesis.

with each member of the PIWI proteins in fly ovaries. The investigation revealed two interesting features regarding piRNA sequences: (1) piRNAs that associate with Aub and Piwi (two members of the PIWI proteins in flies) arise mainly from antisense transcripts of transposons, whereas piRNAs that bind with AGO3 (another member of the PIWI proteins in flies) arise mainly from the sense transcripts of transposons; and (2) piRNAs that associate with Aub and Piwi have mostly uracil (U) at their 5′ ends, whereas piRNAs that bind with AGO3 mostly have adenine (A) at the 10th nucleotide from the heads (**Fig. 3**). piRNAs bound with Aub or Piwi indeed showed complementarities to AGO3-bound piRNAs over the first 10 nucleotides. These findings raised a possibility that PIWI-Slicer activity is involved in the piRNA biogenesis system. Similar supportive data were then obtained from fish and mice, and indicate that the piRNA biogenesis machinery might be highly conserved through evolution. ZILI/ZIWI and MILI/MIWI2 are the PIWI proteins involved in the amplification loop systems in fish and mice, respectively; but, additional factors that are required remain unknown.

The second piRNA biogenesis model is called the nonamplification loop system (Fig. 1), or the primary piRNA processing pathway. Determination of fly piRNA origins showed that piRNAs derived from a particular locus, *flamenco* (*flam*), located on the X chromosome, were exclusively loaded onto Piwi (Fig. 1). *flam* was originally identified as a suppressor of a number of specific transposons. *flam* is expressed only in somatic cells in ovaries, where the expression of Piwi is detectable, but the expression of Aub and AGO3 is absent. These data have strongly suggested that a nonamplification piRNA production system must operate in ovarian somatic cells for Piwi-piRNAs. A similar concept was observed in mice, where pachytene piRNAs loaded onto MIWI did not show the characteristics of the amplification loop system. MIWI is the last and third mouse PIWI protein, whose expression is restricted to pachytene spermatocytes and round spermatids in adult testes (Fig. 2).

In parallel, data have suggested that the primary

piRNA processing system also operates for Aub in fly germline cells (Fig. 1). This means that the aforementioned amplification loop system can be considered as the processing pathway for secondary piRNAs, in which AGO3 and, to a lesser extent, Aub serve as their destination. The same pathway may operate in mouse testes, where it is now believed that piRNAs loaded onto MILI are produced via both the primary and secondary processing pathways.

Function. The functions of piRNAs can be determined by deleting each piRNA-producing locus from

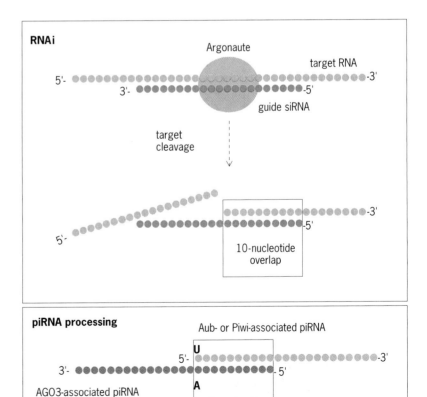

Fig. 3. piRNA associated with Aub, or Piwi, overlaps with piRNA associated with AGO3 through their first 10 nucleotides. Structurally, this RNA pairing shows a strong similarity to the RNA pairing appearing in RNAi (RNA interference), which consists of target RNA cleaved by Argonaute and guide siRNA bound to Argonaute.

a certain chromosome and investigating the resultant phenotypes. To date, the best-studied piRNA locus is *flam* on the *Drosophila* X chromosome. Deletions of *flam* caused upregulation of transposons in ovarian somatic cells, indicating that *flam* is central for the downregulation of selfish (actively self-replicating) transposable elements. These studies demonstrated the functional involvement of piRNAs in silencing such mobile genomic elements.

An alternative strategy is to introduce loss-of-function mutations into genes encoding PIWI proteins. Such genetic studies have been carried out in flies and mice. Loss-of-function mutations of the *piwi* genes caused not only malformation of germline cells, but also upregulation of transposons. Through these studies, the direct involvement of PIWI in controlling transposable elements and the relationship of transposon control and germline maintenance was determined. It was also evident that precise interaction of PIWI proteins with piRNAs is required for transposon silencing and germline survival.

Drosophila Aub is known to silence not only transposons but also protein-coding genes. Target genes of Aub include *stellate* and *vasa*, which encode a casein kinase beta subunit–like protein and an RNA helicase, respectively. Silencing of these two protein-coding genes is male-specific; this is because piRNAs targeting these genes are exclusively expressed in male germline cells. Silencing *stellate* in testes is indispensable for spermatogenesis.

Studies on the *ago3* mutants showed that silencing transposons in the germline requires its protein product; mutant males fail to maintain germline stem cells and mutant females become infertile. As mentioned above, AGO3 prefers to bind with piRNAs originating from sense transposon transcripts. Together, the data suggest that piRNAs in the sense orientation are as important as the antisense piRNAs for maintaining germline development and fertility.

Finally, recent findings from studies in *Drosophila* have strongly supported the idea that piRNAs serve as vectors for epigenetic information. A phenomenon called hybrid dysgenesis can be observed when fruit fly strains with and without certain transposons are crossed. In hybrid dysgenesis, abnormal traits appear in the hybrids, together with greatly elevated rates of genetic mutations and chromosome rearrangements, caused by the mobilization of transposable genetic elements. It was found that the progeny were fertile when an I-element (a type of transposon) was maternally inherited. In contrast, when an I-element was paternally inherited, the progeny showed sterility. The maternal deposition of the PIWI-piRNA complexes acts to prevent transposon mobility in the former case but not in the latter.

For background information *see* GENE; GENETICS; GENOMICS; HYBRID DYSGENESIS; MOLECULAR BIOLOGY; NUCLEOTIDE; PROTEIN; RIBONUCLEIC ACID (RNA); TRANSPOSONS in the McGraw-Hill Encyclopedia of Science & Technology.   Mikiko C. Siomi

Bibliography. A. A. Aravin et al., A piRNA pathway primed by individual transposons is linked to de novo DNA methylation in mice, *Mol. Cell*, 31:785–799, 2008; J. Brennecke et al., An epigenetic role for maternally inherited piRNAs in transposon silencing, *Science*, 322:1387–1392, 2008; J. Brennecke et al., Discrete small RNA-generating loci as master regulators of transposon activity in *Drosophila*, *Cell*, 128:1089–1103, 2007; L. S. Gunawardane et al., A Slicer-mediated mechanism for repeat-associated siRNA 5′ end formation in *Drosophila*, *Science*, 315:1587–1590, 2007; C. Li et al., Collapse of germline piRNAs in the absence of Argonaute3 reveals somatic piRNAs in flies, *Cell*, 137:509–521, 2009; V. V. Vagin et al., A distinct small RNA pathway silences selfish genetic elements in the germline, *Science*, 313:320–324, 2006.

Placozoan genome

Placozoans are small (1–2 mm), disk-shaped animals that were first discovered living on the walls of a saltwater aquarium in the late 1800s. After a long period of diminished interest, these animals were rediscovered in the 1970s amidst hypotheses that they might resemble the earliest animals. The only named species in the phylum Placozoa, *Trichoplax adhaerens*, is one of the simplest known animals, since only four cell types have been described. These cells are organized into an upper epithelium (with cover cells), a lower epithelium (with columnar and secretory cells), and a layer of multinucleate, contractile cells in the middle, but the overall form of the animal is amorphous, resembling a pancake (**Fig. 1**).

To eat, *Trichoplax* uses its lower epithelium as an external gastric surface that takes up food either by phagocytosis or by breaking it down by secreting digestive enzymes. *Trichoplax* has never been observed to reproduce sexually (via the fertilization of egg by sperm); animals in laboratory cultures multiply asexually by the splitting of one animal into two. The focus of many recent studies, the "Grell strain," was originally collected from the Red Sea, but other strains (which might be different species) are found in tropical seas around the world. The animals move by using cilia on the lower epithelium and contractile fiber cells in the middle of the body.

The genome of *Trichoplax adhaerens* was sequenced recently using the whole-genome shotgun strategy (which breaks the genome up into small pieces, followed by sequencing of the pieces and their reassembly into the full genome sequence) and provides a good basis for studies of animal evolution and development. The size of the placozoan genome is approximately 98 million base pairs, among the smallest of animal genomes (comparable to the 97 million base pairs found in the genome of the model nematode species, *Caenorhabditis elegans*). The *Trichoplax* genome is predicted to contain about 11,500 protein coding genes. Nearly 87% of these predicted genes are similar to proteins known from other animals.

Early animal evolution. Although researchers have hypothesized that the earliest animals may have resembled placozoans, the relationship of *Trichoplax* to other animals has been heavily debated. A variety of methods, including the sequence of ribosomal ribonucleic acid (rRNA) and the amino acid sequence of proteins encoded in the nuclear and mitochondrial genomes, have supported conflicting hypotheses. Some analyses place placozoans as cousins of cnidarians, which are a group of animals with saclike body plans that includes corals, sea anemones, and jellyfish. Other analyses have placed *Trichoplax* as a sister group of bilaterians (animals with bilateral symmetry, including vertebrates, fruit flies, and nematodes), whereas still others have positioned placozoans as the earliest branching animal phylum.

The whole genome sequence of *Trichoplax* provides a large amount of data for answering the question of animal relationships, since some of the conflicting hypotheses described above may be a result of the small amount of data being used. The analysis of about 100 genes from fully sequenced animal and nonanimal genomes, including genes made available from the *Trichoplax* genome project, places placozoans as a sister lineage of cnidarians and bilaterians, leaving sponges (the phylum Porifera) as the earliest branching animal group (**Fig. 2**). With the position of *Trichoplax* relative to other animals established, placozoan genes and biology can be studied to understand events in early animal evolution.

Slow-evolving genome. Although the *Trichoplax* genome is 30-fold smaller than the human genome, the intron-exon boundaries (demarcations between regions of genes that are translated into protein and those that are not) of *Trichoplax* genes are similar to their counterparts in the human genome. Considering homologous genes in the human and placozoan genomes, 82% of human introns are also present in *Trichoplax*. This conservation of gene structure is surprising, given that the genes of our closer cousins, fruit flies and nematodes, lack many of these introns. Similarly, the genes of the starlet sea anemone, *Nematostella vectensis*, also share many introns with human genes. The introns that the human genome shares with *Nematostella* and *Trichoplax* were most likely present in the earliest animal genome. The retention of these ancestral introns in the small *Trichoplax* genome is in contrast to the small genomes of fruit flies and nematodes, where the reduction in genome size was accompanied by an accelerated rate of intron loss (in this manner, these latter genomes are faster-evolving than the *Trichoplax* genome).

Another line of evidence for the slow evolution of the *Trichoplax* genome is the organization of genes relative to each other on the human and placozoan genomes. Genes that are present on the same chromosome are referred to as "syntenic" genes (this original definition of the word *synteny* does not require a particular order for the genes). Comparisons of neighboring genes in the *Trichoplax* and human genomes show that many segments of human chromosomes have corresponding segments in the placozoan genome that carry the same genes. Although the order of these genes has been scrambled between the two genomes, it appears that genes that are syntenic in *Trichoplax* are also syntenic in humans. This conserved synteny implies that blocks

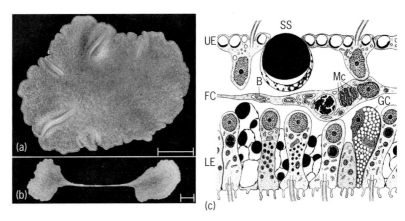

Fig. 1. *Trichoplax* morphology and reproduction. (*a*) Adult *Trichoplax* in laboratory culture (scale bar = 0.2 mm). (*b*) One *Trichoplax* dividing into two animals (scale bar = 0.2 mm). (*c*) Schematic rendering of a transverse section through *Trichoplax* showing the upper epithelium (UE) with cover cells and shiny spheres (SS), the lower epithelium (LE) with columnar and gland cells (GC), and the middle layer with contractile fiber cells (FC). Other abbreviations: B, bacterium in endoplasmic cisterna; Mc, mitochondrial complex. [*Panels a and b courtesy of Ana Signorovitch; panel c originally from K. G. Grell and A. Ruthmann, Placozoa, pp. 13–27, in F. W. Harrison and J. A. Westfall (eds.), Microscopic Anatomy of Invertebrates, vol. 2: Placozoa, Porifera, Cnidaria, and Ctenophora, Wiley-Liss, New York, 1991, and modified by T. Syed and B. Schierwater, The evolution of the Placozoa: A new morphological model, Senckenbergiana Lethaea, 82:259–270, 2002*]

of genes that were neighbors in the earliest animal genome continue to appear on the same segments of chromosomes in extant animals as divergent as humans and placozoans. The fast-evolving genomes of fruit flies and nematodes have lost this synteny, but the sea anemone genome has preserved the ancestral syntenic blocks.

Genes for animal development. Sexual reproduction by the use of sperm and egg is a unifying feature of all animals. The fertilized zygote then unfolds a complex program whereby the dividing cells of an embryo acquire different fates, resulting in the patterning of the adult body plan. Many of the molecular mechanisms involved in embryonic development are shared by different groups of animals. For example, the same

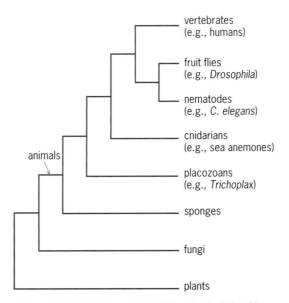

Fig. 2. Phylogenetic tree representing animal relationships. Whole-genome sequence data suggest that placozoans are a sister lineage to cnidarians and bilaterians (vertebrates, flies, and nematodes). Sponges are the earliest branching animal phylum. Fungi and plants are shown here as outgroups (nonanimals).

signal transduction cascades (series of interacting proteins) are used by fruit flies and mice to coordinate the fates of cells during development. Similarly, many of the same DNA-binding proteins that control when and where genes are expressed, called transcription factors, are used to specify the identities of different cells in different bilaterian species. This conservation of molecular mechanisms suggests that the structures that they pattern in diverse species are actually homologous. For example, fruit flies and mice both need to define their anterior-posterior and dorsal-ventral axes, and they do so using the same genes.

Trichoplax adults lack obvious similarity to bilaterian body plans; there is no clear anterior-posterior (head versus tail) axis, and it is unclear if the top versus bottom polarity of the animal is related to the dorsal-ventral (back versus belly) axis of other animals. It is not even known whether the *Trichoplax* life cycle includes embryonic stages similar to those of other animals (rarely occurring dividing cells that resemble embryos have been seen, but they never develop into adult *Trichoplax*). Yet the placozoan genome contains representatives of most of the major known signaling and transcription factor gene families known. However, this is not to say that there are no differences between placozoans and bilaterians in terms of gene content. Of the well-known Hox (homeotic) family of proteins that pattern the anterior-posterior axis of flies and mice, only one Hox-like representative is found in the *Trichoplax* genome. Although some signaling pathways appear to be incomplete in *Trichoplax*, it is striking that an animal with only four known cell types and unknown embryonic development should contain genes used for cell-fate specification in other animals with many more cell types. The findings from the *Trichoplax* genome are similar to those from the sea anemone genome, which also encodes genes known to pattern structures (for example, the mesoderm) that the organism lacks.

Genes for neurons. *Trichoplax* lacks cells that bear an obvious morphological similarity to neurons. However, the genome encodes many genes with neural functions. Voltage-dependent potassium, calcium, and sodium channels and enzymes for neurotransmitter synthesis are predicted in the genome. Although the machinery for vesicle trafficking is present in all eukaryotic organisms, *Trichoplax* appears to have animal-like versions of proteins used for docking vesicles at synapses. The machinery for synthesis, release, and uptake of neurotransmitters, synapse formation, conduction of electrical impulses, and photoreception in the *Trichoplax* genome is most likely functioning in nonneural roles, since placozoans lack neurons.

Prospects. By the measures of intron conservation and synteny of genes, the *Trichoplax* genome appears more similar to the human genome than to the genomes of our closer cousins, such as fruit flies and nematodes. Although the placozoan genome is slow-evolving and may be more similar to the genome of the earliest animal, it is impossible to know if that earliest animal morphologically resembled *Trichoplax*.

Modern placozoans have had just as long to evolve as humans have since the time that the first animals appeared.

The complex gene content of the *Trichoplax* genome, along with similar complexity of the sea anemone genome, suggests that the common ancestor of all animals most likely had a "tool kit" of genes for executing animal development and patterning the adult body plan. However, this genetic tool kit has been elaborated more recently in the evolution of different animal phyla, and further studies are needed to understand how these changes (for example, the expansion of the Hox gene family in cnidarians and vertebrates, which is represented by only one gene in *Trichoplax*) have affected the evolution of animal body plans. Experiments will also be needed to explain the functions of the large repertoire of gene families in an animal as "simple" as *Trichoplax*—the four cell types may be patterned cryptically, at a molecular level, by the many developmental gene families found in the *Trichoplax* genome.

In addition to generating the above questions about the evolution of the animals and their genetic tool kit, the *Trichoplax* genome will further enable the experiments needed to answer those questions.

For background information *see* ANIMAL EVOLUTION; ANIMAL KINGDOM; BILATERIA; CNIDARIA; GENE; GENETIC MAPPING; GENETICS; GENOMICS; PLACOZOA; PORIFERA; PROTEIN; SEA ANEMONE in the McGraw-Hill Encyclopedia of Science & Technology.

Mansi Srivastava; Daniel S. Rokhsar

Bibliography. K. G. Grell and A. Ruthmann, Placozoa, pp. 13–27, in F. W. Harrison and J. A. Westfall (eds.), *Microscopic Anatomy of Invertebrates: vol. 2: Placozoa, Porifera, Cnidaria, and Ctenophora*, Wiley-Liss, New York, 1991; B. Schierwater, My favorite animal, *Trichoplax adhaerens*, *BioEssays*, 27:1294–1302, 2005; B. Schierwater et al., The early ANTP gene repertoire: Insights from the placozoan genome, *PLoS One*, 3:e2457, 2008; M. Srivastava et al., The *Trichoplax* genome and the nature of placozoans, *Nature*, 454:955–960, 2008; T. Syed and B. Schierwater, The evolution of the Placozoa: A new morphological model, *Senckenbergiana Lethaea*, 82:259–270, 2002.

Plant phylogenomics

With an increasing number of plant genomes being sequenced, our understanding of evolutionary processes shaping the gene content and structure of plant genomes is growing rapidly. Comparative analyses of plant genomes have been particularly informative when multispecies comparisons are done within the context of their phylogenetic relationships. For example, one may be able to pinpoint the timing of a gene duplication event if two copies are identified in a pair of closely related species but only one copy is found in a more distantly related species. In this case, one would map gene copy number onto a diagrammatic tree representing relationships among the

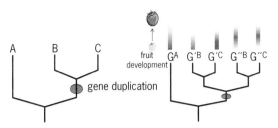

Fig. 1. Gene duplications, orthology, and paralogy. The tree on the left shows phylogenetic relationships among three species (A, B, and C) and the timing of a gene duplication in the common ancestor of species B and C. The tree on the right shows the relationships of genes sampled from species A, B, and C. Following the gene duplication, species B and C have retained both copies. Vertical bars above the gene tree (right) show gene expression through fruit development. The gene G^A is most highly expressed in young fruits of species A, as are genes G$'^B$ and G$'^C$ in species B and C. In contrast, genes G$''^B$ and G$''^C$ are expressed later in fruit development. The most parsimonious explanation for this pattern is that the G^A, G$'^B$, and G$'^C$ share the ancestral expression pattern and there was a shift in expression of G$''$ following duplication in the common ancestor of species B and C. All of the genes shown in the gene tree (right) are related (homologous). However, whereas G$'^B$ and G$'^C$ (or G$''^B$ and G$''^C$) diverged after speciation, G$'^B$ and G$''^C$ (or G$''^B$ and G$'^C$) are separated by speciation and duplication events. Genes sharing common decent through speciation events in the absence of duplication (G$'^B$ and G$'^C$, or G$''^B$ and G$''^C$) are orthologous. In contrast, paralogs (G$'^B$ and G$''^C$, G$'^B$ and G$''^B$, G$'^C$ and G$''^C$, or G$''^B$ and G$'^C$) are related through ancient duplication.

species (a phylogenetic tree) and deduce that the duplication occurred on the branch leading to the most recent common ancestor of the species with two gene copies (**Fig. 1**). Furthermore, one may map the gene expression patterns onto a gene tree in order to determine whether gene function has changed in one or both of the duplicated gene copies (Fig. 1). This is just one example of how phylogenomics—the integration of genomics and phylogenetics—is being used to elucidate the evolution of biodiversity.

The term phylogenomics was coined in the late 1990s to describe how one could use gene trees to predict the function of genes in poorly understood organisms based on their relationship to orthologous genes (Fig. 1) of known function in model organisms. (Genes that share common decent through speciation events in the absence of duplication are orthologous. In contrast, paralogous genes have diverged through ancient duplication.) Later, as the number of microbial genome sequences started to increase, the term also was employed to describe the use of genome-scale sequence data to resolve organismal relationships. Both types of phylogenomic investigation include two essential elements: analyses of genome-scale sequence data and consideration of organismal relationships (that is, phylogeny). In the plant sciences, phylogenomic studies are resolving long-standing phylogenetic questions, and this knowledge is being used to elucidate gene and genome evolution.

Using genome-scale data to resolve relationships. Over the last two decades, researchers have been using gene sequences to estimate relationships among distinct plant groups and develop a classification system based on evolutionary relationships rather than similarities that may arise through con-

vergent evolution. Much progress has been made in this endeavor using a handful of plastid and ribosomal genes. Despite much attention, though, relationships among some major plant lineages have been difficult to resolve. Recently, plant systematists have obtained robust estimates of these relationships using whole plastid genome sequences (**Fig. 2**). For example, there has been growing support for the position of *Amborella trichopoda* [an understory shrub species found only in the cloud forests (rainforests that occur on high mountains in the tropics) of New Caledonia] as the sister to all other flowering plants. However, plant systematists could not rule out the possibility that *Amborella* and the water lilies formed a clade (that is, a limb in the phylogenetic tree) that was sister to the rest of the angiosperms (flowering plants). Phylogenomic analyses using genes extracted from plastid genome sequences indicated that *Amborella* alone is sister to the rest of the angiosperms, and the water lilies are on their own branch that is sister to all other angiosperms except *Amborella* (Fig. 2). These studies also lend strong support for the hypothesis that the two most diverse groups of flowering plants, monocots and eudicots, are more closely related to each other than either is to the magnoliids, which is a group that includes magnolias (*Magnolia*), avocados (*Persea*), and pipevines (*Aristolochia*). Knowledge of these relationships has implications for our understanding of gene and genome evolution throughout the history of flowering plants.

Genomic analysis in the last common ancestor of all living angiosperms. Although there is now consensus for the position of *Amborella* as sister to a

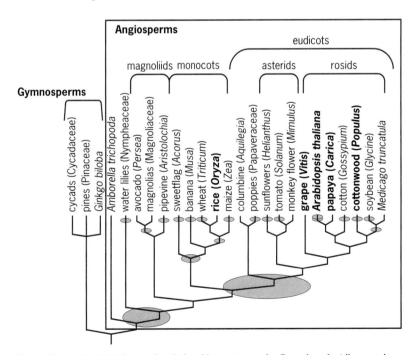

Fig. 2. Resolution of phylogenetic relationships among major flowering plant lineages has improved greatly with the analysis of plastid genome sequences. The tree shows our current understanding of relationships among flowering plants (angiosperms) and the timing of whole genome duplications (ovals) that have been identified through analyses of whole genome sequences (for the species shown in bold type) and transcriptomes (Fig. 3). Note that the timing of more ancient genome duplication events is less certain relative to more recent events, as indicated by larger ovals, encompassing multiple nodes in the phylogenetic tree.

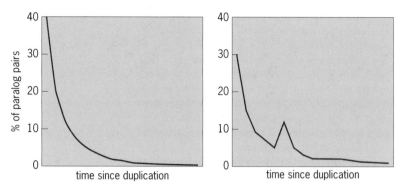

Fig. 3. Ancient whole genome duplications can result in a large proportion of duplicated gene pairs (paralog pairs) with similar divergence times. Expressed genes can be isolated as transcribed messenger ribonucleic acid (mRNA), converted to complementary deoxyribonucleic acid (cDNA), and sequenced en masse. Similar sequence pairs are identified and aligned. The frequency of synonymous nucleotide substitutions is estimated as a proxy for divergence time for each putative paralog pair. The plot on the left shows the distribution of divergence between paralogs in the absence of genome duplication. The plot on the right shows a large number of paralog pairs with similar divergence times, suggesting an ancient genome duplication.

clade containing all other living flowering plants, the branch connecting *Amborella* to the last common ancestor of all extant flowering plants represents just as much time as the branch leading from the last common ancestor to magnolias, maize, tomato, sunflower, grapevine, soybean, and any other plant species (Fig. 2). *Amborella* is not itself ancestral to other angiosperms, but comparisons between the *Amborella* genome and any other flowering plant genome can provide unique insights into characteristics of the last common ancestor of all extant flowering plant lineages.

The complete genome sequence of *Amborella* is yet to be sequenced, but analyses focused on particular gene families have been illuminating. For example, members of the MADS box transcription factor gene family are of great interest because they initiate differentiation of cells, giving rise to distinct floral organs (that is, sepals, petals, pollen-producing stamens, and seed-producing carpels). Phylogenetic analyses of MADS box genes sampled from *Amborella*, other angiosperms, and gymnosperms (seed plants having naked ovules at the time of pollination, and the sister group to angiosperms) indicate gene duplication events in an ancestor of all angiosperms after the divergence of the angiosperm and gymnosperm lineages. This finding has led some investigators to hypothesize that MAD box gene duplication events may have played a role in the origin of the flower. These same phylogenetic analyses have also identified other, lineage-specific duplications that occurred later in angiosperm history.

Comprehensive comparative analyses of genome structure in *Amborella* and other angiosperms will not be possible until the *Amborella* genome has been sequenced, but comparisons of bits of the *Amborella* genome with fully sequenced plant genomes (Fig. 2) suggest that gene order has been conserved in portions of these genomes despite some 150 million years of evolution. Future phylogenomic analyses, including determination of the complete *Amborella* genome sequence, will indicate whether these re-

gions harbor distinctive genes or other features that may reduce the frequency of rearrangements. Moreover, the *Amborella* genome sequence will serve as an evolutionary reference that will help researchers pin down the timing of gene and genome duplications, as well as gene losses and rearrangements, that have driven the divergence of monocot and eudicot genomes.

Phylogenomic analyses indicate genome duplications. The first plant genome to be published was that of *Arabidopsis thaliana* (mouse-ear cress), which is a small annual in the mustard family (Brassicaceae) that has long been a workhorse of plant molecular biology. Its genome is approximately 150 million bases—less than 1/20th the size of the human genome—and organized in only five chromosomes, making it an attractive candidate for whole genome sequencing. Even before the sequence was completed in 2000, though, it became clear that the *Arabidopsis* genome included many blocks with overlapping sets of duplicate genes. This observation was consistent with multiple rounds of ancient genome duplication (polyploidization), with some gene loss and rearrangement following each genome duplication event.

More recently, evidence of ancient genome duplications has been found in the genome sequences of rice, poplar, grapevine, papaya, and sorghum. By comparing duplicated blocks from each of these species within the context of their relationships, researchers are deducing the timing of genome duplications and subsequent gene losses and rearrangements (Fig. 2). For example, two of the three rounds of genome duplication that had been inferred from the *Arabidopsis* genome seem to have occurred within the mustard family lineage after it diverged from the papaya lineage. The timing of the third and oldest genome duplication is still ambiguous, but it clearly occurred before the divergence of the asterid and rosid clades within the eudicots (Fig. 2). Furthermore, analysis of replicated blocks within the grapevine genome suggests that this most ancient event was a triplication. In fact, a common ancestor of *Arabidopsis*, papaya, poplar, grapevine, and tomato was a hexaploid, just as bread wheat (*Triticum aestivum*) is today.

In parallel to the genome duplications described above for the eudicots, comparisons of duplicated blocks in portions of the rice, sorghum, maize, and wheat genomes show evidence of at least two rounds of genome duplication in their common ancestors. Whereas the more recent of these two events seems to have occurred near the time of the origin of the grass family (Poaceae), it is unclear whether the older genome duplication occurred before or after the divergence of the monocots and eudicots (Fig. 2).

In the absence of whole genome sequences for many important flowering plant lineages, researchers are looking for signatures of genome duplication in comparisons of protein-coding genes. It is now routine and relatively inexpensive to sequence thousands of expressed (that is, transcribed) genes from a given species. These expressed genes, comprising the "transcriptome," include duplicates or paralogs (Fig. 1), and the time since each gene duplication can be deduced from estimates of neutral divergence between duplicates in each gene pair. Ancient genome duplications are inferred when many duplicate gene pairs have similar divergence time estimates (**Fig. 3**). Together, analyses of duplicated blocks in sequenced genomes and duplicated genes identified in transcriptome sequences indicate that genome duplications have occurred repeatedly throughout angiosperm history, and these events may have contributed to the diversification of flowering plants (Fig. 2).

Future outlook. These are exciting times in botany. Phylogenomic studies are both improving our understanding of phylogenetic relationships among plant groups and elucidating the processes that shape plant genomes. Beyond identifying ancient genome duplications, phylogenomic investigations of the transcriptome data sets described above are now under way to reconstruct evolutionary relationships among plant lineages and identify innovations associated with gene duplications and interspecific gene transfer. With improvements in genome sequencing and computational technologies, whole genome comparisons should shed even more light on the genetic basis of evolutionary innovations that have fueled the remarkable diversification of flowering plants over the last 150 million years.

For background information *see* CELL PLASTIDS; GENE; GENETICS; GENOMICS; MAGNOLIOPHYTA; PALEOBOTANY; PHYLOGENY; PLANT EVOLUTION; PLANT KINGDOM; PLANT PHYLOGENY; POLYPLOIDY in the McGraw-Hill Encyclopedia of Science & Technology.
Jim Leebens-Mack

Bibliography. L. Cui et al., Widespread genome duplications throughout the history of flowering plants, *Genome Res.*, 16(6):738–749, 2006; R. K. Jansen et al., Analysis of 81 genes from 64 plastid genomes resolves relationships in angiosperms and identifies genome-scale evolutionary patterns, *Proc. Natl. Acad. Sci. USA*, 104(49):19369–19374, 2007; D. E. Soltis et al., Polyploidy and angiosperm diversification, *Am. J. Bot.*, 96(1):336–348, 2009; D. E. Soltis et al., The *Amborella* genome: An evolutionary reference for plant biology, *Genome Biol.*, 9(3):402, 2008; D. E. Soltis et al., The floral genome: An evolutionary history of gene duplication and shifting patterns of gene expression, *Trends Plant Sci.*, 12(8):358–367, 2007.

Polarization lithography

Optical lithography provides a high-productivity, profitable means of making microcircuits on silicon wafers by imaging increasingly smaller features in photoresist. Polarization lithography specifically refers to the use of polarized light for illumination within an optical lithographic system to improve the quality of the imaging and the resolution. In the quest for ever-greater resolution, modern fabrication facilities all use some form of polarization lithography to produce high-technology computer chips. This article discusses basics of polarization lithography and why it is important.

Lithographic resolution is described by Eq. (1), where λ is the laser wavelength, NA is the numerical

$$R = k_1 \frac{\lambda}{\text{NA}} \qquad (1)$$

aperture of the projection lens, and k_1 is a process parameter that is usually about 0.3. Today's advanced lenses can easily image lines and spaces below a resolution of 50 nanometers. This resolution has been achieved by using hypernumerical apertures (NA greater than 1), ArF excimer lasers at a wavelength of 193 nm, and resolution enhancement techniques, such as off-axis illumination and phase masks.

The use of a water-immersion medium to enable lens systems with NA > 1 has extended the product lifetime of using the current 193-nm wavelength by replacing the air between the lens and the wafer with water. The maximum possible numerical aperture using immersion is given by Eq. (2), where n

$$\text{NA} = n \sin \theta \qquad (2)$$

is the index of refraction of the water medium just after the last lens element and before the wafer surface, and θ is the maximum angle, with respect to the

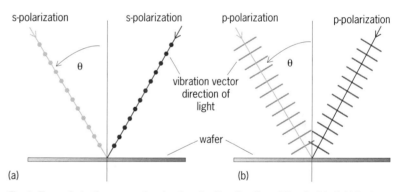

Fig. 1. Two polarization cases showing the vibration direction of the electric field for two light rays incident on a wafer surface with an angle θ. (*a*) With s-polarization, the vibration direction is into the plane of the figure. (*b*) With p-polarization, the vibration direction is parallel to the plane of the figure. In both cases the vibration vector is perpendicular to the ray direction.

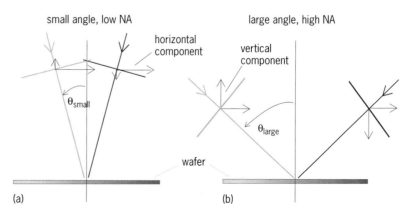

Fig. 2. Two cases of interfering rays with p-polarization. (*a*) Small-angle interference. The vertical component is very small, so its effect on image contrast will be almost negligible. (*b*) Interfering rays with a large angle between them. The vertical component is very large and its effect will lower the image contrast.

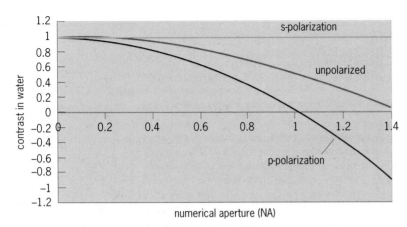

Fig. 3. Contrast calculated in water for two-beam interference as a function of the half-angle between the beams given as the numerical aperture (NA). Contrast for unpolarized light is calculated as the average between s- and p-polarizations.

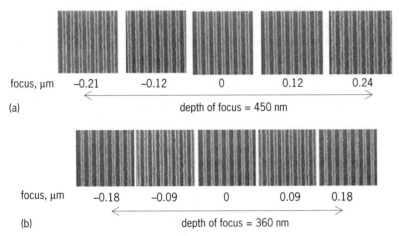

Fig. 4. Scanning electron micrographs (SEMs) of photoresist features in printing 60-nm lines and spaces, comparing (*a*) s-polarized illumination to (*b*) unpolarized illumination. The s-polarized illumination shows increased depth of focus as compared to the unpolarized case, as can be seen by examining the change in line width as the wafer surface is moved through the focus of the lens system. (*Courtesy of Nikon Corporation*)

optical axis, that exits the lens. For water at a wavelength of 193 nm, $n = 1.44$. The maximum angle that can exist in the water medium is close to $70°$, so optical projection systems for lithography have a maximum numerical aperture of 1.35. At this high angle, imaging is determined not just by the NA, but also by the polarization of the light exiting the lens. The type of polarized light has a major influence on image contrast and the resultant resolution.

Polarization and interference. In general terms, the propagation of light can be described with a ray arrow and an electric field wave. The direction of the ray gives the direction of propagation. The wave nature of light gives rise to polarization, that is, the direction of vibration of the light-wave electric field. For the discussion here, this vibration is always perpendicular to the ray direction of the light. Many artificial light sources, including lithography tools at 365-nm wavelengths, emit unpolarized light, where the light consists of many different rays with randomly oriented polarizations. On the other hand, polarized light is that in which the electric field vibrates in a single direction. Many lasers, including excimer lasers, emit polarized or partially polarized light. In

addition, various optical elements can be used to control the amount and type of polarization.

The polarization of light can always be described using only two directions of electric field vibration. These directions are called states of polarization. **Figure 1** shows two cases of light rays incident on a wafer with different states of polarized light. The s-polarization or transverse electric (TE) polarization has a vibration parallel to the wafer surface, while the p-polarization or transverse magnetic (TM) polarization has the vibration vector direction parallel to the plane formed by the two incident rays that are shown in Fig. 1. Unpolarized light can be considered as the average of the s- and p-polarization states.

The use of mainly s-polarization in an optical lithography system can result in higher contrast imaging, as compared to using p-polarization or unpolarized light. This difference can be understood by realizing that the projection lens forms the image by summing all the light at the wafer. The simplest form of this summation is the example of two rays adding and interfering with each other, resulting in bright and dark interference patterns in the photoresist on the wafer surface. The polarization state of the interfering rays determines the contrast of the image. Higher contrast results in improved pattern fabrication.

With s-polarization the two rays have polarization vectors that are parallel to each other; that is, the electric field vibrations are aligned in the same direction and angle. This alignment results in all the light contributing to interference, such that the light and dark patterns have maximum contrast. However, with p-polarization, the two rays have polarization vectors that meet at different angles, and the amplitude must be added by the constituent components. **Figure 2** shows two cases of interfering rays with p-polarization. Figure 2*a* shows a small angle of incidence. The horizontal component of the polarization vector is large compared to the vertical component. In Fig. 2*b*, the angle of incidence is large, and the vertical and horizontal components have approximately equal strengths. Since the vertical components oppose each other, they cancel and reduce the overall contribution to image contrast. This effect becomes larger as the angle of incidence increases. Thus, with p-polarization, large angles degrade image contrast.

Figure 3 shows contrast differences of the various polarization states as a function of the incident angle given in units of numerical aperture. The contrast is calculated in the water-immersion medium. If unpolarized light is assumed, the contrast degrades as the numerical aperture is increased. This is because of the detrimental contribution of p-polarization as θ increases. The contrast for p-polarization becomes negative (inversion of the light and dark areas; that is, areas that were bright on the wafer become dark and areas that were dark become bright) after a numerical aperture of approximately 1.0. This inversion occurs because the incident angles in the water exceed $45°$, causing the vertical components to become larger than the horizontal components. Since the component angles in s-polarization do not change with numerical aperture, there is no negative impact with this polarization. The outcome is that, as lenses increase in numerical aperture, the use of s-polarization to expose detailed circuit patterns becomes necessary to obtain good contrast and resolution. With higher numerical aperture, the effect of p-polarization in unpolarized light can not be ignored.

Polarization in lithographic systems. The use of polarization in optical lithographic systems has been enabled only recently by advances in illumination design. Previously, unpolarized light was used. The current light source used in the latest systems is an ArF excimer laser with 193-nm wavelength, and this light is polarized in one direction. Various optical means are used to shape the polarization emerging from the illuminator in a controlled fashion to obtain the best contrast.

Figure 4 shows an example of the use of s-polarization in an ArF excimer laser lithography system using NA = 0.92 when printing 60-nm lines and spaces. The polarization vector is parallel to the printed lines. The use of polarization increases image contrast by 20%. This improved contrast enables the same optical system to realize an increased range or depth of focus as compared to the unpolarized system. The increase of image contrast not only improves resolution, it also assists in creating a more "uniform circuit pattern" for integrated-circuit production lines. A higher-contrast image decreases sensitivity to errors in illumination that can cause uneven exposure at the wafer. The general outcome is that the use of polarization increases yields in the production output due to increases in the usable ranges of process variables, such as depth of focus.

For background information *see* INTEGRATED CIRCUITS; LASER; MICROLITHOGRAPHY; INTERFERENCE OF WAVES; POLARIZATION OF WAVES; POLARIZED LIGHT in the McGraw-Hill Encyclopedia of Science & Technology.   Tom Milster; Donis G. Flagello

Bibliography. D. Flagello and A. E. Rosenbluth, Lithographic tolerances based on vector diffraction theory, *J. Vac. Sci. Technol. B*, 10(6):2997–3003, 1992; D. Flagello, T. Milster, and A. Rosenbluth, Theory of high-NA imaging in homogeneous thin films, *J. Opt. Soc. Am. A*, 13:53–64, 1996; T. Matsuyama and T. Nakashima, Study of high NA imaging with polarized illumination, *Proc. SPIE*, 5754:1078–1089, 2004; D. Flagello et al., Polarization effects associated with hypernumerical-aperture (>1) lithography, *J. Microlith. Microfab. Microsyst.*, 4:031104, 2005.

Poverty, stress, and cognitive functioning

Social inequalities in income and wealth have a profound effect on the physical and mental health of children. For example, children from low-socioeconomic-status (SES) backgrounds are at greater risk for most forms of childhood morbidities compared to children from higher-SES backgrounds. [Note that the determination of SES is based on varying indices of family income (in relation to federally determined poverty thresholds), parental education (for example, those with college degrees versus those without), and parental occupation (for example, professional versus unskilled labor)]. Impoverished conditions during childhood are associated with poorer adult health, and low childhood SES may be the single most powerful contributor to premature mortality and morbidity worldwide.

Significant relationships have been observed between SES and cognitive ability and between SES and academic achievement in childhood. In fact, SES has a stronger relationship with cognitive performance than does physical health. Children from low-SES backgrounds perform below children from higher-SES backgrounds on tests of intelligence and academic achievement. In addition, SES has been found to have a major effect on language development. For example, one study found that the average vocabulary size of 3-year-old children from families receiving welfare was less than half the size of the average vocabulary of children from higher-SES (professional) families. Low-SES children are also more likely to fail courses, be placed in special education, and drop out of high school compared to high-SES children.

Poverty, cognitive functioning, and the brain. Although intelligence tests and academic achievement reflect cognitive ability, they are not particularly informative about neurocognitive systems or brain regions associated with specific cognitive processes. Recent investigations have employed behavioral tests to parse cognitive function into several relatively independent neurocognitive systems. For example, the prefrontal/executive system refers to executive control processes (decision making, planning, attention, and so on) associated with the prefrontal cortex (PFC). In a study with middle-school children, low-SES children performed below middle-SES children on tests of executive function. In addition, low-SES children had reduced performance on tests of language and memory. In contrast, systems associated with reward processing and visual cognition did not differ between these two groups. These findings indicate that certain neurocognitive systems, such as the prefrontal/executive system, are differentially affected by factors associated with socioeconomic disparity. Although other behavioral studies also found that measures of prefrontal function (for example, alertness and executive attention) are reduced in low- compared to higher-SES children, they provide only indirect measures of brain activity.

Recent studies of socioeconomic disparities in children have employed methods in cognitive neuroscience to measure brain activity directly. For example, by averaging electroencephalogram (EEG) traces that are aligned with a stimulus event, it is possible to extract tiny signals embedded in the EEG known as event-related potentials (ERPs). ERP components provide a precise temporal record of underlying neural activity associated with sensory, motor, or cognitive events. Several studies using ERP methods examined the influence of SES on processes associated

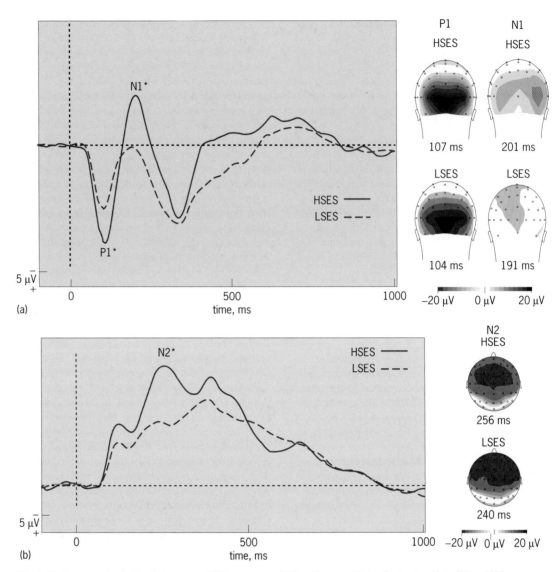

Fig. 1. Socioeconomic studies that measure ERP responses. (*a*) Grand averaged standard extrastriate (P1 and N1) event-related potential (ERP) components for high-SES (HSES) and low-SES (LSES) groups at the PO8 electrode. (Note that the P1 and N1 components are prefrontal-dependent, early latency brain potentials generated in ventral and dorsal extrastriate pathways.) Topographic maps (back view) of peak amplitude times are shown at the right. (*b*) Grand averaged novelty (N2) ERP components for HSES and LSES groups at the Cz electrode. Topographic maps (top view) of peak amplitude times are shown at the right. The asterisk on N1*, P1*, and N2* refers to a significant difference (*p* < .05) in ERP amplitude between the two groups for each of these components.

with prefrontal function, such as attention and inhibition. In one study, low-SES children were found to have reduced novelty and prefrontal-dependent extrastriate ERP responses compared to high-SES children (**Fig. 1**). The PFC is involved in both novelty detection (that is, the involuntary capture of attention by novel events) and the top-down modulation of early visual processing. In another ERP study, low-SES children had difficulty suppressing or inhibiting distracting auditory stimuli compared to middle-SES children. Taken together, the findings from these studies indicate that prefrontal-related ERP responses are reduced in children from low-SES backgrounds.

Another method in cognitive neuroscience is functional magnetic resonance imaging (fMRI). This method detects blood flow changes in the brain while a subject is engaged in cognitive tasks. Because of its greater spatial resolution compared to

ERP techniques, it enables researchers to better identify brain regions associated with these tasks. One study using fMRI examined phonological awareness (the ability to hear and work with the spoken language, which is an essential ability for learning to read) in children (6–9 years old) with below-average reading abilities. A strong positive relationship was found between phonological awareness and activity in the left fusiform gyrus (a region associated with reading ability) in low-SES children. However, this relationship was not found in children from higher-SES backgrounds. This finding may reflect a less typical brain–behavior relationship in higher-SES children, possibly related to the enriched literary environment in which they learn to read. In another fMRI study, 5-year-old children performed a rhyming task during scanning. A positive correlation was found between SES and the degree of hemispheric specialization in the left inferior frontal gyrus (a region involved in

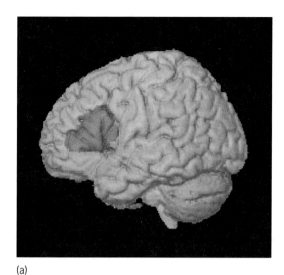

(a)

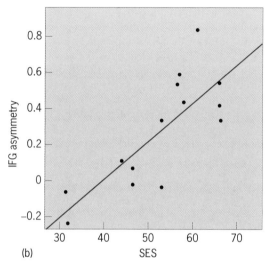

(b)

Fig. 2. Results of an fMRI study. (*a*) The three parts of the
inferior frontal gyrus (IFG): the orbital part, the triangular
part, and the opercular part. (*b*) The correlation between
SES and left-minus-right rhyme task activity in the IFG.
Each dot represents one child.

speech and language processing) [**Fig. 2**]. This finding suggests that there may be a deficit or delay in
the specialization of language function in the left
hemisphere of low-SES children. Although there are
few studies using these methods, the findings are
consistent with behavioral research examining the
influence of SES on neurocognitive systems. In addition, these findings provide more direct evidence of
the relationship between SES disparities and specific
brain regions such as the PFC.

**Influence of environment and stress on brain devel-
opment.** Evidence from animal research has demonstrated that experience can affect brain development. Data from several studies have shown that
environmental complexity can augment brain development, whereas environmental deprivation and
stress can adversely affect behavior, cognitive performance, and the development of specific brain regions, including the PFC. The human PFC may be
particularly sensitive to environmental factors, be-

cause this brain region has a prolonged period of
postnatal development.

A number of factors associated with low-SES rearing environments could influence normal brain development and behavior. These include lead exposure, nutrition, cognitive stimulation, parenting
styles, genetic factors, and stress. For example, low-
SES children often live in cognitively impoverished
environments in which they have limited access
to cognitively stimulating materials and experiences
compared to higher-SES children. In particular, low-
SES children have more experiences with negative
parenting and have fewer positive parent–child interactions (for example, reading, conversation, and
involvement in school activities). Moreover, there is
evidence suggesting that cognitive stimulation and
parental nurturance in early childhood predict later
language and memory functioning, respectively.

Low-SES children also experience greater levels
of stress. For example, children from low-SES backgrounds are exposed to a greater number of chronic
stressors, and they tend to live in more stressful
environments than children from higher-SES backgrounds. In addition, low-SES children have higher
basal levels of the stress hormone cortisol than high-
SES children do. Chronic stress can increase susceptibility to disease, and research has revealed that it
can adversely affect neurological processes, including the suppression of neurogenesis, elevated neurotoxicity, and morphological changes such as dendritic remodeling and decreased hippocampal and
PFC volumes. Evidence from epigenetic research in
animals has also revealed that stressful rearing conditions can reliably predict decreases in cognitive
performance. In addition, animals raised by stressed
mothers are more anxious as adults and, as a consequence, their offspring become more sensitive to
stress. In this manner, susceptibility to stress can
be transmitted across generations. Identifying neurocognitive systems affected by SES disparities could
be helpful in developing intervention programs for
low-SES children.

Intervention. Early-intervention programs have
been shown to improve intellectual development
and academic achievement in low-SES children. In
one study, low-SES children who participated in
a preschool intervention program scored higher
on intelligence and school performance tests than
children who did not participate in this program.
In a similar study, the intelligence quotient (IQ)
of low-SES children who participated in an early-
education program was between 0.5 and 1 standard
deviation higher than that in a low-SES control
group. These findings suggest that early-childhood
intervention programs for low-SES children can
produce persistent positive effects on achievement
and academic success.

Although research focusing on specific neurocognitive systems is relatively new, several intervention programs designed to target such systems have
already been implemented. In one fMRI study, children with reading difficulties showed improved
reading scores and increases in activity of the left

superior temporal gyrus (a region involved in phonological processing) after participating in an intervention program. In another study, reading-impaired children who participated in a study focusing on phonological awareness showed increases in reading fluency compared to a control group (the control group received school-based remedial reading instruction). In addition, these children also showed brain activity that resembled that of a nonimpaired control group. Furthermore, they maintained this pattern of activity for at least one year. Similarly, in a study of 4- and 6-year-old children, investigators found that children who received attention training showed significant improvements in measures of executive attention and intelligence. They also found that these children had more adultlike ERP responses compared to children in a control group.

Very few studies have examined the effects of SES disparities using a targeted intervention approach, such as described above. One such study used a cognitive control curriculum in preschool children to improve executive function skills. Low-SES children who participated in this curriculum showed improved accuracy on tests that measure core aspects of executive function. The goal of future investigations will be to identify neurocognitive systems that are disproportionately affected by SES and to develop targeted interventions for these systems. Studies using methods in cognitive neuroscience will allow us to better characterize and more appropriately target these systems, particularly those that are most vulnerable to SES disparities.

For background information *see* BRAIN; COGNITION; HUMAN ECOLOGY; INFORMATION PROCESSING (PSYCHOLOGY); INTELLIGENCE; LEARNING MECHANISMS; MEMORY; NEUROBIOLOGY; PERCEPTION; STRESS (PSYCHOLOGY) in the McGraw-Hill Encyclopedia of Science & Technology.

Mark M. Kishiyama; Robert T. Knight

Bibliography. R. H. Bradley and R. F. Corwyn, Socioeconomic status and child development, *Annu. Rev. Psychol.*, 53:371–399, 2002; G. J. Duncan, J. Brooks-Gunn, and P. K. Klebanov, Economic deprivation and early childhood development, *Child Dev.*, 65:296–318, 1994; D. A. Hackman and M. J. Farah, Socioeconomic status and the developing brain, *Trends Cogn. Sci.*, 13:65–73, 2009; V. C. McLoyd, Socioeconomic disadvantage and child development, *Am. Psychol.*, 53:185–204, 1998.

Precious element resources

Metals are classic examples of nonrenewable resources, and their extraction from the Earth by mining of ores is not sustainable, in the strict sense of the word. Mining, by definition, depletes the ore reserves. Through mineral processing of the ores and subsequent smelting and refining, the desired metals are isolated for use. Special and precious metals play a key role in modern society, as they are of specific importance for clean technologies and other high-tech equipment. Important applications are information technology (IT), consumer electronics, and sustainable energy production such as photovoltaic (PV), wind turbines, fuel cells, and batteries for hybrid or electric cars (**Fig. 1**). Driving forces for the booming use of these "technology metals" are their extraordinary and sometimes exclusive properties, which make many of them essential components in a broad range of applications. To a large extent, building a more sustainable society with the help of technology depends on sufficient access to technology metals.

The scarcity debate. A discussion on potential metal shortages was triggered in 1972 by the Club of Rome's publication, *The Limits to Growth*. Since the 1970s, a lot has happened specifically with respect to the use of the technology metals. For example, 80% or more of the cumulative mine production of platinum group metals (PGM), gallium (Ga), indium (In), rare-earth elements (REE), and silicon (Si) has occurred over the last 30 years. For most other special metals, more than 50% of their use took place in this period. And for the "ancient metals," gold (Au) and silver (Ag), over 30% of their use can be accounted for from 1978 onward. In many cases, the booming demand, especially from consumer mass applications, has driven metal prices up significantly. For example, the significant increase in demand for platinum (Pt) and palladium (Pd) was mainly caused by automotive catalysts (50% of today's platinum/palladium demand) and electronics (**Fig. 2**). A question that is often raised is how soon precious-element resources will run out, and a related question that is raised occasionally is whether severe shortages of certain metals will occur within the next 10 years.

The current debate takes place between two extremes: resource optimists and resource pessimists. Optimists argue that, in principle, market mechanisms will help to overcome supply shortages. Increased metal prices will lead to new exploration and mining (of so far uneconomic deposits), and technical substitution will be able to replace scarce metals by others with similar properties, or through conservation and innovative technologies. Pessimists start with information about ore resources, compiled by the U.S. Geological Services (USGS), among others, and then divide these numbers by the current and projected annual demand. For some metals, such as indium, this leads to rather short "static lifetimes." We will follow a pragmatic "resource realist" approach, discussing the main parameters and mechanisms that affect metal shortages and what can be done to prevent them.

Dimensions of resource scarcity. Three types of scarcity need to be distinguished, namely, absolute, temporary, and structural resource scarcity, for which an understanding of the primary supply chain is crucial. Absolute scarcity means the depletion of economically mineable ore resources. In this case, all ore deposits of a certain metal, including the ones that have not yet been discovered by exploration, would have been widely mined out, and the total market demand for a metal would exceed the remaining mine production. This would lead first to extreme

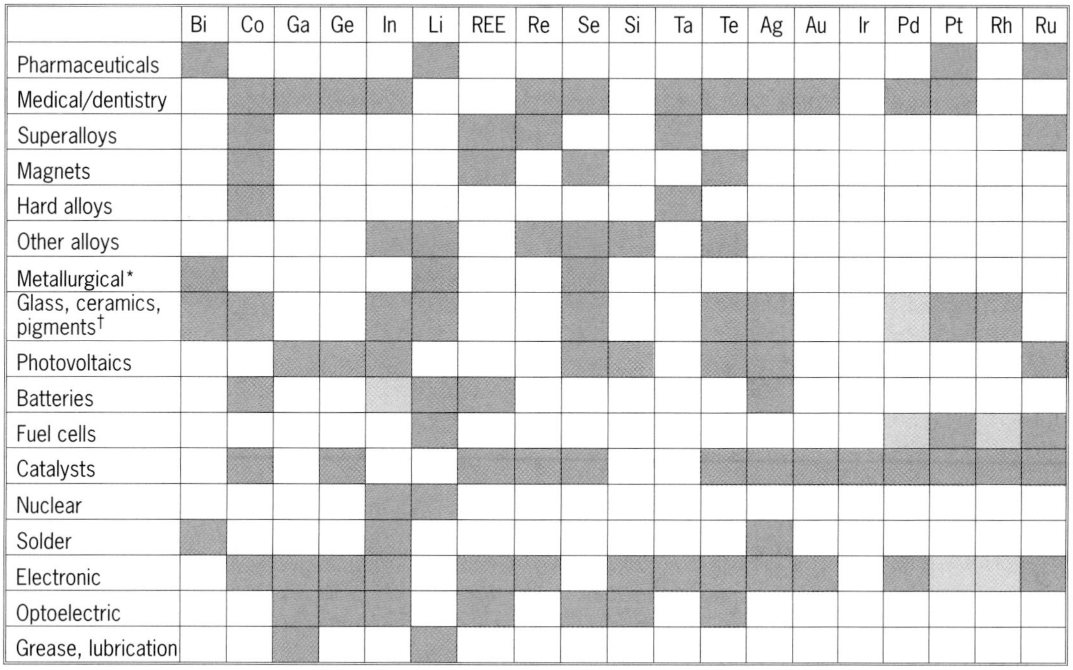

	Bi	Co	Ga	Ge	In	Li	REE	Re	Se	Si	Ta	Te	Ag	Au	Ir	Pd	Pt	Rh	Ru
Pharmaceuticals	▓					▓										▓			▓
Medical/dentistry		▓						▓		▓	▓		▓	▓		▓	▓		
Superalloys		▓						▓			▓								▓
Magnets		▓					▓												
Hard alloys		▓									▓								
Other alloys		▓	▓		▓		▓		▓	▓		▓	▓	▓					
Metallurgical*	▓																		
Glass, ceramics, pigments†	▓				▓				▓								▓		
Photovoltaics			▓	▓	▓				▓	▓		▓							
Batteries		▓				▓	▓												
Fuel cells																▓	▓		
Catalysts		▓		▓			▓								▓	▓	▓	▓	▓
Nuclear										▓									
Solder	▓												▓						
Electronic		▓	▓		▓		▓		▓		▓		▓	▓		▓	▓		
Optoelectric			▓	▓	▓				▓					▓					
Grease, lubrication					▓														

* Additives in smelting, plating,... † Includes indium-tin oxide (ITO) layers on glass.

Fig. 1. Important applications for technology metals.

price increases and finally to forced substitution for that metal (or technology) in certain applications, or would put severe limits on the further distribution of that technology. (In the worst case, a good technology, such as for energy generation, would be endangered because a key metal is not available.) However, within the foreseeable future, such an absolute scarcity is unlikely, and here the arguments of the resource optimists apply. Extremely high prices would make deep-level mining and mining of low-grade deposits, which are currently left aside, economically feasible. Also, it would trigger more exploration, leading to the discovery of new ore bodies. Exploration is very costly and time-consuming. As long as mining companies have enough accessible deposits for the next 20 years, there is not much incentive for them to conduct additional exploration. Accordingly, the data reported by USGS and other geological services do not report the absolute availability of metals on the planet, but rather compile

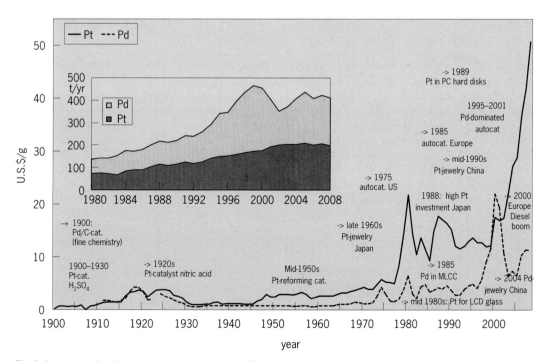

Fig. 2. Long-term development of prices for platinum (Pt) and palladium (Pd) and milestones in application, annual averages until 2008. Prices in U.S. $/g. Inset: Global net demand for platinum and palladium.

the known deposits that can be extracted economically (reserves), or where extraction is expected to be potentially feasible (resources). If exploration and mining efforts extend deeper into the Earth's crust or oceans and cover a wider geographical area, such as into Arctic regions, substantial new metal resources are likely to be accessible. However, this will come with trade-offs, as we will see later in the article.

In contrast, temporary or relative scarcity is a phenomenon that has already been experienced. In this case, for a certain period in time, the metal supply is not able to meet the demand. Reasons are manifold. New technological developments, strong market growth in existing applications, or speculative buying by investors can drive up the demand significantly within a short time, so that mine supply lags behind. Also, the supply can be disrupted by political developments, armed conflicts, natural disasters, or other constraints in the mining countries themselves, within the transport of ore concentrates, or at major smelters/refineries. Temporary scarcities are a major reason for the occasional extreme price volatility in the metal markets. The risk of temporary shortages increases with increasing concentration of the major mines, or smelters, in few and/or unstable regions or in few companies. A low number of applications in which the metal is used also increases the risk. Often, different factors come together and then accelerate development. For instance, in the first quarter of 2008, a soaring demand for PGMs for automotive catalysts and (speculative) investment coincided with a reduced supply from South African mines as a result of shortages of electric power. The prices of platinum and rhodium (Rh) soared to record highs within a short time, as South Africa produces over 75% of the platinum and rhodium supply. Speculation about potential depletion of indium resources started when the sales boom in liquid crystal display (LCD) devices (such as for monitors, televisions, and mobile phones, which use indium-tin oxide as the transparent conductive layer) drove up indium prices significantly. In the future, rapid production growth of thin-film photovoltaics could boost the demand for tellurium (Te), indium, selenium (Se), and gallium. Mass applications of electric vehicles will require large amounts of lithium (Li), cobalt (Co), and some rare-earth elements. And fuel-cell automobiles will need significantly more platinum than is used today in catalytic converters.

Developing and expanding mining and smelting capacities is highly capital-intensive and risky, and takes many years to materialize. Hence, temporary shortages are likely to happen more often in the future.

Structural scarcity is most severe for many technology metals, which often are not mined on their own but occur only as by-products from major or carrier metals. Indium and germanium (Ge), for example, are mainly by-products from zinc (Zn) mining, gallium from aluminum (Al), and selenium and tellurium from copper (Cu) and lead (Pb). The PGMs occur as by-products from nickel and copper mines, and as coupled products in their own mines. Within the PGMs, ruthenium (Ru) and iridium (Ir) are by-products from platinum and palladium. Since the by-product (minor metal) is only a very small fraction of the carrier metal, the usual market mechanisms do not work. Increasing demand will certainly lead to an increasing price for the by-product metal, but as long as the demand for the major metal does not rise correspondingly, mining companies will not produce more, since this would erode the major metal's price. In this respect, the supply of by-product metals is price-inelastic, as even a 10-fold increase in its price usually would not compensate for the negative impact on total revenues when there is an oversupply of the major metal. Moreover, many technology metals are important ingredients for several emerging technologies simultaneously (Fig. 1), so competition among applications becomes likely, and increasing demand from various segments will intensify the pressure on supply. Substitution is not likely to become the solution for many of these metals, since the required functional properties can often be met only by metals from the same metal family. For example, substituting platinum for palladium in catalytic applications will just shift the problem from one temporary/structurally scarce metal to the other. In emerging optoelectronics applications, the crucial metals are Si, Te, Ga, Se, Ge, and In. They can be partially substituted for one another, but this will not mitigate the problem. It can be overcome only by increasing the efficiencies in the primary supply chain (possibly leading to considerable gains) and, above all, by comprehensive recycling efforts, as will be discussed later. Omitting the fact that many technology metals are by-products and ignoring the possibility of structural scarcity are the weak points in the resource optimists' argument.

Independent of whether supply constraints are likely, the impact of the mining of lower-grade ores and from more challenging locations must not be overlooked. It will inevitably lead to increased costs, energy demand, and emissions. It will affect the biosphere (such as the rain forest, Arctic regions, and oceans) and could increase the dependence on certain regions (for example, as a battle for resources). This can put significant constraints on emerging technologies unless effective life-cycle management enables the use of recycled (secondary) metals in the forthcoming years.

Improving recycling. Metals are not consumed. Instead, they are only transferred from one manifestation into another, moving in and between the lithosphere and the "technosphere." Thus, the latter becomes our future renewable resource. Thoroughly extracting "urban mines" is the only sustainable solution to overcome supply disruptions. Metal combinations in products often differ from those in primary deposits, which results in new technological challenges for their efficient recovery. In products such as electronics or catalysts, the precious metals (such as Au, Pt, and Pd) have become the economic drivers for recycling ("paying metals"), while many special metals (such as Se, Te, and In) can be recovered as by-products when state-of-the art treatment and refining operations are used. The very low concentration of technology metals in certain products and dissipation during product use set economic and technical limits in many cases. Technical challenges exist, especially for complex products such as vehicles and computers. Effective recycling requires a well-tuned recycling chain, consisting of different specialized stakeholders, starting with a collection of old products, followed by sorting/dismantling and preprocessing of relevant fractions, and finally recovery of the technology metals. The latter requires sophisticated, large-scale metallurgical operations. For example, Umicore, an integrated smelter-refinery in Antwerp, Belgium, currently recovers and supplies back to the market 7 precious metals as well as 10 base and special metals, and DOWA Eco-System Environmental Management & Recycling, Japan, recovers and recycles 17 different metals.

Recycling technology has made significant progress. Further improvements and an extended range and yield of metals are underway. Designing for sustainability based on a close dialogue between manufacturers and recyclers can further support effective recycling, starting in the design and manufacturing phase and proceeding through a product's life cycle.

The biggest challenges to overcome are the insufficient collection of consumer goods and inefficient handling within the recycling chain. As long as goods are discarded with household waste, stored in basements, or sent to environmentally unsound recycling operations, the total recovery rates will remain disappointingly low, as is the case today for most consumer goods. Legislation can be supportive, but monitoring of the recycling chain along with tight enforcement of the regulations are crucial for success. For example, in spite of a comprehensive European legislative framework [Directive on waste

electrical and electronic equipment (WEEE) and Directive on end-of-life vehicles (ELV)], a significant share of end-of-life computers, cell phones, and cars are currently not recycled properly. Instead, they are (illegally) exported to Asia or Africa under the pretext of "reuse" to circumvent the Basel Convention regulations on the transboundary shipments of waste. The same thing happens in North America and Japan. This leads to a situation in which state-of-the-art, high-financial-investment recycling facilities in industrialized countries are underused because recycling and the associated environmental burden of environmentally unsound treatment are outsourced to the developing world. Except for some inefficient gold and copper recovery, technology metals are lost in such "backyard" recycling processes, and the "urban mine" is wasted irreversibly.

To close the recycling loop for consumer products, new business models need to be introduced that provide strong incentives for returning products at their end of life (EOL). This can include deposit fees on new products; product service systems (PSS), such as leasing; or other approaches. For emerging technologies (such as electric vehicles and photovoltaics), setting up "closed-loop structures" will be essential, and manufacturers that put successful models in place can secure their own supply of technology metals in the future.

In an ideal system, the sustainable use of metals could be achieved by avoiding spillage during each phase of the product life cycle. As shown in **Fig. 3**, such losses occur at various stages, and the specific impact factors need to be analyzed at each stage. Mining and recycling thus need to evolve as a complimentary system, where the primary metals supply is widely used to cover the inevitable life-cycle losses and market growth, and secondary metals from end-of-life products contribute increasingly to

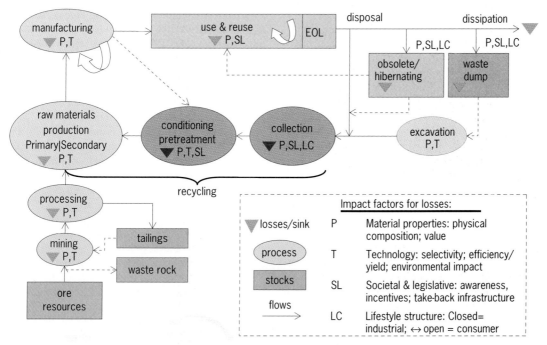

Fig. 3. Life cycle for metals/products and impact factors for losses at various stages.

the basic supply. Effective recycling systems would thus make a significant contribution to conserving natural resources of scarce metals and securing a sufficient supply of technology metals for future generations. It would further mitigate the climate impacts of metal production, which is energy-intensive, especially in the case of precious metals mined from low concentrated ores (for example, gold is mined at 5 g/t from 3000 m underground). The annual mining of 2500 tons of gold worldwide generates some 17,000 t of carbon dioxide (CO_2) per ton of gold produced (based on the ecoinvent version 2.0 database of EMPA/ETH, Zurich), or 42 million tons of CO_2 in total. For PGMs, the ore grade and specific CO_2 impact is of the same magnitude, while copper mining causes 3.5 t CO_2/t Cu, but adds up to 56 million tons CO_2 for production of 16 million tons of Cu annually.

Some mass products are relatively rich metal sources, by comparison. For example, computer motherboards contain approximately 250 g/t of gold, mobile phone handsets 350 g Au/t, and automotive catalytic converters 700 g/t of PGM. If effective collection systems and state-of-the-art recovery processes are used, the secondary metal production from such products requires only a small fraction of the energy and produces less CO_2, compared to mining.

Such products carry a high intrinsic metal value, which makes recycling attractive from an economical point of view as well. The recovery of pure metals from the circuit boards of personal computers costs only about 20% of the intrinsic metal value, leaving sufficient margins to pay for logistics and dismantling. In the case of automobile catalysts, the cost is even less. This similarly applies to large multimetal products, such as cars. For other technology-metal-containing products, such as televisions, audio equipment, and household appliances, the intrinsic metal value is usually not sufficient to pay the total costs of the recycling chain, and incentives by legislation, manufacturers, or distributors are needed to stimulate recycling. However, if the true costs of the landfills and the environmental damage caused by nonrecycling were accounted for, then on a macroeconomic level, proper recycling probably is viable for such products as well. In this sense, efficient recycling of end-of-life products today is insurance for the future. It will prevent/smooth metal price surges and secure a sustainable and affordable supply of metals needed for future products.

For background information *see* CATALYTIC CONVERTER; CONSERVATION OF RESOURCES; ELEMENT (CHEMISTRY); ELEMENTS, GEOCHEMICAL DISTRIBUTION OF; MINERAL RESOURCES; MINING; RARE-EARTH ELEMENTS; RECYCLING TECHNOLOGY in the McGraw-Hill Encyclopedia of Science & Technology.

Christian Hagelüken

Bibliography. R. Gordon, M. Bertram, and T. Graedel, Metal stocks and sustainability, *Proc. Natl. Acad. Sci. USA*, 103(5):1209–1214, 2006; C. Hagelüken and C. E. M. Meskers, Complex life cycles of precious and special metals, in *Linkages of Sustainability*, T. Graedel and E. van der Voet (eds.), Strüngmann Forum Report, vol. 4., MIT Press, Cambridge, Mass. (in preparation), 2009; *NRC 2008*: Minerals, Critical Minerals, and the U.S. Economy, Committee on Critical Mineral Impacts of the U.S. Economy, Committee on Earth Resources, National Research Council, National Academies Press, Washington D.C., 2008; M. A. Reuter et al., *The Metrics of Material and Metal Ecology*, Elsevier, Amsterdam, 2005.

Primate conservation efforts

In the 1960s, the International Union for the Conservation of Nature (IUCN) Red List of Threatened and Endangered Species was created as a conservation tool to identify global extinction risk, while providing species-specific information on distribution, ecology, and threats to persistence. Today, at least 49% of all primate species are identified as threatened with extinction, a number higher than any other mammalian order (**Fig. 1**). In 1993, an assembly was held in Brazil to discuss the crisis of biodiversity loss, and outlined a plan of goals and strategies to help slow the losses. Using this as a guideline, primatologists began a major conservation effort in primate habitat countries. A series of primate action plans for specific geographical regions were created by the IUCN. Integrated Conservation and Development Projects (ICDPs) were funded in specific sites in Africa, South America, and Madagascar. Countrywide policies were developed, such as the Durban Vision in primate-rich Madagascar, in which promises were made by the government to triple the amount of protected areas in 6 years. Because of the more recent attention that climate change has garnered, an international meeting took place in Bali in December 2007. This meeting resulted in the formation of the United Nations Collaborative Programme on Reducing Emissions from Deforestation and Forest Degradation in Developing Countries (UN-REDD Program), which hopes to reduce environmental degradation through funding initiatives to halt tropical forest burning and begin habitat restoration. To increase public awareness of the dire straits of some of our closest relatives, taxon-specific initiatives were also launched, such as "The 25 Most Critically Endangered Primates" list (**Fig. 2***a* and *b*), which was first announced to the media in 2000 and updated every other year since then, and the UN Year of the Gorilla in 2009. All these efforts have helped bring the urgency of preserving primates and their habitats to the attention of governments and the public before it is too late. They have also changed the nature of primate conservation efforts. In addition to hands-on species-specific conservation consisting of evaluating and monitoring, other efforts include the heavy tasks of informing policy, delineating the boundaries of protected areas, gathering information from unexplored and remote regions, and even making difficult decisions about intervention (that is, translocations) in the face of habitat destruction. Primatologists have recently begun to employ new conservation tools that are proving essential for effective, measurable, and rapid results.

Role of primates. As our closest living relatives, nonhuman primates have been used by conservationists as flagship species to bring public attention to the recent large-scale destruction of tropical forests. By saving the natural habitats of primates, it has been possible to save thousands of species of less charismatic (not having popular appeal), but equally endangered, plants and animals in the tropics. Primates have also been targeted as flagship species because they provide many ecosystem services, such as seed dispersal and pollination. A full complement of dispersers and pollinators are essential for tropical forests to continue to reproduce and sustain diversity, and thus maintain the ecosystem health for the plants, animals, and humans that rely upon them. Seven countries have been targeted as primate conservation priorities because of their high primate diversity and threat of extinction: the Democratic Republic of Congo (17.6% threatened), Peru (27%), Brazil (36.4%), Madagascar (40–42%), Nigeria (42.3%), Indonesia (84.1%), and Vietnam (86.4%).

Threats. Some key threats to primate persistence are listed below. After a description of each threat, specific strategies and tools are discussed.

Data deficiency. Many species of primates remain undescribed, often living in small, rarely detected populations in remote regions. Geographic ranges of many species are unknown, and biogeographic limits are not well established. To fully understand how any factor threatens a species, it is imperative to know its abundance and distribution. However, many species have never been studied, and threats to their persistence have not been identified. If such species are in danger of extinction, data deficiency threatens their conservation.

Strategies and tools: Remote-sensing technology (for example, satellite imaging and aerial photography) is used to identify potentially suitable habitats where species might be found, particularly in remote or unexplored regions. Biodiversity surveys were initiated during the 1980s, but have only recently been combined with remote-sensing technology. Ground truthing (whereby technicians gather data in the field to perform verification tasks) using the global positioning system (GPS) helps verify and interpret remote-sensing data, providing more reliable information on land cover and geographic features and barriers that may impact a species' range, or may suggest the presence of isolated or unknown populations. Combined, these tools enable the analysis of floral and faunal spatial patterns, which can be used to model species distributions and prioritize the selection and expansion of protected areas, maximizing their impact in terms of preserving biodiversity, even when it is poorly understood. Such surveys are also augmented by field genetic sampling, making the identification and separation of new species possible. For example, between 1986 and 2006, 55 new species of primates were described in Brazil, 44 in Madagascar, and 11 in Vietnam. Combining geo-

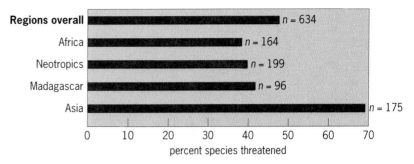

Fig. 1. **Extinction risk of primate species by major primate region.**

graphical information systems analyses with genetic analyses further allows gene flow and ecological barriers to be incorporated into dispersal models, thereby helping to determine optimal linkage of populations for conservation planning.

Deforestation. There are many threats to tropical forests, including commercial exploitation (for example, mining, timber, oil, and beef industries) and local exploitation (for example, subsistence farming, firewood, and construction) [**Fig. 3**]. Mining has become widespread in the tropics, particularly in sensitive primate habitats (Madagascar: titanium, cobalt, nickel, rubies, sapphires, gold; Brazil: rubies, sapphires; Amazon region: gold), and can introduce pollutants and poisons such as mercury into the environment. For example, in Ambatovy, Madagascar, a mining project plans to introduce an open-pit nickel mine and a pipeline for ore slurry in primary forest. Selective logging of tropical hardwoods threatens primate communities in various ways, with population densities of fruit eaters diminishing and those of leaf eaters increasing. The effects of fragmentation range from decreased humidity and drier and poorer quality soil near edges to reduced ranging abilities for primates, which may increase inbreeding. Recent meetings have targeted the burning of tropical forests as the top threat with regard to global climate change and biodiversity loss. Although Madagascar is an extreme case, with approximately 90% of the original forests destroyed by 1990, the rate of deforestation has recently slowed in many regions. However, in the past decade, Kalimantan (the Indonesian territory of Borneo island) has lost over 80% of its forests to palm oil plantations, and many primate-rich regions continue to experience enormous losses.

Strategies: ICDPs were established during the 1980s and 1990s to resolve the competing needs of local people and wildlife by integrating socioeconomic incentives and conservation in rural areas. ICDPs increased the value of protected areas by providing monetary aid for sustainable resource use. For instance, increased ecotourism generated park fees that were used to provide development assistance to surrounding villages. ICDPs also increased opportunities for earning money in the work sector by providing more jobs, many of them in parks. Although ICDPs have been criticized, some attempts have been successful. Between 1997 and 2007, funding and focus shifted to protecting corridors and landscape between protected areas. More recently,

Fig. 2a. Representative species from the 2008 list of the most critically endangered primates. Species represent each major geographic region and taxonomic radiation. (*Courtesy of Stephen Nash*)

emphasis has been placed on restoration ecology, which aims to reverse deforestation through reforestation. With help from media that promote the social value of environmentally conscious consumerism, a "culture of green" has emerged to raise awareness and funds, and put pressure on corporations, thereby benefiting and protecting rainforests where coveted hardwoods grow and primates dwell.

More direct, late-stage approaches to protecting primates from the effects of habitat disturbance must be considered in light of the increasing frequency of destructive practices in primary habitats housing endangered species. For instance, primate translocations and reintroductions are infrequent because they have been met with mixed results, but necessity dictated moving the endangered diademed sifaka (*Propithecus diadema*) in Madagascar (due to mining) and howling monkeys in Brazil (due to dam construction) to intact habitat.

Tools: Perhaps as a natural progression from ICDPs,

Variegated spider monkey
Ateles hybridus

Siau Island tarsier
Tarsius tumpara

Peruvian yellow-tailed woolly monkey
Oreonax flavicauda

Horton Plains slender loris
or
Ceylon Mountain slender loris
Loris tardigradus nycticeboides

Cross River gorilla
Gorilla gorilla diehli

Tonkin snub-nosed monkey
Rhinopithecus avunculus

Pagai pig-tailed snub-nosed monkey
or **simakobu**
Simias concolor

Fig. 2b. Representative species from the 2008 list of the most critically endangered primates. Species represent each major geographic region and taxonomic radiation. (*Courtesy of Stephen Nash*)

research stations have begun to practice not only science-based conservation but also social outreach. For instance, conservation education and health and hygiene programs create a mutualistic relationship between conservationists and those living in sensitive areas surrounding protected areas. Remote-sensing technology is increasingly used by primatologists, with Landsat satellite imagery documenting spatial and temporal changes in deforestation and identifying heavily impacted or critically positioned areas in need of reforestation. Such technology may also be used for routine monitoring of protected areas in proximity to, and potentially impacted by, various forms of land exploitation or encroachment. As a mechanism of carbon offsetting, the corporate sector has begun to finance the restoration of degraded habitat using native species of trees by purchasing "green tags," or renewable energy certificates. Environmental audits are also used to help accurately calculate carbon offsets, determine forests

Fig. 3. Denuded landscape in central Madagascar. (*Courtesy of Stacey Tecot*)

with high potential conservation impact, and provide companies with practical guidelines to becoming more sustainable. Timber company certification for sustainably extracted resources (according to the standards of the Forest Stewardship Council) is being used to combat illegal timber exploitation by verifying points of origin and methods of extraction. Some mining companies have begun to adopt a policy of "net positive impact," whereby they incorporate biodiversity offsetting into their mining program, and illegal mining projects in the Venezuelan Amazon have been challenged in part by indigenous human populations under equal threat.

Captive-based research has recently come to the fore in field-based primate conservation efforts. Conservation genetics has been used to measure heterozygosity in fragmented or disturbed primate populations, perform population viability analyses to determine populations' extinction risks, and even identify difficult-to-follow individuals such as drills in Cameroon through fecal sampling. Such data can be used to identify threatened populations that might need recovery or dispersal assistance (for example, connecting fragments through reforestation), or to collect basic data on group composition and habitat needs. Physiological assessments are just beginning to complement density estimates and behavioral studies to determine the impact of deforestation on primates, thereby establishing a level of human use or habitat quality that is compatible with the reproductive needs of fauna. New techniques such as noninvasive parasite and hormone sampling are being used to identify ecological factors that may pose a challenge and reduce reproductive success. Preliminary research on immunocompetence has determined that selective logging increases the risk of disease transmission in monkeys. Baseline profiles of stress hormone (cortisol) excretion have been constructed for a variety of primate species, and intersite comparisons are just beginning. In the near future, this should lead to the ability to determine if populations are sustainable or corridors are beneficial by monitoring these species in degraded and fragmented forests. Hormone profiles can also be used to identify the least stressful times to translocate animals, and to assess the success of reintroductions. Although early translocation efforts neglected to study

the carrying capacity and social milieu of the forest into which animals were to be transferred, recent successful operations have established criteria before the translocation.

The importance of public awareness as a strategy for combating primate extinction should not be underestimated. With conservation-promoting tools such as media coverage in habitat countries and abroad, goods such as T-shirts for villagers, and biological field guides for tourists, the human community can feel a part of saving their close relatives, and implement simple measures to save habitat and species.

Hunting (bushmeat) and poaching (pet trade). During the past two decades, hunting primates for cash markets has increased exponentially in Africa and Asia. Logging concessions, including building access roads, make primates more accessible to hunters, and human population growth increases the demand for meat. Recently, overfishing off the coast of West Africa has been linked with increases in sales of primate bushmeat (wild-animal meat). However, hunting does not only occur out of nutritional necessity; international networks of gourmet primate dining have been exposed in various cities around the globe, and apes are valued as trophies as well as meat, as can be seen by the sale of gorilla heads and hands. Mothers are often targeted and murdered so that infant primates can be sold as pets.

Strategies: The Convention on International Trade in Endangered Species treaty was agreed upon in 1973 to restrict the trade of endangered species across borders, and is enforced in 175 countries that have signed the treaty. Illegal trade was addressed by TRAFFIC, an IUCN agency that uses legal means to eliminate illegal trade, and the International Primate Protection League, a private organization that has spearheaded investigation and apprehension of illegal traders, and has provided sanctuary to rescued gibbons. A new strategy to combat illegal hunting and poaching is identifying the species and point of origin via genetic analysis so that legal and conservation efforts can focus on areas in crisis. The media also help expose bushmeat networks to public scrutiny.

Tools: In some regions, conservationists tap into hunters' expertise by training them as field assistants and park employees, thus educating those that pose the greatest threat and reducing their impact. Conservation genetics is on the cusp of using species-specific DNA primer sequences to identify bushmeat species, determine the threat to specific species, and monitor bushmeat commerce.

Conclusions. Although there are additional threats to primate persistence, including global climate change, disease transmission between people and nonhuman primates, and the repercussions of political strife, it is imperative to remain optimistic. The tools discussed herein, as well as traditional conservation tools, will lead toward the protection of primate species by helping focus efforts where they will have the greatest positive impact. The most effective new tool is a concerned public and scientific community that will unite in building an overarching strategy

Relationships between conservation tools (columns) and threats to primate persistence (rows)[*]

	Field genetics	Hormones	Parasites and disease	Public outreach	Translocation/ reintroduction	Reforestation	Zoos/sanctuaries	Remote sensing and biodiversity surveys
Data deficiency	Identify hard to study animals; establish group ranges and composition						Support and conduct basic research in the wild and naturalistic captive enclosures on poorly known/ endangered species	Determine where species might be found
Deforestation	Conduct PVAs/ heterozygosity and inbreeding analysis →Reforestation: determine what fragments to connect	Assess effect of deforestation on stress and reproductive physiology →Translocation/ reintroduction: determine best timing; assess success	Assess effect of deforestation on immuno- competence and disease transmission	Provide economic incentives for protecting the forest; teach conservation education	Move animals when destructive practices take place, such as mining, logging, etc.	Carbon offsetting by planting native species to connect fragments		Monitor exploitation; identify areas with high rates of deforestation → Reforestation
Bushmeat/ hunting	Identify bushmeat species		Investigate human and nonhuman primate disease transmission	Train hunters as field assistants, park rangers, etc.				
Climate change		Assess impact of changes in temperature and rainfall						Determine change through time
Pet trade				Train hunters and poachers as field assistants and park rangers			Take in confiscated animals	

[*]Common background shades indicate related tools and strategies.

to reduce these threats through incorporation of policy, field-based science, and social outreach (see **table**).

For background information *see* ADAPTIVE MANAGEMENT; APES; BIODIVERSITY; CONSERVATION OF RESOURCES; ECOLOGICAL MODELING; ECOSYSTEM; ENDANGERED SPECIES; MONKEY; POPULATION ECOLOGY; POPULATION VIABILITY; PRIMATES in the McGraw-Hill Encyclopedia of Science & Technology.
                                    Stacey Tecot; Patricia C. Wright

Bibliography. T. Gillespie, C. Chapman, and E. C. Greiner, Effects of logging on gastrointestinal parasite infections and risk in African primates, *J. Appl. Ecol.*, 42:699–707, 2005; International Union for the Conservation of Nature, *2008 IUCN Red List of Threatened Species*, 2008; J. M. Keay et al., Fecal glucocorticoids and their metabolites as indicators of stress in various mammalian species: A literature review, *J. Zoo Wildl. Med.*, 37(3):234–244, 2006; C. Kremen et al., Aligning conservation priorities across taxa in Madagascar with high-resolution planning tools, *Science*, 320(5873):222–226, 2008; L. K. Marsh (ed.), The nature of fragmentation, in *Primates in Fragments: Ecology and Conservation*, pp. 1–10, Kluwer Academic/Plenum Publishers, New York, 2003; R. A. Mittermeier et al., Primates in peril: The world's 25 most endangered primates, 2006–2008, *Primate Conserv.*, 22:1–40, 2007.

Protein kinase D signaling

Hormones, neurotransmitters, and growth factors stimulate the production of intracellular messengers, including the lipid-derived diacylglycerol (DAG), in their target cells to modify their behavior. One of the prominent intracellular targets of DAG is the protein kinase C (PKC) family. However, the mechanisms by which PKC-mediated signals are decoded by the cell remain incompletely understood.

Protein kinase D1 (PKD1, initially called atypical PKCμ) is the founding member of a novel protein kinase family that includes two additional protein kinases that share extensive overall homology with PKD1, termed PKD2 and PKD3. The modular structure of the members of the PKD family is shown schematically in **Fig. 1**. The three PKD isoforms are now classified as a new protein kinase family, separate from the PKCs. The PKD family occupies a

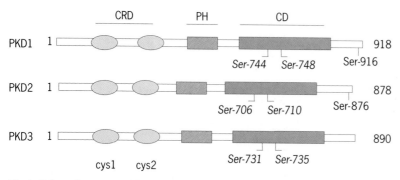

Fig. 1. Schematic representation of the members of the PKD family. Numbers correspond to amino acid positions. Serine residues within the PKD activation loop that became phosphorylated are indicated in italics. CRD = cysteine-rich domain; PH = pleckstrin homology; CD = catalytic domain.

unique position in the signal transduction pathways initiated by DAG and PKC. PKD is not only a direct target of DAG and phorbol esters (highly potent surrogates of DAG), it also lies downstream of PKCs in a novel signal transduction pathway implicated in the regulation of multiple fundamental biological processes.

Model of PKD activation. PKD isolated from multiple cell types or tissues has been seen to exhibit very low catalytic activity, which can be stimulated by phosphatidylserine micelles [colloidal aggregates of a unique number (between 50 and 100) of amphipathic molecules] and either DAG or phorbol esters. Early studies demonstrated that PKD is a phospholipid/DAG-stimulated serine/threonine protein kinase and implied that PKD represents a novel component of the signal transduction initiated by DAG production in their target cells.

Subsequent studies, which aimed to define the regulatory properties of PKD within intact cells, elucidated a mechanism of PKD activation distinct from the direct stimulation of enzyme activity by DAG/phorbol esters plus phospholipids obtained in vitro. Collectively, these studies demonstrated rapid PKC-dependent PKD activation in a broad range of biological systems, but did not exclude the possibility of PKD activation through a PKC-independent mechanism (or mechanisms).

For many protein kinases, catalytic activity is dependent on the phosphorylation of activating residues located in a region of the kinase catalytic domain termed the activation loop or activation segment. Using a number of approaches, two key serine residues in the PKD activation loop, Ser^{744} and Ser^{748} in mouse PKD1 (Fig. 1), were identified. A PKD mutant with both sites altered to nonphosphorylatable alanine was resistant to activation in response to cell stimulation, whereas mutation of Ser^{744} and Ser^{748} to glutamic acid residues, to mimic phosphorylation, generated a constitutively active mutant PKD. The properties of these mutant forms of PKD were consistent with a role of Ser^{744} and Ser^{748} in phosphorylation-dependent activation.

The translocation of signaling protein kinases to different cellular compartments is a fundamental process in the regulation of their activity. PKD1 is present in the cytosol of unstimulated cells, and to a

lesser extent in several intracellular compartments, including the Golgi apparatus and mitochondria, but rapidly translocates from the cytosol to different subcellular compartments in response to receptor activation.

Each translocation step is associated with a particular PKD domain and with rapid and reversible interactions. The first step of PKD translocation is mediated by the cys2 motif of the cysteine-rich domain (CRD) (Fig. 1), which binds to DAG produced at the inner leaflet of the plasma membrane as a result of phospholipase C (PLC) stimulation. In contrast, the cys1 motif recruits PKD to the Golgi apparatus. The second step, that is, reversible translocation from the plasma membrane to the cytosol, requires the phosphorylation of Ser^{744} and Ser^{748} within the activation loop of PKD, leading to its catalytic activation. Active PKD is then imported, via its cys2 motif, into the nucleus, where it transiently accumulates before being exported to the cytosol through a nuclear export pathway that requires the PH domain of PKD. (The PH, or pleckstrin homology, domain is a sequence of about 100 residues involved in intracellular signaling, and which was initially detected in pleckstrin, a major substrate of PKC in platelets.) In addition to the structural determinants present in PKD, other factors, including cell type, stimulus, and scaffolding proteins, also influence its intracellular distribution.

In contrast to PKD1 and PKD2, PKD3 is present in both the cytoplasm and the nucleus of unstimulated cells, but it undergoes rapid and reversible plasma membrane translocation in response to cell stimuli that generate DAG.

A sequential model of PKD1 activation that integrates the spatial and temporal changes in PKD1 localization with its catalytic activity and multisite phosphorylation is illustrated schematically in **Fig. 2**. The salient features are as follows: (1) In nonstimulated cells, PKD1 is in a state of very low kinase catalytic activity (inactive PKD1). It is thought that the CRD and PH domains located in the N-terminal region of PKD1 repress the catalytic activity of the enzyme by intramolecular autoinhibition (inhibition via adoption of a self-imposed latent conformation). Consistent with the repressive function attributed to these domains, deletions or single amino acid substitutions in the PH or CRD domains result in constitutive kinase activity. (2) GPCR (G-protein–coupled receptor) activation leads to a G-protein (Gαq)/PLC–mediated production of DAG at the plasma membrane. The CRD of PKD1 interacts with this second messenger, causing its translocation to the inner plasma membrane, where novel PKCs (for example, PKCε and PKCη) are also translocated in response to DAG generation. Thus, the CRDs of PKC and PKD1 are responsible for mediating proximity between these proteins in membrane microdomains in which DAG is produced. (3) The interaction of the CRD of PKD1 with DAG in the plasma membrane is envisaged to induce conformational changes in PKD1, exposing its two main phosphorylation sites in the kinase activation loop, Ser^{744} and Ser^{748}. Novel PKCε and PKCη, allosterically activated by DAG, are

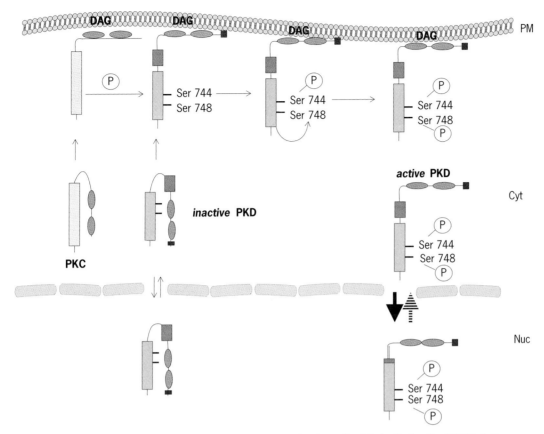

Fig. 2. Model of PKD activation and intracellular distribution regulation. In unstimulated cells, inactive PKD1 is in the cytoplasm. After cell stimulation, phospholipase C (PLC)–mediated hydrolysis of phosphatidylinositol 4,5-biphosphate (PIP$_2$) produces DAG at the plasma membrane, which in turn mediates the translocation of inactive PKDs from the cytosol to that cellular compartment. DAG also recruits, and simultaneously activates, novel PKCs to the plasma membrane, which mediate the transphosphorylation of PKDs on Ser[744] (in mouse PKD1). DAG and PKC-mediated transphosphorylation of PKD act synergistically to promote PKD catalytic activation and autophosphorylation on Ser[748]. Active PKDs then dissociate from the plasma membrane and migrate to the cytosol and subsequently into the nuclei. Upon cessation of agonist-induced cell stimulation, all PKDs return to their steady state, prior to cell stimulation. The scheme represents primarily the rapid activation of PKDs. Potential pathways leading to the inactivation of PKDs were not included for clarity purposes. Cyt, PM, and Nuc denote cytosolic, plasma membrane, and nuclear localization, respectively. Arrow direction and thickness represent directionality of PKDs and differential rates of transport, respectively.

thought to transphosphorylate Ser[744] and partially transphosphorylate Ser[748] in the activation loop of PKD1. Recent studies indicate that Ser[748] also undergoes autophosphorylation [addition of a phosphate to a protein kinase (affecting its activity) by virtue of its own enzymatic activity], even at early times of GPCR stimulation. Phosphorylated Ser[744] and Ser[748] stabilize the activation loop of PKD1 in an active conformation and alleviate the inhibitory influence of the CRD and PH domains (active PKD). (4) Activated (phosphorylated) PKD dissociates from the plasma membrane, moves to the cytosol, and then accumulates transiently in the nucleus. In this manner, activated PKD propagates signals initiated by GPCR activation at the cell surface. This mechanism provides not only for signal amplification but also for added regulatory interfaces that allow the kinetics, duration, and amplitude of PKD activity at different subcellular localizations to be precisely tuned.

Although the model illustrated in Fig. 2 explains the rapid activation of PKD triggered by many stimuli, it should be pointed out that cell stimulation with Gq-coupled agonists (for example, bombesin and vasopressin) initiates a *late* phase of PKC-independent

PKD activation that appears to be driven by PKD autophosphorylation on Ser[748]. More experimental work is needed to define the precise mechanism by which these receptor agonists induce the second phase of PKD activation.

PKD function. The multistep model of activation shown in Fig. 2 suggests that the PKDs are well positioned to regulate membrane, cytoplasmic, and nuclear events. Indeed, it is emerging that the PKDs are implicated in the regulation of a remarkable array of fundamental normal and abnormal biological processes, including cell proliferation, survival, migration, and differentiation; epithelial and neuronal polarity; membrane trafficking; inflammation; cardiac hypertrophy; and cancer. The involvement of PKDs in mediating such a diverse array of normal and abnormal biological activities in different subcellular compartments is likely to depend on the dynamic changes in their spatial and temporal localization, combined with its distinct substrate specificity. It is increasingly evident that a variety of biological responses attributed originally to PKCs are in fact executed by PKDs. Animal models using PKD transgenics or tissue-specific knockouts are emerging and will serve to further clarify the function(s) of PKD isoforms in vivo (knockout gene technology allows experimenters to inactivate specific genes within an organism and determine the effect this has on the functioning of the organism). In this context, it is important to point out that knockout of PKD1 in mice causes embryonic or perinatal lethality with incomplete penetrance (the situation in which an allele that is expected to be expressed is not always expressed).

In view of the multifunctional roles of PKD, the search for physiological substrates and interacting proteins is gathering speed, and already a number of interesting molecules have been identified as PKD targets, including the transcription factor CREB (cyclic AMP–responsive element-binding protein), the tumor suppressor DLC1, the histone deacetylases HDAC5 and HDAC7, the polarity-controlling kinase Par-1, the Ras inhibitor RIN1, and the small heat-shock protein Hsp27. In many cases, PKD-mediated phosphorylation regulates the subcellular localization of the phosphorylated substrate. For example, the phosphorylation of RIN1 on Ser^{351} by PKD induces binding of 14-3-3 proteins (a family of conserved regulatory molecules expressed in all eukaryotic cells, given numerical designations based on column fractionation and electrophoretic mobility), which sequester RIN1 to the cytosol, thereby preventing it from inhibiting the stimulatory interaction between Ras and Raf-1 at the membrane. Similarly, PKD-mediated phosphorylation of HDAC5 and HDAC7 leads to a change in subcellular localization. It is very likely that the same mechanism plays a role in mediating the effects of PKDs on Par-1 and DLC1. An emerging theme is that PKD modulates cell function by altering the subcellular localization of its substrates.

In conclusion, studies on PKD thus far indicate a remarkable diversity of both its signal generation and distribution and its potential for complex regulatory interactions with multiple downstream pathways. It is increasingly apparent that the members of the PKD subfamily are key players in the regulation of cell signaling, organization, migration, inflammation, and normal and abnormal cell proliferation. PKD emerges as a valuable target for development of novel therapeutic approaches in common diseases, including cardiac hypertrophy and cancer.

For background information *see* AMINO ACIDS; BIOCHEMISTRY; CELL (BIOLOGY); CELL ORGANIZATION; ENZYME; PHOSPHOLIPID; PROTEIN; PROTEIN KINASE; SIGNAL TRANSDUCTION in the McGraw-Hill Encyclopedia of Science & Technology.   Enrique Rozengurt

Bibliography. J. Fielitz et al., Requirement of protein kinase D1 for pathological cardiac remodeling, *Proc. Natl. Acad. Sci. USA*, 105:3059–3063, 2008; C. H. Ha et al., Protein kinase D–dependent phosphorylation and nuclear export of histone deacetylase 5 mediates vascular endothelial growth factor–induced gene expression and angiogenesis, *J. Biol. Chem.*, 283:14590–14599, 2008; R. Jacamo et al., Sequential protein kinase C (PKC)–dependent and PKC-independent protein kinase D catalytic activation via Gq-coupled receptors: Differential regulation of activation loop Ser(744) and Ser(748) phosphorylation, *J. Biol. Chem.*, 283:12877–12887, 2008; E. Rozengurt, O. Rey, and R. T. Waldron, Protein kinase D signaling, *J. Biol. Chem.*, 280:13205–13208, 2005.

Pure spin currents

Modern information technologies are based on semiconducting electronic devices, such as the silicon transistor in a computer's central processing unit (CPU), where information is encoded by electric charges and transmitted via charge currents carried by electrons or holes (missing electrons). Since the late 1960s, in accordance with the observation that has become known as Moore's law, semiconducting devices have continued to miniaturize in size, increase in speed, and decrease in cost. Today, a CPU can contain billions of transistors. However, a significant obstacle to this continued downscaling is power dissipation. As individual devices become smaller and the packing density increases, the amount of undesirable heat dissipation becomes increasingly problematic. In fact, the dissipated power density inside modern microprocessors (100 W/cm^2) is nowadays becoming comparable to the power density within a nuclear power reactor (based on the volume of the reactor pool). The difference, of course, is that the active volume of a processor chip is miniscule compared to that of a reactor, so the total power dissipations are vastly different, while the power densities are not.

A novel way to address the increasing power dissipation dilemma is to use the electron's magnetic quantum-mechanical degree of freedom called spin, which is intrinsic to every electron and creates a magnetic moment that can point either parallel ("up") or antiparallel ("down") with an applied magnetic field. In a ferromagnetic material, such as iron, there is an unequal number of spin-up and spin-down electrons. We refer to it as being magnetized. By passing a charge current J_e through such a ferromagnet, the charge carriers become spin polarized, meaning that the charge current is carried by predominantly one spin direction (that is, spin-up in **Fig. 1***a*),

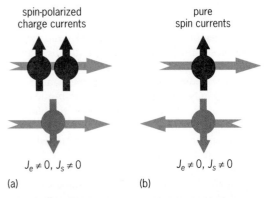

spin-polarized charge currents   pure spin currents

$J_e \neq 0, J_s \neq 0$ $J_e \neq 0, J_s \neq 0$

(a)   (b)

Fig. 1. Schematic for (*a*) a spin-polarized charge current and (*b*) a pure spin current.

resulting in a concomitant spin current J_s. These spin-polarized charge currents can then be used to interact with, or manipulate, the magnetization of a different ferromagnetic component. Since the spin-up and spin-down magnetic states in ferromagnets are inherently stable, this enables the design of nonvolatile electronic devices, meaning ones that do not require energy (or a "boot-up") to retain their information. Such devices can reduce power consumption since they do not require a continuous power supply to refresh their memory, as in today's random access memory (RAM). They go by the name magnetic RAM, or MRAM.

However, the net charge current associated with these spin-polarized charge currents still gives rise to electrical (Joule) heating, similar to unpolarized charge currents. Therefore, the question arises of whether it is possible to use spin currents (the flow of spins) without the drawbacks of charge currents (the flow of electric charge). It is indeed possible to consider pure spin currents, where spin information is transported without a net charge current, as is depicted in Fig. 1b, where the opposite charge currents for the individual spin-up and spin-down directions cancel each other, while the individual spin currents add up due to their antiparallel spin orientations. Interestingly, these pure spin currents are invariant under time reversal, which reverses not only the direction of motion of each electron, but also the spin direction. This symmetry implies that pure spin currents are by their very nature dissipationless. The reason for this is that any dissipation breaks time reversal symmetry, since it increases entropy (disorder) and, therefore, in accordance with the second law of thermodynamics, is related to an increase in time. This inherent lack of dissipation has recently resulted in increased research activity for studying the generation and manipulation of pure spin currents. However, even though pure spin currents are dissipationless, their generation always requires active pumping, which results in power dissipation. In any case, beside their potential impact on information technologies, the investigation of pure spin currents also allows fundamental new insight into spin dynamics and relaxation phenomena.

Nonlocal geometries. Pure spin currents in nonmagnetic conductors were first studied in nonlocal transport devices with electrical spin injection from ferromagnetic contacts. A schematic of such a device is shown in **Fig. 2**. Nonlocal refers to the

fact that voltages are generated in parts of the devices that are outside the direct path of the applied charge currents. In this particular instance, the nonlocal voltages are mediated by pure spin currents. In a ferromagnetic wire, the mobilities for spin-up and spin-down electrons are different, so that any charge current is spin polarized (Fig. 1a). When this spin-polarized charge current is injected from the ferromagnetic contact F1 into a nonmagnetic wire, a net accumulation of one spin direction (that is, spin-up in Fig. 2) is generated in the immediate vicinity of the injection contact. By closing the charge current circuit to the left, all the charge is drained toward the left end of the wire, and the resulting charge current is carried equally by spin-up and spin-down carriers in the nonmagnetic wire (Fig. 2). However, the spin accumulation that originates from the ferromagnetic contact can diffuse equally to the left and right, which results in spin-up diffusing toward the right end of the wire and spin-down diffusing in the opposite direction to maintain charge neutrality. Thus inside of the normal wire, there is a pure spin current without a net charge current on the right side of the injection contact F1.

The lifetime for the spin orientation in the nonmagnetic wire is finite (typically picoseconds in metals and up to several nanoseconds in semiconductors). Therefore, pure spin currents exist only over a finite distance near the injection contact given by the spin diffusion length. A second ferromagnetic contact (F2 in Fig. 2) enables one to map the spatial distribution of the pure spin current by measuring the voltage between the second detection contact F2 and the nonmagnetic wire as a function of the magnetization direction of F2. As the magnetization direction is reversed, a voltage change is generated, which is proportional to the spin accumulation next to the detection contact. These measurements show that the spin orientation can be preserved over distances ranging from several hundred nanometers in many metals to a few tens of micrometers in semiconductors, including silicon. Thus, spin currents are preserved on distances that are comparable or significantly larger than the dimensions of contemporary electronic devices. Therefore, the practical use of pure spin currents becomes plausible, since spin can be essentially preserved within a typical device.

Spin Hall effects. Sometimes it is difficult to integrate ferromagnetic materials with nonmagnetic materials. However, there is also a possibility to generate spin currents from charge currents without the use of any ferromagnetic material. In every material the orbital motion of electrons is coupled to its spin. A consequence of this spin-orbit coupling is that when an electron moves through a wire, depending on the direction of its spin, it scatters with higher probability either to the right or to the left. This spin-dependent scattering gives rise to accumulations of opposite spin orientations at the edges of a wire carrying a charge current (**Fig. 3**). In other words, the charge current generates a transverse, pure spin current. Conversely, a pure spin current generates,

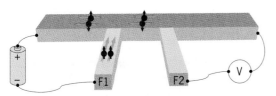

Fig. 2. Pure spin currents through nonlocal electrical spin injection. Spin-up electrons are injected from the ferromagnetic contact F1 into a nonmagnetic wire. While the charge current is drained toward the left, a pure spin current develops in the right part of the nonmagnetic wire. The gray arrows indicate the electron flow direction, that is, negative charge current flow.

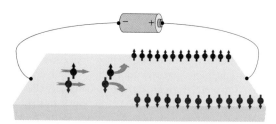

Fig. 3. Spin Hall effect. Spin-dependent scattering separates spin-up and spin-down electrons to opposite edges of a charge current-carrying wire. The gray arrows indicate the electron flow direction, that is, negative charge current flow.

via the same spin-dependent scattering, a transverse charge current. Interestingly, while these spin Hall effects were theoretically predicted in 1971, they were observed experimentally only many years later, in 2004, via magneto-optical detection of the spin accumulation at the edges of current-carrying semiconducting wires.

For practical applications, it is important to understand the efficiency of this transformation from charge to spin currents. Quantifying this efficiency is difficult, and experiments have started only in the past 3 years. So far, reported experimental values vary significantly and range from a maximum of about 10 to less than 0.1% efficiency for converting charge currents into pure spin currents. For now, it remains undecided whether spin Hall effects are merely an academic curiosity or whether they will affect future applications.

Spin pumping. Both electric spin injection and spin Hall effects generate pure spin currents from charge currents. Spin pumping offers an alternative route toward pure spin currents, without the need to apply direct charge currents to the structure. It is possible to excite in a ferromagnetic material (for example, through microwave radiation) ferromagnetic resonances, which are coherent oscillations of the magnetization around its equilibrium orientation. If the ferromagnetic material shares an interface with a nonmagnetic conducting material (**Fig. 4**), then the magnetization dynamics in the ferromagnet gives rise to a time-varying spin-dependent potential at the ferromagnet/normal metal interface. Because of this potential, the transmission and reflection coefficients for electrons impinging the interface will be different for spin-up and spin-down. This difference in transmission coefficients will lead to the genera-

tion of a spin current, which ultimately leads to a spin accumulation diffusing away from the interface. To maintain charge neutrality, there is a backflow of opposite spins, which results in a pure spin current in the nonmagnetic materials carrying spins away from the ferromagnet/normal metal interface. The polarization of this spin current has a direct-current (dc) component parallel and a radio-frequency (rf) component perpendicular to the equilibrium magnetization direction in the ferromagnet. The spin pumping mechanism was originally indirectly observed as an increased damping of ferromagnetic resonance, but recently the concomitant pure spin currents have been directly verified through the inverse spin Hall effect, which transforms a pure spin current into an electrical voltage.

Spin pumping is in many ways the reciprocal effect to spin torque, which occurs when a spin current is injected into a ferromagnet and results in an excitation of magnetization dynamics. This spin-torque effect is currently used in advanced MRAM concepts and in novel tunable microwave generators.

Outlook. It now is well established that pure spin currents can be generated by many different methods and that there are various ways of directly measuring them, as well as their associated spin accumulation. Some of the remaining key questions are how to manipulate spin currents effectively, and whether spin currents themselves can be used to influence the magnetization of ferromagnetic materials. A successful demonstration of these two effects would enable the design of novel logic devices, which ultimately may perform superior to today's conventional semiconducting devices. Apart from that, the investigation of pure spin currents is fundamentally enhancing our understanding of spin-dependent physics in a wide variety of materials.

For background information *see* ELECTRON; ELECTRON SPIN; FERROMAGNETISM; HALL EFFECT; MAGNETIC RESONANCE; METAL; MICROPROCESSOR; SEMICONDUCTOR; SEMICONDUCTOR MEMORIES; SPIN (QUANTUM MECHANICS) in the McGraw-Hill Encyclopedia of Science & Technology.   Axel Hoffmann

Bibliography. A. Hoffmann, Pure spin-currents, *Phys. Stat. Sol. (c)*, 4:4236–4241, 2007; Y. Ji et al., Non-local spin injection in lateral spin valves, *J. Phys. D: Appl. Phys.*, 40:1280–1284, 2007; S. Maekawa (ed.), *Concepts in Spin Electronics*, Oxford University Press, 2006; Y. Tserkovnyak et al., Nonlocal magnetization dynamics in ferromagnetic heterostructures, *Rev. Mod. Phys.*, 77:13751421, 2005; I. Žutić et al., Spintronics: Fundamentals and applications, *Rev. Mod. Phys.*, 76:323–410, 2004.

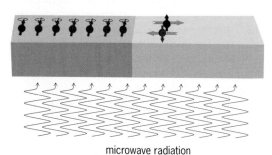

microwave radiation

Fig. 4. Spin pumping. Ferromagnetic resonance excited via microwave radiation in the ferromagnet (left side) generates a pure spin current in an adjacent normal metal (right side).

Pyrocumulonimbus

In the northern hemisphere summer of 1998, smoke from forest fires was detected in the stratosphere, up to 7 km above the troposphere. Initially, the observations of these particles were shrouded in mystery, as there had been no known volcanic eruption. However, these "mystery layers" were abundantly evident in measurements taken by at least three satellite-based and multiple ground-based instruments. Because aerosol layers in the stratosphere have a long residence time and, in the case of volcanic particles, are known to have an impact on climate, there was a compelling need to find their source. Subsequent investigation, using winds to find a common location, pointed to the boreal forest of northwestern Canada, where many large wildfires were active then. Later, a specific and unique form of explosive thunderstorm—pyrocumulonimbus— was discovered to be the "smoking gun," and thus the identification of the stratospheric particles as smoke was confirmed. This discovery is a fundamental challenge to the accepted views of thunderstorm energetics within the atmospheric science community.

Terminology. Pyrocumulonimbus are wildland fire-related convective storms on Earth that have similarities to thunderstorms (cumulonimbus) and certain volcanic eruptions (**Fig. 1**). Meteorologists typically use "Cb" as an abbreviation for cumulonimbus, and hence pyrocumulonimbus is shortened to "pyroCb." The pyroCb is analogous to the Cb in that it is a sufficiently tall cloud column of rising/cooling air that ice crystals form, and hence lightning, precipitation, and sometimes hail occur. Externally, the pyroCb looks much like a Cb, with the characteristic cauliflower texture of the cumulus cloud and a spreading "anvil" cloud where the column reaches the top of the troposphere. However, the pyroCb and the pyrocumulus cloud (less vertically developed and absent ice cloud particles) involve significant amounts of smoke that color the cloud and form distinctive smoky exhaust plumes. The distinction of a pyroCb is that it is caused or significantly augmented by a wildfire. Whereas "regular" Cbs are most often triggered by solar heating of the Earth's surface and subsequently the planetary boundary layer, or frontal lifting, the dominant (and sometimes sole) heat trigger for the pyroCb is the fire. Therefore, the pyroCb is typically anchored to a flaming fire and persists as long as the heat energy release of the fire is sufficient to maintain the high convection column.

Pyroconvection is a phenomenon that has been well known for decades. Forest fire experts have noted in the past a significant transition from lower-intensity to higher-intensity fires associated with "blowup" conditions. Hence, pyroCb can be considered the most energetic and extreme form of wildfire. Fire managers have used terms such as "plume-driven" for pyroconvection and "wind-driven" for other large, uncontrolled types of fire behavior. There are many manifestations of uncontrolled fire behavior; the pyroCb specifically relates to a firestorm that exhibits the violent updrafts leading to cloud, precipitation, glaciation (ice formation) of cloud particles, and other signatures exclusive to cumulonimbus, such as lightning, hail, and tornadoes.

PyroCb plumes. Since the initial discovery of the stratospheric "mystery layers" in 1998, several specific pyroconvective storms have been connected to smoke plumes at unambiguously stratospheric alti-

tude. In addition to the aerosol layers in the stratosphere, the pyroCb has also been connected (with the aid of high-temporal-frequency visible and infrared satellite imagery) to a "day-after" plume that manifests enormous smoke abundance, optical opacity, and altitude (**Fig. 2**). These "day-after" plumes look unlike any "regular" cloud in that they are

(a)

(b)

(c)

Fig. 1. Three types of cumulonimbus: (*a*) pyrocumulonimbus, occurring over the warm fire in northern Arizona on June 25, 2006 (*photograph by Marty Feely*); (*b*) a "regular" cumulonimbus; (*c*) eruption cloud from Redoubt volcano, Alaska, on April 21, 1990 (*photograph by R. Clucas*).

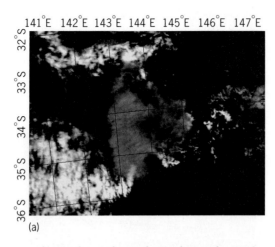

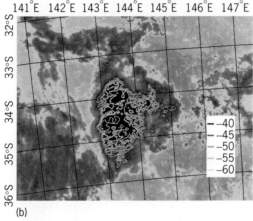

Fig. 2. "Day-after pyroCb" smoke plume in the lower stratosphere on December 18, 2002, in Victoria, Australia: (*a*) true-color visible satellite image, and (*b*) thermal infrared image. Image data taken by NASA's Moderate Resolution Imaging Spectroradiometer.

gray, or smoke-colored, in true-color satellite imagery (instead of the characteristic white of thick water-ice clouds), yet they emit thermal infrared radiation at the same low temperature characteristic of very thick, deep storm clouds. Such "cold" clouds represent a massive concentration of cloud droplets and/or ice crystals at the highest levels within the troposphere, where environmental temperatures are similarly cold. The combination of "grayness" and "cold" within the same cloud does not exist in meteorological clouds, as it is exclusive to a mixture involving cloud water-ice particles and smoke. This "day after" plume is early evidence of the effects of an intense pyroCb. Hours after the pyroconvection ceases, the plume (including some smoke in the stratosphere) is observable in satellite imagery, and the next day it appears as a peculiarly thick, albeit "gray," cloud. Later, after the stratospheric portion of the plume spreads, it is usually seen as the distinctive aerosol layers first observed as mystery clouds in 1998.

Thunderstorm tops. Before 1998, the reigning assumption within the realm of atmospheric science was that thunderstorm convection does not irreversibly penetrate the tropopause, the boundary between the troposphere and the stratosphere. The tropopause represents a sort of "lid" on vertical mixing, in that it is by definition the zone separating the well-mixed lower atmosphere and the stratosphere. The vertical mixing of the lower atmosphere, the troposphere, is enabled by an environmental temperature lapse rate that can give rise to buoyant parcels of air. Moreover, the abundance of tropospheric water vapor also augments buoyancy by virtue of the latent heat of condensation. Buoyantly energetic columns of air meet a strong restraining force near the tropopause, and that suppressive force increases rapidly with altitude in the lowermost stratosphere, where the lapse rate dictates that vertical motions cease, leaving only laminar flow in this very stable environment. Knowledge of the thermal structure of the atmosphere, combined with the typical observation that thunderstorm anvils flatten and spread out near the tropopause, even in the case of severe convection, led to the common belief that cumulonimbus do not punch into the stratosphere. It needs to be noted that observation and research of Cb cloud tops has revealed occasional, small-scale stratospheric penetration, with descriptive terms such as "overshooting top" and "jumping cirrus." The fate of air in these small perturbations is yet uncertain, but it is believed that the air typically collapses back to an equilibrium altitude near the larger anvil cloud.

Volcano versus pyroCb. Volcanic eruptions throughout history are well known to have had sufficient upward force to pollute the stratosphere. For instance, the 1991 eruption of Mount Pinatubo, Luzon, Philippines, deposited material as high as 35 km altitude, double the tropopause height. The Mount Pinatubo event is the most explosive class of volcanic eruption, called plinian eruption. The plinian eruption's vertical column is dominated by what is termed the "convective thrust region," that is, the most climatically relevant volcanic eruptions are primarily a convective phenomenon, not one of ballistics. In that way, and in terms of a heat source near the Earth's surface, the volcanic Cb and pyroCb are quite similar. Historically, the volcano was considered to be the only eruptive force that could inject material from the Earth's surface far across the tropopause. The consequence of this extremity of vertical air transport is that the ejecta remain in the stratosphere and may have lifetimes on the order of years. Climatic consequences can then ensue, such as global cooling of the Earth or chemical perturbations such as ozone destruction. Studies of volcanic eruptions and their effect on climate have been performed in abundance, and much is known about the energetics, dynamics, chemistry, climate impacts, and frequency of these violent events. Much less is known about the pyroCb. The state of the science, in terms of pyroCb, can be considered as being still in its infancy. Skepticism of, and unfamiliarity with, the process is presently being replaced with confirmation of the pyroCb life cycle, its direct injection effect, and the altitude into the stratosphere of its emissions. However, much is still in the frontier of uncertainty. There is no climatology or geography of pyroCb activity, published case studies are still few, and there are no computer climate models that take pyroconvection into account. There is not yet enough of a knowledge base to ascertain whether such a treatment of pyroCb is even warranted.

PyroCb trigger. Pyroconvection is a symptom of extreme fire behavior. Buoyant air motions in the free troposphere (that is, the altitudes above the blanket of air closest to the Earth called the planetary boundary layer) are caused by "hot" thermals—parcels of relatively warm air surrounded by cooler air. These thermals rise with a speed and to a vertical extent consistent with the heat difference between the thermal and the surrounding atmospheric environment. The pyroconvective thermal is directly related to the intensity of the flaming fire front. That intensity is related to the concept of "energy release rate (ERR)." ERR is an integral of flaming-front size (measured in meters), fuel concentration or loading (measured in consumable mass per area), and rate of fire-front spread (in meters per second). Thus, ERR increases with the amount of fuel being consumed. (Fuel availability depends primarily on moisture content and the rate of advance of the fire.) One can see that when intense flaming is the mode of burning, a fire in sufficiently dense biomass spreading at speed can generate an ERR resulting in a thermal that rises into the free troposphere far enough to cool and form a cloud (**Fig. 3**). The convective dynamic includes an excitation of wind flowing in to fill the void created by the buoyant "pyrobubble" and this wind further advancing the fire (through direct spread and "spotting," where flaming debris is launched and deposited downwind). In some circumstances, the atmospheric and fuel ingredients combine to generate ERRs that transform pyrocumulus into the more extreme pyroCb.

Recent investigations into specific pyroCb storms strongly indicate that a unique additive to pyroCb intensity may be crucial to the storm's effectiveness in penetrating the tropopause. The enormous concentration of smoke aerosols emitted by the fire "seeds" the buoyant air parcel with superabundances of efficient cloud condensation nuclei. These extraordinary abundances effectively limit cloud-particle growth and thereby inhibit the formation of precipitation-size cloud droplets. This happens because the water vapor in the parcel cooling to its condensation/freezing temperature has an unusually large inventory of nucleation sites for cloud particle formation, and the condensing/subliming water spreads equally among many small, nascent droplets. Normal precipitation processes require some large droplets or ice crystals to be mixed in with the overall cloud-particle population. The abnormal pyroconvective air parcel thus inhibits precipitation formation. Even when intense pyroconvection and pyroCb develop, large cloud droplets and precipitation are suppressed, and as a result the main drag on cumulonimbus intensity is obviated. Consequently, the pyroCb gains an intensity "advantage" over regular Cb, with updrafts—unencumbered by heavy precipitation—that achieve unique speed/momentum. Evidence for this unique momentum is the unusually great vertical extent of the pyroCb injection. Thus, the fire directly triggers

Fig. 3. Extreme crown fire in May 1986 in northwestern Ontario, Canada. Helicopter view of a pyroCb-generating, high-intensity crown fire in the Canadian boreal forest. This fire was spreading at 2 km/h and generating an ERR of about 50,000 kW/m of fire front. (*Photograph by B. Stocks*)

pyroCb formation via its sensible heat and smoke emissions. Still uncertain is the relative importance of heat energy, peculiar cloud-droplet formation (also known as cloud microphysics), air-moisture content, and environmental meteorological forcing in the pyroCb-extremity equation.

Though pyroCb understanding is still in its infancy, some striking facts are at hand, facts that fortify a call for deeper exploration leading to a climatology and atmospheric modeling of this phenomenon and its effects. For instance, a pyroCb in Australia's capital, Canberra, in January 2003, destroyed hundreds of homes, took several lives, and polluted the stratosphere with smoke that was detectable for weeks and spread around the southern hemisphere. As long ago as 1950, a firestorm in western Canada injected a pall of smoke, observed by aircraft in the lowermost stratosphere, that darkened midday skies as far away as Washington, D.C., and caused eyewitness observations of a blue sun and moon as far away as Europe. A single, 3-h-long pyroCb in Alberta, Canada, in 2001 injected an amount of smoke aerosol into the stratosphere estimated to be on the order of 5–36% of the entire northern hemispheric stratosphere's "background" aerosol load. The theorized "nuclear winter" that would result from large-scale conflagrations after a military nuclear conflict is based on the idea that the hypothesized hemispheric smoke in the stratosphere would sufficiently reduce incoming solar radiation to cool the Earth and shorten growing seasons. Thus, the discovery of stratospheric smoke is a compelling signal that environmental impact consistent with theory is part of the pyroCb dynamic.

For background information *see* AEROSOL; CLIMATOLOGY; CLOUD PHYSICS; FOREST FIRE; SATELLITE METEOROLOGY; STRATOSPHERE; THUNDERSTORM; TROPOPAUSE; TROPOSPHERE; VOLCANO; VOLCANOLOGY in the McGraw-Hill Encyclopedia of Science & Technology.   Michael Fromm; Brian Stocks

Bibliography. M. Fromm et al., New directions: Eruptive transport to the stratosphere: Add fire-convection to volcanoes, *Atmos. Environ.*, 38(1):163-165, 2004; M. Fromm et al., Observations of boreal forest fire smoke in the stratosphere by POAM III, SAGE II, and lidar in 1998, *Geophys. Res. Lett.*, 27:1407-1410, 2000; S. J. Pyne, *Awful Splendour: A Fire History of Canada*, University of Washington Press, 2007; D. Rosenfeld et al., The Chisholm firestorm: Observed microstructure, precipitation and lightning activity of a pyro-cumulonimbus, *Atmos. Chem. Phys.*, 7:645-659, 2007.

Raccoon rabies in urban settings

Rabies is one of the oldest and deadliest zoonotic diseases, killing tens of thousands of people worldwide each year. It is a viral infection typically transmitted to people via bites from infected animals, especially bats and carnivores or domestic mammals. The disease has no cure, but pre- and postexposure prophylaxes are available; therefore, human deaths due to rabies are relatively rare in North America. Because of aggressive vaccination programs for pets, most rabies cases in developed countries such as the United States come from wildlife species, and current rabies management programs are focused on the challenging goal of reducing infection within wildlife [in contrast to reducing infection within domestic species (especially domestic dogs), as is the case in underdeveloped countries].

The rabies virus, a single-stranded ribonucleic acid (RNA) virus, has been classified into several different strains, or genotypes. Each strain is typically host-specific, with the raccoon rabies variant (RRV) being of greatest management interest in North America. Whereas most rabies strains show little geographic variation and are found primarily in rural areas, RRV has undergone a major expansion with significant effects on human health agencies. Historically, RRV was apparently restricted to the southeastern United States, specifically northern Florida and southern Georgia, where it was enzootic (affecting animals in a limited geographic region) [**Fig. 1**]. In the late 1970s, RRV-infected raccoons were translocated north to a Virginia location that contained a rabies-naïve raccoon population. Over the next 20 years, RRV spread along the Atlantic Coast from Alabama to Maine, reaching southern Ontario in 2002. Management of

Raccoon response to urbanization		
Population parameter	Urban versus rural	Magnitude
Population density	Increase	High
Annual survival	Increase	Moderate
Home range size	Decrease	High
Home range aggregation/overlap	Increase	High
Contact rate	Presumed increase	??

RRV infection is especially challenging in urban landscapes because of the urban ecology of the raccoon host.

Ecology of urban raccoons. The raccoon is a medium-sized terrestrial mammal that is distributed across most of North America. The species has a generalized natural history, including a highly omnivorous diet, and considerable flexibility in habitat use. These traits combine to produce a highly opportunistic exploiter of the landscape, including metropolitan areas. Toward the latter part of the twentieth century, raccoon populations experienced substantial increases across most parts of the United States, and their range has continued to expand northward. The raccoon is the most abundant native carnivore species in metropolitan areas across the eastern half of the United States.

Raccoons respond strongly to anthropogenic resources characteristic of cities, including food and den sites (see **table**). Indeed, raccoons are quick to take advantage of novel food sources, such as refuse and pet food, to the point of habituation. Not surprisingly, the highest recorded densities for raccoon populations have most commonly come from urban populations, often exceeding 100 raccoons/km^2 (typically 40-120 raccoons/km^2). However, most of the reported densities have come from urban populations residing in high-quality habitat patches, infused with human-related foods, with relatively little information on raccoon populations residing completely within the developed matrix. This knowledge gap has been addressed recently, and density estimates reported from mark–recapture grids in developed landscapes within the Chicago metropolitan area indicate ranges of 2–10 raccoons/km^2, which is consistent with many rural areas with poor habitat. Thus, the variability of urban raccoon densities reflects the diverse characteristics of metropolitan landscapes.

The extremely high population densities in many parts of the urban landscape appear to be a consequence of relatively high survival rates of adults combined with high fecundity of urban raccoons. Although roadkill is a major cause of mortality for raccoons, traditional radio tracking has shown that they adjust to traffic patterns in urban areas to such a degree that their survival (80%) remains as high as, or higher than, that of rural populations with much lower traffic volumes. For example, urban raccoons have responded to increases in traffic by choosing not to cross major thoroughfares. It is likely that their nocturnal activity reduces their exposure to peak traffic volumes in cities.

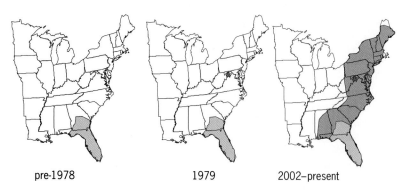

pre-1978　　　　1979　　　　2002–present

Fig. 1. Spatial spread of raccoon rabies variant (RRV) in the United States.

Anthropogenic resources have a profound influence on the movements and sociospatial system of raccoons. In urban areas, in which human-related foods are available in a predictable fashion, such as refuse in garbage cans in parks, raccoons reduce their home range size, and the local population becomes spatially aggregated with high levels of home range overlap among individuals. For example, mean home range sizes for female raccoons in the Chicago metropolitan area ranged from 21 to 52 hectares (ha) [1 ha = 10,000 m^2] across seasons, which were considerably smaller than the averages of 71–182 ha for rural female home ranges. Furthermore, radio tracking showed that activity centers remained stable across seasons for urban raccoons, reflecting the relatively predictable distribution of anthropogenic foods, whereas activity centers shifted for rural raccoons searching for ephemeral and unpredictable food types.

The social system of raccoons also contributes to high densities in urban areas. Raccoons exhibit a certain degree of social tolerance, especially with aggregated resources (food and dens), and do not exhibit territorialism in the traditional sense. Indeed, raccoons have often been observed feeding together in loose aggregations where food is abundant. Thus, large numbers of raccoons can live in small patches across the urban landscape. Intuitively, contact rates among urban raccoons should be high, especially where anthropogenic resources are highly aggregated. However, contact rates for free-ranging raccoons have not been quantified until recently using newly developed technology.

Contact rates. Contact rate of the host population is a crucial metric for understanding the dynamics of transmissible diseases, including the development of predictive rabies models. Despite its importance, contact rate remains difficult to estimate for secretive, nocturnal animals such as raccoons. Therefore, proximity-detecting collars have been recently developed in response to the need to model rabies in raccoons.

The proximity-detector system is a modification of the traditional very high frequency (VHF) radio collars commonly used for wildlife studies. The modified system basically consists of multiple radio collars that communicate with each other over a short-range (ultrahigh frequency, UHF) radio data link (**Figs. 2 and 3**). Each transmitter broadcasts a unique identification (ID) code over a UHF channel at 1.5-s intervals. When not broadcasting its ID code, each radio collar also functions as a detector that "listens" for other ID codes by sampling the UHF channel. A "contact" record begins when a proximity-detecting collar receives the UHF signal of another collar within a pre-programmed detection distance, and continues until the collar fails to detect the signal of the other for a period exceeding a user-defined "separation time." Once separation time is exceeded without reception of the signal and the contact is considered terminated, data are logged onboard in nonvolatile memory (that is, memory that keeps its contents even if power is absent). Each contact record consists of

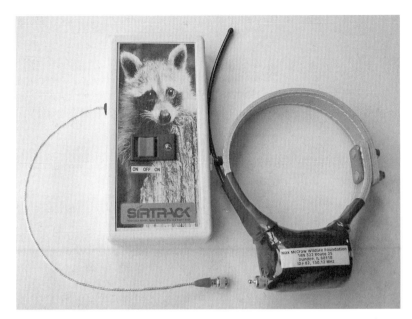

Fig. 2. A proximity-detector collar and interface unit (*Sirtrack, Ltd., Havelock North, New Zealand*) for raccoons.

five fields: data entry number (record number), ID number of the contacted collar, date, time that contact was initiated (h:min:s), and contact duration in seconds. Each collar is theoretically capable of simultaneously and accurately recording the duration of contacts with up to eight others at any given time. There is currently no remote download; the radio collar must be recovered to download data to a computer via an interface unit (Fig. 2).

Contact rates of a free-ranging urban raccoon population were monitored by fitting proximity-detecting collars on a local population of raccoons in the Chicago metropolitan area during 2004–2005. The UHF signal strength was set at the lowest level, resulting in a detection distance of 1–1.5 m (3.3–5 ft) and a separation time of 45 s. Of 32 raccoons

Fig. 3. Raccoon fitted with a proximity-detecting collar.

monitored for a complete annual period, more than 77,500 contacts lasting more than 1 s were recorded. The mean number of contacts recorded per day ranged from 0 to 96 for individual pairs, and the duration of each contact ranged from 1 s to 576 min. The overall pattern that emerged was that contacts were extensive within the local population throughout the year and across dyad (pairing) types, and the greatest social distance (as determined by social networking analysis) between all pairs of individuals was only one raccoon. These data are the only estimates of contact rates for free-ranging raccoons without the use of feeding stations, and the degree to which contact rates are elevated in urban versus rural populations of raccoons is unknown. Proximity-detecting collars are currently in use worldwide to estimate contact rates for other species such as white-tailed deer, bighorn sheep, and brush-tailed possums.

Control strategies. Attempts to control spread of RRV within raccoon populations have focused on combined strategies of population reduction and immunization at the local level. In either case, the strategy is to reduce the number of susceptible hosts, either by lethal means or by increasing herd immunity through the use of vaccines distributed via oral baits. The large-scale oral baiting program mimics successful fox rabies programs conducted in parts of Europe and consists of large-scale distribution of oral baits, usually by airplane or helicopter, over prescribed drop zones in rural areas. Currently, oral baiting barriers have been established along the western front of the disease, from the Ohio–Pennsylvania border southward along the Appalachian ridge through Alabama. Baits are dropped from the air along this line annually. Each bait typically consists of a fish meal polymer enclosing a sachet containing a vaccinia-rabies glycoprotein (V-RG) recombinant vaccine. Raccoons that chew or ingest the bait/vaccine develop antibodies in 2–3 weeks, thereby becoming immune to infection.

Urban centers present unique challenges to these rabies-control strategies. These include habitat fragmentation; complex patterns of property ownership; human social and political pressures (which include opposition to lethal removal); large numbers of domestic dogs, cats, and other nontarget species that may consume baits; and intentional and unintentional relocations of raccoons. Large-scale aerial drops of baits are limited in cities, so oral bait programs are usually implemented by hand or vehicle. Thus, the distribution of numbers of baits sufficient to achieve herd immunity is difficult. A recent program in Toronto, Canada, used extremely high bait densities (200 and 400/km^2) to reach bait acceptance levels of 74 and 82% for a local raccoon density of only 20 individuals/km^2. The Toronto experience suggests that baiting levels must be 2.5–5 times higher in cities than in comparable rural areas to achieve similar vaccination rates.

Finally, it should be remembered that RRV becomes enzootic in an urban area, and eradication via oral baits may not completely eliminate the risk of transmission, because the vaccine is not effective in raccoons incubating the disease or that are clinically affected. Rabies has been documented cycling in a raccoon population in Washington, D.C., 9 years after the initial epizootic (epidemic outbreak of disease), suggesting that RRV may remain in an urban population for many years once it becomes established.

For background information *see* ANIMAL VIRUS; BIOLOGICALS; DISEASE; DISEASE ECOLOGY; EPIDEMIOLOGY; PUBLIC HEALTH; RABIES; RACCOON; VACCINATION; VIRUS; ZOONOSES in the McGraw-Hill Encyclopedia of Science & Technology.   Stanley D. Gehrt

Bibliography. C. K. Bozek, S. Prange, and S. D. Gehrt, The influence of anthropogenic resources on multi-scale habitat selection by raccoons, *Urban Ecosyst.*, 10:413–425, 2007; S. Prange, S. D. Gehrt, and E. P. Wiggers, Demographic factors contributing to high raccoon densities in urban landscapes, *J. Wildl. Manag.*, 67:324–333, 2003; S. Prange et al., New radiocollars for the detection of proximity between individuals, *Wildl. Soc. Bull.*, 34:1333–1344, 2006; P. C. Ramey et al., Oral rabies vaccination of a northern Ohio raccoon population: Relevance of population density and prebait serology, *J. Wildl. Dis.*, 44:553–568, 2006; R. C. Rosatte et al., Raccoon density and movements in areas after population reduction to control rabies, *J. Wildl. Manag.*, 71:2373–2378, 2007; R. C. Rosatte, R. R. Tinline, and D. H. Johnston, Rabies control in wild carnivores, pp. 595–634, in A. Jackson and W. Wunner (eds.), *Rabies*, 2d ed., Academic Press, San Diego, 2007; C. A. Russell et al., Predictive spatial dynamics and strategic planning for raccoon rabies emergence in Ohio, *PLoS Biol.*, 3:e88, 2005.

Radio-frequency MEMS

Radio-frequency micro-electro-mechanical systems (RF MEMS) is a technology aimed at the realization of high-quality components to enable superior circuits (that is, capable of operation over wider frequency bandwidths and exhibiting lower losses) for wireless communications systems. RF MEMS adds to conventional integrated circuit technology two main fabrication techniques, namely, surface micromachining and bulk micromachining, to reduce the extraneous capacitive and resistive effects introduced by the substrate/wafer by virtue of the fact that the components are naturally embedded in it. Because of their extraneous nature, these effects are called parasitic; they are parasites to the desired components being implemented. Surface micromachining reduces component parasitics by suspending them over the substrate. Bulk micromachining, on the other hand, reduces parasitics by removing the substrate in a core region underneath the components. This decoupling from the substrate results in fully three-dimensional components, such as capacitors, varactors, inductors, resonators, switches, and transmission lines, that not only exhibit a minimum of substrate parasitics, but that may also be caused to undergo mechanical motion upon the application of electrostatic

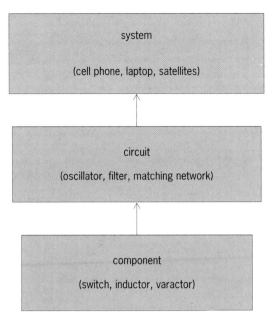

Fig. 1. Elements of RF MEMS.

or magnetic forces. This latter feature is very valuable for producing reconfigurable circuits and systems that may adapt to operating conditions to maintain optimum performance. The elements of RF MEMS may be construed as depicted in **Fig. 1**.

RF MEMS is poised to revolutionize wireless communications because the components it is capable of producing exhibit performance levels that surpass those germane to conventional integrated circuits. These higher levels of performance, in turn, provide the necessary degrees of freedom to enable designers to achieve better-performing wireless systems, from cell phones, to wireless-enabled laptops, to communications satellites.

The key aspects of conventional components improved by RF MEMS include the loss incurred by a signal in traversing the component (the so-called insertion loss, IL), the range of frequencies over which the component may be employed (the so-called bandwidth), and the component's ability to reconfigure.

Advantages of low insertion loss and wide bandwidth. In portable applications, lower-insertion-loss components will ultimately reduce battery current draw; this will extend battery life and reduce the frequency of battery recharging events. Less power-consuming circuits are desirable because the power saved may be employed for new functions. This would be manifest, for instance, in extending the functional capabilities of a cell phone to include multiple FM radios, television sets, computers, and more, all in one unit and with extended battery life.

The observed power consumption (for example, battery current draw) in a communications system is rooted in both direct and indirect dissipation in circuit components. Direct power dissipation is determined by the amount of resistance the component presents to the signal current flowing through it. Since, to achieve a certain system performance, certain levels of signal power and amplitude must be maintained throughout its architecture, it is nec-

essary to compensate for the signal lost by introducing amplifiers to boost it. But the added amplifiers consume battery current, so the ultimate impact of a lossy component is additional power consumption.

Indirect power dissipation, on the other hand, is determined by the fact that, at the signal frequencies usually employed, a certain amount of the signal intended to go through a component is coupled to the ground plane (and sometimes to other components) [**Fig. 2**]. Thus, less of the input signal to the component reaches its output and, again, further amplification may be required to compensate for this loss, with the concomitant battery current draw.

Another deleterious effect of signal coupling to the ground plane is the narrowing of the component bandwidth. As indicated in Fig. 2, capacitive coupling to the ground, exemplified by C_{sub}, shunts the component's output signal. Since, for a given substrate capacitance, C, the impedance, $|Z_C| = 1/(2\pi f C)$, is inversely proportional to the frequency, f, higher signal frequencies experience greater shunting to ground, thus limiting a series component's maximum frequency of operation. Similar effects are present in all components implemented in conventional integrated circuit processes.

Key components enhanced by RF MEMS. The building blocks of communications circuits may be divided into passive and active components. Active components, which are not dealt with by RF MEMS, include transistors, employed for signal amplification. Passive components of import to communications systems include transmission lines, capacitors, inductors, varactors, and switches.

Transmission lines. Transmission lines are utilized to interconnect components in a radio-frequency (RF) circuit. As such, they should introduce negligible insertion loss and not limit signal bandwidth. Transmission lines have been improved with RF MEMS by removing the bulk from underneath them (the air-suspension approach). This has been effective in

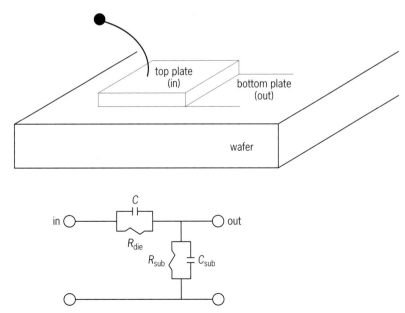

Fig. 2. Sketch of parallel-plate capacitor on a wafer.

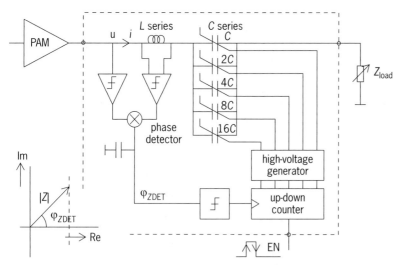

Fig. 3. Block diagram of the adaptive series-*LC* matching module. It compensates the reactive part of the load impedance by controlling the detected phase ϕ_{Z_DET} of the matched impedance to approximately zero. (*From A. van Bezoojien et al., RF-MEMS based adaptive antenna matching module, 2007 IEEE Radio Frequency Integrated Circuits Symposium, pp. 573–576, IEEE Press, 2007,* © *2007 IEEE*)

reducing the insertion loss by up to 65% and increasing the bandwidth by a factor of 5.

Capacitors. These components are employed in matching networks and for dc blocking, and their performance is captured by two parameters, namely, quality factor, Q (the higher the Q, the less lossy the capacitor), and self-resonance frequency. (This is the maximum frequency of operation; beyond it the capacitor behaves inductively.) These have been improved, respectively, by a factor of 10 upon removal of the substrate underneath them.

Varactors. These components are variable capacitors employed for tuning and reconfigurability purposes. They may adopt a finite set of capacitor values, such as in a two-state capacitor, or a continuous range of capacitor values. In addition to Q and self-resonance frequency, varactor linearity is a very important performance parameter. Conventional integrated circuit varactors derive from *pn*-junction diodes, which are intrinsically nonlinear. As a result, they tend to add harmonic frequencies to the signal being processed. RF MEMS varactors, on the other hand, contain no nonlinear elements. Thus, the signal spectral purity is preserved. The typical varactor performance at cell phone frequencies includes Q between 14 and 23, and a tuning range of 1:1.35 for a control voltage range of 2.7 V. In some designs the tuning range has been extended to 1:6 for a control voltage of 10 V.

Inductors. These components are employed as high-impedance chokes and in matching networks, and their performance is also captured by their Q and self-resonance frequency. These have been improved, respectively, by factors of 10 and 3 by removing the substrate underneath them.

Switches. These components are employed for signal routing, and their performance is captured by their insertion loss and isolation (ISO). The isolation measures the degree to which the input signal is coupled to the switch's output, in particular, the ratio of the output power to the input power expressed in decibels (dB) when the switch is set to the open-

circuit state. The typical insertion loss and isolation of RF MEMS switches are 0.1 and 50 dB, respectively, in the dc-to-4-GHz frequency range.

Wireless system reconfigurability. RF MEMS-based components are ideal for implementing reconfigurable circuit functions. These include, in particular, impedance-matching networks (IMN). An example of a system to reconfigure an impedance-matching network between the power amplifier and the antenna in a cell phone handset is shown (**Fig. 3**).

The problem addressed is that of compensating the antenna detuning due to body-proximity-induced changes in the reactive part of the antenna impedance. These effects manifest themselves as changes in radiated power due to mismatch in the impedance presented to the output power amplifier (PAM). The approach involves inserting, at the output of the power amplifier, an adaptive series *LC* impedance-matching network. In this impedance-matching network the inductor is fixed, but the capacitor is variable, implemented by a parallel combination of binary-weighted (5-bit) RF MEMS capacitors exhibiting a tuning range between 1 and 15 pF.

The process to effect IMN tuning is as follows. When a mismatch exists, the phase shift between the input current and voltage to the *LC* network, ϕ_{Z_DET}, is nonzero. Then, to compensate for reactive changes in the antenna impedance, the value of ϕ_{Z_DET}, is fed to an up-down counter, which drives a high-voltage generator (HVG) which, in turn, drives the RF MEMS capacitors to such values as are necessary to drive ϕ_{Z_DET} toward zero. Finally, a value of ϕ_{Z_DET} equal to zero signifies an impedance-matched condition has been achieved. Since cell phones carry only a 5-V battery, and the RF MEMS capacitors require an activation voltage of 60 V, the implementation includes a charge pump HVG to transform 5 V to 60 V.

The performance of the reconfigurable impedance-matching network was successfully demonstrated for various antenna load impedances, thus validating the concept. The significance of this example is that it demonstrates the power of RF MEMS in enabling the automatic adaptability of an impedance-matching network to transform reactive antenna loads into real loads. As a by-product, this new adaptive capability is expected to make isolators redundant and to enable the utilization of smaller-size antennas. In addition, the scheme will improve performance in terms of power dissipation and overall signal reception and transmission.

For background information *see* CAPACITOR; COMMUNICATIONS SATELLITE; IMPEDANCE MATCHING; INDUCTOR; MICRO-ELECTRO-MECHANICAL SYSTEMS (MEMS); MOBILE COMMMUNICATIONS; TRANSMISSION LINES; VARACTOR in the McGraw-Hill Encyclopedia of Science & Technology.   Héctor J. De Los Santos

Bibliography. H. J. De Los Santos, *Introduction to Microelectromechanical (MEM) Microwave Systems*, 2d ed., Artech House, Norwood, MA, 2004; G. M. Rebeiz, *RF MEMS: Theory, Design, and Technology*, John Wiley & Sons, 2003; A. van Bezoojien et al., RF-MEMS based adaptive antenna matching module, *2007 IEEE Radio Frequency Integrated Circuits Symposium*, pp. 573–576, IEEE Press, 2007.

Satellite detection of thunderstorm intensity

Despite the availability of an advanced system of geostationary satellites covering the United States, the detection of severe or violent thunderstorms has been dependent mainly on radar and storm spotters, when available. In Europe and Africa, radar and spotter networks are less developed or nonexistent. The question arises, to what extent can satellites be used for detecting severe weather phenomena? Research into this question started with the first satellite observations more than 40 years ago in the United States. This paper reviews the principal findings from this research, with some current applications of the satellite data. The emphasis will be on inferring the intensity or severity of existing thunderstorms, which includes the occurrence of damaging winds, hail, and tornadoes. The estimation of rainfall from satellite imagery and the use of the data in numerical weather prediction are important topics, but beyond the scope of this paper. However, storm initiation and intensification will be covered briefly.

First pictures. In 1960, the first pictures of thunderstorms were transmitted to Earth from the *TIROS-1* (Television and Infrared Observation Satellite) low-orbiting satellite. These were exciting times, as much was to be learned from extensive pictures of the clouds from above. The first published studies reported on the visible characteristics of thunderstorms. For example, in 1963, L. F. Whitney reported observations that the storm clouds were "conspicuous and distinctive" and sometimes had sharp, scalloped edges. He also noted that multiple anvils were combined into a single, highly reflective cloud top. Moreover, he noted that these clouds covered a much larger area than was indicated by the radar echoes and locations of sferics (lightning) associated with the storms.

One of the first studies to propose an index of severity based on cloud characteristics was by R. J. Boucher in 1967. Based on 17 cases from *TIROS IV–VII* (1962–1964), the diameter of the cirrus shield was suggested to be an index of storm severity. It was noted that those with diameters of less than about 100 km were rarely severe, while those with diameters of greater than 250 km often had severe weather. It was also observed that penetrating tops (above the anvil) were not always associated with severe weather.

Some of the first ideas for using satellite imagery in severe weather forecasting and warnings were presented by J. F. W. Purdom in 1971. The appearance of squall lines as a wedge-shaped cloud, narrowing to the south, was noted. Methods to precisely locate the polar and subtropical jet stream, the thermal ridge, and vertical wind shear, which are factors in severe thunderstorm development, were illustrated using *NOAA-1* (National Oceanic and Atmospheric Administration) satellite imagery. He also proposed several ideas on how geostationary imagery from the Applications Technology Satellite *ATS-3* could be used to improve warnings. It was noted that developing squall lines could be detected earlier than from radar,

and that specific regions of convective clusters could be identified as being under threat of severe weather (often the southern portion). Finally, the sequence of images (as frequent as 11 min) allowed measurements of the rate of anvil growth.

The idea of anticipating thunderstorm initiation from the location of boundaries and other mesoscale features in high-resolution geostationary imagery (**Fig. 1**) was suggested by J. F. W. Purdom in 1976. The effect of terrain (coastlines, rivers, and lakes) on the formation of convective clouds was shown. Precise location of storm outflow boundaries could be made from high-resolution GOES (Geostationary Operational Environmental Satellites). Given favorable environmental conditions (stability and shear), intersecting or merging of boundaries was observed to coincide with new storm initiation or intensification. Some possible mechanisms for such intensification were suggested by R. A. Maddox and coinvestigators in 1980. The utility of locating thunderstorm outflow boundaries and other mesoscale cloud features has been one of the most powerful applications of satellite imagery in severe storm forecasting to date.

Exploration of infrared (IR) imagery. One of the earlier studies of thunderstorm anvil structure, which made use of the infrared imagery, was that of C. E. Anderson in 1979. Observations of a few select severe storms suggested anticyclonic rotation and spiral bands with similarity to hurricanes. The cirrus plumes extended downwind, but to the right of the upper-level flow. The significance of these characteristics for storm severity was not determined. In a latter study, R. F. Adler and coinvestigators reported in 1981 a cyclonic spiral cloud shadow in visible imagery, possibly associated with a mesocyclone observed from Doppler radar.

An early study of overshooting (penetrating) cloud tops from geostationary satellites was reported by D. W. McCann in 1979. An example is shown in **Fig. 2**. (The collapse of overshooting tops had been previously observed by high-altitude aircraft and possibly linked to tornado formation by T. T. Fujita and later T. A. Umenhofer.) In this study, the collapse appeared to be related to enhanced low-level outflow from the storm. In some cases, this coincided with acceleration of the gust front, strong surface outflow (bow echoes), and occlusion of a mesolow, which sometimes precedes tornado genesis. This suggested a possible utility in anticipating strong surface winds; however, it was emphasized that the collapsing top was not a direct cause of the tornado.

In 1980, R. A. Maddox identified a unique class of convective storm systems on the basis of the infrared observations of GOES. Frequenting the midwestern United States in summer, these systems, termed mesoscale convective complexes (MCCs), had unique characteristics in the satellite imagery. They were observed to have long lifetimes as compared to individual storms (>6 h) and extensive, cold cloud shields (>100,000 km^2 at <−32°C), and they were nearly circular in shape (eccentricity >0.7). These weather systems, which produce a wide variety of severe weather and heavy rain, are still difficult to forecast. They occur with weak

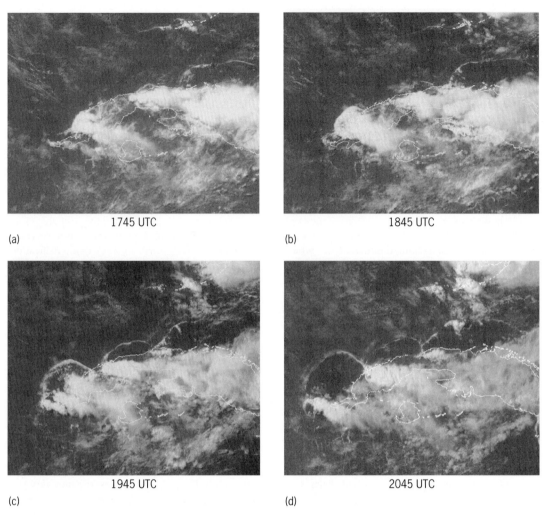

Fig. 1. Visible imagery from *GOES-12* on June 8, 2007, showing development of outflow boundaries from existing storms and new storm formation where these expanding boundaries intersect.

upper-level forcing and appear to be driven in part by low-level warm air advection.

Growth rates. One of the first attempts to use geostationary imagery to relate rates of cloud-top

Fig. 2. Example of overshooting cloud tops and plumes: Tornadic thunderstorms in Kansas and Oklahoma from *GOES-12* visible imagery, May 5, 2007, 00:45 Coordinated Universal Time (UTC).

cooling and anvil expansion to storm severity was conducted by R. F. Adler and D. D. Fenn (1979). IR imagery taken at 5-min intervals from the NASA Synchronous Satellite (*SMS-2*) was used to compare a set of severe and nonsevere storms from a single day; the severe storms exhibited more rapid growth and colder cloud-top temperatures. Storm-top divergence and vertical velocity inferred from rates of cooling and anvil expansion were twice as large for the severe storms. Tornadoes occurred after rapid expansion in several of the storms. It was suggested that a 30-min lead time in anticipating severe weather might be possible. Limitations to the measurements were that existing anvil clouds may have obscured new storm growth and that storm heights estimated from cloud-top temperature appeared to be underestimated when compared to radar.

These ideas were investigated further by R. F. Adler and coinvestigators (1985). Using imagery from *SMS-2* and NOAA's *GOES-1* at 5-min intervals, an index of thunderstorm intensity was developed. The index was computed as the maximum updraft intensity inferred from cloud-top ascent rates and expansion rates of contours of cloud-top temperature. Based on storms on four different days, the index could distinguish the most severe from nonsevere storms when tested on independent data. The difficulty of identifying growing cloud tops because of existing anvils was reaffirmed in the study. In addition, cold cloud tops sometimes appeared to be too warm because of instrument resolution. The error in estimating height from cloud-top temperature was recognized again.

In 1980, D. W. Reynolds reported the relationship of cloud-top temperature to the occurrence of hail. It was expected that a more direct relationship might exist between inferred updraft intensity and hail than with other types of severe weather. For example, tornadoes depend on other factors, such as low-level wind shear. The storms studied developed large, cold cloud tops and were long-lasting (3–5 h). The onset of hail was related to rapid cloud growth and the cloud tops becoming colder than the tropopause. Maximum hail occurred when the cloud-top temperature reached a minimum (and the difference with the tropopause temperature was largest). A visual enhancement of the infrared imagery was proposed to identify clouds colder than the tropopause.

Enhanced-V signature. From infrared imagery, a distinctive warm spot embedded at the top of the anvil cloud was observed with some severe thunderstorms by P. Mills and E. Astling (1977). Hypotheses for the local, warm region of the cloud included greater emissivity, mixing of warm stratospheric air, and subsidence. When visually enhancing the imagery to view small temperature differences within the cloud top, D. W. McCann in 1983 reported on an area of cold cloud-top temperature adjacent to the warm spot in the shape of the letter "V." The orientation of the V was related to the upper-level wind, with the open end pointing downwind. The signature was referred to as an "enhanced V" but is often shaped more like the letter "U" (**Fig. 3**).

McCann used IR imagery at 30-min intervals from a large number of cases (~900) to investigate the relation between the V-shaped pattern of cloud-top temperature and severe weather occurrence. Most storms with an enhanced V exhibited some form of severe weather. The appearance of the V usually preceded the severe weather—on the average, 30 minutes earlier. The false alarm rate (FAR) from warning on the basis of the V was about 0.3, similar to those of current radar methods based on mesocyclone or three-dimensional (3D) reflectivity structure, as reported by P. D. Polger and coinvestigators in 1994. Many severe storms did not have an enhanced V, leading to a relatively low probability of detection (POD), 0.25, which is much less than is possible with radar. The enhanced-V occurrence was linked to the combination of strong upper-level wind and penetrating tops (intense convection).

The causes of the warm spot feature were studied further by R. F. Adler and R. A. Mack (1986). Using stereoscopic observations from a pair of GOES satellites (1979) and a simple updraft model, they identified three classes of storms. In essence, the cold-warm couplet was a manifestation of the wind shear, which separated the top of the updraft from sinking air downwind. The offset between the coldest and highest points is the result of mixing of progressively warmer air as the overshooting top continued

to ascend. The effects of the temperature gradient, mixing, and shear above the updraft's equilibrium level explained the uncertainty in estimating cloud-top height from radiometric temperature noted in previous studies.

The cloud-top temperature structure observed from above severe storms was investigated further by G. M. Heymsfield and coinvestigators (1983) and G. M. Heymsfield and R. H. Blackmer (1988). Similar to other studies, warm areas were observed 10–20 km downwind from the updraft. These formed near the time of tropopause penetration and moved with the storm motion. A subsidence mechanism was again proposed. In some cases, a distant warm area was observed 50–75 km downwind. These areas moved with the upper-level winds. In the rapid growth stages, the coldest tops were collocated with the radar echoes. At later stages, these were sometimes displaced from the echo cores. It was concluded that (1) the width of the V was related to the distance between the cold-warm couplet, and (2) the temperature difference between the couplet was related to the amount of overshoot of the penetrating top. Ingredients for the V were believed to be intense updrafts, overshooting tops, and wind shear near the tropopause. Limited resolution of the satellite imagery, unknown ice-crystal and temperature structure, complexity of multicell storms, and simplified 3D models were identified as factors limiting progress in studies such as this.

A 2007 study by J. C. Brunner and coinvestigators examined storms with enhanced-V signatures using

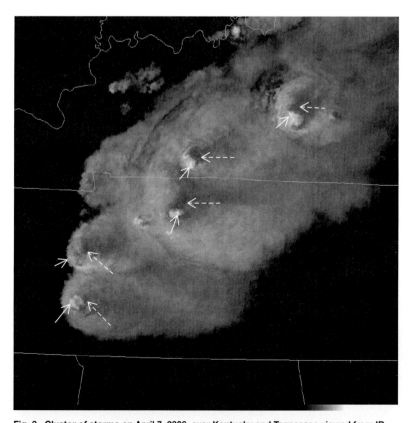

Fig. 3. Cluster of storms on April 7, 2006, over Kentucky and Tennessee viewed from IR imagery from MODIS on the AQUA satellite. Brightest areas are coldest (<−70°C). Five enhanced-V signatures with their cold spots (solid arrows) and warm spots (dashed arrows) are located above thunderstorm cores likely to have strong updrafts.

higher-resolution (1-km) imagery from polar-orbiting satellites. In addition to obtaining results similar to McCann's earlier study, they found the following to be useful indicators that a thunderstorm was severe: (1) the temperature difference between the cold and warm spots (exceeding 15°C) and (2) wind speeds near cloud tops, exceeding 50 kt (knots). The wind speeds are available from observed displacement of cloud and features in water vapor imagery from R. M. Rabin and coinvestigators (2004).

Plumes and short-wave infrared reflectance. Short-wave infrared imagery from the Advanced Very High Resolution Radiometer (AVHRR) has been observed to contain unusual features above some intense thunderstorms in Europe. The AVHRR is aboard the NOAA series of low-orbiting satellites, which limited the observations to fixed times. Daytime images from the AVHRR 3.7-μm band showed regions of enhanced solar reflectivity at or above the anvil surface. Occasionally, the areas were in the shape of plumes (Fig. 2), extending from a single point near a penetrating top to long distances downwind over a portion of the anvil, as reported by M. Setvak and C. A. Doswell in 1991. The occurrence of hail with some of these storms raised the question of a practical use of 3.7-μm observations in issuing severe weather warnings.

The time evolution of highly reflective areas above storm tops, studied with *GOES-8*, 3.9-μm imager data, was reported by M. Setvak and coinvestigators in 2003. These areas had time scales ranging from minutes to hours, and sizes from individual pixels to entire anvils. In several cases, there was a link to the location and initiation time of mesocyclones as observed from Doppler radar. In one of the cases, the highly reflective spot moved downwind with the speed of the upper-level flow, while in others the spot remained above the mesocyclone. A general relationship between these features and severe weather has yet to be identified.

Stratospheric water vapor above deep convective clouds has been identified from Meteosat (EUMETSAT) observations in the infrared window and water-vapor absorption bands by J. Schmetz and coinvestigators (1997). It was demonstrated that the equivalent brightness temperature in the water vapor can exceed that in the infrared window by several degrees because of stratospheric water vapor above the cloud top. The use of the difference in brightness temperature between the two bands to monitor convective cloud growth and areas of tropospheric-stratospheric exchange has been proposed.

Research by D. Rosenfeld and co-investigators, reported in 2003, has identified characteristic differences in the microphysical structure of prestorm clouds. Using the 3.7-μm band of AVHRR, hydrometeor size at the cloud top has been estimated for cloud clusters at differing stages of growth. It was found that the relation between the effective radius of hydrometeors and the cloud-top temperature, and the cloud-top temperature of glaciation are uniquely different for severe storms. Strong updrafts in severe storms result in relatively limited growth of hydrometeors. Hence, hydrometeor size increases more slowly as the cloud top grows, and glaciation occurs at a colder cloud-top temperature than in clouds with weaker updrafts. They proposed that hydrometeor size and glaciation temperature of developing cumulus inferred from satellite might be used to help forecast storm severity.

Radar advances in storm detection. There is little doubt that forecasters have made extensive use of satellite imagery in many aspects of severe weather forecasting. However, the extent to which many of the findings previously cited have been incorporated into the real world of forecasting and warning appears limited. One reason for this was the advent of Doppler radar in the United States.

While research on thunderstorm cloud-top structure and growth rates from geostationary satellites was flourishing, the generation of mesocyclones and tornadoes was being studied with Doppler radar, as reported by R. A. Brown and coinvestigators in 1978. The radar's ability to probe the internal wind and hydrometeor structure of storms made it an obvious tool for the detection of hail, high winds, and tornadoes. The implementation of a national network of Doppler radars across the United States took place in the 1980s–1990s. Automated algorithms for detecting mesocyclones, tornadoes, hail, and cell tracking have been used by the National Weather Service (NWS) to aid in a rapid, uniform scheme for issuing warnings. At the same time, the measurement of cloud-top cooling, growth rates, and other measures of storm intensification from satellite have yet to be implemented for automated, routine use. In addition, the identification of features such as the enhanced V and cold-warm couplets has yet to be automated for extended testing.

Current applications. Early operational use of satellite data was primarily subjective, as most forecast offices did not have access to digital data and were limited by the resolution of imagery. As reported by Molenar and coinvestigators in 2000, programs such as the RAMM Advanced Meteorological Satellite Demonstration and Interpretation System (RAMSDIS) have advanced the availability of higher-quality digital data since the 1990s. However, many forecast offices today still lack the means to implement automated schemes using satellite data. An Advanced Weather Interactive Processing System (AWIPS) is now bringing digital data to the forecast offices and may include analyses from satellite data in the future. The *GOES-R* Proving Ground program is setting the stage for future implementation of automated algorithms for nowcasting thunderstorms by the National Weather Service.

There are several current attempts to apply some of the early concepts to the routine nowcasting of storms. These do not specifically address identification of current storm severity, but rather initiation and intensification. A few examples are cited here.

In 2003, R. D. Roberts and S. Rutledge reported the development of a technique for the nowcasting of storm initiation and growth using both GOES and Doppler radar data (WSR-88D). The technique is based on observed cloud growth rates over extended periods. The onset of storm development depends on surface convergence features such as gust fronts, using aspects of the earlier work of Purdom. In addition, cloud tops reaching subfreezing altitudes and rapid cooling of these clouds are also important factors. Similar to the findings of Adler and others, intensity has been related to the rate of cooling at these early stages of development. The development of radar echoes aloft can be anticipated 30 min in advance, as shown by J. R. Mecikalski and K. M. Bedka in 2006. The concepts of this algorithm are being incorporated into the auto-nowcast system under development at the National Center for Atmospheric Research. The auto-nowcast system uses fuzzy logic to combine predictor information from radar, satellite, surface observations, profilers, and forecast models to provide 0–1-h nowcasts of storm location. Testing has indicated improvement over extrapolation and persistence.

A Web-based system for the automated monitoring of convective systems has been developed by R. M. Rabin and T. Whittaker (2005). It identifies and tracks cloud clusters from GOES satellite and WSR-88D radar based on user-selected thresholds of cloud-top temperature and radar reflectivity. Time trends of size, cloud-top temperature, and reflectivity are displayed. After defining inflow regions to the storm, the user may obtain time trends of the environmental parameters in a storm relative frame of reference. These parameters include thermodynamic stability and wind shear from a forecast analysis. The usefulness of these time trends in monitoring the evolution of mesoscale convective systems is being explored.

Another automated system for nowcasting convective storms using Meteosat Second Generation (MSG) satellite data was developed by C. Morel and coinvestigators (2002). Adaptive thresholding of cloud-top temperature is used to discriminate convective systems. It tracks and monitors growth and decay of convective systems and time trends of associated lightning.

More recent techniques for (1) tracking the initiation and growth of storms and (2) detection of overshooting tops and accompanying warm spots (reported by K. M. Bedka and coinvestigators in 2009), are being tested on current GOES data for eventual use with the next generation of U.S. GOES satellites. These future satellites, *GOES-R* and beyond, will have enhanced resolution and spectral bands similar to MSG with image intervals of 5 min or less. In addition, a lightning detector capable of monitoring time trends of total lightning for the first time on a 10-km scale will be included. The *GOES-R* Proving Ground program was started in 2009 to test and improve techniques that use GOES data for short-term forecasting of thunderstorm initiation and intensity with input from scientists and forecasters. It is envisioned that these techniques will be used by forecasters after launch of the *GOES-R* satellites around 2015.

Conclusions. There has been a long history of research into the development of severe thunderstorms from satellite data. From measurements of cloud-top cooling and expansion, there is valuable information on storm initiation and intensification. A serious limitation of the satellite visible and infrared observations is that they capture only the structure of the storm tops. The internal storm structure and subsequent new surface development are often masked by the upper-cloud layer. For fully developed storms, inferred processes involve the interaction of the cloud top with upper-level winds. Although the location and strength of the major updrafts can be estimated from information on the overshooting tops, features such as cold-warm couplets, rates of expansion and cloud-top cooling, and processes in the lower altitudes of the storms cannot always be routinely monitored. For example, the wind flow in these lower regions is critical to tornado generation and other strong surface winds.

Other limitations have included spatial and temporal resolution. While low-orbiting satellites appear to have sufficient resolution to capture most significant cloud-top features in both the visible and infrared regions, they lack adequate temporal coverage. Geostationary satellites still have less than optimal spatial resolution; however, current (MSG) and future (*GOES-R*) sensors have horizontal resolutions approaching those of earlier low-orbiting data. Data from these sensors should prove valuable in providing additional information on the nowcasting of thunderstorms.

For background information *see* CLOUD PHYSICS; DOPPLER RADAR; LIGHTNING; MESOMETEOROLOGY; METEOROLOGICAL SATELLITES; SATELLITE METEOROLOGY; SFERICS; SQUALL LINE; STORM; STORM DETECTION; THUNDERSTORM; WEATHER FORECASTING AND PREDICTION in the McGraw-Hill Encyclopedia of Science & Technology.　　　Robert M. Rabin; Jason Brunner; Scott Bachmeier

Bibliography. R. F. Adler and D. D. Fenn, Thunderstorm intensity as determined from satellite data, *J. Appl. Meteorol.*, 18:502–517, 1979; R. F. Adler and R. A. Mack, Thunderstorm cloud top dynamics as inferred from satellite observations and a cloud top parcel model, *J. Atmos. Sci.*, 43:1945–1960, 1986; R. F. Adler, D. D. Fenn, and D. A. Moore, Spiral feature observed at top of rotating thunderstorm, *Mon. Weather Rev.*, 109:1124–1129, 1981; R. F. Adler, M. J. Markus, and D. D. Fenn, Detection of severe Midwest thunderstorms using geosynchronous satellite data, *Mon. Weather Rev.*, 113:769–781, 1985; C. E. Anderson, Anvil outflow patterns as indicators of tornadic thunderstorms, *Preprints 11th Conf. Severe Local Storms*, Kansas City, American Meteorological Society, pp. 481–485, 1979; K. M. Bedka et al., Objective satellite-based overshooting top detection using infrared window channel brightness temperature gradients, *J. Appl. Meteorol. Climatol.*, submitted May 2009; R. J. Boucher, Relationships between the size of satellite-observed cirrus shields and the severity of thunderstorm complexes, *J. Appl. Meteorol.*, 6:564–572, 1967; R. A. Brown, L. R. Lemon, and

D. W. Burgess, Tornado detection by pulsed Doppler radar, *Mon. Weather Rev.*, 106:29–38, 1978; J. C. Brunner et al., A quantitative analysis of the enhanced-V feature in relation to severe weather, *Weather Forecast.*, 22:853–872, 2007; T. T. Fujita, Proposed mechanism of tornado formation from rotating thunderstorm, *Preprints 8th Conf. Severe Local Storms*, Denver, American Meteorological Society, pp. 191–196, 1973; G. M. Heymsfield, R. H. Blackmer, and S. Schotz, Upper-level structure of Oklahoma tornadic storms on 2 May 1979. I: Radar and satellite observations, *J. Atmos. Sci.*, 40:1740–1755, 1983; G. M. Heymsfield and R. H. Blackmer, Satellite-observed characteristics of Midwest severe thunderstorm anvils, *Mon. Weather Rev.*, 116:2200–2223, 1988; R. A. Maddox, L. R. Hoxit, and C. F. Chappell, A study of tornadic thunderstorm interactions with thermal boundaries, *Mon. Weather Rev.*, 108:322–336, 1980; R. A. Maddox, Mesoscale convective complexes, *Bull. Am. Meteorol. Soc.*, 61:1374–1387, 1980; D. W. McCann, On overshooting-collapsing thunderstorm tops, *Preprints 11th Conf. Severe Local Storms*, Kansas City, American Meteorological Society, pp. 427–432, 1979; D. W. McCann, The enhanced-V: A satellite observable severe storm signature, *Mon. Weather Rev.*, 111:888–894, 1983; J. R. Mecikalski and K. M. Bedka, Forecasting convective initiation by monitoring the evolution of moving cumulus in daytime GOES imagery, *Mon. Weather Rev.*, 134:49–78, 2006; P. Mills and E. Astling, Detection of tropopause penetrations by intense convection with GOES enhanced infrared imagery, *Preprints 10th Conf. Severe Local Storms*, American Meteorological Society, Omaha, Nebraska, pp. 61–64, 1977; D. A. Molenar, K. J. Schrab, and J. F. W. Purdom, RAMSDIS Contributions to NOAA satellite data utilization, *Bull. Am. Meteorol. Soc.*, 81:1019–1030, 2000; C. Morel, S. Senesi, and F. Antones, Building upon SAF-NWC products: Use of the rapid developing thunderstorms (RDT) product, *2002 EUMETSAT User Conference*, Dublin, Ireland, 2002; C. Mueller et al., NCAR auto-nowcast system, *Weather Forecast.*, 18:545–561, 2003; P. D. Polge et al., National Weather Service warning performance based on the WSR-88D, *Bull. Am. Meteorol. Soc.*, 75:203–214, 1994; J. W. F. Purdom, Satellite imagery and severe weather warnings, *Preprints 7th Conf. Severe Local Storms*, Kansas City, American Meteorological Society, pp. 120–137, 1971; J. W. F. Purdom, Some uses of high-resolution GOES imagery in the mesoscale forecasting of convection and its behavior, *Mon. Weather Rev.*, 104:1474–1483, 1976; R. M. Rabin, Detecting winds aloft from water vapor satellite imagery in the vicinity of storms, *Weather*, 59:251–257, 2004; R. M. Rabin and T. Whittaker, Tool for storm analysis using multiple data sets, *Advances in Visual Computing: First International Symposium, ISVC 2005*, Lake Tahoe, Nevada, December 5–7, 2005, edited by G. Bebis et al., *Proceedings*, Springer, pp. 571–578, 2005; D. W. Reynolds, Observations of damaging hailstorms from geosynchronous satellite digital data, *Mon. Weather Rev.*, 108:337–348, 1980; R. D. Roberts and S. Rutledge, Nowcasting storm initiation and growth using *GOES-8* and WSR-88D data, *Weather Forecast.*, 18:562–584, 2003; D. Rosenfeld et al., Satellite detection of severe convective storm by their retrieved vertical profiles of cloud particle effective radius and thermodynamic phase, *J. Geophys. Res.*, vol. 113, D04208, 2008; J. Schmetz et al., Monitoring deep convection and convective overshooting with METEOSAT, *Adv. Space Res.*, 19(3):433–441, 1997; M. Setvak and C. A. Doswell, The AVHRR Channel 3 cloud top reflectivity of convective storms, *Mon. Weather Rev.*, 119:841–847, 1991; M. Setvak et al., Satellite observations of convective storm tops in the 1.6, 3.7 and 3.9 μm spectral bands, *Atmos. Res.*, 67–68:607–627, 2003; T. A. Umenhofer, Overshooting top behavior of three tornado-producing thunderstorms, *Preprints 9th Conf. Severe Local Storms*, Norman, Okla., American Meteorological Society, pp. 96–99, 1975; L. F. Whitney, Severe storm clouds as seen from TIROS, *J. Appl. Meteorol.*, 2:501–507, 1963.

Schizophrenia and prenatal infection

Schizophrenia is a chronic psychotic disorder that is characterized by the presence of hallucinations, delusions, disorganized thinking and behavior, and deficits in information processing. Schizophrenia affects approximately 1% of the population, with a typical onset in adolescence or later. The causes of the disorder are still largely unknown. There is an approximately 50% concordance rate among identical twins—meaning that if one twin has the disorder, there is a 50% chance that the other twin will also have it. This finding suggests that, while genetics plays a substantial role, there is a significant contribution from environmental factors. Schizophrenia is believed to be a neurodevelopmental disorder, in which disruptions in programming by prenatal conditions can have a significant role. The prenatal period is the most significant and sensitive window for brain development in human life; therefore, an interruption in brain maturation during this critical time can have irreversible and lasting effects. Increasing evidence indicates that infection during pregnancy could be one such prenatal condition. The correlation between prenatal infection and the later development of schizophrenia is one of the most compelling pieces of epidemiological evidence regarding the cause of schizophrenia. This connection may assume particular importance in the future prevention of this disorder, given that many infections are themselves preventable.

Findings on infection and schizophrenia. The initial interest in the relationship between infection and schizophrenia was a result, in part, of a well-replicated correlation between winter and spring births and schizophrenia. This finding led investigators to test whether naturally occurring events in populations, such as influenza epidemics, were associated with schizophrenia. These *ecologic studies* indicated an association between influenza epidemics and risk of producing offspring with schizophrenia,

Examples of studies on prenatal infection and schizophrenia		
Study design	Description of association	Comments
Ecologic studies		
Ecologic study of relationship between 1957 influenza epidemic in Finland and risk of schizophrenia	Significant association found for second trimester of pregnancy	Risk of schizophrenia was compared with risk of other diagnoses rather than with healthy controls
Ecologic study of relationship between prenatal exposure to poliovirus and risk of schizophrenia in Finland	Significant association found for fifth month of pregnancy	Measure of exposure was only an approximation of incidence of poliovirus infection in pregnant women
Birth cohort studies		
Case-control study that measured influenza antibody levels from archived maternal serum from Kaiser patient records in Alameda County, CA, for offspring born from 1959 to 1967	Threefold increased risk of schizophrenia after influenza exposure in the first half of pregnancy; sevenfold increased risk from first-trimester exposure	Study used an indirect measure for influenza, but this was well validated
Cohort study of children prenatally exposed to 1964 rubella epidemic in New York City	Rubella-exposed birth cohort had a markedly increased risk of schizophrenia (20.4%)	By population estimates, the risk of schizophrenia would be increased 15 times
Study comparing healthy controls with patients using data from a national population and a national neonatal biobank of blood samples in Denmark	Increased risk of schizophrenia with neonatal seropositivity for toxoplasma	Sera taken from infants in the first week of life

NOTE: There have been over 50 studies to date; for purposes of brevity, only a few examples are given in the table.

suggesting that risk of schizophrenia may be related to infection during pregnancy.

This promising evidence called for a more refined research design, specifically, one that would assess individuals rather than populations. *Cohort studies* have the power to confirm population findings with more precise data, including individual diagnosis of infection. Ideally, in such studies, the assessment of infection and the diagnosis of schizophrenia would be based on documented medical and clinical information spanning from early pregnancy to the time of expected onset of schizophrenia in the offspring. The use of biomarkers (biological indicators found in the blood) to confirm infection is a powerful tool for increasing the validity and strength of these studies. In one birth cohort study, analyses of blood stored during pregnancy were used to confirm infection; follow-up revealed a greater likelihood of schizophrenia for offspring whose mothers had influenza infection during the first and second trimesters of pregnancy compared to those who were unexposed in pregnancy.

Additional cohort studies have provided evidence that other maternal infections are associated with schizophrenia in the offspring. These infections include rubella, respiratory infection, toxoplasmosis (infection by the protozoan *Toxoplasma gondii*, manifested clinically in severe cases by jaundice and by liver and spleen enlargement), and possibly herpes simplex virus. There is also evidence of elevated levels of maternal cytokines, which are robust markers or intercellular signals of a maternal immune response to an infection. Some infections have a greater effect on schizophrenia than others; for example, rubella infection is related to a nearly 20-fold increase in risk, compared to a 3-fold increased risk for influenza (as discussed above). A limited number of examples are given in the **table**.

Mechanisms. Investigations into the causes of correlations between in utero infection and schizophrenia are still in the theoretical stage, and many factors are being considered. Some of these factors include interactions between genes and the infections, the detrimental effects of infection on the physiology of the brain, and the route of transmission of the infection.

Because the majority of mothers who have infection in early pregnancy do not produce offspring with schizophrenia and, as mentioned earlier, only half of identical twins with schizophrenia have a co-twin affected with the disorder, it is likely that both genetic predisposition for schizophrenia and infection are needed for the onset of schizophrenia among cases exposed to infection. This genetic vulnerability is therefore dependent on prenatal infection for schizophrenia to become manifest.

Although the causes of schizophrenia are not well understood, the neurophysiologic outcomes are relatively well documented. Among the many abnormalities in the physiology of a brain affected with schizophrenia, functioning of the hippocampus (involved in inhibition and certain kinds of memory) and certain regions of the brain cortex (involved in attention, awareness, and thought) are most aberrant. Other cardinal abnormalities include the enlargement of fluid-filled cavities in the brain called ventricles (**Fig. 1**). One mechanism by which the aforementioned areas of the brain could be affected by infection during prenatal brain development involves disruption of cell proliferation and neuronal migration, that is, the process by which cells are created, specified, and targeted to the appropriate regions of the brain. This initial disruption may cause continued disruptions in future brain development from childhood to early adulthood. Finally, another key abnormality in schizophrenia is hyperactivity of

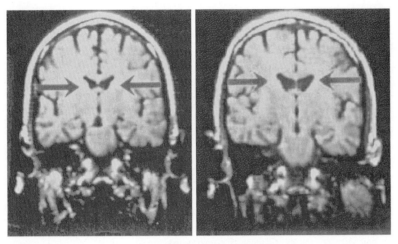

Fig. 1. Magnetic resonance images of identical twins. Ventricles of healthy brain (left) and enlarged ventricles of a brain affected with schizophrenia (right). (*Image courtesy of Dr. Daniel R. Weinberger, Clinical Studies Section, Clinical Brain Disorders Branch, National Institute of Mental Health*)

a specific protein called dopamine, which is a neurotransmitter that facilitates communication between neurons; other brain chemicals have also been implicated.

Maternal infection may lead to aberrant effects in the fetal brain by two modes of transmission. The first involves passage of the infection across the placenta and into the fetal brain through the immature blood–brain barrier; in this case, the infection itself causes damage directly. The second involves indirect transmission. This refers to an effect produced by the mother's physiological response to infection, which results in a disruption to fetal brain tissue. Cytokines are a type of signaling molecule released following many different types of infections that are known to disrupt brain development in a fetus in infected pregnant mothers. Cytokine effects provide evidence for the theory that a common mechanism is responsible for all infectious damage to the fetal brain, regardless of the individual type of infection. Another indirect mechanism—autoantibodies—produced in response to infection can attack the fetal brain tissue instead of attacking the infection. These and other mechanisms are depicted in **Fig. 2**.

Implications. Public health knowledge of maternal and reproductive health is improving as epidemiologic studies become more sophisticated.

This is being made possible in part by research models that include rigorously measured biological indicators of maternal infection, larger samples of cases, additional data to better understand the mechanisms that give rise to schizophrenia, and genetic material to reveal the role of interactions between infections and genes in this disorder. These developments will help increase awareness, treatment, and prevention of maternal infections, potentially decreasing the incidence rate of schizophrenia in the coming years. Despite accumulating evidence, though, no definitive causes for schizophrenia have yet been identified. Even in the most ambitious current studies, the majority of exposed mothers did not produce offspring with schizophrenia. This suggests that infection does not operate alone to increase schizophrenia risk, but rather that additional susceptibility factors (for example, genetic mutations) are also necessary. Nonetheless, in view of the fact that maternal infection causes many additional known diseases in the offspring, more rigorous adherence to current recommended precautions to minimize exposure to potentially harmful infections and other risk factors during pregnancy is strongly advocated.

For background information *see* BRAIN; CYTOKINE; DOPAMINE; EPIDEMIOLOGY; INFECTION; INFLUENZA; NEUROBIOLOGY; PREGNANCY; PREGNANCY DISORDERS; PRENATAL DIAGNOSIS; PSYCHOSIS; SCHIZOPHRENIA in the McGraw-Hill Encyclopedia of Science & Technology.　　　Havalyn Logan Arader; Alan S. Brown

Bibliography. A. S. Brown, Prenatal infection as a risk factor for schizophrenia, *Schizophr. Bull.*, 32(2):200–202, 2006; A. S. Brown et al., Elevated maternal interleukin-8 levels and risk of schizophrenia in adult offspring, *Am. J. Psychiat.*, 161(5):889–895, 2004; I. I. Gottesman and S. O. Moldin, Schizophrenia genetics at the millennium: Cautious optimism, *Clin. Genet.*, 52(5):404–407, 1997; S. R. Hirsch and D. R. Weinberger, *Schizophrenia*, Blackwell Science, Malden, MA, 2003; K. S. Kendler and C. A. Prescott, *Genes, Environment, and Psychopathology: Understanding the Causes of Psychiatric and Substance Use Disorders*, Guilford Press, New York, 2006.

Science on the International Space Station

The year 2008 marked a significant milestone in the construction of the *International Space Station (ISS)* with the addition of two new laboratory complexes intended to greatly expand the resources available on-orbit for scientific experimentation and observation. Both the Kibo complex built by the Japan Aerospace Exploration Agency (JAXA) and the Columbus Orbiting Facility (COF) produced by the European Space Agency (ESA) will add to the existing capabilities of the U.S. Destiny laboratory module to provide a platform for investigations in such diverse fields as astrophysics, biology, material sciences, and Earth climate observation. With these additions the *ISS* is now finally able to realize its original vision as a long-duration, general-purpose laboratory

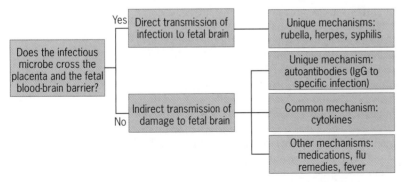

Fig. 2. Mechanisms by which maternal infection may alter fetal brain development.

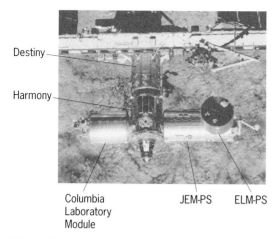

Destiny

Harmony

Columbia   JEM-PS   ELM-PS
Laboratory
Module

Fig. 1. *ISS* international partner laboratory modules.
(*Courtesy of NASA*)

for performing scientific experiments requiring human interaction in a space environment for the benefit and advancement of humankind.

Figure 1 shows the configuration of the *ISS* laboratory modules at the end of space shuttle mission STS-124 in June 2008. At the bottom center is the Harmony connecting node with the Destiny laboratory located attached above it. The ESA Columbus facility is located on the left side, and on the right is the Japanese Experiment Module-Pressurized Module (JEM-PM) with the Experiment Logistics Module-Pressurized Section (ELM-PS) berthed on top.

Columbus Laboratory Module. Launched aboard space shuttle mission STS-122 on February 7, 2008, the Columbus Laboratory Module is ESA's primary contribution to *ISS*-based scientific research and marks their first return to crewed space since the completion of the Spacelab program in the late 1990s. On flight day four, Columbus was unberthed from the orbiter's payload bay and attached to the starboard port on the *ISS* Harmony node. Power, data, and cooling lines were connected, and the laboratory was successfully activated the following day. Weighing 10.3 metric tons (22,700 lb) and with a pressurized volume of 75 m^3 (2650 ft^3), this facility is the smallest of the *ISS* laboratories and is capable of accommodating up to 16 equipment racks arranged as four rows of four racks each around the circumference of the main cylinder section. To reduce costs and ensure high reliability, the basic design of the laboratory is based on the structures built for the ESA-supplied Multi Purpose Logistics Modules (MPLMs). [The MPLMs are large, reusable pressurized modules used by the shuttle to ferry up to 7.5 tons (16,500 lb) of scientific experiments, consumable supplies, and spare parts to the *ISS*. ESA built three MPLMs for NASA, and they have flown a total of eight missions since 2001.] The Columbus laboratory was designed for a life span of at least 10 years (**Fig. 2**).

Of the 16 rack locations in Columbus, 10 are outfitted for accommodation of International Standard Payload Racks (ISPRs), which provides a set of common interfaces for power, cooling, video, and high-rate data. Each ISPR provides a framework for housing payloads and experiments with a mass of up to 700 kg (1550 lb). At the time of launch, Columbus was outfitted with five experiment racks that were activated shortly after installation of the laboratory on the *ISS*:

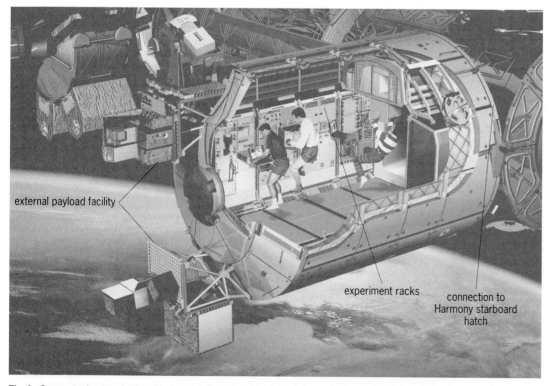

external payload facility

experiment racks

connection to Harmony starboard hatch

Fig. 2. Conceptual cutaway, showing interior of the Columbus Laboratory Module. (*Courtesy of ESA*)

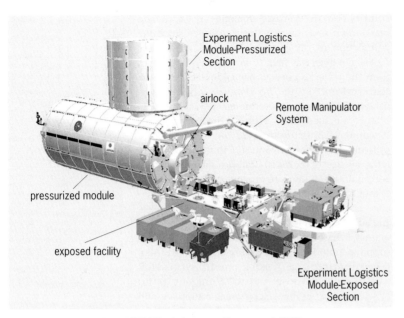

Experiment Logistics
Module-Pressurized
Section

airlock

Remote Manipulator
System

pressurized module

exposed facility

Experiment Logistics
Module-Exposed
Section

Fig. 3. **Components of the JAXA Kibo Laboratory.** (*Courtesy of JAXA*)

1. Biolab, a facility supporting experiments on microorganisms, cell and tissue cultures, and small plants and animals.

2. The Fluid Science Laboratory, a facility for studying the complex behavior of fluids in microgravity with the purpose of exploring improvements in energy production and propulsion efficiency.

3. The European Physiology Modules Facility, a platform for performing experiments in human physiology in a space environment including bone loss, circulation, respiration, and immune system response.

4. The European Drawer Rack, providing a flexible experiment carrier for a large variety of scientific disciplines housed in standard-sized drawers and lockers.

5. The European Transport Carrier, which accommodates transfer and stowage of experiment instruments and samples housed in the standard *ISS* Cargo Transfer Bags.

In addition to the experiment racks inside the pressurized module, ESA has equipped the Columbus module with an External Payload Facility (EPF), which provides four locations for accommodation of instruments that are required to be exposed directly to the environment of space. Consisting of a structural framework mounted on the module's end cone, it provides similar power, command, and data interfaces to externally mounted equipment that is available to ISPRs. Two science payloads were launched on a separate Integrated Cargo Carrier and installed by astronauts on the EPF during the third spacewalk of the STS-122 mission:

1. The European Technology Exposure Facility (EuTEF), a programmable, fully automated, multiuser facility with modular and flexible accommodation for a variety of technology payloads. EuTEF is specifically designed to facilitate the rapid turnaround of experiments and will contain nine different instruments for its initial configuration on orbit. The EuTEF was returned to Earth during the STS-128 space shuttle mission in September 2009.

2. A Sun-monitoring (SOLAR) facility. Located on the side of the EPF directly opposite the Earth, SOLAR is an instrument platform mounted on a set of gimbals such that it can be aimed at any point on the Sun. It was originally launched with a set of three spectrophotometers for measuring and analyzing radiation emitted by the Sun from the near infrared to the extreme ultraviolet portions of the spectrum.

Once initial activation of the Columbus laboratory was completed, the work of coordinating the operation of the facility systems and conducting scientific experiments between the science community, flight crew, and ISS Mission Control was assumed by ESA's Columbus Control Center (Col-CC). The center is situated at the German Aerospace Center (DLR) facility in Oberpfaffenhofen, near Munich, Germany, and serves as the direct link to the Columbus laboratory for on-orbit operations.

Kibo. Japan's contribution to the *ISS*, and the culmination of 23 years of design and development, the Kibo laboratory complex represents Japan's first human-rated space facility and is the largest research module on the *ISS*. Kibo, which means "hope" in Japanese, is built and operated by the Japan Aerospace Exploration Agency and is composed of five major components that required three separate space shuttle missions to transport them to the *ISS* (**Fig. 3**).

Experiment Logistics Module–Pressurized Section (ELM-PS). This module provides transportation and storage space for experiments, instruments, samples, spare parts, and consumables used on-board Kibo. The ELM-PS can be berthed to either the ISS Harmony connecting node or to a port on the main Kibo laboratory segment allowing astronauts to transfer equipment between modules in a shirt-sleeve environment. Capable of transporting up to eight standard *ISS* equipment racks, the ELM-PS was the first piece of the Kibo complex launched into orbit aboard space shuttle STS-123 on March 11, 2008. For that mission it was loaded with five Kibo subsystem racks, two experiment racks, and one stowage rack. These were later relocated into the Kibo module after it was installed on the *ISS* during the following mission. The ELM-PS was originally intended to be returned to Earth in the orbiter's payload bay so that it could be flown multiple times. However, with the upcoming termination of the space shuttle program in 2010, the decision was made to leave the ELM-PS installed on the JEM Pressurized Module to serve as a permanent storage facility.

Japanese Experiment Module–Pressurized Module (JEM-PM). At 11.2 m (37 ft) in length, 4.4 m (14.5 ft) in diameter, and weighing almost 16 metric tons (35,000 lb) when fully loaded with equipment racks, the JEM-PM is about the size of a large tour bus. It can accommodate up to 23 racks of which 10 can be International Standard Payload Racks. An active Common Berthing Mechanism (CBM) is located on the top side of the module for attaching the ELM-PS. A small airlock is built into one end of the module through which experiments or spare parts can be transferred between the JEM pressurized module and the Exposed Facility. Two large windows just above the airlock allow astronauts to view the Exposed Facility for observation of unpressurized payloads and monitoring operation of the JEM-RMS. The JEM-PM was launched into orbit on June 14, 2008, as the primary payload on mission STS-124, and is berthed on the port side of the *ISS* Harmony module.

Japanese Experiment Module–Remote Manipulator System (JEM-RMS). This is a pair of robotic arms designed to manipulate experiments and perform maintenance tasks on the Kibo Exposed Facility. Both arms have six independent joints patterned after those of the human arm to provide maximum dexterity. At 10 m (33 ft) in length, the Main Arm is primarily used for grappling and positioning large objects and was integrated into the JEM-PM at the Kennedy Space Center. The Small Fine Arm was launched along with the Exposed Facility and is capable of more precise movements. The JEM-RMS arms are operated by the flight crew from a control console in the Kibo laboratory aided by remote television cameras mounted on the ends of each arm.

Japanese Experiment Module–Exposed Facility (JEM-EF). This is a multipurpose experiment platform on which various scientific experiments can be attached for exposure to the microgravity and vacuum environment of space. Since its installation on the end of the JEM-PM module, up to twelve separate payloads can be mounted on the platform using automated attachment mechanisms called Equipment Exchange Units. These are used to supply standard services to experiments such as power, data, and cooling, and provide a means by which payloads can be easily changed out on-orbit.

Experiment Logistics Module–Exposed Section (ELM-ES). This is a logistics carrier designed to be launched and returned on board the space shuttle. It can be used to transport up to three Exposed Facility payloads to the *ISS*. The ELM-ES will be attached to the EF while transferring payloads to the facility in order to provide the best access for robotics transfer operations with Kibo's robotic arm. Both the JEM-EF and ELM-ES were successfully delivered to the space station as a part of shuttle flight STS-127 in July, 2009. The following three payloads were carried on the ELM-ES and transferred to the EF during this mission, and the empty ELM-ES was returned to Earth:

1. The Monitor of All-Sky X-Ray Image (MAXI), a pair of cameras to scan the x-ray spectrum of the entire sky visible from the EF every 90 min.

2. The Space Environment Data Acquisition Equipment-Attached Payload (SEDA-AP), to measure the space environment (neutrons, plasma, heavy ions, high-energy light particles, atomic oxygen, and cosmic dust) in the station's orbit.

3. The Inter-orbit Communication System–Exposed Facility (ICS-EF), the Kibo-specific communications system for uplink and downlink data, images, and voice data between Kibo and the JAXA's Tsukuba Space Center.

All Kibo operations are monitored and controlled from the Mission Control Room of the Space Station Integration and Promotion Center at the Tsukuba Space Center in Japan. The JAXA Flight Control Team works in close cooperation with NASA's Mission Control Center and Payload Operation Integration Center for monitoring and controlling Kibo systems and Japanese experiments onboard Kibo, developing and implementing operation plans, and supporting launch preparations.

In preparation for continuing operation of the Kibo complex once the space shuttle program has ended, JAXA has been developing their own resupply vehicle to be launched on the Japanese H-IIB booster called the H-II Transfer Vehicle (HTV). This will have both a Pressurized Logistics Carrier that will be docked to the zenith port of the *ISS* Harmony module, and an Unpressurized Logistics Carrier containing Exposed Pallets that can be mounted either to the JEM-EF or to attachment points on the U.S. segment. Once they have finished their intended use, the PLCs, ULCs, and EPs would be jettisoned and burned up during atmospheric reentry. The first flight of the HTV was launched on September 11, 2009.

ISS achieves 6-person crew capability. Science research aboard the *ISS* was really able to get fully underway after delivery of the fourth and final set of solar arrays during the STS-119 space shuttle mission, launched on March 15, 2009. The addition of the Starboard 6 (S6) truss segment increased the electrical generation capacity of the space station to 120 kW and doubled the amount of power available for science payloads to 30 kW. This cleared the way for the launch of the Expedition 20 crew aboard a Russian Soyuz spacecraft from the Baikonur Cosmodrome in Kazakhstan on May 27, 2009. Enlarging the complement of full-time astronauts from three to six finally allowed for sufficient crew time to be devoted to conducting experiments rather than be limited strictly to operating and maintaining the *ISS* systems. Expansion of research facilities aboard the space station is expected to continue in 2010 with the widely anticipated launch of the Russian Multipurpose Laboratory Module (MLM). This additional capability is expected to supplement the Columbus and Kibo facilities for ushering in an unprecedented era of international cooperation in space-based scientific research.

For background information *see* ROCKET PROPULSION; SPACE BIOLOGY; SPACE POWER SYSTEMS; SPACE SHUTTLE; SPACE STATION; SPACE TECHNOLOGY; WEIGHTLESSNESS in the McGraw-Hill Encyclopedia of Science & Technology.   Michael Peacock

Bibliography. D. Barry and K. Wa, Kibo, *Air and Space Magazine*, May 1, 2008; F. Morring, Columbus lab attached to space station, *Aviat. Week Space Tech.*, February 11, 2008; I. Oei and C. Mirram, Europe's human research experiments integration on the International Space Station, *Acta Astron.*, 63(7–10):1126–1136, October-November 2008; N. Peter and R Delmotte, Overview of global space activities in 2007/2008, *Acta Astron.*,

65(3–4):295–307, August-September 2009; J. A. Robinson et al., International Space Station research—Accomplishments and pathways for exploration and fundamental research, AIAA 2008-799, *46th AIAA Aerospace Sciences Meeting*, Reno, Nevada, January 7–10, 2008.

Ship waves in the atmosphere

Waves in the atmosphere and oceans can have a wide variety of configurations and propagation characteristics, but one of the most distinctive is the so-called atmospheric ship-wave pattern that is commonly observed in the lee (downwind side) of flow over and around isolated obstacles. An example of several atmospheric ship waves in a high-resolution visible satellite image is shown in **Fig. 1**. These waves were generated by air flow over the South Sandwich Island chain in the South Atlantic, which extends roughly 240 mi (400 km) in the north–south direction. The same general structure of the lee wave pattern behind each island is obvious; each pattern is confined to a wedge or V-shaped region, with maximum amplitudes near the legs of the V with little attenuation downstream.

Water ship waves. The term ship wave derives from analogy to the pattern of waves behind a moving ship. These waves travel with the ship; that is, they are stationary in a coordinate frame moving with the ship. In relatively deep water, this wave pattern consists of two sets of waves: a transverse wave set, with waves aligned approximately normal to the direction of motion of the ship; and a diverging wave set that appears to extend out radially behind the ship at an angle to the direction of motion of the ship (**Fig. 2**). This wave pattern was first successfully explained in 1891 by Lord Kelvin (William Thompson) and is therefore often referred to as the Kelvin ship-wave pattern. Kelvin's analysis correctly predicted all the essential features of the observed deep-water ship-wave pattern, namely, the existence of both diverging and transverse wave sets that are confined to a fixed angle or wedge of half-width approximately $19.5°$. (A schematic of the pattern of wave crests is shown in **Fig. 3**.) The maximum wave amplitudes occur along the wedge line where the transverse and diverging waves coalesce, and decrease exponentially away from and outside of the wedge line. The amplitudes also depend on the shape of the ship; wide ships produce larger-amplitude transverse waves, whereas narrow ships produce larger-amplitude diverging waves. The diverging wave length is highly spatially dependent, but the transverse wave length is nearly constant and depends only on the ship speed. Physically, both waves are free-surface gravity waves; that is, gravity is the restoring force for the wave motion. By changing the reference frame to move with the ship, it is apparent that the same wave pattern will be generated for uniform fluid flow moving over an isolated obstacle, such as a rock in a stream.

These arguments must be modified somewhat for ships traveling on water of finite depth. In that case, the diverging gravity wave set, similar to the Mach cone in supersonic flow, is still present, because the waves can always align themselves at some angle to remain stationary, but the transverse wave set can only remain stationary if the ship speed is less than the maximum gravity-wave phase speed (that is, $U < \sqrt{gH}$, where U is the ship speed, g is gravity, and H is the water depth). Thus, the wave pattern for a fast-moving ship in shallow water (where the Froude number, $F = U/\sqrt{gH} > 1$) consists only of diverging waves. In cases where $F < 1$, both diverging and transverse waves are present but the wedge angle containing them is variable, increasing from $19.5°$ for $F = 0$ to $90°$ for $F = 1$. Also, the transverse wavelength no longer depends only on ship speed, but now depends on both U and H. All these features can be seen for the various boats moving at different speeds in Fig. 2. Some have only a diverging wave pattern; others have both transverse and diverging waves, and the wedge angle is different for different boats. For example, the very fast moving jet ski in the

Fig. 1. Example of atmospheric ship waves over the South Sandwich Islands in the South Atlantic, as observed by the Moderate-resolution Imaging Spectroradiometer (MODIS) instrument on NASA's *Aqua* satellite, on January 27, 2004. (*NASA, http://visibleearth.nasa.gov*)

center-right of Fig. 2 produces only diverging waves and is relatively wide compared to the other slower moving ships.

Atmospheric ship waves. Atmospheric ship waves in the lee of isolated obstacles (also called lee waves) have properties entirely analogous to finite-depth water ship waves. The atmospheric ship-wave pattern is more complicated, in part because the lee waves are generated in a continuously stratified atmosphere, as opposed to a ship wave, where the wave pattern is confined only to regions near the surface. In the atmosphere, the role of finite depth in regulating the ship-wave pattern on the water surface is replaced by the vertical distribution of wind (U) and stability (N, proportional to the vertical temperature gradient) in the vicinity of the obstacle. In particular, transverse waves may or may not exist, depending on the vertical structure of the ratio of the wind to stability (the so-called Scorer parameter, $\ell^2 = N^2/U^2$) in the vicinity of the obstacle. In general, gravity waves produced by flow over topography can propagate both horizontally and vertically, losing energy and decreasing in amplitude as they propagate away from the obstacle. However, when ℓ^2 decreases sufficiently rapidly with height, vertical propagation is inhibited and the waves are "trapped" or "ducted" in finite layers. These conditions allow the formation of transverse waves, which decay very slowly downstream from the forcing obstacle. Also analogous to finite-depth water ship waves, atmospheric transverse waves are favored when the flow is over wide-breadth mountain ranges, rather than over isolated narrow-breadth obstacles.

These lee-wave structures may be visible from the ground or from space when the water vapor concentration (humidity) is sufficient to allow cloud formation in the rising parts of the wave pattern. However, lee waves can be present even when the humidity is too low to form clouds in the wave crests. **Figure 4** shows an example of a predominately transverse lee wave pattern in dry air as observed in satellite infrared imagery centered on a water-vapor emission band.

Because of the complex nature of atmospheric temperature and wind structures and terrain forcing, three-dimensional atmospheric ship-wave patterns can be extremely varied. For example, in Fig. 1, even though the meteorology in the vicinity of each island is similar, the wave patterns behind each island have obvious differences, especially in the structure of the diverging wave components. Further, the vertical structure of the atmosphere can be highly variable, leading to the possibility of multiple trapping layers, resulting in different transverse wave amplitudes and wavelengths at different altitudes. In addition, if the wind direction changes with height, the wave pattern may rotate accordingly. The shape of the obstacle and the orientation of the wind relative to the obstacle also affect the wave pattern. If the obstacle height is large enough, or, more precisely, if the internal Froude number $F = U/Nh$ is small enough (where h is the obstacle height), the flow in the lee can become dominated by shed vortices (often

Fig. 2. Photograph taken by the author of ship-generated waves produced by several small boats in Long Beach, California, harbor.

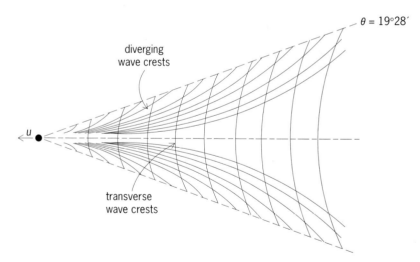

Fig. 3. Schematic of Kelvin's classical analysis of deep-water ship waves behind a ship (dot) moving at a constant speed U. Diverging and transverse wave patterns are confined to a wedge of half-width $\theta = 19°28'$.

called Kármán vortex streets), or turbulent wakes, destroying the coherence of the wave pattern. **Figure 5** shows an example of patterns produced by two islands in the Crozet Island archipelago in the southern Indian Ocean. The westernmost island is smaller and produces a familiar ship-wave pattern, while the taller island to the east (with a lower internal F) produces a string of vertically oriented vortices. Another complication is that ship-wave patterns produced by individual obstacles may overlap and interfere (this can be seen in both Figs. 1 and

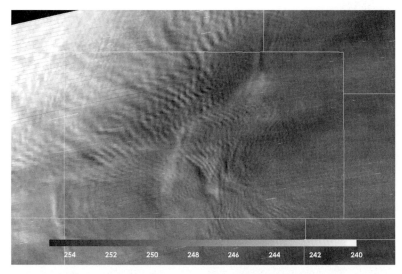

Fig. 4. Example of lee waves over Colorado, observed in the water-vapor channel of the MODIS instrument on October 3, 2005, 1945 UTC. (*Courtesy Kris Bedka, CIMSS, University of Wisconsin-Madison*)

Fig. 5. Atmospheric ship-wave pattern and wake vortices behind two islands in the Crozet Island Archipelago in the southern Indian Ocean, as observed in MODIS visible satellite imagery on November 4, 2004. (*NASA, http://visibleearth.nasa.gov*)

4). In regions of constructive interference, wave amplitudes may become large enough to initiate wave breaking and turbulence.

The ship-wave pattern is generally possible in any system in which the wave propagation is dispersive (that is, the wave speed depends on the wavelength) and there is a point-source disturbance. As discussed above, this is the case for free-surface water ship waves and atmospheric gravity waves forced by flow over isolated obstacles (such as an island or an overshooting convective tower atop a thunderstorm) in the atmosphere and oceans, and for inertio-gravity waves (where rotational effects are significant). It is also the case for lightning-generated "whistlers" (audio-frequency electromagnetic waves that propagate along the Earth's magnetic field lines).

For background information *see* ATMOSPHERE; FROUDE NUMBER; KÁRMÁN VORTEX STREET; VORTEX; WAKE FLOW in the McGraw-Hill Encyclopedia of Science & Technology. Robert Sharman

Bibliography. R. A. Houze, Jr., *Cloud Dynamics*, Academic Press, 1993; J. Lighthill, *Waves in Fluids*, Cambridge University Press, 1978; Y.-L. Lin, *Mesoscale Dynamics*, Cambridge University Press, 2007; R. S. Scorer, *Dynamics of Meteorology and Climate*, John Wiley & Sons, 1997.

Silk and the earliest spiders

Spiders are familiar animals whose intricate webs have been marveled at by humans for thousands of years. Silk production is not unique to spiders; indeed, silk is made by arthropods as diverse as silkworms, the spectacular glowworms of New Zealand, and other arachnids including pseudoscorpions. All spiders produce silk, and the possession of silk glands in the abdomen (opisthosoma) is a characteristic feature of the arachnid order Araneae. Other characters of spiders include venom glands in the forepart of the body (prosoma), which emerge from a pore in the cheliceral fang, and the presence of hairless fangs separates spiders from closely related arachnids such as Amblypygi (tailless whip scorpions). During spider evolution, not all of these characters appeared at the same time. For example, the most primitive living spiders, members of the suborder Mesothelae, lack venom glands and hence have no pore in the fang. All other spiders (the suborder Opisthothelae) possess venom glands, except rarely when they have been secondarily lost. New evidence from 380-million-year-old fossil spiderlike animals shows that they lacked venom glands. However, like true spiders, they had silk glands and spigots. The weaving apparatus was different, though, and thus we can speculate on how the silk was used. Arachnologists can now see more clearly the evolutionary pathway of silk production and silk use in early arachnids.

Silk. Spider silk consists of proteins together with many different organic and inorganic components, including neurotransmitter peptides, glycoproteins, lipids, sugars, phosphates, calcium, potassium, and sulfur. There are many different kinds of spider silks, each with its own mode of formation and function, even within a single species. Simple dragline silk is composed of a liquid-crystal protein complex and is remarkable for its strength and elasticity; it is produced constantly by running and jumping spiders, and is the main structural silk in webs. Cribellate silk (produced from the cribellum, which is a specialized, flattened spinning organ) has an outer layer of extremely fine strands that can entrap insect legs and hairs like burrs on sheep's wool. The capture spiral of araneoid orb webs (such as those of garden spiders) is covered with glue droplets to which insects adhere. Other silks are modified for use in lining burrows, entrance doors, trip lines, egg sacs, sleeping bags, sperm webs, parachuting or ballooning,

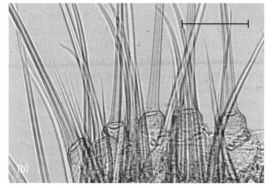

Fig. 1. Spinnerets and spigots of extant mesothele spiders. (a) Anterior lateral spinneret of *Liphistius* showing spigots. **(b)** Spigots of *Heptathela*. Scale bar = 50 μm.

wrapping prey, hibernating and molting chambers, retreats, and nursery webs for spiderlings.

Different types of glands produce different types of silk; for example, in araneoid orb weavers, one type of gland produces dragline silk, another provides the core of the threads of the capture spiral, and the glue coating is supplied by a third gland type. Spider silk glands consist of a chamber that leads to a duct that emerges to the outside through a spigot. Spigots are hollow hairs (setae), often modified in some way. Simple spigots (**Fig. 1**) appear different from normal setae by their bell-shaped bases. In modern spiders, the spigots are found on spinnerets, which are modified opisthosomal appendages. One exception to this is the row of spigots (called fustules) along the anterior edge of the epigastric furrow (the posterior edge of opisthosomal segment 2) in adult male spiders. This silk, from the epigastric glands, is used to make a tiny web on which the male spider deposits his sperm before charging his palps with it to inseminate the female.

The spinnerets of spiders are unusual in being opisthosomal appendages, which all other arachnids have lost. The spinnerets can be shown embryologically to represent homologs of biramous appendages (that is, appendages having two branches) of opisthosomal segments 4 and 5. The primitive chelicerate *Limulus* shows biramous appendages in this position in the form of a segmented median branch and a lateral branch with a plate covering lamellate gills. The most primitive spiders, the mesotheles, have a maximum of eight spinnerets, with two pairs

on opisthosomal segments 4 and 5, called anterior and posterior medians and laterals; even in living mesotheles, though, the anterior medians are nonfunctional. The anterior medians, at least, are missing in mygalomorph opisthotheles (tarantulas, and funnel-web and bird-eating spiders), and they are either modified to a cribellum or absent in araneomorph opisthotheles.

Webs. Most people tend to think of spiders as using silk to make webs for prey capture, but the more primitive kinds of spiders, in the suborder Mesothelae, live in burrows. They use silk for wrapping eggs, lining the burrow entrance, weaving the entrance door, and making radiating trip wires that the spider uses to detect passing prey. Mesotheles are known as fossils from as old as the Pennsylvanian period [ca. 300 million years ago (mya)], by which time insects had already taken flight, and the aerial prey-capture web is thought to have developed to follow the insects into the air to harvest this vast food resource. There are many different kinds of prey-capture webs. For example, a sheet web, as used by the Agelenidae (funnel-web spiders), consists of a dense meshwork of fine strands stretched out from a retreat inside a bush. This kind of web captures grasshoppers and other jumping insects most effectively. When the prey struggles, the vibrations are transmitted to the spider, which runs out and immobilizes the insect. Orb webs are constructed in gaps in vegetation through which insects might fly. Orb webs can use cribellate silk or silk with glue (ecribellate). Uloboridae (the hackled-band orb weavers) make cribellate orb webs and araneoids make the ecribellate type, but both catch prey in the same way. An insect flying through what appears to be an empty space hits the web and sticks to it, and its flight velocity is quickly dampened by the nature of the structural silk. The spider senses the struggling insect, emerges from its retreat, and immobilizes the prey by injecting venom and/or wrapping it in silk. One of the problems with a static prey-capture web is that predators, such as small birds or spider-hunting wasps, develop a search image for a web and its associated occupant, regardless of how the spider attempts to conceal itself. Consequently, many spider families have abandoned the prey-capture web and stalk their prey by sight or other senses. This is especially noticeable in the most diverse of all spider families today, the Salticidae (jumping spiders), whose greatly enlarged anterior median eyes enable them to pinpoint prey and jump on it with great accuracy.

Origin and evolution of silk. Although it is clear that silk has been employed by spiders for many diverse uses, several theories have been put forward regarding its origin and early evolution. One idea is that it was used first for lining burrows to prevent their collapse, as is done by modern mesotheles. Similarly, a mesothele dashes out of its burrow to capture prey and then needs to be able to find its way back, so a silken trail would be useful for homing, which could later evolve into a trip wire. Silk is used in many ways during reproduction, for example, in sperm webs

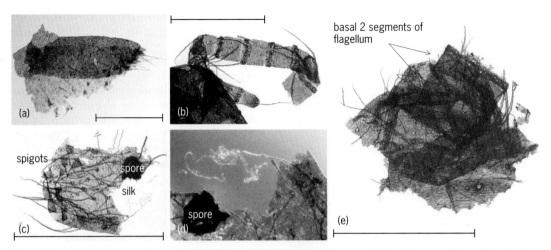

Fig. 2. *Attercopus fimbriunguis*, Devonian of New York. (*a*) First-described "spinneret"; the darkness of the cuticle reflects the number of layers, so the fragment is folded over twice. (*b*) Flagellar structure with 12 segments (including the possible distalmost one) from the original Gilboa locality; segments show distal collars and setae. (*c*) Piece of cuticle from a corner of the opisthosomal ventral plate showing setae, spigots, and a possible silk strand. (*d*) Close-up of panel *c* showing the possible silk strand emerging from the spigot shaft. (*e*) Two flagellar segments emerging from the posterior part of the opisthosoma. Scale bars = 0.5 mm.

and egg sacs, which has led to suggestions that this was its primary function.

Of course, it is unlikely that the silk glands, spigots, and spinnerets appeared together during evolution, so which came first? It is apparent from the discovery of fossils showing spigots arranged along the posterior edge of an opisthosomal plate (**Fig.** 2*c*) that the silk glands and spigots developed before the spinnerets appeared. Therefore, it seems likely that the genetic mechanism for the production of a biramous appendage, which had been suppressed in the ancestors of Araneae, was switched back on when the advantage of having spigots on appendages—the maneuverability for weaving webs—became selected for. A strand of material apparently issuing from a spigot (Fig. 2*d*) may be the oldest fossil occurrence of silk, although it would be difficult to analyze chemically. Silken strands occur alongside the earliest spiders in amber of Cretaceous age (ca. 140 mya).

Early fossil arachnids. In the 1980s, paleobotanists macerating Devonian (ca. 380 mya) shales from Gilboa, New York, for fossil plant cuticles found tiny fragments of animals: the oldest land animals from North America. One specimen appeared to be a nearly complete spider spinneret bearing about 20 spigots (Fig. 2*a*). On the basis of the simple spigot type and the lack of tartipores (vestigial spigots from earlier molts), the fossil spinneret was compared most closely with posterior median spinnerets of mesotheles. The distinctiveness of the cuticle enabled association of the spinneret with remains previously referred tentatively to another kind of arachnid: a trigonotarbid. Detailed study of this material resulted in a fuller description of the animal, *Attercopus fimbriunguis*, which was at that time thought to be the oldest known spider. The appendicular morphology of *Attercopus*, but little of the body, is now known in great detail. A number of other arachnids, including other trigonotarbids, were described alongside *Attercopus* from Gilboa,

including some enigmatic fragments that resembled multisegmented flagella (Fig. 2*b*).

Later collections made in the 1990s at South Mountain, New York, yielded more *Attercopus* material, including some specimens with spigots. With more specimens available for study, it became clear that the original spinneret was actually a rolled-up piece of cuticle and that the spigots were arranged in two rows along the edge of plates. These are ventral opisthosomal plates, features that occur in relatives of Araneae, such as trigonotarbids and Amblypygi, but which have been lost in true spiders.

The fossil record of spiders is sparse. For example, they were unknown from strata of the Permian period (ca. 300–250 mya) until 2005. Then, *Permarachne*, from the Urals of Russia, was described as an unusual mesothele because it appeared to have a long structure emerging from the posterior of its opisthosoma: an anal flagellum (**Fig. 3**). Moreover, while the main part of the fossil showed the ventral side, there were plates on the opisthosoma. Because nothing like it was known from mesothele spiders, the flagellum was interpreted as a long spinneret (the other spinnerets were presumed to be absent in the fossil) and the opisthosomal plates were presumed to be dorsal.

Finally, comparing *Permarachne* with the new interpretation of *Attercopus*, it became clear that forcing these animals into a modern definition of spiders did not work. One specimen proved that the structures found with *Attercopus* are really anal flagella, as is the supposed spinneret of *Permarachne*, and the plates on *Permarachne* are ventral, not dorsal (Fig. 2*e*). Thus, these animals form the basis of the new arachnid order Uraraneida ("tailed spiders"). The spigot location in *Attercopus* suggests that the original use of silk in these protospiders was to produce sheets, perhaps used as burrow linings, as homing trails, or to cover egg masses. The evolution of spinnerets, which would allow more precise weaving of webs, by some mutation that switched back on

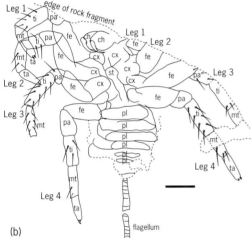

Fig. 3. *Permarachne*, Permian of Russia. (*a*) *Permarachne* in rock matrix. (*b*) Explanatory drawing of panel *a*. Abbreviations: ch, chelicera; cx, coxa; fe, femur; mt, metatarsus; pa, patella; pl, ventral plate; st, sternum; ta, tarsus; ti, tibia. Scale bar = 1 mm.

the genes for opisthosomal appendages, occurred in their relatives, that is, the true spiders.

For background information *see* ARACHNIDA; ARANEAE; ARTHROPODA; FOSSIL; INSECTA; NATURAL FIBER; ORGANIC EVOLUTION; PREDATOR-PREY INTER-ACTIONS; SILK; SPIDER SILK in the McGraw-Hill Encyclopedia of Science & Technology.   Paul A. Selden

Bibliography. K. Y. Eskov and P. A. Selden, First record of spiders from the Permian period (Araneae: Mesothelae), *Bull. Br. Arachnol. Soc.*, 13:111–116, 2005; P. A. Selden, W. A. Shear, and P. M. Bonamo, A spider and other arachnids from the Devonian of New York, and reinterpretations of Devonian Araneae, *Palaeontology*, 34:241–281, 1991; P. A. Selden, W. A. Shear, and M. D. Sutton, Fossil evidence for the origin of spider spinnerets, and a proposed arachnid order, *Proc. Natl. Acad. Sci. USA*, 105:20781–20785, 2008; W. A. Shear (ed.), *Spiders: Webs, Behavior and Evolution*, Stanford University Press, 1986; F. Vollrath and P. A. Selden, The role of behavior in the evolution of spiders, silks, and webs, *Annu. Rev. Ecol. Evol. Syst.*, 38:819–846, 2007.

Sirtuins

Aging is influenced by complex interactions between genetic and environmental parameters. Sirtuins are a family of nicotinamide adenine dinucleotide (NAD)–dependent protein deacetylases (which remove acetyl groups from protein molecules) and adenosine diphosphate (ADP)–ribosyltransferases (which catalyze the transfer of the ADP-ribose moiety from NAD^+ onto specific substrates) that are thought to play an important role in modulating these interactions. Mammalian sirtuins have been recently implicated in a variety of aging-related processes, and it has been suggested that these enzymes may prove to be effective therapeutic targets for treating age-associated diseases in humans.

Function. Sirtuins are named for the founding member of the family, that is, the budding yeast Silent Information Regulator 2, Sir2. Sir2 is a histone deacetylase that promotes the formation of silenced chromatin at three cellular loci in yeast: the telomeres (the ends of chromosomes), the silent mating loci, and the ribosomal DNA (rDNA). [Note that chromatin is the deoxyribonucleoprotein complex forming the major portion of the nuclear material and of chromosomes; histones are a group of positively charged proteins that aid in compaction of the chromatin.] Modulation of the histone acetylation state is an important mechanism for regulating gene expression. Chromatin with highly acetylated histones tends to be permissive for messenger ribonucleic acid (mRNA) transcription, whereas removal of histone acetyl groups reduces transcription by creating a more compact chromatin structure. Yeast cells that lack Sir2 have defects associated with loss of transcriptional silencing at these loci, including sterility, reduced telomere length, genomic instability at the ribosomal DNA, and shortened life span.

The enzymatic activity of Sir2 differs from non-sirtuin deacetylases in that the reaction uses NAD^+ as a substrate and produces 2′-O-acetyl-ADP-ribose and nicotinamide as products (see **illustration**). Nicotinamide and NADH (the reduced or hydrogenated form of NAD) are both inhibitors of sirtuin activity and are believed to be biologically relevant regulators in vivo. Based on this unique catalytic mechanism, Sir2 is defined as a class III histone deacetylase. Sir2 also has ADP-ribosyltransferase activity, in which the ADP-ribose moiety of NAD^+ is covalently joined to the substrate protein, but this activity is believed to be relatively low in vivo. Most sirtuins studied thus far appear to have the capacity to catalyze both reactions under appropriate conditions in vitro.

The sirtuin family is highly conserved from yeast to mammals. Sirtuins have been classified into homology groups (I, II, III, and IV) based on their protein sequence similarity and presumed evolutionary divergence (see **table**). Yeast have five sirtuin proteins, that is, Sir2 and four homologs of Sir2, Hst1–4, all of which are group I sirtuins. The nematode *Caenorhabditis elegans* has four sirtuins, the fruit fly *Drosophila melanogaster* has five sirtuins, and both mice and humans have seven sirtuins

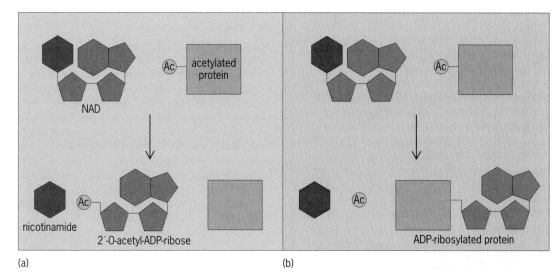

Catalytic activities of sirtuins. Sirtuins catalyze two distinct enzymatic reactions: (*a*) protein deacetylation and (*b*) ADP-ribosylation. Both reactions remove an acetyl group from a substrate protein while consuming NAD and producing nicotinamide. Sirtuin-dependent deacetylation results in the formation of 2′-O-acetyl-ADP-ribose and deacetylated substrate protein, whereas ADP-ribosylation covalently attaches ADP-ribose to the substrate protein at the former acetylation site.

Sirtuin homology groups by organism		
Class	Organism	Sirtuin
Ia	Yeast	Sir2
	Yeast	Hst1
	Nematode	SIR-2.1
	Fly	Sir2
	Mouse/Human	SIRT1
Ib	Yeast	Hst2
	Fly	Sirt2
	Mouse/Human	SIRT2
	Mouse/Human	SIRT3
Ic	Yeast	Hst3
	Yeast	Hst4
II	Nematode	SIR-2.2
	Nematode	SIR-2.3
	Fly	Sirt4
	Mouse/Human	SIRT4
III	Mouse/Human	SIRT5
IVa	Nematode	SIR-2.4
	Fly	dSirt6
	Mouse/Human	SIRT6
IVb	Fly	dSirt7
	Mouse/Human	SIRT7

(SIRT1–7). The mammalian sirtuins span all four homology groups. All sirtuins examined thus far have a similar enzymatic mechanism to that of Sir2; however, many sirtuins have multiple nonhistone substrates, and a few are thought to act primarily as mono-ADP-ribosyltransferases rather than as protein deacetylases (see illustration).

Role in aging. The importance of sirtuins in aging was first demonstrated in yeast, where overexpression of Sir2 caused an approximately 35% increase in mother-cell replicative life span, which is defined by the number of daughter cells produced before senescence. The mechanism by which Sir2 promotes longevity in yeast is believed to be via repression of recombination within the ribosomal DNA, which leads to the formation of extrachromosomal ribosomal DNA circles. These circles are known to accumulate with age in yeast mother cells and cause senescence. Subsequent studies in nematodes and flies have shown that increased activity of SIR-2.1 or Sir2 also increased life span in these multicellular organisms. Interestingly, ribosomal DNA circles do not appear to cause aging in either nematodes or flies, suggesting that sirtuins have evolved to promote longevity via different mechanisms in different species. There has been speculation, though, that yeast Sir2 may also promote longevity by a mechanism different from repression of ribosomal DNA circles, leaving open the possibility of a conserved molecular mechanism by which sirtuins modulate aging in simple eukaryotes. So far, only class Ia sirtuins have been shown to play a significant role in aging.

Dietary restriction, defined as reduced nutrient availability in the absence of malnutrition, increases life span in a variety of organisms, including yeast, nematodes, flies, mice, and rats. Based on their NAD-dependent catalytic activity, it has been proposed that sirtuins have evolved to serve as conserved metabolic sensors linking nutritional availability and aging. If this is the case, then this hypothesis may account for the apparently different mechanisms by which sirtuins slow aging in different species. However, the evidence supporting the hypothesis that sirtuins are activated by dietary restriction is mixed, with some studies reporting induction of sirtuins and others failing to observe any change in sirtuin expression or activity. This hypothesis has also been challenged in both yeast and nematodes, where dietary restriction has been shown to increase life span primarily by sirtuin-independent pathways. Hence, despite extensive experimental examination, it remains unclear whether sirtuins play a direct role in life-span extension from dietary restriction.

Mammalian SIRT1. Given the conserved longevity-promoting role of Sir2 orthologs in invertebrate species, there is substantial interest in understanding whether SIRT1 also influences aging in mammals. To investigate this, knockout gene technology has been employed, which inactivates a specific gene in a laboratory organism in order to study gene function. SIRT1 knockout mice have a shortened life span and reduced insulin sensitivity but do not show general features of accelerated aging. It has also been reported that SIRT1 knockout mice have reduced levels of oxidative damage in the brain, which is not predicted in a model of accelerated aging. Several transgenic SIRT1 mouse models have also been generated and examined for aging-associated phenotypes. Mice that overexpress SIRT1 specifically in pancreatic β cells show increased glucose-stimulated insulin secretion and improved glucose tolerance, and broad overexpression of SIRT1 is reported to confer protection against metabolic defects associated with a high-fat diet. Brain-specific overexpression of SIRT1 confers protection against neurodegeneration in a mouse model of Alzheimer's disease. Thus far, however, it has not been demonstrated that overexpression or activation of SIRT1 (or other sirtuins) is sufficient to increase life span in mice.

The idea that SIRT1 might function as a key regulator of cellular survival has been bolstered by several studies linking SIRT1 to important metabolic, stress resistance, and cell death (apoptosis) pathways. SIRT1 has been reported to deacetylate and regulate more than a dozen different protein targets, many of which have been implicated in longevity control in invertebrate organisms or in human disease. For example, SIRT1 is reported to promote enhanced stress resistance in mammalian cells by deacetylating various transcription factors that have homologs seen to promote longevity in *C. elegans*. The tumor suppressor p53 is another SIRT1 target that has been suggested to play a role in aging and age-associated disease and clearly regulates cell survival and cancer progression. Additional SIRT1 targets are also likely to contribute to the effects of SIRT1 in balancing cell death and survival in response to various types of damage.

SIRT1-mediated control of metabolism and energy expenditure has also been suggested as an important potential mechanism by which sirtuins might influence aging and disease in mammals. As mentioned above, SIRT1 knockout and transgenic mice have altered insulin sensitivity and glucose homeostasis. Additionally, SIRT1 is reported to inhibit adipogenesis (the formation of fat or fatty tissue) by deacetylating the peroxisome proliferator–activated receptor γ (PPAR-γ, which is a member of the nuclear hormone receptor superfamily of transcription factors and which responds to specific factors by altering gene expression), and SIRT1 regulation of the PPAR-γ coactivator PGC-1α is thought to modulate energy production and mitochondrial biogenesis, perhaps in response to nutrient availability. Regulation of PGC-1α by SIRT1 has been proposed as one mechanism by which SIRT1 could mediate metabolic effects associated with dietary restriction; however, as mentioned previously, the importance of SIRT1 in dietary restriction has yet to be firmly established.

Pharmacology. Much excitement has been generated recently regarding the therapeutic potential of sirtuin-activating compounds as "dietary restriction mimetics," which are hypothetical compounds capable of providing the health benefits of dietary restriction without requiring reduced caloric intake. The best-characterized sirtuin activator is resveratrol, which is a polyphenolic compound found in red wine. Resveratrol activates Sir2 orthologs in vitro in a substrate-specific manner and has been reported to confer phenotypes consistent with activation of SIRT1 in vivo. Resveratrol was initially reported to increase life span in yeast, nematodes, flies, and a short-lived species of fish (the turquoise killifish, *Nothobranchius furzeri*). Subsequent studies failed to reproduce life-span extension from resveratrol in yeast and flies, though, and showed sirtuin-independent life-span extension in nematodes. Resveratrol has also been reported to enhance survival and performance of mice fed a high-fat diet, which has led to speculation that dietary supplementation with resveratrol might have similar effects in humans. Somewhat disappointingly, however, a recent report showed that supplementation with resveratrol failed to enhance survival in mice fed a normal diet. Second-generation sirtuin activators are currently under development, some of which are structurally unrelated to resveratrol. Initial reports indicate that these compounds may provide similar protection against metabolic defects associated with a high-fat diet in mice.

The effects of pharmacological activation of sirtuins in humans remain unknown. If activation of SIRT1 slows aging in people, then sirtuin-activating compounds may be efficacious against a wide variety of age-associated diseases. Concern has been raised, however, that activation of SIRT1 may also promote cancer, based in part on its function as a negative regulator of the tumor suppressor p53. Sirtuin-activating compounds are now being tested in human clinical trials for safety and, if they are found to be safe, will likely be examined for efficacy against metabolic disease.

Future outlook. In the decade since overexpression of Sir2 was first shown to slow aging in yeast, much has been learned about the sirtuin family of enzymes. Based on their conserved longevity-promoting function in invertebrate organisms, it is reasonable to speculate that mammalian sirtuins play a similar role. Going forward, it will be important to untangle the complex interactions of SIRT1 with multiple important regulators of cell survival. There is also much to be learned about the functions of the other six mammalian sirtuins, which have been relatively uncharacterized. For now, there is reason to be optimistic that sirtuin activators might have significant health benefits in humans, and it will be of interest to follow their progression through clinical trials.

For background information *see* AGING; APOPTOSIS; CANCER (MEDICINE); CELL SENESCENCE;

CHROMOSOME; DEOXYRIBONUCLEIC ACID (DNA); ENZYME; ONCOLOGY; PROTEIN; TRANSCRIPTION; YEAST in the McGraw-Hill Encyclopedia of Science & Technology.　　　　　　Matt Kaeberlein

Bibliography. N. Dali-Youcef et al., Sirtuins: The "magnificent seven," function, metabolism and longevity, *Ann. Med.*, 39:335–345, 2007; L. Guarente, Mitochondria—A nexus for aging, calorie restriction, and sirtuins?, *Cell*, 132:171–176, 2008; M. Kaeberlein, The ongoing saga of sirtuins and aging, *Cell Metab.*, 8:4–5, 2008; M. Kaeberlein and R. W. Powers III, Sir2 and calorie restriction in yeast: A skeptical perspective, *Ageing Res. Rev.*, 6:128–140, 2007.

Sleep-dependent memory processing

The functions of sleep remain largely unknown, which is a surprising fact given the vast amount of time that this state takes from our lives. One of the most exciting hypotheses is that sleep contributes importantly to the brain plasticity that underlies learning and memory. Over the last decade, an abundance of evidence has emerged supporting

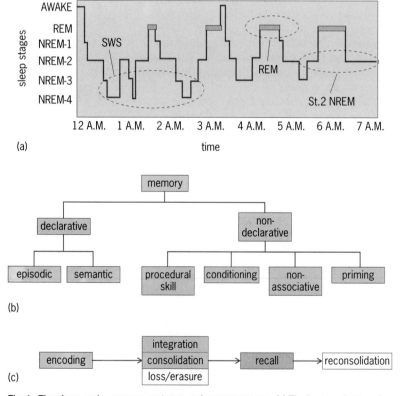

(a)

(b)

(c)

Fig. 1. The sleep cycle, memory systems, and memory stages. (a) The human sleep cycle. Across the night, rapid eye movement (REM) sleep and non-REM (NREM) sleep cycle every 90 minutes in an ultradian manner (that is, occurring in cycles that are repeated often), while the ratio of NREM to REM sleep shifts. During the first half of the night, NREM stages 3 and 4 slow-wave sleep (SWS) dominate, whereas stage 2 NREM and REM sleep prevail in the latter half of the night. (b) Memory systems. Human memory is most commonly divided into declarative forms, including episodic and semantic memory, and nondeclarative forms, including an array of different types (for example, procedural skill memory). (c) Developing stages of memory. Following the initial encoding of a memory, several ensuing stages are proposed, beginning with consolidation and including integration of the memory representation and even the erasure of memory. Also, following later recall, the memory representation is believed to become unstable once again, requiring periods of reconsolidation.

this role of sleep in what is becoming known as sleep-dependent memory processing. However, the question of whether sleep affects memory has proved to be a complex scientific challenge. Three core components lie at the heart of this conundrum: (1) memory systems, (2) memory stages, and (3) sleep stages (**Fig. 1**).

While it is often used as a unitary term, *memory* is not a single entity. Human memory involves a variety of different forms and corresponding locations in the brain, including conscious fact-based memory (for example, textbook learning) and more nonconscious skill memories (for example, learning to ride a bicycle). Just as memory is not monolithic, neither are the processes that create, sustain, and modify it. Instead, memories evolve in several distinct stages over time. We first form or "encode" a memory by engaging with an object or performing an action, leading to the development of a "memory representation" within the brain. Yet following encoding, memories continue to evolve through additional stages, including long-term "consolidation," "association/integration" with existing knowledge, and even the "erasure" of knowledge. Finally, sleep itself cannot be treated homogeneously, being divided into non-rapid eye movement (NREM) sleep and rapid eye movement (REM) sleep. NREM sleep has been further divided into substages 1 through 4, increasing in their depth, whereas REM sleep is a time of intense neural and mental activity (dreaming), with dramatic changes in brain chemistry and electrical activity accompanying these sleep cycles.

The issue, then, is to understand how these different stages of sleep influence the different stages of learning in different systems of memory. Based on a selection of recent cognitive and neuroimaging studies, three important results have emerged: (1) the essential need for sleep before learning so as to prepare the human brain for initial memory formation, (2) the critical need for sleep after learning for the subsequent plastic consolidation of memory, and (3) evidence that sleep not only strengthens individual memories, but actually integrates and flexibly associates them with one another—perhaps, the basis of human creativity.

Sleep before learning. While there is good evidence for the benefit of sleep after learning (see below), studies also indicate that there is a critical need for sleep *before* learning, that is, before the attempted formation of new memories. For example, recent studies have shown that subjects who are prevented from sleeping for one night demonstrated an inability to learn and acquire new facts such as lists of words or a series of pictures. This deficit ranges from 20% to 40% relative to participants who obtained a full night of sleep. These studies have also examined exactly what is failing within the brain and causing this memory deficit by placing participants inside a magnetic resonance imaging (MRI) brain scanner during the initial learning phase. Accompanying these performance impairments under conditions of sleep deprivation, and relative to a control group that slept, there was a selective deficit identified in the

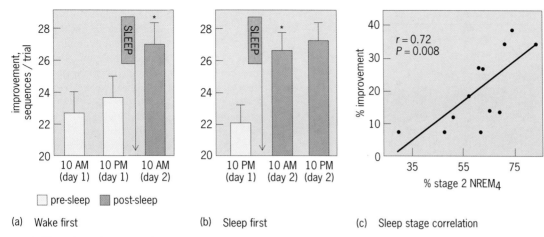

Fig. 2. Sleep-dependent motor skill learning. (*a*) Wake first. After morning training (10 A.M., clear bar), subjects showed no significant change in performance when tested after 12 hours of wake time (10 P.M., clear bar). However, when tested again following a night of sleep (10 A.M., solid bar), performance had improved significantly. (*b*) Sleep first. After evening training (10 P.M., clear bar), subjects displayed significant performance improvements just 12 hours after training following a night of sleep (10 A.M., solid bar), yet showed no further significant change in performance following an additional 12 hours of wake time (10 P.M., solid bar). (*c*) The amount of overnight improvement on the motor skill task correlated with the percentage of stage 2 NREM sleep in the last (fourth) quarter of the night (stage 2 NREM$_4$). Asterisks indicate significant improvement relative to training, and error bars indicate standard error of the mean.

hippocampus, which is a brain structure known to be critical for learning new fact-based information. Therefore, sleep deprivation appears to specifically impair those brain structures that are critical for the initial recording of new memories.

Sleep after learning consolidates memories. In addition to the requirement of sleep before learning, numerous studies have now described a proactive benefit of sleep *after* learning in the subsequent strengthening of newly formed memories. A collection of studies indicate that, for procedural skill memories, sleep (both REM and NREM, depending on the memory type) will not only stabilize memories, but actually facilitate their enhancement. As a consequence, memory performance following sleep is even better than that before sleep (**Fig. 2**). Moreover, these "offline" enhancements are associated with a neuroplastic reorganization of these memory representations within the brain. Such findings suggest that sleep may remodel the neural anatomy of skill memories at night, potentially transferring them into more efficient storage locations. As a consequence, this can improve access and hence recall the next day.

In addition to improving procedural skill memories, sleep has also been implicated in the consolidation of episodic declarative (fact-based) memories. However, these findings indicate that sleep may stabilize fact-based memories rather than necessarily enhance them over time. For example, a collection of reports have demonstrated that, for the learning of lists of word pairs, deep NREM slow-wave sleep (SWS; stages 3 and 4 combined) plays a particularly important role in memory solidification.

That this memory benefit is causal, rather than simply correlational, was recently demonstrated by manipulating NREM SWS at night. Following the learning of a word pair list, a technique called direct current stimulation was used to boost the strength of the slow brain waves of NREM SWS in the pre-

frontal cortex (a region that, like the hippocampus, is involved in declarative fact-based memory). Direct current stimulation not only increased the strength of slow-brain-wave oscillations during this stage of sleep, but also enhanced next-day word pair retention, suggesting a critical role for SWS neurophysiology in the offline consolidation of episodic facts.

Exactly how this type of sleep may be stabilizing memories is still unclear. One exciting possibility is that the signature of nerve cell firing that occurred during learning while awake may be reactivated or "replayed" during NREM SWS at night, potentially strengthening the connections pertinent to the memory representation, or even transferring them to different storage locations. Alternatively, rather than strengthening the necessary neural connections, NREM SWS sleep may instead weaken non-necessary connections. Either one or both of these processes would ultimately lead to a more efficient stored memory trace within the brain.

Sleep to integrate and associate. As critical as consolidation is to later memory performance, the association and integration of new experiences into preexisting networks of knowledge is equally, if not more, important. The resulting creation of associative webs of information offers numerous and powerful advantages. Indeed, the end goal of sleep-dependent memory processing may not be the enhancement of individual memories in isolation, but instead their integration into a common schema, thereby facilitating the development of universal concepts, which is a process that forms the basis of generalized knowledge and even creativity. Exciting new studies have confirmed the many anecdotal reports of sleep facilitating creative insights, which may be more strongly associated with the REM stage (in which most dreaming occurs).

In one study, participants initially learned five individual object memory pairs (A > B, B > C, C > D, D > E, E > F). Unknown to the subjects, the pairs

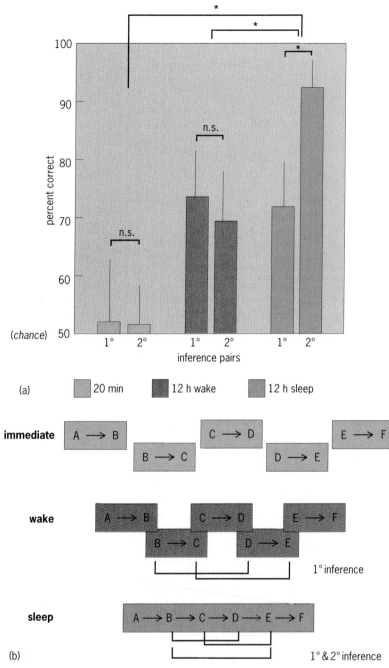

(a)

20 min	12 h wake	12 h sleep

Fig. 3. Sleep-dependent integration of human relational memory. (a) Delayed inference (associative) memory performance (% correct) in a relational memory task following different offline delays. Immediate testing after just a 20-min offline delay demonstrated a lack of any inferential ability, resulting in chance performance on both one-degree (first-order) and two-degree (second-order) associative judgments. Following a more extended 12-h delay, across the day (wake group), performance was significantly above chance across both the one- and two-degree inference judgments. However, following an equivalent 12-h offline delay, but containing a night of sleep (sleep group), significantly better performance was expressed on the more distant two-degree inference judgment compared with the one-degree judgment. (b) A conceptual model of the effects of sleep on memory integration. Immediately after learning, the representation of each premise is constituted as the choice of one item over another (A > B, etc.), and these premises are isolated from one another despite having overlapping elements. After a 12-h period with no sleep, the premise representations are partially integrated by their overlapping elements, sufficient to support first-order transitive inferences. However, following a 12-h offline period with sleep, the premise representations are fully interleaved, supporting both first- and second-order transitive inferences. Asterisks indicate significance; n.s. = not significant.

contained an embedded hierarchy (A > B > C > D > E > F). Following an offline delay of 20 min, 12 h across the day, or 12 h containing a night of sleep, knowledge of this associative hierarchy was tested by examining relational judgments for novel "inference" pairs, separated either by one degree of associative distance (B > D and C > E pairs) or by two degrees of associative distance (B > E pair). Subjects in the 20-min group that were tested soon after learning showed no evidence of inferential ability, performing at chance levels. In contrast, the 12-h groups displayed highly significant relational memory development. Most remarkable, however, if the 12-h period contained a night of sleep, a near 25% advantage in relational memory was seen for the most distantly connected inferential judgment (the B > E pair; **Fig. 3**). Together, these findings demonstrate that human memory integration takes time to develop, requiring slow, offline associative processes. Furthermore, sleep appears to preferentially facilitate this integration by enhancing hierarchical memory binding and biasing the development of the most distant/weak associative links among related, yet separate, memory items.

In another striking demonstration of sleep-inspired insight, participants were tested on a mathematical "number reduction task." During this task, participants analyzed and worked through a series of eight-digit string problems, using specific addition rules. Following initial training, after various periods of waking or sleeping, subjects returned for a further series of trials. When retested after a night of sleep, subjects solved the task, using this "standard" procedure, at a 16.5% faster rate. In contrast, subjects who did not sleep prior to retesting averaged less than a 6% improvement. However, hidden in the construction of the task was a much simpler way to solve the problem. On every trial, the last three response digits (for example, "4-1-9") were the mirror image of the preceding three (that is, "9-1-4"). As a result, the second response digit in the string always provided the answer to the problem, and subjects using such "insight" could stop after producing the second response digit—a shortcut to solving the problem. Most dramatically, nearly 60% of the subjects who slept for a night between training and retesting discovered this shortcut the following morning. In contrast, no more than 25% of the subjects in any of four different control groups that did not sleep had this insight. Thus, sleeping after exposure to the problem more than doubled the likelihood of solving it.

Such evidence suggests that sleep serves a meta-level role in memory processing that moves far beyond the consolidation and strengthening of individual memories, and instead aims to intelligently assimilate and generalize these details offline. In doing so, sleep may offer the ability to test and build common informational schemas of knowledge, providing increasingly accurate statistical predictions about the world and allowing for the discovery of novel, even creative, next-day solution insights.

In conclusion, it appears that one potential function of sleep, in combination with awake daytime learning, is to provide the symbiotic support and coordination of the encoding, consolidation, and integration of our memories, the ultimate aim of which may be to create a generalized catalogue of stored knowledge that does not rely on the verbose retention of all previously learned facts. It is perhaps therefore no coincidence that we are never told to stay awake on a problem, but instead to sleep on it.

For background information *see* BRAIN; COGNITION; INFORMATION PROCESSING (PSYCHOLOGY); LEARNING MECHANISMS; MEMORY; NEUROBIOLOGY; PROBLEM SOLVING (PSYCHOLOGY); PSYCHOLOGY; SLEEP AND DREAMING in the McGraw-Hill Encyclopedia of Science & Technology.   Matthew P. Walker

Bibliography. L. Marshall and J. Born, The contribution of sleep to hippocampus-dependent memory consolidation, *Trends Cognit. Sci.*, 11(10):442–450, 2007; B. Rasch and J. Born, Maintaining memories by reactivation, *Curr. Opin. Neurobiol.*, 17(6):698–703, 2007; G. Tononi and C. Cirelli, Sleep function and synaptic homeostasis, *Sleep Med. Rev.*, 10(1):49–62, 2006; M. P. Walker, The role of sleep in cognition and emotion, *Ann. New York Acad. Sci.*, 1156:168–197, 2009; M. P. Walker and R. Stickgold, Sleep, memory, and plasticity, *Annu. Rev. Psychol.*, 10(57):139–166, 2006.

Smart grid

In 1940, only 10% of the energy consumption in North America was used to produce electricity. By 1970, this had risen to 25%, and by 2002, it was 40%. The North American electric power grid plays a critical role in the economy and society, and has been hailed by the National Academy of Engineering as the twentieth century's engineering innovation most beneficial to civilization.

Electric power grids were once considered to be loosely interconnected networks of largely local systems. Now, these grids increasingly host large-scale, long-distance wheeling (movement of wholesale power) from one region or company to another. Likewise, currently, the connection of distributed resources, primarily small generators, is growing rapidly. In terms of the sheer number of nodes, as well as the variety of sources, controls, and loads, electric power grids are among the most complex networks.

Defining a smart grid. Smart grids consist of computer, communication, sensing, and control technology that are used in parallel with an electric power grid. Smart grids enhance electric power delivery reliability, minimize electric energy cost to consumers, and facilitate the interconnection of new generating sources to the grid.

Grid operators monitor how the system is being affected by the "environment" and keep transmission lines within their operating limits, while also maintaining an instantaneous balance between available generation (supply) and loads (demand). In recent decades, generation and transmission capacity was not sufficiently increased; thus, electric power grids have been operating with lower reserve margins, and thus, are closer to the edge of stability. With fewer margins for error, grid operators often have to make quick decisions under considerable stress.

In the future, emerging issues involving electric power grids include integration and management of renewable resources and "microgrids"; use and management of the integrated infrastructure with an overlaid sensor network, secure communications, and intelligent software agents; active control of high-voltage devices; development of new business strategies for a deregulated energy market; and system stability, reliability, robustness, and efficiency in a competitive marketplace and carbon-constrained world (**Fig. 1**).

The goal of smart grids is to transform current electric power infrastructures so that they will become energy-delivery, computer, and communications networks with unprecedented reliability, robustness, efficiency, and quality. This will require addressing challenges and developing tools, techniques, and integrated probabilistic risk assessment and impact analysis for wide-area sensing and control for digital-quality infrastructure. These include sensors, communication, and data management, as well as improved state estimation, monitoring, and simulation, all of which are linked to intelligent controllers. This would provide improved protection and discrete-event control.

For example, in numerous major power outages, narrowly programmed protection devices have sometimes contributed to worsening the severity and impact of an outage. This is due to the operation of these devices since they typically perform a simple on/off logic function, which acts locally while destabilizing a larger regional interconnection. With its millions of relays, controls, and other components, the parameter settings and structures of the protection devices and controllers in the electricity infrastructure are crucial to the operation of the grid. It is analogous to the expression "for want of a horseshoe nail. . . the kingdom was lost," in other words, relying on the dependability of an inexpensive (25¢) chip and narrow control logic can have a major impact on the operation and protection of a multibillion-dollar machine.

There is a need to coordinate the protection actions of such relays and controllers with each other to achieve overall stability; a single controller or relay cannot do it all. They are often tuned for worst cases; therefore control action may become excessive from a systemwide perspective. On the other hand, they may be tuned for best case, and then the control action may not be adequate. This calls for coordinating protection and control. Neither device, using its local signal, can by itself stabilize a system, but by building the devices to have intelligent agent abilities, multiple agents, each using its local signal, can stabilize the overall system.

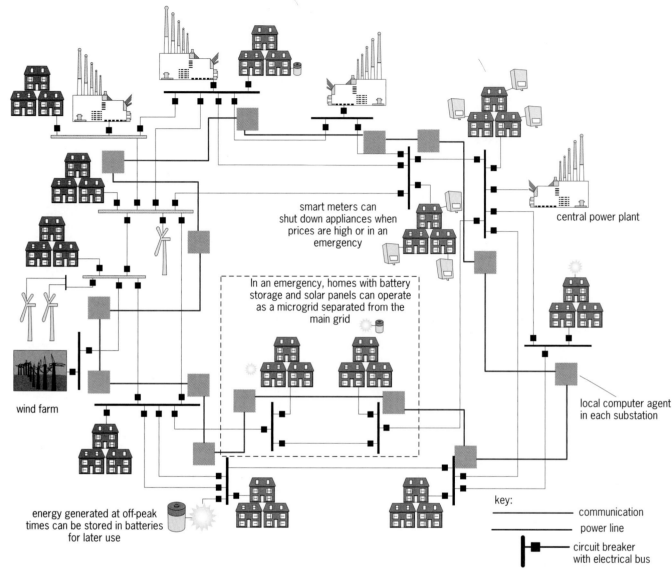

smart meters can shut down appliances when prices are high or in an emergency

In an emergency, homes with battery storage and solar panels can operate as a microgrid separated from the main grid

central power plant

local computer agent in each substation

wind farm

energy generated at off-peak times can be stored in batteries for later use

key:

communication

power line

circuit breaker with electrical bus

Fig. 1. The smart grid, a vision of the future where intelligent software agents and communications can manage and restore the system. (*Drawn by Laurie E. Miller; used by permission*)

It is important to note that the key elements and principles of operation for interconnected power systems were established in the first half of the twentieth century, prior to the emergence of extensive computer and communication networks. Computation is now heavily used in all levels of the power network—for planning and optimization, fast local control of equipment, and processing of field data. But coordination across the network occurs on a slower time scale. Some coordination occurs under computer control, but most of it is still based on telephone calls between system operators at the utility control centers, even during emergencies.

Building a smart grid. Efforts in the area of building a smart grid have developed, among other things, a new vision for integrated sensing, communications, protection, and control of the power grid. Pertinent issues involve the development of protection and control devices for centralized versus decentralized control. Other issues involve adaptive operation and robustness when dealing with various destabilizers.

Instead of performing real-time societal tests, which can be disruptive, extensive simulations of devices and operating policies in the context of the whole system have been performed. The prediction of unintended consequences of designs and policies provide a greater understanding of how this technology might fit into the continental grid, as well as guidance for effective deployment and operation.

The equipment used to build a smart grid consists of communications, computers, sensors, and control devices. The communications infrastructure enables the passing of information between all components of the power system, between individual customers and central computers, and between buyers and sellers of electric power. The computers operate independently and process information as intelligent agents. Traditionally, computers have been available in system control centers and power plants. The smart grid extends computer capability to each component and each customer of the power system. Sensors measure power system parameters such as voltage, current, temperature, and power flows. The sensors feed this information to the closest independent agent computers that decide where to send it or what actions to take. Control devices at customer locations can turn appliances on and off, as well as change the temperature set point of heating and cooling systems.

Advanced technology now in development (and under consideration) holds the promise of meeting the electricity needs of a robust digital economy. The architecture for this new technology framework is evolving through early research on concepts and the necessary enabling platforms. This architectural framework envisions an integrated, self-healing, electronically controlled electricity supply system of extreme resiliency and responsiveness—one that is fully capable of responding in real time to the billions of decisions made by consumers and their increasingly sophisticated agents. The potential exists to create an electricity system that provides the same efficiency, precision, and interconnectivity as the billions of microprocessors that it will power.

Benefits to the existing power system. Revolutionary developments, in both information technology and material science and engineering, promise significant improvement in the security, reliability, efficiency, and cost effectiveness of all critical infrastructures. Having an intelligent computer agent built into each circuit breaker, transformer, and capacitor in the power system will allow a complete database to be built and maintained easily and automatically. It will allow central computers to know exactly what each component's parameters are and how that component is performing when parameters are measured and sent into the smart grid. This will allow electric companies to monitor and maintain all components (**Fig. 2**) much more closely than at present. At this time, control over the flows on a power grid is done primarily by operators in a central location, by adjusting generator outputs and switching lines in and out (**Fig. 3**). When overloads occur it is almost impossible to force flow to go over differing paths. With a smart grid the loads themselves can be adjusted temporarily to give a much tighter control over the flows.

One problem that is very difficult for power grid companies to deal with is locating, isolating, and fixing failed components. A smart grid that uses intelligent agents (**Fig. 4**), placed at many locations, makes it easier to monitor and locate the failure of a component, thus enabling repair crews to go directly to the component to replace it. In addition, a smart grid allows for components to be switched out and, if the necessary alternate circuits are available, the customers affected by the switch could be supplied through alternate means. Sometimes a failed component forces the power system's own protective devices to open circuit breakers. This could lead to large pieces of the system being disconnected from the main grid. A smart grid should be able to sense when this happens and confirm the exact location of the disconnections so that reconnections can proceed quickly.

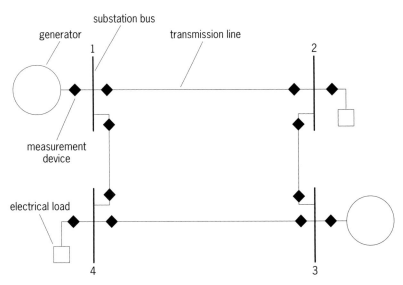

Fig. 2. Power system components. (*Copyright © Bruce F. Wollenberg, Massoud Amin*)

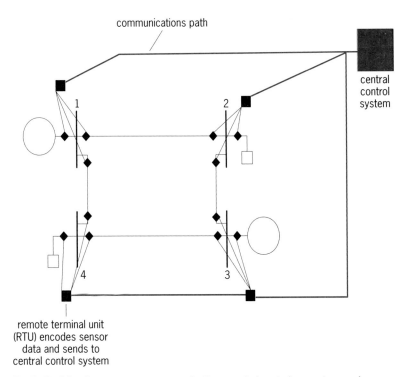

Fig. 3. Traditional power system communications, central control computers, and sensors. (*Copyright © Bruce F. Wollenberg, Massoud Amin*)

The fact that the power grid would have thousands of independent computer agents located throughout its power plants, substations, and customer connections means that very large computational tasks could be carried out by these agents as a "distributed computer" system. Present power system computer tasks are done in central locations where data are brought in, sorted into a database, and then run through algorithms to analyze the system. By distributing this task throughout the smart grid, these tasks can be done much faster and more reliably.

Benefits to consumers. Since power systems were deregulated in the late 1990s, the price for delivery

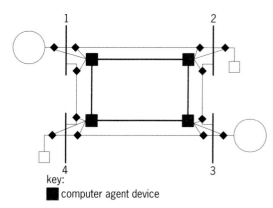

Fig. 4. Power system with smart grid. (*Copyright* © *Bruce F. Wollenberg, Massoud Amin*)

to a consumer has been calculated on an hourly basis. At present, only wholesale buyers receive pricing information and make use of it to adjust purchases. The economics of operating a power system become greatest when the customers receive pricing information and respond to it by lowering load in high-price hours, and then using power more heavily during lower-priced periods. Customers can obtain a decrease in their electric bills if they are allowed to observe pricing changes and schedule the use of electric power accordingly. However, this requires an hourly observation of the price and hourly adjustments of the power consumption. A smart grid would terminate at the customer's location in what has been called a "smart meter" that would measure not only energy consumption but also the prices that were being paid each hour. By coupling the smart meter with so-called smart appliances that could be signaled by a wireless signal in the house or business, the appliances could be scheduled automatically by algorithms in the smart meter. When prices change significantly, the use of electricity could also be changed to keep the customer's costs to a minimum. Such systems would allow the grid to "flatten out" the load curve (that is, make usage more evenly distributed throughout a 24-h period), so that peaks would be lower. At the same time, this would eliminate the need for additional equipment to be installed in the grid, simply to supply power for short periods during peaks.

Electric and pluggable hybrid electric vehicle (PHEV) automobiles would benefit greatly from smart metering, which could schedule and adjust the rate of charging to minimize costs. If the PHEV had the ability to send power back into the grid from its batteries, it could be used as a storage device. The charging and discharging of a PHEV can be regulated for the benefit of the consumer and the power system. Smart metering could then schedule when the customer bought and sold energy to minimize costs or even maximize profits. Similarly customers with large heating or cooling systems could buy and store thermal energy efficiently to minimize costs.

Integration of renewable energy devices. The biggest drawback to the use of renewable energy systems, such as wind generators or solar panel gen-

eration, is the inability to schedule wind and solar energy. These sources need to be coordinated with the power system as they come online or go offline. If large electric storage systems are available, then the renewable energy can be stored until needed by the electric grid. The advent of PHEVs offers the power grid a means of unprecedented ability to control frequency and make adjustments as intermittent generation sources (such as wind or solar energy) come online or go offline. If PHEVs are available in large numbers, the rate of battery charge or discharge can be scheduled to coordinate with variable amounts of wind and solar energy.

A smart grid would make it possible to "island" parts of the electric system from the main grid when the main grid prices are very high or when it is blacked out. The island would need to be managed by intelligent agents until it is reconnected to the grid.

Future challenges. Some of the unanswered questions relating to the smart grid pertain to who will build it. If electric companies build the smart grid, it may reduce their income. If consumers pay for the consumer end of the smart grid, it is critical to know what components of the smart grid they will be financing and how these components will be installed. At this time, the overall cost-benefit analysis (that is, what are the major savings and costs) of a smart grid is not known. The U.S. electric, computer, communication, and electronics industries need to begin to develop standards for communication, computer message structure, and sensing and control device interfaces before the smart grid can be built.

For background information *see* COMPUTER; CONTROL SYSTEMS; DIGITAL CONTROL; DISTRIBUTED SYSTEMS (COMPUTERS); DISTRIBUTED SYSTEMS (CONTROL SYSTEMS); ELECTRIC POWER GENERATION; ELECTRIC POWER SYSTEMS; ELECTRIC VEHICLE; ENERGY STORAGE; SOLAR ENERGY; WIND POWER in the McGraw-Hill Encyclopedia of Science & Technology.

S. Massoud Amin; Bruce F. Wollenberg

Bibliography. S. M. Amin, For the good of the grid: Toward increased efficiencies and integration of renewable resources for future electric power networks, *IEEE Power Energy Mag.*, 6(6):48–59, November/December 2008; S. M. Amin, Toward self-healing energy infrastructure systems, *IEEE Compu. Appl. Power*, 14(1):20–28, January 2001; S. M. Amin, and B. F. Wollenberg, Toward a smart grid: Power delivery for the 21st century, *IEEE Power Energy Mag.*, 3(5):34–41, Sept/Oct. 2005; United States Department of Energy, *The Smart Grid: An Introduction*, 2008.

Soy protein adhesives

In the quest to manufacture and use building materials that are more environmentally friendly, soy adhesives can be an important component (**Fig. 1**). Trees fix and store carbon dioxide in the atmosphere. After the trees are harvested, machinery converts the wood into strands, which are then bonded together with adhesives to form strandboard, used in constructing long-lasting houses. Soybeans fix both carbon dioxide and nitrogen and can be converted into flour that can be made into a waterborne adhesive (glue) for bonding wood strands together. Although soy and other protein adhesives were used for centuries, fossil fuel–based adhesives have generally replaced biobased adhesives because they are more cost-effective and durable. However, new soy technology and an interest in low formaldehyde emissions for interior wood products are providing an impetus for a resurgence of biobased adhesives.

History. For millennia, humankind has used natural adhesives to make useful products. Since the early Egyptians used adhesives to bond wood veneer onto furniture, bonding wood has become common in applications from furniture assembly to roof beams. Traditionally, adhesives were based on proteins from animals (hoof, hide, blood, milk, and fish scales) and plants (soybeans). Among the wood products developed using protein adhesives were interior plywood (soybean adhesives) and glued laminated timber (glulam) beams (casein adhesives from milk). However, synthetic adhesives have generally displaced these protein adhesives.

Fossil fuel–based adhesives came into prominence in the 20th century and have generally displaced biobased adhesives. Compared with many natural products, fossil fuels are relatively easy to convert into well-defined, uniform, specifically designed polymers. These synthetic polymers also provide greatly improved water resistance compared with standard biobased adhesives. Synthetic adhesives' resistance to heat, moisture, and decay helped to increase the wood product market for structural and exterior wood products. In recent years, however, volatile fossil fuel prices, a desire for renewable products, and the need to minimize formaldehyde emissions have led to opportunities for biobased adhesives—in particular, soybean-based ones.

Soybeans are an abundant agricultural crop with high oil and protein contents. The two main soy products are oil and meal. The oil is used for food and chemical applications, with alkyd (oil-based) paints being an example of the latter. Newer uses of the oil fraction are biodiesel fuel, and polyurethane and epoxide components. Soybean meal has mainly been used for animal feed and, to a limited extent, human food products. The meal ground into flour is the feedstock for making adhesives for wood bonding. The traditional method of dispersing the soy flour in water used aqueous caustic (sodium hydroxide). However, this type of adhesive had poor dispersion stability and poor water resistance after bonding.

Improved adhesives. To improve protein adhesives, it is important to realize that protein adhesives are different from other adhesives in two major ways. First, in chemical terms, most adhesives, whether synthetic or biobased, are polymers made from one or a few monomers, usually with similar functionality. However, proteins are made from more than 20 amino acid monomers with side chains contain-

Fig. 1. Conversion of plants to houses. The left side illustrates the conversion of the soybeans to flour and then to liquid adhesive, which is used to bond the strands that come from the trees shown on the right side. The strands and adhesive make strandboard used in housing construction.

ing a variety of functional groups. Second, morphologically, adhesives generally can be well described by understanding their primary and secondary structures. However, soy proteins also have tertiary and quaternary structures. Their polypeptide backbone not only influences the secondary structures of crystalline α-helix and β-sheet regions (which are distinct protein structural configurations), but also influences the specific folding that forms the tertiary structures. Understanding how conditions alter these tertiary structures is an active research area for fields ranging from enzymatic activity to prion (proteinaceous infectious particle)–related diseases. The protein folding influences polypeptide agglomerates, which is the quaternary structure.

Recent research on soy adhesives has focused on both dispersing and curing proteins. To form a satisfactory bond, soy flour with about 50% protein needs to be dispersed in water at high concentrations. Then, during the bonding process, the adhesive needs to be cured to provide water resistance for the bonded assembly.

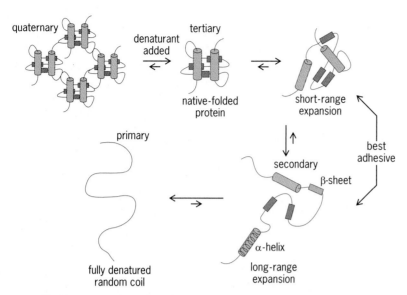

Fig. 2. Alteration (protein denaturation) of native soy protein structure to improve adhesive properties. Dispersion of soy flour can break apart the native quaternary state into individual polypeptides folded into their native tertiary structure. This tertiary structure can be opened into short-range and then long-range expansion, leaving the secondary structure intact. Further disruption of α-helices and β-sheets provides a normal polymer chain. The best adhesion has been found when the tertiary structure is disrupted, but disruption of the secondary structure is limited.

The complex structure of soy proteins makes them more difficult to disperse or solubilize than most adhesives. Most soy proteins are complex polypeptide agglomerates referred to as the quaternary protein structure (**Fig. 2**). Breaking apart these agglomerates yields the individual polypeptides, still in their ball-shaped tertiary structure. Depending upon conditions, the ionic, hydrogen bond, disulfide, and hydrophobic associations holding the tertiary structure together may be opened to allow short- or long-range expansion, leaving the secondary structure of α-helices and β-sheets intact. Retention of some secondary structure is important to good adhesive performance. Dispersions tend to be unstable because proteins tend to associate with each other over time. These associations can have a dramatic effect upon viscosity, which is important for adhesive distribution and adhesion of the proteins.

The curing process improves wet performance by decreasing the protein's water affinity and solubility. Given the variety of side-chain functional groups (amines, carboxylic acids, phenols, and thiols), many different chemicals can react with the proteins to cross-link the polypeptide chains. Research in recent years has focused on a variety of curing chemicals, including formaldehyde, phenol-formaldehyde, glutaraldehyde, and poly(amidoamine)-epoxy resin.

Currently, soy adhesives pass full exterior water exposure tests only when combined with fossil fuel-based adhesives. In such systems, soy proteins can lower the cost and improve the adhesives' performance, but the adhesives still depend upon a synthetic resin for the high moisture resistance. When soy is the main adhesive component, the resulting products are well suited for indoor applications. The absence of added formaldehyde with these adhesives eliminates the problem of urea-formaldehyde adhesives breaking down under hot and humid conditions.

Further opportunities for improving soy adhesives lie in increasing bond strength under wet conditions, increasing the protein concentration in the adhesive, and decreasing the cure time and temperature.

For background information *see* ADHESIVE; ADHESIVE BONDING; FORMALDEHYDE; POLYMER; PROTEIN; SOYBEAN; STRUCTURAL MATERIALS; WOOD COMPOSITES; WOOD ENGINEERING DESIGN; WOOD PRODUCTS; WOOD PROPERTIES in the McGraw-Hill Encyclopedia of Science & Technology.   Charles R. Frihart

Bibliography. C. R. Frihart, Adhesive groups and how they relate to the durability of bonded wood, *J. Adhes. Sci. Tech.*, 23:601–617, 2009; A. L. Lambuth, Chap. 20: Protein adhesives for wood, in A. Pizzi and K. L. Mittal (eds.), *Handbook of Adhesive Technology*, 2d ed., Marcel Dekker, New York, 2003; H. Lee et al., Single-molecule mechanics of mussel adhesion, *Proc. Natl. Acad. Sci. USA*, 103(35):12999–13003, 2006; K. Li, Formaldehyde-free lignocellulosic adhesives and composites made from the adhesives, U.S. Patent No. 7,252,735, 2007; G. E. Means and R. E. Feeney, Chemical modifications of proteins: A review, *J. Food Biochem.*, 22(5):399–426, 1998; R. H. Pain, *Mechanisms of Protein Folding*, 2d ed., Oxford University Press, Oxford, 2000; C. L. Pearson, Chap. 21: Animal glues and adhesives, in A. Pizzi and K. L. Mittal (eds.), *Handbook of Adhesive Technology*, 2d ed., Marcel Dekker, New York, 2003; J. Wescott and M. Birkeland, Stable adhesives from urea-denatured soy flour, WIPO Patent Application WO/2008/011455, 2008; R. P. Wool and X. S. Sun, *Bio-Based Polymers and Composites*, Elsevier/Academic Press, San Diego, 2005.

Space flight, 2008

Space flight in 2008 saw several significant events and accomplishments. Four space shuttle missions were flown successfully. All of the international partner *International Space Station* (*ISS*) pressurized elements have now been placed in orbit. SpaceX Corporation flew the first commercially developed rocket to orbit. The National Aeronautics and Space Administration (NASA), in conjunction with the European Space Agency (ESA), continued to explore Mars and Saturn. China demonstrated its first spacewalk, and India launched its first spacecraft to the Moon. NASA also celebrated its 50th anniversary in 2008.

Human spaceflight. NASA's *Atlantis* STS-122 crew began the first space shuttle mission of 2008 on February 7 and arrived at the *ISS* on February 9. Crewmembers added the ESA Columbus laboratory to the station, increasing the orbital outpost's scientific capabilities. On the mission, NASA astronaut Steve Frick commanded a crew of six, including Pilot Alan Poindexter and Mission Specialists Leland Melvin, Rex Walheim, Stanley Love, and Hans Schlegel of ESA. The mission delivered ESA astronaut

Leopold Eyharts to the *ISS*, and returned Daniel Tani from the *ISS* to Earth.

The space shuttle *Endeavour* STS-123 crew began its mission March 11 for a 16-day mission to the *ISS*. Aboard the shuttle were Commander Dominic Gorie, Pilot Gregory H. Johnson and Mission Specialists Robert Behnken, Mike Foreman, Rick Linnehan, Garrett Reisman, and Takao Doi, a Japan Aerospace Exploration Agency astronaut. The astronauts delivered the Japanese logistics module-pressurized section (JLP), the first pressurized component of the Japan Aerospace Exploration Agency's Kibo laboratory. The crew of *Endeavour* also delivered the final element of the station's mobile servicing system, the Canadian-built Dextre, also known as the special-purpose dextrous manipulator.

On March 10, *Jules Verne*, the first of the ESA automated transfer vehicles (ATV), a new series of autonomous spaceships designed to resupply the station with supplies and equipment and reboost the *ISS*, was successfully launched into low-Earth orbit by an Ariane 5 rocket. The ATV is automated and can remain in orbit for an extended period of time and then dock with the station. On September 30, *Jules Verne* successfully completed its 6-month logistics mission with a controlled destructive reentry over a completely uninhabited area of the South Pacific.

On April 8, Commander Sergei Volkov, Flight Engineer Oleg Kononenko, and South Korean Guest Cosmonaut So-Yeon Yi launched on *Soyuz TMA-12* to the *ISS*. Yi was South Korea's first space traveler. Then, on April 19, *ISS* Commander Peggy Whitson, Soyuz Commander Yuri Malenchenko, and Yi returned to Earth in a Soyuz spacecraft that had been docked at the station. During reentry they experienced a controlled, ballistic descent. Yi had to be hospitalized several days for back pain following the landing. Faulty explosive bolts caused the Soyuz equipment module to remain attached to the crew section as the craft reentered the Earth's atmosphere. Atmospheric drag eventually caused separation and disaster was averted.

During *ISS Expedition 16*, Whitson, Malenchenko, and rotating crewmembers Clay Anderson, Tani, Eyharts, and *Expedition 17* astronaut Reisman added many new pieces to the station. Modules added to the station during this time included the Harmony connecting module, the European Columbus research lab, and a Japanese logistics module. A set of solar arrays was also relocated. This activity set the stage for completing all the required elements to allow the station to begin more complex scientific research and support six permanent crewmembers onboard the facility.

On May 31, space shuttle *Discovery* headed for orbit on the STS-124 mission and arrived at the station June 2, delivering the Japanese Pressurized Module, the second pressurized component of the Japan Aerospace Exploration Agency's Kibo laboratory. Discovery landed at Kennedy Space Center on June 14. The Commander was Mark Kelly, Navy Commander. Kenneth T. Ham served as the pilot. Mission Specialists were Karen L. Nyberg, Ronald J. Garan, Jr., Michael E. Fossum, and Japan Aerospace Exploration Agency (JAXA) astronaut Akihiko Hoshide. Astronaut Gregory E. Chamitoff flew to the station on this mission, taking Reisman's place.

The year 2008 marked a major milestone for the *ISS* project. Space shuttle mission STS-124 was a landmark event as all the major partners' pressurized elements have now been launched into Earth orbit. These elements will advance the international collaboration, which involves 16 separate countries and will allow the station to become a functional research lab. *See* SCIENCE ON THE INTERNATIONAL SPACE STATION.

On September 25, 2008, China launched its third crewed spacecraft with three astronauts onboard to attempt the country's first-ever space walk. The spaceship *Shenzhou*-7 blasted off on a Long March II-F carrier rocket from the Jiuquan Satellite Launch Center in the northwestern Gansu province. Zhai Zhigang, a 41-year-old astronaut, successfully completed the spacewalk and waved to a national audience during a live broadcast. The three Chinese astronauts landed safely back on Earth after a 68-h voyage.

On October 12, 2008, Richard Garriott, son of former NASA astronaut Owen Garriott, and his crew successfully launched aboard a Soyuz TMA spacecraft from the Baikonur Cosmodrome in Kazakhstan en route to the *ISS*. Garriott joined the *Expedition 18* crew, which included NASA astronaut Michael Fincke and Russian cosmonaut Yuri Lonchakov. This historic mission marked Garriott as the world's first second-generation astronaut.

On October 24, 2008, Volkov and Kononenko of the 17th *ISS* crew along with Garriott landed on the steppes of Kazakhstan at 11:37 p.m. EDT, after more than 6 months in space. All three were reported to be in good condition.

On November 16, 2008, STS-126/*Endeavour* docked smoothly at the *ISS* pressurized mating adapter-2. *Endeavour* had launched on November 14. Sandra Magnus, who arrived aboard *Endeavour*, replaced Chamitoff as *Expedition 18* Flight Engineer. The seven-member crew of *Endeavour* returned to Earth on November 30, landing at Edwards Air Force Base in California. Mission managers had waived off landing at the Kennedy Space Center in Florida, the shuttle's primary landing site, after thunderstorms and strong winds prevented *Endeavour* from attempting either of the two opportunities for Kennedy. Besides Magnus, veteran space flier Christopher J. Ferguson commanded the STS-126 mission, Eric A. Boe served as the Pilot, and the Mission Specialists were Stephen G. Bowen, Robert S. Kimbrough, Heidemarie M. Stefanyshyn-Piper, and Donald R. Pettit. This was the fourth and final space shuttle mission for 2008.

In November, nations around the world celebrated a milestone in space exploration, the 10th birthday the *ISS*. Now the largest spacecraft ever built, the orbital assembly of the space station began with the launch from Kazakhstan of its first component,

Zarya, on November 20, 1998, marking the start of an international construction project of unprecedented complexity and sophistication.

In December, NASA awarded two contracts for commercial cargo resupply services to the *ISS*: one to Orbital Sciences Corporation of Dulles, Virginia, and the other to Space Exploration Technologies (SpaceX) of Hawthorne, California. At the time of award, NASA has ordered eight flights valued at about $1.9 billion from Orbital and 12 flights valued at about $1.6 billion from SpaceX. These contracts began January 1, 2009, and are effective through December 31, 2016. The contracts each call for the delivery of a minimum of 20 metric tons of cargo to the space station. These are the first commercial contracts awarded by NASA for human space-flight-related activities.

Robotic solar system exploration. On January 14, NASA's Mercury Surface, Space Environment, Geochemistry, and Ranging (*MESSENGER*) spacecraft flew by Mercury and imaged much of the surface not previously seen by a spacecraft. *MESSENGER* is only the second spacecraft to have ever visited Mercury. *Mariner 10* flew past Mercury in 1974–1975. Because of geometry during each of its encounters, *Mariner 10* was able to image less than half the planet.

MESSENGER successfully completed its second flyby of Mercury on October 6, unveiling another 30% of Mercury's surface that previously had never before been seen. About 95% of the surface of Mercury has now been observed. The results indicate that Mercury's surface is overall more ancient and heavily cratered than the Moon or Mars. They also indicate that volcanism played a more important role in shaping Mercury than previously thought. Also observed were younger volcanic plains that lie between impact craters.

NASA's *Cassini* spacecraft made a very close flyby of Saturn's moon Enceladus on March 12. The spacecraft saw geysers, much like Old Faithful, erupting from giant fractures on the south pole of Enceladus. *Cassini* sampled the erupting plumes and identified water-ice, dust, and gas.

On June 30, *Cassini* completed its 4-year primary mission and began a 2-year extended mission. Among other things, *Cassini* revealed the Earth-like world of Saturn's moon Titan and showed the potential habitability of another moon, Enceladus, during its primary mission.

Cassini also confirmed in 2008 that at least one other body in our solar system has a liquid surface lake. A lakelike feature was observed in the south polar region of Titan which was 235 km (146 mi) long. This lake contains liquid hydrocarbons, and liquid ethane was positively identified.

NASA's *Phoenix* spacecraft made an unprecedented landing in the northern polar region of Mars on May 25. Its mission was to examine the possibility of frozen water within reach of the lander's robotic arm. In July, it was announced that water was found in a soil sample. *Phoenix* operated successfully on Mars for more than 5 months before the winter decline in sunlight received on its solar cells finally would no longer allow the lander's batteries to be recharged. *Phoenix*'s cameras beamed back more than 25,000 pictures of Mars, including photos taken with the first space-operational atomic force microscope. It also observed snow falling in the Martian air and provided daily weather observations from Mars's north polar region. *See* PHOENIX MARS MISSION.

The compact reconnaissance imaging spectrometer for Mars (CRISM) and other instruments onboard NASA's *Mars Reconnaissance Orbiter* (*MRO*) also observed evidence for water on Mars from orbit in 2008. *Mars Reconnaissance Orbiter*'s instruments showed that vast regions of the ancient highlands of Mars contain clay minerals, which form only in the presence of water. The *Mars Reconnaissance Orbiter* has also revealed vast Martian glaciers of water ice under protective blankets of rocky debris at much lower latitudes than any ice previously identified on the planet. Scientists analyzed data from the spacecraft's ground-penetrating radar and found buried glaciers extending for tens of kilometers from the edges of mountains and cliffs.

The Mars Exploration rovers *Spirit* and *Opportunity* approached their fifth-year anniversary on Mars in 2008. The two rovers have made important discoveries about historically wet and violent environments on ancient Mars. They have returned a quarter-million images and driven more than 21 km (13 mi), sending back more than 36 gigabytes of data.

The Mars Express mission, operated by ESA, returned the highest-resolution three-dimensional map of the surface of the Martian moon Phobos. The images were collected in July during a flyby only 100 km (62 mi) from the center of the tiny moon. This was the first of a series of eight flybys at distances ranging between 4500 and 93 km (2800 and 58 mi), which provided detailed scientific information about the density, structure, and chemical make-up of Phobos. *Mars Express* has been orbiting the red planet since December 2003.

After a very successful 17-year-long mission, the *Ulysses* spacecraft finally ran out of power and could no longer send data back to Earth in 2008. The mission, a joint effort of NASA and ESA, sent back data about activity in the polar regions of the Sun. The spacecraft was originally designed to last 5 years but far surpassed all expectations. Some of *Ulysses'* discoveries include that the Sun's magnetic field allows particles to transfer between high and low solar-system inclinations or latitudes. This is significant for future human exploration of the solar system, as astronauts will have to be shielded from hazardous particles from a variety of latitudes not previously anticipated. It also showed that fast solar wind particles from the magnetic poles of the Sun dominate the overall solar wind for much of the solar cycle.

On October 22, 2008, the Indian Space Research Organisation's Polar Satellite Launch Vehicle, *PSLV-C11*, successfully launched the 1380-kg (3042-lb) *Chandrayaan-1* spacecraft on a trajectory to the Moon. The spacecraft successfully entered lunar orbit on November 8. This was India's first spacecraft outside of Earth orbit. India collaborated with ESA, the United States, and Bulgaria for scientific instruments on the mission.

Astronomy and Earth science. On May 14, NASA announced the *Chandra X-ray Observatory* and the National Radio Astronomy Observatory's Very Large Array observed the most recent supernova to occur in the Milky Way. The supernova explosion occurred about 140 years ago. The observations will help in understanding how a supernova generates heavy elements. The supernova was obscured by gas and dust since it occurred in the center of the galaxy. Only x-ray and radio observations were able to detect it.

NASA launched the *Gamma-ray Large Area Telescope* (*GLAST*) on a Delta 2 rocket on June 11, 2008. The telescope, a partnership of the United States, France, Germany, Japan, Sweden, and Italy, is the most powerful instrument ever put into space for observing the universe in high-energy gamma rays. After reaching orbit, the telescope was renamed *Fermi* for Italian scientist Enrico Fermi. Soon after becoming operational, the telescope discovered the first-ever gamma-ray pulsar within a supernova remnant known as CTA 1, located about 4600 light-years away from Earth in the constellation Cepheus. *See* FERMI GAMMA-RAY SPACE TELESCOPE.

The NASA-French space agency (CNES) collaborative satellite *Ocean Surface Topography Mission/Jason 2* (*OSTM/Jason 2*) launched from Vandenberg Air Force Base in California on June 20 on a Delta II rocket. The satellite will map sea-surface height, which will enable more accurate weather, ocean, and climate forecasts. The data will feed into models that predict weather and climate changes and activity. The *Jason 1* satellite is already in orbit, and the new *OSTM/Jason 2* observations will be used in conjunction with observations that continue with *Jason 1*. Less than 1 month later *OSTM/Jason 2* had produced its first complete maps of the global ocean surface topography, surface wave heights, and wind speed.

In July, NASA announced that the Time History of Events and Macroscale Interactions during Substorms (*THEMIS*) mission discovered releases of magnetic energy one-third of the way to the Moon that power substorms that cause sudden brightening and rapid movements of the aurora borealis, also called the northern lights. Substorms often accompany intense space storms, caused by solar activity, and can disrupt radio communications and Global Positioning System signals and cause power outages. A better understanding of substorms will allow models to be developed to predict activity intensity. *THEMIS* is made up of five identical, small satellites and was launched in February 2007. The satellites line up once every 4 days along the Equator and take observations synchronized with ground observatories. Each ground station uses a magnetometer and a camera pointed upward to determine where and when a substorm will begin. Instruments measure the light from particles flowing along the Earth's magnetic field and the electrical currents these particles generate.

The NASA/ESA *Hubble Space Telescope* celebrated its 100,000th orbit of the Earth in August with images of the Tarantula nebula near the star cluster NGC 2074. This region of space is 170,000 light-years away, and is one of the most active star-forming regions in our Local Group of galaxies.

NASA's *Swift* satellite found the most distant gamma-ray burst ever detected on September 19. An exploding star 12.8 billion light-years away caused the explosion, designated GRB 080913.

In November, the *Hubble Space Telescope* took the first visible-light photograph of a planet circling another star. The planet, no more than three times Jupiter's mass, orbits the bright southern star Fomalhaut, located 25 light years away in the constellation Piscis Australis. The telescope observed a faint point of light 2.9×10^6 km (1.8×10^6 mi) inside a ring of debris and dust around the star. Previous observations had indicated that the ring was off center around the star. *Hubble* detected that the asymmetry was due to gravitational shepherding by the planet.

On December 10, *Hubble* detected carbon dioxide in the atmosphere of a planet orbiting another star. This is an important step in finding the chemical biomarkers of life as we know it on other worlds. The Jupiter-sized planet, HD 189733b, is too hot for life. However, these new *Hubble* observations prove that the basic chemistry for life can be measured on planets orbiting other stars. Organic compounds can also be a by-product of life and their detection on an Earth-like planet may one day provide evidence for life beyond Earth.

Other activities. On February 20, 2008, a missile launched from a Navy cruiser soared 209 km (130 mi) above the Pacific Ocean and smashed into a dying and potentially deadly U.S. spy satellite. The *USS Lake Erie*, armed with an SM-3 missile designed to knock down incoming missiles (not orbiting satellites), launched the attack, hitting the satellite about 3 min later as the spacecraft traveled in polar orbit at more than 17,000 mi/h (27,360 km/h).

In March 2008, noted author and visionary Sir Arthur C. Clarke died at age of 90 in Sri Lanka. He is credited with proposing the idea of using satellites in geostationary orbit for telecommunications. His science fiction writings also influenced generations of potential scientists and engineers.

On April 28, India's Polar Satellite Launch Vehicle, *PSLV-C9*, successfully launched the 690-kg (1520-lb) Indian remote-sensing satellite *CARTOSAT-2A*, the 83-kg (183-lb) Indian Mini Satellite (*IMS-1*), and eight nanosatellites for international customers into a 637-km (396-mi) polar Sun-synchronous orbit. *PSLV-C9*, in its "core alone" configuration, launched 10 satellites, with a total weight of about 820 kg (1810 lb).

The third flight of the Space Exploration Technologies Corporation (SpaceX) Falcon 1 launch vehicle occurred on August 2, 2008. A problem occurred with stage separation, causing the stages to be held together and the rocket never made it to orbit.

Launches to Earth orbit and beyond in 2008		
Country of launch	Attempts	Successful
Russia	27	26
United States	15	14
China	11	11
Europe	6	6
Russian-Ukrainian Zenit-3SL	5	5
India	3	3
Japan	1	1
Iran	1	0
Total	69	66

September 28, 2008, the fourth Falcon 1 launch vehicle was successfully launched and achieved Earth orbit. With this key milestone, Falcon 1 becomes the first privately developed liquid-fuel rocket to orbit the Earth.

NASA's Constellation program to develop a human space vehicle to replace the space shuttle and to return humans to the Moon by 2020 progressed in 2008, with engineering test hardware being built for a series of launch-pad abort and vehicle dynamics tests. Successful tests of the launch abort system and shuttle-derived solid-rocket motors for the Ares I vehicle occurred in 2008. Major hardware was also completed and delivered to Florida for the 2009 Ares I-X flight test, including the new forward skirt extension and forward skirt. Also completed were numerous Ares Stage 1 recovery tests with new parachutes that will be used on the Ares I-X flight test.

Launch summary. The year 2008 was a relatively busy one for launches (see **table**). There was a significant increase in commercial launches over 2007. Of the 69 total launch attempts to Earth orbit and beyond, 28 were commercial.

Russia led with 27 launch attempts, a post-Soviet era record. Eleven Russian launches were commercial, while the United States had six commercial launches. Russia used seven different launch vehicles (Soyuz, Proton M, Proton K, Kosmos 3M, Dnepr, Molniya, and Rokot, in order of frequency of use).

The United States had 15 launches in 2008, with one failure (the SpaceX Falcon 1). Of these, nine were noncommercial, including four space shuttle missions, two U.S. Department of Defense (DoD) related payloads, and three NASA missions. The Delta II rocket launched five payloads, and the Atlas V, Falcon I, and Pegasus XL each launched two payloads.

All European launches were done with the Ariane 5 launch vehicle from their equatorial launch site in French Guiana. Five launches were commercial and the sixth was the *Jules Verne* space station resupply demonstration mission.

China's 11 launches included two for geostationary communications satellites and nine Chinese government missions. India's launches included their first lunar mission (*Chandrayaan-1*) as well as a military radar satellite for Israel and an India military mission that also deployed a series of cubesat nanosatellites. The Japanese launch was a government satellite for wide-band Internet access.

Iran attempted its first orbital launch in 2008 using the Safir two-stage rocket. The payload, believed to be a dummy satellite, did not achieve orbit.

For background information *see* AURORA; CHANDRA X-RAY OBSERVATORY; COMMUNICATIONS SATELLITE; EXTRASOLAR PLANETS; GALAXY, EXTERNAL; GAMMA-RAY ASTRONOMY; GAMMA-RAY BURSTS; HUBBLE SPACE TELESCOPE; MARS; MERCURY (PLANET); MOON; REMOTE SENSING; SATELLITE (ASTRONOMY); SATURN; SCIENTIFIC AND APPLICATIONS SATELLITES; SOLAR WIND; SPACE FLIGHT; SPACE STATION; SPACE TECHNOLOGY; SUN; X-RAY ASTRONOMY in the McGraw-Hill Encyclopedia of Science & Technology.

Donald Platt

Bibliography. *Aviation Week & Space Technology*, various 2008 issues; *Commercial Space Transportation: 2008 Year In Review*, Federal Aviation Administration, January 2009; ESA Press Releases 2008; NASA Public Affairs Office News Releases, 2008.

Stimulating innate immunity as cancer therapy

In contrast to machines, organisms exchange parts continuously. Body cells die, whereas others divide and proliferate. It is one of the great wonders to observe an organism, starting from a single cell, reaching its final architecture, and maintaining this architecture for the rest of its life, with all pieces in permanent flow exchanging several million cells per minute in humans.

Cells could not build up the architecture without communicating with each other and exerting mutual control. The control of cell division is at the core of a functioning organism and is checked by several mechanisms acting from inside and outside the cell. Although proliferation is controlled on multiple layers, it can become defective when multiple mutations within the genome of a cell accumulate over time. Usually, the result of mutations is the suicide (apoptosis, or programmed cell death) of the cell. Sometimes, though, when mutations affecting this suicide mechanism combine with mutations of division control mechanisms, uncontrolled proliferation can result, ending in a so-called neoplasm or tumor. Malignant neoplasms can lead to fatal organ failure. Fortunately, there is another important player that helps in the protection from neoplastic threat: the immune system. Billions of cells of the immune system patrol through blood vessels, lymphatic vessels, and even dense tissue to check if anything is out of order, for intruders such as bacteria and viruses, or for defective cells that need to be eliminated. For a long time, it was assumed that cancer cells, in contrast to pathogens, "look" too similar to healthy cells to be detected by immune cells; however, it is now known that this is not the case. Tumor tissue very often is surrounded and invaded by tumor-infiltrating lymphocytes (TILs), that is, by immune cells. Thus, detection does not seem to be the problem here. Why then is eradication common with pathogens and yet appears to be rare for tumors?

Human body "weakness." A long-held notion among oncologists was that the human body is too weak to reject palpable neoplasms, and without exception will need help from drugs, surgery, or radiation, since neoplasms usually grow and spread in clinical routine. Cancer so often is fatal, and cancer immunotherapy, so far, has not been very successful. A closer look, though, reveals that the simple picture of the human body's incapability to deal with cancer is not quite correct. Some neoplasms such as melanoma and neoblastoma are known to disappear after a while with considerable frequency. It is striking that survival rates of many cancer forms correlate with the number of TILs: the more TILs found in a tumor, the longer is the survival. Thus, TILs are not just observers; some of these immune cells actually must be able to slow neoplastic growth or, perhaps, even be able to kill some malignant cells. The weapons, in principle, are there.

Spontaneous regression. The killing of cancer cells up to the complete eradication of large tumors can occur. This is proven by the occasional observation of a spontaneous regression. Sometimes, spontaneous regressions lead to a shrinkage of a neoplasm that grows again later; sometimes, they can lead to cure. About 1000 cases of spontaneous regression have been reported in the scientific literature. One must assume a much larger number of unreported cases in addition because physicians did not report their observations in the literature, patients did not reappear in the clinic, the case could not be documented well, or other reasons prevented publication. Still, spontaneous regressions were considered to be very rare events, unexplainable and thus without the potential to aid in the treatment of cancer. This was the case until November 2008, when, in a carefully designed study, it was shown that the rate of spontaneous regressions from breast cancer in women must be on the order of 20%, an astonishing rate, suggesting that the role of the immune system in cancer rejection was vastly underestimated. In other words, the effort required to stimulate the immune system against cancer cells might be much smaller than initially thought. Putting this high rate of spontaneous regression into perspective means that an important detail was probably missed in cancer immunotherapy so far.

Lowered cancer risk after infection. How much then is there to learn from spontaneous regressions? A close look at published cases from 2001 showed that a large number of cases were preceded by an acute feverish infection (28–80% of cases, with a putative surplus number of reported regressions with unreported infections). One must always be cautious not to overinterpret connections in time as causal connections, but this correlation is puzzling. Could feverish infections impact established neoplasms? If this were true, developing neoplasms should be impacted even more since the number of cancer cells to be battled is orders of magnitude lower in microscopic foci than in established cancers. According to

this logic, people with a personal history of feverish infections should be less likely to develop cancer. Strikingly, this seems to be the case. More precisely, it is possible to find epidemiological studies supporting the hypothesis of cancer-protecting effects of infections, but these studies were published over many years and not connected and discussed with regard to spontaneous regressions until recently. The literature on spontaneous regression and the literature on cancer epidemiology were isolated islets of knowledge.

Treatment using bacterial extracts. There is a third line of evidence giving credence to the hypothesis of the positive effects of infections. More than 140 years ago, the German physician W. Busch observed that neoplasms can shrink after an erysipelas infection, an infection of the epidermis caused by *Streptococcus pyogenes* bacteria, which is today rare and easily managed by antibiotics, but common and often life-threatening those days. In 1868, Busch infected a female neck-cancer patient with streptococci, after which the huge tumor shrank rapidly. In the 1890s, the New York surgeon William Coley continued this work by infecting patients not with dangerous live streptococci but with heat-sterilized streptococcus extracts. Until his death in 1936, Coley (along with contemporaries) treated hundreds of patients, with mixed results. He achieved miraculous cures in some cases of inoperable, late-stage cancer, but failed in others. His daughter, Helen Coley-Nauts, later meticulously investigated her father's patient records. From her publications, it is possible to draw the conclusion that results were better when high fever was induced by the bacterial extract, and when the extract was administered close to the tumor rather than systemically. Overall, Coley's success rate, judged by 5-year survival, was similar to present-day therapies.

The Nobel Prize in Physiology or Medicine for 2008 was awarded to Harald zur Hausen in Heidelberg, Germany, for his observation that chronic papilloma virus infection can cause cervical cancer. It is estimated that 15–20% of all cancers are caused by chronic infections. However, the observations sketched above suggest the opposite: that acute infections can protect from and even cure cancer. What is going on here? The fundamental difference is "chronic" versus "acute," which may cause opposite outcomes.

Role of the innate immune system. What may be the common denominator of spontaneous regressions after infection, of protection from cancer after infection, and of tumor lysis after streptococcal infection? One hypothesis is that pathogen-associated molecular pattern molecules (PAMPs) could be the answer. PAMPs are a diverse class of molecules such as LPS (lipopolysaccharide from bacterial cell walls), flagellin (a protein found in bacterial propellers), dsRNA (double-stranded ribonucleic acid found in viruses), CpG-DNA (deoxyribonucleic acid characteristic for bacteria), zymosan (a substance found in infectious fungi), and many others, which are found in pathogens only and not generated by human tissues. PAMPs are known to stimulate the innate arm of the immune system even in tiny amounts. The innate arm is the evolutionarily older, up to recently largely unexplored part of the immune system, while almost all efforts in cancer immunotherapy were aimed at stimulating the adaptive arm, which can generate antibodies and killer T-cells. As is now known, when under infection, the adaptive arm is not independent but needs proper enhancement by the innate arm. Both parts of the immune system are tightly interconnected. Cancer cells do not generate PAMPs. Could this be the reason why the immune reaction usually is too weak and can become strong enough, at least occasionally, when a cancer patient gets an acute feverish infection?

PAMPs in clinical trials. Some PAMPs have been tested in clinical trials, without much success. However, these trials did not take into account lessons derived from the old experiments and from epidemiological findings. Previous PAMP trials were guided by pharmaceutical magic-bullet thinking: the idea of a cancer toxin or an immune stimulant wiping all cancer cells in a blow. Single patented PAMP substances were tested, whereas pathogens upon normal infection release a cocktail of PAMPs. PAMPs were tested on patients with compromised immune systems following radiation and chemotherapy. Fever, in clinical trials of novel drugs, is regarded as an adverse event and treated with antipyretics (fever-reducing agents) or by lowering drug dose. While this is advisable for drugs interfering with metabolism, the opposite can be adequate when immune stimulation is the goal. PAMP treatment was stopped when the disease progressed. In contrast, Coley applied bacterial extracts over weeks and months, and it sometimes took very long until the first signs of improvement appeared. Whereas the adaptive immune system can generate memory cells, the innate arm has no memory. It must be stimulated again and again.

Future outlook. It is imperative to combine the lessons derived from experiments done over a century ago with today's improved understanding of the innate immune system. PAMP therapy should be started before surgery, when cancer antigen load is high. Multiple PAMPs should be combined into a cocktail. PAMPs should be applied as close to the neoplasm as possible. Fever should not be suppressed, but viewed as an enhancer able to support healing. Last, PAMP immunotherapy should be tested on the most promising patients, that is, those with the best immune status and the most "immunogenic" forms of cancer like melanoma, sarcoma, and neuroblastoma.

For background information *see* CANCER (MEDICINE); CELL DIVISION; CELLULAR IMMUNOLOGY; CHEMOTHERAPY AND OTHER ANTINEOPLASTIC DRUGS; IMMUNOLOGICAL DEFICIENCY; IMMUNOLOGY; IMMUNOTHERAPY; INFECTION; MUTATION; ONCOLOGY; TUMOR in the McGraw-Hill Encyclopedia of Science & Technology. Uwe Hobohm

Bibliography. S. L. Christian and L. Palmer, An apparent recovery from multiple sarcoma with involvement of both bone and soft parts treated by toxin of erysipelas and *Bacillus prodigiosus*, *Am. J. Surg.*, 43:188–197, 1928; H. C. Coley-Nauts, G. A. A. Fowler, and F. H. Bogatko, A review of the influence of bacterial infection and of bacterial products (Coley's toxins) on malignant tumors in man, *Acta Med. Scand.*, 145:1–105, 1953; U. Hobohm, Fever and cancer in perspective, *Canc. Immunol. Immunother.*, 50:391–396, 2001; U. Hobohm, Fever therapy revisited, *Br. J. Canc.*, 92:421–425, 2005; U. Hobohm, J. Grange, and J. Stanford, Pathogen associated molecular pattern in cancer immunotherapy, *Crit. Rev. Immunol.*, 28:95–107, 2008; P. Matzinger, An innate sense of danger, *Semin. Immunol.*, 10:399–415, 1998; P.-H. Zahl, J. Maehlen, and H. G. Welch, The natural history of invasive breast cancers detected by screening mammography, *Arch. Int. Med.*, 168:2311–2316, 2008.

Stimuli-responsive polymers

Materials that are capable of responding to internal or external stimuli, such as temperature, pH, ionic strength, concentration gradients, electric field, and light, have been of interest for a number of years. Responses include changes in polymer morphology, surface characteristics, and solubility, as well as self-assembly. Potential applications include changes of surface wettability, dimensions, opacity, color, electromagnetic signature, and bioactivity, to name some. Because of the tremendous interest in new materials and their importance for novel technologies, research themes involving stimuli responsiveness have accelerated.

In a quest for "smart" materials, scientists have pursued new avenues to generate molecular designs "invisible" to the naked eye that would result in notable visual changes, similar to flipping an electric switch that turns electricity "on" and "off" to illuminate or darken a room. However, the design and synthesis of materials capable of turning on or off on demand requires orchestrated efforts. To accomplish these challenging tasks in materials, an electrical switch needs to be converted to a much smaller entity that will function at molecular bond (angstrom), nanometer, or micrometer scales. However, individual events at these scales usually will not result in visible changes. In an effort to actually "see" these responses, higher-order events must take place that collectively respond to internal or external stimuli. An illustrative example of a multilevel event is the response of the eye to illumination. This effective photoswitching vision device enables fairly complex processes to be governed by the cis-trans isomerization of the retinal molecule embedded in rhodopson helices, where a small dimensional change amplifies a series of events leading to larger dimensional and chemical changes, with the generation of an electrical signal transmitted to the brain and a vision event. For these events to occur, interconnected molecular segments must respond in an organized fashion.

Solutions. In polymeric solutions, stimuli responsiveness is significantly simpler because spatial

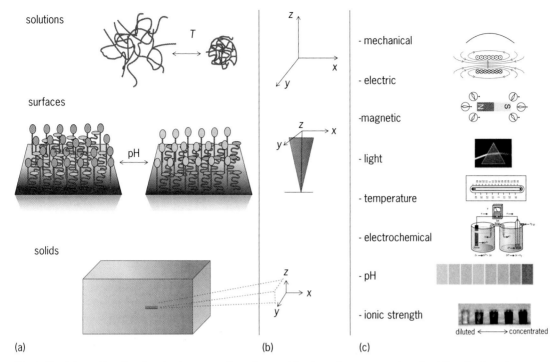

solutions

T

surfaces

pH

solids

- mechanical

- electric

-magnetic

- light

- temperature

- electrochemical

- pH

- ionic strength

diluted ⟷ concentrated

(a)   (b)   (c)

Fig. 1. Spatial considerations in stimulus-responsive polymers: (*a*) schematic representation of mobility of polymeric segments in solutions, on surfaces, and solids of stimuli-responsive polymers; (*b*) spatial restrictions of mobility in *xyz* directions in solutions, at surfaces, and in solids; (*c*) examples of physical and chemical stimuli in polymers.

availability is achieved by the replacement of solvent molecules, which energetically requires overcoming the Brownian motion of the solvent molecules. This is shown schematically in **Fig. 1***a*. One can easily envision a macromolecule that, upon temperature changes, will exhibit conformational changes and collapse. Several examples of polymers that exhibit dimensional changes as a function of temperature are manifested by using chemical segments that exhibit lower critical solution temperature (LCST). Copolymer solutions of poly(N-isopropylacrylamide) (PNIPAm), poly(N-vinylcaprolactam) (PVCL), 2-dimethylaminoethyl methacrylate (DMAUMA), and poly(2-carboxyisopropylacrylamide) are known for their temperature responsiveness and expand below LCST. There are other amphiphilic copolymers, such as poly(ethylene oxide)-poly(propylene oxide) and poly(ethylene oxide)-poly(1,2-butylene oxide) that also show temperature responsiveness but do not exhibit LCST.

Another common stimulus in polymeric solutions is a change in proton concentration or pH change. Solubility changes upon pH variation in polyelectrolytes containing ionizable pendant groups at the polymer backbone are manifested by a variation of the hydrodynamic volume due to the expansion or crumbling of the polymer chain. Acrylic acid polymers, such as poly(methacrylic acid) and poly(2-propyl acrylic acid), are examples of pH-responsive species that extend the backbone as a result of the absorbance of protons by a pendant carboxylic group at basic pH values. The presence of amine groups along a polymer backbone results in polybases that release the captured proton with a change in pH. Block copolymers that carry a common core block, such as 2-(diethylamino)ethyl methacrylate (DEAEMA), and one of the coronal blocks, 2-(dimethylamino)ethyl methacrylate (DMAEMA) and poly(ethylene oxide) (PEO), are other examples.

Surfaces. Polymeric segments anchored at surfaces exhibit another level of spatial freedom. Figure 1*b* shows the surface response to pH changes, in which H$^+$ concentration levels above the surface induce extension and collapse of polymeric segments that are sensitive to pH. Mobility restrictions come from anchoring one end of the segment at the surface and, consequently, limited mobility at the other end. Polymers in the form of brushes, thin films, and colloids with sophisticated architectures and high levels of precision have been developed, in which the formation of reconstructable surfaces by spontaneously assembled interfacial layers is of particular interest. Polymer surfaces may respond to the surroundings changes in a variety of ways. If surface polymeric chains are capable of rearranging themselves as a result of environmental changes by adopting new conformations, the chemical makeup and molecular weight, and consequently, solute–solvent interactions, will play crucial roles in surfaces being stimuli-responsive. Several studies have examined the dynamic nature of polymer surfaces and their reconstruction dynamics, which is associated with switching the external medium or changing the chemical makeup of the surface itself. Evaluation of the wettability of oxidized polyethylene after various environmental exposures and temperature variations and the formation of surfaces with broad lyophobicity ranges and various end groups, ranging from carboxylic acid to fluorine-containing moieties, have been reported.

Another tunable response was prepared by chemically grafting poly(vinylmethyl siloxane) [PVMS] networks with alkanethiols bearing a hydrophilic end group (-COOH or -OH) through azobisisobutyronitrile (AIBN)-mediated thiol-ene reactions, which demonstrated that the responsive nature of PVMS-S-$(CH_2)_{11}$OH can be fine-tuned by varying the temperature, where a faster response was obtained at temperatures above the melting point of the -S-$(CH_2)_{11}$OH moiety. Polyelectrolyte and zwitterionic (dipolar) polymer brushes that respond to ionic strength and pH by large conformational changes also have been reported with useful stimuli-responsive properties, including changes in surface wettability, permeability, mechanical, adhesive, adsorptive, and optical properties.

Solids. Although it is relatively easy to achieve stimuli-responsiveness in solutions and on surfaces, solids represent another level of difficulty. Figure 1*b* schematically illustrates spatial restrictions in solids, which are quite severe and network-dependent. As was pointed out earlier, the presence of free volume accomplished by designing the networks with desirable glass transition temperatures (T_g) is one approach to provide sufficient space for segmental rearrangements. The importance of the T_g in the network rearrangements in photomechanical processes was appreciated when azobenzene polymers were exposed to light intensity gradients, which resulted in surface deformations, with the original surface topography recovered by heating the films above the T_g. Photosensitive polymers typically respond by conformational cis–trans transformations of their backbone upon exposure to ultraviolet (UV) and visible light. Photosensitive azobenzene chromophores expand and contract, whereas species such as bis(4-(dimethylamino)phenyl)(4-vinylphenyl)methylleucocyanide swell as a result of the ionization of cyanide groups. Because azopoly-

mers offer superior properties and are easily processed, they find use as actuators capable of responding to external stimuli by deforming their shape. Among naturally derived species, chitosan is a copolymer of N-acetyglucosamine and glucosamine obtained by deacetylation of chitin extracted from the shells of shrimp, lobster, crab, or squid. Two functional groups (-NH_2 and -OH) on its backbone makes chitosan pH-responsive when carboxyl, amine, and allyl (-$CH_2HC=CH_2$) groups are present.

The grand challenge is to create mechanically stable solid networks that exhibit restricted volume changes. Because spatial limitations resulting from chemical or physical connectivity of polymers often restricts molecular motions, the presence of molecular-size "voids," often referred to as free volume, facilitates the opportunity for designing stimuli-responsive polymeric networks. To achieve responsiveness, polymer networks must contain responsive components as well as a sufficient space for rearrangements, resulting from stimuli. One approach used to create additional controllable free volume is to copolymerize stimuli-responsive monomers with monomers that exhibit film-forming properties. This concept was used in the synthesis of n-(DL)-(1-hydroxymethyl) propylmethacrylamide/n-butyl acrylate (n-DL-HMPMA/nBA) colloidal particles, which, upon coalescence, formed films. The temperature responsiveness can be maintained by the n-DL-HMPMA monomer, whereas the lower T_g of nBA provides sufficient free volume for spatial rearrangements.

These attributes can be measured by differential scanning calorimetry (DSC) or dynamic mechanical analysis (DMA) as a function of temperature if the stimulus is the temperature, which will be reflected in the glass transition (T_g), temperature, as well as stimuli-responsive transitions (T_{SR}). When 2-(N,N-dimethylamino) ethyl methacrylate (DMAEMA) and n-butyl acrylate (nBA) monomers were copolymerized into colloidal dispersions and allowed to coalesce to form solid continuous films, in addition to T_g, composition-sensitive endothermic stimuli-responsive transitions T_{SR} are observed. The T_{SR} transitions change with the composition of stimuli-responsive component of the copolymer, temperature, and the rate of temperature changes. Based on the experimental data, the relationship

$$1/T_{SR} = w_1/T_{binary} + w_2/T_{form}$$

$$\text{or} \quad 1/T_{SR} = w_1(1/T_{binary} - 1/T_{form}) + 1/T_{form}$$

was established, where T_{SR} is the temperature of the stimuli-responsive transition, T_{binary} is the temperature of stimuli-responsive homopolymer in a binary polymer–water equilibrium, w_1 and w_2 ($w_2 = 1 - w_1$) are weight fractions of each component of the copolymer, and T_{form} is the film formation temperature. This relationship is shown in **Fig. 2** and allows predictions of the T_{SR} transitions in stimuli-responsive solid copolymers.

Self-repair. Stimuli responsiveness is also manifested by the ability of polymer networks to

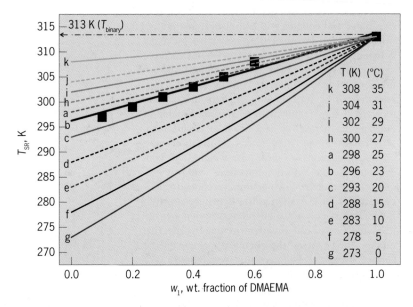

	T (K)	(°C)
k	308	35
j	304	31
i	302	29
h	300	27
a	298	25
b	296	23
c	293	20
d	288	15
e	283	10
f	278	5
g	273	0

Fig. 2. Experimental T_{SR} values obtained from DSC measurements and predicted T_{SR} using $1/T_{SR} = w_1/T_{binary} + w_2/T_{form}$ for different T values plotted as a function of w_1. (*Reproduced with permission from M. W. Urban, Progr. Polymer Sci., in press, 2009*)

self-repair. Approaches include the physical incorporation of encapsulated fluids, which, upon mechanical damage, release crosslinking agents, and the incorporation of chemical entities within networks, which, upon thermal exposure, electromagnetic radiation, or oscillating magnetic field result in self-repairing. Although physical incorporation of mechanically fragile capsules or hollow fibers introduces an inherent lack of mechanical integrity to the networks, the presence of self-repairing components covalently incorporated into the networks not only provides mechanical stability but also controls the self-repair process.

For background information *see* COLLOID; COPOLYMER; FILM (CHEMISTRY); GLASS TRANSITION; MACROMOLECULAR ENGINEERING; PH; POLYMER; SUPRAMOLECULAR CHEMISTRY in the McGraw-Hill Encyclopedia of Science & Technology.

Marek W. Urban

Bibliography. X. Chen et al., A thermally remendable cross-linked polymeric material, *Science*, 295(5560):1698–1702, 2002; C. Corton and M. H. Urban, Repairing polymers using oscillating magnetic field, *Adv. Mater.*, 2009, in press; J. A. Crowe-Willoughby and J. Genzer, Formation and properties of responsive siloxane-based polymeric surfaces with tunable surface reconstruction kinetics, *Adv. Func. Mater.*, 19(3):460–469, 2009; B. Ghosh and M. W. Urban, Self-repairing oxetane-substituted chitosan polyurethane networks, *Science*, 323(5920):1458–1460, 2009; F. Liu and M. W. Urban, New thermal transitions in stimuli-responsive polymers, *Macromolecules*, 42(6):2161–2167, 2009; M. W. Urban, Stratification, stimuli-responsiveness, self-healing, and signaling in polymer networks, *Prog. Polym. Sci.*, 34(8):679–687, 2009; S. R. White et al., Autonomic healing of polymer composites, *Nature*, 409:794–797, 2001.

Strong-interaction theories based on gauge/gravity duality

One of the triumphs of theoretical physics of the twentieth century was the development of quantum electrodynamics (QED), the fundamental theory of electrons and photons. QED describes not only all of electrodynamics, atomic physics, and chemistry with extraordinary precision, but also the basic properties of the electron itself. For example, the electron g factor is correctly predicted by QED to 10 significant figures. The corresponding problem in particle and nuclear physics is to be able to describe the structure and properties of hadrons, such as protons and neutrons, in terms of their fundamental constituents, quarks and gluons. High-energy experiments, such as the deep inelastic electron–proton scattering pioneered at the Stanford Linear Accelerator Center (SLAC), which revealed the quark structure of the proton, have shown that the basic elementary interactions of quarks and gluons are well described by a remarkable generalization of QED called quantum chromodynamics (QCD).

Quantum chromodynamics. In QCD, quarks and gluons interact with each other via a new type of charge called "color." However, unlike the photons of QED, gluons also interact with each other, and these "non-Abelian" gauge couplings lead to color confinement, that is, the impossibility of free colored quarks and gluons.

The strong couplings and confinement of quarks and gluons into hadrons in QCD make the calculation of hadronic properties such as hadron masses and their interactions a much more difficult problem than solving the equations of QED. The most successful theoretical approach thus far has been to employ very large numerical simulations on advanced computers using an approximate representation of QCD called lattice gauge theory. *See* LATTICE QUANTUM CHROMODYNAMICS.

AdS/CFT correspondence. The AdS/CFT correspondence introduced by Juan Maldacena (also called gauge-gravity duality) between string theories of gravity in anti-de Sitter (AdS) space and conformal (scale-invariant) gauge field theories (CFT) in physical space-time has created a completely new set of tools for studying the dynamics of strongly coupled quantum field theories such as QCD. In effect, the strong interactions of quarks and gluons are represented by a simpler semiclassical (without quantum effects such as particle creation and annihilation) gravity theory in higher dimensions. Although a perfect string-theory dual of QCD is not yet known, the AdS/CFT correspondence has already provided many new and remarkable insights into QCD, including color confinement, and quantitative predictions for the meson and baryon spectra and the wave functions that describe the structure of hadrons at a fundamental level. The AdS/CFT correspondence has also been applied to the behavior of quark and gluon matter at extreme conditions, such as high temperatures and pressure, thus providing insight into phenomena now being studied experimentally in relativistic heavy-ion collisions at Brookhaven National Laboratory (BNL). For example, AdS/CFT predicts a very low ratio of viscosity to entropy density for the quark–gluon plasma phenomena seen at BNL, in apparent agreement with experiment.

Geometry of AdS space. The constant radial distance r between any point on a sphere and its center is given by the Euclid (300 BC) formula $r^2 = x^2 + y^2 + z^2$. In the special theory of relativity, an invariant space-time distance interval σ also involves the time dimension as measured by the speed of light c: $\sigma^2 = x^2 + y^2 + z^2 - c^2 t^2$. Anti-de Sitter space has one extra mathematical space dimension labeled u. Unlike ordinary space, AdS space is intrinsically curved; consequently, the distance between neighboring points ds also depends on a "warp" factor R/u, where R is the radius of curvature of AdS space. This is represented mathematically by a "metric" formula (1).

$$(ds)^2 = \frac{R^2}{u^2}[(dx)^2 + (dy)^2 + (dz)^2 - c^2(dt)^2 - (du)^2] \quad (1)$$

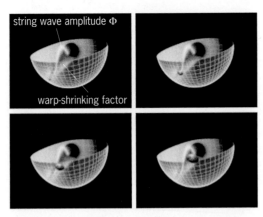

Fig. 1. Geometry of AdS space and the size of a proton. Different values of the fifth-dimensional AdS coordinate u (the radial coordinate of the sphere) correspond to different scales at which the proton is examined. The outer sphere corresponds to events at very short distances in the usual four-dimensional space; that is, the size of the proton (blob with three quarks inside), as seen by a four-dimensional observer, shrinks to zero near the $u = 0$ boundary of AdS space. Large-distance confinement physics corresponds to the inner sphere at $u = u_0$. To help visualize the five dimensions in the figure we may think in two tangent spatial dimensions x and y, with origin of coordinates in the small proton at the outer sphere, and a fifth dimension coordinate u pointing inward to the center of the sphere. The string wave amplitude Φ is illustrated by the upper stream in each picture. The warp-shrinking factor is represented by the lower stream.

In AdS space the size of an object as seen by an observer in four dimensions depends on its location inside the higher-dimensional space (**Fig. 1**). Technically, AdS space is a space of negative curvature with a four-dimensional space-time boundary. Due to the warp factor R/u in the metric, the object shrinks near the boundary of AdS space at $u = 0$ represented by the outer sphere in Fig. 1. The picture is reminiscent of a hologram in the two-dimensional surface of a sphere, depicting objects located in the three-dimensional interior volume of the sphere. Thus, gauge–gravity dual theories can be termed "holographic theories." The holographic description requires that the quantum theory defined in the four-dimensional boundary "surface" is equivalent to the gravity theory defined in the "volume" at the "surface" interior (Fig. 1).

Changes of scale. It follows from Eq. (1) that the interval ds remains invariant if one simultaneously changes the length and time scales of ordinary space-time by an arbitrary factor λ ($x \to \lambda x, y \to \lambda y, z \to \lambda z, t \to \lambda t$), with a corresponding change of length scale in the fifth dimension ($u \to \lambda u$). Thus changes in the length scale in the physical four-dimensional space-time world are equivalent to a change of scale in the fifth dimension.

Toward a dual theory of QCD. The original application of the AdS/CFT correspondence by Maldacena connected a gravity theory in AdS space to a particular type of supersymmetric theory of colored quarks and gluons in physical space-time. In this example, one can compute physical observables in a strongly coupled gauge theory—such as QCD—in terms of a classical gravity theory. However, the supersymmetric theory studied by Maldacena is "conformal";

that is, it has neither an inherent length dimension nor color confinement. Moreover, in a supersymmetric theory, every particle has a partner of the same mass but with different intrinsic spin. Such a theory is thus very different from QCD, the confining theory of quarks and gluons that underlies the real world of hadrons and nuclei.

Nonetheless, in a limited domain there is a "conformal window," where AdS/CFT can be applied. Thus, in effect, one can use a theory of gravity in AdS space to characterize a strongly interacting theory of quarks and gluons in physical space-time.

AdS/QCD. It is possible to modify the AdS/CFT theory to incorporate quark and gluon confinement. The simplest method is to introduce a boundary at a point $u = u_0$ in the fifth dimension of AdS space, so that the quarks and gluons are restricted to propagate inside the domain $0 < u < u_0$. This is effectively a model in which quarks are permanently confined inside a "bag" of given radius, in the AdS fifth dimension, but with the important feature that Lorentz invariance is preserved. Because the change in the length scale in physical space is equivalent to a change of scale in the fifth dimension, this also restricts the separation between quarks and gluons in physical space-time to a finite domain. Moreover, the value of u_0 introduces a unit of mass, $\Lambda_{QCD} = \hbar c/u_0$, which characterizes the mass spectrum of the quark and gluon bound states such as the proton ($\hbar$ is Planck's constant divided by 2π). Inside the bag, the quarks and gluons propagate freely, but their configuration depends on the different energy scales at which the proton is examined. For example, in a very high energy collision experiment the three quarks in the proton are close together near the boundary of AdS space at a small value of u; conversely, a large-size proton, with far-separated quarks as perceived in a low-energy experiment, has been pulled by the gravitational field in AdS space up to the largest size allowed by confinement, as limited by the inner sphere in Fig. 1.

In the "bottom-up" approach, known as AdS/QCD, an effective gravitational theory is constructed that encodes salient properties of the QCD dual theory, such as color confinement and chiral symmetry breaking. It has been observed that one can modify the metric of AdS space with a harmonic oscillator confining potential, $U = \kappa^4 u^2$, to reproduce the observed linear behavior in the hadronic spectrum, the well-known "Regge trajectories." In this "soft-wall" model, the value of κ breaks conformal invariance and sets the mass scale for the hadronic spectrum.

Dirac's amazing idea. One of the most useful theoretical tools for describing bound states is "light-front quantization," inspired by an insightful paper by Paul A. M. Dirac in 1949. Dirac showed that there are extraordinary advantages for relativistic theories if one replaces ordinary time t with the time marked by the front of a light wave τ. For example, when one takes a flash photograph, one obtains an image at specific values of τ, not t. Light-front quantization is the ideal framework to describe the structure of hadrons

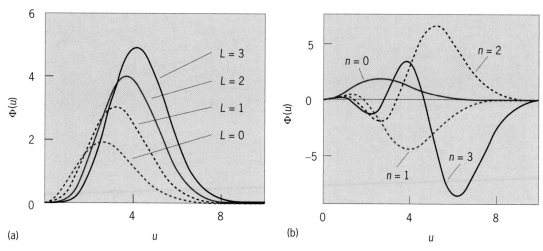

Fig. 2. Meson wave functions in AdS space in the soft-wall holographic model of confinement: (a) orbital modes ($n = 0$) and (b) radial modes ($L = 0$). Constituent quark and antiquark fly away from each other as the orbital and radial quantum numbers increase.

in terms of their quark and gluon degrees of freedom, because wave functions defined at fixed "light-front" time τ are independent of the total momentum of the state. In contrast, wave functions defined at fixed "instant" time t depend in a very complicated way on the total momentum. Even worse, the computation of even the simplest processes, such as form factors, requires one to take into account processes in which currents arise spontaneously from the vacuum. In contrast, the simple structure of the light-front vacuum allows an unambiguous definition of the quark and gluon content of a hadron and their wave functions. The light-front wave functions of relativistic bound states thus can provide a description of the structure and internal dynamics of hadronic states in terms of their fundamental constituents.

Light-front holography. There is a remarkable and direct connection between AdS space and the light-front formalism, called "light-front holography." This procedure allows information of the wave amplitude $\Phi(u)$, that propagates in AdS space to be precisely mapped to the light-front wave functions of hadrons in physical space-time in terms of a specific light-front variable ζ, which measures the separation of the quark and gluonic constituents within the hadron, independent of its total momentum.

Schrödinger equation for hadrons. We can also use "light-front holography" to transform the bound-state equations for the wave function in AdS space to a corresponding bound-state equation in physical space at fixed light-front time τ. The resulting light-front equation is similar to the celebrated Schrödinger radial wave equation at fixed t, which describes the quantum-mechanical structure of atomic systems. Internal orbital angular momentum L and its effect on quark kinetic energy play an explicit role. Thus, by using the AdS/CFT correspondence, one obtains a relativistic wave equation applicable to hadron physics, where the light-front coordinate ζ has the role of the radial variable r of the nonrelativistic theory. The solutions of the wave equation determine the mass spectrum M and the wave functions

$\Phi(\zeta)$ of hadrons. The light-front Schrödinger wave equation for a pion in holographic QCD is Eq. (2),

$$\left[-\frac{d^2}{d\zeta^2} - \frac{1 - 4L^2}{4\zeta^2} + U(\zeta) \right] \Phi(\zeta) = M^2 \Phi(\zeta) \quad (2)$$

where the vast complexity of the QCD interactions among constituents is summed up in the addition of the effective potential $U(\zeta)$, which is then modeled to enforce confinement. For example, in the soft-wall model the potential is $U = \kappa^4 \zeta^2 + 2\kappa^2(J - 1)$, where J is the total angular momentum of the hadron. The corresponding wave functions of a pion describe the probability distribution of its constituents for the different orbital and radial states. The separation of the constituent quark and antiquark in AdS space gets larger as the orbital angular momentum increases. Radial excitations are also located deeper inside AdS space (**Fig. 2**).

Hadronic spectrum. Thus AdS/CFT and light-front holography provide a quantum-mechanical wave equation formalism for hadron physics. The soft-wall model, in particular, appears to provide a very useful first approximation to QCD. The solutions of the light-front equation determine the masses of the hadrons, given the total internal spin S, the orbital angular momenta L of the constituents, and the index n, the number of nodes of the wave function in ζ. For example, if the total quark spin S is zero, the meson bound-state spectrum follows the quadratic form $M^2 = 4\kappa^2(n + L)$. The pion, with $n = 0$ and $L = 0$, is massless for zero quark mass, in agreement with general arguments based on chiral symmetry. If the total spin of the constituents is $S = 1$, the corresponding mass formula for the orbital and radial spectrum of the ρ and ω vector mesons is $M^2 = 4\kappa^2(n + L + \frac{1}{2})$. The states are aligned along linear Regge trajectories (**Fig. 3**).

The light-front wave functions describe the quark and gluon composition of the hadrons in close analogy to the way positronium is described as an electron–positron bound state by the Schrödinger

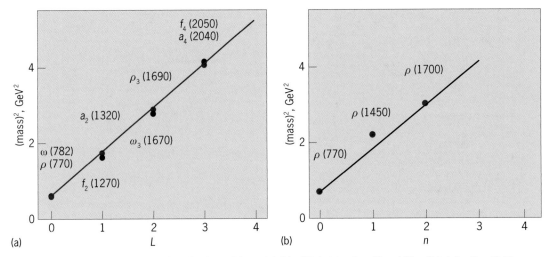

Fig. 3. Vector meson states in the soft-wall holographic model: (*a*) orbital states ($n = 0$) and (*b*) radial states ($L = 0$). The value of the mass scale is $\kappa = 0.54$ GeV. Experimental values are marked by dots.

equation. The hadron spectra and their wave functions are obtained for general spin and orbital angular momentum. The hadronic form factors predicted by the theory—which give the probability that a hadron remains intact in a collision—agree well with experiment. The nucleon has both *S*- and *P*-wave components, allowing one to compute both Dirac (spin-conserving) and Pauli (spin-flip) form factors. Even the spectrum of glueballs—the bound states of two or more gluons—is predicted within this framework. *See* EXPERIMENTAL SEARCH FOR GLUONIC HADRONS.

In conclusion, a significant theoretical advance in recent years has been the application of the AdS/CFT correspondence to study the dynamics of strongly coupled quantum field theories in terms of classical gravity in a higher-dimensional space. Extensions of the AdS/CFT method that have been recently developed by theorists have led to new analytical insights into the confining dynamics of QCD, a concept that is difficult to realize using other methods. The AdS/QCD model, together with light-front holography, provides a semiclassical first approximation to strongly coupled QCD. The model does not account for particle creation and absorption, and thus it is expected to break down at very short distances where such relativistic quantum effects become important. However, the model can be systematically improved, for example, by using AdS/QCD solutions as the basis functions to compute higher-order corrections to the full theory at fixed light-front time.

For background information *see* DUALITY (PHYSICS); ELEMENTARY PARTICLE; GAUGE THEORY; GLUONS; HOLOGRAPHY; NONRELATIVISTIC QUANTUM THEORY; POSITRONIUM; QUANTUM CHROMODYNAMICS; QUANTUM ELECTRODYNAMICS; QUANTUM FIELD THEORY; QUANTUM MECHANICS; QUARK-GLUON PLASMA; QUARKS; RELATIVITY; SUPERSTRING THEORY; SUPERSYMMETRY in the McGraw-Hill Encyclopedia of Science & Technology.

[This research was supported by U.S. Department of Energy Contract DE-AC02-76SF00515.]

Stanley J. Brodsky; Guy F. de Téramond

Bibliography. S. J. Brodsky and G. F. de Téramond, Hadronic spectra and light-front wavefunctions in holographic QCD, *Phys. Rev. Lett.*, 96:201601, 2006; L. Da Rold and A. Pomarol, Chiral symmetry breaking from five-dimensional spaces, *Nucl. Phys. B*, 721:79–97, 2005; G. F. de Téramond and S. J. Brodsky, The hadronic spectrum of a holographic dual of QCD, *Phys. Rev. Lett.*, 94:201601, 2005; G. F. de Téramond and S. J. Brodsky, Light-front holography: A first approximation to QCD, *Phys. Rev. Lett.*, 102:081601, 2009; J. Erlich et al., QCD and a holographic model of hadrons, *Phys. Rev. Lett.*, 95:261602, 2005; N. Evans, The gravity of hadrons, *Phys. World*, 18:26–27, 2005; A. Karch et al., Linear confinement and AdS/QCD, *Phys. Rev. D*, 74:015005, 2006; J. Maldacena, The illusion of gravity, *Sci. Amer.*, 293(5):56–63, November 2005; J. Polchinski and M. J. Strassler, Hard scattering and gauge/string duality, *Phys. Rev. Lett.*, 88:031601, 2002.

Structural design of high-rise towers

Recent trends and developments in creating leading skyscraper architecture (generally greater than 40 or 50 stories) are attributable to scientific and technological advances during planning, design, construction, operations, and maintenance phases. Architecturally coordinated functional, aesthetic, environmental, and economic requirements are combined with progressive architectural engineering systems to produce efficient high-rise towers for human occupancy and use, as permitted by zoning, fire and life-safety, energy, and building codes and standards.

Essential transitioning integration of mechanical, electrical, electronic, plumbing, civil, and structural engineering design capabilities in conjunction with innovative contemporary sculptural forms present many challenges to the project team. However, more advanced computer software computational and visualization techniques and programmable mathematical engineering calculations using photoelastic analysis (PEA), finite element analysis (FEA), load and resistant factor design (LRFD), and seismic response modification factors design (SRMFD) have enhanced the projection of environmental simulation studies. These include virtual preconstruction walk-through and fly-through space-time four-dimensional (4D) modeling as well as building information modeling (BIM) considered five-dimensional (5D) and automated shop-drawing production in conjunction with integrated project delivery (IPD) to facilitate critical cybernetic decision-making processes.

High-rise towers. Despite the devastating structural collapse of the World Trade Center (WTC) towers in Lower Manhattan in New York City on September 11, 2001, skyscrapers continue to competitively soar higher, globally (**Fig. 1a**).

The Physics Factbook™ developed an asymptotic power curve of graphed high-rise tower clusters, which predicted the recent remarkable ascension toward greater record-breaking heights within diminishing time spans. This is because of factors such as (1) symbolic corporate identity, (2) increasing land costs, (3) computerized advances in elastic, inelastic, and nonlinear dynamic structural analyses and design, as well as materials science, computer-aided design drawings, and computer-aided manufacturing (CADD/CAM), (4) introduction of new design-build risk management, cost- and quality-control methodologies, (5) economic progress in construction technologies, including "fast track" critical-path scheduling, remote radio controls, virtual-to-real robotics, and modularized prefabrication of component parts, (6) voluntary leadership in energy and environmental design (LEED) and "Energy Star" concerted efforts to significantly reduce fossil fuel consumption and focus on sustainable solar, wind, and ocean wave power sources to arrive at self-generation of electrical power and net-zero carbon emissions via high-rise building towers, (7) compliance with environmental systems sustainability and zoning regulations, (8) trends toward performance-oriented versus prescriptive codes, and (9) improvements in design and construction standards because of increased project and occupancy safety records.

Figure 1b shows the structural height versus rank of the top 100 tallest high-rise buildings in the world above 122 m (400 ft). This graph shows a fairly smooth curve for about 80% of the towers less than 305 m (1000 ft) in height in the greater density range. Beyond that, the smaller groups of plotted points are randomly spaced and irregular in their deviation from the power curve, illustrating the sporadically accelerating planning, design, and construction impetus provided by the disastrous September 11 event as well as incentives provided since 2000 on the verge of astounding worldwide economic and continuing population growth.

Nevertheless, considering the infill provided since the initial 2002 power-curve plot, with the abundance of new, taller high-rise tower construction in Dubai, Shanghai, and elsewhere because of progressive architecture and structural design advances, it appears that with slight modifications to the base criterion threshold at 198 m (650 ft), the smoothness of the power-curve equation is readily achiev-

able. However, the prior benchmark of 80%, noted above, would be reduced to about 75%, and the number and density of tallest high-rise towers in the database remarkably increased to the top 200 or 250. In this replotting process, the asymptotic curve gap graphed would increase by about 5% in width near its crescendo and allow for greater high-rise tower heights in a dramatically revised forecast (Fig. 1b).

Council on Tall Buildings and Urban Habitat. The Council on Tall Buildings and Urban Habitat (CTBUH) is an international organization of architects, structural engineers, city planners, and construction industry representatives. Originally, their standard for the height to the architectural top of tall buildings was measured from the sidewalk level of the main entrance to the structural roof or framed skylight. The CTBUH altered this standard so that the structural top now includes spires, which are architectural elements, but not television antennas, radio antennas, or flag poles.

The change was centered on the controversy of the 442-m-tall (1450-ft) Sears Tower (110 stories), completed in 1974 in Chicago (Fig. 1a), breaking the 1972 WTC towers (also at 110 stories) record of 415 and 417 m (1362 and 1368 ft). The Sears Tower antenna at 527 m (1730 ft) exceeded the height of the WTC, North Tower, antenna by only 1 m (3.28 ft). Prior to that, the tallest record was maintained for 41 years by the Empire State Building, constructed in 1931, at 381 m (1250 ft).

After the CTBUH issued the structural top/spire rule, another dispute was caused by the sparsely occupied rivaling (mixed-use) Petronas twin high-rise towers—88 stories each—constructed in Kuala Lumpur, Malaysia, in 1998 at 452 m (1483 ft) to the top of their spires.

The controversy subsided in 2004 with the completion of the Taipei 101 Building, as it became the world's tallest skyscraper at only 101 stories, but with a structural top height of 509 m (1671 ft) and 614 m (2015 ft) with its antennas (Fig. 1a).

The predecessor tallest structure, the Eiffel Tower (in 1889) at 305 m (1000 ft), designed by civil and structural bridge engineer Gustave Eiffel, is intermittently occupied by tourists, restaurant patrons, service personnel, and tower administrators and is considered a nonoccupied wrought-iron observation tower by CTBUH. Therefore, it is not included in their official listings. Nevertheless, Eiffel's innovative tower probably forecasted the eventual iconic shape and form of occupied ultra-high-rise tower structures, incorporating super "octet" (octahedral-tetrahedral) space-frames, via the ingenious use of the "Phoenix (lattice) column" for space-columns, -beams, and -diagonal braces to ultimately create "green" parklike apertures and multilevel atria.

Structural design trends. Architects, structural engineers, building owners, and skyscraper developers are increasingly producing avant-garde forms. Their shapes are not limited to the traditional Cartesian coordinate (x, y, z) geometry of past edifices. Instead, recent trends portray innovative sculptural forms constructed of polygonal and curvilinear floor plates.

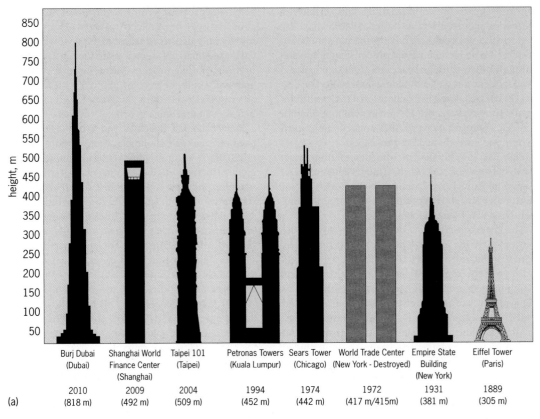

(a)

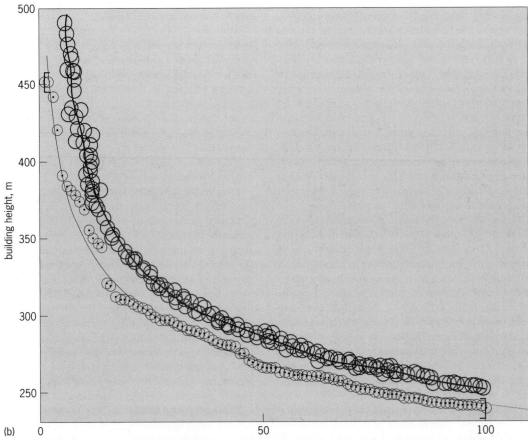

(b)

Fig. 1. **Tall building structures worldwide. Construction completion dates (right to left) range from 1889 to 2010 (***a***) Council on Tall Buildings and Urban Habitat (CTBUH) architectural heights. Although the Eiffel Tower is not a CTBUH tall building, it is strategically significant as a visionary precipitator of structural skyscrapers in the historical past as well as a progenitor of futuristic forms emerging for very tall and very densely populated city structures. The destroyed World Trade Center Twin Towers are included because of their prominent CTBUH significance. (***b***) Height versus rank (right to left) of top 100 buildings. The curves indicate relatively current (small circles) as well as projected indefinite future trends (large circles). (***Modified from The Physics Factbook***™)**

Tapered cross sections with inclined exterior glass-walled structures, as in architect Santiago Calatrava's occupied spiraling tower in Malmo, Sweden (2005), and the proposed 124-story, 610-m-tall (2000-ft) Chicago Spire, culminate in virtually roofless pinnacles containing multistory sky lobbies, boardroom suites, and restaurants, capped with observation decks and spires for communications antennae and aircraft beacons.

Soaring high-rise towers depend on high-speed elevator systems and multistory escalators. The vertical cross-section configuration of stacked elevator banks, where most accumulate at lower floor levels and only a few service the uppermost tower floors, is often reflected in the stepped tapering exterior forms of high-rise towers and in net leasable (efficiency-factored) floor-plate areas.

Contemporary high-rise towers continue to be designed as untethered structural cantilevers (for example, flagpoles) anchored in the earth via a variety of sophisticated load-bearing foundation systems to resist sliding, rotating, leaning, and overturning. Normally, lateral resistance to horizontal and downdraft vortex-shedding wind forces govern over dynamic seismic vibrations caused by sinusoidal and compressive earthquake wave frequencies, in compliance with the structure's natural period and other structural integrity requirements of building codes and structural design standards.

Since constructed high-rise tower origins in the late nineteenth and early twentieth centuries, their structural design has increasingly depended on counterbalancing the predominant compressive gravity loads of normally centralized cores that traditionally house the clustered elevators, restrooms, mechanical and electrical rooms, and code-required fire exit stairways, with the tensile lateral resistance provided primarily by exposed structural framing facade systems. Basically included are fundamental structural design elements of triangulation, involving diagonal braces or trusses within the rectilinear or trapezoidal exterior frames as expressed architecturally in (1) the John Hancock Center in Chicago, designed by famed pioneering structural engineer Fazlur Kahn ("father" of the modern skyscraper) of Skidmore, Owings and Merrill, (2) the Bank of China in Hong Kong, designed by leading modern architect I. M. Pei, and (3) the recently completed Hearst Tower in New York City, designed by London architect Sir Norman Foster, a proponent of the half-mile-high mixed-use conical Millennium Tower diamond-grid tube anchored in Tokyo Bay (**Fig. 2**).

Structural engineering designers are constantly challenged to balance ductility (flexibility) with rigidity (stiffness) in creating "soft" and "hard" structural members and connections, which take into consideration safety factors that often exceed the required maximum loadings as well as design criteria involving foreshortening, deflections, elongation, bonding, cracking, twisting, shear lag, and shearing potential, with approved code-required fire-resistance ratings to avoid progressive collapse, loss of life, and severe property damage.

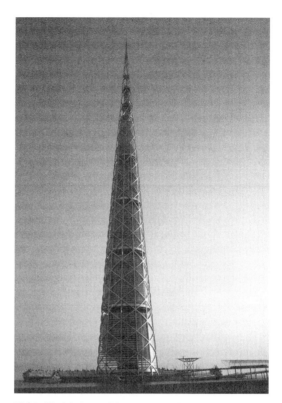

Fig. 2. Model photo of the proposed Millennium Tower in Tokyo Bay. Note the decreasing widths of the elevator and fire exit stairway core with increasing tower height. (*Courtesy Foster + Partners, http://www.fosterandpartners.com*)

With increasing trends toward optimizing high-rise tower heights, the aspect ratio (CTBUH heights over width at base), as a prevailing structural design criterion (Fig. 1a), has climbed dramatically, from less than 3.5:1 at the Empire State Building to 6.7:1 at the WTC to 7.3:1 at Taipei 101, and is projected to be 7.6:1 at Burj (Tower) Dubai (**Fig. 3**). However, leading architects and structural engineers are designing even more slender, visually dynamic (as in simulated seismic-motion) towers, with several under construction globally that not only approach (Chicago Spire design) but also exceed an aspect ratio of 10:1, like the "The Illinois" mile-high tower proposed in 1956 by Frank Lloyd Wright. This is made possible by using high-strength carbon-fiber-reinforced concrete (least cost and most rigid) construction, prefabricated high-strength welded and bolted structural steel connections, and composite concrete-steel design specifications. Tubular frame designs, as initiated by Fazlur Kahn, with inherent "bundled tube" concepts (Sears Tower, now called Willis Tower) in conjunction with core, outrigger, and exterior envelope belt trusses, amalgamate the structure's natural periods of varying sizes and tube heights and resist excessive tower sway with integrated damping devices to prevent disastrous collisions between tubular segments and calamitous overturning. Application of these advanced structural design principles culminate in the world's tallest, Burj Dubai (Fig. 3), however, without the merits of 1-acre column-less floor areas, as at the WTC,

Fig. 3. Burj Dubai. (*Courtesy Emaar Properties*)

X-Seed 4000 skyscraper form for Tokyo (inspired by Mount Fuji), with a height of 4000 m (13,123 ft) and a 6-km-wide (3.7-mi) base with a very low 0.67:1 aspect ratio.

With advanced computerization, computational and environmental simulation techniques, a combination of the tensegrity (tension plus integrity) structural design principles introduced by Fuller-Sadao for large-scale mixed-use high-rise tower applications, the powerful demonstrable envisioning concepts of leading American architect-planner-engineer Frank Lloyd Wright for treelike high-rise tower structures, such as Johnson Wax (1936 and 1944) and Price (1952), and the ecological concepts of leading international architects Paolo Soleri, Ken Yeang, and William McDonough in all probability are bound to lead to structural design innovations forecasted periodically. Of The Illinois mile-high tower, Wright openly declared, "Although this skyscraper may be considered too expensive to build right now, in the future we cannot afford not to build it."

For background information *see* ARCHITECTURAL ENGINEERING; BUILDINGS; STRUCTURAL ANALYSIS; STRUCTURE (ENGINEERING); STRUCTURAL MECHANICS

because of the implementation of compartmented reinforced concrete load-bearing structural shear walls.

Outlook. Currently being contemplated for ultra-high-rise tower structures for human occupancy in the 0.5- to over 2-mi (0.8- to 3.2-km) range are articulated "octet" super-space-frame structures designed to withstand the squared wind-velocity design pressures and cubed wind-energy power accumulations achieved with incremental increases in height. These include lateral wind-relieving green landscaped apertures and multilevel parklike atria within tubular structures architecturally designed to accommodate "city within a city" populations of 100,000 to 1,000,000, as with the unbuilt floating terraced Tetrahedral City designed for Tokyo Bay and New York Harbor by architects and engineers Buckminster Fuller and Shoji Sadao in 1968.

A noticeable simultaneous trend in the last few years for super-high-rise tower proposals located outside the urban grid, in alignment with grandiose city-in-a-park concepts of famed twentieth-century architect Le Corbusier, is the lowering of the typically ascending aspect ratio noted above. These proposals include (1) the 1000-m-tall (3280-ft) Sky City 1000 (**Fig. 4**) on land or water by Takenaka Construction Co. of Japan (2.5 to 1), (2) the 2-mi-high (3.2-km) Ultima Tower by (Eugene) Tsui Design and Research, Inc., incorporating draped steel suspension cables with a 1-mi (1.6-km) base diameter (2 to 1); and (3) the 800-story asymptotic-conical

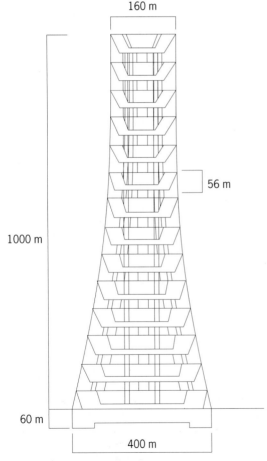

Fig. 4. Proposed Sky City 1000, with 14 levels and 800 hectares of floor area for offices, residences, commercial spaces, and so on.

in the McGraw-Hill Encyclopedia of Science & Technology.   Andrew Charles Yanoviak

Bibliography. G. Binder, *101 of the World's Tallest Buildings*, Images Publishing Group, Australia, 2006; W. F. Chen and E. M. Liu, *Principles of Structural Design*, CRC Press, 2006; K. M. Hays and D. Miller (eds.), *Buckminster Fuller: Starting with the Universe*, Yale University Press, 2008; A. Lepik, *Skyscrapers*, Prestel Publishing USA, 2004; M. Wells, *Skyscrapers Structure and Design*, Yale University Press, 2005.

Supercapacitors

Supercapacitors are energy storage devices. They are charged from an external power source, such as a wall socket, and store the energy for later use. As an energy storage system, they are compatible with alternative, or clean, energy sources, such as windmills or solar panels. For example, a supercapacitor may collect and store the energy produced by solar panels during the day and provide back this energy at night. Supercapacitors are also called electrochemical capacitors and ultracapacitors. In certain cases, they may be called double-layer capacitors.

Double-layer capacitors. Charge storage in a double-layer capacitor is similar to that in a typical parallel-plate capacitor, where charge is placed on one of two parallel metal plates and charge of the same magnitude but opposite sign is placed on the other metal plate. In double-layer capacitors, the two metal plates are replaced by an electrode/electrolyte boundary. The electrode is an electronically conducting solid, usually a metal. The electrolyte is, typically, a solution containing many ions which is ionically conducting and electronically insulating, meaning that the ions can travel through the electrolyte (and therefore carry the current) but electrons cannot pass through the electrolyte. The surface of the electrode is charged. Ions of the opposite charge are attracted electrostatically from the electrolyte and reside in the region of the electrolyte just outside the electrode surface, exactly balancing the charge placed on the electrode surface (**Fig. 1**). There is a double layer of charge on each electrode/electrolyte boundary, where one layer is the charge on the electrode surface and the other layer is the charged ions in the electrolyte. A similar result will occur on the other electrode in the supercapacitor, but with charges of opposite sign on the electrode surface (Fig. 1). Supercapacitors which store their charge in this way are also called double-layer capacitors.

Typically, double-layer capacitors use high-surface-area carbon electrodes, because carbon is relatively inexpensive and widely available in a number of forms, such as carbon powders, carbon cloths, and carbon nanotubes. The electrolyte for these capacitors can be water-based (aqueous) or can contain no water (nonaqueous). The most common aqueous electrolytes are sulfuric acid (H_2SO_4) and potassium hydroxide (KOH). Aqueous electrolytes will break down (forming oxygen or hydrogen gas) when the double-layer capacitor voltage exceeds approximately 1 V. Conversely, the nonaqueous electrolytes are typically acetonitrile or propylene carbonate. Nonaqueous electrolytes have a larger voltage window than aqueous electrolytes, typically between 2.5 and 4 V. Having a large voltage window leads to greater energy storage capacity in the supercapacitor. However, nonaqueous electrolytes require more stringent preparation methods, because the electrolytes must be completely dry and are flammable, which may require extra handling considerations.

Because the charge storage mechanism of the double-layer capacitor does not require any phase changes or changes to the material during charging/discharging, these supercapacitors are capable of achieving hundreds of thousands or even a million cycles without significant degradation in their energy/charge storage characteristics. This is a major advantage, particularly for applications in which many cycles are desired.

The time required to form and unmake the double layer is very short (approximately 10^{-8} s), meaning that the charge is accessible at a high rate (high current) and that the supercapacitor can be charged and discharged very rapidly. One important performance characteristic for energy storage systems is the power of which they are capable, where power is the product of the voltage and the current that can be drawn at that voltage. The short formation/deformation time of the double layer in the supercapacitor means that these systems are theoretically able to provide high power. This is important in high-power applications.

Storing the charge in the double layer also leads to one of the major disadvantages of double-layer capacitors, that being low energy. Energy is the product of the charge and the voltage at which the charge can be withdrawn. High energy means longer times between charging. For double-layer capacitors, the stored energy is relatively low. Stored energy is also described as energy density, the energy available

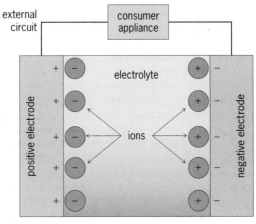

Fig. 1. **Charged double-layer capacitor.**

from the system per volume, and as specific energy, the energy per system mass.

The charge stored per surface atom in a double layer can be calculated from experimental capacitance values, where the capacitance for a metal is approximately 30 μF/cm^2 of accessible surface area. Given that there are approximately 10^{15} atoms/cm^2 in a typical surface, that one electron holds 1.6×10^{-19} C of charge, and that in a double-layer capacitor with a water-based electrolyte the practical voltage window is approximately 1 V, the charge per surface atom can be estimated as

$$\left(\frac{30\,\mu F}{cm^2}\right)\left(\frac{cm^2}{10^{15}\,atoms}\right)\left(\frac{e^-}{1.6\times10^{-19}\,C}\right)(1V) \quad (1)$$

$$= 0.19\ e^-\ \text{per surface atom}$$

Thus, less than one electron is stored per surface atom in the double layer. This can be compared with a Faradaic reaction, in which one or more electrons can be stored per surface atom as the atom is oxidized or reduced. The double-layer capacitor stores less charge than a system such as a battery, which uses a Faradaic reaction, which leads to an inherently lower energy for the double-layer system.

Supercapacitor energy can be increased by replacing the aqueous electrolyte used in the above equation with nonaqueous electrolytes, which have up to a four times higher practical voltage window, thereby increasing the number of electrons that can be stored per surface atom. Though, for some nonaqueous electrolytes the capacitance per surface area may be smaller than in aqueous electrolytes. Many commercial supercapacitors presently available use nonaqueous electrolytes to provide higher energy densities.

Comparison of double-layer capacitors to batteries. Supercapacitors are similar to batteries in that they are energy storage devices that must be charged from an external power source. In double-layer capacitors, the charge storage mechanism is very different than that in batteries, resulting in different cycle lives, power and energy densities, and potential profiles during charging/discharging.

In a battery, the charge is stored in a Faradaic reaction, in which one species loses electrons (is oxidized) and another species gains those electrons (and is reduced). This oxidation and reduction is often coupled with a change of phase of the species. In a battery, changes in phase and chemical composition are one of the modes that often lead to failure over time (multiple cycles), and this is why only a few thousand cycles are usually possible before the battery loses its ability to store charge, or fails altogether. Because there is very little physical change going on in the supercapacitor, several hundred thousand or even a million cycles are easily achievable. This longer cycle life means that supercapacitors need to be replaced less often than batteries, which is ideal for systems which need to be charged often or for systems in locations that are difficult to access.

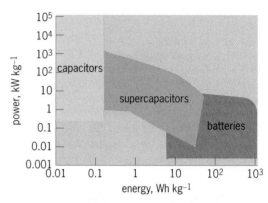

Fig. 2. Comparison of the specific energy and power of supercapacitors, batteries, and parallel-plate capacitors.

The time required to form or break down the double layer is very short (10^{-8} s), much shorter than the rate of many Faradaic reactions (10^{-4}–10^{-2} s). This shorter time may lead to higher power capabilities for the double-layer capacitor, compared to a battery (**Fig. 2**). However, double-layer capacitors cannot store as much charge as batteries, with double-layer capacitors storing 0.19 electrons per surface atom and the Faradaic reaction in a battery allowing for one or more electrons to be stored per surface atom. So, the energy of a double-layer capacitor is inherently smaller than that of a battery (Fig. 2).

Supercapacitors based on pseudocapacitance. There is a second type of supercapacitor that stores its charge based on a Faradaic reaction which exhibits pseudocapacitance. The term pseudocapacitance describes a Faradaic reaction which behaves electrochemically like a capacitor. Essentially, this means that the electrochemistry of the Faradaic reaction is so fast that the reaction is always at equilibrium throughout the given potential region. Three criteria need to be met for a reaction to be considered pseudocapacitive. First, the cyclic voltammogram (a plot of the current that flows as potential is cycled linearly between two potentials) must be a mirror image of itself above and below the zero current line (**Fig. 3**). Ideally, this current should be constant throughout the potential range. Capacitance, the amount of charge

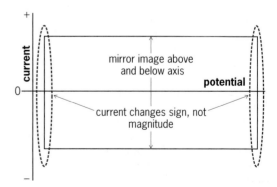

Fig. 3. Cyclic voltammogram of idealized pseudocapacitive supercapacitor electrode.

stored per applied potential, can be determined by dividing the current by the sweep rate (the rate of the potential change in the cyclic voltammogram), so a constant current indicates that the capacitance is also constant throughout that region and the same amount of charge can be stored or is available at each voltage to which the capacitor is charged or discharged.

Second, at the potential extremes of the cyclic voltammogram, the current should change sign but have a small or no change in magnitude (Fig. 3). This means that charge can be put onto the supercapacitor up to the maximum voltage and that this charge can be taken off at this voltage, so it is not necessary for the voltage to fall significantly before the charge can be recovered and used to power an appliance.

Third, as the sweep rate is increased, the current of the reaction, at a given potential, should also increase linearly. If these criteria are met, then the Faradaic reaction is said to be at equilibrium throughout the window, and the reaction is considered to be pseudo-capacitive and acceptable for use in a supercapacitor.

Pseudocapacitance generally arises in one of three ways. (1) It can be exhibited by systems in which there is a species which deposits onto the electrode surface and the amount of coverage is related directly to the potential applied. At each applied potential there is a certain characteristic surface coverage, and when a particular potential is applied to the electrode, the film deposits or dissolves essentially instantaneously to reach the characteristic surface coverage; the system is then at equilibrium. This is exhibited by hydrogen on platinum electrodes during underpotential deposition. (2) A similar situation may occur with the intercalation of some species into the lattice of the electrode material. Again, for each applied potential, there is some characteristic amount of intercalation which is related directly to the potential applied. This type of pseudocapacitive electrochemistry is exhibited by the intercalation of Li into TiS_2. (3) Some supercapacitors are based on the oxidation/reduction of the electrode material itself. For these supercapacitors, each potential has a characteristic ratio between the oxidized and reduced species, and when a particular potential is applied there is an instantaneous conversion of species in the electrode to achieve the required ratio. There are several examples of supercapacitors based on these types of pseudocapacitive materials, such as ruthenium oxide and manganese oxide supercapacitors. These electrode materials provide high energy densities (typically, 10–60 Wh/kg, depending on supercapacitor configuration, electrolyte, and metal oxide); however, these materials tend to be quite heavy and expensive.

Because the reaction responsible for pseudocapacitance is a true Faradaic reaction, it is able to store multiple electrons per surface atom as that atom undergoes oxidation or reduction. Thus, supercapacitors based on pseudocapacitance store significantly more charge and have significantly higher energies (10–60 Wh/kg) than double-layer capacitors (typically, 5–10 Wh/kg).

Self-discharge in supercapacitors. One concern for supercapacitor manufacturers is self-discharge, the loss of voltage experienced by a supercapacitor as it sits unused in a charged state. In other words, a charged supercapacitor does not hold its charge indefinitely; it loses charge steadily with storage time. The different types of supercapacitors described above have different self-discharge mechanisms. Some of these mechanisms have been identified, such as electrolyte decomposition or discharge due to the oxidation/reduction of an electrolyte impurity (for example, $Fe^{2+/3+}$), whereas others are still under study. The rate of self-discharge limits the applications for which supercapacitors can be used. For instance, a supercapacitor which loses its charge within 7 days could not be used in place of a lead–acid battery in a motor vehicle. Other applications are not so sensitive to self-discharge, such as those in which the supercapacitor is attached to a power supply all the time and is used only in high-power situations (such as in the camera flash described below) or when the power supply fails (such as in a power backup system). Nevertheless, manufacturers and research scientists are attempting to identify the self-discharge mechanisms in the hope that self-discharge can be minimized, which would lead to a broader range of applications.

Effect of surface area on supercapacitors. The amount of charge that can be stored in a supercapacitor is based on the amount of material in the electrode (for supercapacitors based on pseudocapacitance) or on the surface area of the electrode (for double-layer capacitors). Therefore, increasing the surface area is generally considered to be desirable if large energy densities are required. It also is desirable for the supercapacitor to remain relatively small, particularly for applications such as portable electronics. To increase the surface area, a porous electrode material is generally used. Decreasing the pore diameter and increasing the number of pores in the electrode increases the surface area and therefore increases the charge storage, leading to higher energy densities.

To charge the supercapacitor, ions must pass through the pores to counter the charge that is placed on the electrode surface. As ions move through the electrolyte, they encounter resistance to their movement, commonly called solution resistance. The dimensions of the pores have a significant effect on solution resistance, similar to the resistance to electron movement seen in wires. As the pore increases in length, there is a higher solution resistance; and as the pore narrows, the solution resistance increases. The practical outcome is that for long, small pores, the tip of the pore will charge rapidly, because ions do not have difficulty reaching these areas. However, the ions will require longer times to reach areas deep in the pore, and these areas will charge more slowly than the areas at the tip of the pore. Therefore, if long, small pores are used in the supercapacitor electrode to increase the energy, the charge stored deep in the pores is accessible only at relatively slow rates. This lowers the power of the supercapacitor. Recall that one of the main advantages of the supercapacitor is high power. By increasing the energy with the pores, this advantage is lost.

Many research groups are attempting to design supercapacitor electrodes which may have high surface areas to provide high energy densities and wide pores for easier access and therefore higher power densities. Several methods are used to accomplish this, including using carbon nanotubes and templating carbon with silica spheres. Generally, all of these methods are designed to produce materials, typically carbons, that have large pores (mesopores) through which the ions can pass easily, and in the walls of these pores there are very small pores (micropores) which provide the high surface area (**Fig. 4**). In this way, the ions that need to move into these very small pores to counter the charge will have large "highways" through which to pass most of the distance and need to travel only very short distances in the very small pores. This provides both high power and high energy.

Applications. Supercapacitor applications tend to require either high power capabilities or a long cycle life. Supercapacitors are being used in place of lead-acid batteries in fleet vehicles (for example, delivery vans), where they are required to turn engines on and off many times a day. The high cyclability of the supercapacitor ensures that it does not wear out as quickly as a lead-acid battery in this situation. Additionally, supercapacitors can be purchased which provide high power, as needed for vehicle stereos and the flash for cell phone cameras. The high power required for these two situations (loud bass beats from a stereo or strong flash for the camera) is more than can be supplied by the battery. Thus, the supercapacitor supports the battery by providing the power needed during these specific situations. Supercapacitors are being used in cordless electric hand tools, where the high power capabilities allow

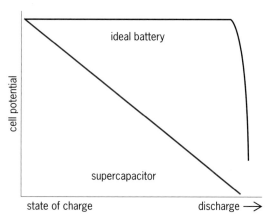

Fig. 5. Potential profile of supercapacitor and ideal battery during discharging.

for very rapid recharges and the long cycle life means the power supply of the power tool need never be replaced.

The potential profile exhibited by supercapacitors during charging and discharging is characterized as a linear drop in potential during supercapacitor discharge or use (**Fig. 5**). This is unlike the potential profile of batteries, which, ideally, will produce the same potential throughout their discharge until right before the battery dies. The linear drop in potential can be a hindrance in using supercapacitors in applications where a constant potential is required to run the appliance. For other applications, however, it may be important to know how charged the supercapacitor is, and the linear drop in potential with use can provide a method of state-of-charge identification.

It is unlikely that one charge storage system, such as the supercapacitor, battery, or fuel cell, will be ideal for all of our consumer applications and will win out over the others. More likely, all three systems (as well as nonelectrochemical systems, such as flywheels) will work in tandem to supply consumers' needs. Each storage system has benefits and disadvantages which will limit its possible applications. It is likely that for some consumer appliances, a combination of two or even all three of these storage systems will be required.

For background information *see* BATTERY; CAPACITANCE; ELECTRIC VEHICLE; ELECTROCHEMISTRY; ELECTRODE; ELECTROLYTE; ELECTROLYTIC CONDUCTANCE; ENERGY STORAGE; OXIDATION–REDUCTION; SOLID-STATE BATTERY in the McGraw-Hill Encyclopedia of Science & Technology.   Heather Andreas

Bibliography. B. E. Conway, *Electrochemical Capacitors: Scientific Fundamentals and Technological Applications*, Kluwer Academic/Plenum Publishers, 1999; B. E. Conway, Transition from "supercapacitor" to "battery" behavior in electrochemical energy storage, *J. Electrochem. Soc.*, 138(6):1539-1548, 1991; E. Frackowiak, Carbon materials for supercapacitor application, *Phys. Chem. Chem. Phys.*, 9:1774-1785, 2007; M. Winter and R. J. Brodd, What are batteries, fuel cells, and supercapacitors?, *Chem. Rev.*, 104:4245-4269, 2004.

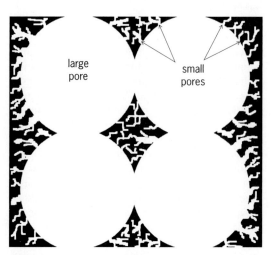

Fig. 4. Templated carbons containing large and small pores.

Supernova 1987A

Supernova 1987A (SN1987A) is the brightest supernova to be observed since Johannes Kepler observed a supernova in 1604 and the first to be observed in every band of the electromagnetic spectrum. It was first detected on February 24, 1987, at 0530 Universal Time in the Large Magellanic Cloud near the bright star-forming nebula 30 Doradus.

Two independent experiments detected a flash of neutrinos from SN1987A that occurred about 2 h before the discovery of the optical outburst. The observed properties of the neutrino flash [energy $\sim 3 \times 10^{46}$ J, temperature $\sim 4.6 \times 10^{10}$ K (the characteristic temperature of the Fermi–Dirac distribution function of the neutrinos, which reflects the temperature of their source), and decay time scale ~ 4 s] were in remarkably good agreement with predictions from theoretical models in which a degenerate iron core collapsed to form a neutron star.

At discovery, SN1987A had a visual magnitude of approximately $V = 5$, which made it bright enough to be seen with the unaided eye. Early ultraviolet spectra showed that the photosphere was expanding with velocities exceeding 30,000 km s^{-1} (19,000 mi s^{-1}). The supernova brightened by an additional factor of 6 until it reached maximum light, $V = 3$, about 3 months after outburst. After that, the supernova debris became transparent to visible light and faded rapidly, decaying with time t approximately as exp ($-t$ /111.3 days). This decay rate, and the observation of characteristic gamma-ray emission lines at energies of 847 and 1238 keV (wavelengths of 1.48 and 1.00 pm), confirmed the hypothesis that the supernova light was dominated at early times by the radioactive decay of 0.07 solar masses of cobalt-56 (^{56}Co) that was produced during the supernova explosion. The integrated light resulting from this decay was about 0.7×10^{42} J.

As the supernova debris expanded, it cooled rapidly. By about 500 days after the outburst, the temperature dropped below 3000 K (5000°F), and carbon monoxide (CO) molecules and dust grains formed in the interior. The inner radioactive debris is expanding with a velocity of about 3000 km s^{-1} (1900 mi s^{-1}). Today, its light is dominated by the radioactive decay of titanium-44 (^{44}Ti). Although the debris has faded by a factor of about 10^7 since maximum (**Fig. 1**), it can still be imaged by the *Hubble Space Telescope* (**Fig. 2**). With a present temperature of less than 100 K ($-280°$F), the inner debris is perhaps the coldest optically emitting source known to astronomers. Most of its luminosity (about 200 Suns) emerges in the far-infrared (10–100-μm) band.

Circumstellar matter. SN1987A is surrounded by a bright ring that appears to be elliptical in shape but is actually a circular equatorial ring of radius approximately 0.67 light-years that is inclined at 45° with respect to the plane of the sky (Fig. 2). Besides this equatorial ring, SN1987A is surrounded by an even more remarkable pair of outer loops that are not concentric with the supernova. It is believed that these loops are actually also circular rings with approxi-

mately the same polar axis as the inner ring, but not coplanar with the inner ring. If so, the outer loops have physical radii of about 1.5 light-years and lie on planes approximately parallel to the plane of the inner ring but displaced by about 1.3 light-years.

The approximate bipolar symmetry of the ring system suggests that the supernova progenitor star was once a binary system and that the circumstellar gas was ejected when the two stars merged. The measured expansion velocity of the rings indicates that they were ejected by the supernova progenitor system some 20,000 years before it exploded.

These rings began to glow in emission lines when their gas was illuminated and heated by ultraviolet radiation and soft x-rays produced by the supernova during the first day after the explosion. From the fading rate of the equatorial ring, it is inferred that the glowing gas in the equatorial ring has atomic densities in the range $(0.6$–$3.3) \times 10^4$ cm^{-3} and mass of about 0.04 solar masses.

Actually, the glowing gas in the rings represents only a small fraction of the circumstellar gas that was

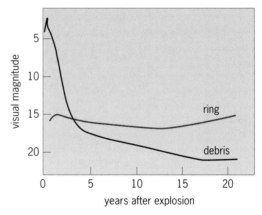

Fig. 1. Light curves of the debris of supernova 1987A and of its inner circumstellar ring.

Fig. 2. Supernova 1987A and its triple ring system 20 years after the explosion, as observed by the *Hubble Space Telescope*. The faint outer loops are barely visible. The faint amorphous nebulosity at the center is the inner debris of the supernova explosion, which is glowing by virtue of the radioactive decay of ^{44}Ti. The two bright stars are unrelated to the supernova. (*NASA; ESA; Space Telescope Science Institute; P. Challis; R. Kirshner, Harvard-Smithsonian Center for Astrophysics*)

ejected by the supernova progenitor system. During the first few years after the explosion, astronomers also detected "light echoes"—rapidly varying nebulosity beyond the rings produced by reflection of supernova light off dust grains in the circumstellar gas. The gas responsible for the light echoes was not ionized by the supernova and is now invisible. Its morphology and mass are uncertain but probably comprise a few solar masses.

From supernova to supernova remnant. From observations of the absorption lines during the first few months, one can infer that the supernova debris has kinetic energy of about 1.5×10^{44} J, or roughly 200 times the integrated optical luminosity. This energy will be converted into radiation as a result of the impact of the debris with the circumstellar matter, a process that will take place over decades or centuries after the explosion.

As the supernova debris expands, it drives a blast wave through the circumstellar matter, while at the same time a "reverse shock" propagates backwards into the expanding debris. The shocked gas between the blast wave and the reverse shock is heated to temperatures greater than 10^7 K. Such gas radiates primarily in the soft x-ray band (about 0.5–2 keV or 2.5–0.6 nm). In addition, such shock waves accelerate electrons to relativistic energies and compress magnetic fields, resulting in synchrotron radiation that can be observed at radio wavelengths.

Evidence for interaction of the supernova debris with circumstellar matter first appeared about 3.5 years after the supernova explosion in the form of radio synchrotron emission and soft x-rays. Images of the radio emission showed that it was coming from an annulus having a radius of about two-thirds of the inner circumstellar ring, indicating that the source had expanded with velocities exceeding 30,000 km s^{-1} (19,000 mi s^{-1}). The expansion velocity abruptly decelerated to about 4000 km s^{-1} (2500 mi s^{-1}) after the emission appeared, indicating that the blast wave had encountered a sudden increase in density of the circumstellar gas inside the equatorial ring. Thereafter, the radio and x-ray luminosities continued to increase steadily, and both the radio and x-ray images continued to expand with velocities of about 4000 km s^{-1} (2500 mi s^{-1}).

The equatorial ring was predicted to brighten rapidly when the blast wave reached it, about a decade after the explosion. The first indication of this event occurred in July 1997, when the *Hubble Space Telescope* detected a rapidly brightening "hotspot" on optical images of the inner ring. Thereafter, several more such hotspots appeared, so that today the ring is almost completely encircled by some 30 such hotspots (Fig. 2). Most of the hotspots are still unresolved by the *Hubble Space Telescope* and have not merged with each other. Spectral line profiles indicate that the hotspots are being crushed by shocks having velocities of about 200 km s^{-1} (120 mi s^{-1}).

The integrated optical light of SN1987A is now dominated by the light from the hotspots in the inner ring (Fig. 1). The appearance of the hotspots marks the transition of SN1987A from its supernova phase

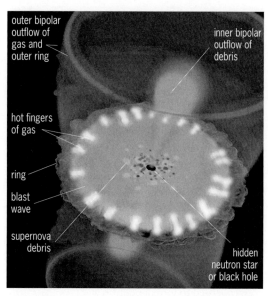

Fig. 3. Schematic model of SN1987A. (*NASA; ESA; A. Feild, Space Telescope Science Institute*)

into its supernova remnant phase, when its radiation is dominated by impact of the supernova debris with circumstellar matter rather than by radioactivity from the interior.

At about the same time as the hotspots began to appear, the x-rays from SN1987A began to brighten at an accelerating rate. The x-ray image is an elliptical ring that resembles the distribution of hotspots on the optical ring. The radial expansion velocity of the x-ray image decreased from about 4000 km s^{-1} (2500 mi s^{-1}) to about 1700 km s^{-1} (1050 mi s^{-1}). From 1997 to 2009, the x-ray luminosity of SN1987A increased by a factor of about 100.

Figure 3 illustrates the hydrodynamics of this event. The hotspots are produced when the supernova blast wave overtakes "fingers" of dense gas protruding radially inward from the inner equatorial ring. The blast wave suddenly slows down as it transmits into the dense gas in the fingers. The shocked gas behind the transmitted shock is sufficiently dense to radiate optically, manifested in the hotspots. The x-ray emission is dominated by emission from lower-density shocked gas near the fingers.

There are many unresolved questions regarding the model shown in Fig. 3. The inner debris appears to be expanding preferentially along the polar axis. We still lack a detailed understanding of the distribution of gas beyond the circumstellar rings and of the presupernova hydrodynamics that shaped the circumstellar ring system and the fingers.

Nature of compact object. Perhaps the outstanding mystery of SN1987A is the absence of any evidence (since the initial neutrino flash) for a compact object at its center. The bolometric luminosity of the supernova debris (that is, the total energy of the emitted electromagnetic radiation, mostly infrared) is now about 200 Suns, and a brighter compact object could not have escaped detection. If the compact object is a neutron star, as is expected, it must be very faint—perhaps because it has an anomalously low magnetic field or spin rate—and it can accrete no more than 10^{-11} solar masses per year.

Future behavior. The remnant of SN1987A will continue to brighten in the radio, infrared, optical, ultraviolet, and x-ray bands and will remain bright for many decades. Within another decade, the hotspots will merge as the blast wave overtakes the entire inner circumstellar ring and begins to interact with the gas beyond the ring. The number of ionizing photons from the shock interaction will exceed the initial flash of the supernova. These photons will cause heretofore unseen circumstellar matter to become visible, providing a new opportunity to understand the complex history of mass loss from the progenitor star. The ionizing radiation from the shock interaction will also shine inward, causing newly synthesized elements in the inner supernova debris to glow, well before they reach the reverse shock.

Large ground-based telescopes equipped with adaptive optics and the *James Webb Space Telescope* will provide infrared images and spectra of SN1987A with unprecedented angular resolution and sensitivity, providing a much clearer view of the shock interactions and the distribution of the glowing inner debris, and an opportunity to search more deeply for infrared emission from a central compact object. The Atacama Large Millimeter Array (ALMA) will provide images of SN1987A at millimeter wavelengths having angular resolution superior to optical images obtained by the *Hubble Space Telescope*. ALMA images of nonthermal continuum radiation will provide a detailed view of the acceleration of relativistic particles in the circumstellar shocks. They will also enable us to push the search for millimeter emission from a compact object to much deeper limits than are currently possible. Finally, ALMA may be able to image CO emission from the cold circumstellar gas beyond the visible ring system, providing vital clues to the hydrodynamic processes by which the supernova progenitor ejected this gas.

For background information, *see* GAMMA-RAY ASTRONOMY; HUBBLE SPACE TELESCOPE; INFRARED ASTRONOMY; INTERSTELLAR MATTER; NEBULA; NEUTRINO; NEUTRINO ASTRONOMY; RADIO ASTRONOMY; RADIO TELESCOPE; SHOCK WAVE; SUBMILLIMETER ASTRONOMY; SUPERNOVA; X-RAY ASTRONOMY in the McGraw-Hill Encyclopedia of Science & Technology.
Richard McCray

Bibliography. W. D. Arnett et al., Supernova 1987A, *Annu. Rev. Astron. Astrophys.*, 27:629–700, 1989; R. McCray, Supernova 1987A revisited, *Annu. Rev. Astron. Astrophys.*, 31:175–216, 1993; K. Weiler, S. Immler, and R. McCray (eds.), *Supernova 1987A: 20 Years After*, American Institute of Physics, 2008.

Sustainability and printing inks

Large retailers now insist on packaging made from sustainable sources with less waste and lower cost. Their first contacts for more sustainable packaging were the package printers, who, in turn, contacted substrate and ink manufacturers, pushing the need to suppliers further up the chain. One piece of this large puzzle, and perhaps the most visible piece, is the wide range of printing inks used for packaging. As this change propagates through the entire product manufacturing chain, there is increasing interest in sustainable inks for all types of printing, and not just in packaging.

Many terms have been used to describe sustainability, including green, environmentally friendly, Earth-friendly, and "eco." Sustainability also includes social and business issues. Simply put, if a product requires raw materials that are declining in availability, is made by a process that is not cost-competitive (often due to excess waste), is manufactured by a company that will not survive, or cannot be processed cleanly at the end of its life, then that product is not sustainable. The process for defining these issues is a life-cycle analysis of the entire process for the making, using, and disposing of a product.

Life-cycle analyses are very difficult. Standardized processes for creating them are still being developed. In their absence, ink makers tend to judge sustainability using four major factors: (1) the amount of renewable (plant-derived) material; (2) the environmental emissions (both in manufacturing and on press), usually the volatile organic compound (VOC) content; (3) the energy requirements to make, ship, print, and dry the inks; and (4) the end-of-life issues, such as recycling, biodegradability, and compostability. These factors are balanced with cost and performance as well as business proceeds.

To understand ink sustainability, these four criteria must be applied to the four major ink types: solvent, water, oil, and radiation-curable [ultraviolet/electron beam (UV/EB)]. In the larger scope of printing, the major printing technologies (lithography, flexography, gravure, screen, and digital) also must be considered. The **table** shows a 4-by-4 matrix of sustainability factors versus ink technology.

Plant-derived content. All plant-based (sometimes called vegetable-based) raw materials come from crops planted, cultivated, harvested, and processed for the purpose of making ink. Because ink raw materials are not made from plants harvested in the wild, it is assumed, in the absence of life-cycle analyses, that farming, harvesting, and processing of these crops are equally sustainable and that the sustainability of all plant-based ink raw materials can be weighted equally.

Plant-based ink raw materials are of four main types: (1) those derived from plant fats, such as linseed and soybean oils, their fatty acids, and derivatives; (2) those derived from pine rosin, which comes mainly from the pulping operations for plant fiber used in papermaking; (3) plant proteins, which usually are extracted from biomass residue after processing to remove plant fats; and (4) cellulose derivatives, usually made from cotton and/or paper pulp and often used in solvent-based inks.

Oil-based inks. Of the four major ink types, oil-based lithographic inks can have the highest levels of plant-based content. Usually these levels range between 30

Sustainability factors versus ink technology				
	Ink technologies			
	Solvent	Water	Oil	UV/EB
Plant content	20–40%	<5%	30–70%	<10%
Emissions	40–60%	3–10%	0–20%	0–2%
Energy demand	High	High	Low	Medium
End-of-life	Allow reuse or degradation of underlying substrate			

and 70%, depending on the limitations imposed by print performance, printing technology, and cost.

Solvent-based inks. Solvent-based inks for flexography and gravure generally have the second-highest level of plant-derived content. Polyamide resins derived from plant fatty acids and cellulose derivatives, such as nitrocellulose, ethyl cellulose, and ethyl hydroxyethyl cellulose, are widely used in solvent-based inks. Some solvent-based inks contain high levels of pine rosin–derived materials. Also, some of the ethanol-based solvents, such as ethyl acetate, can be made from plant-based ethanol, but there is much debate whether corn-based ethanol production is itself a sustainable process. Levels of plant-derived materials in solvent-based inks often range from 20 to 40%.

UV/EB inks. In the past, UV/EB inks often contained essentially no plant-based materials. In recent years, however, many companies began using vegetable-derived materials in these inks. The average level of plant-sourced material in UV/EB inks is probably 2 to 5%, but up to 15% seems possible.

Water-based inks. Water-based inks usually contain no plant-derived material. Occasionally, some protein-based ink systems for flexography are available, but they are niche products and do not generally offer any performance advantages.

Emissions: VOC content. UV/EB products are the big winners here. UV- and EB-curable inks usually show measured VOC levels in the 0–2% range. Because the experimental error for the VOC test is ±2%, UV/EB inks are usually described as having essentially zero VOC content.

VOC levels in aqueous inks tend to be quite low (3–10%), although they can be higher. The concern is that the drying process for water-based inks requires heat to drive off the volatiles. Thus, it can reasonably be assumed that 100% of the VOC content is emitted into the air. Because most of the mass being evaporated is water, generally there are no emission controls and all the organics (VOCs) also are released into the air.

Oil-based inks for sheet-fed lithographic printing often contain between 0 and 20% VOCs, mostly in the form of hydrocarbon mineral oils, which are just barely volatile enough to qualify as VOCs. Although all of these oils might vaporize during the entire life cycle of the printed product, it is a slow-release process. Conversely, oil-based inks for heatset lithographic printing contain 30–40% hydrocarbon oils, which are more volatile and designed to be removed in a gas-fired drying oven at the end of the printing press. In heatset printing, there is usually an incineration process that reduces the VOC emis-

sion. However, the resulting carbon dioxide emissions contribute to the overall carbon footprint.

Solvent-based inks generally are the worst in terms of VOC emissions. These inks have VOC levels between 40 and 60%. On smaller printing equipment, nearly all of this is emitted directly into the air and clearly represents the worst-case scenario for VOC emissions from ink. On larger presses, such as publication gravure presses, much of the evaporated solvent is captured, reclaimed, and returned to the ink manufacturer for reuse, making this a closed-loop, high-volume recycling program with low VOC emissions.

Energy requirements. All four ink types require roughly the same amount of energy to manufacture, with one caveat. For acceptable shelf life, UV/EB inks need more delicate handling and thus their manufacturing areas generally require UV-free lighting (nonfluorescent) and a level of air conditioning in the summer. Temperatures in the manufacturing area should be kept under 80°F (27°C). Other ink chemistries do not need air conditioning in the manufacturing area. Shipping and print drying also have significant energy requirements.

Shipping. Burning diesel fuel to ship inks in trucks when essentially 100% of the ink will remain on the substrate (that is, everything you ship, you use) is the baseline standard. This is the case for UV/EB inks and nearly the case for oil-based sheetfed lithographic inks. Water-based, solvent-based, and oil-based heat-set inks are roughly 50% solids, meaning that only half the amount of shipped material remains on the substrate. The remaining part (mostly water or solvent) is evaporated. This is substantially less efficient than shipping 100% solids inks. Large-volume inks, such as some news inks and solvent publication gravure inks, are shipped by rail, an economical and energy-efficient method. These large-volume inks are stored in tanks and require no primary packaging to handle them.

Drying. Oil-based sheetfed inks are the best here. There is no postprinting drying equipment. Instead, the printed sheets sit stacked on pallets in a warehouse for a couple of days until dry. There are some environmental costs for operating the warehouse and for the floor space, but oil-based sheet-fed still has the lowest energy cost.

UV/EB printing requires either UV curing lamps or an electron-beam unit on the press. Both of these require electricity to operate. The UV lamps generate a fair amount of heat that must be managed. The EB unit requires an oxygen-free (nitrogen-inerted) curing zone, and this adds some cost. Nitrogen is not consumed but is released back into the air and is quite sustainable.

Water-based, solvent-based, and heatset oil-based printing inks require active drying systems involving heating and air flow. These are the most energy-intensive inks to print and dry.

End-of-life issues. Despite marketing claims to the contrary, no printed ink of any technology is recyclable. No one takes waste print and reclaims the cyan, magenta, yellow, and black inks for reuse. In the context of printing, recycling applies only to the underlying substrate. All of the ink technologies allow this.

Although there are no good data to prove it, there is no reason to think that printed ink from any of these ink technologies will biodegrade or be compostable. They are all based on high-molecular-weight polymers, and many are converted to highly crosslinked films during drying. As a result, they are quite stable. However, the inks represent a very small amount of the weight of the printed article, usually 1–3%. As with recycling, biodegradation and composting of printed materials is normally a substrate issue, with no evidence that the ink positively or negatively affects the degradation of any substrate.

For background information *see* CONSERVATION OF RESOURCES; ENVIRONMENTAL MANAGEMENT; INK; PAPER; PRINTING; RECYCLING TECHNOLOGY in the McGraw-Hill Encyclopedia of Science & Technology.

Don P. Duncan

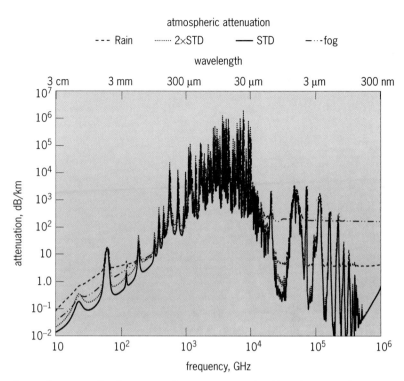

Attenuation across the electromagnetic spectrum at sea level based on currently accepted models. Rain = 4 mm/hr, fog = 100 meter visibility, STD (standard atmosphere) = 7.5 gm/m³ water vapor, and 2 × STD = 15 gm/m³ water vapor. (*Courtesy of H. B. Wallace, MMW Concepts LLC*)

Terahertz imaging

The terahertz (THz) region of the electromagnetic spectrum, broadly defined, comprises the range of frequencies from 100 GHz (10^{11} Hz) to 10 THz (10^{13} Hz), corresponding to free-space wavelengths from 3 mm to 30 micrometers. This radiation, sometimes called "T-rays," bridges the frequency gap between microwaves and long-wavelength infrared waves, and historically was of interest primarily to astronomers probing the cosmic microwave background, and fusion physicists diagnosing plasmas. This difficult transition region between radio frequencies and optics has become much more accessible in recent years with the development of new sources and detectors, spawning tantalizing potential applications for terahertz imaging in areas as diverse as hazardous-materials detection and security, nondestructive testing, medical imaging, and astronomy. As is the case in other wavelength regions, imaging is the natural way to retrieve spatial information contained in a terahertz scene.

Much of the excitement about terahertz imaging results from the fact that most nonmetallic and non-polar substances are relatively transparent to terahertz radiation. Therefore, terahertz imaging systems can be used to see through materials such as clothing, cardboard, polystyrene foam, or leather for nondestructive inspection or security purposes. Many explosives, drugs, or chemical and biological agents have distinctive spectral features in the terahertz regime that may be useful for identification of concealed substances. Imaging at terahertz frequencies has shown in vivo surface features and depth information of skin cancers useful in guiding the surgical removal of tumors and has potential use for real-time imaging of other tumors during surgery. Finally, terahertz radiation poses no health risks to people because it is inherently nonionizing. Realizing these benefits in widespread applications will require substantial further development in detectors, sources, and instruments. Broadband or monochromatic terahertz images may be useful for some applications; but more generally, spectral information about each pixel in the scene may be required.

Atmospheric transmission. As seen in the **illustration**, absorption by water vapor in the atmosphere is the primary source of terahertz attenuation at sea level. The transmission is characterized by windows separated by water vapor absorption peaks. The windows are rather broad below 0.3 THz (corresponding to wavelengths greater than 1 mm) and become relatively narrower at higher frequencies. Importantly, the attenuation in the window regions increases as the frequency increases to about 3 THz (corresponding to a wavelength of 100 μm). Even under the best atmospheric conditions, the range at which objects can be imaged is limited to no greater than about 10 m (33 ft) for frequencies above 1 THz (corresponding to wavelengths less than 300 μm). In addition, the attenuation varies greatly depending on humidity. Accounting for atmospheric effects will complicate calibrated measurements and spectroscopic measurements.

Imaging approaches. Many potential implementations are possible for a terahertz imaging system, depending on the application. First, the illumination geometry can be arranged in either reflection or transmission modes. The transmission mode is effective for thin samples, while the reflection mode can scan samples that are virtually opaque to the terahertz radiation. Thus the transmission mode could be useful for screening mail, while the reflection mode would be more appropriate for screening people for hidden explosives.

Second, the imaging system could be scanning or staring. A scanning system can use a tightly focused low-power source and a single detector or a small number of detectors to gradually build up an image, whereas a staring system would illuminate an entire scene simultaneously and the scene would be recorded with a large-format sensor (a focal-plane array) in the same way that a modern electronic camera records an image. Staring sensors are preferred for real-time imaging because of the inherently parallel nature of detection; however, large-format focal-plane arrays for terahertz application are still in development.

Third, an imaging system can be configured to image either a broad band of frequencies, which could enhance the sensitivity, or to image over a narrow frequency range. Typically, in the terahertz spectral region, narrow-band imaging is accomplished by using a narrow-band source. Spectroscopic imaging can be done if the source frequency can be scanned.

Sources of terahertz radiation. While objects near room temperature naturally emit some terahertz radiation, as part of the blackbody thermal radiation described by Planck's law, practical terahertz imaging will probably require active illumination by a terahertz source. Widespread application and commercialization of terahertz technology will require efficient, high-power, compact sources. Even the best available compact sources for terahertz radiation are weak, producing only about 1 mW of average power, in contrast to most other spectral regions where compact sources producing on the order of kilowatts are available. Terahertz radiation sources being considered for imaging applications are generally laser-based and are fundamentally of two types: short-pulsed laser sources for broadband or continuous-wave (CW) sources for narrow-band terahertz generation.

Infrared-pumped gas terahertz lasers have existed for many years, but their application for imaging is limited because of cost, complexity, and size. Alternatively, significant CW terahertz radiation can be generated using difference-frequency generation in a nonlinear optical crystal using infrared or visible pump lasers. Difference-frequency generation is a nonlinear optical process where two beams at frequencies ω_1 and ω_2 interact in a nonlinear medium to generate radiation at a frequency $\omega_3 = \omega_1 - \omega_2$. The terahertz radiation ω_3 can be swept throughout the terahertz region by slightly tuning the pump lasers.

Solid-state quantum cascade lasers (QCLs) are being rapidly developed as compact solid-state terahertz sources, although cryogenic cooling is still required for devices that directly generate terahertz radiation. The quantum cascade structure relies on radiative intersubband or interminiband transitions of electrons between energy levels defined in the conduction band of a semiconductor heterostructure, a band-gap-engineered structure that is typically implemented in III–V semiconductor alloys. Recently, a device was reported that demonstrated room-temperature terahertz generation near 5 THz (corresponding to a wavelength of 60 μm) with a dual-wavelength midinfrared QCL employing intracavity difference-frequency generation.

Photoconductive dipole antennas can be used to generate broadband terahertz pulses. These sources illuminate a semiconductor with a short ($\sim$100 femtosecond) laser pulse to cause a brief current to flow across a biased dipole antenna patterned on the semiconductor. The sharp rise and fall time of the current flow result in a pulse of terahertz radiation emitted by the antenna.

Optical rectification can also be used to generate terahertz pulses. Here, a high-intensity femtosecond laser pulse illuminates a nonlinear crystal that rectifies the high-frequency oscillations of the pulse, leaving only the envelope of the laser signal. Because the envelope lasts only a few hundred femtoseconds, it contains a broad range of frequencies in the terahertz regime.

Detectors. Heterodyne detectors have been the most common terahertz detectors because high-resolution spectroscopy has been the most common application. In heterodyne detection, the terahertz signal is mixed with a reference frequency, typically in a photodiode, and the difference frequency is amplified and analyzed, as done for conventional heterodyne radio receivers. The lower difference frequency is more easily amplified and manipulated with more conventional radio-frequency electronics. Spectroscopy at terahertz frequencies can be done, for example, by tuning the reference frequency and amplifying only a narrow, constant difference frequency, much like tuning a radio to various stations.

For imaging, however, arrays of direct detectors are most desirable, and these detectors should be sensitive over a fairly broad terahertz spectrum for the best sensitivity. Large focal-plane arrays optimized for the infrared and visible spectral regions are readily available; however, such arrays optimized for terahertz frequencies do not yet exist.

The most sensitive terahertz detectors require cryogenic cooling; most commonly these are helium-cooled bolometers of many varieties including silicon, germanium, indium antimonide (InSb), and superconducting transition edge bolometers. These have been fabricated recently into sensitive small arrays by micromachining techniques. The requirement for helium cooling precludes many commercial uses of terahertz imaging.

Large-format uncooled microbolometer arrays have been extensively developed for long-wavelength infrared applications ($\sim$10 μm wavelength). Several preliminary studies have demonstrated terahertz imaging at 2.5 THz (118.8 μm wavelength) and 4.3 THz (70 μm wavelength) using these arrays, albeit with low sensitivity and with pixel sizes that were too small (46.25 μm pixel pitch). Several groups are working to integrate microantennas with microbolometers to provide inexpensive imaging arrays with better sensitivity.

Outlook. Terahertz imaging has demonstrated tantalizing results in areas such as medical imaging, hazardous materials detection, security, and astronomy. The realization of practical and widespread terahertz imaging systems will require substantial improvements in lower-cost and higher-power sources that can operate near room temperature. In conjunction with this requirement, further development of room-temperature, large-format terahertz focal-plane-array detectors is necessary to take advantage of the substantial parallelism of an array to do real-time imaging.

For background information *see* BOLOMETER; HEAT RADIATION; HETERODYNE PRINCIPLE; LASER; NONDESTRUCTIVE EVALUATION; NONLINEAR OPTICAL DEVICES; SEMICONDUCTOR HETEROSTRUCTURES; SUBMILLIMETER ASTRONOMY; SUBMILLIMETER-WAVE TECHNOLOGY in the McGraw-Hill Encyclopedia of Science & Technology.

Maryn G. Stapelbroek; Eustace L. Dereniak

Bibliography. J. F. Federici et al., THz imaging and sensing for security applications—explosives, weapons and drugs, *Semicond. Sci. Technol.*, 20:S266–S280, 2005; K. Humphreys et al., Medical applications of terahertz imaging: A review of current technology and potential applications in biomedical engineering, pp. 1302–1305, *26th Annual International Conference of the IEEE Engineering in Medicine and Biology Society, 2004, IEMBS '04*, September 1–5, 2004; A. W. M. Lee et al., Real-time imaging using a 4.3 THz quantum cascade laser and a 320 × 240 microbolometer focal-plane array, *IEEE Photon. Tech. Lett.*, 18:1415–1417, 2006; M. J. Rosker and H. B. Wallace, Imaging through the atmosphere at terahertz frequencies, pp. 773–776, in *Proc. IEEE MTT-S International Microwave Symposium*, edited by A. Mortazawi et al., Honolulu, Hawaii, June 3–8, IEEE, 2007.

The role of metal ions in DNA damage

The cell has many vital parts. The mitochondria are the power plants that provide energy (adenosine triphosphate) for cellular processes, and the ribosomes are factories that make all the proteins and enzymes required for cellular functioning. DNA is an indispensible component of cells, since it contains the genetic instructions for cellular functioning, protein synthesis, and heredity. If DNA is damaged, it can be fixed or replaced by DNA repair enzymes, but if DNA damage is not corrected, cell mutation or death can occur. Over time, this damage can accumulate and lead to diseases such as cancer, Alzheimer's, Parkinson's, and cardiovascular disease.

In addition to DNA in the nucleus, which contains the cellular genetic code, DNA is also found in mitochondria. In mitochondria, the oxygen we breathe is converted to water, and the resulting energy from this respiration is used for cellular processes. Damage to mitochondrial and nuclear DNA can occur from outside sources, such as chemical toxins or ultraviolet light. Most frequently, DNA damage results from by-products of the respiration process. Some oxygen compounds formed during respiration are highly reactive and can damage cellular components, including DNA. The most reactive and damaging of these oxygen species are formed when a stable oxygen-containing molecule gains an electron, creating compounds with an unpaired electron, called radicals, which are extremely reactive in order to pair up or lose the extra electron. The most reactive of these radical species, and the one that is most damaging to DNA, is the hydroxyl radical (represented by $\cdot$OH; the dot refers to the unpaired electron). Superoxide ($O_2^-\cdot$) is a radical that forms when oxygen (O_2) gains an electron. However, oxygen species do not have to be radicals to be reactive. Hydrogen peroxide (H_2O_2) is an example of a nonradical reactive oxygen species that is a by-product of respiration in the mitochondria. Collectively, the oxygen compounds that contribute to DNA damage are called reactive oxygen species. Some reactive oxygen species, such as hydroxyl radicals, can damage DNA directly, while others, such as hydrogen peroxide and superoxide, must be converted into DNA-damaging hydroxyl radicals by metals.

Metal ions can generate DNA-damaging reactive oxygen species. Metals such as iron and copper are required for proper cell functioning. Because these two metals readily give up electrons, they can also generate reactive oxygen species in cells. It has been known for some time that iron (Fe^{2+}) and copper (Cu^+) react with stable oxygen species to form damaging radical species, including the hydroxyl radical. For example, although hydrogen peroxide does not by itself damage DNA, iron or copper can react with hydrogen peroxide to generate the reactive hydroxyl radical [reaction (1)]. In fact, DNA damage

$$H_2O_2 + Fe^{2+} \quad \text{or} \quad Cu^+ \rightarrow \cdot OH + OH^- + Fe^{3+} \quad \text{or} \quad Cu^{2+} \tag{1}$$

from the iron-generated hydroxyl radical is the primary cause of cell death under conditions of oxidative stress. Reactions (1)–(3) show the generation of damaging reactive oxygen species: superoxide ($O_2^-\cdot$) [reaction (2)] and hydroxyl radical ($\cdot$OH) by

$$O_2 + Fe^{2+} \quad \text{or} \quad Cu^+ \rightarrow O_2^-\cdot + Fe^{3+} \quad \text{or} \quad Cu^{2+} \tag{2}$$

copper and iron [reaction (1)]. Once formed, superoxide reacts to form hydrogen peroxide (H_2O_2) [reaction (3)].

$$2O_2^-\cdot + 2H^+ \rightarrow H_2O_2 + O_2 \tag{3}$$

Thus, the oxygen that we breathe to live, combined with the metals required for cell functioning, generates DNA-damaging reactive oxygen species.

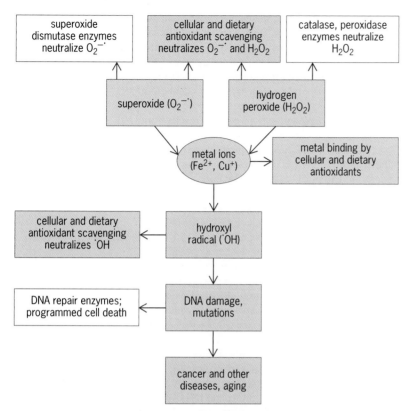

A chart depicting the role of metals such as iron and copper (oval) in the formation of potentially harmful reactive oxygen species (O_2^{-}, H_2O_2, and $\cdot OH$), and the cellular enzymes and processes (white boxes) that control reactive oxygen species and DNA damage. Cellular and dietary antioxidants can act to either scavenge and neutralize reactive oxygen species or bind metal ions to prevent hydroxyl radical formation and damage.

Because reactive oxygen species are continually being produced, cells must constantly protect themselves from DNA damage. One of the many ways in which cells accomplish this task is by producing enzymes such as superoxide dismutases, catalases, and peroxidases that break down superoxide and hydrogen peroxide (see **illustration**). Another strategy is the use of antioxidants, which are small molecules that prevent oxidation of other molecules, including DNA. Cellular antioxidants can either scavenge reactive oxygen species by reacting with and neutralizing them or prevent the formation of reactive oxygen species altogether.

Cellular and dietary antioxidants. Antioxidants—both those found naturally in cells and those obtained from the diet—prevent cellular damage from reactive oxygen species. It is the accumulation of damage to DNA and other cellular components that is linked to cancer, aging, and cardiovascular and neurodegenerative diseases (see illustration). Some DNA damage can be repaired by enzymes in the cell. But if DNA damage is too extensive, the cell is no longer viable and will initiate programmed cell death (apoptosis) in response to the damage. If natural antioxidant defenses are high, DNA damage is minimized and disease risk is lowered. It is becoming clearer that hydroxyl radical damage to the ends of DNA in chromosomes (telomeres) is a major cause of aging. In addition to long-term DNA damage, natural cellular antioxidants such as glutathione are increas-ingly depleted as people age or experience harmful stress. There are many causes of this stress, including disease, habits such as smoking, environmental exposure to toxic compounds, and some drug treatments such as those used in chemotherapy. All of these conditions can upset the balance between the formation of reactive oxygen species and the cellular defenses that protect against them.

Dietary antioxidants are of particular interest, because antioxidants obtained from food may be able to augment natural antioxidant defenses. Many types of dietary antioxidants are known, including vitamins A, C, and E; selenium; carotenoids such as beta-carotene; and flavinoids, which include polyphenol compounds. All fruits and vegetables have antioxidants called polyphenols (plants use these antioxidant compounds to prevent fungal and bacterial infection), as do teas, dark chocolate, and red wine. It has been known for many years that people who eat a diet rich in fruits and vegetables (foods that contain high levels of antioxidants) lower their risk of cancer and of cardiovascular and neurodegenerative diseases. This lowered risk is believed to be a result of dietary antioxidants supporting cellular antioxidant defenses to prevent damage from reactive oxygen species. However, a tremendous amount of research is currently in progress to determine how antioxidants prevent cellular damage by reactive oxygen species. Since we do not know exactly how antioxidants prevent DNA damage, it is difficult to determine which of the thousands of dietary antioxidant compounds might be the most effective. To make the situation more complex, dietary antioxidants may act best in combination, which may be why a balanced diet can lower disease risk.

Vitamins A, C, and E are also considered antioxidants, and much research is being done to determine if these vitamins can help to treat or prevent diseases caused by reactive oxygen species damage. Selenium is a mineral that boosts cellular antioxidant defenses through its incorporation into glutathione peroxidase enzymes that decompose hydrogen peroxide and by forming small-molecule antioxidant compounds. However, these vitamins and minerals are also harmful in large doses, which is another example of the complexity of antioxidants. Some foods, such as green tea and pomegranates, have high concentrations of dietary antioxidant compounds. Measuring the antioxidant content of foods is not sufficient for determining their ability to prevent reactive oxygen species damage, because these measurements do not determine how much of these compounds is absorbed (bioavailable) from foods. Thus, dietary antioxidants must be both bioavailable and capable of preventing DNA damage by reactive oxygen species. Since foods and supplements often contain many different antioxidant compounds, it is also difficult for researchers to determine which compounds, or combinations of compounds, have beneficial health effects.

Mechanisms for antioxidant prevention of DNA damage. There are several ways in which cellular and dietary antioxidants can prevent DNA damage from reactive oxygen species. The most-studied antioxidant mechanism is the ability to scavenge and neutralize reactive oxygen species (see illustration). This scavenging ability is an important aspect of antioxidant activity, but it does not take into account the role of metals in generating reactive oxygen species. In addition, scavenging very reactive species such as hydroxyl radicals is difficult because they react very quickly and damage cellular components before the antioxidants can neutralize them. Recently, researchers have focused on antioxidant interactions with metals such as iron and copper that generate reactive oxygen species. Many classes of cellular and dietary antioxidants are capable of metal binding and may prevent DNA damage by preventing metals from forming or releasing damaging reactive oxygen species (see illustration). This metal-binding mechanism prevents damage at the source instead of depending on the antioxidant finding and neutralizing reactive oxygen species before they cause damage.

The study of oxidative damage and how antioxidants prevent it is extremely complex. Many thousands of antioxidants are known, and their interactions with each other and with various cellular components are a rapidly evolving topic of research. This research will take years because of the number of variables that must be investigated (antioxidant combinations, different interactions with cellular components) and the need to prove antioxidant efficacy for disease prevention in animal models and long-term clinical trials. For now, a significant gulf exists between the dietary studies that have established the value of an antioxidant-rich diet in disease prevention and the scientific studies that are working toward understanding why this is true. The recently identified metal-binding mechanism for antioxidant activity is extremely important, because unlike many other processes that repair oxidative damage or neutralize damaging radicals, this mechanism may prevent damaging hydroxyl radicals from forming, highlighting an important new research area. Additional research is also required to determine specifically how DNA and other cellular damage results in neurodegenerative and cardiovascular diseases and cancer. Ultimately, a better understanding of how damage caused by reactive oxygen species contributes to disease and the mechanisms for antioxidant activity to prevent this damage will lead to improved dietary recommendations for antioxidant-rich foods, as well as the development of antioxidant drugs or supplements to treat and prevent disease.

For background information *see* ANTIOXIDANT; CELL (BIOLOGY); COPPER; DEOXYRIBONUCLEIC ACID (DNA); ENZYME; FREE RADICAL; IRON; MITOCHONDRIA; OXYGEN TOXICITY; SUPEROXIDE CHEMISTRY in the McGraw-Hill Encyclopedia of Science & Technology.   Julia L. Brumaghim

Bibliography. M. S. Fernandez-Panchon et al., Antioxidant activity of phenolic compounds: From in vitro results to in vivo evidence, *Crit. Rev. Food Sci. Nutr.*, 48:649–671, 2008; N. R. Perron and J. L. Brumaghim, A review of the antioxidant mechanisms of polyphenol compounds related to iron binding, *Cell Biochem. Biophys.*, 53:75–100, 2009; M. Singh et al., Challenges for research on polyphenols from foods in Alzheimer's disease: Bioavailability, metabolism, and cellular and molecular mechanisms, *J. Agr. Food Chem.*, 56:4855–4873, 2008; M. Valko, H. Morris, and M. T. D. Cronin, Metals, toxicity and oxidative stress, *Curr. Med. Chem.*, 12:1161–1208, 2005; M. Valko et al., Free radicals and antioxidants in normal physiological functions and human disease, *Int. J. Biochem. Cell Biol.*, 39:44–84, 2007.

Three-dimensional measurement of river turbulence

Approaches to understanding the mechanical processes shaping river channels have historically been limited by the challenge of collecting data capable of delineating the complexity reflected in the fluid flow of water. At the same time, accurately capturing this complexity is critical for research in the fields of fluvial geomorphology, hydrology, and environmental engineering. The study of fluid dynamics within the context of river environments reveals that fundamental properties, such as velocity, are highly variable over small spatial and temporal scales. The characteristics and behavior of flowing water determine how streams influence environments, such as the shape and structure of the channel, the erosive power exerted by the water, and the amount and character of the materials the river can transport. At watershed scales, time-averaged measurements of velocity in the downstream direction may provide reasonable estimates of reach-averaged stream power and the ability of the flowing water to perform mechanical work. However, where there is a need for higher-resolution measures of velocity and stream power to calculate bed shear stress or stream competence, define hydraulic microhabitat conditions, or generate deterministic models of stream channel adjustment, small-scale variations in the directionality and magnitude of flow vectors should be taken into account.

Measurement of fluid properties in river environments can be very challenging because flow is often moving not only in the downstream direction, but also vertically in the flow column and horizontally in a cross-stream direction. The complexity of these flow patterns in all three dimensions is greatly influenced by the configuration of the river bed and the presence of any obstructions within the channel. While it is typical in many applications to generate a single value for stream velocity at a cross section, that value does not accurately reflect the instantaneous downstream velocity at every location along the cross section, let alone water movement in the vertical or horizontal direction.

Flowing water is characterized by two different types of movement: laminar and turbulent flow. When simple, highly viscous fluids move slowly in one direction, the velocity profile is mainly determined by the internal resistance of the material and the way in which momentum is transferred between successive layers at the molecular scale. In this case, friction at the bed creates enough resistance to inhibit movement, and the shear stress created between successive molecular layers of the fluid generates a velocity gradient that is directly proportional to the distance from the source of friction. In open-channel environments, the laminar sublayer is a very small portion of the total flow column.

Under most circumstances, water has a low dynamic viscosity, which means that internal resistance via viscous forces at the molecular scale is less important than the transfer of momentum via larger coherent structures, called turbulent eddies, moving within the flow column. Eddy viscosity is a more efficient means of transferring momentum and results in a less pronounced vertical velocity gradient. Turbulent flow occurs above the laminar sublayer in the boundary layer, which dominates the flow column. Three-dimensional (3D) flow dominates in streams where turbulent eddies transfer momentum within the boundary layer (the zone affected by fluid interactions with a frictional surface) or at depths sufficient to be peripheral to the influence of the boundary layer. Above the boundary layer lies the fully turbulent free-stream layer, where vertical mixing and the absence of boundary effects result in no velocity gradient.

Measuring turbulence. Detailed measurements of flow velocity can be used to characterize the structure of turbulent flow. Patterns in the u (streamwise), v (vertical), and w (cross-stream) components are used to characterize different turbulence statistics. A wide variety of methods for delineating turbulence characteristics have been used in the laboratory and in field studies, including Pitot tubes, propeller current meters, ultrasonic current meters, hot-film and hot-wire anemometers, and laser Doppler velocimeters. Other visualization techniques, allowing for the direct examination of the entire 3D flow field, include the use of tracers such as dyes, hydrogen bubbles, or tracer particles. The two most widely used instruments in the field measurement of 3D flow conditions in shallow-river flows are the electromagnetic current meter (ECM) and the acoustic Doppler velocimeter (ADV). Both instruments are robust and accurate and allow the user to collect high-frequency, high-resolution velocity data under a wide range of field conditions. Both require an Eulerian approach, in which the instrument itself remains fixed in position within the stream flow and measures a moving target. Both instruments are capable of measuring turbulence statistics, such as the mean standard deviation, skewness, and quadrant analysis of Reynolds shear stress. Thorough measurements can be used to reconstruct the velocity across the area sampled, which, in effect, provides a generalized picture of the turbulent structure.

Acoustic Doppler velocimeters emit sonic pulses and measure the change in frequency, known as the Doppler shift, as a portion of the signal bounces off suspended sediment, air bubbles, or other particles that scatter sound and is reflected back to the instrument sensor. Flow, carrying scattering particles away from the sensor, generates a Doppler shift to a lower frequency proportional to the rate of movement between the two. This shift in frequency and propagation delay can be used to calculate instantaneous velocities, which are averaged over time to obtain the mean velocity. In order to derive flow in the streamwise, cross-stream, and vertical vectors, there must be multiple sonic pulses pointed in different directions. To measure 3D velocity, at least three acoustic beams are required. Common problems associated with ADVs are that they are susceptible to low-quality signals and can improperly estimate the signal variance in the presence of noise.

The electromagnetic current meter works by inserting a probe that generates an electric field into the flow column. This is based on the Faraday principle of electromagnetic induction, which states that the voltage induced across a conductor as it moves through a magnetic field is proportional to the velocity of the conductor. The difference in voltage between two electrodes indicates the average velocity. In order for ECM to produce accurate measurements, it must be properly calibrated to the water velocity. ECM is also very sensitive to electrical interference and faulty grounding. In a direct comparison of ECM and ADV data collection in riverine environments, B. J. MacVicar and coworkers indicated that for most commonly used turbulence measures, there is little difference between the two instruments. Because ECM samples flow only once, ECM may perform better under highly turbulent conditions, such as when the ADV reliance on Doppler shift results in an increase in noise and a decrease in correlated measurements.

Outlook. Recent advances in the sampling rate and accuracy of instrumentation available for open-channel field investigations and a corresponding increase in computing power have led to an enormous growth in our ability to detect 3D flow and test theories of the role that turbulence plays in shaping river channels, dissipating energy, and creating microhabitat conditions. Further use of these tools and techniques in river studies increases the potential for advance in the science of fluvial geomorphology and river mechanics and our ability to understand the complex dynamics behind instantaneous flow variation and the development of coherent turbulent structures.

For background information *see* FLOW MEASUREMENT; FLUID-FLOW PRINCIPLES; FLUID MECHANICS; FLUVIAL EROSION LANDFORMS; FRICTION; GEOMORPHOLOGY; HYDROLOGY; LAMINAR FLOW; RIVER; TURBULENT FLOW in the McGraw-Hill Encyclopedia of Science & Technology.   Claire Ruffing; Michael Urban

Bibliography. T. Buffin-Bélanger and A. G. Roy, 1 min in the life of a river: Selecting the optimal record length for the measurement of turbulence in fluvial boundary layers, *Geomorphology*, 68:77–94, 2005; B. J. MacVicar et al., Measuring water velocity in highly turbulent flows: Field tests of an electromagnetic current meter (ECM) and an acoustic Doppler velocimeter (ADV), *Earth Surf. Proc. Land.*,

32:1412–1432, 2007; M. Muste et al., Practical aspects of ADCP data use for quantification of mean river flow characteristics, pt. II: fixed-vessel measurements, *Flow Meas. Instrum.*, 15:17–24, 2004; I. Nezu, Open-channel flow turbulence and its research prospect in the 21st century, *J. Hydraul. Eng.*, 131(4):229–246, 2005; A. G. Roy et al., Size, shape and dynamics of large-scale turbulent flow structures in a gravel-bed river, *J. Fluid Mech.*, 500:1–27, 2004; A. Tominaga and I. Nezu, Turbulent structure in compound open-channel flows, *J. Hydraul. Eng.*, 117(1):21–41, 1991.

Tools to assess community-based cumulative risk and exposures

Multiple agents and stressors can interact in a given community to adversely affect human and ecological conditions. A cumulative risk assessment (CRA) analyzes, characterizes, and potentially quantifies the effects from multiple stressors, which include chemical agents (for example, benzene) as well as biological (for example, vector-borne illness), physical (for example, housing characteristics), and psychological (for example, socioeconomic) ones. The distinguishing feature of a CRA is an analysis of combined effects and interactions.

In risk assessment, the term "community-based" indicates a focus on a given population. A community may be defined by geophysical boundaries such as a watershed, by geopolitical limits such as county or state borders, or by socioeconomic criteria within a defined geographic boundary. These assessments may also include community participation in project formulation and implementation, such as by providing test kits to residents to take measurements of environmental pollutants.

Cumulative assessment challenges. Single-chemical risk assessments typically involve the process of hazard identification, dose-response, exposure assessments, and risk characterization. These source-based assessments are typically performed in the context of regulatory requirements or isolated actions, such as the issuance of an air permit for an industrial facility. In contrast, population-based assessments focus on identifying persons exposed, determining the chemicals or stressors to which they were exposed, and characterizing risks. This approach is more effective for implementing risk-reduction strategies based on public or ecological health.

Single chemicals often have various sources or exposure pathways, which describe full source-to-receptor paths for a particular agent. Aggregate exposure is exposure to a single stressor across all sources, pathways, routes, and time, such as a pesticide. Cumulative risk is the combined risk from aggregate exposures to multiple agents or stressors. Aggregate exposure can be calculated with deterministic approaches that combine single-point exposure values for various pathways, or with probabilistic approaches that produce a distribution of exposure values. The latter produce a better idea of the sources of uncertainty and variability, which can lead to a clearer picture of where additional data are needed.

A cumulative assessment can be quantitative, such as a cancer risk assessment of inner-city teenagers exposed to urban air pollutants, or more qualitative in nature, such as by drawing from various sources of information to visually display potential risk factors. Multiple stressors, sources, dose-response relationships, and health effects may be involved. Stressors may be chemical, biological, radiological, physical, or psychological and may be related to a range of health effects. Multistressor, multieffect cumulative assessments are considerably more complex methodologically than aggregate or single-chemical assessments. Indoor sources of pollution are an example of multiple stressors that can impact health (see **table**).

Publicly available tools. A community-based cumulative risk assessment (CBCRA) typically begins with a characterization of initiating factors, which prompt decisions to undertake a CBCRA and may include multiple pollutant sources within a community, increases in illness in the population, or elevated chemical concentrations either in the environment or in humans (such as found in blood or urine samples) (**Fig. 1**, Initiating factors). The next steps involve defining the relevant population and study area, generating a list of environmental stressors related to the initiating factors, and identifying links between exposures and vulnerabilities within the population. Several factors may be examined throughout this process, including population vulnerabilities, public health information, toxicological and epidemiological data, exposure pathways, differential exposures, and contact with environmental media and pollutant sources (Fig. 1, Data elements).

Indoor air pollutants and typical sources	
Pollutant	Source
Combustion gases—CO and NO	Combustion—furnace, cooking stove, etc.
Volatile organic compounds	Outgassing of building materials
Formaldehyde and other aldehydes and carbonyls	Outgassing of pressed wood and insulation foam
Pesticides	Household products
Particulate matter	Combustion
Biological agents—molds, spores, dander	Contaminated ventilation systems, ceiling tile and wallboard, pets
Environmental tobacco smoke	Smoking in building
Radon	Infiltration from soil beneath structure
Asbestos	Construction coatings, tile, insulation

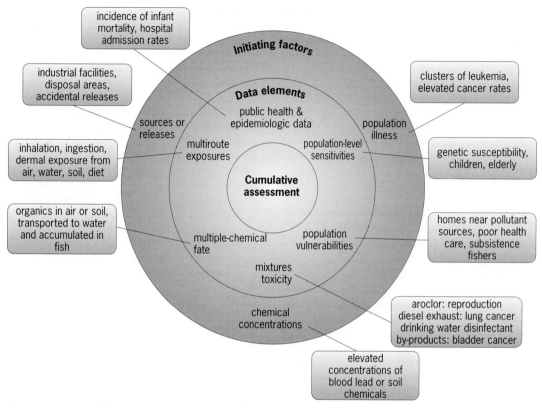

Fig. 1. Examples of CRA initiating factors and data elements. (*Adapted from U.S. EPA, http://oaspub.epa.gov/eims/ eimscomm.getfile?p_download_id=474337*)

One of the primary differences between communities is in their patterns of exposure. While emission source and dose-response characteristics are common across communities, susceptibility and vulnerability differ. Tools that isolate exposure routes and pathways for a given community and then incorporate toxicity information will lead to a better characterization of risk. A number of tools are publicly available that provide information on initiating factors and data elements, including Web-based mapping tools, databases, guidance documents, and exposure models. Several types of measurement test kits are also available for a variety of chemically related stressors.

Current assessment tools. Many tools developed for general risk assessments can be applied to CRAs. They fall into one of four categories based on which aspect of risk assessment they inform: (1) planning, scoping, and problem formulation, including stakeholder involvement, (2) contaminant fate and transport and subsequent exposure, (3) toxicity evaluation, and (4) characterization of risk and uncertainty, and presentation of results. Various types of tools are included in these categories. Guidance documents and facility or air quality web-based mapping tools fall under the first category. Computer models often inform the second category, in addition to monitoring methods and databases of information. The third includes toxicity databases, interaction profiles and regression models for meta-analysis of toxicology data. Probabilistic approaches (for example., Monte Carlo methods) and geospatial analysis tools (for ex-

ample, geographic information systems) address the fourth category.

To assess differences between community-specific exposures, the second category of tools should address certain exposure-related questions: (1) How are people exposed to multiple chemicals? (2) In which media, at what levels, where and when? (3) What are the intensity and duration of these exposures? (4) Are there uniquely susceptible or vulnerable subpopulations? Exposure models that incorporate human activity patterns and pollutant concentration fields begin to address many of these questions; however, they are not widely available, and generally require specific inputs and technical expertise to operate. Exposure metrics, such as proximity to a pollutant sources, provide screening level assessments and could prove to be more transferable across communities, but generally lack the precise concentration estimates necessary for dose-response relationships.

Emerging tools. Additionally, emerging scientific tools are also being applied to better understand environmental risks, especially with respect to multichemical toxicity. These tools coincide closely with recent advances in high-performance computing, and in genomics research and chemical structure-biological activity relationships. They address processes that occur within a physiological system after exposure to an agent or stressor. Biomarkers are a product of physiological processes that occur after exposure to an agent or stressor. Some biomarkers reflect actual exposure concentrations, such as total blood-lead levels for lead exposure, whereas other contaminants may be better reflected by measuring chemical by-products that result from the metabolism and detoxification process.

Computational toxicology and research in genomics, proteomics, metabolomics, and metabonomics (the "omics") have the potential to address multichemical toxicity at the molecular and cellular level. This precludes the necessity for whole-animal toxicity testing, and addresses risks based on specific molecular changes. These techniques provide high-throughput assessments and facilitate research on the effects of multiple chemicals on various physiological systems. Three examples within these fields include quantitative structure-activity relationships (QSAR), physiologically based toxicokinetic (PBTK) models, and in-vitro toxicity pathways.

QSAR assumes a sufficiently strong relationship between chemical structure and biological activity that toxicities of minimally tested compounds can be estimated from those of better-known compounds with similar structures. QSAR can relate the physiochemical properties of a given chemical to its lowest observed adverse effect level (LOAEL), effective concentration (EC50), and carcinogenicity. An extension of QSAR is the virulence factor-activity relationship (VFAR), which extends the application to biological agents.

PBTK models describe the transport and metabolism of a chemical entering the body, and estimate and predict internal doses for organs, tissues, or groups of both (**Fig. 2**). PBTK models incorporate parameters such as partition coefficients, organ and body volumes, blood flows, ventilation rates, absorption rates, clearance, and metabolic transformation rates. PBTK models effectively act as in-silico (computer-simulated) mimics of body tissues, organs, or systems.

The omics relate to advances in mapping and evaluating changes in the human genome. Those described in this article specifically refer to research on full DNA sequences, structures and functions of proteins, metabolites (chemical fingerprints) left behind from specific chemical processes, and the quantitative measurement of metabolic responses to pathophysiological stimuli or genetic modification. For risk assessments, the omics can relate perturbations in in-vitro biocellular pathways to chemical stressors, such as activation of specific genes by arsenic exposure.

Moving forward. Community-based cumulative risk assessments include two key components: one is involvement of community stakeholders, and another is the cumulative risk assessment. This article focuses on developing tools for communities and researchers to perform a CRA. One of the defining features that differentiate communities from one another is their differences in exposure to chemical agents. While toxicity and dose-response remain fixed across communities, exposure will determine a community's susceptibility to an agent or stressor. Each community may also have different demograph-

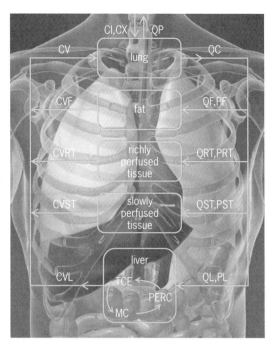

Fig. 2. Ternary-mixture PBPK model representation for humans. Fat (F), richly perfused (RT), slowly perfused (ST), and liver (L) tissue groups are characterized with their volumes, perfusion rates (QF, QRT, QST, and QL), partition coefficients (PF, PRT, PST, and PL), and concentrations of venous blood effluents (CVF, CVRT, CVST, and CVL). CV is the mixed venous blood concentration, QC and QP are cardiac output and pulmonary ventilation, and CI and CX are inhaled and exhaled air concentrations. (*I. D. Dobrev, M. E. Andersen, and R. S. H. Yang, In silico toxicology: Simulating interaction thresholds for human exposure to mixtures of trichloroethylene, tetrachloroethylene and 1,1,1-trichlorothane, Environ. Health Perspect., 110(10):1031–1039, 2002*)

ics of especially vulnerable populations, such as children or the elderly.

CBCRAs represent real-world exposure scenarios, in which community members are exposed to a wide range of chemical, biological, physical and psychological stressors. The confluence of exposure assessment tools with ones that address multichemical toxicities, and their application by and within communities, represents the current multidisciplinary drive towards tool development. Availability to community-based researchers and transferability across different communities are two logistical aspects that also need to be overcome. CBCRAs are becoming more common and widely available tools would support this momentum and increase their effectiveness in decreasing exposure and subsequent risk.

[Disclaimer: The U.S. Environmental Protection Agency through the Office of Research and Development funded and managed some of the research described here. The present article has been subjected to the Agency's administrative review and has been approved for publication.]

For background information *see* AIR POLLUTION, INDOOR; CHEMOMETRICS; ENVIRONMENTAL ENGINEERING; ENVIRONMENTAL TOXICOLOGY; HAZARDOUS WASTE; MONTE CARLO METHOD; MUTAGENS AND CARCINOGENS; RISK ASSESSMENT AND

MANAGEMENT; TOXICOLOGY in the McGraw-Hill Encyclopedia of Science & Technology.

Timothy M. Barzyk

Bibliography. T. M. Barzyk et al., Tools available to communities for conducting cumulative exposure and risk assessments, *J. Expo. Sci. Environ. Epidemiol.*, 2009; R. C. Fry et al., Activation of inflammation/NF-kappaB signaling in infants born to arsenic-exposed mothers, *PLoS Genetics*, 3(11):e207, 2007; M. Johnson et al., A participant-based approach to indoor/outdoor air monitoring in community health studies, *J. Expo. Sci. Environ. Epidemiol.*, 19(5):492–501, 2009; M. Medina-Vera et al., An overview of measurement tools available to communities for conducting exposure and cumulative risk assessments, *J. Expo. Sci. Environ. Epidemiol.*, 2009; National Research Council, *Toxicity Testing in the 21st Century: A Vision and a Strategy*, National Academy of Science, Washington, D.C., 2007; S. N. Sax et al., A cancer risk assessment of inner-city teenagers living in New York City and Los Angeles, *Environ. Health Perspect.*, 114(10):1558–1566, 2006; U.S. Environmental Protection Agency, *Concepts, Methods, and Data Sources for Cumulative Health Risk Assessment of Multiple Chemicals, Exposures and Effects: A Resource Document*, National Center for Environmental Assessment, EPA/600/R-06/013F, 2007a; U.S. Environmental Protection Agency, Framework for cumulative risk assessment, *Risk Assessment Forum*, Washington, D.C., EPA/630/P-02/001F, 2003; U.S. Environmental Protection Agency, Organophosphorus cumulative risk assessment—2006 update, 2006; U.S. Environmental Protection Agency, Science and decisions: Advancing risk assessment, National Research Council, Committee on Improving risk Analysis Approaches Used by the U.S. EPA, 2008; U.S. Environmental Protection Agency, *The U.S. Environmental Protection Agency's Strategic Plan for Evaluating the Toxicity of Chemicals*, Office of the Science Advisor, Science Policy Council, EPA/100/K-09/001, 2009; U.S. Environmental Protection Agency, *QSAR/VFAR Workshop Summary Report*, Office of Research and Development, National Homeland Security Research Center, EPA/600/R-07/095, 2007b; V. G. Zartarian and B. D. Schultz, The EPA's human exposure research program for assessing cumulative risk in communities, *J. Expo. Sci. Environ. Epidemiol.*, 2009.

Tractor-trailer truck aerodynamics

Commercial trucks in the United States vary significantly in size and load-carrying capacity. These vehicles are classified by the United States Department of Transportation (DOT) based upon their gross vehicle weight rating (GVWR). The classes range from 1 to 8, with categories of light-duty (1–3), medium-duty (4–6), and heavy-duty (7–8). The 2007 statistics for combination trucks on the highway show that there are over 2 million trucks, each traveling an average of 65,000 mi per year (105,000 km per year) and consuming 12,800 gal (48,600 L) of fuel per year to give an average fuel economy of 5.1 mi/gal (46 L/100 km).

Class 8 tractor-trailers consume roughly 12–13% of the total United States petroleum usage (21 million barrels/day or 3.3×10^6 m³/day). At highway speeds, that is, 65 mi/h (105 km/h), a class 8 tractor-trailer uses 53% of the usable energy produced by the vehicle engine to overcome aerodynamic drag, while rolling resistance consumes about 32% of the usable energy (**Fig. 1**). Notably, a 2% reduction in the aerodynamic drag of tractor-trailers translates into 285 million gallons (1.1 billion liters) of diesel fuel saved per year, which would also greatly reduce greenhouse gas emissions. This provides a significant motivation in the modern trucking industry to pursue aerodynamic drag reduction technologies.

History. In the 1930s, the Labatt Brewing Company developed a very successful streamlined tractor-trailer, designed by Count Alexis de Sakhnoffsky, to provide a higher cruising speed and a larger load-carrying capacity (**Fig. 2**). Unlike the other trucks on the road, which had a maximum speed of about 30 mi/h (48 km/h), the Labatt Streamliner could travel up to 50 mi/h (80 km/h) and carry a 50% larger load. From 1936 to 1955, the Labatt Streamliner was used throughout Canadian cities, and in 1939, it won a Best Design award at the World's Fair in New York. Over the next several decades, the design of tractor-trailers for transportation of commodities became focused on providing a maximum hauling capability, which resulted in the modern square-edged trailer shape with its dimensions (maximum width, height, and length) dictated by regulations. During the late 1970s and early 1980s, a significant effort was given to the improvement of tractor-trailer aerodynamics in order to reduce fuel consumption. This effort involved reducing tractor and trailer aerodynamic drag by studying the influence of tractor-trailer reshaping and streamlining through the use of aftermarket drag-reducing devices. The majority of this work was done experimentally using various types of wind tunnels, which allowed for a deeper understanding of the aerodynamics of tractor-trailers.

Aerodynamic drag of a tractor-trailer. The field of aerodynamics investigates the motion of air as it interacts with a moving object. Understanding the motion of air, commonly referred to as the flow field, around an object enables the calculation of the forces and moments acting on the object. Typical flow field properties include velocity, pressure, density, and temperature. The conservation equations for mass,

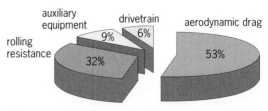

Fig. 1. Energy breakdown for tractor-trailers.

momentum, and energy can be defined and used to solve for the flow-field properties. The drag, or air resistance, refers to the forces that oppose the relative motion of an object through the air.

For the flow field around a tractor-trailer, drag can be divided into two categories: pressure drag and skin friction drag (**Fig. 3**). Pressure drag arises because of the pressure difference on the front and rear of the vehicle and it is strongly influenced by the shape of the vehicle. Vehicles with a larger cross section have a higher pressure drag than slender vehicles. Skin friction drag arises from the friction, or shear, of the air moving against the surface of the vehicle. Skin friction is directly related to the total surface area of the vehicle that is in contact with the air and to surface roughness. It is caused by viscous forces in a thin region, called the boundary layer, near the surface of the vehicle where the air speed rapidly changes to the vehicle speed. At the front of the vehicle, the boundary layer is usually laminar and relatively thin, but it becomes turbulent and thicker toward the rear of the vehicle. The thickness of the boundary layer at the base of a 53-ft (16-m) trailer is roughly 8–10 in. (20–25 cm).

The total vehicle drag is defined by the equation below, where C_D is the vehicle drag coefficient, A is

$$\text{drag} = C_D \times A \times \tfrac{1}{2}\rho U^2$$

the vehicle cross-sectional area, ρ is the air density, and U is the vehicle speed. For a typical tractor-trailer, the drag coefficient varies widely between 0.5 and 0.8, depending on the vehicle shape and configuration, and the cross-sectional area is roughly 108 ft^2 (10 m^2). It is clear from the above equation that the total drag can be lowered through streamlining the vehicle (decreasing C_D), reducing the cross-sectional area of the vehicle, and decreasing the vehicle speed. The first two approaches in drag reduction can be achieved by either modifying the geometry of the vehicle or by flow conditioning. The flow conditioning works by means of air injection at high or low speeds into the flow field around the vehicle to reproduce a flow field caused by a more streamlined body. Since the drag is a function of speed squared, a significant aerodynamic drag reduction can also be achieved simply by reducing the vehicle speed.

Figure 3 shows a typical tractor-trailer configuration where the total drag is primarily caused by the pressure difference between the front of the tractor and the base of the trailer. The contribution of the skin friction drag is more significant on the trailer body where a large surface area is in contact with the air. For a common tractor-trailer, roughly 66% of the total drag comes from the tractor, 16% from the trailer body, and 18% from the trailer axle and wheel assembly. The skin friction drag is negligible on the tractor and the trailer axle and wheel assembly, but it contributes 26% of the drag on the trailer body where surfaces are smooth.

There are several critical aerodynamic flow regions around a typical tractor-trailer (**Fig. 4**) that

Fig. 2. A 1947 Labatt Streamliner. (*Courtesy of Labatt USA*)

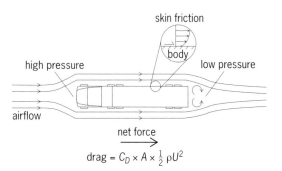

Fig. 3. Sources of aerodynamic drag on a tractor-trailer.

can significantly influence the overall vehicle drag. These flow regions include the trailer base, trailer underbody, trailer axle and wheel assembly, tractor-trailer gap, flow around the tractor body, and flow through the engine. The drag produced by these critical flow regions can be treated by existing drag-reducing add-on devices, such as trailer boat-tails, skirts, and gap sealers. The boat-tails are made of three or four flat plates that are attached to the trailer base and are angled slightly inward (**Fig. 5**). These plates turn the flow field more sharply into the trailer wake, which results in an increase in pressure on the trailer base and a reduction in the drag coefficient by about 7–14% for a vehicle with a baseline C_D of 0.65. The trailer skirts (Fig. 5) are flat plates that extend beneath the trailer and span the distance between the rear tractor wheels and the trailer wheels. These skirts function by shielding the trailer wheels and axles from crosswinds, thereby streamlining the trailer underbody and reducing C_D by about 5–7%. Some designs also include skirts that cover the trailer wheels and extend to the trailer

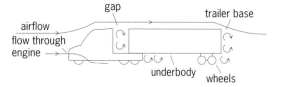

Fig. 4. Critical aerodynamic flow regions contributing to the overall drag on a tractor-trailer.

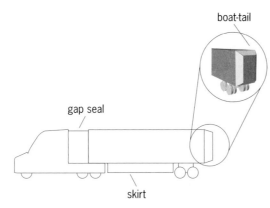

Fig. 5. Trailer boat-tail, skirt, and gap sealer used for tractor-trailer aerodynamic drag reduction.

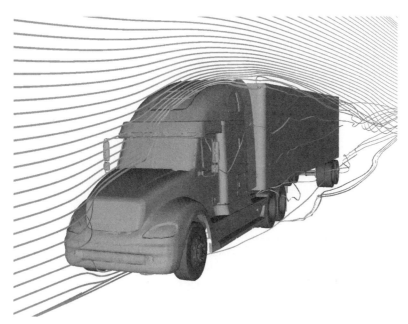

Fig. 6. Instantaneous velocity streamlines over a modern class 8 tractor-trailer in a crosswind predicted using computational fluid dynamics (CFD). The streamlines highlight the complex flow structures that exist around the vehicle.

base. Gap sealers (Fig. 5) are devices that reduce the amount of crossflow in the tractor-trailer gap, a phenomenon which occurs when the tractor-trailer is operating in a crosswind and a portion of the flow field is drawn into the gap. Some designs of gap sealers consist of rigid plates that span a portion of the gap. Other designs have a flexible membrane, similar to a roll blind, which extends from the back of the tractor to the front of the trailer. Wind tunnel tests demonstrate that these gap sealers can reduce C_D by about 1–3%.

Despite the effectiveness of trailer boat-tails, skirts, and gap sealers, these devices have not been accepted on a wide-scale basis throughout the tractor-trailer industry for reasons of cost and liability. For example, the trailer must be modified to accommodate drag-reducing devices, and since there are more trailers than tractors, a greater initial investment is needed by fleet companies. Additionally, the boat-tails may interfere with loading operations and the side skirts may restrict access to the trailer underside

where brake inspections must be performed. A lack of durability of the gap sealers has prevented them from withstanding the wear and tear of normal operations. However, with increasing fuel costs and efficiency standards, there is a greater motivation today to decrease the fuel usage of tractor-trailers through aerodynamic drag reduction techniques. For these drag-reducing devices to gain the acceptance of the tractor-trailer industry, not only must they be further developed to maximize their aerodynamic drag reduction performance, they also must provide practical designs for fleet operations.

While tractor-trailer aerodynamic studies have historically depended solely on model or full-scale wind tunnel testing, one of the tools currently being employed to design operationally minded drag-reduction devices is computational fluid dynamics (CFD). This technique provides detailed flow-field information, such as the three-dimensional, time-dependent velocity field about the tractor-trailer (**Fig. 6**). Increased computing power and improved modeling capabilities will allow CFD to provide a deeper understanding of the critical aerodynamic flow regions and to offer a means of rapidly assessing the performance of future drag-reducing concepts and designs for tractor-trailers.

For background information *see* AERODYNAMIC FORCE; AERODYNAMICS; BOUNDARY-LAYER FLOW; COMPUTATIONAL FLUID DYNAMICS; FLUID MECHANICS; GAS DYNAMICS; STREAMLINING; TRUCK; WIND TUNNEL in the McGraw-Hill Encyclopedia of Science & Technology.   Kambiz Salari; Jason Ortega

Bibliography. K. R. Cooper, *Commercial Vehicle Aerodynamic Drag Reduction: Historical Perspective as a Guide, UEF Symposium on The Aerodynamics of Heavy Vehicles: Trucks, Buses and Trains*, Asilomar Conference Center, Monterey-Pacific Grove, California, Dec. 2–6, 2002; K. R. Cooper, *Truck Aerodynamics Reborn—Lessons from the Past*, SAE Paper 2003-01-3376, 2003; J. Leuschen and K. R. Cooper, *Full-Scale Wind Tunnel Tests of Production and Prototype, Second-Generation Aerodynamic Drag Reducing Devices for Tractor-Trailers*, SAE Paper 06CV-222, 2006; United States Department of Transportation, Federal Highway Administration, *Highway Statistics*, 2007; United States Department of Transportation, *Transportation Energy Data Book*, Edition 26, 2007.

Transportation efficiency and smart growth

Personal transportation is a major component of greenhouse gas emissions as well as household expenses in the United States, accounting for about 18% of each. Both vehicle miles traveled and car ownership, the major determinants of both emissions and household costs, have risen continuously since the 1940s and at far higher rates than general population growth. Yet, it is only relatively recently that the effects of land use and transportation infrastructure investments, on the one hand, and pricing policies for roads, parking, auto insurance, fuel, and vehicles, on the other hand, have begun to attract significant researcher attention.

Smart growth is defined by its proponents as a type of human settlement pattern that includes a wide variety of physical, social, and economic aspects, ranging from density of development and availability of transit services, to diversity of incomes and housing types, availability of parks and recreation, and protection of sensitive habitats. Many of these criteria do not relate directly to energy consumption or efficiency.

In this article, we focus on location efficiency—the extent to which certain types of infrastructure and pricing policies provide desired levels of mobility and accessibility at lower levels of car ownership and with fewer vehicle miles traveled, thereby saving energy. Location efficiency is an important component of smart growth. It reduces the consequences and costs of travel, just as building energy efficiency reduces the consequences and costs of comfort in homes and workplaces.

Location efficiency studies. Conventional analyses of energy policy have tended to overlook the contributions of location efficiency. For example, the "National Energy Strategy of 1991–1992" spends 13 pages discussing transportation energy use and ways to save energy, listing fuel economy, vehicle scrappage, alternative fuels, and consumer education. In contrast, the report spent only about a half page, none of it quantitative, on location efficiency. The 2001 National Energy Policy document failed to address location efficiency at all.

These studies were typical of the prevailing view at the time, which is still heard in policy debates in 2009, that engineering-based improvements in efficiency, such as higher fuel economy in cars or less energy intensity in fuels, are acceptable subjects for energy policy discussion; whereas economic and social-science–based changes in policy, such as allowing higher densities for residential developments, providing more transit, or pricing transportation services to reflect incremental costs, are not suitable subjects for analysis.

The importance of location efficiency can be seen by examining the trends in vehicle miles traveled. Up until 2005, vehicle miles traveled (VMT) in the United States rose at an essentially constant rate, indistinguishable from the rate of increase before 1973. The pre-1973 increase could, at least intuitively, be explained by decreases in the real cost of cars and fuel, coupled with increases in real median family incomes and the expansion and completion of the interstate highway system. However, the persistence of the VMT increase post-1973, when income growth essentially came to a halt, suggests that a purely economic explanation for the increase in VMT would be strained.

Since about 1990, an increasing number of research papers have looked at data from a variety of perspectives and find consistent patterns of public policy decisions that have a strong effect on car ownership and VMT per car.

Statistical studies generally have found strong correlations between compact residential development (high number of dwelling units per unit land area) and reductions in travel. These studies have also found that high provision of mass-transit services lowers VMT. Interestingly, this result applies not only to those who use transit, but also to the average area household. Adequate transit is often found in, and encourages the development of, neighborhoods with a variety of services close to urban cores, which are accessible by car, transit, bicycle, and on foot, lowering automobile VMT.

The referenced studies in published overviews of location efficiency are hard to compare directly to each other because of issues of data availability. For example, some studies define residential housing density at a relatively large scale, which can obscure some of the influence of this variable. For example, the metropolitan area of New York City is less dense than the metropolitan area of Los Angeles. Nevertheless, the finer-scaled look at density shows that the large reductions in driving in the five boroughs of New York City far more than compensate for the higher amount of driving in suburbs of New York, which are more sprawling than those of Los Angeles. Conversely, defining density over too small a land-use area would suggest that a pair of eight-story buildings surrounded by low-density housing should somehow have an effect in reducing VMT. In addition, some studies reflect the results of travel surveys, while others rely on more accurate methods, such as odometer readings or (more recently) GPS tracking of cars.

An additional caution in comparing studies is that most of the influencers of location efficiency are highly multicollinear; that is, when one parameter, such as density, is higher, the other parameters, such as transit access, mixed use, proximity to the center, and cost of parking, are also higher. It is difficult to separate out the effects of these variables, and many studies do not even attempt to do so. There is a strong level of comparability in the results, which gets stronger as one looks in detail at the strengths and limitations of the underlying studies. Some of the original research may find a particular variable to be very important, while others find it to be less important (or even statistically insignificant). Nonetheless, trends with respect to VMT are consistent. One does not find direct conflicts in which one study will find a positive influence of a particular variable, while another finds a negative influence.

Government policy. Government policy is at the heart of market decisions on the main influencers of location efficiency. Density is regulated at the local level almost everywhere in the United States. With few exceptions, localities limit the maximum density that can be built. These limitations typically are at a level far below that needed to achieve significant savings from location efficiency. Parking is similarly regulated. It is common for municipalities or counties to require minimum amounts of parking, both for residential and for commercial development. These regulations also cut in the direction of reducing location efficiency.

Transportation infrastructure is provided almost exclusively by governments, with the exception being most off-street parking. Governments also build and operate mass-transit systems as well as build, maintain, and police roads and highways. The predominant public policy trend since the end of World War II has been to encourage the dispersal of land uses to individual, single-use sites, which are connected by a transportation network designed almost exclusively for private automobile travel.

Pricing policies. Pricing policies can also have a major effect on location efficiency. Many of the real costs of auto travel are either externalized (pollution, public health, road safety), subsidized through public policy (road construction, undervalued or free parking), or inaccurately reflected to consumers largely due to tradition (auto insurance). Studies repeatedly have shown that drivers reduce VMT as these "hidden" costs are revealed to them through effective pricing policies. Pricing programs can be designed to pay for the provision of transportation infrastructure, to reduce driving and congestion, or both.

Road, cordon, or congestion pricing. Drivers reduce VMT to avoid direct charges that reflect the incremental costs they cause (that is, delay) through their addition to traffic.

Parking pricing. Some estimate that more than 85% of all parking in the United States is free, although parking is not free to build, operate, or maintain. This serves as a direct subsidy to car travel, while also creating parking shortages. Revealing the true cost of parking to drivers has been shown to reduce VMT.

Pay-as-you-drive insurance. Auto insurance premiums are generally not based on how many miles someone drives, but rather on a car's model year. The total risk pool, then, has low-mileage drivers subsidizing high-mileage drivers, with few drivers able to reduce premiums by driving less. Pay-as-you-drive insurance directly ties an insurance premium to the number of miles driven. Transforming auto insurance from what is essentially a "fixed" cost of auto ownership to a "variable" cost of ownership (like gasoline) has been predicted to decrease VMT by up to 8%.

Energy prices. Notwithstanding the high level of predicted effectiveness of these pricing policies, the first such policy that is usually discussed is energy prices. Recent evidence suggests that this potentially is not a particularly important variable.

A recent study found that the short-term price elasticity of gasoline was only 3–8% (best estimate 4%), based on data following the most recent spike in gasoline prices. Price elasticity is the measure of how much the demand for a "good" changes in response to price. A price elasticity of 0.10 means that if price goes up, say, 20%, then demand goes down by 2% (that is, by 10% of 20%). Thus, a price elasticity of 4% is very low. It is not surprising that the price elasticity of gasoline is low, because gas represents only about 15% of the total costs of owning and operating a car, so doubling gas prices only increases driving cost by about 15%. Interestingly, even this effect may be overstated; while consumption declined as gasoline climbed to over $4/gal, the subsequent decline in gasoline prices to about $2/gal occurred simultaneously with continuing decreases in VMT.

Summary and recommendations. Considerable research has been done, leading to consistent results in identifying the main influencers of location efficiency. These factors have been shown to allow predictions of how changes in public policy could reduce car ownership and vehicle miles traveled at a net economic benefit, resulting in VMT reductions of 10–40% in 20–30 years. This estimate is based primarily on policies affecting transportation infrastructure and land use. Including the effects of pricing, it would put the reduction at the high end of the range, or possibly even higher.

With the exception of a few cities, such as Portland, Oregon, many of these general relationships have not yet been incorporated into local land-use and transportation planning models. For example, municipalities continue to base their off-street parking requirements on observations of single-use, suburban locations that lack transit. Better efforts to use location efficiency and project context to determine necessary parking would facilitate the kind of development shown to reduce household VMT. California has led the way by requiring improvements in their planning models to reflect the research discussed here.

Further research can offer even more insight. Virtually no research has been done on the effect of commercial land uses on auto use or on the effect of any of the location efficiency variables on freight transportation. While it is plausible to assume that the ton-miles of freight transportation for intrametropolitan trips should decline more or less in proportion to the decline in automobile vehicle miles traveled, there is no empirical data to validate or falsify this assumption. We also lack sufficient empirical analysis of the effects of telecommuting, flexible work schedules, or other increasingly popular work arrangements on VMT.

For background information *see* CIVIL ENGINEERING; ENVIRONMENTAL MANAGEMENT; LAND-USE PLANNING; TRANSPORTATION ENGINEERING; VALUE ENGINEERING in the McGraw-Hill Encyclopedia of Science & Technology.   David B. Goldstein; Justin Horner

Bibliography. F. K. Benfield et al., *Solving Sprawl*, Natural Resources Defense Council, 2001; J. E. Bordoff and P. J. Noel, *Pay-As-You-Drive Auto Insurance: A Simple Way to Reduce Driving-Related Harms and Increase Equity*, Brookings Institution, July, 2008; Congressional Budget Office, *Using Pricing to Reduce Traffic Congestion*, CBO, March 2009; R. Ewing et al., *Growing Cooler: The Evidence on Urban Development and Climate Change*, Urban Land Institute, 2008; T. A. Litman, *Evaluating Transportation Land Use Impacts*, Victoria Transport Policy Institute, 2009; D. Shoup, *The High Cost of Free Parking*, Planners Press, 2005.

Ultrasonic density measurement

Density is an important bulk property of any material, and its measurement is very important in many disciplines such as material characterization and quality or process control. Ultrasonic density measurement is an attractive technique for rapid non-destructive evaluation of a material's density.

Conventional techniques and their limitations. Conventional ultrasonic density measurements can be carried out by determining the time of flight of an ultrasonic wave between an emitting and a receiving transducer; this is a measure of the speed of sound in the material. Because the speed of sound in any medium is dependent on only the bulk modulus (compressibility) and the density of the medium, the density of the medium can be determined if the bulk modulus is known and the speed of sound is measured. In particular, the speed of sound in the material, c, is given by Eq. (1), where K is the bulk

$$c = \sqrt{\frac{K}{\rho}} \qquad (1)$$

modulus and ρ is the density of the material, and therefore the density is given by Eq. (2).

$$\rho = \frac{K}{c^2} \qquad (2)$$

Another approach to ultrasonic density measurement involves determining the amplitude of the reflection of an ultrasonic wave by an interface between a known material and an unknown material. The reflection coefficient (the ratio of the amplitude of the reflected wave to the amplitude of the incident wave) depends on the ratio of acoustic impedances of the known and unknown materials. The acoustic impedance Z is defined by Eq. (3). The reflection

$$Z = \rho c \qquad (3)$$

coefficient RC for normally incident waves (traveling at $90°$ to the plane of the interface) is given by Eq. (4), where Z_1 is the impedance of material 1 in

$$RC = \frac{Z_1 - Z_2}{Z_1 + Z_2} \qquad (4)$$

which the wave is traveling and Z_2 is the impedance of material 2 that forms the interface with material 1 from which the wave is to be reflected. This formula shows that if the reflection coefficient is used to deduce the density of the material from which the wave is reflected, some further information has to be known about material 2, namely, its speed of sound.

A disadvantage of the ultrasonic determination of density using the time-of-flight and the reflection-coefficient methods is the need to know the compressibility of the medium in order to calculate the density from the measurement results. A further disadvantage is the need for accurate positioning of the sending and receiving transducers relative to the sample. Especially for fluid samples, this requirement

means that the measurement has to take place in a carefully designed test cell of tightly controlled dimensions. For accurate measurements, it is also important to ensure that the transducers are mounted with parallel surfaces. These requirements can be impractical if measurements have to be carried out in the field.

Therefore, recently, several ultrasonic "dipstick" techniques have been developed in order to measure fluid properties ultrasonically. More generally, several techniques to measure fluid properties using guided waves have been developed. As early as 1989, a circular torsional waveguide sensor was developed to measure viscosity, and the possibility of measuring density using a waveguide of noncircular cross section was also considered. Recently, a technique was presented for accurately determining fluid densities by measuring the change in velocity of torsional waves in waveguides. Techniques to measure fluid bulk velocity by means of dipsticks also exist. Dipstick techniques for fluid bulk measurements do not suffer from the geometric restriction problems that limit test cells, because their shape predefines the geometry and direction of wave propagation. The remainder of this article describes ultrasonic fluid density measurements using dipsticks in more detail.

Guided-wave dipstick techniques. The idea of the guided-wave dipstick technique is that an ultrasonic wave that propagates in a solid structure can sense the presence and nature of the adjacent fluids. When a torsional wave pulse propagates along a waveguide submerged in a fluid, it interacts at the boundary with surrounding fluid (**Fig. 1**). As a result, the boundary layer of the fluid is alternately accelerated and decelerated. If the waveguide has a noncircular cross section, normal forces are exerted on the surrounding fluid, and fluid will be displaced as the cross section rotates back and forth. This mechanism effectively adds some of the mass of the fluid to the waveguide and changes its inertia. The change in inertia is reflected in a change in the velocity of the torsional wave in the waveguide. Hence, by measuring the speed of propagation of the torsional wave, the density of the fluid can be estimated.

In 1986, an approximate theory was suggested to relate the speed of the torsional wave to the density of the surrounding fluid. A two-dimensional, inviscid flow field of the fluid was calculated in this theory. However, the accuracy of the approximate inversion of the measurements to infer the density of the fluid was compromised because of the complexity of the wave behavior associated with the noncircular cross-sectional shape.

A semianalytical finite element (SAFE) model can be used to accurately predict wave propagation along a solid bar with a noncircular cross section immersed in a fluid. It therefore provides a more precise inverse model relating the group velocity of the torsional wave and the density of the surrounding fluid. This method uses a finite element representation of the cross section of the waveguide, thereby allowing its shape to be specified arbitrarily, together

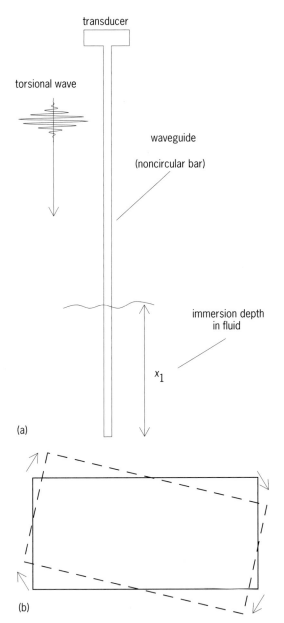

(a)

(b)

Fig. 1. Torsional dipstick sensor for density measurements. (a) Schematic diagram. (b) Cross section (rectangular in this case), showing deformation due to the torsional wave.

with an analysis into harmonics of motion along the propagation direction. An absorbing region, which has the same mass density as the fluid but with increasing damping properties with distance from the central axis, is used in the modeling to absorb waves radiating away from the waveguide, so that the fluid is considered to have infinite extent (**Fig. 2a**). The solution to the SAFE model is obtained by solving an eigenequation (characteristic equation) and finding the wave numbers for the propagation modes at a chosen value of frequency.

Typical results. Figure 2b shows the modeling results of the torsional mode on an aluminum bar with rectangular cross section (1.1 × 2.2 mm) immersed in alcohol at a frequency of 70 kHz. An enlargement of the bar and nearby fluid is shown in the figure. The radial stress (with respect to the center of the

bar) and pressure in the fluid are displayed as a gray scale, and the displacement in the fluid and the cross section of the bar are plotted by arrows. It can be clearly seen that the fluid is displaced by the movement of the cross section; thus it is to be expected that the propagating speed of the torsional mode along the bar should be influenced by the fluid.

Using the SAFE method, the group velocity of the torsional mode propagating in the aluminum rectangular bar can be calculated as a function of the density of the fluid. A typical result is shown in **Fig. 3**, where the prediction of the previously mentioned approximate theory is also plotted. This prediction is based on Eq. (5), where ρ_s and ρ_f are

$$\frac{C_a - C_f}{C_a} = 1 - \left(1 + \frac{\rho_f I_f}{\rho_s I_s}\right)^{-1/2} \tag{5}$$

the densities of the solid and adjacent fluid, respectively; C_a and C_f are the group velocities of the torsional mode of the bar in air and immersed in the fluid, respectively; and I_s and I_f are the polar moments of inertia of the solid waveguide only and the solid waveguide when it is immersed in a liquid, respectively. The polar moment of inertia of a body is a measure of its ability to resist torsion. The larger a body's polar moment of inertia, the smaller will be the twist experienced by that body for a

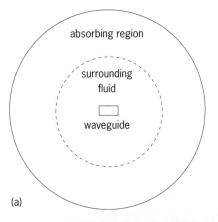

(a)

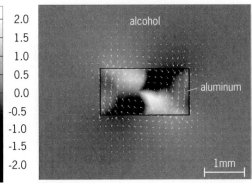

(b)

Fig. 2. Semianalytical finite element (SAFE) modeling of a solid waveguide immersed in an infinite fluid. (a) Schematic diagram. (b) Typical results. Gray scale indicates the magnitude of radial stresses in the bar and the pressure in the liquid; the arrows indicate the displacements in the fluid.

constant level of torque applied to the body. The immersion of the waveguide in a fluid will alter the polar moment of inertia of the waveguide by adding additional mass to the waveguide. The new polar moment of inertial of the immersed waveguide (I_f) will depend on the polar moment of inertia of the waveguide (I_s), its shape, and the fluid properties of the liquid adjacent to the waveguide. As shown by Eq. (5), the change in the polar moment of inertia will also cause a change in the propagation of torsional waves in the waveguide. The quantity I_f can be predicted using the fluid properties and a first-order approximation of the two-dimensional flow field of the fluid around the waveguide.

However, several researchers have reported the inaccuracy of this prediction. The deviation of the measured velocity of torsional waves from the predictions was larger than 20%. With SAFE modeling, the group velocity of the torsional mode along the bar immersed in a fluid can be calculated accurately, and thus a precise inverse model can be provided to relate the density of the fluid to the group velocity of the torsional wave; this SAFE method prediction is also shown in Fig. 3. Experiments were conducted by measuring fluid samples with densities ranging from 0.8 to 1.1 g/cm³, with a 5-cycle tone burst signal at 70 kHz. It can be seen that the SAFE method predictions agree very well with the measurements, and that they represent a substantial improvement over the approximate model.

Sensor optimization. With the SAFE method, one can easily design the geometry of the cross section and the material properties of the bar so that the sensitivity of the dipstick can be optimized. In the SAFE calculation, it was found that $(C_a - C_f)/C_a$ varies almost linearly with the fluid density, and therefore the sensitivity can be presented by the slopes of the lines. **Figure** 4a compares the sensitivities of aluminum dipsticks of rectangular, elliptical, diamond-shaped, and hollow rectangular cross sections with the same aspect ratio. The results show that the diamond-shaped cross section outperforms the elliptical one, and the elliptical one has better sensitivity

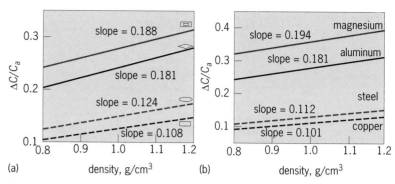

Fig. 4. Comparisons of measurement sensitivities of different dipsticks. (a) Comparison of aluminum dipsticks with different cross sections, indicated by symbols at right. (b) Comparison of dipsticks with diamond-shaped cross sections (1-mm and 3-mm axes), made of different materials indicated by labels at right. $\Delta C = C_a - C_f$.

than the one with rectangular cross section. The hollow rectangular waveguide has sensitivity similar to that of the diamond-shaped cross section with the same aspect ratio. Figure 4b compares the sensitivities of dipsticks with diamond-shaped cross sections (with axes of 1 and 3 mm) that are made of copper, steel, aluminum, and magnesium. It can be seen that the measurement becomes more sensitive when the density of the solid bar is closer to that of the fluid.

In conclusion, there are several ways to measure density ultrasonically. Conventional measurements use the time-of-flight and reflection-coefficient methods. However, in these methods the liquid compressibility has to be known and measurements have to be carried out in a test cell. For rapid field measurements, dipstick sensors can be an alternative that does not require accurately machined test cells. A further advantage of dipsticks is the separation of the fragile transducer element from the measurement region, so that fluids in harsh environments (high temperature, high radiation, corrosive, and so forth) can be tested. A semianalytical finite element model can function as an accurate inverse model to determine the density of the fluid from the change in the group velocity of the torsional mode. This model also enables the optimization of the dipstick sensor by changing the material of the dipstick and the geometry of the cross section.

For background information *see* ACOUSTIC IMPEDANCE; DENSITY; FINITE ELEMENT METHOD; MOMENT OF INERTIA; NONDESTRUCTIVE EVALUATION; SOUND; ULTRASONICS; WAVEGUIDE in the McGraw-Hill Encyclopedia of Science & Technology.

Frederic B. Cegla; Zheng Fan; Michael J. S. Lowe

Bibliography. H. H. Bau, Torsional wave sensor—A theory, *Trans. ASME. J. Appl. Mech.*, 53:846–848, 1986; Z. Fan et al., Torsional waves propagation along a waveguide of arbitrary cross section immersed in a perfect fluid, *J. Acoust. Soc. Am.*, 124:2002–2010, 2008; K. F. Graff, *Wave Motion in Elastic Solids*, Oxford University Press, 1975, reprint, Dover, 1991; J. O. Kim and H. H. Bau, On line, real-time densimeter—Theory and optimization, *J. Acoust. Soc. Am.*, 85:432–439, 1989.

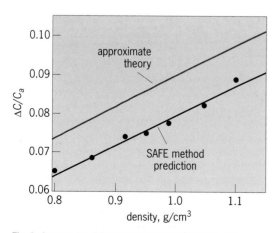

Fig. 3. Inverse model relating the group velocity of the torsional mode and the density of the fluid. The predictions of the model are compared with experimental measurements, indicated by data points. $\Delta C = C_a - C_f$.

Umami taste receptor

Umami is one of the five basic taste qualities that humans can detect, along with sweet, bitter, salty, and sour. It was discovered in 1908 by Kikunae Ikeda, a professor of chemistry at Tokyo Imperial University, while doing research on the strong flavor in seaweed broth. However, umami has only recently been accepted as a basic taste. Umami in Japanese means roughly "delicious." It is often used to describe the "meaty" or "savory" flavor common to products such as seafood, meat, cheese, and mushrooms. The primary umami tastant (any chemical that stimulates the sensory cells in a taste bud) is L-glutamate, which is a naturally occurring amino acid found in abundance in protein-rich foods. A secondary umami tastant is the structurally similar amino acid, aspartate. The most unique characteristic of umami taste is the synergy between glutamate and purinic ribonucleotides such as inosine 5′-monophosphate (IMP). IMP is an "enhancer" of umami taste; that is, it does not taste as umami at submillimolar concentration,

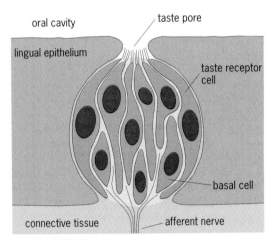

Fig. 1. Schematic drawing of a taste bud.

oral cavity
taste pore
lingual epithelium
taste receptor cell
basal cell
connective tissue
afferent nerve

but can strongly potentiate the umami taste of glutamate. Recent progress in the molecular biology of the umami taste receptor has revealed the molecular mechanism of the synergy between the two umami ligands (that is, molecules with an affinity to bind to the umami receptor). This novel mechanism provides new insights into allosteric modulation (in which modulators bind to regulatory sites distinct from the active site on the enzyme or protein, resulting in conformational changes that may profoundly influence enzyme or protein function) of G protein–coupled receptors (GPCRs). [G proteins are guanosine 5′-triphosphate (GTP)–binding proteins. The GPCRs are cell surface receptors that, when activated by the binding of a ligand, in turn activate a cytosolic G protein molecule, initiating a cascade of reactions effecting the intracellular response to the extracellular signal (the ligand).]

Perception of taste. Taste perception is mediated by taste receptor cells. In mammals, taste receptor cells are assembled into specialized structures called taste buds (**Fig. 1**), which are distributed on the surface of the tongue, along the soft palate, and in the epithelium of the pharynx and epiglottis. The taste receptor cells are connected to the taste center in the brain through afferent nerves at their bases and are exposed to the oral cavity at their apical tips, where the taste receptors are located. Activation of the receptors by tastants triggers a series of signal transduction steps in the cell, which leads to neurotransmitter release and the signal being sent to the brain through the afferent nerves. Different taste qualities are mediated by different taste receptors, which are located in separate groups of taste cells, such as sweet taste cells, bitter taste cells, umami taste cells, and so on, allowing the brain to differentiate the different qualities.

The umami taste receptor is a heterodimer (a protein consisting of two nonidentical subunits) of T1R1 and T1R3 (distinct GPCR members) and is closely related to the sweet taste receptor, which is a heterodimer of T1R2 and T1R3. The two taste receptors share a common subunit, T1R3. The T1R family of taste receptors belongs to the class-C type of GPCRs, featuring a large N-terminal extracellular fragment followed by a seven transmembrane domain (TM) [**Fig. 2**]. The large N-terminal fragments of the T1Rs are composed of the Venus flytrap domain (VFT) and a small cysteine-rich domain (CRD). The VFT domain forms the ligand-binding pocket and structurally resembles the trapping structure of the carnivorous Venus flytrap plant. It consists of two globular subdomains connected by a hinge region. This bilobed structure can form open or closed conformations. Closure of the VFT domain, induced by glutamate, leads to activation of the receptor.

Experimental findings and modeling. Functional assays were developed for the T1R1/T1R3 receptor in experimental embryonic cells expressing a promiscuous G protein (that is, having the ability to interact with a wide range of GPCRs). The in vitro activities of the human T1R1/T1R3 receptor correlate nicely with umami taste. The heteromeric receptor is highly

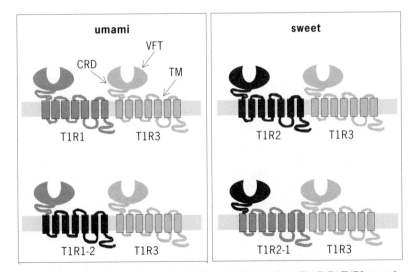

umami
VFT
CRD
TM
T1R1   T1R3

sweet
T1R2   T1R3

T1R1-2   T1R3

T1R2-1   T1R3

Fig. 2. Schematics of umami, sweet, and chimeric T1R receptors. The T1R1/T1R3 umami taste receptor and the T1R1-2/T1R3 chimeric receptor recognize umami ligands, whereas the T1R2/T1R3 sweet taste receptor and the T1R2-1/T1R3 chimeric receptor recognize sweet ligands. Abbreviations: CRD, cysteine-rich domain; TM, transmembrane domain; VFT, Venus flytrap domain.

selective for umami stimuli, recognizing only the umami-tasting glutamate and aspartate among the 20 natural L-amino acids. The receptor responded to glutamate with an EC_{50} (effective dose 50, the amount required to produce a response in 50% of the maximal activity) closely matching the umami detection threshold. More importantly, the glutamate-induced activity was strongly potentiated by IMP. The same functional assay for human T1R1/T1R3 was adopted for high-throughput screening of synthetic libraries. Multiple novel chemical classes of umami-tasting compounds have been identified in this process, further validating the role of T1Rs as human umami taste receptors. Behavioral and physiological studies using knockout mice also have confirmed the role of T1Rs in rodent amino acid taste.

The molecular mechanism of umami synergy has been recently revealed using a combination of chimeric receptors, mutagenesis (formation of a genetic mutation), and molecular modeling approaches. Since the umami and sweet taste receptors recognize different taste stimuli, the primary ligand-binding sites should reside on the unique subunits T1R1 and T1R2, and not on the shared subunit T1R3. Chimeric receptors with the T1R1 N-terminal domain and T1R2 transmembrane domain (T1R1-2) displayed essentially the same ligand specificity as the umami taste receptor, and the activity can be enhanced by IMP. Conversely, chimeric receptors with the T1R2 N-terminal domain and T1R1 transmembrane domain (T1R2-1) displayed the same ligand specificity as the sweet taste receptor (Fig. 2). These data thus indicate that the T1R1 N-terminal domain is critical for binding of IMP as well as glutamate.

Mutagenesis analysis further defined the binding site of glutamate and IMP to be on the VFT domain of T1R1. Thirty-eight residues in the VFT domain of T1R1 were mutated individually, and the mutants were tested for their response to glutamate and IMP. Among the 38 residues, 4 were found to be essential for glutamate recognition, and another set of 4 residues were found to be critical for IMP activity (**Fig. 3**), suggesting that glutamate and IMP occupy different parts of the space within the VFT domain.

Molecular modeling of the T1R1 VFT based on the crystal structure of metabotropic glutamate receptors (mGluRs, which respond to glutamate, an important neurotransmitter, by activating proteins inside nerve cells that affect cell metabolism) revealed the relative positions of the 8 critical residues. The 4 residues important for glutamate binding are located near the hinge region, whereas the 4 residues important for IMP are located near the opening or "lips" of the bilobed structure. A cluster of positively charged residues has been found near the opening of the T1R1 VFT, whereas the negatively charged phosphate group of IMP is important for its umami enhancement activity. Indeed, 3 of the 4 residues that are important for IMP are positively charged, allowing for interaction with the phosphate group of IMP through a salt bridge. A molecular model has been constructed, where both glutamate and IMP are positioned in the cleft of the T1R1 VFT domain (Fig. 3).

hT1R1	Activities		
	Glutamate	IMP	S807
WT	+	+	+
H71A	+	−	+
R277A	+	−	+
S306A	+	−	+
H308A	+	−	+
S172A	−	+	+
D192A	−	+	+
Y220A	−	+	+
E301A	−	+	+

(a)

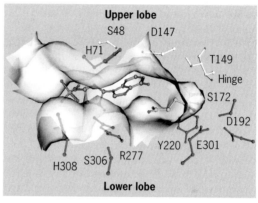

(b)

Fig. 3. Glutamate recognition and IMP activity. (*a*) Critical residues for glutamate and inosine 5′-monophosphate (IMP) recognition, and (*b*) schematics of the molecular model to explain the synergy between glutamate and IMP. S807 is a control umami agonist that targets the transmembrane domain of T1R1. Glutamate is located deep inside the VFT domain near the hinge region, whereas IMP is located close to the opening of the VFT domain.

Glutamate binds close to the hinge region of the VFT domain and induces closure of the lobes, activating the receptor. IMP binds close to the opening of the VFT domain and coordinates the positively charged residues from both sides of the bilobed structure, thereby stabilizing the closed conformation and enhancing the activity of the receptor.

Based on the model, the positively charged residues at the opening of the T1R1 VFT could be repulsive to each other and could destabilize the closed conformation in the absence of IMP. Attempts were made to reverse the positive charge on one side of the opening so as to create an artificial salt bridge between the two lobes and help stabilize the closed conformation. The resultant mutant was expected to have higher activity than the wild-type receptor, as if it were already enhanced by IMP. Indeed, one such mutant (H308A) did show higher affinity and efficacy for glutamate than the wild-type umami receptor (Fig. 3), offering a nice validation of the model.

GPCR modulators. There are two types of GPCR modulators based on their binding sites. The orthosteric modulators work on the same binding site as the natural ligand of the receptor, whereas the allosteric modulators work on different sites. In recent years, researchers have become interested in allosteric modulators of GPCRs as novel therapeutic agents. In many cases, allosteric modulators have advantages over orthosteric ones. For instance, the family of mGluRs shares the same ligand and highly conserved ligand-binding domain, making it very difficult to develop selective orthosteric modulators for an individual receptor. Allosteric modulators, on the other hand, could target the less conserved portion of the receptors and achieve subtype selectivity. Many synthetic allosteric modulators for class-C GPCRs have been developed over the years, although all of them target the transmembrane domains (as opposed to the N-terminal domains).

IMP is a naturally occurring allosteric modulator of the umami taste receptor, as it occupies a different part of the VFT binding pocket when compared to glutamate. It is the only reported example of an allosteric modulator working on the VFT domain, representing a novel mechanism of enhancing class-C GPCRs. Recently, the same mechanism has been found to apply to the sweet taste receptor. High-throughput screening using human sweet taste receptors has identified a series of positive allosteric modulators. These sweet taste enhancers are being developed to reduce dietary caloric intake. Using the same approach of chimeric receptors, mutagenesis, and molecular modeling, initial data have indicated that these sweet taste enhancers target the VFT domain of T1R2 and follow a very similar mechanism as IMP. Understanding the mechanism for enhancement of T1R taste receptors could thus stimulate new ideas to develop allosteric modulators of other class-C GPCRs.

For background information *see* AMINO ACIDS; CHEMICAL SENSES; LIGAND; MONOSODIUM GLUTAMATE; NEUROBIOLOGY; PROTEIN; SENSATION; SENSE ORGAN; SPICE AND FLAVORING; TASTE; TONGUE in the McGraw-Hill Encyclopedia of Science & Technology.

Xiaodong Li

Bibliography. J. Chandrashekar et al., The receptors and cells for mammalian taste, *Nature*, 444:288–294, 2006; X. Li et al., Human receptors for sweet and umami taste, *Proc. Natl. Acad. Sci. USA*, 99:4692–4696, 2002; H. Xu et al., Different functional roles of T1R subunits in the heteromeric taste receptors, *Proc. Natl. Acad. Sci. USA*, 101:14258–14263, 2004; F. Zhang et al., Molecular mechanism for the umami taste synergism, *Proc. Natl. Acad. Sci. USA*, 105:20930–20934, 2008.

Vaccines for schistosomiasis

Schistosomiasis, caused by trematode blood flukes of the genus *Schistosoma*, is the most important human helminth disease in terms of morbidity and mortality. The three major species infecting humans are *Schistosoma mansoni* (which occurs in much of sub-Saharan Africa, areas of South America,

the Caribbean, Egypt, and the Arabian peninsula), *S. haematobium* [present in much of sub-Saharan Africa, Egypt, Sudan, the Maghreb (northwestern Africa), and the Arabian peninsula], and *S. japonicum* (endemic to southern China and the Philippines, with small foci in Indonesia). *Schistosoma haematobium* infections cause fibrosis, stricturing, and calcification of the urinary tract, whereas the other two species have well-described associations with chronic hepatic and intestinal fibrosis and their attendant consequences. Despite the existence of a highly effective antischistosome drug, praziquantel (PZQ), schistosomiasis is spreading into new areas, and, although it is the cornerstone of current control programs, PZQ chemotherapy does have limitations. Mass treatment does not prevent reinfection, and there is increasing concern about the development of parasite resistance to PZQ. Consequently, vaccine strategies represent an essential component for future control of schistosomiasis. An improved understanding of the immune response to schistosome infection, both in animal models and in humans, suggests that development of an effective vaccine is possible, although this goal has yet to be achieved.

Life-cycle features and transmission. Unlike other trematodes, schistosomes are dioecious (that is, they have separate sexes), with the adults having a cylindrical body featuring two terminal suckers, a complex tegument (outer covering or layer), a blind digestive tract (having a single opening through which food enters and undigested waste is expelled), and reproductive organs. The body of the male forms a groove or gynecophoric channel, in which it holds the longer and thinner female. Schistosomes are transmitted through freshwater containing free-swimming larval forms of the parasite called cercariae. These utilize an elastase proteolytic enzyme produced in the head region to penetrate the skin of humans or, in the case of *S. japonicum*, other mammalian hosts (domestic livestock such as water buffalo, pigs, sheep, and dogs). The cercariae shed their bifurcated tails and enter capillaries and lymphatic vessels on route to the lungs. After several days, the young worms or schistosomula migrate to the portal venous system, where they mature and unite. These worm pairs then migrate to their ultimate vascular bed—the superior mesenteric veins (*S. mansoni*), the inferior mesenteric and superior hemorrhoidal veins (*S. japonicum*), or the vesical plexus and veins draining the ureters (*S. haematobium*). Egg production commences 4–6 weeks after infection and continues for the life of the worm, which can be up to 15 years in the definitive mammalian host. Eggs are deposited in the vein lumen. The females produce hundreds (*S. mansoni*; *S. haematobium*) to thousands (*S. japonicum*) of eggs per day. Eggs pass into the host tissues and then many journey through the intestinal or bladder mucosa and are shed in the urine or feces. The life cycle is completed when the eggs hatch, releasing miracidia (ciliated larvae) that, in turn, infect specific freshwater snails (*Biomphalaria* sp. are infected by

TABLE 1. *Schistosoma mansoni* proteins that correlate with resistance in human studies or have shown vaccine efficacy in animal models

Protein or cDNA	Location in adult worm	Identity	Protective vaccine in mice	Protective role in humans
Sm-TSP-2 (tetraspanin D)	Tegument apical membrane	Tetraspanin integral membrane protein	++ worms (recombinant protein) ++ eggs (recombinant protein)	Yes
Sm-TSP-1	Tegument apical membrane	Tetraspanin integral membrane protein	+ worms (recombinant protein) ++ eggs (recombinant protein)	No
Sm-29	Tegument apical membrane	Unknown but has C-terminal transmembrane domain	++ worms (recombinant protein)	Yes
Sm-23	Tegument apical membrane	Tetraspanin integral membrane protein	+ worms (multiantigenic peptide) + worms (plasmid DNA)	Yes
Sm-p80	Associated with tegument inner membrane	Calpain (neutral cysteine protease)	+ worms (plasmid DNA) ++ worms (plasmid DNA including cytokines)	Not determined
Sm-14	Whole body, cytosolic	Fatty acid–binding protein	++ (recombinant protein)	Yes
Sm-28-GST	Whole body	Glutathione-S-transferase	+ worms (recombinant protein) ++ eggs (recombinant protein)	Yes
Sm-28-TPI	Tegument of newly transformed schistosomula	Triose phosphate isomerase	+ worms (transfer of anti-TPI mAb)	Yes
Sm-97 (paramyosin)	Tegument of schistosomula and musculature of adults	Paramyosin	+ worms (recombinant and native proteins)	Yes
CT-SOD	Tegument and gut epithelia	Cytosolic Cu-Zn superoxide dismutase	++ worms (plasmid DNA)	Not determined

SOURCE: Adapted with permission from D. P. McManus and A. Loukas, Current status of vaccines for schistosomiasis, *Clin. Microbiol. Rev.*, 21:225–242, 2008.

S. mansoni; *Bulinus* sp. by *S. haematobium*; and *Oncomelania* sp. by *S. japonicum*). After two generations of primary and then daughter sporocysts within the snail, asexually produced cercariae are released.

Immune response in schistosomiasis. Most chronic morbidity in schistosomiasis is due to the host's immune response, which is directed against schistosome eggs trapped in tissues. The trapped eggs secrete a range of molecules leading to a marked CD4+ (T-helper) cell programmed granulomatous inflammation involving eosinophils, monocytes, and lymphocytes. Granulomas are also characterized by collagen deposition. Severe hepatic periportal fibrosis (Symmers' fibrosis) occurs in the intestinal schistosomes. Much of the morbidity is attributable directly to the deposition of connective tissue elements in affected tissues. In mice, a predominantly T-helper-1 (Th1) reaction in the early stages of infection shifts to an egg-induced T-helper-2 (Th2)–biased profile, and imbalances between these responses lead to severe lesions. Similar regulatory control could be at the basis of fibrotic pathology in humans, although this has not yet been fully established.

Strategies for vaccine development. Vaccination against schistosomes can be targeted toward the prevention of infection and/or to a reduction in egg production. The failure to develop efficacious schistosome vaccines can be attributed in part to the complex immuno-evasive strategies used by schistosomes to avoid elimination from their intravascular environment. Nevertheless, convincing arguments still support the likelihood that effective vaccines can be developed. First, irradiated cercariae induce high levels of protection in experimental animals and additional immunizations boost this level further;

second, endemic human populations develop varying degrees of resistance, both naturally and drug-induced; and third, veterinary antihelminth recombinant vaccines against cestode platyhelminths have been successfully developed and applied in practice. Considerable efforts have been aimed at identifying relevant schistosome antigens that may be involved in inducing protective immune responses, with a view to developing a recombinant protein, synthetic peptide, or DNA vaccine.

Status of vaccine development. Despite the discovery of numerous potentially promising vaccine antigens from *S. mansoni* and, to a lesser extent, *S. haematobium*, only one vaccine has entered clinical trials: Bilhvax, or 28-kilodalton (28-kDa) glutathione-S-transferase, from *S. haematobium*. Data for some of the most promising *S. mansoni* vaccine antigens discovered in the last 10 years, as well as those that were independently tested in the mid-1990s, are summarized in **Table 1**. The leading *S. japonicum* vaccine candidates are shown in **Table 2** and, as with *S. mansoni*, the majority are membrane proteins, muscle components, or enzymes. Vaccine development against *S. mansoni* and *S. haematobium* necessitates the use of clinical vaccines for human application. The zoonotic transmission of schistosomiasis japonica allows for a complementary approach for *S. japonicum* involving the development and deployment of a transmission-blocking veterinary vaccine.

New antigen discovery. The current *Schistosoma* vaccine candidates may prove not to be the most effective. It is important to identify new target antigens and to explore alternative vaccination strategies to improve vaccine efficacy. Mining and functional annotation of the greatly expanded

TABLE 2. Lead *S. japonicum* vaccine candidates that have shown efficacy in the mouse model and reservoir hosts of schistosomiasis japonica

Antigen (native or recombinant protein or DNA plasmid)	Abbreviation	Size, kDa	Stage expressed	Biological function	Worm burden reduction, %*	
					Mice	Other hosts
Paramyosin	Sj-97	97	Schistosomula, adults	Contractile protein + others	20–86	17–60 (buffalo/cattle/pigs/sheep)
Triose phosphate isomerase	Sj-TPI	28	All stages	Enzyme	21–33	42–60 (buffalo/pigs)
23-kDa integral membrane protein	Sj-23	23	Adults	Membrane protein	27–35	0–59 (water buffalo/cattle/sheep)
Aspartic protease	Sj-ASP	46	All stages	Digestion of hemoglobin	21–40	
Calpain large subunit	Sj-calpain	80	All stages	Protease	40–41	
28-kDa glutathione-S-transferase	Sj-28GST	28	All stages	Enzyme	0–35	16–69 (water buffalo/cattle/sheep)
26-kDa glutathione-S-transferase	Sj-26GST	26	All stages	Enzyme	24–30	25–62 (water buffalo/cattle/pigs/sheep)
Signaling protein 14-3-3	Sj-14-3-3	30	All stages?	Molecular chaperone	26–32	
Fatty acid–binding protein	Sj-14	14	All stages?	Binds fatty acids	34–49	32–59 (rats/sheep)
Serpin	Sj-serpin	45	Adults	Serine proteinase inhibitor	36	
Very-low-density lipoprotein–binding protein	Sj-SVLBP	20	Adult males	Binds lipoproteins	34	
Ferritin	Sj-Fer	450	All stages?	Iron storage	35	

*Egg reduction (in feces and/or liver) was also recorded with many of the candidates. When evaluated, reduced egg-hatching capacity of *S. japonicum* eggs into viable miracidia occurred with some vaccines. Cocktails of several of the leading candidates have also been tested.
SOURCE: Adapted with permission from D. P. McManus and A. Loukas, Current status of vaccines for schistosomiasis, *Clin. Microbiol. Rev.*, 21:225–242, 2008.

S. mansoni and *S. japonicum* transcriptomes and their accessibility through public databases, in combination with postgenomics technologies, including DNA microarray profiling, proteomics, glycomics, and immunomics, has the potential to identify a new generation of potential vaccine target molecules that may induce greater potency than the current candidate schistosome antigens.

Whereas it is the schistosome external surface or tegument on which many researchers have focused their efforts, it is those few tegument proteins—that is, the tegument plasma membrane proteins—that are truly exposed to the host immune system in a live worm that are likely to be a major focus for future vaccinology effort. Where they have been investigated, membrane spanning proteins of the tegument have shown great promise, for example, the tetraspanins and Sm-29 (Table 1). This subset of exposed proteins, which present extracellular regions of various sizes, represent excellent targets for future vaccine design and development.

Vaccine formulation. Extracellular vaccine candidates need to be expressed in bacteria or eukaryotic expression systems. Many of the selected targets are likely to require processing through the endoplasmic reticulum by virtue of their expression sites in the parasite (that is, secreted or anchored in the tegument), and this may prove challenging. An additional important consideration is that antigen identification and successful protective results are of little value if good manufacturing practice (GMP) cannot be applied for scaling up of production of any vaccine candidate.

The selection of a suitable adjuvant and delivery system to aid in the stimulation of the appropriate immune response is a critical step in the path to the development and employment of successful anti-

schistosome vaccines, and a number of approaches have been tested with some success. Traditional approaches have seen the use of Freund's adjuvants (a group of water-oil emulsions containing killed microorganisms that enhance antigenicity) when antigens are first being assessed as vaccines in the mouse model. It must be remembered, however, that Freund's complete adjuvant, although it has been the mainstay of immunological adjuvants in research for decades, is not suitable for human application because it can produce a number of undesirable side effects, including the formation of local inflammatory lesions at the site of the injection, which can result in chronic granulomas and abscesses. Once efficacy has been proven with Freund's adjuvants, other adjuvants, particularly those that are licensed (or have the potential to be licensed) for human use, should be used to formulate an antigen.

Conclusions. Although major challenges exist, there is considerable optimism that effective schistosome vaccines will be developed. Integrated genomic and proteomic studies provide a method to select new vaccine antigens. The apical membrane proteins expressed on the surface of the schistosomulum and adult worm are logical vaccine targets on which to focus. In addition, there are messenger ribonucleic acids (mRNAs) encoding novel, putatively secreted proteins without known homologs that are lodged in the tegument membrane, and these have yet to be explored. Indeed, there are very few descriptions of schistosomiasis vaccine trials with proteins that are completely unique to schistosomes and that do not share sequence identity with any other proteins. It is important to emphasize that antischistosome vaccines, when developed and deployed, will not be a panacea. They need to be regarded as one component, albeit a very important one, of integrated schistosomiasis control programs that complement existing strategies, including chemotherapy and health education. Although it is debatable, PZQ resistance is either here or on the horizon, and the need for vaccines is now more pressing than ever.

For background information *see* ANTIGEN; CELLULAR IMMUNOLOGY; DRUG RESISTANCE; EPIDEMIOLOGY; IMMUNITY; IMMUNOLOGY; MEDICAL PARASITOLOGY; PUBLIC HEALTH; SCHISTOSOMIASIS; TREMATODA; VACCINATION in the McGraw-Hill Encyclopedia of Science & Technology.   Donald P. McManus

Bibliography. R. Bergquist, J. Utzinger, and D. P. McManus, Trick or treat: The role of vaccines in integrated schistosomiasis control, *PLoS Negl. Trop. Dis.*, 2:e244, 2008; A. G. Capron et al., Schistosomes: The road from host-parasite interactions to vaccines in clinical trials, *Trends Parasitol.*, 21:143–149, 2005; A. Loukas, M. Tran, and M. S. Pearson, Schistosome membrane proteins as vaccines, *Int. J. Parasitol.*, 37:257–263, 2007; D. P. McManus and A. Loukas, Current status of vaccines for schistosomiasis, *Clin. Microbiol. Rev.*, 21:225–242, 2008; E. J. Pearce and A. S. MacDonald, The immunobiology of schistosomiasis, *Nat. Rev. Immunol.*, 2:499–511, 2002.

Variable data printing

Variable data printing (VDP) is a dynamic form of printing in which text and images in a printed piece are changed based on the specific characteristics of the reader—these changes are triggered by information in a database. They may relate the printed piece to someone's age, gender, occupation, and personal interests. The piece can address multiple demographic points to target the message to the reader. In essence, the printed piece is created uniquely for the individual (**Fig. 1**). Studies have shown that if the message is properly targeted to the interests of the recipients, there will be a higher response to the offer.

Variable data printing is already used in many applications. The most common use is in direct mail. Variable data are used to market new products, create brand loyalty, promote unique sales or services, and target individuals with specific messages.

Over 40 years ago, variable data printing was used for invoices, statements, and bills, with the individual's mailing address being displayed using a window envelope. In the early 1980s, variable data started to be used for marketing materials. This was enabled by the introduction of electrophotographic digital printers from Xerox that imaged directly from electronic files. The result was black text being added to preprinted pieces such as a company's stationery with a color letterhead or variable text being added to a brochure that had one or two panels left blank for personalized information. This was the start of one-to-one personalized marketing with variable data printing. However, these machines were slow, and personalization was expensive to produce.

In the past 20 years, the software tools used to manage and move data have improved, and adding graphic images as variable elements in a document has become practical. Additionally, the cost of high-quality digital color printing has decreased, making the option of using VDP more affordable. An important breakthrough came in the early 1990s with the introduction of high-quality color digital electrophotographic presses. Digital printing technology has advanced steadily, with higher quality, increased run speeds, and reduced toner costs. As an example of the state of the art today, the Xerox iGen4 produces 120 color impressions per minute on stock up to 14.33×22.5 in. (36×57 cm) [**Fig. 2**].

Furthermore, digital color printing by the inkjet process is improving in speed and quality and might rival the electrophotographic process for the VDP market in the near future. New inkjet digital printers introduced in 2009 run at speeds of 1000 ft/min (305 m/min). However, the quality is currently not as high as that of the digital electrophotographic presses, like the iGen4.

Marketing services. Traditionally, most marketing materials are printed by the thousands, with each sheet being identical (static printing). With variable data printing, multiple versions are produced, targeting groups of people or individuals. With today's technology, variable data printing is a marketing tool that is available to companies of all sizes, ranging from the local pizzeria to a large equipment manufacturer to a national realty firm. Variable data (VD) as a marketing tool is not confined to printed materials but is also used for e-mail and Web-page advertising.

The effectiveness of VDP compared to static printing has not been authoritatively quantified. Descriptions of improved response rates can be confusing when companies refer to gains as percentages of baseline responses. For example, a company might report a 34% gain when the baseline response rate of 2% increased to 2.68% when VDP was used. If the same VDP mailing campaign showed an actual response of 8%, this would be reported as a 300% increase. The author recommends reporting this improvement as a 6% increase over a 2% baseline, since it provides a more realistic perception of the additional response.

There are case studies available from Kodak, Xerox, Canon, Ricoh, and Xeikon showing increased response rates when VDP was used. Controlled studies have been conducted by academic institutions showing significant increases in response rates with the use of VDP as well.

Two controlled studies have been run by academic institutions showing significant increases in response rate (Romano, 1999; Gilbert, 2003). While there may not be a clear benchmark on the increased value of variable data printing, hundreds of case studies available from manufacturers and trade associations consistently show the increased value of VDP.

The Romano study dealt with the automotive industry and tracked the increase in sales with brochures changing from black and white to color, and each option was done with and without variable

Fig. 1. Two variable data samples are shown. The sample on the left is for a hiking boot company, where the image changes for gender and the boots change for individuals. On the right, the images change for gender, and names are used in several different places.

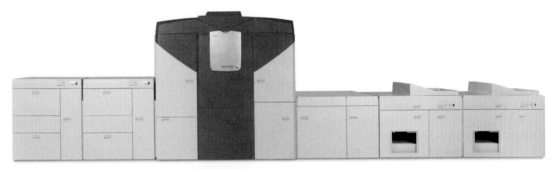

Fig. 2. Critical changes in hardware have advanced the wider use of variable data printing with high-speed/high-quality digital color devices. Shown here is a Xerox iGen Color Digital Press. (*Courtesy of Xerox Corporation*)

data. In the Gilbert study, four completely different industries were tracked and each showed positive increases when personalization was used.

In general, the highly effective uses of VDP are differentiated from the marginally effective uses by the appropriate use of variable data and the creative design of the piece. A successful campaign is tailored to be personal to the recipient, with a relevant offer delivered at the optimum time.

Personalizing the message. The best way for a company to personalize promotional pieces is by tracking and storing customers' purchasing histories, including times of purchases and methods of payment. Patterns develop over time. For example, when someone rents a DVD from an online store, the store may

suggest additional DVDs based on the rental patterns of other people that ordered the same movie. Analyzing the behavior of the entire consumer base and predicting what customers will want becomes more statistically powerful as the size of the customer base increases.

A critical element in implementing VDP is the software that enables variable text and graphics to be inserted into a document template. There are three general types of VDP software—one is a plug-in for prepress page assembly programs, such as Adobe InDesign or QuarkXPress; the second type works within a PDF workflow; and the third type is an independent software program. An example of the first type is uDirect from XMPie, which works with

Adobe's InDesign. An example of the second type is Fusion Pro from Printable, which is a plug-in for Adobe Acrobat. An example of the third type is Persona from PageFlex, which is a stand-alone software application.

Data: the key to successful VDP. The activities of consumers, such as purchases, are increasingly being recorded. Some companies monitor these activities nationwide. There are also companies that receive and analyze every newspaper published, recording information like birth announcements. They sell this information to advertising, marketing, and sales services across the country. Still other companies combine the information that can be obtained from public records, such as real estate records, automobile registrations, census data, and survey data, with consumer lists from mail-order companies and other data sources. It is common for companies to exchange lists. There are companies that compile lists of characteristics for consumers and businesses and track as many as 450 different attributes. USAData, ACCUDATA, and InfoUSA are three national companies that make data available for creating targeted direct marketing.

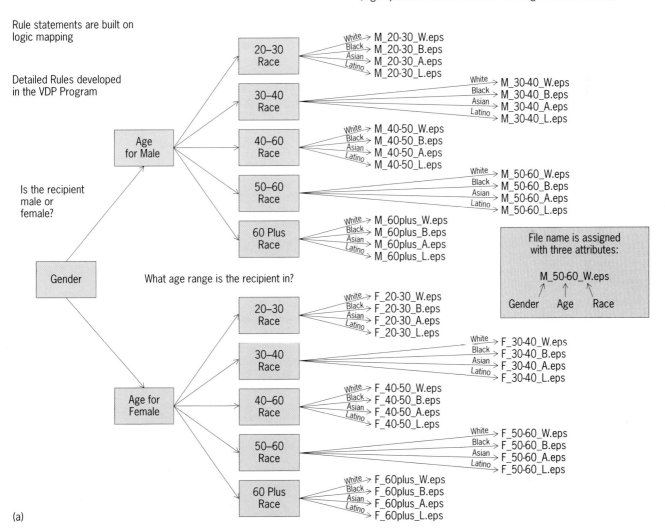

File name is assigned in the database and the field points to the right image to use.

Record No	First Name	Last Name	Address	City	ST	Zip	M/F	Photo Choice	Date Towed	Age	Race	Miles Towed	Cost
45831	Jane	Doe	4 Apache Drive	Danbury	CT	06804	F	F_40-50_W.eps	9/1/2009	40-50	Black	12	$50
45832	John	Sample	307 Monaco Circle	Clemson	SC	29631	M	M_50-60_W.eps	8/13/2009	50-60	White	125	$200
45833	Maria	Sanchez	3947 North Lake Drive	Milwaukee	WI	53211	F	F_20-30_B.eps	9/2/2009	20-30	Latino	52	$150

(b)

Fig. 3. Typically the image selection is based on (a) a logic statement working with various fields in a database or (b) the specific photograph assigned in the database.

Advertisers, ad agencies, and printers all have access to consumer data. They can mine this information to determine relevant images for a personalized brochure, giving an individual an offer that is more likely to create a response. The strategy is to evoke an emotional response that keeps the recipient reading. The longer someone looks at a brochure, the more likely that person is to purchase or donate. Even seeing their name in print or how long they have been a customer makes consumers more likely to respond to an advertisement. Personalizing the information brings relevance to the piece, causing the consumer to identify better with the offer.

Linking the data to the design. Writing rule statements for placing variable data is different for the various VDP software applications. Some applications use scripting languages and allow users to adjust the parameters for rule statements. Rule statements typically are structured as "if-then" statements. Software either will work within a single database, using rules to combine fields and customize the data, or will combine information from multiple databases. Regardless of the approach, the outcomes of the rules need to be considered in the planning and design phases of the variable data piece (**Fig. 3**).

For example, assume that a company tracks the people whose cars were towed in the last month who were not members of a motor club. The goal is to promote membership in a motor club to these people. The collected data include names, addresses, towing dates, cost and distance of towing, and type of car. Assume that the tow truck drivers also recorded the approximate ages, genders, and ethnicities of the customers. A brochure could now be produced where the cover image is a picture of a person of the appropriate gender, race, and age sitting on a curb next to a car of the appropriate type waiting for towing service. Text blocks could be included based on the number of miles the car was towed and the cost of the towing. In producing this brochure, one needs to consider the increased cost related to the number of pictures necessary for all the contingencies. With 2 gender options, 4 race options, and 5 age groups, 40 different photographs would be needed. The assumption is that the recipient will relate better to someone of the appropriate age, race, and gender on the cover of the brochure. Mapping out the plan for writing the rules requires working with logic statements (Fig. 3).

In this example, the designer would specify a set of rule statements about gender, age, and race

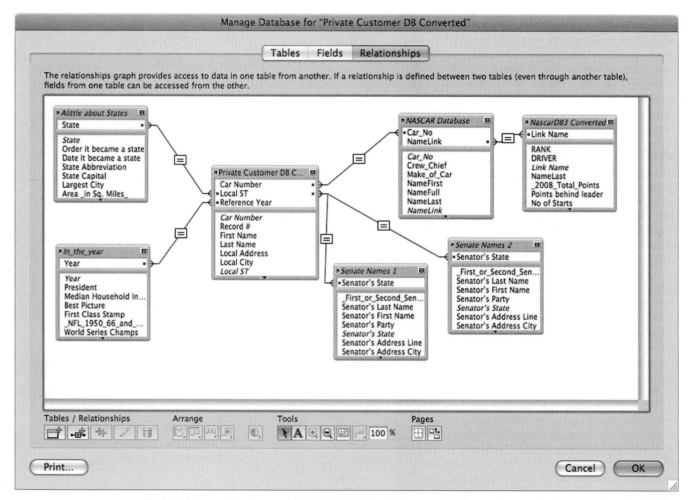

Fig. 4. **Databases drive the personalization of the text and graphic images in the variable data piece. Combining multiple databases allows for triggered variable data that are linked to key data in the main database.**

that identifies the corresponding file saved on the hard drive to be inserted into the file (Fig. 3*a*). A second approach is to reference the file name of the photo directly in the database, and the software then inserts the correct image. The software references the name of the photo from the database and inserts the correct image (Fig. 3*b*).

Once the relevant information is identified for a customer, data from multiple databases can be linked to create personalized messages. In **Fig. 4**, the main database is in the center, and the subordinate databases with relational information are linked to it. The database on the upper left of Fig. 4 has information about the state in which the person lives. The database in the lower left has historical data for various topics for every year from 1950 to the present. The creative use of the data is the designer's responsibility. For example, if the designer knows the year in which someone was born, was married, or graduated from college, facts from that year can be included to increase the recipient's interest in the message.

Going green with print. In the printing industry, variable data printing is being promoted as being a more environmentally friendly and sustainable production process. Fewer total printed pieces are produced, and, therefore, postage costs are reduced with VDP. Producing fewer printed pieces implies that less paper, ink/toner, production time, and energy are used. It is also significant that variable data are used in personalizing e-mail and smart phone messages, Web pages, and other electronic messages, thus reducing the amount of printed material produced.

Outlook. VDP is an effective means of advertising and marketing. It provides a service in helping consumers make better choices by providing them with relevant information. The next time you print a boarding pass at home before leaving for the airport, the boarding pass might contain weather information for your destination city at the top of the page, then your boarding pass, followed by advertisements for restaurants, rental cars, and hotels at the bottom. As the process becomes more sophisticated, the specific restaurant, rental car, and hotel ads on the boarding pass will be targeted to the dietary, automotive, and lodging preferences of each individual. This type of marketing will become so commonplace that consumers will take it for granted.

For background information *see* DATABASE MANAGEMENT SYSTEM; ELECTRONIC MAIL; INK; INKJET PRINTING; INTERNET; LOGIC; PHOTOCOPYING PROCESSES; PRINTING; WORLD WIDE WEB in the McGraw-Hill Encyclopedia of Science & Technology.

John Leininger

Bibliography. P. Bennett, *The Handbook for Digital Printing and Variable-Data Printing*, PIA/GATF Press, Pittsburgh, PA, 2006; D. Broudy and F. Romano, *Direct Mail Responses—Digital White Paper No. 1*, Printing Industries of America/Digital Printing Council, Pittsburgh, PA, 1999; D. Gilbert, *Conventional Direct Mailing versus Variable Data, Personalized Printing: A Comparative Analysis of Cost-per-Response Effectiveness*, Clemson University, Clemson, S.C., 2003; A. Hughes, *Strategic Database Marketing*, 3d ed., McGraw-Hill, New York, 2006; F. Romano, *Variable Data Printing*, Independent Graphics, New York, 2001.

Volcano seismicity in the laboratory

There are some 600 volcanoes on the Earth that are known to have erupted in historical time, with nearly 500 million people currently living on an edifice or nearby. Improved understanding of volcanic mechanisms is therefore a central goal in volcano-tectonic research and hazard mitigation. Although sophisticated techniques are available for monitoring volcanoes, there is still no universally accepted quantitative physical model for determining whether or not a sequence of precursory phenomena will end in an eruption or for forecasting the time or the type of eruption, such as relatively benign effusive volcanism (for example, Mauna Loa, Hawaii), devastating and explosive plinian eruption (for example, Vesuvius, AD 79), or flank collapse (for example, St. Helens, 1980). With the advent of modern portable broadband seismology, GPS (Global Positioning System), and laser/satellite surveying, seismicity and ground deformation are now the most common types of monitoring technology, complementing more traditional geochemical indicators in assessing volcanic unrest. Seismic monitoring has been used with great success to forecast and analyze eruptive episodes in numerous settings, and is arguably the most important method. Although our understanding of the processes driving these observations has increased substantially, a key part of the forecasting challenge is similar to that experienced with earthquakes; often, very little data can be observed before a catastrophic main eruption, and so novel and innovative statistical strategies are required to arrive at a failure forecast. Central to these strategies is the application of fundamental rock mechanics in assessing the failure of the rock mass (analyzed as the seismic event rate), whether driven by tectonic stresses or volcanically driven fluid pressures.

In general, two distinct types of seismicity are measured on volcanoes. Volcano-tectonic (VT) earthquakes are generated by deformation and faulting of the rock. Ideally, as with natural earthquakes, a link between the local stresses and the failure strength of the rock could be used to ascertain how close the rock is to failure, and when this failure might occur. Unfortunately, to assess the strength of a rock mass such as a volcanic edifice (**Fig. 1**), it is necessary to know, a priori, the strength of the rock at depth (that is, under the conditions of interest), information that is impossible to directly measure in the field. To link the fundamental micromechanical and petrological processes to seismicity data, a number of laboratory studies have recently concentrated on the failure strength of representative volcanic rocks, with the aim of improving the link between the observed seismic rate and rock failure. Laboratory

Fig. 1. Photograph taken by the author of the Stromboli volcano, Italy. The sloping feature on the left-hand side of the edifice is called the Sciara del Fuoco, the site of a flank collapse. A sample of volcanic basalt (inset) is shown emerging from a steel pressure vessel after a laboratory test. Although the scales are very different, the same fundamental physical processes are at work.

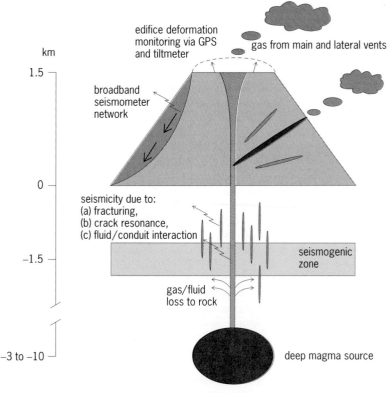

Fig. 2. Conceptual plumbing systems under an active volcano, illustrating the diverse and complex processes and features.

by fluid movement though conduits, cracks, and the associated fault damage zone, known in volcanology as low-frequency (LF) harmonic events. Such events were first identified in the early 1990s and can be compared to the resonance of air through an organ pipe, generating a characteristic frequency and tone. Studies of these earthquakes have greatly elucidated our knowledge of subsurface volcanic-tectonic processes, even suggesting, for a time, a potential route to accurate forecasting methodologies.

Advances in the use of experimental rock physics as a laboratory tool. The volcanic plumbing system (**Fig. 2**) is one of enormous complexity and diversity. A deep magma chamber provides a source of heat and molten (and partially molten) rock, creating and interacting with various fractures, faults, and damage zones above it. Further toward the surface, interaction with groundwater provides another source of complexity. The edifice is composed of a wide variety of volcanic rocks, ranging from basalt and andesite through rhyolite and assorted different types of volcanic tuffs, depending on the specific volcano (magma viscosity). Even though the physical conditions vary hugely, all of these conditions of temperature and pressure can be simulated in the laboratory using a variety of specialized apparatus, the design of which has improved enormously in recent years with the advent of new and modern materials, sensors, and electronics. For work at relatively low temperatures (<300°C), a "triaxial" deformation apparatus is commonly used, where the name is derived from the simulation of three principal tectonic forces, with the maximum load direction (provided by a hydraulic ram) used to deform and fracture a specimen loaded inside the vessel (Fig. 1). For a rock density of 2500 kg/m^3, pressure increases by 25 MPa per 1 km. Most typical triaxial apparatus can easily attain pressure equivalent to 10 km, permitting research into volcano-tectonic processes, especially in the "seismogenic zone." This is a depth interval where fluids, pressures, and forces preferentially collide to produce seismicity. Using such equipment, it is now relatively easy to measure seismicity (AE), using piezoelectric sensors, such as lead zirconate titanate (PZT), to generate a three-dimensional (3D) image of the evolving fracture through time. At higher temperatures (<1200°C), the fragility of the PZT precludes direct attachment to the specimen; insulating rods are needed, and so only two sensors can be fitted. Therefore, a trade-off exists between the image and the desired conditions. Even so, the novel application of piezoelectric sensors has greatly elucidated the picture of volcanic processes through direct laboratory simulation.

The key to successfully reproducing volcanic processes lies in the separation of the two key physical mechanisms: the fracturing and faulting of the edifice that ultimately provides conduits for eruption, and the more subtle processes involving fluid movement leading to resonance in the fractures, the conduits, or the fluid itself. In terms of forecasting strategies, the statistical procedures favored recently can now be tested and verified in the laboratory at

seismicity, a well-known analogue to tectonic seismicity and better known in experimental studies as acoustic emission (AE), has shown that the field scale processes, such as earthquakes and volcanism, are controlled by the same fundamental physics as can be controlled in the laboratory. The second key type of seismicity is specific to volcanoes and is generated

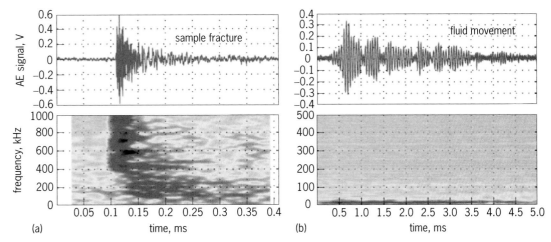

Fig. 3. Comparison of laboratory seismicity (*a*) from faulting and (*b*) from rapid fluid movement through the same faulted and damaged zone. The lower panels illustrate the power (dark shade) and frequency content (vertical axis) associated with the waveforms, showing dramatically different characteristics. (*Adapted from P. M. Benson et al., Laboratory simulation of volcano seismicity, Science, 322:249–252, 2008; reprinted with permission from AAAS*)

elevated temperature and pressure to test their accuracy. Such models assess how close a rock mass is to failure by calculating the inverse seismicity rate with time, which tends to zero as failure (eruption) is approached. This method has been applied successfully to both field data and laboratory data. Ideally, the method is applied to the trend in minima in inverse rate rather than the entire catalog; however, accurate forecasts can be made using the whole catalog when these minima cannot be distinguished.

Simulating volcano seismicity: Comparisons and pitfalls. A side-by-side comparison of seismicity recorded during the two stages of fracturing and fluid flow shows a very obvious change in frequency and power content (**Fig. 3**). A typical laboratory high-frequency (HF) event has considerable power in the 400–800-kHz frequency range, and dies off quickly with time. Conversely, LF events—induced by venting the pore fluid (water) via the top part of the apparatus, which has the effect of isolating the HF generation mechanism (faulting) from crack, conduit, and fluid resonance—show a signal with virtually no power present at frequencies above 20 kHz. The power is essentially monochromatic along the waveform. These observations are directly analogous to the observed VT seismicity and LF resonance on active volcanoes. Although the support for experimental work is high, the issue of how to scale the approximately 10-cm-size samples to kilometer-scale volcanoes is ever present. Fortunately, a simple inverse relationship exists between the length scales involved (cm to km) and the frequency seen in the observations, as the kHz frequencies measured in the laboratory scale to the 1–2 Hz measured in the field by the same proportion. A similar approach can be used with other physical parameters, such as viscosity. Although this is very much a first-order and simplified approach, the scale invariance seen from these analyses now allows researchers to address the scaling issue with confidence, whether they are dealing with magma, water, or hydrothermal fluid generally. Additional investigation of coupled mechanical/

fluid volcano-tectonic mechanisms, which are directly applicable to the field setting, can be derived from the advanced analysis of the waveforms from the suite of sensors surrounding the laboratory sample. This permits the calculation of the type of event involved, such as an explosion, implosion, tensile fracture, or shear fracture (for example, a tectonic earthquake). Such analyses, known as moment tensor inversion, are commonly used by seismologists to determine relative plate motion after earthquakes. Taking all of these methods together, recent work in the laboratory with multisensor configurations has confirmed what volcanologists have suspected for a long time—that low-frequency seismicity is generated by fluid movement, resonance in fractures, and the interaction at the fluid/rock boundary. Laboratory work both at ambient and at high temperature is therefore playing a key part in improving the physical basis behind forecast models.

For background information *see* ACOUSTIC EMISSION; EARTHQUAKE; MAGMA; ROCK MECHANICS; SEISMOLOGY; VOLCANO; VOLCANOLOGY in the McGraw-Hill Encyclopedia of Science & Technology.

Philip M. Benson

Bibliography. K. Aki and P. G. Richards, *Quantitative Seismology: Theory and Methods*, Freeman & Co., San Francisco, 1980; S. R. McNutt, Seismic monitoring and eruption forecasting of volcanoes; a review of the state-of-the-art and case histories, in R. Scarpa and R. Tilling (eds.), *Monitoring and Mitigation of Volcanoes*, Springer, New York, 1996; M. Rosi et al., *Volcanoes*, Firefly, Toronto, 2003; C. H. Scholz, *The Mechanics of Earthquakes and Faulting*, 2d ed., Cambridge University Press, Cambridge, U.K., 2002.

Walking on water

Walking on water is a skill that has evolved independently many times during the course of evolutionary history, allowing a minority of nature's denizens to forage on the water surface and better avoid predators. Over 1200 species of insects and spiders are capable of walking on water, as are several larger creatures, such as some birds, lizards, and dolphins. While the weight of water walkers is supported by one of two means, a variety of ingenious propulsion mechanisms have evolved.

Resting on the water surface. Despite having a density slightly higher than that of water, and so being incapable of floating on the surface by virtue of buoyancy, small water walkers, such as insects and spiders, can reside at rest on the free surface, with their weight supported by the surface tension (**Fig. 1a**). This property of an air-water surface, which has its origins in the intermolecular forces between polar water molecules, makes the water surface behave like a trampoline, resisting surface deflection and enabling it to bear weight. In order to avoid falling through the surface and then facing the daunting task of crossing it from below, a feat that typically requires that they generate a force comparable to 100 times their body weight, water-walking arthropods are covered by a dense matt of waxy hairs

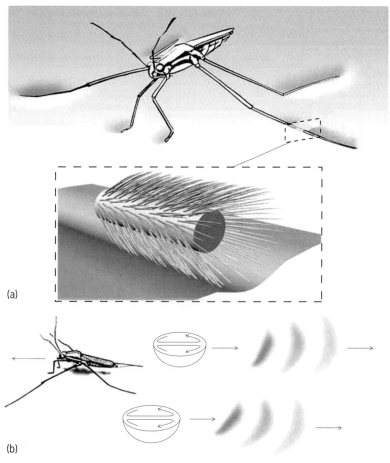

(a)

(b)

Fig. 1. The water strider, one of the most common water-walking insects. (a) At rest, the strider deforms the water surface and is supported by the surface tension. Inset shows the hair layer on the driving legs of the water strider. The leg is a hairy brush with tilted, flexible hairs. The resulting leg surface is water-repellent, and so prevents the strider from sinking through the interface; moreover, its directionality enhances the strider's propulsive efficiency. (b) To propel itself, the strider drives its central pair of legs in a rowing motion, generating a field of rearward-propagating capillary waves in addition to a pair of jets that roll up into a pair of dipolar vortices. (D. L. Hu, B. Chan, and J. W. M. Bush, *The hydrodynamics of water strider locomotion, Nature, 424:663–666, 2003*)

(Fig. 1a inset). By increasing the effective surface area of their bodies and thus the energetic cost of wetting, their hairy coat ensures their water repellency, thus allowing them to survive impacts with raindrops or momentary submersion as may arise from their interaction with a predator or a breaking wave. On the body cavity, a dense mat of hair ensures that water does not penetrate the spiracles through which they breathe, allowing some insects to breathe underwater for extended periods, others indefinitely.

Locomotion of water-walking arthropods. The dynamic role of the hair cover of water-walking arthropods has recently been recognized. On the driving legs, flexible grooved hairs point toward the leg tips (Fig. 1a inset); the resulting directional anisotropy ensures that the driving legs behave as traditional cross-country skis. As the creature strikes the free surface, either in the specialized rowing motion of the water strider and fisher spider or in the alternating tripod gait common to most terrestrial insects, the contact forces between the driving leg and the water are maximized. Conversely, these contact forces are minimized during the gliding phase and when the creature extracts its leg from the interface. This directional anisotropy is apparent when one watches a water strider on a flowing stream: the striders may reside at near rest if they are facing upstream; however, if they turn to face downstream, they are rapidly swept in that direction.

To the unaided eye, the only visible manifestation of the locomotion of most water-walking arthropods is a field of rearward-propagating surface waves generated by the driving stroke. Flow visualization studies demonstrate that these waves are generally accompanied by a pair of dipolar vortices (Fig. 1b). These studies indicate that the great majority of water-walking arthropods rely primarily on surface tension for weight support, and momentum transfer via subsurface vortices for their forward propulsion. Propulsion by momentum transfer via coherent vortical structures is a characteristic feature of biolocomotion through fluids, common to both flying birds and swimming fish. Compared to these creatures, water-walking creatures are relatively efficient in that they generate a propulsive force by striking the water surface, but are resisted primarily by air drag.

Less common propulsion mechanisms do not require a leg strike, but instead rely on manipulation of the water surface, either the surface tension or the surface geometry, and so are referred to as quasistatic (**Fig. 2**). As an emergency escape mechanism, certain insects release a surface active chemical, typically a lipid, in their wake; the resulting surface tension gradient propels them along the free surface for a limited distance (Fig. 2a). Some water-walking insects are too slow and weak to glide up or leap over the upward-sloping menisci that adjoin floating objects or emergent vegetation, and so are unable to escape the water surface to land, as is sometimes necessary in order to lay eggs or avoid predators. In order to do so, some species have developed an

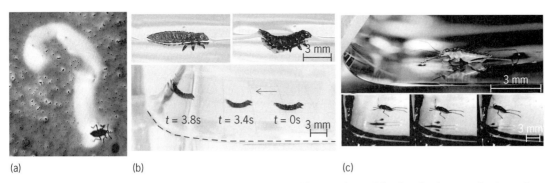

Fig. 2. Quasistatic water-walking techniques. (*a*) *Microvelia* utilizes Marangoni propulsion: by releasing a small volume of a surfactant, specifically a lipid, the insect generates a surface tension gradient that propels it forward. The surface divergence generated by the surfactant is evident in the clearing of blue dye from the free surface. Some insects can propel themselves against gravity, along upward-sloping menisci, simply by deflecting the free surface. (*b*) Wetting insects such as Collembola are circumscribed by a contact line with the water surface, and so may deform the free surface by arching their backs. Doing so propels them up the meniscus, from right to left. (*c*) Nonwetting insects such as *Mesovelia* climb menisci by clasping the free surface with their wetting unguis and pulling upward.

Fig. 3. Large water walkers. (*a*) Basilisk lizard (*photo courtesy of Joe McDonald*). (*b*) Western Grebe (*photo from the feature film "Winged Migration"*). (*c*) Tail-walking dolphin (*photo courtesy of Fran Hackett, New York Aquarium*). (*d*) Humans, who can walk on water only by using water-walking flotation skis such as those conceived by Leonardo da Vinci (*Leonardo da Vinci, Codex Atlanticus, folio 26, 1475–1480, found in Il Codice Atlantico di Leonardo da Vinci nella biblioteca Ambrosiana di Milano, Hoepli, Milan, 1894–1904*).

ingenious meniscus-climbing technique that relies on the attractive force between like-signed menisci, the force responsible for the formation of bubble rafts atop a glass of champagne. Wetting insects such as the beetle larvae are circumscribed by a contact line, and can propel themselves up a meniscus simply by arching their back to match its curvature (Fig. 2*b*). Water-walking insects, which are predominantly hydrophobic, clasp the surface with retractable hydrophilic claws on their front and rear pairs of legs, thereby generating a lateral force that draws them upward (Fig. 2*c*). The principal propulsive force arises from the menisci on the front legs. The rear legs simply balance torques on the creature, while the central pair of legs pushes downward in order to support the creature's weight.

Large water walkers. Creatures too large to rely on surface tension for weight support cannot generally reside at rest on the free surface (**Fig. 3**). Clark's grebe, a shorebird, sprints across the water surface as part of its mating ritual. The most impressive water walker is perhaps the basilisk lizard, and the largest water-walking creature is the tail-walking dolphin. Each of these creatures relies on dynamic weight support, striking the free surface so as to generate reaction forces that ultimately bear the creature's weight. The basilisk lizard also uses hydrostatic pressure for propulsion, pushing off the back of the cavity generated by the leg strike, thereby generating downward, rearward-propagating vortices that ac-

count for both weight support and forward propulsion. If humans were to walk on water, we would have to master a similar technique; however, we would have to run twice as fast as we can, and generate 15 times as much power. As it is, we are incapable of walking on water without floatation devices, such as those envisaged by Leonardo da Vinci (Fig. 3*d*) and employed by fifteenth-century ninjas.

Water-walking and microfluidic devices. Inspired by their natural counterparts, a number of water-walking devices of varying degrees of sophistication have been developed, and make clear the relative importance of the various anatomical adaptations of water walkers. Perhaps most importantly, the world of water-walking insects is dominated by surface tension, and so can serve to inform the design of microfluidic devices. For example, just as the rough, waxy surface of the lotus leaf has inspired the development of water-repellent surfaces commonly used in corrosion-resistant and self-cleaning surfaces, the anisotropic form of the hairy coating of water-walking insects has inspired the design of novel unidirectional superhydrophobic surfaces that may find application in directional draining and directed fluid transport in microfluidic devices.

For background information, *see* COLLEMBOLA; HEMIPTERA; INTERFACE OF PHASES; MICROFLUIDICS; PODICIPEDIFORMES; SURFACE TENSION; VORTEX in the McGraw-Hill Encyclopedia of Science & Technology.

John W. M. Bush; David L. Hu

Bibliography. J. W. M. Bush and D. L. Hu, Walking on water: Biolocomotion at the interface, *Annu. Rev. Fluid Mech.*, 38:339–369, 2006; J. W. M. Bush, D. L. Hu, and M. Prakash, The integument of water-walking arthropods: Form and function, *Adv. Insect Physiol.*, 34:117–192, 2008; S. T. Hsieh and G. V. Lauder, Running on water: Three-dimensional force generation by basilisk lizards, *Proc. Nat. Acad. Sci.*, 101:16784–16788, 2004; D. L. Hu et al., Water-walking devices, *Exp. Fluid.*, 43:769–778, 2007; D. L. Hu and J. W. M. Bush, Meniscus-climbing insects, *Nature*, 437:733–736, 2005; D. L. Hu, B. Chan, and J. W. M. Bush, The hydrodynamics of water strider locomotion, *Nature*, 424:663–666, 2003.

Wave processes and shoreline change

Shorelines are among the most strikingly dynamic geological features on the planet. Winds, waves, and tides are constantly reshaping the coast, moving the shoreline back and forth. Where waves have caused significant deposition along the coast, the shore consists primarily of sand and gravel. Of social, financial, and often sentimental value, the sandy coasts of the world increasingly are being developed despite the ongoing changes that endanger structures and infrastructure built near the shore. Increased rates of sea-level rise over the last century already threaten many developed coasts; these risks will only increase over the coming decades and centuries with predicted increases in the rate of sea-level rise. Coastal scientists endeavor to understand how coasts change. Predicting future coastal changes requires an understanding of how and why coasts have both eroded and accumulated over time. Although beaches and sandy coastlines, such as barrier islands, are currently mostly being eroded or are moving landward, these shores were originally formed through depositional processes. Coastal land-forms can store information about the environmental conditions in the past and can help us better understand how the world's coasts could behave in the future. Until recently, the cause of several types of coastline shapes—regularly spaced coastal undulations and landforms—has been poorly understood.

Transformation of waves approaching a coast. Water waves, which are generated offshore by winds blowing across the ocean surface, are the dominant environmental force that drives change along sandy coastlines. As waves approach the shallower regions near the coast, the limited depth of the water affects their passage. This shoaling in shallower water makes waves slow down. As waves slow, conservation of energy tends to make them become taller. Eventually, as waves continue to slow (shorten) and increase in height, they become oversteep and break. The region where waves break and dissipate their energy is called the surf zone (**Fig. 1**). The large amounts of energy and momentum delivered by waves in the surf zone causes this region to be dynamic, characterized by strong currents that can transport significant quantities of sand. Another change occurs to waves if they approach the shore at an angle: As they slow down, they also refract, reducing the angle between wave crests and the shoreline.

Cross-shore sediment transport and shoreline change. Changes to a shoreline can best be understood by using a simplified framework that separates the coast into two component directions: alongshore (in a direction walking along the coast) and cross-shore (in a direction swimming directly offshore). Over a period of years, there can be significant cross-shore changes to a coast. Storms (with large waves) tend to be the most influential, as they typically transport large amounts of sand offshore during a relatively short period of time (the duration of the storm). However, between storms and during months or years of relatively calmer wave conditions, waves tend to return this sediment from offshore back into shallower regions and the beach. Although coastal changes from storms can be significant, the offshore and onshore movements of sand tend to cancel out over time periods of years and longer.

Transport of sediment along the shore and coastline change. The alongshore transport of sediment (sand) by breaking waves constitutes one of the most persistent and effective forces for moving sediment on the Earth. Waves breaking at an angle to the coast drive a current along the shore, within the surf zone. This current, combined with the mobilization of sediment by wave breaking itself, creates a drift of sediment along the coast, confined to the surf zone and moving in the alongshore direction that the incoming waves were traveling (Fig. 1). The quantity of sediment transported alongshore is affected by the height of breaking waves and the angle between waves and the shoreline (**Fig. 2a**).

Although it may seem that this movement of sediment from one section of coast to another could itself be a cause of coastal erosion, the presence of this sediment "conveyor" does not necessarily mean that the shoreline location will be changing. At any position

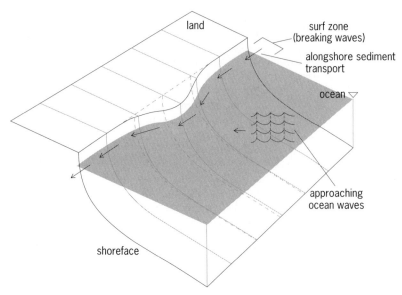

Fig. 1. Waves approach the shore and break in the surf zone. Alongshore sediment transport is driven by waves approaching the shore at an oblique angle.

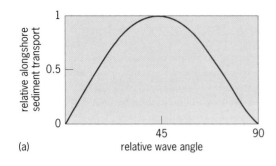

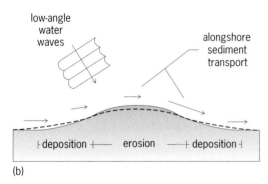

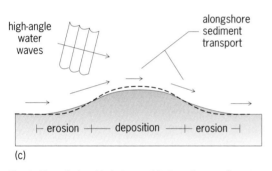

Fig. 2. The relationship between (*a*) alongshore sediment transport and the angle between waves and the coastline. (*b*) Low-angle waves tend to smooth the shape of a coast. (*c*) High-angle waves tend to make bumps along a coast grow.

Coastline instability. If the coast is not straight, the relative angle between waves and the local coast will change, moving along the shoreline. Accordingly, the quantity of sediment being moved alongshore will not be constant along the coast. When wave crests have a small angle compared to the coast (with a "low" angle), any map-view bump in the shoreline will flatten and disappear over time (Fig. 2*b*).

Because waves break at relatively small angles to the coast (<10°), it was previously thought that alongshore sediment transport would only flatten, or diffuse, shoreline bumps. Recent research, however, has indicated something surprising: The important angle for determining the changes in sediment transport is not the angle of wave breaking (which is generally small), but the angle waves make to the coast before shoaling and refracting (an angle that can be much larger than the breaking angle). The relationship between sediment transport and wave angle has a maximum (for angles of about 45°). If waves approach the coast at angles greater than this maximum, these "high-angle" waves will cause bumps along the coast to grow, not flatten (Fig. 2*c*); a coast exposed to these high-angle waves is unstable.

Formation of capes and "flying spits." When a sandy coast is exposed to mostly low-angle waves, its evolution is relatively simple: It will flatten over time. Because most of the world's coasts experience primarily low-angle waves, they are straight. However, research using numerical simulations of coastal evolution has revealed that when coasts are influenced more by high-angle than by low-angle waves, their evolution can be far more interesting. The instability in coastal shape causes any preexisting coastline bumps to grow. Many bumps growing along a

along the coast, just as sand is being moved down-current, new sediment arrives from up-current. If the shoreline is straight and there are no structures blocking this flow, the amount of sediment arriving tends to equal the amount leaving, and the coast does not move. If the coast is curved or if there is a blockage to the alongshore current, the amount of sand arriving at one location will not equal the amount leaving, and the coast will either erode (more leaving than entering) or grow into the ocean (more arriving than leaving).

The fact that shoreline change is driven not by alongshore sediment transport itself, but rather by alongshore changes in the amount of sediment transported, is illustrated by many of the engineering structures that have been placed along coasts to reduce alongshore currents. A blockage or even a "speed bump" to the alongshore current placed along a coast will cause a beach to grow up-current of the structure; however, down-current regions will not be receiving this sediment, and a structure will cause erosion in neighboring regions.

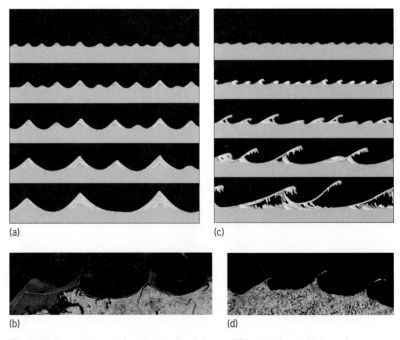

Fig. 3. High-angle coast (*a*) computer simulation and (*b*) example of a high-angle coast with symmetrical wave approach (Carolina coast, southeastern United States). (*c*) Computer simulation and (*d*) natural example of high-angle coast exposed to high-angle waves (Azov coast, Ukraine).

coast interact with one another, resulting in a phenomenon whereby smaller bumps become part of larger ones, and eventually the coast attains a shape with semiregular undulations (**Fig. 3***a, c*). Similar regularly undulating coasts can be seen around the globe in locations where high-angle waves dominate (Fig. 3*b, d*). If more waves approach the coast from one direction than another, the resulting asymmetry in alongshore sediment transport causes an asymmetry in shoreline features, and instead of shoreline capes, a series of offshore-extending sandy deposits, or "flying spits," develops.

This process, where regularly spaced features appear from a generally smooth coast, is an example of self-organization in nature. These examples illustrate an important, and often surprising, aspect of the formation of many patterns in the natural world: The size (or wavelength) of self-organized patterns can arise from dynamics of the system itself, and can be independent of a scale determined exclusively from the external driving forces. The large-scale periodicity in the coast can be many orders of magnitude larger than any wavelength associated with the waves responsible for their formation.

Segmentation of elongate water bodies. Although the dominance of high-angle waves is more the exception than the rule on the Earth's open ocean coast, enclosed, elongate water bodies (such as ponds or lagoons with sandy coasts) are expected to have more high-angle than low-angle waves. Even if winds blow equally from all directions, the larger fetch, or distance across which winds blow, along the long axis of the water body results in larger waves than when the wind blows across the narrow parts of the water body. This causes a predominance of high-angle waves along the long coast of the water body. Numerical simulations of the evolution of a wave-dominated elongate water body show initial behavior similar to that of an open coast: Capes arise where waves approach symmetrically, and flying spits develop where there is an asymmetry in wave attack (**Fig. 4***a*). Once these capes and spits extend almost halfway across the water body, a new dynamic arises and the shoreline undulations attract each other across the water body by affecting the wave climates felt on the opposite side. Eventually, the elongate water body segments into smaller, rounder, stable ponds. These numerical simulations suggest a formation mechanism for enigmatic series of water bodies found across the surface of the Earth (Fig. 4*b*).

Summary. Along a sandy coast—a coast where beaches consist of loose sand and gravel—wave processes control the large-scale shape of the coast through the alongshore transport of sediment. Changes in the shoreline shape over long times are dominated by down-coast changes in the sediment transport conveyer moving alongshore, typically due to changes in the orientation of the coast. Although alongshore sediment transport often smoothes a

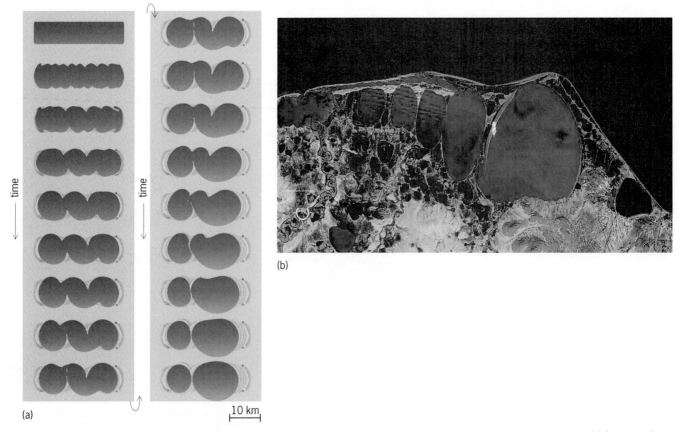

(a)   (b)   time   time   10 km

Fig. 4. Computer simulation of (*a*) the segmentation of an elongate water body by wind-generated waves. (*b*) Segmented water bodies Laguna Val'karkynmangkak, Russian Arctic coast.

coast, when waves dominantly approach from large angles, the coastline is unstable and periodic features such as capes and flying spits emerge. The dominance of high-angle waves is expected in elongate water bodies, and wave processes tend to segment them into series of round ponds or lakes.

For background information *see* COASTAL ENGINEERING; COASTAL LANDFORMS; MARINE SEDIMENTS; NEARSHORE PROCESSES; OCEAN WAVES in the McGraw-Hill Encyclopedia of Science & Technology.

Andrew Ashton; Brad Murray

Bibliography. A. Ashton, A. B. Murray, and O. Arnoult, Formation of coastline features by large-scale instabilities induced by high-angle waves, *Nature*, 414:296–300, 2001; A. D. Ashton et al., Fetch-limited self-organization of elongate water bodies, *Geology*, 37(2):187–190, 2009; R. A. Davis, *The Evolving Coast*, Scientific American Library, NY, 1996; R. A. Davis and D. Fitzgerald, Beaches and Coasts, Blackwell Publishing, Malden, MA, 2004; P. D. Komar, *Beach Processes and Sedimentation*, Prentice Hall, Upper Saddle River, NJ, 1997; C. D. Woodruffe, *Coasts: Form, Process and Evolution*, Cambridge University Press, Cambridge, U.K., 2003.

White-nose syndrome of bats

White-nose syndrome (WNS) is causing major mortality in bats of the northeastern United States, particularly little brown bats (*Myotis lucifugus*) and the federally endangered Indiana bat (*Myotis sodalis*). This syndrome was first documented by a photograph taken in Howe Cave, Schoharie County, New York, on February 16, 2006, and was observed in a few bats in 2006 by the New York State Department of Environmental Conservation (NYDEC). However, in the winter of 2006–2007, an estimated 9000–11,000 bats died in four caves in New York. Many of the dead bats had white fungus growing around their nose (**Fig. 1**), so the ailment was termed "white-nose syndrome." Since then, WNS has been brought to the attention of the research community, and investigations have been undertaken to determine the cause of WNS and to find the means to control WNS.

Bat arousals during hibernation. In the winter of 2007–2008, as many as 500,000 bats may have died of WNS. Many more have died in the winter of 2008–2009. The cause of the syndrome is not yet known, but the bats actually die from a lack of body fat that normally supports them through the winter. Under normal conditions, little brown bats put on about 2 g of fat prior to hibernation, and it is this fat that sustains them through the winter. However, they arouse about once every other week during the hibernation period. The reasons for these arousals are not well understood, but the bats may fly around inside the hibernaculum (winter shelter) or even outside, although they do not feed. These arousals may occur to allow the bats to drink, urinate, or dispose of wastes; to allow their immune systems to operate; or simply to exercise. These arousals are costly, using up to 75–80% of their stored fat and leaving perhaps only half a gram of fat to get the bats through the winter. However, bats with WNS have body weights that are very low for this particular time of year, and they do not have enough fat to last them through the winter. Therefore, many are arousing and flying outside, often hanging on buildings or elsewhere. Normally, when an outsider enters a bat cave in winter, some of the bats arouse and begin flying about. However, bats in WNS-affected caves show very little response; they seem to have very little energy and thus rarely arouse from disturbance. Mortality appears to approach 100% in some of the WNS-affected caves. In the summer of 2008, dead and dying bats were reported in buildings in New York,

Fig. 1. Little brown bats with white-nose syndrome. (*Photo by Nancy Heaslip, New York Department of Environmental Conservation*)

Vermont, Connecticut, Massachusetts, Pennsylvania, and New Hampshire.

Possible causes of WNS. What is the cause of WNS? First, there is the much publicized white nose caused by fungi, but it is now known that several different fungi are involved, including one that grows at low temperatures. The white fungus is not limited to the nose and may be found on wing and tail membranes or in the fur. The fungus that grows at low temperatures (*Geomyces* sp.) is suspected by many biologists to be the cause of WNS, and this possibility is being investigated at various research facilities. If this fungus should prove to be the cause, it might act by irritating the bats, causing them to arouse from hibernation more often than is normal. This might cause them to use an excessive amount of fat. Also, the fungus may grow on the wing membrane and can cause severe damage there, resulting in tissue decay that can result in an inability to fly and forage. Other researchers are also addressing the problem. Some are studying changes in body composition during the prehibernation period to determine whether the bats are entering hibernation with inadequate fat reserves to survive the winter. Others are using temperature-sensitive radio transmitters to determine whether WNS bats hibernate normally or arouse more frequently than normal bats. A number of researchers at the Indiana State University Center for Research and Conservation of North American Bats are investigating chitinase-producing bacteria in the digestive tracts of bats afflicted with WNS. Normal bats have numerous bacteria of several species that digest tiny pieces of chitin remaining in the digestive tracts from late fall feeding. This pro-

Fig. 3. Dead bats at the base of a tree outside an anthracite mine in Carbondale, PA, in February 2009. Approximately 300 dead bats around six trees were found on this visit. (*Photo by Kevin Wenner, Pennsylvania Game Commission*)

cess can produce additional energy for the bats, although it is not yet known whether it is enough to be significant. These bacteria are much less abundant both in species and in number in WNS bats. Other work seeks to gather information on weights and temperatures of bats during their hibernation period.

Documentation. WNS was documented in four states before 2008: New York, Vermont, Massachusetts, and Connecticut. There also was some evidence that it occurred in Pennsylvania. In 2008–2009, Pennsylvania was confirmed as a location for WNS, and it also was confirmed in New Jersey, West Virginia, and Virginia, making a total of eight confirmed states at present (**Figs. 2** and **3**).

The species of bats most heavily affected are those forming large clusters, the little brown bat and the Indiana bat. A few northern bats (*Myotis septentrionalis*) and small-footed bats (*Myotis leibii*) have been found with WNS. These last two species are generally difficult to find in hibernacula because they hibernate in cracks and under rocks. A few large brown bats (*Eptesicus fuscus*) and eastern pipistrelles (*Perimyotis subflavus*) have also been found with visible signs of WNS. Thus, all species occurring in caves in New York and New England have been found to be affected.

Since WNS has expanded into caves of West Virginia, additional bats are threatened, including the federally endangered gray bat (*Myotis grisescens*) and the eastern subspecies of Townsend's big-eared bat (*Corynorhinus townsendii virginianus*). Furthermore, it is only about 300 mi (480 km) to the large wintering population of *Myotis sodalis* in southern Indiana, thereby posing another potential threat to this bat species, and WNS continues to spread (**Fig. 3**). Hopefully, the causation or means of containment can be found before WNS reaches the great Indiana bat hibernacula in Indiana and Kentucky. With the cause still in doubt, no cure is known. However, a possible means of slowing WNS has been suggested, and this entails adding small heaters at selected spots in affected caves. These spots could serve as warm "refuges" for when the bats arouse as

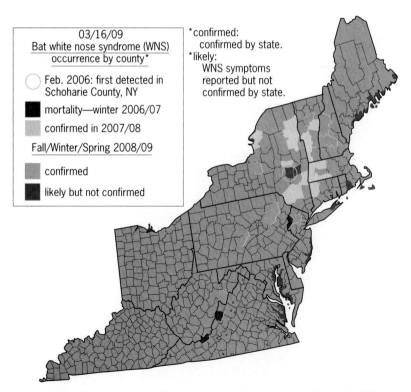

03/16/09
Bat white nose syndrome (WNS)
occurrence by county*

○ Feb. 2006: first detected in Schoharie County, NY

■ mortality—winter 2006/07

▨ confirmed in 2007/08

Fall/Winter/Spring 2008/09

▨ confirmed

■ likely but not confirmed

*confirmed:
    confirmed by state.
*likely:
    WNS symptoms
    reported but not
    confirmed by state.

Fig. 2. Confirmed occurrence of white-nose syndrome by county as of March 16, 2009. (*Map by Cal Buchkoski*)

they normally do about every 2 weeks during hibernation. These arousals cost the bats a great deal of energy, but the bats would lose much less fat if they would fly to the "warm spots" when aroused. This approach is now being tested, but has generated a great deal of interest.

Why should there be any concern about the bats though? The major reasons are related to their food habits. In tropical areas, they pollinate plants and disperse seeds. Bats in the Eastern United States feed on huge numbers of flying insects, including numerous agricultural and garden pests, many of which carry plant and animal diseases. Bats are the major predator on nocturnal flying insects, thereby helping to maintain the balance of nature.

For background information *see* CAVE; CHIROPTERA; DISEASE; DISEASE ECOLOGY; ECOSYSTEM; ENERGY METABOLISM; EPIDEMIOLOGY; FUNGAL ECOLOGY; FUNGI; HIBERNATION AND ESTIVATION in the McGraw-Hill Encyclopedia of Science & Technology.

John O. Whitaker, Jr.

Bibliography.   D. S. Blehert et al., Bat white-nose syndrome: An emerging fungal pathogen?, *Science*, 323:227, 2009; J. G. Boyles and C. K. R. Willis, Could localized warm areas inside cold caves reduce mortality of hibernating bats affected by white-nose syndrome?, *Front. Ecol. Environ.*, in press, 2009; A. Hicks, White-nose syndrome: Background and current status, New York State Department of Environmental Conservation, 2008; J. D. Reichard, Wing-damage index used for characterizing wind condition of bats affected by white-nose syndrome, Center for Ecology and Conservation Biology, Department of Biology, Boston University, 2008; WNS Science Strategy Group, Questions, observations, hypotheses, predictions, and research needs for addressing effects of white-nose syndrome (WNS) in hibernating bats, *Science Strategy Meeting Synopsis*, 2008.

Wide-Area Augmentation System (WAAS)

The Global Positioning System (GPS) is used in a wide variety of both civil and military applications. These applications affect users in everyday activities, from communications and power generation to precisely determining one's location. The GPS is also increasingly being used in aircraft guidance. However, the operational GPS system was not designed to meet the stringent requirements for bringing aircraft safely within close proximity of other objects, and thus the GPS by itself does not sufficiently guarantee the required safety-of-life performance at all times. In particular, safety-of-life performance requires that any error in the reported position of an aircraft must be strictly limited at all times. Therefore, aviation authorities around the world, including the Federal Aviation Administration (FAA) in the United States, have defined augmentation systems to monitor the performance of the GPS and provide timely alerts to users when significant errors may be present.

These augmentation systems are independent of the GPS; they monitor GPS performance continuously. Most important, they detect GPS faults in real time and warn pilots within seconds. Such assistance is needed because the GPS ground control system may not detect and report faults for tens of minutes or longer. The fault-detection alternatives include Aircraft-Based Augmentation Systems (ABASs), Ground-Based Augmentation Systems (GBASs), and Satellite-Based Augmentation Systems (SBASs). This article focuses entirely on SBASs. More specifically, it focuses on the Wide-Area Augmentation System (WAAS), which is the SBAS for North America and was the first operational SBAS. SBASs are also being deployed in Europe (EGNOS), Japan (MSAS), and India (GAGAN).

Architecture.   The WAAS was originally commissioned in July 2003. It currently has 38 reference stations spread across the United States, Canada, and Mexico (**Fig. 1**). Each reference station includes three GPS receivers. Three receivers are included so that a fault in a single receiver can be readily detected.

The receivers are dual frequency. Every second, they make pseudorange measurements and carrier-phase measurements at the GPS L1 and L2 frequencies. The L1 and L2 frequencies are 1575.42 and 1227.60 MHz, respectively. These measurements, together with L-band carrier-to-noise (power) ratio measurements, are brought back to the WAAS master stations over redundant land lines. The WAAS has three master stations so that availability is assured should there be failure at one, and redundancy is provided during WAAS upgrades.

Each master station processes the dual frequency measurements arriving from all of the fielded GPS receivers to create differential corrections for the GPS measurements made by aircraft within the service area. These corrections are intended to compensate for ionospheric delay along the path from the satellites to the airborne receiver, as discussed below. The master stations also process the GPS measurements to bound the potential range errors that will remain after the differential corrections are applied. To begin this process, each master station screens the measurements for bad data and estimates an upper bound on the multipath and noise errors present in each measurement.

The master station then uses the dual frequency measurements to generate the corrections for the ionospheric delays mentioned above. The ionosphere is dispersive, and so the ionospheric delay at L1 is different from the delay at L2. More specifically, the observed delay is inversely proportional to the square of the GPS carrier frequency. The WAAS ground system leverages this relationship to estimate the ionospheric delay at all of the vertices in an ionospheric grid. This grid is 5° by 5° in latitude and longitude south of 60° North. It is less dense over Alaska and northern Canada. As with the single-frequency ionospheric delay model embedded in the software of civil GPS receivers, the WAAS ionospheric corrections model the ionosphere as though it were a

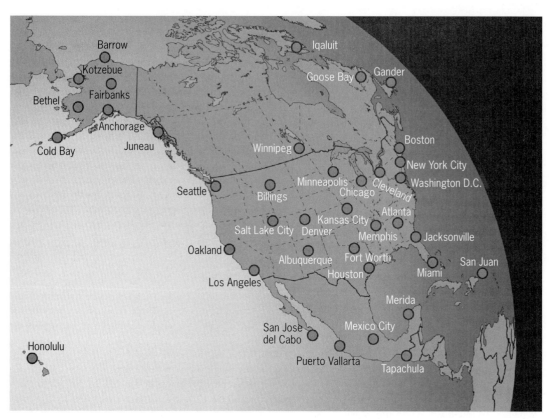

Fig. 1. WAAS reference station locations.

thin shell existing at 350 km (217 mi) above the surface of the Earth. The line of sight between the receiver and the satellite penetrates this layer at a point labeled the ionospheric pierce point. The density of the reference-station network is determined by the spatial decorrelation of the ionospheric delay. If the ionosphere were spatially smoother, then fewer reference stations would be required to cover the service area. If the ionosphere had steeper gradients in latitude and longitude, then a greater number would be required.

Unfortunately, typical avionics receivers cannot make use of the L2 signal because it lies in a non-aviation portion of the radio spectrum. Therefore they cannot determine the amount of ionospheric delay affecting their own signals—hence the need for the ground system to estimate the ionospheric delays that affect the L1-only signals and send the above-described grid of ionospheric delay estimates to the airborne user. Over the next decade, the GPS will begin to broadcast a second signal for civil aviation (L5). After that time, new avionics will be able to use both frequencies to compute the ionospheric delay directly in the aircraft, because the L5 frequency does fall within an aviation band. A dual-frequency user will not require the ground to send ionospheric corrections.

Each master station also generates a set of corrections for certain data in the clock and ephemeris message broadcast by each GPS satellite in view of the reference network. These corrections contain four elements per satellite. One element cor-

rects the satellite clock, and the other three elements are corrections for the three dimensions of satellite position. The corrections are generated from the pseudorange measurements after the ionospheric contribution has been removed and errors due to multipath and tropospheric delay have been minimized.

While the master-station processing to generate the ionospheric and satellite-specific corrections is sophisticated, the greater challenge is to bound the position errors that will remain after the corrections are applied. Each master station provides bounds for the residual errors in the ionospheric corrections called grid ionospheric vertical errors (GIVEs). The GIVE bounds the ionospheric correction error for a given point in the grid when the error is projected into the vertical. Each master station also bounds the impact of the satellite-specific error after correction, and these bounds are called user differential range errors (UDREs). They bound the projection of the satellite clock and location errors when projected onto the line of sight to the worst-case location in the coverage area.

Together these quantities are combined with bounds on the user ranging measurement errors to formulate a final bound on position error. The GIVEs and UDREs assure that the calculated upper bound for the position error always exceeds the actual value of the position error. However, the UDREs and GIVEs cannot be made too large, because the assured bound must be small enough to meet the requirements of the intended operation.

The master station packs the ionospheric corrections, satellite-specific corrections, and associated bounds into the WAAS message stream. This message stream is uplinked to the geostationary satellites (**Fig. 2**). These satellites simply shift the uplink signal frequency and broadcast the message to users everywhere in the geostationary footprint using the same L1 frequency as the GPS. At present, the WAAS broadcasts the differential corrections and integrity information from the geostationary satellites *Telesat Anik F1R* and *PanAmSat Galaxy XV* (**Fig. 3**). The WAAS signals are similar to the GPS signals in design. Further, the WAAS signals are synchronized to GPS time and so can be used for ranging, just as though they were provided by another GPS satellite. Because the satellites are geostationary, they are always in view of the users in the coverage area and greatly enhance availability and continuity of accurate position fixes.

Performance. For an aircraft en route at its full cruising altitude, the requirements on navigation performance are measured in hundreds of meters. However, they become more restrictive as the aircraft comes closer to the ground. Currently, the most demanding level of performance provided by any GPS-based system allows the aircraft to operate safely within 60 m (200 ft) of the ground even when visibility is limited. The requirements for this type of operation are accuracy to better than 4 m (13 ft) 95% of the time; less than a 1-in-10^7 chance that the error is greater than 35 m (115 ft) without the pilot being alerted within 6 s; less than a 1-in-10^5 chance of not being able to complete the operation once initiated; and availability of the above greater than 99% of the time.

Every quarter, the FAA publishes a performance analysis report for the WAAS covering the previous 3 months. This report identifies the availability of service for different flight modes as a function of location. It also reports on WAAS outages and integrity. As expected, there has not been a single integrity fault observed at any location since the commissioning of WAAS. **Figure 4** shows the availability of the vertically guided service averaged over the 3-month period from January 1 through March 31, 2009. As can be seen, the target value of 99% availability was provided to nearly all of the United States and a large portion of Canada and Mexico.

The GPS and WAAS enable flight crews to choose fuel-efficient routes rather than follow indirect routes based on the location of terrestrial radio navigation aids. The WAAS further enables vertically guided approaches that allow the aircraft to come within 60 m (200 ft) of the ground before the pilot must be able to see the runway environment visually. These capabilities enhance aviation safety and also allow for more efficient, continuous descent approaches. The WAAS does this without requiring additional infrastructure to be installed at the airport. Currently, more than 1600 vertically guided procedures are published for use within the United States. Many of these procedures are at airports that had no previous instrument guidance.

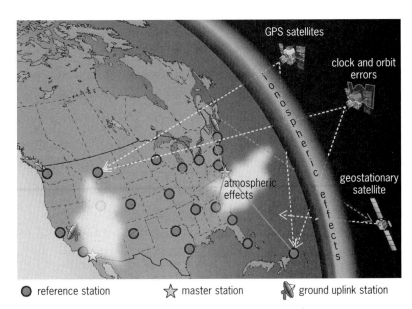

reference station   ☆ master station   ⚡ ground uplink station

Fig. 2. WAAS architecture.

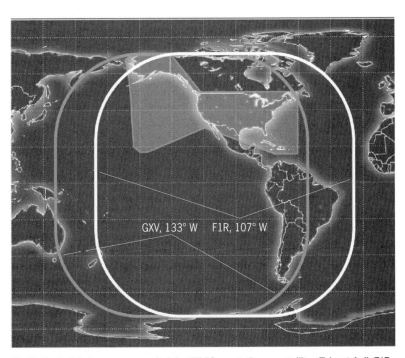

Fig. 3. Footprints (coverage areas) of the WAAS geostationary satellites *Telesat Anik F1R* **(F1R), located at 107° West longitude, and** *PanAmSat Galaxy XV* **(GXV), located at 133° West longitude. (***Federal Aviation Administration, Global Navigation Satellite System (GNSS) Program Office***)**

The WAAS adheres to an international standard that other SBAS providers also follow. Any certified receiver that works with the WAAS is compatible with the other SBASs around the world. MSAS was certified for operation in Japan in 2007. Europe is planning to bring EGNOS into service in 2010. With service extensions to South America and Africa, it will not be long before an aircraft with an SBAS receiver can achieve vertical guidance across much of the globe.

For background information *see* IONOSPHERE; SATELLITE (SPACECRAFT); SATELLITE NAVIGATION SYS-

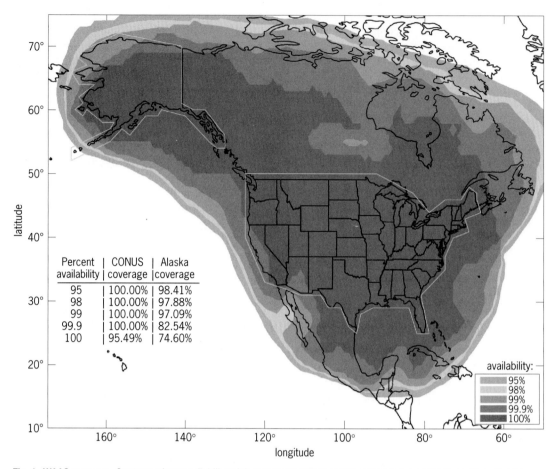

Percent availability	CONUS coverage	Alaska coverage
95	100.00%	98.41%
98	100.00%	97.88%
99	100.00%	97.09%
99.9	100.00%	82.54%
100	95.49%	74.60%

Fig. 4. WAAS coverage. Contours show availability of the vertically guided service averaged over the 3-month period from January 1 through March 31, 2009. CONUS = continental United States. (*Federal Aviation Administration/William J. Hughes Technical Center, Wide-Area Augmentation System Performance Analysis Report #28, January 1 to March 31, 2009, April 2009*)

TEMS in the McGraw-Hill Encyclopedia of Science & Technology.
                                                Todd Walter

Bibliography. S. Datta-Barua et al., Verification of low latitude ionosphere effects on WAAS during October 2003 geomagnetic storm, *Proceedings of ION Annual Meeting*, Boston, MA, June, 2005; Federal Aviation Administration/William J. Hughes Technical Center, *Wide-Area Augmentation System Performance Analysis Report #28, January 1 to March 31, 2009*, April 2009; D. Lawrence et al., Wide Area Augmentation System (WAAS) program status, *Proceedings of ION GNSS 2007*, Fort Worth, TX, September 2007; T. Schempp et al., WAAS performance improvements as a result of WAAS expansion, *Proceedings of ION GNSS 2006*, Fort Worth, TX, September 2006; T. Walter et al., Integrity lessons from the WAAS Integrity Performance Panel (WIPP), *Proceedings of ION NTM*, San Diego, January 2003.

Nobel Prizes for 2009

The Nobel Prizes for 2009 included the following awards for scientific disciplines.

Chemistry. The chemistry prize was awarded to Ada E. Yonath of the Weizmann Institute of Science in Rehovot, Israel, Thomas A. Steitz of Yale University, and Venkatraman Ramakrishnan of the MRC Laboratory of Molecular Biology in Cambridge, UK, for their elucidation of the structure and function of the ribosome.

The ribosome is a structure in the cytoplasm of the cell that uses genetic information, exported from the nucleus as ribonucleic acid (RNA) and translated into another format, to synthesize protein. Every protein used in the structure and function of organisms, from bacteria to plants and animals, is assembled by the ribosome. The ribosome itself is a complex particle composed of protein molecules and RNA. These combine to form what are called the small and a large subunit—the large subunit being roughly twice the molecular weight of the small subunit—which are arranged around and move along the segment of messenger RNA (mRNA) to be translated.

Their work allowed the modern understanding of that function. Specifically, the ribosome reads information on a segment of mRNA, which is a copy of a DNA sequence (gene) that encodes for—or carries the instructions to construct—a specific protein. Translating these instructions, the ribosome joins amino acid pairs to make peptides, and arranges these peptides in a specific sequence to build a polypeptide chain that, when folded, becomes a functional protein molecule.

Yonath, Steitz, and Ramakrishnan used x-ray crystallography to determine the ribosome's structure and how it functions at the molecular level. The technique involves crystallizing a molecule and directing x-rays at the resulting crystal to map the arrangement of its atoms and generate a three-dimensional (3D) depiction of its structure. In 1980, Yonath generated the first low-resolution structure of the ribosome's large subunit from the bacterium *Geobacillus stearothermophilus*, a heat-tolerant bacterium she selected for its hardiness. Further studies would also employ so-called extremophile bacterial species known for their tolerance to a variety of stressors with the expectation that they might remain stable under the experimental regime. Over the next 20 years, Yonath, Steitz, and Ramakrishnan made iterative improvements in obtaining higher-quality x-ray crystallographic structures. In August 2000, Steitz reported the first high-resolution (atom-level) structure of the large subunit from *Haloarcula marismortui* (a halophilic, or salt-tolerant, bacterium). This was followed in September 2000 by Yonath and Ramakrishnan separately reporting high-resolution structures of the small subunit from *Thermus thermophilus* (an extremely thermophilic bacterium).

Ramakrishnan went on to discover that the ribosome's small subunit employs a "molecular ruler" to measure the fit between a section of mRNA and its complementary match (transport RNA) twice, and thus double-check that the correct amino acids are incorporated into the growing peptide chain. This system explains why the ribosome so rarely makes mistakes in translating nucleotides into proteins. Further work by Steitz has shown that the large subunit is responsible for peptide bond formation between amino acids by imaging structures of the large subunit during the bond-forming process. In 2001, Yonath was able to produce a high-resolution structure of the large subunit from *Deinococcus radiodurans* (a radiation-resistant bacterium). This proved useful for studying how antibiotics disrupt the function of bacterial ribosomes. In addition, Yonath, Steitz, and Ramakrishnan have produced 3D structures that show the functional sites at which antibiotics are able to disable the ribosome's protein production in bacteria. This in turn has contributed to the understanding of the growing problem of antibiotic resistance and paves the way for the development of new antibiotics.

For background information *see* ANTIBIOTIC; BIOCHEMISTRY; CELL (BIOLOGY); DEOXYRIBONUCLEIC ACID (DNA); GENETIC CODE; MOLECULAR BIOLOGY; PROTEIN; RIBONUCLEIC ACID (RNA); RIBOSOMES; X-RAY CRYSTALLOGRAPHY in the McGraw-Hill Encyclopedia of Science & Technology.

Physiology or medicine. The prize in physiology or medicine was awarded jointly to Elizabeth H. Blackburn of the University of California, San Francisco, Jack W. Szostak of Harvard Medical School, and Carol W. Greider of Johns Hopkins University School of Medicine, for their elucidation of how chromosomes are protected by telomeres and the discovery of the enzyme telomerase.

Telomeres act as caps at the ends of the chromosomes, which are constructed of the tightly packed double helix of the DNA with associated proteins. While broken arms of a chromosome are unstable at the break and can unravel or fuse with another chromosome, telomeres essentially seal off the end and prevent random fusion events.

Telomeres also allow the complete DNA sequence to be copied and prevent its degradation during cell division. Each telomere provides a landing place for the DNA polymerase primer at the end of the chromatid being replicated. Without it, the primer would cover the end of the DNA to be transcribed, preventing synthesis of that section. The resulting gap in the code would cause a small piece of DNA at each end of the chromatid to be lost in every round of cell division, slowly eroding the molecule. Most of the DNA molecule is composed of sequences of nucleotides, each of which encodes a specific gene, so any loss could have negative consequences. The telomere, in contrast, is composed of a single short sequence of nucleotides repeated multiple times. With these repeats making up the end of each chromatid, only a segment of the expendable telomere is lost with each cell division.

Telomerase is an enzyme that rebuilds the telomeres after replication. The exceptions include most fetal tissues, reproductive cells in adult males, inflammatory cells, proliferative cells of renewal tissues such as the epithelial tissue and bone marrow, and most tumor cells. It is generally active only while cells remain undifferentiated. Once a cell has taken on a specific form and role, telomerase activity during subsequent divisions ceases; the telomere begins to erode and the cell line begins to age. This characteristic has implications for the understanding of aging and of cancer, which can be related to heightened telomerase activity resulting in cell immortality and uncontrolled cell division.

Decades prior to the work of Blackburn, Szostak, and Greider, researchers had observed that the ends of the chromosomes, already called telomeres, differed from the rest of the DNA and seemed to prevent the chromosomes from attaching to each other. In the early 1970s repeats were found to exist near the ends of the chromosomes, but the true character of telomeres remained hidden.

The prize was awarded for the sequence of discoveries that followed, leading to the understanding of the composition, function, and importance of telomeres and the existence, composition, and action of telomerase.

It was Elizabeth Blackburn who was able to produce a detailed map of the ends of the chromosomes of a unicellular aquatic organism called *Tetrahymena*. Although she found many repeats of the DNA sequence CCCCAA at the tips, she could ascribe no specific function to them.

While at a conference in 1980 to present her findings, Blackburn crossed paths with Jack Szostak who had been studying the effect of the insertion of a linear segment of DNA called a plasmid into a yeast cell. Inside the yeast the DNA acted like a minichromosome, but over the course of several cell divisions it degraded. The two researchers decided to combine these DNA segments of interest from two very different organisms, splicing the repeats onto the plasmid. In a paper published in 1982 they reported that the *Tetrahymena* telomeres prevented the degradation of the plasmids inside yeast cells, suggesting it was a very old and widely shared mechanism. The means of telomere formation, however, remained a mystery.

Greider became involved as a graduate student in Blackburn's laboratory. The two suspected that an unknown enzyme might be responsible for synthesizing the telomere. Analyzing *Tetrahymena* cell extracts she had developed, Greider found signs of enzymatic activity on Christmas day 1984.

Following this work, Grieder and Blackburn purified and named the enzyme telomerase, demonstrating that it was a ribonucleprotein—composed of protein and RNA. They speculated in the paper describing the enzyme that the RNA portion might contain the familiar repeat in order to act as "an internal guide sequence" for determining primer recognition or synthesis of the telomeres. They later confirmed this mechanism while continuing to work together after Grieder became an independent scientist. Telomerase binds to telomeres using the internal complementary template of the RNA component, and synthesizes new telomere DNA with the protein component via reverse transcriptase.

Szostak and his lab meanwhile identified mutations in yeast that caused a gradual erosion of the telomeres, cell senescence (premature aging), and the eventual halt of cell division. Blackburn's group found the same effect in mutated *Tetrahymena*. The opposite occurred with the addition of telomerase—cell damage was prevented and delay of senescence delayed—an effect which Greider's group then reproduced in human cell lines. Not only have the discoveries of these three researchers illuminated the mechanism of cell division, they have spawned intense study in the fields of aging of cells and of whole organisms, as well as in cancer research once telomerase activity was found to be heightened in some cancer cells. In addition, the work has shed light on certain diseases of the skin, lungs, and blood cells now understood to result from telomerase defects.

For background information *See* ANIMAL GROWTH; CELL SENESCENCE AND AGING; CHROMOSOME; SYNTHETIC CHROMOSOME in the McGraw-Hill Encyclopedia of Science & Technology.

Physics. The physics prize was divided, with half going to Charles K. Kao of the Standard Telecommunication Laboratories, Harlow, United Kingdom, and the Chinese University of Hong Kong, China, for his pioneering work on the transmission of light in fibers for optical communication; and half shared by Willard S. Boyle and George E. Smith of the Bell Laboratories, Murray Hill, New Jersey, for their invention of an imaging semiconductor circuit, the

charge-coupled device (CCD). The prize honored the important role the winners have played in the development of modern information technology. Kao developed the theoretical basis and vision for low-loss optical fibers, which have revolutionized telecommunications. Boyle and Smith's invention is used in the ubiquitous digital camera, which has revolutionized photography, but has also enabled advances in medical imaging technology, astronomical instrumentation, and other scientific applications.

Low-loss optical fibers. The transmission of light through media such as glass, or even through water jets, had been attempted by differing means and for differing purposes for centuries. However, it was not until the 1950s that light transmission through glass fibers for applications such as imaging (for example, in gastroscopes) became possible. It was found that glass fibers clad with glass of slightly lower index of refraction (and with a protective plastic coating) could guide light by the principle of total internal reflection. If light strikes a boundary with a medium having a lower index of refraction, some of the light will be bent ("refracted") as it passes into the new medium, but some will be internally reflected. If the angle of the incident light exceeds a particular angle—the critical angle—depending on the refractive indices of the materials, total internal reflection occurs. The theory was worked out in detail, and by 1960 industrial production of fiber optics-based devices had been achieved.

The use of fiber optics for communication was a tempting objective, given that the short wavelength of visible light in the electromagnetic spectrum meant that a tremendous amount of information could, in theory, be transmitted along a single fiber, promising much greater efficiency than radio-based or coaxial cable-based transmission. Moreover, other enabling technologies, such as the laser, were being realized at about the same time. One type of optical fiber, the so-called single-mode fiber, is theoretically most efficient in conjunction with laser light sources in transmitting light over long distances at high bit rates. However, the fibers in use in early fiber optics placed severe practical limitations on long-distance transmission because of attenuation rates of about 1000 decibels per kilometer. That is, 99% of light would be lost in only 20 m (66 ft). Kao and coworkers provided the breakthrough by studying not only the physical properties but also the materials properties of the glass fibers. They concluded that much of the observed attenuation was due to absorption and scattering of light—that is, to impurities—not to theoretical limitations, and in an article in 1966 stated that glass of high purity should be capable of long-distance transmission "with important potential as a new form of communication medium." Further experimental and theoretical work confirmed the suitability of the single-mode design but also pointed to fused silica as having the requisite purity. Four years later at Corning Glass Works, the production of such fibers was achieved using a process known as chemical vapor deposition. Over the years, fibers of even greater efficiency have been manufactured, so that today attenuations of less than 0.2 dB/km (less than 5% light loss per kilometer) are typical. The "important potential" envisioned by Kao and coworkers has been realized, as long-distance telephone and data communications (including the Internet) are based mainly on fiber-optic technology in conjunction with other modern technologies such as light-emitting diodes and light amplifiers.

Charge-coupled device (CCD). The CCD invented by Boyle and Smith is now used primarily to produce high-quality images in electronic form. The first papers describing both the theory and their experimental results were published in 1970. CCDs take advantage of the sensitivity of silicon to light in the ultraviolet (wavelengths around 400 nm) to near-infrared (1100 nm) part of the electromagnetic spectrum. Light impinging on the surface of silicon creates electron-hole pairs, with the number of pairs related to the wavelength (photon energy), intensity (number of photons), and duration (length of exposure) of the incident light. In the CCD, the silicon surface is organized in an array of picture elements—pixels—upon which the image is focused and which collect the electrons produced by the light hitting them. The charge collected on the pixels is read out periodically, with the amount per pixel representing the intensity of the incident light on it in the preceding time frame. It thus becomes possible to reconstruct the image electronically from the charge collected and location of the elements in the pixel array.

Image sensors based on CCDs may be categorized as linear or area sensors. Linear image sensors, as the name implies, have their pixels arranged along a central axis and require relative motion between the image and the pixels such that the object is scanned and electronically reconstructed one line at a time. These sensors found use in scanners and photocopiers, but with increasing speed and sensitivity they are part of sophisticated applications such as industrial inspection systems and remote sensing by means of landform scanning from satellites and airplanes.

Area image sensors consist of two-dimensional arrays of pixels connected with collection and output circuitry. As such, they do not require relative motion between the image source and CCD. The high resolution and sophisticated processing circuitry of modern CCDs have enabled digital photography to virtually replace the use of chemically processed films. They are used in television cameras and camcorders as well, and in a variety of medical imaging devices. Although CCDs are not directly sensitive to x-rays, a layer of scintillating material over the sensor allows x-rays to be visualized in dentistry by the visible light emitted by a phosphor struck by the radiation. The amount of x-rays required to produce an image is less than with conventional film, and the image is immediately available on a nearby computer. In astronomy, CCDs have played an important role in visualizing distant objects by means of sensitive cameras in conjunction with powerful telescopes.

The digital imaging technology launched by Boyle and Smith's invention has combined with the dramatic advances in digital computing and telecommunications to fundamentally change and greatly expand the uses of photography both at the professional and consumer levels.

For background information *see* ASTRONOMICAL IMAGING; CHARGE-COUPLED DEVICES; FIBER-OPTIC CIRCUIT; OPTICAL COMMUNICATIONS; OPTICAL FIBERS; OPTICAL MATERIALS in the McGraw-Hill Encyclopedia of Science & Technology.

Contributors

Contributors

The affiliation of each Yearbook contributor is given, followed by the title of his or her article. An article title with the notation "coauthored" indicates that two or more authors jointly prepared an article or section.

A

Aksan, Dr. Nusret. *University of Pisa, Italy.* NATURAL CIRCULATION IN NUCLEAR SYSTEMS.

Akturan, Dr. Riza. *Engineering, Sirius XM Radio, Monmouth Junction, New Jersey.* PICOCELLS AND FEMTOCELLS.

Amin, Prof. S. Massoud. *Department of Electrical and Computer Engineering, University of Minnesota, Minneapolis.* SMART GRID—coauthored.

Andreas, Dr. Heather. *Department of Chemistry, Dalhousie University, Halifax, Nova Scotia, Canada.* SUPERCAPACITORS.

Arader, Ms. Havalyn Logan. *Department of Psychiatry, Columbia University Medical Center, New York, New York.* SCHIZOPHRENIA AND PRENATAL INFECTION—coauthored.

Arkhipova, Dr. Irina R. *Josephine Bay Paul Center for Comparative Molecular Biology and Evolution, Marine Biological Laboratory, Woods Hole, Massachusetts.* BDELLOID ROTIFERS—coauthored.

Arora, Dr. Rajeev. *Department of Horticulture, Iowa State University, Ames.* FREEZING TOLERANCE AND COLD ACCLIMATION IN PLANTS.

Ashton, Dr. Andrew. *Woods Hole Oceanographic Institution, Woods Hole, Massachusetts.* WAVE PROCESSES AND SHORELINE CHANGE—coauthored.

Assmann, Dr. Sarah M. *Department of Biology, Pennsylvania State University, University Park.* GUARD CELLS—coauthored.

B

Bachmeier, Scott. *Cooperative Institute for Meteorological Satellite Studies, University of Wisconsin-Madison.* SATELLITE DETECTION OF THUNDERSTORM INTENSITY—coauthored.

Balakotaiah, Dr. Vemuri. *Department of Chemical and Biomolecular Engineering, University of Houston, Texas.* CATALYTIC AFTERTREATMENT OF NOx FROM DIESEL EXHAUST—coauthored.

Barsley, Dr. Robert E. *Louisiana State University Health Sciences Center School of Dentistry, New Orleans.* FORENSIC DENTISTRY—coauthored.

Barzyk, Dr. Timothy M. *U.S. Environmental Protection Agency, Research Triangle Park, North Carolina.* TOOLS TO ASSESS COMMUNITY -BASED CUMULATIVE RISK AND EXPOSURES.

Benson, Dr. Philip M. *Rock and Ice Physics Laboratory, Department of Earth Sciences, University College London, United Kingdom.* VOLCANO SEISMICITY IN THE LABORATORY.

Berchtold, Dr. Nicole C. *Institute of Brain Aging and Dementia, University of California, Irvine.* EXERCISE AND COGNITIVE FUNCTIONING.

Berg, Roger W. *DENSO International America, Inc., Vista, California.* INTERVEHICLE COMMUNICATIONS—coauthored.

Betts, Jonathan B. *Los Alamos National Laboratory-National High Magnetic Field Laboratory, Los Alamos, New Mexico.* DIGITAL ULTRASONICS FOR MATERIALS SCIENCE—coauthored.

Beveridge, Prof. Christine A. *ARC Centre of Excellence for Integrative Legume Research, School of Biological Sciences, University of Queensland, Brisbane, Australia.* CONTROL OF SHOOT BRANCHING IN PLANTS—coauthored.

Bird, Prof. Michael. *School of Earth and Environmental Sciences, James Cook University, Cairns, Australia.* BAT GUANO: RECORD OF CLIMATE CHANGE —coauthored.

Bobik, Dr. Alexander. *Cell Biology Laboratory, Baker IDI Heart and Diabetes Institute, Melbourne, Victoria, Australia.* APOLIPOPROTEIN C-III.

Böddeker, Bert. *DENSO AUTOMOTIVE Deutschland GmbH, Eching, Germany.* INTERVEHICLE COMMUNICATIONS—coauthored.

Bowers, Prof. Philip L. *Department of Mathematics, Florida State University, Tallahassee.* DISCRETE ANALYTIC FUNCTIONS.

Brewer, Dr. Philip B. *ARC Centre of Excellence for Integrative Legume Research, School of Biological Sciences, University of Queensland, Brisbane, Australia.* CONTROL OF SHOOT BRANCHING IN PLANTS—coauthored.

Brodsky, Prof. Stanley J. *SLAC National Accelerator Laboratory, Stanford University, Stanford, California.* STRONG-INTERACTION THEORIES BASED ON GAUGE/GRAVITY DUALITY—coauthored.

Brophy, Dr. John. *Jet Propulsion Laboratory, California Institute of Technology, Pasadena, California.* DAWN ION PROPULSION SYSTEM.

Brown, Prof. Alan S. *Department of Psychiatry, Columbia University Medical Center, New York, New York.* SCHIZOPHRENIA AND PRENATAL INFECTION—coauthored.

Brumaghim, Dr. Julia L. *Department of Chemistry, Clemson University, South Carolina.* THE ROLE OF METAL IONS IN DNA DAMAGE.

Brunner, Jason. *Cooperative Institute for Meteorological Satellite Studies, University of Wisconsin-Madison.* SATELLITE DETECTION OF THUNDERSTORM INTENSITY —coauthored.

Bush, Dr. John W. M. *Department of Mathematics, Massachusetts Institute of Technology, Cambridge.* WALKING ON WATER—coauthored.

C

Cegla, Dr. Frederic B. *Department of Mechanical Engineering, Imperial College London, United Kingdom.* ULTRASONIC DENSITY MEASUREMENT—coauthored.

Chen, Dr. An. *Technology Research Group, Advanced Micro Devices, Sunnyvale, California.* BEYOND CMOS TECHNOLOGY.

Choi, Prof. Young B. *Department of MIS & CIS, College of Business, Bloomsburg University of Pennsylvania.* GREEN COMPUTING—coauthored.

Cole, Christopher. *Finisar Corporation, Sunnyvale, California.* OPTICAL ETHERNET.

Cooper, Ms. Lisa Noelle. *Department of Anatomy and Neurobiology, Northeastern Ohio Universities College of Medicine, Rootstown, Ohio.* INDOHYUS: THE ORIGIN OF WHALES—coauthored.

Cronin, Dr. Thomas W. *Department of Biological Sciences, University of Maryland, Baltimore.* COLOR VISION IN MANTIS SHRIMPS.

Culver, Dr. Joseph P. *Department of Radiology, Washington University School of Medicine, St. Louis, Missouri.* NONINVASIVE DIFFUSE OPTICS FOR BRAIN MAPPING—coauthored.

D

Dalal, Dr. Yamini. *Laboratory of Receptor Biology and Gene Expression, National Cancer Institute, Bethesda, Maryland.* ALTERNATIVE NUCLEOSOMAL STRUCTURE.

Dash, Dr. Philip R. *School of Biological Sciences, University of Reading, Berkshire, United Kingdom.* DEATH RECEPTORS.

Davies, Dr. T. Jonathan. *National Center for Ecological Analysis and Synthesis (NCEAS), University of California, Santa Barbara.* PHYLOGENETICS: PREDICTING RARITY AND ENDANGERMENT.

Davis, Dr. Burtron H. *Center for Applied Energy Research, Lexington, Kentucky.* FISCHER-TROPSCH SYNTHESIS.

De Los Santos, Dr. Héctor J. *NanoMEMS Research, LLC, Irvine, California.* RADIO-FREQUENCY MEMS.

DeLuca, Dr. Jennifer G. *Department of Biochemistry and Molecular Biology, Colorado State University, Fort Collins.* MITOSIS AND THE SPINDLE ASSEMBLY CHECKPOINT.

Dempsey, Prof. John P. *Center for Offshore Research and Engineering, National University of Singapore.* ARCTIC ENGINEERING—coauthored.

Dereniak, Prof. Eustace L. *College of Optical Sciences, University of Arizona, Tucson.* TERAHERTZ IMAGING—coauthored.

de Téramond, Prof. Guy F. *University of Costa Rica, San José.* STRONG-INTERACTION THEORIES BASED ON GAUGE/GRAVITY DUALITY—coauthored.

Dever, Timothy. *NASA Glenn Research Center, Cleveland, Ohio.* CONICAL BEARINGLESS MOTOR-GENERATORS—coauthored.

Dilcher, Dr. David L. *Paleobotany and Palynology Laboratory, Florida Museum of Natural History, Gainesville, Florida.* COEVOLUTION BETWEEN FLOWERING PLANTS AND INSECT POLLINATORS—coauthored.

Doxastakis, Dr. Manolis. *Department of Chemical and Biomolecular Engineering, University of Houston, Texas.* MOLECULAR MODELING OF POLYMERS AND BIOMOLECULES.

Duncan, Dr. Don P. *Wikoff Color Corporation, Fort Mill, South Carolina.* SUSTAINABILITY AND PRINTING INKS.

E

Ehrsson, Dr. Henrik. *Department of Neuroscience, Karolinska Institutet, Stockholm, Sweden.* BODY SELF-PERCEPTION—coauthored.

Elliott, Prof. Stephen J. *Institute of Sound and Vibration Research, University of Southampton, Highfield, Southampton, United Kingdom.* ACTIVE NOISE CONTROL IN VEHICLES—coauthored.

Eschenfelder, Paul. *Avion Corporation, Spring, Texas.* AIRPORT WILDLIFE HAZARD CONTROL.

Evans, Dr. David C. *Department of Natural History, Royal Ontario Museum, Toronto, Canada.* HADROSAURID (DUCK-BILLED) DINOSAURS.

F

Fan, Zheng. *Department of Mechanical Engineering, Imperial College London, United Kingdom.* ULTRASONIC DENSITY MEASUREMENT—coauthored.

Fanelli, Dr. Victor. *Los Alamos National Laboratory-National High Magnetic Field Laboratory, New Mexico.* DIGITAL ULTRASONICS FOR MATERIALS SCIENCE—coauthored.

Fino, Dr. Debora. *Department of Materials Science and Chemical Engineering, Politecnico di Torino, Italy.* DIESEL PARTICULATE FILTERS.

Flagello, Dr. Donis G. *Nikon Research Corporation of America, Oro Valley, Arizona.* POLARIZATION LITHOGRAPHY—coauthored.

Forman, Dr. William R. *Smithsonian Astrophysical Observatory, Cambridge, Massachusetts.* COSMIC ACCELERATION AND GALAXY CLUSTER GROWTH—coauthored.

Frihart, Dr. Charles R. *USDA Forest Service, Forest Products Laboratory, Madison, Wisconsin.* SOY PROTEIN ADHESIVES.

Frohlich, Dr. Michael W. *Jodrell Laboratories, Royal Botanic Gardens, Kew, Surrey, United Kingdom.* ORIGIN OF THE FLOWERING PLANTS.

Fromm, Dr. Michael. *U.S. Naval Research Laboratory, Washington, District of Columbia.* PYROCUMULONIMBUS—coauthored.

G

Gade, Dr. Alexandra. *National Superconducting Cyclotron Laboratory, Michigan State University, East Lansing.* NEUTRON-RICH ATOMIC NUCLEI.

Garcia-Barriocanal, Dr. Javier. *Universidad Complutense, Madrid, Spain.* ENHANCED IONIC CONDUCTIVITY IN OXIDE HETEROSTRUCTURES—coauthored.

Gehrt, Dr. Stanley D. *School of Environment and Natural Resources, Ohio State University, Columbus.* RACCOON RABIES IN URBAN SETTINGS.

Gertsch, Dr. Leslie. *Rock Mechanics and Explosives Research Center and the Department of Geological Science and Engineering, Missouri University of Science and Technology, Rolla.* LUNAR AND PLANETARY MINING TECHNOLOGY—coauthored.

Gilleo, Dr. Ken. *ET-Trends LLC, West Greenwich, Rhode Island.* CELL PHONE CAMERAS.

Gilliland, Dr. Laura Ullrich. *Department of Biology, Pennsylvania State University, University Park.* GUARD CELLS—coauthored.

Gladyshev, Dr. Eugene A. *Department of Molecular and Cellular Biology, Harvard University, Cambridge, Massachusetts.* BDELLOID ROTIFERS—coauthored.

Goldstein, Dr. David B. *National Resources Defense Council, San Francisco, California.* TRANSPORTATION EFFICIENCY AND SMART GROWTH—coauthored.

Gronewold, Dr. Andrew. *National Exposure Research Laboratory, U.S. Environmental Protection Agency, Research Triangle Park, North Carolina.* APPLICATIONS OF BAYES' THEOREM FOR PREDICTING ENVIRONMENTAL DAMAGE—coauthored.

Grunze, Prof. Heinz. *Department of Psychiatry, Institute of Neuroscience, Newcastle University, United Kingdom.* BIPOLAR DISORDER.

Gunatilaka, Dr. A. A. Leslie. *Southwest Center for Natural Products Research and Commercialization, University of Arizona, Tucson.* FUNGAL SECONDARY METABOLITES.

Gunnell, Dr. Gregg F. *Museum of Paleontology, University of Michigan, Ann Arbor.* ORIGIN AND EVOLUTION OF ECHOLOCATION IN BATS.

H

Hagelüken, Dr. Christian. *Umicore, Precious Metals Refining, Hanau, Germany.* PRECIOUS ELEMENT RESOURCES.

Hall, Dr. Brian K. *Department of Biology, Dalhousie University, Halifax, Nova Scotia, Canada.* DEVELOPMENT AND EVOLUTION.

Harding, Dr. Alice K. *Astrophysics Science Division, NASA Goddard Space Flight Center, Greenbelt, Maryland.* FERMI GAMMA-RAY SPACE TELESCOPE—coauthored.

Harley, Prof. John P. *Department of Biological Sciences, Eastern Kentucky University, Richmond.* MICROBIAL SURVIVAL MECHANISMS.

Harold, Dr. Michael P. *Department of Chemical and Biomolecular Engineering, University of Houston, Texas.* CATALYTIC AFTERTREATMENT OF NOₓ FROM DIESEL EXHAUST—coauthored.

Harrison, Prof. T. Mark. *Institute of Geophysics and Planetary Physics, University of California, Los Angeles.* ANCIENT ZIRCONS PROVIDE A NEW PICTURE OF EARLY EARTH.

Herbst, Dr. Jan F. *Materials and Processes Laboratory, General Motors Research and Development Center, Warren, Michigan.* HYDROGEN-POWERED CARS—coauthored.

Hibbett, Dr. David S. *Department of Biology, Clark University, Worcester, Massachusetts.* PHYLOGENETIC CLASSIFICATION OF FUNGI.

Hobohm, Dr. Uwe. *Department of Bioinformatics, University of Applied Sciences, Giessen, Germany.* STIMULATING INNATE IMMUNITY AS CANCER THERAPY.

Hoffmann, Dr. Axel. *Materials Science Division, Argonne National Laboratory, Argonne, Illinois.* PURE SPIN CURRENTS.

Holt, Prof. William. *Department of Geosciences, Stony Brook University, New York.* EARTHSCOPE: OBSERVATORIES AND FINDINGS.

Horner, Justin. *National Resources Defense Council, San Francisco, California.* TRANSPORTATION EFFICIENCY AND SMART GROWTH—coauthored.

Houck, Max. *Director of the Forensic Science Initiative, West Virginia University, Morgantown.* FORENSIC SCIENCE EDUCATION.

Hu, Dr. David L. *School of Mechanical Engineering, Georgia Institute of Technology, Atlanta.* WALKING ON WATER—coauthored.

I

Ibach, Dr. Rebecca E. *USDA Forest Service, Forest Products Laboratory, Madison, Wisconsin.* DURABILITY OF WOOD-PLASTIC COMPOSITE LUMBER.

Iborra, Dr. Enrique. *Universidad Politecnica—ETSIT, Madrid, Spain.* ENHANCED IONIC CONDUCTIVITY IN OXIDE HETEROSTRUCTURES—coauthored.

J

Jansen, Ralph. *NASA Glenn Research Center, Cleveland, Ohio.* CONICAL BEARINGLESS MOTOR-GENERATORS—coauthored.

Jarzen, Dr. David M. *Paleobotany and Palynology Laboratory, Florida Museum of Natural History, Gainesville, Florida.* COEVOLUTION BETWEEN FLOWERING PLANTS AND INSECT POLLINATORS—coauthored.

K

Kaeberlein, Dr. Matt. *Department of Pathology, University of Washington, Seattle.* SIRTUINS.

Kageyama, Dr. Ryoichiro. *Institute for Virus Research, Kyoto University, Japan.* DEVELOPMENTAL TIMING AND OSCILLATING GENE EXPRESSION—coauthored.

Kapovich, Prof. Michael. *Department of Mathematics, University of California, Davis.* GEOMETRIZATION THEOREM.

Karr, Dr. Jesse W. *Department of Chemistry, Boston University, Massachusetts.* METAL IONS IN NEURODEGENERATIVE DISEASES OF PROTEIN MISFOLDING—coauthored.

Kascak, Peter. *College of Engineering, Mechanical, Industrial, and Manufacturing Engineering Department, The University of Toledo, Ohio.* CONICAL BEARINGLESS MOTOR-GENERATORS—coauthored.

Keith, Dr. Brian. *Abramson Family Cancer Research Institute, University of Pennsylvania School of Medicine, Philadelphia.* OXYGEN SENSING IN METAZOANS.

Kishiyama, Dr. Mark M. *Helen Wills Neuroscience Institute, University of California, Berkeley.* POVERTY, STRESS, AND COGNITIVE FUNCTIONING—coauthored.

Klein, Dr. Donald A. *Department of Microbiology, Immunology, and Pathology, Colorado State University, Fort Collins.* MICROBIAL INTERACTIONS.

Knight, Dr. Robert T. *Evan Rauch Professor of Neuroscience and Helen Wills Neuroscience Institute, University of California, Berkeley.* POVERTY, STRESS, AND COGNITIVE FUNCTIONING—coauthored.

L

Leebens-Mack, Dr. Jim. *Department of Plant Biology, University of Georgia, Athens.* PLANT PHYLOGENOMICS.

Leininger, Dr. John. *Department of Graphic Communications, Clemson University, South Carolina.* VARIABLE DATA PRINTING.

Leinmüller, Tim. *DENSO AUTOMOTIVE Deutschland GmbH, Eching, Germany.* INTERVEHICLE COMMUNICATIONS—coauthored.

Leon, Dr. Carlos. *Universidad Complutense, Madrid, Spain.* ENHANCED IONIC CONDUCTIVITY IN OXIDE HETEROSTRUCTURES—coauthored.

Li, Dr. Xiaodong. *Senomyx, Incorporated, San Diego, California.* UMAMI TASTE RECEPTOR.

Liew, Prof. Jat-Yuen Richard. *Department of Civil Engineering, National University of Singapore.* DEPLOYABLE SYSTEMS—coauthored.

Lima, Dr. Nelson. *Centro de Engenharia Biológica, Micoteca da Universidade do Minho, Braga, Portugal.* FUNGI AND FUNGAL TOXINS AS WEAPONS—coauthored.

Lowe, Prof. Michael J. S. *Department of Mechanical Engineering, Imperial College London, London, United Kingdom.* ULTRASONIC DENSITY MEASUREMENT—coauthored.

Lycett, Dr. Stephen J. *Department of Anthropology, University of Kent, Canterbury, United Kingdom.* MOVIUS LINE—coauthored.

Lynn, Dr. Jeffrey W. *NIST Center for Neutron Research, Gaithersburg, Maryland.* IRON-BASED SUPERCONDUCTORS.

M

Mackenzie, Dr. Paul B. *Theoretical Physics, Fermilab, Batavia, Illinois.* LATTICE QUANTUM CHROMODYNAMICS.

MacLatchy, Dr. Laura. *Department of Anthropology, University of Michigan, Ann Arbor.* MOROTOPITHECUS.

Mahony, Dr. James B. *Department of Pathology and Molecular Medicine, Regional Virology and Chlamydiology Laboratory, St. Joseph's Healthcare Hamilton, East Hamilton, Ontario, Canada.* DETECTION OF RESPIRATORY VIRUSES.

Marshall, Prof. Peter W. *Center for Offshore Research and Engineering, National University of Singapore.* ARCTIC ENGINEERING—coauthored.

Mason, Prof. William H. *Department of Aerospace and Ocean Engineering, Virginia Polytechnic Institute & State University, Blacksburg.* AIRPLANE WING DESIGN.

Mata-Toledo, Dr. Ramon A. *Professor of Computer Science, James Madison University, Harrisonburg, Virginia.* GREEN COMPUTING—coauthored.

McCray, Prof. Richard. *Department of Astrophysical and Planetary Sciences, University of Colorado, Boulder.* SUPERNOVA 1987A.

McFarlane, Dr. Donald. *Joint Science: Biology, The Claremont Colleges, California.* BAT GUANO: RECORD OF CLIMATE CHANGE—coauthored.

McManus, Dr. Donald P. *Molecular Parasitology Laboratory, Division of Infectious Diseases, Queensland Institute of Medical Research, Brisbane, Australia.* VACCINES FOR SCHISTOSOMIASIS.

McPherson, Dr. Brian J. *Department of Civil and Environmental Engineering, University of Utah, Salt Lake City.* CARBON CAPTURE AND STORAGE.

Meyer, Prof. Curtis A. *Department of Physics, Carnegie-Mellon University, Pittsburgh, Pennsylvania.* EXPERIMENTAL SEARCH FOR GLUONIC HADRONS.

Meyer, Prof. Michael D. *School of Civil and Environmental Engineering, Georgia Institute of Technology, Atlanta.* MARINE TRANSPORTATION AND THE ENVIRONMENT.

Meyerhoff, Prof. Robert. *Department of Mathematics, Boston College, Chestnut Hill, Massachusetts.* HYPERBOLIC 3-MANIFOLDS—IN PART.

Migliori, Dr. Albert. *Los Alamos National Laboratory-National High Magnetic Field Laboratory, New Mexico.* DIGITAL ULTRASONICS FOR MATERIALS SCIENCE—coauthored.

Miller, Dr. Mikel M. *Air Force Research Laboratory (AFRL/RW), Eglin Air Force Base, Florida.* ANIMAL NAVIGATION—coauthored.

Milster, Prof. Tom D. *College of Optical Sciences, University of Arizona, Tucson.* POLARIZATION LITHOGRAPHY—coauthored.

Miserez, Dr. Ali. *Materials Department and Marine Science Institute, University of California, Santa Barbara.* HUMBOLDT SQUID BEAK BIOMIMETICS—coauthored.

Mohammadi, Dr. Farhad. *Advanced Cerametrics, Inc., Lambertville, New Jersey.* PIEZOELECTRIC FIBER COMPOSITES.

Mumma, Dr. Michael J. *Solar System Exploration Division, NASA Goddard Space Flight Center, Greenbelt, Maryland.* METHANE ON MARS.

Murray, Dr. Brad. *Division of Earth and Ocean Sciences, Duke University, Durham, North Carolina.* WAVE PROCESSES AND SHORELINE CHANGE—COAUTHORED.

N

Niwa, Dr. Yasutaka. *Institute for Virus Research, Kyoto University, Japan.* DEVELOPMENTAL TIMING AND OSCILLATING GENE EXPRESSION—coauthored.

Norton, Dr. Christopher J. *Department of Anthropology, University of Hawaii, Honolulu.* MOVIUS LINE—coauthored.

O

Omodei, Dr. Nicola. *Istituto Nazionale di Fisica Nucleare, Sezione di Pisa, Italy.* FERMI GAMMA-RAY SPACE TELESCOPE—coauthored.

Ortega, Dr. Jason. *Science & Technology, Lawrence Livermore National Laboratory, Livermore, California.* TRACTOR-TRAILER TRUCK AERODYNAMICS—coauthored.

P

Palmer, Prof. Andrew C. *Center for Offshore Research and Engineering, National University of Singapore.* ARCTIC ENGINEERING—coauthored.

Paterson, Dr. Robert Russell Monteith. *Centro de Engenharia Biológica, Micoteca da Universidade do Minho, Braga, Portugal.* FUNGI AND FUNGAL TOXINS AS WEAPONS—coauthored.

Paun, Dr. Ovidiu. *Molecular Systematics Section, Jodrell Laboratory, Royal Botanic Gardens, Kew, Richmond, United Kingdom.* EPIGENETICS AND PLANT EVOLUTION.

Peacock, Mike. *Florida Institute of Technology, Merritt Island.* SCIENCE ON THE INTERNATIONAL SPACE STATION.

Pennycook, Dr. Stephen J. *Materials Science & Technology Division, Oak Ridge National Laboratory, Tennessee.* ENHANCED IONIC CONDUCTIVITY IN OXIDE HETEROSTRUCTURES—coauthored.

Petkova, Dr. Valeria I. *Department of Neuroscience, Karolinska Institutet, Stockholm, Sweden.* BODY SELF-PERCEPTION—coauthored.

Pfeiffer, Dr. Karl-Peter. *Department of Medical Statistics, Informatics, and Health Economics, Innsbruck Medical University, Austria.* ELECTRONIC MEDICAL RECORD.

Pierce, Dr. Marcia M. *Department of Biological Sciences, Eastern Kentucky University, Richmond.* NEW DRUG TO CONTROL TUBERCULOSIS; NEW MALARIA VACCINE.

Pinkerton, Dr. Frederick E. *Materials and Processes Laboratory, General Motors Research and Development Center, Warren, Michigan.* HYDROGEN-POWERED CARS—coauthored.

Platt, Dr. Donald. *Micro Aerospace Solutions, Inc., Melbourne, Florida.* SPACE FLIGHT, 2008.

Politis, Dr. Gustavo G. *CONICET, Facultad de Ciencias Sociales, Universidad Nacional del Centro de la Provincia de Buenos Aires, Argentina.* EARLIEST HUMANS IN THE AMERICAS.

Preziosi, Dr. Richard. *Faculty of Life Sciences, University of Manchester, United Kingdom.* COMMUNITY GENETICS—coauthored.

R

Rabin, Dr. Robert M. *NOAA/National Severe Storms Lab, Norman, Oklahoma and Cooperative Institute for Meteorological Satellite Studies, University of Wisconsin-Madison.* SATELLITE DETECTION OF THUNDERSTORM INTENSITY — coauthored.

Rando, Dr. Oliver J. *Department of Biochemistry and Molecular Pharmacology, University of Massachusetts Medical School, Worcester.* HISTONE MODIFICATIONS, CHROMATIN STRUCTURE, AND GENE EXPRESSION.

Raskar, Dr. Ramesh. *MIT Media Lab, Cambridge, Massachusetts.* COMPUTATIONAL PHOTOGRAPHY.

Raubal, Dr. Martin. *Department of Geography, University of California, Santa Barbara.* LOCATION-BASED DECISION SUPPORT.

Ren, Dr. Tianying. *Oregon Hearing Research Center, Department of Otolaryngology and Head and Neck Surgery, Oregon Health & Science University, Portland.* COCHLEAR WAVE PROPAGATION.

Rivera-Calzada, Dr. Alberto. *Universidad Complutense, Madrid, Spain.* ENHANCED IONIC CONDUCTIVITY IN OXIDE HETEROSTRUCTURES—coauthored.

Rokhsar, Dr. Daniel S. *Center for Integrative Genomics and Department of Molecular and Cell Biology, University of California, Berkeley.* PLACOZOAN GENOME—coauthored.

Rosas, Dr. Antonio. *Department of Paleobiology, Museo Nacional de Ciencias Naturales, Consejo Superior de Investigaciones Científicas (CSIC), Madrid, Spain.* ATAPUERCA FOSSIL HOMININS.

Rosenkrans, Dr. Wayne A., Jr. *Chairman and President, Personalized Medicine Coalition, Washington, D.C.* PERSONALIZED MEDICINE.

Rostami, Dr. Jamal. *Department of Energy and Mineral Engineering, Pennsylvania State University, University Park.* LUNAR AND PLANETARY MINING TECHNOLOGY—coauthored.

Rowell, Dr. Roger M. *Professor Emeritus, Department of Biological Systems Engineering, University of Wisconsin, Madison.* HARDENING OF WOOD.

Rowntree, Dr. Jennifer. *Faculty of Life Sciences, University of Manchester, United Kingdom.* COMMUNITY GENETICS—coauthored.

Royer, Danielle. *Department of Anthropology, Stony Brook University, New York.* HUMAN FOSSILS FROM OMO KIBISH.

Rozengurt, Dr. Enrique. *Division of Digestive Diseases and CURE: Digestive Diseases Research Center, David Geffen School of Medicine, University of California at Los Angeles.* PROTEIN KINASE D SIGNALING.

Ruffing, Claire. *Department of Geography, University of Missouri-Columbia.* THREE-DIMENSIONAL MEASUREMENT OF RIVER TURBULENCE—coauthored.

Rutkowski, Dr. Adam J. *Air Force Research Laboratory (AFRL/RW), Eglin Air Force Base, Florida.* ANIMAL NAVIGATION—coauthored.

Ryan, Dr. Kevin. *Department of Chemistry, City College of New York (CUNY).* MOLECULAR SHAPE AND THE SENSE OF SMELL—coauthored.

Ryan, Dr. Xiaozhou P. *Department of Chemistry, City College of New York (CUNY).* MOLECULAR SHAPE AND THE SENSE OF SMELL—coauthored.

S

Salari, Dr. Kambiz. *Computation, Lawrence Livermore National Laboratory, Livermore, California.* TRACTOR-TRAILER TRUCK AERODYNAMICS—coauthored.

Sánchez-Dehesa, Prof. José. *Ingenieria Electronica [Department of Electronic Engineering], Universidad Politecnica de Valencia [Polytechnic University of Valencia], Spain.* ACOUSTIC CLOAKING—coauthored.

Santamaria, Dr. Jacobo. *Universidad Complutense, Madrid, Spain.* ENHANCED IONIC CONDUCTIVITY IN OXIDE HETEROSTRUCTURES—coauthored.

Schaller, Dr. Emily L. *Institute for Astronomy, University of Hawaii, Honolulu.* LARGE KUIPER BELT OBJECTS.

Schlachter, Dr. Alfred S. *Advanced Light Source, Lawrence Berkeley National Laboratory, Berkeley, California.* NATURAL-GAS-POWERED VEHICLES.

Schmidt, Robert K. *DENSO AUTOMOTIVE Deutschland GmbH, Eching, Germany.* INTERVEHICLE COMMUNICATIONS—coauthored.

Sefrioui, Dr. Zouhair. *Universidad Complutense, Madrid, Spain.* ENHANCED IONIC CONDUCTIVITY IN OXIDE HETEROSTRUCTURES—coauthored.

Selden, Dr. Paul A. *Paleontological Institute, University of Kansas, Lawrence.* SILK AND THE EARLIEST SPIDERS.

Senn, Prof. David R. *University of Texas Health Science Center at San Antonio, Center for Education and Research in Forensics.* FORENSIC DENTISTRY—coauthored.

Sharman, Dr. Robert. *National Center for Atmospheric Research (NCAR) Research Applications Laboratory, Boulder, Colorado.* SHIP WAVES IN THE ATMOSPHERE.

Sharoni, Dr. Amos. *Department of Physics, University of California-San Diego, La Jolla, California.* AVALANCHES AND PHASE TRANSITIONS.

Shimojo, Dr. Hiromi. *Institute for Virus Research, Kyoto University, Japan.* DEVELOPMENTAL TIMING AND OSCILLATING GENE EXPRESSION—coauthored.

Shpyrko, Dr. Oleg G. *Department of Physics, University of California-San Diego, La Jolla, California.* LOCAL STRUCTURAL PROBES.

Siomi, Dr. Mikiko C. *Department of Molecular Biology, Keio University School of Medicine, Tokyo, Japan.* PIRNAS (PIWI-INTERACTING RNAS).

Smith, Peter H. *Lunar and Planetary Laboratory, University of Arizona, Tucson.* PHOENIX MARS MISSION.

Spink, Dr. John. *Assistant Professor, School of Criminal Justice, East Lansing.* BRAND SECURITY IN PACKAGING.

Srivastava, Dr. Mansi. *Center for Integrative Genomics and Department of Molecular and Cell Biology, University of California, Berkeley.* PLACOZOAN GENOME—coauthored.

Stapelbroek, Prof. Maryn G. *College of Optical Sciences, University of Arizona, Tucson.* TERAHERTZ IMAGING—coauthored.

Stocks, Brian. *Wildfire Investigations Ltd., Sault Ste. Marie, Ontario, Canada.* PYROCUMULONIMBUS —coauthored.

Storm, Dr. Peter A. *Department of Mathematics, University of Pennsylvania, Philadelphia.* HYPERBOLIC 3-MANIFOLDS—IN PART.

Stothers, Dr. Ivan M. *St. John's Innovation Centre, Cambridge, United Kingdom.* ACTIVE NOISE CONTROL IN VEHICLES—coauthored.

Suzuki, Tadao. *DENSO Corporation, Japan, Kariya, Aichi, Japan.* INTERVEHICLE COMMUNICATIONS—coauthored.

Suzuki, Dr. Yoko. *Los Alamos National Laboratory - National High Magnetic Field Laboratory, Los Alamos, New Mexico.* DIGITAL ULTRASONICS FOR MATERIALS SCIENCE—coauthored.

Szalai, Dr. Veronika. *Department of Chemistry and Biochemistry, University of Maryland, Baltimore County.* METAL IONS IN NEURODEGENERATIVE DISEASES OF PROTEIN MISFOLDING—coauthored.

T

Tecot, Dr. Stacey. *Department of Anthropology and Institute for the Conservation of Tropical Environments, Stony Brook University, New York.* PRIMATE CONSERVATION EFFORTS—coauthored.

Teller, Prof. Seth. *Computer Science and Artificial Intelligence Laboratory, Massachusetts Institute of Technology, Cambridge.* AUTONOMOUS PASSENGER VEHICLES.

Thewissen, Dr. J. G. M. *Department of Anatomy and Neurobiology, Northeastern Ohio Universities College of Medicine, Rootstown, Ohio.* INDOHYUS: THE ORIGIN OF WHALES—coauthored.

Thomas, Dr. Howard. *Institute of Biological, Environmental, and Rural Sciences, Aberystwyth University, Ceredigion, United Kingdom.* LEAF SENESCENCE AND AUTUMN LEAF COLORATION.

Torrent, Dr. Daniel. *Ingenieria Electronica [Department of Electronic Engineering], Universidad Politecnica de Valencia [Polytechnic University of Valencia], Spain.* ACOUSTIC CLOAKING—coauthored.

U

Urban, Dr. Marek W. *School of Polymers and High Performance Materials, The University of Southern Mississippi, Hattiesburg.* STIMULI-RESPONSIVE POLYMERS.

Urban, Dr. Michael. *Department of Geography, University of Missouri-Columbia.* THREE-DIMENSIONAL MEASUREMENT OF RIVER TURBULENCE—coauthored.

V

Vallero, Dr. Daniel. *Adjunct Professor of Engineering Ethics, Pratt School of Engineering, Duke University, Durham, North Carolina.* APPLICATIONS OF

Bayes' theorem for predicting environmental damage—coauthored; Biotechnology and the environment.

vanEngelsdorp, Dr. Dennis. *Department of Entomology, Pennsylvania State University, University Park.* Colony collapse disorder.

Varela, Dr. Maria. *Materials Science & Technology Division, Oak Ridge National Laboratory, Tennessee.* Enhanced ionic conductivity in oxide heterostructures—coauthored.

Verani, Dr. Cláudio N. *Department of Chemistry, Wayne State University, Detroit Michigan.* Films of metal-containing surfactants.

Vikhlinin, Dr. Alexey. *Smithsonian Astrophysical Observatory, Cambridge, Massachusetts.* Cosmic acceleration and galaxy cluster growth—coauthored.

Vu, Dr. Khac Kien. *Department of Civil Engineering, National University of Singapore.* Deployable systems—coauthored.

W

Waite, Dr. J. Herbert. *Department of Molecular, Cellular and Developmental Biology, University of California, Santa Barbara.* Humboldt squid beak biomimetics—coauthored.

Walker, Dr. Matthew P. *Department of Psychology, University of California, Berkeley.* Sleep-dependent memory processing.

Walter, Dr. Todd. *Department of Aeronautics and Astronautics, Stanford University, Stanford, California.* Wide-Area Augmentation System (WAAS).

Weese, Dr. J. Scott. *Department of Pathobiology, Ontario Veterinary College, University of Guelph, Ontario, Canada.* Methicillin-resistant Staphylococcus aureus in the horse.

Whitaker, Dr. John O., Jr. *Department of Biology, Center for North American Bat Research and Conservation, Indiana State University, Terre Haute.* White-nose syndrome of bats.

White, Mr. Brian R. *Department of Radiology, Washington University School of Medicine, St. Louis, Missouri.* Noninvasive diffuse optics for brain mapping—coauthored.

Winger, Prof. Jeff Allen. *Department of Physics and Astronomy, Mississippi State University, Mississippi State.* Beta-delayed neutron emission.

Winograd, Dr. Nicholas. *Department of Chemistry, Pennsylvania State University, University Park.* Cluster ion mass spectrometry.

Wollenberg, Prof. Bruce F. *Department of Electrical and Computer Engineering, University of Minnesota, Minneapolis.* Smart grid—coauthored.

Wright, Dr. Patricia C. *Department of Anthropology and Institute for the Conservation of Tropical Environments, Stony Brook University, New York.* Primate conservation efforts—coauthored.

Wurster, Dr. Christopher. *School of Geography and Geosciences, University of St Andrews, United Kingdom.* Bat guano: record of climate change—coauthored.

X

Xiao, Dr. Shuhai. *Department of Geosciences, Virginia Polytechnic Institute and State University, Blacksburg.* Evolutionary patterns of the Ediacara biota.

Y

Yanoviak, Andrew Charles. *University of Hawaii, Honolulu.* Structural design of high-rise towers.

Z

Zvegintsev, Dr. Valerie. *Institute of Theoretical and Applied Mechanics, Siberian Branch of the Russian Academy of Sciences, Novosibirsk, Russian Federation.* Hypersonic test facilities.

Index

Index

Asterisks indicate page references to article titles.